W9-ANN-519

ADVANCES IN CHEMICAL PHYSICS

VOLUME XCVIII

EDITORIAL BOARD

Advances in CHEMICAL PHYSICS

Edited by

I. PRIGOGINE

Center for Studies in
Statistical Mechanics and
Complex Systems
The University of Texas
Austin, Texas
and
International Solvay
Institutes, Université
Libre de Bruxelles
Brussels, Belgium

and

STUART A. RICE

Department of Chemistry
and
The James Franck Institute
The University of Chicago
Chicago, Illinois

VOLUME XCVIII

AN INTERSCIENCE® PUBLICATION
JOHN WILEY & SONS, INC.
NEW YORK • CHICHESTER • WEINHEIM • BRISBANE • SINGAPORE • TORONTO

This text is printed on acid-free paper.

An Interscience® Publication

Library of Congress Catalog Number: 58-9935

ISBN 0-471-16285-X

Printed in the United States of America

10 9 8 7 6 5 4 3 2 1

CONTRIBUTORS TO VOLUME XCVIII

JOHN CURRO, Sandia National Laboratories, Albuquerque, New Mexico.

J. L. DORMANN, Laboratoire de Magnétisme et d'Optique de Versailles, Université de Versailles-Saint Quentin, Versailles, France.

T. ERBER, Department of Physics and Department of Mathematics, Illinois Institute of Technology, Chicago, Illinois.

D. FIORANI, Istituto di Chimica dei Materiali, CNR, Area della Ricerca di Roma, Monteretondo Stazione, Italia.

S. HAVRILIAK, JR., Rohm and Haas Research, Bristol Research Park, Bristol, Pennsylvania.

S. J. HAVRILIAK, Havriliak Software Development Co., Huntingdon Valley, Pennsylvania.

G. M. HOCKNEY, Theoretical Physics Department, Fermi National Accelerator Laboratory, Batavia, Illinois.

KENNETH S. SCHWEIZER, Department of Materials Science and Engineering, University of Illinois at Urbana-Champaign, Urbana, Illinois.

E. TRONC, Laboratoire de Chimie de la Matière Condensée, Université Pierre et Marie Curie, Paris Cedex 05, France.

STANISLAW URBAN, Institute of Physics, Jagellonian University, Kraków, Poland.

ALBERT WÜRLINGER, Physical Chemistry II, Ruhr-University, Bochum, Germany.

INTRODUCTION

Few of us can any longer keep up with the flood of scientific literature, even in specialized subfields. Any attempt to do more and be broadly educated with respect to a large domain of science has the appearance of tilting at windmills. Yet the synthesis of ideas drawn from different subjects into new, powerful, general concepts is as valuable as ever, and the desire to remain educated persists in all scientists. This series, *Advances in Chemical Physics*, is devoted to helping the reader obtain general information about a wide variety of topics in chemical physics, a field that we interpret very broadly. Our intent is to have experts present comprehensive analyses of subjects of interest and to encourage the expression of individual points of view. We hope that this approach to the presentation of an overview of a subject will both stimulate new research and serve as a personalized learning text for beginners in a field.

I. PRIGOGINE
STUART A. RICE

CONTENTS

ADVANCES IN CHEMICAL PHYSICS

VOLUME XCVIII

INTEGRAL EQUATION THEORIES OF THE STRUCTURE, THERMODYNAMICS, AND PHASE TRANSITIONS OF POLYMER FLUIDS

KENNETH S. SCHWEIZER

Departments of Materials Science and Engineering, Chemistry, and Frederick Seitz Materials Research Laboratory, University of Illinois, 1304 West Green Street, Urbana, Illinois 61801

JOHN G. CURRO

Sandia National Laboratories, Albuquerque, New Mexico 87185

CONTENTS

Advances in Chemical Physics, Volume XCVIII, Edited by I. Prigogine and Stuart A. Rice.
ISBN 0-471-16285-X

I. INTRODUCTION AND SCOPE

Condensed polymeric fluids exhibit a rich and complex set of experimental phenomena associated with the combined influences of local, system-specific monomer structure and global connectivity and flexibility. Such behavior is of both fundamental and practical interest. Early pioneering theoretical work focused largely on simple lattice models that invoked severe simplifications of both molecular structure and statistical mechanics.[1] More recently, remarkable progress has been made in describing relatively long wavelength structure and properties by employing scaling and renormalization group approaches inspired by analogies with critical phenomena, as well as self-consistent field methods.[2–4] However, these modern continuous-space approaches have restricted ranges of applicability (e.g., long chains, low and moderate densities), and generally address only the generic qualitative behavior of macromolecular systems from a polymer physics point of view. System-specific chemical structure features are lumped into fitting constants, or *prefactors*, and local fluid structure is not addressed. Thus, the *a priori* predictive capacity of such approaches is generally modest or nonexistent. In contrast, for simple atomic (and colloidal) and small-molecule fluids much theoretical progress for both structural and thermodynamic properties has been made over the past two to three decades based on continuous-space *integral equation* methods.[5] Such microscopic approaches are nonperturbative in interaction potentials and density (though generally "uncontrolled"), and can treat the physical consequences of the local molecular structure and intermolecular forces over a wide range of thermodynamic state conditions.

The purpose of this chapter is to summarize *some* recent progress toward developing liquid-state theories of macromolecular systems. The attractiveness of such a theoretical approach is its ability to describe structure and correlations on all length scales, thereby allowing a *quantitative* treatment of both the universal polymer physics aspects and the molecule-specific questions of great interest to chemists, materials scientists, and engineers. We shall focus on one particular integral equation approach, the Polymer Reference Interaction Site Model (PRISM) theory,[6] first proposed by us in 1987. PRISM is a macromolecular extension of the pioneering RISM theory of Chandler and Andersen.[7,8] We must emphasize that over the past few years many different liquid-state integral equation theories have been developed and applied (see Section X), and this activity is growing rapidly within the chemical physics community. We will only briefly address these developments here since a detailed survey and comparison of the emerging integral equation methods is the subject of a future review.[9] Moreover, we note that PRISM theory is presently by far the most developed and widely applied polymer integral equation approach.

A first review of the PRISM approach was written two and one half years ago, and was primarily intended for the polymer science community.[10] The present article will emphasize the most recent theoretical developments from a more liquid-state, chemical physics perspective. The detailed scientific issues of interest to polymer scientists that have motivated many of the PRISM developments and applications will be mentioned only briefly. Moreover, the often subtle and important question of the choice of molecular model that is adequate for a particular physical problem will not be emphasized here. However, examples of the influence of molecular structure simplification, or *coarse graining*, on physical predictions will be given throughout the chapter.

We have attempted to touch on all the fundamental ideas associated with the wide range of *homogeneous phase* phenomena and systems treated to date with the PRISM approach, supplemented with examples of specific applications. Liquid–liquid phase separation and copolymer self-assembly will also be discussed. However, the treatment of spatially *in*homogeneous systems and/or first-order phase transitions by combining PRISM theory with modern polyatomic density functional methods will not be discussed. Significant progress in this direction has been recently achieved (see Section XI), including problems such as the structure of polymers near surfaces and interfaces, confined chain molecule and polymer fluids, macromolecules in porous media, melt crystallization, block copolymer microphase separation, and liquid–vapor phase transitions.

A major goal of this chapter is to summarize the essential modifications of RISM integral equation theory[7,8] required to accurately treat condensed phases of flexible macromolecules. There appears to be three broad aspects which are worth enumerating explicitly.

1. *Many Coupled Integral Equations.* Consider a one-component fluid consisting of macromolecules each of which is composed of N elementary "sites" (degree of polymerization). In general (except for cyclic ring polymers) there will be of order N^2 coupled nonlinear integral equations describing the site–site intermolecular pair correlation functions. Since N is of the order 10^2–10^5 for polymers, this leads to an intractable numerical problem. Thus, approximation schemes must be constructed that result in a tractable number of coupled equations for suitably defined "averaged" correlation functions. In general, the physically appropriate simplifications depend on the global macromolecular architecture (e.g., chain, rigid rod, star-branched, etc.).

2. *Closure Approximation.* Many integral equation approaches, including RISM and PRISM, introduce an effective or renormalized site–site interaction potential called the *direct correlation function.*[5–8] The fundamental statistical mechanical approximations are made for this quantity by relating it to the bare intermolecular potential, thermodynamic state, and inter- and intramolecular pair correlation functions. Such relations are known as the *closure approximation.* Even for atomic fluids, the most useful closures depend on the form of the intermolecular potential, temperature, and density.[5] In particular, following the classic van der Waals idea, harsh repulsive forces are generally treated differently than slowly varying potentials such as the attractive branch of the Lennard-Jones interaction or Coulombic interactions.[5,8,11] A fundamental question is how to construct closures for macromolecular systems that yield reliable thermodynamic and structural predictions on *all* length

scales. As a preview of subsequent sections, we have found that the standard site–site Percus–Yevick closure[5,7] appears to work as well for the structure of repulsive force *(athermal)* macromolecular melts and mixtures as it does for small, rigid molecule fluids such as benzene and carbon tetrachloride.[8] However, a qualitatively correct treatment of the effect of attractive forces on structure is much more difficult for polymers.

3. *Self-Consistent Treatment of Intramolecular and Intermolecular Correlations.* Most macromolecules are conformationally flexible. Thus, the question of intermolecular packing and intramolecular structure are coupled and in principle must be solved for self-consistently.[8,12] This is true even for small flexible molecules such as the *n*-alkanes,[12] but the polymer problem is more complex due to "long range" (in chemical sequence) N-dependent intramolecular excluded volume effects, which can be progressively "screened" as the polymer density is increased.[2–4] The development of a complete ab initio theory of packing and conformation within the RISM formalism requires tractable schemes for both constructing a medium-induced solvation potential and the self-consistent mathematical solution of the resulting complicated *effective* single-chain problem.

In addition to the three general aspects given, there remains the *thermodynamic inconsistency* problem inherent to *all* integral equation approaches formulated at the level of pair correlations and not the partition function.[5] A basic question is how much the integral equation predictions for thermodynamic properties and phase boundaries differ according to the route chosen (e.g., free energy charging, compressibility) as a function of macromolecular size and other system-specific variables. It appears this thermodynamic inconsistency problem worsens as the polymers become larger, both for melt properties such as the pressure and isothermal compressibility, and for phase boundaries of multicomponent fluids. This problem has motivated the development of novel *molecular closure* approximations described in Section VI.

As true for atomic and small molecule fluids, the most unambiguous test of the accuracy of PRISM theory is via comparison with exact computer simulations of the same theoretical model. Unfortunately, simulation of long-chain, high-density polymer fluids in continuous space remains extremely time consuming even with modern supercomputers. However, significant progress has been made recently and some results have been obtained for both polymer melts and binary mixtures, which serve as benchmarks to test approximate theories. Direct comparisons of PRISM theory with experimental wide-angle scattering and selected thermodynamic property measurements on one-component hydrocarbon polymer liquids will also be presented.

In macromolecular fluids several interesting physical aspects and questions arise that have no analog in atomic and small-molecule systems. There are two broad issues worthy of explicit enumeration.

A. What is the role of macromolecule degree of polymerization N and global architecture on equilibrium properties? For one-component fluids, the effect of N on intrinsic thermodynamic properties, *local* packing, and collective scattering patterns saturates rather quickly within a given architectural class (e.g., linear chains) and is only quantitatively affected by global architecture. However, intermolecular structure on the macromolecular scale is always strongly influenced by N and depends explicitly on polymer global architecture (chain, ring, rod, star-branched, polymeric fractal). In macroscopically phase-separating polymer mixtures and microphase-separating block copolymers, long wavelength concentration fluctuations and phase boundaries are strongly affected by the degree of polymerization.

B. How does the level of chemical structure detail (or degree of coarse graining) retained in a theoretical model impact physical property predictions? This is a subtle question the answer to which depends on many factors including the nature of the physical phenomenon of interest and the level of accuracy deemed acceptable. One broad goal of our research is to use PRISM theory to investigate a particular problem with a range of different single-polymer models varying from the most coarse-grained "Gaussian thread" model commonly employed in field-theoretic studies[2-4] to atomistically realistic descriptions such as the rotational isomeric state (RIS) model.[13] In this way, the influence of the system-specific molecular structure details on questions such as scattering patterns and phase equilibria can be systematically established, and adequate "minimalist" models for a particular question can hopefully be deduced. Examples of such an approach will be given throughout the chapter. Progress in understanding such chemical issues is essential in order to use integral equation methods as an interpretative and predictive tool in materials chemistry and polymer science and engineering.

The question of the relationship between PRISM theory predictions and heavily coarse-grained scaling and field-theoretic approaches[2-4] is an interesting one, particularly to the polymer physics community. For some problems good qualitative agreement is found, and PRISM theory can be viewed as a microscopic derivation of results obtained by hueristic scaling ansatzes or heavily coarse-grained incompressible field theories. However, for other problems, such as correlation effects and phase separation in polymer blends and block copolymers, qualitative differences emerge. These reflect the differences in realism of the models adopted and/or the basic statistical mechanical approximations employed. We do not discuss these issues directly here, but shall refer the reader to the original literature where appropriate.

An attractive virtue of PRISM theory is the ability to derive *analytic* solutions for many problems *if* the most idealized Gaussian thread chain model of polymer structure is adopted. The relation between the analytic results and numerical PRISM predictions for more chemically realistic models provides considerable insight into the question of what aspects of molecular structure are important for particular bulk properties and phenomena. Moreover, it is at the Gaussian thread level that connections between liquid-state theory and scaling and field-theoretic approaches are most naturally established. Thus, throughout the chapter analytic thread PRISM results are presented and discussed in conjunction with the corresponding numerical studies of more realistic polymer models.

II. PRISM THEORY: BASIC ASPECTS

The integral equation approach to simple classical liquids was pioneered by Kirkwood and many others.[5,14] Considerable progress was made initially in the application of integral equation theory to simple monatomic liquids.[5,15] The most accurate theories for simple liquids are based on the well-known Ornstein–Zernike equation that defines the direct correlation function $C(r)$ in terms of fluid density and the radial distribution function $g(r) = 1 + h(r)$. Pioneering work was done in the 1960s and early 1970s.[5] For *dense* simple liquids with strongly repulsive and weak attractive interactions, the Percus–Yevick (PY) approximation[5,16] gives remarkably accurate results when compared to computer simulation and x-ray scattering experiments on monatomic liquids. The PY approximation can be viewed as a closure that approximately relates the direct correlation function of the radial distribution function, interatomic potential, and temperature. This closure, together with the Ornstein–Zernike equation, leads to a nonlinear integral equation for the radial distribution function $g(r)$ of a monatomic liquid. Theoretical treatment of the structural consequences of attractive forces at moderate and low densities is far more difficult even for simple fluids.[5] This area remains active in order to get better quantitative and thermodynamically consistent theories,[5,17] a better description of nonclassical critical phenomena,[18] and also to correctly treat situations where the intermolecular interactions are complex such as in colloidal suspensions.[19]

In the 1970s Chandler, Andersen, and co-workers initiated the pioneering extension of atomic integral equation concepts to molecular liquids based on the Reference Interaction Site Model, or RISM, theory.[7,8] This work, and other theoretical approaches based on interaction site models, has been reviewed in several places.[8,20] In RISM theory each molecule is subdivided into bonded spherically symmetric *interaction sites.* For small molecules (e.g., nitrogen, benzene, carbon tetrachloride) the definition of such sites is essentially obvious based on the chemical view of a molecule as a bonded collection of elementary units or functional groups. The liquid structure can be characterized by a matrix of site–site intermolecular pair correlation or radial distribution functions $g_{\alpha\gamma}(r)$ defined according to[7,8]

$$\tilde{\rho}^2 g_{\alpha\gamma}(r) = \left\langle \sum_{i \neq j=1}^{M} \delta(\mathbf{r}_i^\alpha)\delta(\mathbf{r} - \mathbf{r}_j^\gamma) \right\rangle \tag{2.1}$$

for a fluid of M molecules. In Eq. (2.1) $\tilde{\rho}$ is the number density of molecules and $\mathbf{r}_i^\alpha$ specifies the position of site α on molecule i. In RISM theory Chandler and Andersen generalized the Ornstein–Zernike equation of monatomic liquids to molecular liquids in a manner that includes intramolecular as well as intermolecular correlations.[7] Physically, the key idea is that intramolecular chemical bonding constraints, which describe the molecular shape of rigid molecules, strongly influence *inter*molecular packing. Based on heuristic arguments, Chandler and Andersen then employed a PY-type closure for the direct correlation functions in analogy with the monatomic case.[7,8] The resulting set of nonlinear integral equations can be solved numerically for the intermolecular pair correlation functions.[8,21]

Chandler and co-workers successfully applied this RISM formalism to describe the structure of rigid diatomic and polyatomic molecular liquids.[8,21] The generalization of the RISM theory to treat flexible molecules was initiated by Chandler and Pratt[12] in the late 1970s, and extensively applied to short alkane liquids[12] and the hydrophobic effect.[22] By combining the RISM methodology for a single flexible *ring molecule* (in imaginary time) with the Feynman path integral formulation of quantum mechanics, Chandler and co-workers have recently developed microscopic theories of quantum processes in fluids focusing particularly on the solvated electron problem.[23,24]

Beginning in 1987, we and our co-workers have extended and widely applied the RISM concepts to the case of flexible polymer solutions and melts,[6,25,26] polymer mixtures or blends,[27–32] and block copolymers.[33] We generically refer to this work as polymer RISM, or PRISM, theory.[10] The connection of the elementary aspects of PRISM theory with the quantum electron work has been discussed.[34]

The earliest version of PRISM theory rests on two very simple ideas that allow the circumvention of difficult computational and conceptual problems inherent to flexible macromolecular systems: points 1 and 2 enumerated in the Introduction. The first technical simplification applies to linear polymers when the degree of polymerization N is large. In this case one can, to a good approximation, take each of the monomers along the chain backbone as equivalent. At the most fundamental level, this corresponds to assuming the site–site direct correlation functions are independent of where monomers are located along the chain. This "preaveraging of end effects" approximation results in a reduced theory for *chain-averaged* site–site pair correlation functions[6] such as $g(r) = N^{-2}\sum_{\alpha\gamma}^{N} g_{\alpha\gamma}(r)$. Such a simplification, $g_{\alpha\gamma}(r) = g(r)$, would be exact for cyclic ring polymers. Of course, this approach represents a loss of detailed structural information, and interesting questions such as the packing of chain ends cannot be addressed. Tractable schemes to go beyond the equivalent monomer approximation have been proposed,[6] but to our knowledge not implemented. Numerical RISM studies on short linear molecules (propane, butane) suggest the preaveraging approximation is very accurate for the chain-averaged pair correlations[35] even when $N = 3$ or 4.

Within the equivalent monomer approximation scheme, each monomer in the linear chain is constructed from one or more spherically symmetric interaction sites A, B, C, and so forth. The generalized Ornstein–Zernike-like matrix equations of Chandler and Andersen[7] can be conveniently written in Fourier transform space in the general form

$$\hat{\underset{\sim}{H}}(k) = \hat{\underset{\sim}{\Omega}}(k)\hat{\underset{\sim}{C}}(k)[\hat{\underset{\sim}{\Omega}}(k) + \hat{\underset{\sim}{H}}(k)] \tag{2.2a}$$

where the caret denotes Fourier transformation with wave vector k. In real space one obtains

$$\underset{\sim}{H}(r) = \int d\mathbf{r}' \int d\mathbf{r}'' \underset{\sim}{\Omega}(|\mathbf{r} - \mathbf{r}'|)\underset{\sim}{C}(|\mathbf{r}' - \mathbf{r}''|)[\underset{\sim}{\Omega}(\mathbf{r}'') + \underset{\sim}{H}(\mathbf{r}'')] \tag{2.2b}$$

The first set of terms on the right-hand side of Eq. (2.2a,b) describes all possible site–site correlation pathways *between a pair of tagged molecules*. In the low-molecular-density limit only these contributions survive. The second set of terms describes all correlation pathways between two sites on a pair of molecules, which are mediated by one or more *different* molecules. The matrix multiplications in Eqs. (2.2) run over ν-independent sites A, B, C, . . . and $C_{\alpha\gamma}(r)$ is the $\nu \times \nu$ matrix of direct correlation functions. Because of symmetry there are $\nu(\nu + 1)/2$ independent

Ornstein–Zernike equations for the total correlation functions $H_{\alpha\beta}(r)$

$$H_{\alpha\gamma}(r) = \rho_\alpha \rho_\gamma h_{\alpha\gamma}(r) = \rho_\alpha \rho_\gamma [g_{\alpha\gamma}(r) - 1] \tag{2.3}$$

where $\rho_\alpha \equiv \tilde{\rho} N_\alpha$ is the density of sites of type α, and N_α is the number of sites of type α per chain.

When the generalized Ornstein–Zernike-like or PRISM matrix Eq. (2.2) is applied to flexible macromolecules, a *conformational preaveraging* assumption is employed by replacing the instantaneous, N-body intramolecular structure of the flexible chain by its ensemble-averaged pair correlation function description.[6,8,12,24] Thus, all information concerning the intramolecular structure of polymer chains is contained in the functions $\Omega_{\alpha\gamma}(r)$ defined as

$$\Omega_{\alpha\gamma}(r) = \tilde{\rho} \sum_{i \in \alpha, j \in \gamma} \omega_{ij}(r) \tag{2.4}$$

where $\omega_{ij}(r)$ is the normalized probability density between two sites i and j on the *same* molecule. In Fourier transform space $\hat{\Omega}_{\alpha\gamma}(k)$ can be identified as the single-chain partial structure factors.

The generalized Ornstein–Zernike-like equations in Eq. (2.2) define $\nu(\nu + 1)/2$-independent direct correlation functions. In order to have a solvable system of equations, additional approximate "closure relations" are required. This is the critical step, since the RISM or PRISM equations are really just defining relations for the site–site direct correlation functions. The most accurate closure is system-specific and is a question of enduring interest. In our original work on *dense one-component repulsive force* liquids, we followed Chandler and Andersen by adopting the approximate site–site PY closure[7,8]

$$C_{\alpha\gamma}(r) \cong \{1 - \exp[\beta v_{\alpha\gamma}(r)]\} g_{\alpha\gamma}(r) \tag{2.5a}$$

where $v_{\alpha\gamma}(r)$ is a spherically symmetric, *repulsive* interaction potential between sites α and γ, and $\beta = 1/k_B T$ at temperature T where k_B is Boltzmann's constant. For hard sphere interaction between sites, the PY closure reduces to the particularly simple form

$$\begin{aligned} g_{\alpha\gamma}(r) &= 0 \qquad r < d_{\alpha\gamma} \\ C_{\alpha\gamma}(r) &\cong 0 \qquad r > d_{\alpha\gamma} \end{aligned} \tag{2.5b}$$

which is equivalent to the so-called mean spherical approximation (MSA).[5,8,11] The condition inside the distance of closest approach $d_{\alpha\gamma}$ is an exact statement reflecting the impenetrability of hard spheres. The second condition, in which the direct correlation functions are approximated as zero outside the hard core, exploits the standard idea of Ornstein and Zernike that the direct correlation function is spatially short range. For atomic fluids, Eq. (2.5b) can be derived by established graph-theoretical partial summations and other functional methods.[5] However, for interaction site molecular fluids the PY closure is argued to be useful based on analogies with atomic fluids and heuristic physical concepts.[7,8] The lack of a rigorous interaction site cluster series basis for Eq. (2.5b) has led to RISM theory being described as a "diagrammatically improper" theory.[8]

Equations (2.3)–(2.5) lead to $\nu(\nu + 1)/2$-coupled integral equations that make up the polymer RISM theory in its simplest form appropriate for dense, repulsive force polymer fluids. The integral equations can be solved numerically using a variety of standard techniques.[5,8]

Alternative closure approximations for the repulsive force fluid have been investigated and will be briefly commented on in subsequent sections. Based on the idea that the atomiclike closures are useful by analogy for molecular fluids, there are several alternatives to the PY or MSA for hard core fluids. These include the hypernetted chain (HNC) approximation[5,20,36]

$$c_{\alpha\gamma}(r) = h_{\alpha\gamma}(r) - \ln[1 + h_{\alpha\gamma}(r)] \qquad r > d_{\alpha\gamma} \tag{2.6}$$

and the Martytnov–Sarkisov (MS) closure[5,37]

$$c_{\alpha\gamma}(r) = h_{\alpha\gamma}(r) - (\tfrac{1}{2})(\{1 + \ln[1 + h_{\alpha\gamma}(r)]\}^2 - 1) \qquad r > d_{\alpha\gamma} \tag{2.7}$$

Numerical studies of chain molecule fluids have also been carried out by Yethiraj[38] using the considerably more complicated "diagrammatically proper" formulation of RISM theory due to Chandler et al.[39,40] Novel, even more complicated closures have been recently proposed by several workers,[41,42] but numerical predictions for polymer fluids have not been established.

Appropriate closures for describing the influence of attractive forces on polymer liquid structure is a much more subtle and difficult problem than the repulsive force or hard core fluid case. We defer discussing this aspect until Section VI.

In our application of PRISM theory to flexible polymer systems, one expects that the *intra*molecular structure, represented by Eq. (2.4), depends on the *inter*molecular structure specified in Eq. (2.3) and vice versa.[6,8,12] Thus, in a rigorous calculation the intramolecular and intermolecular structure must be determined in a *self-consistent* manner leading to problem 2 mentioned in the Introduction. This problem represents a major conceptual difficulty and might be thought to be especially formidable for large macromolecules. The self-consistent issue for flexible molecules was originally addressed by Chandler and Pratt in both a formal diagrammatic manner and in the context of tractable approximation schemes formulated for short-chain molecules.[8,12] For macromolecules, several new theories for performing such self-consistent structural calculations have been formulated and applied,[43–47] which will be discussed in Sections VIII and IX.

A simple, zeroth approximation to the self-consistent problem *for dense one-component polymer melts* can be invoked as suggested by our earliest PRISM work.[6] Subsequent structural self-consistent calculations (see Section VII), as well as computer simulations and experiments, suggest that to a good approximation one can avoid (*under appropriate conditions*) the self-consistency complication by exploiting Flory's "ideality hypothesis". Flory[1,48] argued many years ago that in a high-density melt of strongly interpenetrating chains, the "long-range" *intra*molecular excluded volume interactions that lead to chain expansion in a dilute solution in a good solvent are "screened out" or "cancelled" by the *compressive* intermolecular interactions between chains embedded in a nearly incompressible fluid. *At the level of a single chain*, the net result is a cancellation of repulsive bare intrachain interactions by the attractive, "solvent-mediated" interactions. Thus, in a dense, one-component melt the chains act as a "theta or ideal solvent" for themselves in the sense that the chain radius-of-gyration obeys the maximum entropy ideal random walk scaling law: $R_g \propto N^{1/2}$. The "prefactor" in this scaling relation can be computed based on an atomistic single chain model[13] which *ignores* interactions among monomers beyond close neighbours. Neutron scattering experiments[49] and computer simulations[50,51] on polymer melts have demonstrated the accuracy of Flory's conjecture. This approximation provides an enormous simplification because the intramolecular correlations in Eq. (2.4) can then be calculated from a separate single-chain computation in which long-range (in chemical sequence) interactions along the chain backbone are set to zero. A wide variety of single-chain models are available,[13] and thus the connection between polymer structural features and bulk properties and phenomena can be systematically investigated.

It is important to note that in calculation of the intramolecular structure factors for input into PRISM theory, one can include as much (or as little) chemical detail regarding the molecular architecture as desired. For questions regarding intermolecular packing on relatively long length scales (e.g., the so-called correlation hole regime[2] corresponding to intermolecular separations of several monomer diameters and larger), the local monomeric structure is not important, and one can use a coarse-grained description of the polymer chain structure.[2-4,13] In this case a Gaussian, freely jointed, or semiflexible chain model for $\hat{\Omega}_{\alpha\beta}(k)$ would suffice. Such coarse-grained models are also useful for investigating general trends which transcend the fine details of specific polymer molecules.

On the other extreme, in order to make specific quantitatively accurate predictions for thermodynamic properties and the details of local packing, we anticipate that the local monomeric structure is important. For the often subtle question of macromolecular mixture miscibility and copolymer self-assembly, it is often unclear a priori what level of chemical detail is adequate. For such problems, one may employ a model that includes the effects of constant bond lengths, bond angles, and rotational potentials such as the rotational isomeric state model.[13] Inclusion of these local details into $\hat{\Omega}_{\alpha\beta}(k)$ is feasible but requires significantly more numerical effort. With modern workstations, a tractable option is to perform a *single-chain* simulation to provide a chemically realistic input to PRISM. Thus, PRISM theory is versatile in its ability to make predictions about intermolecular packing on both local (monomeric) and global (radius of gyration) length scales, as a functional of intramolecular architecture. In this chapter we will describe PRISM applications that include the entire range of local chemical detail.

III. STRUCTURE AND THERMODYNAMICS OF DENSE MELTS

Pure one-component polymer liquids, or *melts,* are in one sense the simplest case since the single-chain conformation is nearly "ideal." However, there remains the question of the influence of local chemical architecture on melt structure, thermodynamic properties, and physical phenomena (e.g., wide-angle scattering, crystallization). In the context of PRISM theory, the question is on what length scale, or degree of coarse-graining, is an "interaction site" defined? Since there does not exist a rigorous renormalization group type scheme to "integrate out" degrees of freedom and chemical details, the practical approach is to study families of models of variable levels of realism.[52] Figure 1 illustrates this process schematically in the context of an industrially important class of saturated hydrocarbon polymers (polyolefins). Three general levels of chain models are illustrated. (1) An atomistic level where there may be multiple symmetry-inequivalent sites within a monomer repeat unit. For polyolefins, sites may be a methylene, methyl, or methyne group for which angstrom-level structure is explicitly accounted. (2) A *single-site*

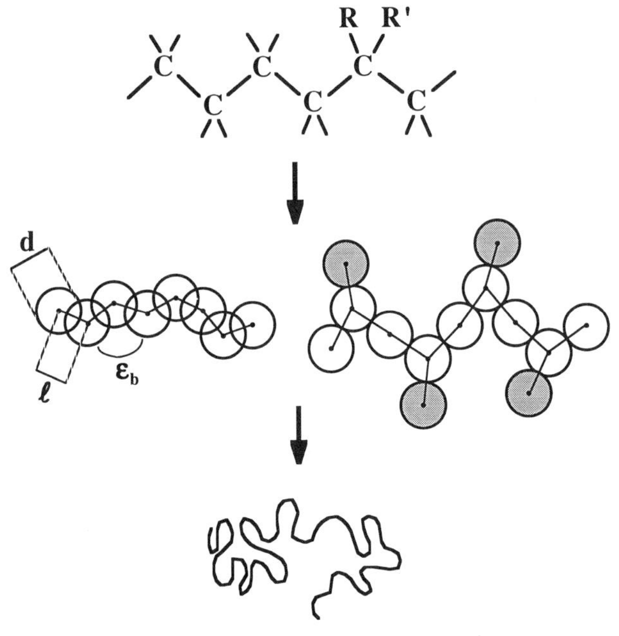

Figure 1. Schematic representation of three levels of chain models considered and the coarse-graining procedure. The top level is an atomistic model of polyolefins. The second level shows two intermediate models: site overlapping semiflexible chain (with bending energy ε_b) and freely jointed branched chain. The bottom level is the Gaussian thread chain.

intermediate-level "semiflexible chain" (SFC) model or multiple-site "freely jointed" branched chain. Such models correspond to a modest degree of coarse graining. (3) The extreme, heavily coarse-grained *Gaussian thread model* where the polymer is crudely treated as a thin, fully flexible, ideal random walk space curve. It is at this level that self-consistent field theoretic approaches describe polymer structure.[2–4]

In the next two sections we consider melt structure, as embodied in the intermolecular site–site radial distribution functions and the total structure factor describing collective density fluctuations in Fourier space, as a function of degree of coarse graining. Possible *mappings*, which relate the different chain models, are briefly mentioned.[52] Purely repulsive (generally hard-core) interchain site–site potentials are employed corresponding to an athermal melt situation. At high liquid densities, structure is expected to be dominated by such purely steric packing forces.[5,11] Use of the structural information to compute thermodynamic properties is addressed in Section C.

A. Single-Site Homopolymers

Consider first linear polymers composed of identical spherical sites that interact intermolecularly via a pair decomposable site–site hard-core potential of diameter d. The dimensionless reduced fluid density is $\rho_m d^3$, where $\rho_m = N\rho$ is the site number density.

1. Gaussian Thread Chains

At the most coarse-grained level the polymer is described as an ideal random walk *on all length scales*. The intramolecular structure factor matrix is Gaussian and given by[2–4]: $\hat{\omega}_{\alpha\gamma}(k) = \exp(-k^2\sigma^2|\alpha - \gamma|/6)$, where σ is the so-called statistical segment length. Physically, it represents a length scale beyond which real chain units are orientationally uncorrelated. The mean-square end-to-end distance R and radius-of-gyration R_g are given by $\langle R^2 \rangle = N\sigma^2$, and $R_g = R/\sqrt{6}$, respectively, where N is the number of statistical segments. The single-chain structure factor $\hat{\omega}(k) = N^{-1} \Sigma_{\alpha,\gamma}^{N} \hat{\omega}_{\alpha\gamma}(k)$ is easily computed in closed form. Numerically obtained PRISM predictions of $g(r)$ and dimensionless collective density fluctuation structure factor, $\hat{S}(k) = \hat{\omega}(k) + \rho_m \hat{h}(k)$, for such a model have been represented for a wide range of N and reduced densities.[6,25,30] Gaussian ring polymers have also been studied.[6,25] As expected physically, *for large N* only minor structural differences between ring and chain melts are found *on* macromolecular length scales, and identical behavior is predicted for the local region of $g(r)$.

A further model simplification, corresponding to a type of "continuum limit" (commonly employed in field-theoretic approaches[2–4] in the large N regime), can be introduced in order to obtain analytic results that capture all the essential physical features of the Gaussian chain model.[25,30] The single-chain structure factor is approximated by a Lorentzian[3]

$$\hat{\omega}(k) = \frac{1}{(k^2\sigma^2/12) + N^{-1}} \tag{3.1}$$

This form neglects the self-scattering term appropriate for the $k\sigma \rightarrow \infty$ regime, but which is irrelevant in a continuum-of-sites description. Equation (3.1) very accurately describes the exact Gaussian $\hat{\omega}(k)$ for the $k\sigma < 1$ regime of interest in a continuum model. In particular, it exactly reproduces the $k = 0$ value and the "self-similar" intermediate scaling regime, $\hat{\omega}(k) = 12(k\sigma)^{-2}$ for $R_g^{-1} \ll k \ll \sigma^{-1}$. In real space, this self-similar behavior corresponds to power law, or critical-like, correlations, $\omega(r) \propto r^{-1}$. This is a polymeric effect associated with the ideal random walk chain statistics on intermediate length scales, and is widely exploited in the "scaling theory" approach to polymer physics problems.[2–4] The second simplification is to take the "thread" limit, corresponding to $d \to 0$ and $\rho_m \to \infty$ such that the reduced fluid density is finite and nonzero. This simplification reduces the hard-core impenetrability constraint to a point condition, $g(r=0)=0$. Thus, within the PY closure approximation the site–site direct correlation function reduces to a delta-function form: $C(r) = C_0\delta(\mathbf{r})$, where $C_0 = \hat{C}(k=0)$ is a parameter to be determined by application of the PRISM integral equation and the core exclusion condition.[25,30]

The resulting PRISM integral equation is analytically solvable for the Gaussian thread model. The structural predictions are[25,30]

$$g(r) - 1 = \frac{3}{\pi \rho_m \sigma^2 r}\left[\exp\left(\frac{-r}{\xi_\rho}\right) - \exp\left(\frac{-r}{\xi_c}\right)\right] \tag{3.2}$$

$$\begin{aligned}\hat{S}(k) &\equiv \hat{\omega}(k) + \rho_m \hat{h}(k) \\ &= \frac{12(\xi_\rho/\sigma)^2}{1+(k\xi_\rho)^2}\end{aligned} \tag{3.3}$$

The fundamental length scales are the density screening length, ξ_ρ, given by

$$\xi_\rho^{-1} = \frac{\pi \rho_m \sigma^2}{3} + \xi_c^{-1} \tag{3.4}$$

which controls the *local* packing of threads, and the "correlation hole" length scale $\xi_c = R_g/\sqrt{2}$. Equation (3.2) shows that the correlated part of $g(r)$ consists of a local and macromolecular contribution. "Negative" correlation is predicted on all length scales, that is, $g(r) < 1$ for all r, and simple liquidlike solvation shells are entirely absent. Remarkably, these general features survive qualitatively in more chemically realistic, even atomistic, models of polymer structure due to thermal conformational disorder and destructive interference between the packing consequences of multiple local length scales (see Section II.B). For the simple thread model the local contribution to $g(r)$ is directly related to $\hat{S}(0) = 12(\xi_\rho/\sigma)^2$ and hence the isothermal compressibility, κ, via the thermodynamic relation $\hat{S}(0) = \rho_m k_B T \kappa$. The simple Yukawa forms in Eqs. (3.2) and (3.3) are a consequence of the technical simplifications invoked by the Gaussian thread model. Hence, the precise details of $g(r)$ in the local region will change as more chemically realistic models are employed.

The depth of the *local* correlation hole is predicted to be controlled by a so-called packing length $(\rho_m \sigma^2)^{-1}$. This quantity is invariant to arbitrary redefinition of a coarse-grained segment (or regrouping of real monomers). Under melt conditions and for normal temperatures (T = 250–500 K), the packing length falls typically in the range of 1.7–5.5 Å for a very wide class of semiflexible polymers.[52,53]

The predicted power law relation (for large N) of Eq. (3.4) between the density screening length and ρ_m is in excellent agreement with experiments, scaling arguments, and field theories for dense solutions but *not* melts.[52,54] However, under many solution conditions the "ideality" approximation breaks down and the effective statistical segment length, and hence R_g, acquires a polymer concentration dependence. This aspect has been incorporated by using the fully self-consistent version of PRISM (see Section VIII), or more simply by combining field-theoretic and/or scaling predictions[2,3] for single-chain size (e.g., $\sigma \propto \rho^{-1/8}$ in good solvents) with the PRISM analysis of intermolecular packing.[54]

The second contribution to $g(r)$ in Eq. (3.2) is called the correlation hole effect by deGennes[2] and is associated with the longer wavelength universal aspects of chain connectivity and interchain repulsive forces. On intermediate length scales it has a critical power law form due to chain conformation self-similarity, and this simple analytic form remains an excellent representation even for chemically realistic models when intersite separations exceed an intrinsic (N-independent) distance of the order of 3–5 site diameters.[25]

The dimensionless collective structure factor, $\hat{S}(k)$ in Eq. (3.3), is of a purely decaying, or "diffusive," form; no large angle peaks (which must be present in real dense fluids) are present. Again, this is a consequence of the idealized Gaussian thread model, although the diffusive form is in general accord with experiments (in the $k\sigma < 1$ regime) and field-theoretic predictions for moderately concentrated ("semidilute") polymer solutions.[2,3]

The analytic Gaussian thread model has been generalized to approximately treat nonzero chain thickness ($d \neq 0$) in a simple average manner.[30] This generalization is called the *Gaussian string model,* and results in a $g(r)$ and $\hat{S}(k)$ of the same form as Eq. (3.2) but the density–density screening does not obey Eq. (3.4). For long chains all the basic structural and thermodynamic features remain the same as the thread model, although the contact value, $g(d)$, is now nonzero, and this has important implications for particular physical problems. The Gaussian string model has been shown to be generally in remarkable agreement with numerical PRISM predictions for discrete, nonzero thickness Gaussian chains.[30] This agreement suggests that a type of "self-averaging" process occurs in polymer fluids, that is, the essential part of $C(r)$ is its long wavelength, integrated strength $C_0 = \hat{C}(k=0)$.

2. *Semiflexible Chain Models*

The most fundamental aspects of real polymer structure are: (a) nonzero chain thickness, (b) semiflexibility, that is, a system-specific and thermodynamic-state-dependent tendency for chain bending or coiling due to rotational isomerism, and (c) an overall size strongly correlated with the degree of polymerization. As displayed in Figure 1, the discrete SFC model includes these features by introducing (a) a site diameter d; (b) a local bending energy ε_b, which controls the "chain persistence length," defined as $\xi_p = \Sigma_\alpha^N \langle \mathbf{l}_1 \cdot \mathbf{l}_\alpha \rangle / l_b$, which for large N is given by $\xi_p = l_b / \{1 + \langle \cos(\theta) \rangle\}$ where l_b is the magnitude of the nearest neighbor rigid bond length; and (c) a degree of polymerization, N, which determines the overall size $\langle R^2 \rangle = l_b(2\xi_p - l_b)N$ (for large N). The ratio l_b/d controls the amount of exposed surface area available for interchain site–site interactions or packing.

Summarizing, the structure of a fluid of *hard-core* SFC polymers is characterized by four dimentionless variables, which can be chosen to be $\rho_m d^3$ (reduced density) or total site packing fraction η, $\Gamma = \xi_p/d$ (chain aspect ratio), l_b/d, and degree of polymerization, N. Novel approximate, but accurate and computationally convenient, numerical procedures have been developed by Honnell et al.[55] for the calculation of the single-chain structure factor, $\hat{\omega}(k)$, of the SFC model. In the polymer field jargon, the SFC model is the discrete, finite thickness generalization of the "worm-like chain" or "Koyama" model,[4,56] which interpolates between the rigid rod and ideal random walk chain models. The approximate calculation of $\hat{\omega}(k)$ is based on a cumulant expansion and rigorous evaluation of the second and fourth moments of $r_{\alpha\gamma}$, or equivalently $\langle \cos(\theta) \rangle$ and $\langle \cos^2(\theta) \rangle$. In addition, it is possible to exactly compute the next nearest neighbor correlations,[43] $\omega_{\alpha,\alpha+2}(r)$, and this extension is generally adopted and accounts for the most local part of the intramolecular excluded volume interactions.

Details of the rotational potentials, chemical bond lengths, bond angles, and nonspherical monomar structure are ignored in the SFC model and thus can only be mimicked by judicious choice of SFC model parameters. However, it has been recently demonstrated by Schweizer et al.[52] that by appropriate choice of SFC parameters both the single-chain and interchain packing of real polymer liquids can be reproduced to surprising accuracy by the SFC model. Although inherently nonunique, specific procedures for "mapping" a real polymer onto the SFC have been formulated and successfully applied. Here we present a few

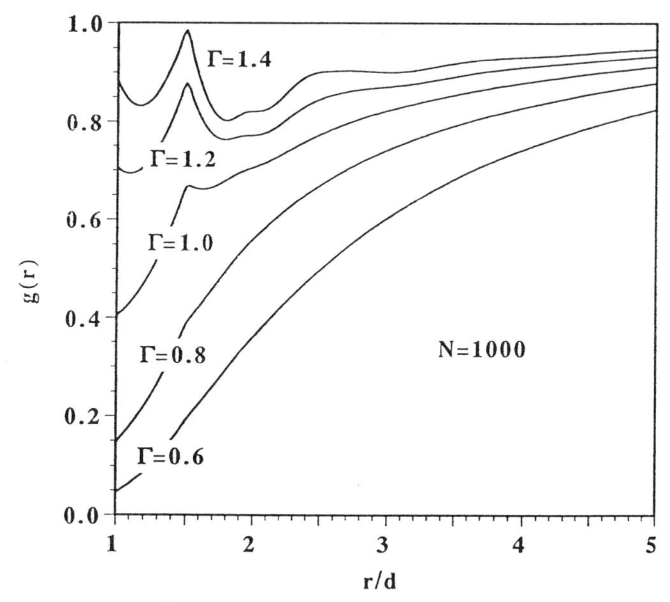

Figure 2. PRISM results[52] for the site–site $g(r)$ of a $N = 1000$, $\eta = 0.5$ SFC model liquid and several experimentally relevant choices of chain aspect ratio Γ.

representative results and refer the reader to the original literature for the details.[52]

Figure 2 shows the predicted $g(r)$ at meltlike density for $N = 1000$ repeat unit chains, $l_b/d = 0.5$, and a range of aspect ratios (of order unity) relevant to typical flexible polymers of experimental interest.[52] For these cases $g(r)$ is relatively featureless and slowly varying in rough accord with the simple Gaussian thread model behavior. There is both a local and global correlation hole, and $g(r) < 1$ for all r. These features are in qualitive accord with atomistic calculations. Moreover, the form of $g(r)$ appears to be a remarkably good, coarse-grained representation of the site-averaged correlations predicted by atomistic PRISM theory (see Sections III.A.4 and III.B) and atomistic simulations.[57–59] Such agreement is *not* because real polymers are Gaussian on all length scales as assumed by the thread model. Rather, it is the multitude of local chemical lengths, and thermal conformational disorder associated with chains composed of real monomers, which frustrates the development of well-defined solvation shells and positive correlation in $g(r)$.[52]

Another important structural feature in Figure 2 is that the local correlation hole is very sensitive to aspect ratio. As expected physically, it deepens as the chain becomes more flexible and less able to efficiently pack with neighboring polymers. This feature has important consequences for thermodynamic properties (e.g., cohesive energy density) and the miscibility of polymer mixtures.[52]

As the chain aspect ratio is significantly increased above $\Gamma \cong 1.4$, and/or the accessible site surface area is enhanced by increasing l_b/d, more well-defined solvation shells develop and "positive correlation" $[g(r) > 1]$ occurs. The extreme limit is the rigid rod polymer. The predicted packing of such models (not shown) begins to resemble a smeared version of the $g(r)$ of simple atomic liquids, particularly in the "tangent" SFC limit[55] where $l_b = d$. Thus, such a tangent model appears to be a *poorer* coarse-grained representation (relative to the $l_b/d < 1$ SFC models) of the $g(r)$ of real polymer fluids.

The tangent SFC model with chain aspect ratio of $\Gamma \cong 1.4$ has been studied in recent large-scale molecular dynamics and Monte Carlo simulations of dense melts by several groups.[51,60] Comparisons of PRISM theory predictions with these benchmark simulations has shown agreement at roughly the same level obtained for atomic and small molecule (RISM) liquids (e.g., 10–20% errors at contact and much better as r increases). This is significant since it shows that the standard site–site PY closure for hard-core fluids suffers no obvious loss of accuracy as the chains become longer. For linear chain solutions and melts interacting via pure hard-core potentials the site–site PY closure appears to be the most accurate closure for $g(r)$ as judged by overall comparison with computer simulations.[60] At the highest meltlike densities, Yethiraj and Schweizer[61] have shown that PRISM theory with the PY, HNC, and MS closures yield similar results for $g(r)$. However, the PY closure is the most robust since under certain density and chain length conditions the HNC, MS, and diagramatically proper closures[38–40] can fail to converge and/or result in extremely poor descriptions of collective structure and density fluctuations especially on long length scales. Theoretical arguments for this have been suggested.[61] Very recent simulation studies by Yethiraj[62] of hard-core fluids composed of rather stiff chains of larger aspect ratios (roughly $\Gamma > 2$) reveals that PRISM theory predicts the local (near contact) behavior of $g(r)$ very well, but longer range aspects associated with liquid layering and solvation shell structure is not accurately described.

An example of a comparison by Honnell et al.[55] of PRISM theory with molecular dynamics simulations[51] are shown in Figure 3. Details of the model are given elsewhere.[51,55] Briefly, a meltlike density was studied for $N = 50$–150 unit chains. The linear polymers were modelled as freely jointed beads with a purely repulsive, shifted Lennard-Jones interaction between all segment pairs. The corresponding chain aspect ratio is $\Gamma \cong 1.4$. PRISM theory with the PY closure (plus a standard correction

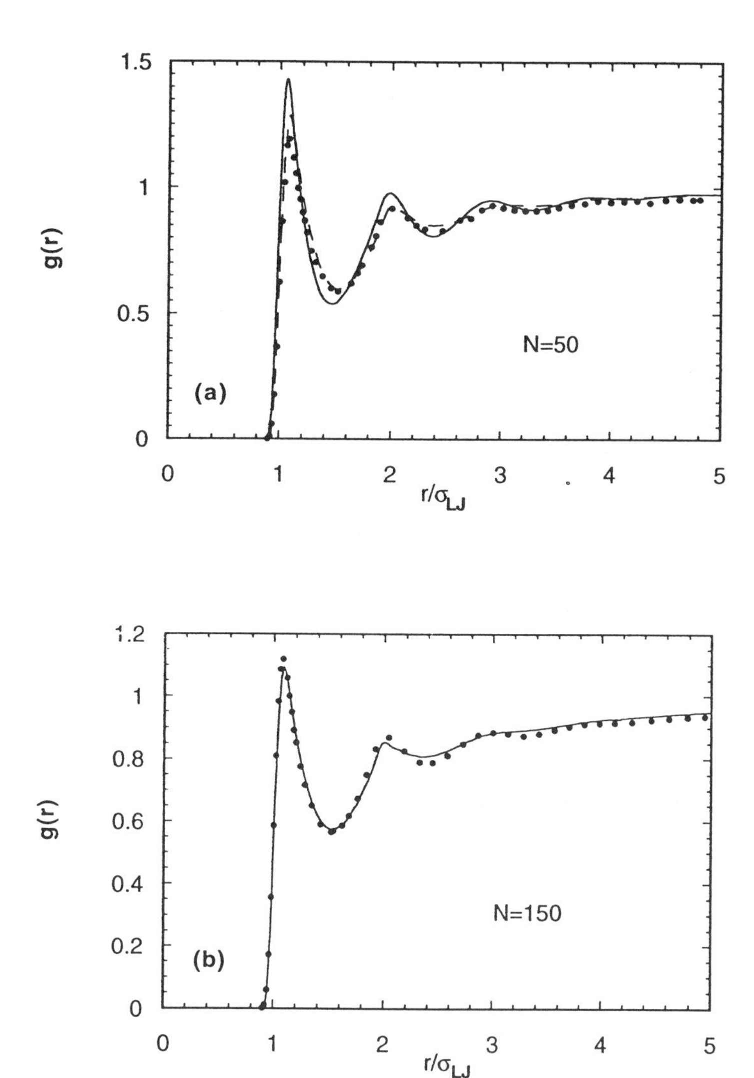

Figure 3. Intermolecular site–site radial distribution function for soft-core, repulsive Lennard-Jones chain liquids at $\eta = 0.464$ and (a) $N = 50$ and (b) $N = 150$. The circles are the molecular dynamics simulation results of Grest and Kremer.[51] Curves are PRISM predictions[55] based on the exact $\hat{\omega}(k)$ (solid line for $N = 50$) and the SFC model (dashed curve for $N = 50$ and solid curve for $N = 150$). Distances are scaled by the Lennard-Jones site diameter.

for repulsive force softness) was applied for two choices of the intramolecular structure factor $\hat{\omega}(k)$: (i) an SFC chain model with bending energy chosen to reproduce the simulated value of the chain end-to-end distance and (ii) the exact simulation result for the *single-chain* quantity $\hat{\omega}(k)$. The second approach involves the fewest statistical mechanical approximations and provides a precise check on the accuracy of the PRISM/PY theory for interchain packing. Errors of the size of 15% are found at small separations, which become much smaller as r increases. Calculation (i) is in the best agreement with the $N = 50$ simulation. This is partially fortuitous, that is, errors in PRISM theory and errors in the approximate calculation of $\hat{\omega}(k)$ have largely cancelled to yield nearly perfect agreement for $N = 50$, 100 (not shown), and 150. Calculations for a simpler ideal freely jointed chain model have also been performed[51,55] (not shown here) and are in the poorest agreement with simulation since this fully flexible ideal model ignores the very local intrachain excluded volume interactions between monomers separated by two bonds. Thus, chain size is *underestimated* leading to the strong underestimation of $g(r)$ locally.

The trends of $g(r)$ with decreasing fluid density are qualitatively similar to decreasing aspect ratio at melt density.[55] Single-chain conformational entropy becomes increasingly more important relative to interchain packing entropy as the fluid becomes more dilute, resulting in a $g(r)$ that is less structured with a much deeper *local* correlation hole. Many examples have been given in the literature.[6,25,55] However, as true for RISM theory of simple molecules,[8] the quantitative accuracy of PRISM theory is reduced as the fluid density is lowered even if the exact, simulated $\hat{\omega}(k)$ is employed.[60.61] Similarly, at fixed density and aspect ratio, the chains pack more poorly [less solvation shell structure and deeper hole locally in $g(r)$] as N is increased.[6,25,55] However, a stable long-chain limit is approached in the local region of $g(r)$, and this occurs more quickly as the density and/or chain aspect ratio is increased.

3. *Atomistic Models*

The structurally simplest polymer, and one of the most commercially important, is polyethylene. It consists of a linear chain of CH_2 units, which we model as single spherical sites in the *single-site* homopolymer spirit. There exist well-developed ideal rotational isomeric state chain models[13] where the bond rotational degrees of freedom are represented as discrete *trans* and *gauche* isomers. Numerical calculation of the required single-chain structure factor can be achieved via Monte Carlo simulation or using the recently developed computationally convenient approximate methods of McCoy and co-workers.[63]

The predictions of PRISM theory for melts of $(—CH_2—)_N$ chains based on purely hard-core intermolecular potentials have been numerically obtained, including a systematic study of the *n*-alkanes by Honnell et al. ($N = 4–20$).[64,65] Detailed comparisons with wide-angle x-ray scattering measurements of Narten and Harbenschuss have been carried out, and excellent agreement between theory and experiment has been demonstrated.[64–66] This agreement has motivated theoretical extensions that employ PRISM theory of the liquid as input to treat melt thermodynamic properties (PVT equation-of-state, compressibility, thermal expansion coefficient),[67] and the development and application of novel polymeric density functional theories of crystallization[68] and polymer near surfaces and interfaces. For the strongly first-order crystallization transition, it has been found by McCoy et al.[68] that the atomistically realistic description of polyethylene is required for a proper description of the phase transition.[68] This is not surprising since crystallization is a phenomenon exquisitely sensitive to local molecular structure and packing.

Figure 4 shows the predicted melt methylene–methylene $g(r)$ for a range of temperatures.[52] The experimental density and a temperature-

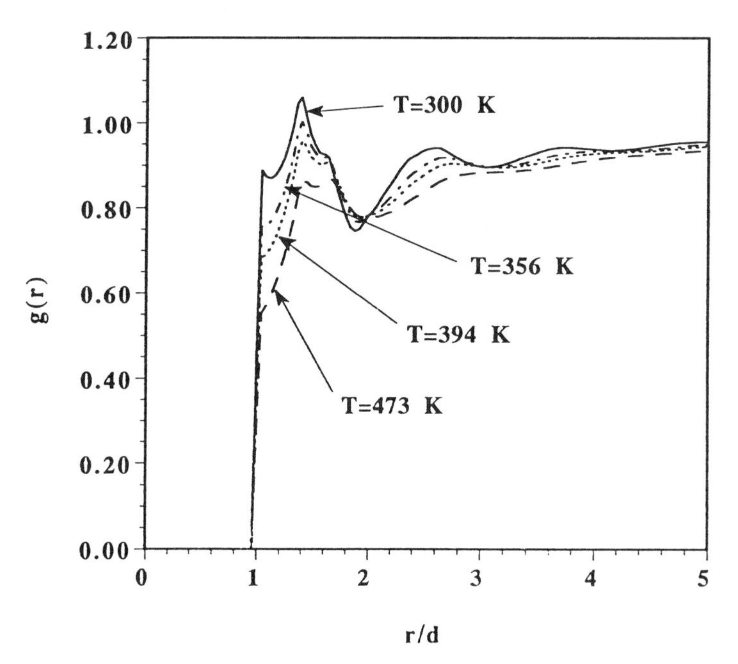

Figure 4. PRISM predictions for the site–site radial distribution function of an $N = 1000$ RIS model of a hard-core polyethylene melt at various temperatures.[52] Distances are scaled by the effective (T-dependent) hard-core diameter.

dependent hard-core diameter associated with the repulsive Lennard-Jones methylene interactions (computed according to the standard Barker–Henderson procedure)[5,11] has been employed in the calculations. Although there is some fine local structural details for $r < 20$ Å, they are rather weak and *g(r)* crosses over to the long-range correlation hole form for larger separations [$h(r) \propto r^{-1}$]. Random behavior corresponding to $g(r) = 1$ is attained only when $r > R_g$. As the temperature is raised, the reduction of local packing efficiency occurs due to the lower liquid density and enhanced conformational disorder (more twisted *gauche* ± conformers). Comparison of the predicted dimensionless collective structure factor, $\hat{S}(k)$, with wide-angle scattering data is shown in Figure 5.[64,66] Excellent agreement is obtained for the chemically sensible value of 3.9 Å for the methylene hard-core diameter. Comparable agreement between theory and experiment has been found for the entire alkane series.[65]

The most significant feature of Figure 5 is the strong first diffraction peak or "amorphous halo," which is influenced by *both* inter- and intramolecular pair correlations.[64–66] The very large angle scattering reflects single-chain correlations that are *input* to the theory. Agreement of the theoretical prediction for the collective structure factor at $k = 0$ with the measured data point is partially fortuitous since the attractive

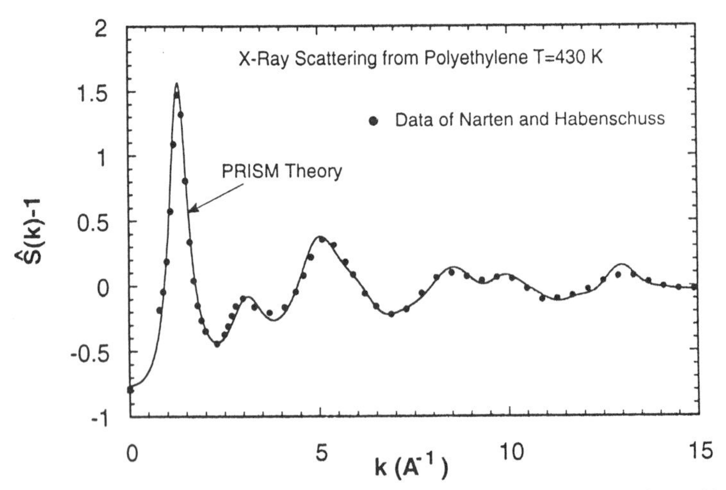

Figure 5. Dimensionless total structure versus absolute wave vector for a $N = 6429$ polyethylene melt just above its melting temperature.[64,66] The solid circles are the x-ray scattering data, and the line is the PRISM prediction based on a hard-core model with $d_{CH_2} = 3.9$ Å. The solid square at $k = 0$ represents the experimental value based on the measured isothermal compressibility and liquid density.

intermolecular interactions present in the real fluid have not been included in the calculation of this thermodynamic property.

The broad message of all the atomistic PRISM studies of the linear hydrocarbons is that the theory is capable of an essentially quantitive, ab initio description of melt structure for the structurally simplest case of $(\text{—CH}_2\text{—})_N$.

4. *Coarse Graining and Relationship of Different Chain Models*

The collective density fluctuation melt structure factor, $\hat{S}(k)$, has been computed for a wide range of single-site models and chain parameters.[6,25,55,64–66,68] There are two primary packing-related features of interest: the zero-angle scattering $\hat{S}(0)$ and the first strong diffraction peak. These basic density correlation features are qualitatively the same for all chain models since they are not intrinsically of polymeric origin. However, clear differences exist with rather well-defined trends. For example, at fixed fluid packing fraction, chain length, and chain persistence length both the inverse zero-angle scattering amplitude and the amplitude and sharpness of the amorphous halo *increase* as the monomer structural model includes more local structural features.[55,64–66] Within the SFC model,[55] these features also increase as the chain aspect ratio increases and/or N decreases since local packing is enhanced, although a saturation behavior occurs for sufficiently large Γ and/or N.

The ability to construct coarse-grained models in such a way as to mimic, or reproduce, selected properties of real polymers or atomistic computations is a goal of both computational and conceptual value. Recently, some progress has been made in this direction using PRISM theory.[52] Briefly, in its minimalist implementation the key ideas employed to *select* the parameters of the coarse-grained model are as follows: (1) require an identical aspect ratio as the atomistic model or real experimental polymer: this parameter has been argued to be the primary one in determining the average interchain packing efficiency. (2) Set N equal to the degree of polymerization on a monomer basis. (3) Choose the reduced density such that $\hat{S}(0)$ is equal to the experimental value; part of the motivation here is the intriguing direct connection between $\hat{S}(0)$ and the *local g(r)* suggested by the Gaussian thread analytic results and empirically verified for more realistic chain model studies. In the initial studies based on the SFC model, a purely hard-core interaction has been employed with a common value of d and $l_b/d = 0.5$.

The results of this approach as applied to polyethylene are shown in Figure 6. Remarkable agreement between the atomistic model *g(r)* and the SFC *g(r)* is found. Moreover, even the Gaussian thread result seems reasonable as an "interpolation" through the atomistic *g(r)*. For inte-

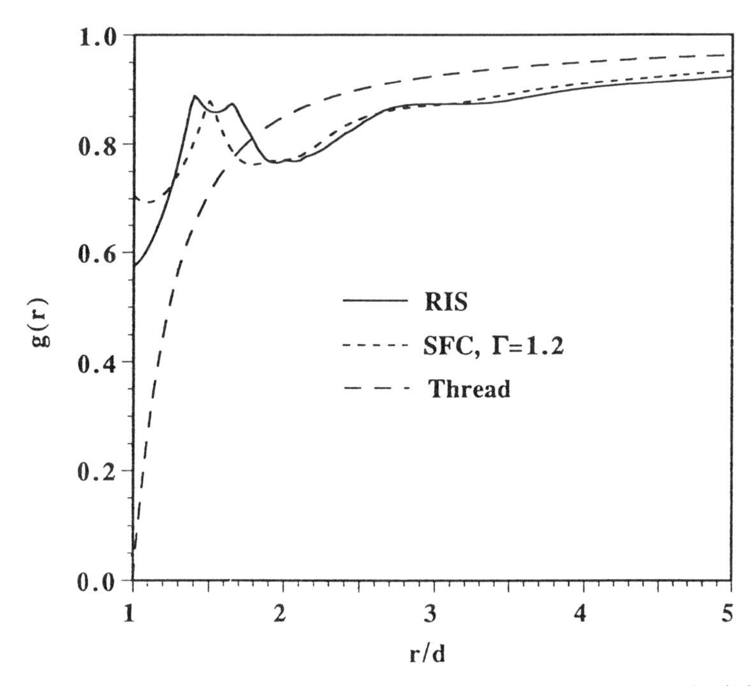

Figure 6. Predicted interchain radial distribution function for a hard-core polyethylene melt described by three single-chain models: atomistic RIS at 430 K, overlapping ($l/d = 0.5$) SFC model with appropriately chosen aspect ratio and site number density (see text), and the Gaussian thread model (shifted horizontally to align the hard core diameter with the value of $r/d = 1$).

grated thermodynamic quantities, such as the cohesive energy density associated with intermolecular attractive forces, close agreement is found between all three approaches.[52]

Generalization of this mapping scheme to polymers of more complex monomer structure, such as polypropylene, also yield promising results[52] for the *chain-averaged* carbon–carbon radial distribution $g_{av}(r)$. Although there will surely be systems and phenomena where such a "preaveraging" of chemical structure detail will incur significant (and perhaps fatal) errors, such a mapping scheme allows one to construct and study coarse-grained SFC models for a very large number of materials. Thus, this approach has significant potential for making PRISM theory a "molecular design tool" in the sense that many possible material systems can be quickly studied based on input of a small amount of conformational and related information. This approach has been recently implemented by Schweizer and co-workers with considerable success for understanding and predicting melt solubility parameters and polyolefin blend miscibility.[52,69]

B. Multiple-Site Vinyl Polymers

In order to capture the nonspherical nature of monomers for polymers more complicated than polyethylene, one can use additional independent sites to build the monomer structure. An example is shown in Figure 7 for a vinyl polymer. Note that the sites can be overlapping to maintain the correct bond lengths, angles, and steric volume of the atoms or groups of atoms making up each site. We use a united atom scheme[26] to construct a vinyl monomer from three independent sites where site A represents a CH_2 group, site B represents a CH group, and site C depicts a side chain substituent. PRISM theory in Eqs. (2.3)–(2.5) now yields six integral equations for the six independent radial distribution functions $g_{AA}(r)$, $g_{BB}(r)$, $g_{CC}(r)$, $g_{AB}(r)$, $g_{AC}(r)$, and $g_{BC}(r)$, which characterize the intermolecular packing.

As a first approximation[26] one can model the vinyl polymer as a freely jointed, tangent hard sphere chain as depicted on the second line of Figure 1. Thus each bond (of fixed length) is completely flexible with each site, including the side group site C, acting as a universal joint. Invoking the Flory ideality hypothesis, the intramolecular structure functions $\hat{\Omega}_{\alpha\gamma}(k)$ in Eq. (2.4) become[26]

$$\hat{\Omega}_{\alpha\gamma}(k) = \tilde{\rho} \sum_{i\in\alpha} \sum_{j\in\gamma} \left[\frac{\sin(kd_{\alpha\gamma})}{kd_{\alpha\gamma}} \right]^{m(i,j)} \tag{3.5}$$

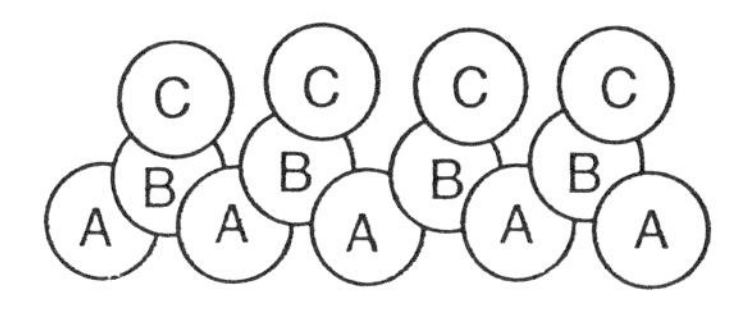

Vinyl Chain

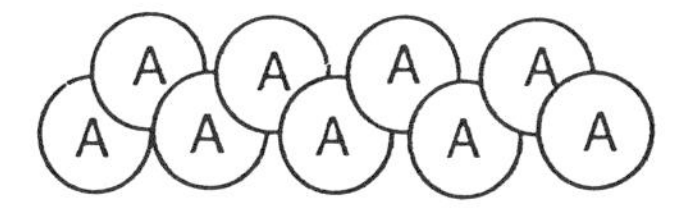

Polyethylene

Figure 7. Schematic representation of a three-site model for describing vinyl polymers contrasted with a one-site model for polyethylene.

where $m(i, j)$ is the number of bonds between a pair of sites i and j on the same chain. The summations in Eq. (3.5) can be performed in a straightforward manner and are detailed in Ref. 26.

In reality it is known from computer simulation[51] that the intramolecular excluded volume is not completely screened out in a polymer melt, even at high density. Overall the chains will exhibit ideal scaling with $R_g \sim N^{1/2}$ characteristic of a chain with no long-range repulsions, but the chain expands locally due to intramolecular overlaps. This is confirmed in self-consistent calculations[43–47] as discussed in Section VIII. In order to quantitatively compare PRISM calculations for the intermolecular structure with computer simulations, it is necessary to compensate for this local chain overlap. This can be accomplished by using the intramolecular structure functions $\hat{\Omega}_{\alpha\gamma}(k)$ obtained from the full many-chain simulation. Alternatively one can compute $\hat{\Omega}_{\alpha\gamma}(k)$ from a single-chain calculation or simulation in which only local, short-range repulsions are included. Figure 8 shows selected components of the intramolecular structure factor matrix for vinyl chains of 33 monomers obtained from Monte Carlo simulations of Yethiraj and co-workers.[70] The points are from the full, many-chain simulation, whereas the curves were obtained from a single-chain Monte Carlo simulations of Yethiraj and co-workers. The points are from the full, many-chain Monte Carlo simulation in

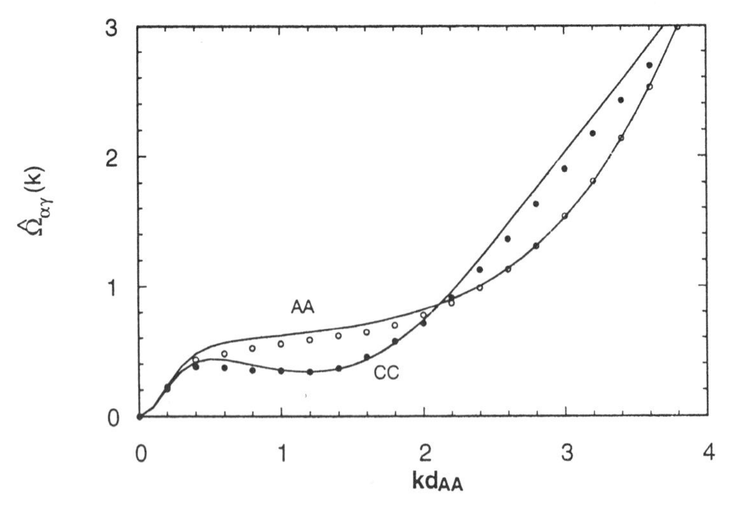

Figure 8. Predictions for two diagonal partial structure factors of vinyl chain melts of 33 monomers. The points are from the multiple-chain Monte Carlo simulations of Yethiraj and co-workers.[70] The curves are from single-chain simulations[70] in which repulsive interactions between sites separated more than 2 bonds are screened (set to zero). The BB structure factors are similar to AA and were omitted for clarity.

which sites separated by two bonds are prohibited from overlapping. Interactions between sites separated by more than two bonds are set to zero based on the physical expectation that long-range excluded volume is screened under melt conditions. It can be seen from this figure that the intramolecular structure of a chain, with local repulsions only is an excellent approximation of a chain in a melt.

Using these results for $\hat{\Omega}_{\alpha\gamma}(k)$ as input to PRISM theory, the six coupled equations resulting from Eqs. (2.2) and (2.5) can be solved numerically using a straightforward Picard iteration scheme.[26] The intermolecular packing of vinyl chains of 33 monomers is compared with the simulations of Yethiraj et al.[70] in Figure 9. While the agreement is not quantitative, it can be seen that the PRISM theory certainly captures the essential features of the intermolecular packing. The comparison with simulation in Figure 9 was carried out at a packing fraction characteristic of a concentrated solution ($\eta = 0.35$). We anticipate that the agreement between PRISM and theory would improve as the packing fraction increases to $\eta \sim 0.5$ characteristic of a neat polymer melt.

It is instructive to examine the details of the six intermolecular radial distribution functions in Figure 9. Note that on long length scales ($R_g \sim 3$) all the $g_{\alpha\gamma}(r)$ are essentially identical in the correlation hole regime. This verifies that the local monomer architecture does not affect the packing on intermediate and long length scales. On the other hand, on short length scales near contact, significant local packing differences are seen between the different types of sites making up each monomer. We observe from Figure 9a that $g_{\mathrm{CC}}(r)$ is much larger than all the other local correlations. This is a consequence of the fact that the C sites are situated on the outside of the chain and hence can easily approach each other near contact.

By contrast, $g_{\mathrm{BB}}(r)$ is small near contact because of screening effects. The B site is located on the chain backbone underneath the C groups and therefore is strongly shielded by the surrounding sites. These qualitative screening ideas[26,70] can be carried further to explain the relative order of all the radial distribution functions near contact: $g_{\mathrm{CC}}(d) > g_{\mathrm{AC}}(d) > g_{\mathrm{BC}}(d) > g_{\mathrm{AA}}(d) > g_{\mathrm{AB}}(d) > g_{\mathrm{BB}}(d)$. Not surprisingly the local packing, characterized by the six different $g_{\alpha\gamma}(r)$, is a sensitive function of the detailed monomeric structure. For example $g_{\mathrm{CC}}(r)$ is seen[26,70] to systematically increase near contact when the hard-core diameter of the C site is increased. For typical nonpolar van der Waals interactions the attractive interactions between sites are spatially short range. For this reason one expects the local intermolecular packing details are important in determining the thermodynamic properties (e.g., cohesive energy) of the polymer liquid.

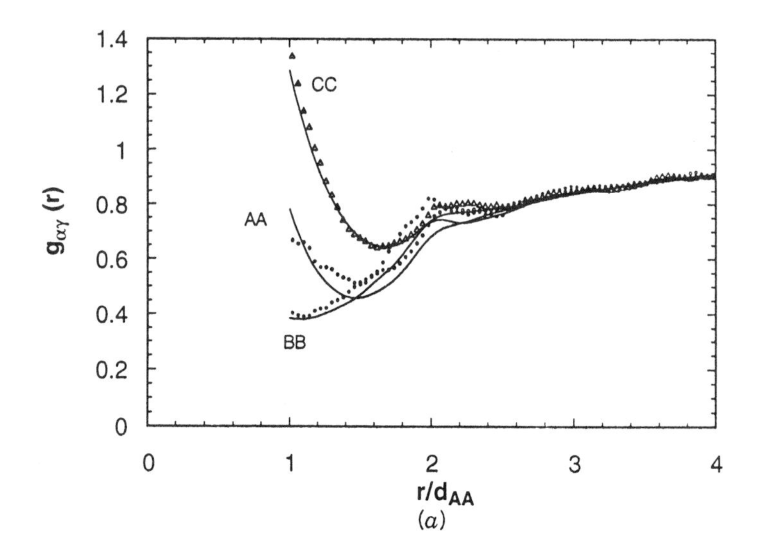

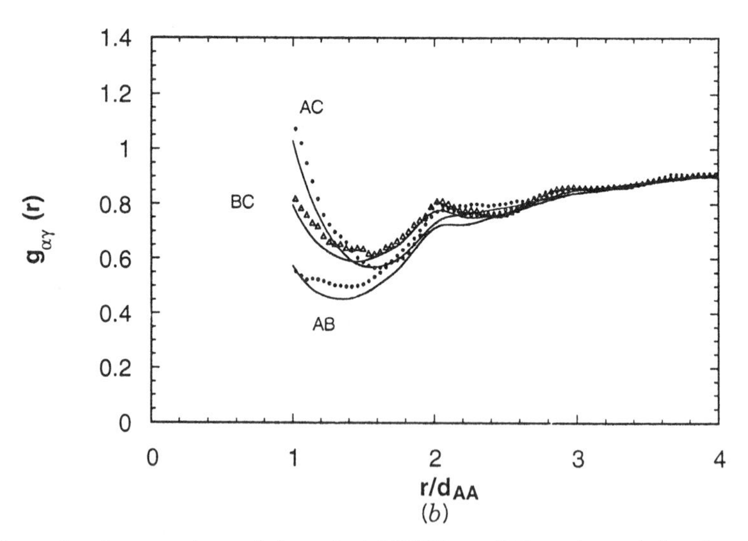

Figure 9. A comparison of theoretical PRISM predictions (curves) for the radial distribution functions with Monte Carlo simulations (points).[70] The simulations were performed on vinyl chain melts of $N = 33$ monomers at a packing fraction of 0.35. Note the shielding effects at short distances. (a) The diagonal components AA, BB, and CC of the correlation functions. (b) The off-diagonal components AB, AC, and BC of the correlation functions.

The structure of a polymer melt can be probed by x-ray or neutron scattering experiments. The density of scattering $I(k)$ is given by

$$I(k) = \sum_{\alpha\gamma} b_\alpha b_\gamma \hat{S}_{\alpha\gamma}(k) \tag{3.6}$$

where b_α is a scattering cross section of species α, and the $\hat{S}_{\alpha\gamma}(k)$ are the partial structure factors making up a structure factor matrix defined analogously to Eq. (3.3)

$$\begin{aligned}\hat{\underset{\sim}{S}}(k) &\equiv \hat{\underset{\sim}{\Omega}}(k) + \hat{\underset{\sim}{H}}(k) \\ &= (\underset{\sim}{1} - \hat{\underset{\sim}{\Omega}}(k)\hat{\underset{\sim}{C}}(k))^{-1}\hat{\underset{\sim}{\Omega}}(k)\end{aligned} \tag{3.7}$$

The second equality in Eq. (3.7) follows from Eq. (2.2). The summations in Eq. (3.6) run over the ν-independent sites making up the monomer. Assuming for the moment that the scattering cross sections of each site are equal, then the scattering intensity of a three-site vinyl polymer melt is proportional to the average structure factor defined according to

$$\hat{S}_{\text{av}}(k) = \frac{1}{9} \sum_{\alpha\gamma} \hat{S}_{\alpha\gamma}(k) \tag{3.8}$$

$\hat{S}_{\text{av}}(k)$ is plotted in Figure 10 for a tangent hard-sphere, freely jointed

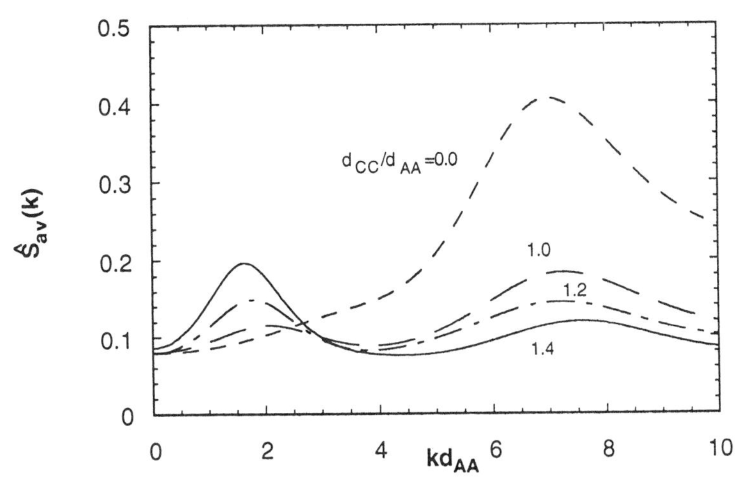

Figure 10. The average total structure factor calculated from PRISM theory for a melt of freely jointed, vinyl chains having $N = 50$ monomers.[26] Each curve was computed for different diameters of the C (side chain) site. The dashed curve ($d_c = 0.0$) refers to a one site, polyethylene-like chain. Note the emergence of a low angle prepeak at $kd_{\text{AA}} \sim 1.5$–2.0 as the size of the side group is increased.

chain melt[26] for various diameters d_{CC} of the side group site labeled as C. Note that when there is no side chain group ($d_{CC} = 0$) the vinyl chain reduces to a one-site, freely jointed polyethylene-type chain. In this case the main structural feature at low wave vector is seen from Figure 10 to occur at $kd_{AA} \approx 7$ corresponding to the nearest neighbor distance ($\approx 2\pi/k$). When a side group substituent is added, however, a new peak grows in below $kd_{AA} \approx 2$. This low-angle peak grows in intensity and shifts to smaller wave vectors as the size of the side group increases. Curiously, this low-angle feature corresponds to packing distances in real space of approximately three hard-core diameters. Such a "prepeak" has been reported from some vinyl polymer melts such as polystyrene;[71] however, recent x-ray scattering measurements[72] on isotactic polypropylene show no indication of a prepeak. Examination[26] of the partial structure factors reveals that the prepeak is arising from relatively long-range interchain correlations between backbone carbon centers that are modulated by the presence of the side groups.

It should be emphasized that Figure 10 was calculated for an idealized freely jointed chain melt in which the site diameter and bond length are the same. In order to make quantitative contact with experiments, it is necessary to more faithfully represent the monomer architecture through the intramolecular functions $\hat{\Omega}_{\alpha\gamma}(k)$. A model that captures more of the local chemical structure of real polymer chains is the well-known rotational isomeric state model.[13] In order to mimic a chain in a theta solvent or a melt, intramolecular repulsions are included between sites separated by less than or equal to four bonds (the "pentane effect").[13]

Detailed PRISM calculations were performed by Rajasekaran et al.[32] on the stereochemically regular *isotactic* polypropylene (i-PP) of $N = 200$ monomers employing the rotational isomeric state model of Suter and Flory[73] to compute the required $\hat{\Omega}_{\alpha\gamma}(k)$. The characteristic ratio of a linear chain is defined as $C_\infty = \langle R^2 \rangle / N_b l_{CC}^2$, where N_b is the number of backbone carbon–carbon bonds of length l_{CC}. According to the Suter–Flory rotational isomeric state calculation for i-PP, $C_\infty \cong 4.0$ at 473 K. SANS measurements,[74] however, indicate that $C_\infty = 6.2$ for i-PP in the melt state. In order to compensate for this discrepancy in chain dimensions, the rotational state energies (or equivalently the temperature $T = 286$ K) was rescaled to obtain the experimental C_∞. Although the moments of the distribution can be computed in closed form, the single-chain structure functions cannot be computed conveniently for the rotational isomeric state model. Thus, a Monte Carlo simulation of a single chain was employed to obtain the six functions $\hat{\Omega}_{\alpha\gamma}(k)$.

The six intermolecular radial distribution functions for i-PP were then deduced from PRISM calculations using the single-chain simulation

11a, and the off-diagonal components are given elsewhere.[32] It can be seen that the i-PP correlation functions are qualitatively similar to the idealized chain results in Figure 9. An interesting feature of the BB radial distribution function for the atomistically realistic model is that because of shielding effects, and the added constraints of the local chain architecture, $g_{\mathrm{BB}}(r)$ approaches zero at a distance greater than the $d_{\mathrm{BB}} = 3.9$ Å hard-core diameter. In other words, because of local steric constraints,

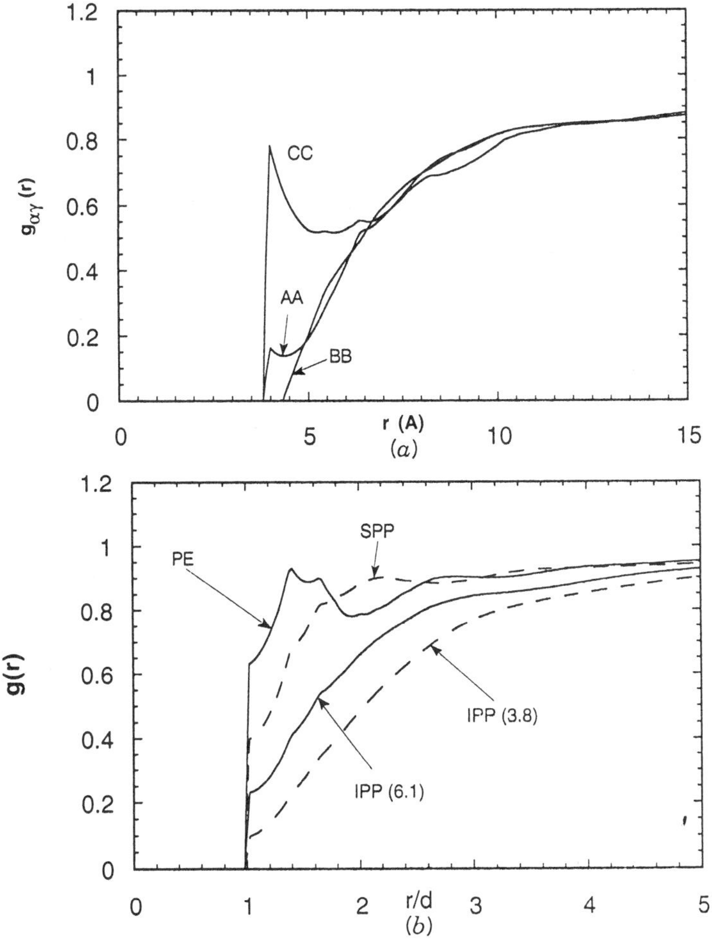

Figure 11. PRISM predictions for hard-core atomistic RIS models of polyolefins. (a) The three diagonal radial distribution functions of isotactic polypropylene.[32] (b) A comparison of chain *averaged* site–site radial distribution functions at 473 K for $N_b = 400$ models of polyethylene, isotactic polypropylene, and syndiotactic polypropylene.[52] The characteristic ratio, C_∞, of the RIS models employed for PP are shown in parentheses.

the CH sites of different chains are restricted from coming into direct contact.

The resulting x-ray scattering pattern of the i-PP melt was found to be in some disagreement with the wide-angle scattering measurements of Londano et al.[72] in the region $k \approx 1\ \text{Å}^{-1}$. This led Curro and co-workers[75] to introduce further realism into the single-chain structure by performing single-chain Monte Carlo simulations in which the internal rotational angles ϕ were allowed to vary in a continuous manner. This additional realism in the local chain architecture leads to excellent agreement with scattering measurements on i-PP melts, as discussed elsewhere.[75]

A comparison at *fixed* liquid packing fraction of the *chain-averaged* carbon–carbon radial distribution function, $g_{av}(r)$, for polyethylene ($C_\infty \cong 7$), isotactic polypropylene (with two values of C_∞), and syndiotactic polypropylene (s-PP) are shown in Figure 11.[52] The polymers i-PP and s-PP are regular multiple-site homopolymers of different stereochemistry and significantly different characteristic ratios were used in the calculation ($C_\infty \cong 10.5$ for s-PP). The clear differences among all the systems demonstrates the sensitivity of local packing in polymer melts to monomer shape, stereochemistry or tacticity, and backbone stiffness. Finally, as discussed in depth elsewhere, the $l_b/d = 0.5$ SFC model of section III.A.2 reproduces the structural variations quite well based on the effective aspect ratio mapping idea.[52]

C. Thermodynamics

Having determined the structure of the polymer liquid, it is in principle possible to compute most thermodynamic properties of interest.[5,8] Whereas the structure or radial distribution functions at liquid density are primarily controlled by the repulsive part of the intersite potentials, thermodynamic quantities will also be sensitive to the attractive potentials. In the case of a one-component melt, thermodynamic quantities of interest include the pressure P, isothermal compressibility κ, and the internal or cohesive energy U. Since in general one theoretically knows $g(r)$ only approximately, the thermodynamic properties derived from structure will be approximate. Moreover, integral equation theory leads to *thermodynamically inconsistent* results in the sense that the predictions depend on the particular thermodynamic route used to relate the thermodynamic quantity to the structure.[5,8]

1. *Equation of State*

Thermodynamic inconsistency is particularly apparent for the pressure of polymer fluids.[67] There are at least three routes that relate the pressure to the structure. Perhaps the easiest method to implement is the so-called compressibility route:

$$\frac{P}{k_B T} = \int_0^{\rho_m} \frac{d\rho}{\hat{S}_{av}(0)} \tag{3.9a}$$

where $\rho_m k_B T \kappa = \hat{S}_{av}(0)$. The analog of the virial route of monatomic liquids for the pressure of a molecular liquid is the *free energy charging formula.* For hard-core potentials one can write[5,8]

$$\frac{F - F_0}{V k_B T} = 2\pi\rho_m^2 \sum_{\alpha\gamma} d_{\alpha\gamma}^3 \int_0^1 g_{\alpha\gamma}^{(\lambda)}(\lambda d_{\alpha\gamma})\lambda^2 \, d\lambda \tag{3.9b}$$

where $g_{\alpha\gamma}^{(\lambda)}(\lambda d_{\alpha\gamma})$ is the contact radial distribution function for a fluid composed of molecules with hard-core sites of diameter $\lambda d_{\alpha\gamma}$; F and F_0 are the Helmholtz free energies of the fluid of interest and a corresponding ideal gas, respectively. In Eq. (3.9b) the hard-core diameter is turned on as the "charging parameter" λ changes from 0 to 1. The pressure follows from differentiation of F with respect to volume.

Yethiraj and co-workers[67] calculated the hard-core contribution to the equation-of-state of polyethylene by various thermodynamic routes using PRISM theory. It can be seen from the results plotted in Figure 12 that very large differences are found between the compressibility and charging routes. Qualitatively similar results are seen for ethane and n-butane;[67] however, the thermodynamic inconsistency appears to increase significantly with N. One contributing reason for this large, N-dependent thermodynamic inconsistency is traceable to the fact that RISM theory,[7,8] unlike the PY theory[5] for atomic liquids, is not exact in the low-density limit. Both Eqs. (3.9a) and (3.9b) effectively integrate the structure of the fluid over the complete range of density. A route that uses only structural information at liquidlike density, where RISM theory is accurate, might produce better results.

Based on an argument by Percus,[76] another route to the pressure was proposed by Dickman and Hall[77] making us of the density of sites in a nonuniform molecular liquid at a hard wall $\rho_w(0)$:

$$P = \rho_w(0) k_B T \tag{3.9c}$$

Yethiraj and Hall[78] developed a "wall PRISM" theory to compute $\rho_w(0)$. The wall PRISM theory prediction for hard sphere polyethylene is also shown in Figure 12. It can be seen that pressures intermediate between the charging and compressibility route predictions are found.

Also plotted in Figure 12 is the predictions of a continuous space Flory–Huggins type of approach: the generalized Flory dimer (GFD)

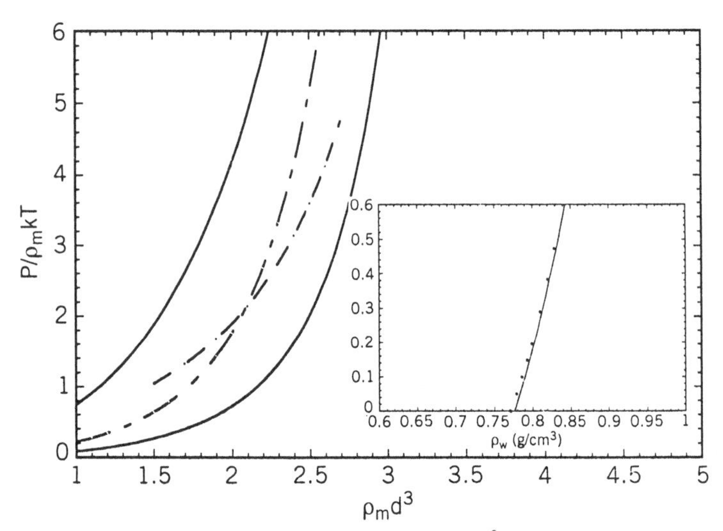

Figure 12. The hard-sphere equation-of-state ($d = 3.9$ Å) as a function of reduced fluid density computed[67] for polyethylene at $T = 430$ K and $N = 6429$ by various thermodynamic routes: free energy (upper solid), compressibility (lower solid), wall (dashed), and GFD (short/long dash). The inset includes attractions by perturbation theory using the GFD curve as the reference system; the points represent experimental results.[83]

equation of state of Dickman et al.[79] suitably modified by Yethiraj and co-workers to rotational isomeric state chains of overlapping sites.[67] Monte Carlo studies[79] have documented the accuracy of the GFD equation of state for chain molecule liquids at high densities. It can be seen from Figure 12 that the wall PRISM and GFD predictions for hard-core polyethylene chains are in reasonable accord at densities $\rho_m d^3 \sim 2$ characteristic of polyethylene melts. It can be seen from the slopes of the curves, however, that the isothermal compressibility from wall PRISM calculations is somewhat higher than from the GFD model. Recently, density functional theory has been combined with PRISM theory by McCoy and co-workers[80] as an alternative to wall-PRISM in computing the density profile of sites near a wall. This approach avoids the wall PRISM assumption of intramolecular chain ideality in the vicinity of the wall. Preliminary calculations[80] for the pressure by this approach are in close agreement with corresponding GFD calculations. Direct comparisons of PRISM predictions for the equation of state of hard-core *n*-alkane fluids with simulations have been recently carried out by Yethiraj,[81a] and for the interchain structure by Dodd and Theodorou.[81b]

Although the intermolecular packing of a dense polymeric melt is generally believed to be controlled by the repulsive part of the potential, attractions will have a strong influence on the equation of state. The conceptually and computationally simplest way to incorporate the effects of attractions on the pressure is through thermodynamic perturbation theory[5] about a hard-core reference system. For convenience we employ the Barker–Henderson[5,82] version of perturbation theory in which the intersite Lennard-Jones potential $v(r)$ is divided into a repulsive branch $v_0(r)$ and an attractive branch $v_a(r)$:

$$\begin{aligned} v(r) &= 4\varepsilon[(\sigma/r)^{12} - (\sigma/r)^6] \\ v_0(r) &= v(r) \qquad r \leq \sigma \\ &= 0 \qquad r > \sigma \\ v_a(r) &= 0 \qquad r \leq \sigma \\ &= v(r) \qquad r > \sigma \end{aligned} \tag{3.10}$$

The Helmholtz free energy can then be written to first order as

$$\frac{F}{k_B T} = \frac{F_{HS}}{k_B T} + \frac{1}{2}\rho_m^2 \int \frac{v_a(r)}{k_B T} g_0(r)\, d\mathbf{r} + \cdots \tag{3.11}$$

where F_{HS} and $g_0(r)$ are the free energy and intermolecular radial distribution function of the corresponding hard core reference system. Equation (3.11) is written for a single-site monomer but is easily generalized to monomers consisting of multiple sites. The pressure is then found by differentiation of Eq. (3.11) with respect to volume. The optimum diameter d of the hard-core reference system is given by

$$d = \int_0^\infty \{1 - \exp[-v_0(r)/k_B T]\}\, dr \tag{3.12}$$

Any of the four hard-core equation-of-state curves in Figure 12 could be used in conjunction with Eq. (3.11) to obtain the pressure of a polyethylene melt at any desired temperature. The method that appears to be the most accurate is to use the GFD equation of state for the reference system and PRISM theory for the structure $g_0(r)$ of the reference system. The results[67] of this procedure are shown in the inset of Figure 12 along with experimental PVT data of Olabisi and Simha.[83] In calculating the polyethylene pressure curve, the hard-core diameter d was maintained at 3.9 Å in order to be consistent with x-ray scattering measurements[64,66] on polyethylene at 430 K. The Lennard-Jones well depth parameter was then adjusted in order to *fit* the experimental data; this procedure yields $\varepsilon/k_B = 38.7$ K, which fixes $\sigma = 4.36$ Å according to

Eq. (3.12). It can be seen from the inset in Figure 12 that excellent agreement is obtained with experiment. However, this approach is not completely satisfying since PRISM theory is not used for the equation of state of the reference system. On the other hand, the simulation studies of Yethiraj and Hall have shown that the use of the PRISM theory $g_0(r)$ leads to an accurate prediction of the attractive potential contribution to the pressure *within* a perturbative HTA framework.[84]

Reasonable ab initio results have also been obtained for the thermal expansion coefficient and isothermal compressibility $[\hat{S}(k=0)]$ of Polyethylene melts.[52,67] The latter was computed using the experimental T-dependent density and the assumed dominance of soft repulsive forces. The resulting $\hat{S}(0)$ was found to be roughly 20% larger than the experimental values, although excellent agreement was obtained for the relative temperature dependence[52] over the entire experimental range of $T = 380$–525 K.

Finally, analytic predictions for the osmotic pressure of polymers in good and theta solvents can be derived based on the Gaussian thread model, PRISM theory, and the compressibility route.[30] The qualitative form of the prediction for large N is[54] $\beta P \propto (\rho\sigma^2)^3$, which scales as ρ^3 for theta solvents and $\rho^{9/4}$ for good solvents. Remarkably, these power laws are in complete agreement with the predictions of scaling and field-theoretic approaches and also agree with experimental measurements in semidilute polymer solutions.[2–4]

2. *Melt Solubility Parameters*

The internal or cohesive energy density U is also a useful thermodynamic parameter for polymer melts. It is defined as[85]

$$U = \frac{1}{\nu^2} \sum_{\alpha\gamma} \rho_\alpha \rho_\gamma \int v^a_{\alpha\gamma}(r) g_{\alpha\gamma}(r)\, d\mathbf{r} \tag{3.13a}$$

where the integration is carried out over the attractive branch of the potential in Eq. (3.10). In first-order perturbation theory the radial distribution function is approximated by its reference hard-core melt value $g^0_{\alpha\gamma}(r)$ thereby yielding

$$U \cong \frac{1}{\nu^2} \sum_{\alpha\gamma} \rho_\alpha \rho_\gamma \int v^a_{\alpha\gamma}(r) g^0_{\alpha\gamma}(r)\, d\mathbf{r} \tag{3.13b}$$

In the absence of correlations, or the random mixing limit, the radial distribution functions are all unity. In this "mean-field" limit the cohesive energy density reduces to $U_{\mathrm{MF}} = \frac{-32}{9} \pi \varepsilon \rho_m^2 \sigma^3$ for the Lennard-Jones potential. Note that the simple quadratic scaling of U with density is as expected for a mean field theory which accounts only for the number of (random) pairwise interactions.

The melt solubility parameter δ can be computed from[85]

$$\delta = \sqrt{-U} \tag{3.14}$$

For small molecule liquids δ can be measured directly from the heat of vaporization. For polymer melts the solubility parameter can only be indirectly estimated from solubility data in various solvents,[86] group contribution tables,[87] or model-dependent fitting of PVT measurements.[88]

The cohesive energy density for polyethylene as a function of temperature has been computed from Eq. (3.13b).[52] In agreement with experiment, $|U|$ was found to decrease nearly linearly with temperature.[89] This trend arises from the fact that the correlation hole of polyethylene deepens as the temperature is increased due to the combined effects of decreased density, and an increase in the number of *gauche* states of the polyethylene (PE) chain backbone (see Fig. 4). The magnitude of the predicted solubility parameter is in good agreement with experimentally inferred values for polyethylene in the range 15–19 $(\mathrm{J/cm^3})^{1/2}$ based on PVT measurements[45] and group contributions.[87]

The cohesive energy of isotactic polypropylene has also been calculated from Eq. (3.13a) using the intermolecular structural information in Figure 11 for the hard-core system. The united atom Lennard-Jones potentials of Jorgensen and co-workers[90] for CH_2, CH, and CH_3 groups with $\sigma \cong \sigma_{\alpha\gamma}$ were employed, along with appropriate values for the CH_2 group well depth energy $\varepsilon \equiv \varepsilon_{AA}$ and relative values for the CH_3 and CH groups of

$$\lambda_1 = \sqrt{\varepsilon_{CC}/\varepsilon_{AA}} \qquad \lambda_2 = \sqrt{\varepsilon_{BB}/\varepsilon_{AA}} \tag{3.15}$$

Based on the Jorgensen parameters,[90] the relative solubility parameter is predicted[52] to be $\delta_{PE}/\delta_{i\text{-}PP} = 1.24$. If as a simplifying approximation one sets $\lambda_1 = \lambda_2 = 1$, then $\delta_{PE}/\delta_{i\text{-}PP} = 1.26$ is predicted.[52] These predicted ratios are in good agreement with $\delta_{PE}/\delta_{i\text{-}PP} = 1.20$ inferred by Rodgers et al.[87] from equation-of-state data for isotactic polypropylene. For the case of syndiotactic polypropylene one finds[52] that $\delta_{PE}/\delta_{s\text{-}PP} = 1.07$ under the assumption $\lambda_1 = \lambda_2 = 1$. Thus, polypropylene is predicted to have a smaller solubility parameter than polyethylene because of differences in packing on local length scales (see Fig. 11b).

Schweizer and co-workers[52] have estimated the cohesive energy of a range of polymers of varying chain architecture using both the single-site semiflexible chain model and the analytic Gaussian thread model. Model calculations of the reduced solubility parameter are shown in Figure 13. As discussed in Section II.B.2, for experimental applications a system-specific effective aspect ratio was employed to map the semiflexible chain model to a particular polymer of interest. As described in detail elsewhere,[52] the relative solubility parameters computed with the SFC approach are in good agreement with the atomistic values quoted above for both polyethylene and the various tacticities of polypropylene. Moreover, predictions for many other hydrocarbon polymer melts have also been made and compared with experimental solubility parameters.[87,88,91] Good agreement is found that provide a simple understanding of how polymer structure influences melt solubility parameters.[52]

The Gaussian thread model, in conjunction with a Yukawa form for the attractive interchain potential of spatial range a:

$$v(r) = -\varepsilon a \frac{e^{-r/a}}{r} \qquad \varepsilon > 0 \tag{3.16}$$

yields a simple analytic expression for the reduced solubility parameter. The result can be written in several alternative, but equivalent, forms[52]

$$\tilde{\delta} \equiv \frac{\delta}{\delta_{\mathrm{MF}}} = \frac{1}{\sqrt{1 + \dfrac{3}{\pi \rho_m \sigma^2 a}}} = \frac{1}{\sqrt{1 + \dfrac{\sigma}{a}\sqrt{\dfrac{\hat{S}(0)}{12}}}} = \frac{1}{\sqrt{1 + \dfrac{d}{2a\eta}\Gamma^{-2}}} \tag{3.17}$$

This form is plotted in Figure 13 for two experimentally relevant choices[52] of the parameter $d/2a\eta$. The prediction of a direct connection between polymer density, aspect ratio [or packing length $(\rho_m \sigma^2)^{-1}$], and spatial range of the attractive potential is intuitively reasonable. Equation (3.17) has been shown by Lohse to provide an excellent representation of experimentally deduced solubility parameters of polyolefin melts.[92]

As a cautionary remark, we note that significant quantitative differences between the SFC and Gaussian thread model predictions are evident in Figure 13. These differences are not surprising, and reflect the poorer local packing of Gaussian threads relative to semiflexible chains. These differences also highlight the potential subtleties of the proposed mapping schemes, that is, the need to separately "calibrate" the different coarse-grained model parameters against experimental data or atomistic PRISM computations.[52]

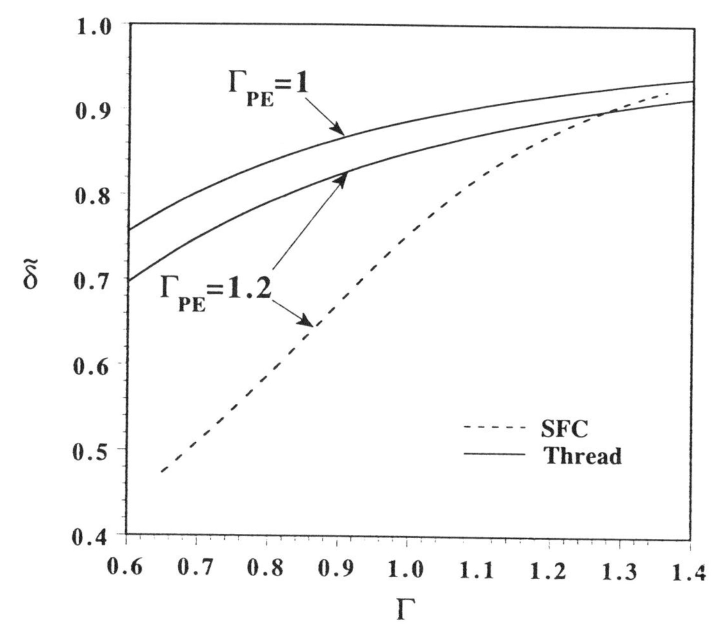

Figure 13. Reduced solubility parameter as a function of chain aspect ratio for the $l/d = 0.5$ SFC model and the analytic Gaussian thread model.[52] Predictions based on two choices of polyethylene aspect ratio at 430 K are shown. The liquid density is determined by the calibration procedure discussed in Ref. 52.

IV. ATHERMAL POLYMER BLENDS

Mixtures of polymers, or *blends*, are of major scientific and materials engineering interest.[93–95] Moreover, the phase behavior of high polymer blends is very subtle due to the enormous reduction of the ideal (combinatorial) entropy of mixing due to chain connectivity.[1] *Athermal* blends are mixtures of two or more polymer components for which the heat of mixing is zero. An example of such a blend is one in which all intermolecular site–site interactions are entirely repulsive hard core in nature. Although athermal blends do *not* exist in reality, their behavior is important from a theoretical point of view. Based on studies of atomic[5,11] and small molecule liquids[8] the structure of one-component liquids at high density is believed to be primarily determined by the repulsive part of the potential. This suggests that a useful strategy for describing general polymer blend thermodynamics might be to treat the attractive interactions by a perturbation expansion about an athermal reference system. Thus the problem of determining the intermolecular packing in athermal polymer blends is a fundamental one and forms the basis of the simplest conceivable general blend theory. In addition, the role of excess entropic effects on mixing, and possible athermal phase separation, are questions of basic statistical mechanical interest.

The well-known mean-field incompressible Flory–Huggins theory[1] of polymer mixtures assumes random mixing of polymer repeat units. However, it has been demonstrated that the radial distribution functions $g_{\alpha\gamma}(r)$ of polymer melts are sensitive to the details of the polymer architecture on short length scales. Hence, one expects that in polymer mixtures the radial distribution functions will likewise depend on the intramolecular structure of the components, and that the packing will not be random. Since by definition the heat of mixing is zero for an athermal blend, Flory–Huggins theory predicts athermal mixtures are ideal solutions that exhibit complete miscibility.

In this section we examine athermal binary mixtures using PRISM theory. Tests of both the structural and thermodynamic predictions of PRISM theory with the PY closure against large-scale computer simulations are discussed in Section IV.A. Atomistic level PRISM calculations are presented in Section IV.B, and the possibility of nonlocal entropy-driven phase separation is discussed in Section IV.C at the SFC model level. Section IV.D presents analytic predictions based on the idealized Gaussian thread model. The limitations of overly coarse-grained chain models for treating athermal polymer blends are briefly discussed.

A. Comparison with Computer Simulations

An important question is whether PRISM theory can predict the packing in athermal blends with the same good accuracy found for one-component melts. To address this question Stevenson and co-workers[96] performed molecular dynamics simulations on binary, repulsive force blends of 50 unit chains at a liquidlike packing fraction of $\eta = 0.465$. The monomeric interactions were very similar to earlier one-component melt simulations[51] which served as benchmark tests of melt PRISM theory. Nonbonded pairs of sites (both on the same and different chains) were taken to interact via shifted, purely repulsive Lennard-Jones potentials. These repulsive potentials were adjusted so that the effective hard site diameters, obtained from Eq. (3.12), were $d_{AA} = 1.015$ and $d_{BB} = 1.215$ for the chains of type A or B, respectively. Chain connectivity was maintained using an intramolecular FENE potential[51] between bonded sites on the same chain. The resulting chain model has nearly constant bond lengths that are nearly equal to the effective hard-core site diameter.

The three intermolecular radial distribution functions $g_{AA}(r)$, $g_{AB}(r)$, $g_{BB}(r)$ in the blend were obtained from the simulation as a function of the concentration. Corrections were made for finite size effects in the simulation.[96] Likewise, the intramolecular structure factors $\hat{\Omega}_{AA}(k)$ and $\hat{\Omega}_{BB}(k)$ were obtained from the simulation and used as input for PRISM calculations on the athermal blend. In the PRISM calculation the PY closure of Eq. (2.5a) was used for the same soft repulsive potentials as in the simulation. A comparison between the results from the molecular dynamics simulation and PRISM theory is shown in Figure 14 for the case of volume fraction of A chains $\phi = 0.368$. Although deviations are seen at small distances, overall the agreement is quite good and comparable to similar studies[51,60] done earlier on one-component polymer melts (see Fig. 3). Similar agreement was found at other blend concentrations.[96]

The simulation of Stevenson and co-workers allow a direct test of the random mixing approximation. Strictly speaking, at the structural level the random mixing approximation in its polymeric Flory–Huggins form[1] implies that all the radial distribution functions in the mixture are identically unity, $g_{\alpha\gamma}(r) = 1$. As can be seen from Figure 14 this is obviously a poor approximation. A less restrictive definition of random mixing might be that the packing is the same for both species in the blend; in other words all the $g_{\alpha\gamma}(r)$ are the same. We can probe this approximation by defining an excess correlation function $\Delta g(r)$ based on

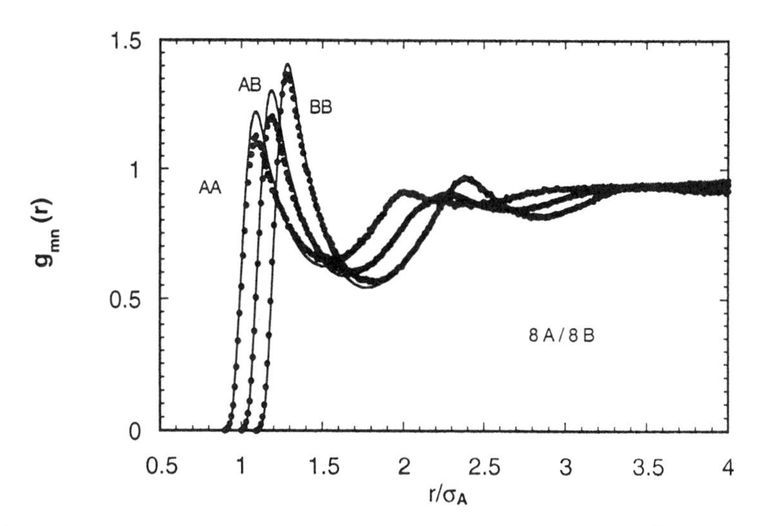

Figure 14. A comparison between PRISM theory and MD simulations[96] for the radial distribution functions in an athermal blend of 50 unit chains. The composition was maintained at $\phi = 0.368$. The points are the simulations and the curves are the PRISM predictions.

the differences in radial distribution functions between species.

$$\Delta g(r) = g_{AA}(r) + g_{BB}(r) - 2g_{AB}(r) \tag{4.1}$$

This function quantifies the tendency for "pairing" or physical clustering of like species. Figure 15 depicts $\Delta g(r)$ for the $\phi = 0.368$ case obtained from the simulation. Significant departures from random mixing, caused by local differences in monomer size, are evident at short length scales. Remarkably, PRISM theory is able to capture these subtle packing effects as seen by the solid line in Figure 15.

Another important question regarding the structure of athermal blends is whether the single-chain conformation changes with composition. In the molecular dynamics simulations,[96] small changes (at most 10–15%) were observed in $\hat{\Omega}_{AA}(k)$ and in the mean-square end-to-end distance $\langle R^2 \rangle$ of the chains in the blend. However, such changes are barely within the statistical error of the simulation. The collective partial structure factors were also monitored in the simulation and no evidence for incipient phase separation was detected in this athermal mixture.[96]

Benchmark Monte Carlo simulations of a different class of athermal polymer mixtures have recently been carried out by Weinhold et al.[97] An equimolar ($\phi = 0.5$), constant-volume binary blend was considered. The polymers were modeled as semiflexible, tangent bead chains of equal degrees of polymerization, N, interacting via a purely hard-core potential

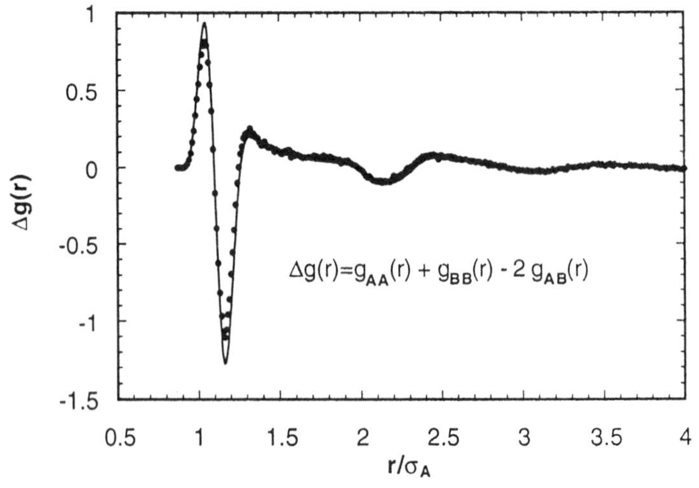

Figure 15. The nonrandom mixing $\Delta g(r)$ in an athermal blend of composition ϕ = 0.368. The points are the simulations and the curves are the PRISM predictions.[96]

of the same diameter for all sites. The reduced fluid density was $\rho_m d^3 = 0.65$ ($\eta \cong 0.34$) representative of a concentrated solution. The only difference between the A and B species was the local chain bending energy, which controls the chain aspect ratio. A statistical segment length was defined as $a = (6R_g^2/N)^{1/2}$, and an aspect ratio as $\Gamma = a/d$. The A ("flexible") chain aspect ratio was fixed at $\Gamma_f = 1.5$, and the B ("stiff") chain aspect ratio was varied over the wide range of $\Gamma_s = 1.7$–4.2. Various model blends were studied for $N = 20$ with the *stiffness asymmetry* of each characterized by the variable $\xi = 2(\Gamma_s - \Gamma_f)/(\Gamma_s + \Gamma_f)$. Besides providing exact results to test PRISM theory for athermal blends, these studies were motivated by scientific questions related to the relevance of a purely entropic *packing frustration* mechanism for phase separation in chemically similar polymer blends such as the saturated polyolefins.[98–100]

Here representative examples are given of the PRISM/simulation comparisons and the main conclusions are summarized.[97] The PRISM calculations were carried out assuming conformational ideality, and the required single-chain structure factor was computed using the discrete, tangent site Koyama model[55] adjusted such that it reproduced the radius of gyration of the chains in the *one-component melt*. To within the statistical errors of the simulations, *no* changes in single-chain conformation on going from the melt to the equimolar mixture were detected.

The agreement of PRISM theory with simulations for the three blend pair correlation functions, $g_{ss}(r)$, $g_{ff}(r)$, and $g_{fs}(r)$, was found to be typical of prior studies of dense melts, that is, errors of roughly 10–20% close to contact, but much more accurate as r increases. More importantly, the *relative* form of the three $g(r)$'s, and their changes with increasing blend stiffness asymmetry, were accurately described by PRISM theory.[97] One simulation was carried out for $N = 200$ and $\xi = 0.28$. No loss of accuracy of PRISM for structural properties was found as N was increased from 20 to 200.

Blend thermodynamic properties were also computed. Comparisons of PRISM theory and simulations for the partial excess (interaction entropic) free energies of mixing per site, $\Delta F_{\text{mix},i}$, are listed in Table I for $N = 20$ and 200. The PRISM results are based on the free energy charging route expressions, which for $\phi = \frac{1}{2}$ are given by[97]

$$\beta\, \Delta F_{A,\text{mix}} = \frac{\pi \rho d^3}{2} \int_0^1 d\lambda\, \lambda^2 [g^{\lambda}_{\text{AA}}(\lambda d^+) + g^{\lambda}_{\text{BB}}(\lambda d^+) - 2 g^{\lambda}_{\text{AB}}(\lambda d^+)] \quad (4.2)$$

for the A species. The B-species expression is obtained by interchanging A and B labels in Eq. (4.2). Surprisingly good agreement between theory and simulation is obtained for all stiffness asymmetries and both values of

TABLE I
Comparison of Theory and Simulation Values of Excess Partial Free Energy of Mixing Changes as Defined in Text[97] [a]

	Simulation		Theory		
ξ	$\beta\,\Delta F_s$	$\beta\,\Delta F_f$	$\beta\,\Delta F_s$	$\beta\,\Delta F_f$	$\beta\,\Delta F_{exc}$
0.024	—	—	−0.0063	0.0063	0.00003
0.14	−0.022	0.020	−0.0175	0.0179	0.0004
0.28	−0.032	0.031	−0.0276	0.0296	0.0017
$N = 200$	−0.065	0.069	−0.0539	0.0549	0.0010
0.41	−0.036	0.035	−0.0306	0.0337	0.0031
0.53	−0.038	0.036	—	—	—

[a] Theory results are computed based on free energy charging method. The statistical uncertainties of the simulation values is roughly ±0.005. Subscripts s and f denote stiff and flexible components. The flexible statistical segment length, a_f, is fixed at 1.50 ± 0.01 and the stiffness asymmetry variable is $\xi = 2(a_s - a_f)/(a_s + a_f)$. $\beta\,\Delta F_{exc}$ is the total excess free energy of mixing predicted by PRISM theory; to within statistical uncertainty this quantity is found to be zero in the simulations for all cases shown. All results shown are for $N = 20$, except for the one $N = 200$ case.

N. Note that the stiff and flexible excess free energy of mixing are opposite signs, which implies that the flexible (stiff) chain is destablized (stabilized) upon transfer from the melt to the blend. Such behavior reflects local packing differences and equation-of-state effects,[93,101] which cannot be described within an incompressible theory as often employed in polymer science.[2] Note, however, the *total* net excess free energy was extremely small in all cases [even much smaller than the ideal entropy of mixing per segment $= N^{-1}\ln(2)$].

Thus, for the short and moderately long chains studied the athermal stiffness blend behaves as a *nearly ideal mixture in a thermodynamic sense* even though there are significant differences in segmental packing among the different species consistent with the molecular dynamics (MD) simulations described above.[96] PRISM computations of an effective interaction (or "chi" in polymer science) parameter based on the free energy or compressibility route have also been shown to be in surprisingly good agreement with simulation[97] and are very small, thus supporting the above conclusions.

We note that there are hints in Table I, and structural fluctuation quantities such as $\Delta g(r)$ discussed elsewhere,[97] that as N or stiffness mismatch increase the excess free energy of mixing also increases and the blend is less stable. Thus, the possibility of entropy-driven phase separation due to packing frustration of dissimilar flexibility chains as N increases beyond 200 remains open based on the simulation studies of Ref. 97.

Summarizing, the major conclusion of this section is that PRISM theory provides an excellent description of the structure and (constant volume) free energy of mixing of high-density athermal polymer blends composed of the short and modest chain length molecules presently accessible to computer simulation. This has motivated the application of the theory to experimentally relevant situations such as long chain ($N \approx 10^3$–10^4) and chemically realistic atomistic models.

B. Multiple-Site Homopolymer Blends

It was demonstrated in Section III.B for one-component melts that subtle screening effects resulted from the packing of nonspherical monomers. It is natural to expect that similar screening effects would also be operable in athermal blends of vinyl polymers. In order to probe this aspect at a *chemically realistic* level, Rajasekaran and co-workers[32] studied an athermal mixture of polyethylene and isotactic polypropylene. The chains in this mixture were modeled as illustrated in Figure 7 with three sites (A, B, C) making up a polypropylene monomer, and a single D site representing the CH_2 group of polyethylene. Application of Eqs. (2.2)–(2.5) lead to a set of 10 coupled integral equations that were solved numerically using standard Picard iteration techniques.

Assuming that the intramolecular structure of the chains in the athermal blend is independent of composition, then the elements of the 4×4 $\hat{\Omega}_{\alpha\gamma}(k)$ intramolecular matrix for the blend are already available from the corresponding one-component melt intramolecular structure functions. The $\hat{\Omega}_{\alpha\gamma}(k)$ for α, $\gamma = $ A, B, C were obtained from Monte Carlo simulations of a single, rotational isomeric state chain using the parameterization of Suter and Flory[73] discussed in Section III.B. Likewise, $\hat{\Omega}_{DD}(k)$ was obtained from the RIS calculations of Honnell and co-workers for polyethylene.[63,64]

Figure 16 depicts the intermolecular packing in the athermal PE/i-PP blend for chains of 200 *monomers* at a volume fraction of polyethylene sites of $\phi = 0.50$. Although there are 10 independent correlations, only the diagonal components are shown. The radial distributions in the blend are qualitatively similar to those in the one-component melt (see Fig. 11). However, as demonstrated elsewhere,[32] the detailed structure is found to be significantly composition dependent. For example, the local peaks in the polyethylene $g_{DD}(r)$ increase monotonically in magnitude as more polypropylene is added to the mixture (by roughly 30% at $\phi = 0.1$ relative to the pure PE melt). This suggests a tendency of the polyethylene to cluster in the mixture as a result of unfavorable cross correlations between the PE and i-PP chains. Despite this clustering

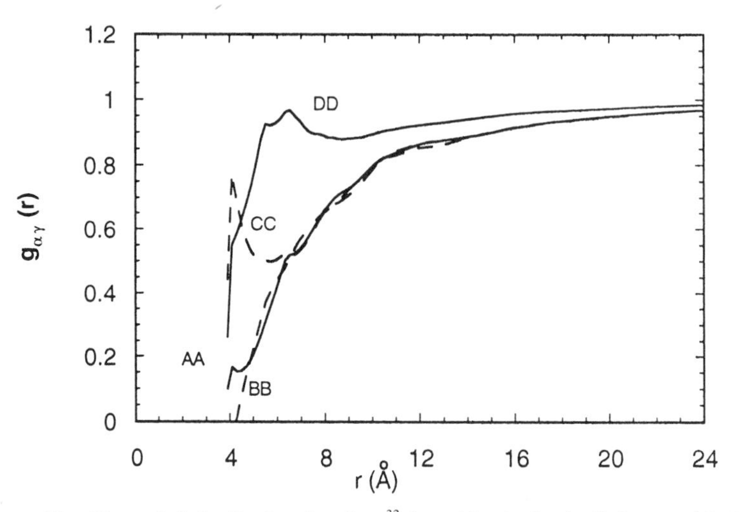

Figure 16. The radial distribution functions[32] for a blend of polyethylene and isotactic polypropylene ($N_{PE} = N_{PP} = 200$) at a volume fraction of polyethylene of $f = 0.5$. The four diagonal correlations are shown.

tendency, no thermodynamic evidence was found for macroscopic phase separation in the athermal mixture at any composition.

From a knowledge of the radial distribution functions in the blend as a function of composition, one can obtain the various thermodynamic state functions by applying the analysis of Kirkwood and Buff.[102] For the present case the following relationship for the entropy of mixing ΔS_{mix} can be derived[32]

$$\left(\frac{\partial^2 \Delta S_{\text{mix}}}{\partial \phi^2}\right)_{TP} = \frac{-k_B(\eta/\nu)^2 V}{9\phi^2 \hat{S}_{AA}(0) - 6\phi(1-\phi)\hat{S}_{AD}(0) + (1-\phi)^2 \hat{S}_{DD}(0)} \tag{4.3}$$

where η is the overall blend packing fraction and ν is the volume of one of the sites (assumed to be all equal). If incompressibility ($\kappa \to 0$) and random mixing [$g_{\alpha\gamma}(r) = 1$] are enforced, then the continuous-space analog of the Flory–Huggins, or ideal, solution relationship is obtained[32]

$$\left(\frac{\partial^2 \Delta S_{\text{mix}}}{\partial \phi^2}\right)_{TV} = -k_B\left(\frac{\eta}{\nu}\right)V\left[\frac{1}{2N_{PE}\phi} + \frac{1}{3N_{PP}(1-\phi)}\right] \tag{4.4}$$

Calculations by Rajasekaran and co-workers[32] of the entropy of mixing derivative for the PE/PP blend using Eq. (4.3) predict a *smaller* entropy stabilization of the mixture, by approximately a factor of 2, relative to ideal solution behavior.[32] Thus, excess entropic effects do destabilize this PE/i-PP blend relative to the ideal solution behavior, but no athermal phase separation is found even for this very structurally asymmetric case. Experimentally, an equimolar PE/PP mixture is *immiscible* for all accessible temperatures and values of N_{mon} *far less* than 200. Thus, explanation of the experimental behavior requires consideration of thermal effects (attractive forces) as discussed in Section V.

It should be mentioned that Eq. (4.3) is only one of several possible thermodynamic routes to the entropy of mixing in the athermal blend. Another possible route is through the "charging formula" of Chandler[8] used earlier in Eq. (3.9b) for the one-component polymer melt.

C. Semiflexible Blends and Entropy-Driven Phase Segregation

Motivated by both scientific questions related to the origin of phase separation in saturated polyolefin alloys,[97–100] and the basic statistical mechanical question of entropy-driven phase segregation, Singh and Schweizer[100] have carried out a detailed numerical PRISM study of the structure and phase behavior of the binary athermal "stiffness" blend discussed in Section IV.A. A wide range of chain aspect ratios of the tangent SFC chain, fluid density, blend composition, and ratio of A and B site diameters were investigated. Liquid–liquid spinodal phase separation is defined as when all the partial collective structures at zero wave vector, $\hat{S}_{\mathrm{MM}'}(k=0)$, simultaneously diverge. This condition is precisely given as

$$0 = 1 - \rho_{\mathrm{A}} N_{\mathrm{A}} C_{\mathrm{AA}} - \rho_{\mathrm{B}} N_{\mathrm{B}} C_{\mathrm{BB}} + \rho_{\mathrm{A}} \rho_{\mathrm{B}} N_{\mathrm{A}} N_{\mathrm{B}} (C_{\mathrm{AA}} C_{\mathrm{BB}} - C_{\mathrm{AB}}^2) \quad (4.5)$$

where $C_{\mathrm{MM}'} = \hat{C}_{\mathrm{MM}'}(0)$. For the cases of interest here $N_{\mathrm{A}} = N_{\mathrm{B}} = N$, $\rho_{\mathrm{A}} = \phi\rho$, $\rho_{\mathrm{B}} = (1-\phi)\rho$, and ρ is the total site number density of the binary blend.

The possibility of entropy-driven phase separation in purely hard-core fluids has been of considerable recent interest experimentally, theoretically, and via computer simulations. Systems studied include binary mixtures of spheres (or colloids) of different diameters,[103] mixtures of large colloidal spheres and flexible polymers,[104] mixtures of colloidal spheres and rods,[105] and a polymer/small molecule solvent mixture under *infinite dilution* conditions (here an athermal conformational "coil-to-globule" transition can occur).[106] For the latter three problems, PRISM theory could be applied, but to the best of our knowledge has not. The first problem is an old one solved analytically using PY integral equation theory by Lebowitz and Rowlinson.[107] *No liquid–liquid phase separation was found, that is, the hard sphere mixture is completely miscible.* Recent simulations and experiments suggest this is not true for *highly* size-asymmetric cases,[103] and modifications of the atomic closures have been proposed to account for the observed phase separation.[108] For the polymer problem, we are also using the site–site PY closure. However, entropy-driven phase separation, if it occurs, is associated with the difficulty of packing chains of different stiffness or bending rigidity. In analogy with liquid crystal systems,[109] one might expect phase separation if the "packing frustration" is sufficiently great (although for mixtures of rods and coils segregation into an isotropic and nematic phase generally occurs). Relevant variables for such an entropy-driven phenomenon would include the individual chain aspect ratios, overall packing fraction, and degree of polymerization.

Here only a few of the highlights of the extensive study[100] will be mentioned for the simplest case of equal A and B site hard-core diameters, and an equimolar mixture of chains of N sites each ($\phi = \frac{1}{2}$). The A chain aspect ratio is fixed at approximately 1.3, which is representative of a polymer such as polyethylene.[52] Thus, the structural asymmetry variable $\xi = 2(\Gamma_B - \Gamma_A)/(\Gamma_B + \Gamma_A) < 0$ for most experimental polyolefin mixtures since unbranched polyethylene generally has the largest aspect ratio of the saturated hydrocarbon polymers.[52,97,100] Figure 17 shows spinodal phase boundaries based on Eq. (4.5) for two reduced densities representative of a dense melt and concentrated solution (as studied by simulation).[97] There are several important features.[100]

1. *No* phase separation is found for a stiffness asymmetry variable less than roughly 0.4, or for low values of N and any value of stiffness mismatch (consistent with simulation).[97] Since experimentally one expects for hydrocarbon polymers that $\xi < 0$ (since polyethylene has a high aspect ratio), this result suggests that a purely entropy-driven mechanism cannot account for the facile tendency of polyolefins to demix. Moreover, for chain parameter values typical of most semiflexible polymers of interest, the excess entropic effects appear small and much weaker than enthalpy related considerations associated with *local* packing differences between species (see Section V).

2. Phase separation can occur at *large enough N* under the appropriate conditions. It seems clear that since large N is required, the predicted phase separation is driven by spatially long range, or nonlocal, aspects of polymer connectivity and excluded volume interactions. *Thus, nonlocal entropy-driven phase segregation requires a large enough N and sufficient absolute stiffness and aspect ratio mismatch of the polymer backbones.* This deduction seems natural in that packing frustration is created locally

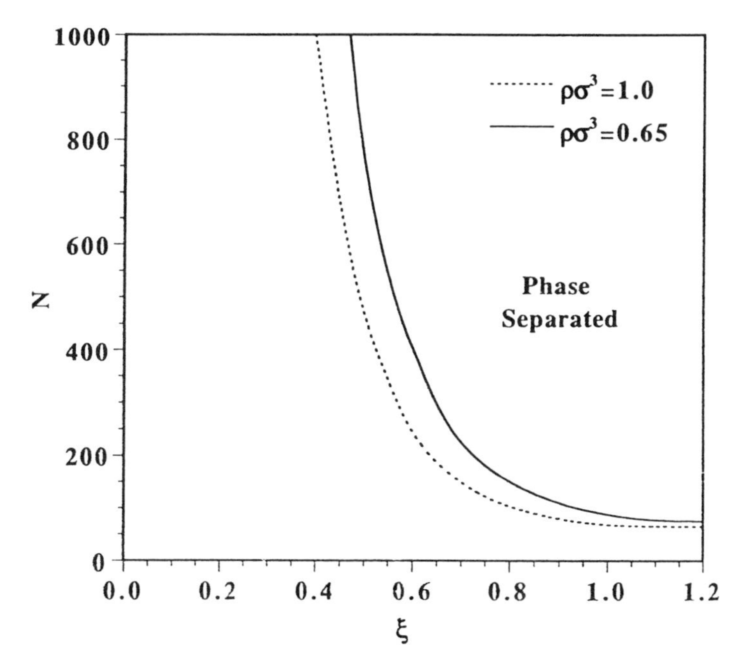

Figure 17. Athermal spinodal phase diagram obtained using PRISM theory[100] and the compressibility route for two reduced fluid densities ($d = \sigma$) and a flexible chain aspect ratio of 1.3. The structural asymmetry variable $\xi = 2(\Gamma_B - \Gamma_A)/(\Gamma_B + \Gamma_A)$.

and then propagated to macomolecular length scales via a backbone stiffness dependent chain connectivity mechanism. This scenario is further reinforced by the fact that PRISM theory studies to date have found *no* athermal phase separation for highly flexible chain models such as Gaussian or freely jointed models[27–29] (see also Section IV.D), which lack completely the local rodlike stiffness on length scales comparable and shorter than the chain persistence length.

3. The effect of fluid density is relatively weak, but phase separation is enhanced at higher density consistent with intuition.

A structural interpretation of the predicted nonlocal entropy-driven phase separation can be deduced by examining the spatially resolved pair correlation function measure of incompatibility or clustering, $\Delta g(r)$ defined in Eq. (4.1). Figure 18a shows this function for the case of fixed chain aspect ratios and several values of N and fluid packing fraction. As the chains become longer, local clustering of like segments is enhanced but tends to "saturate" at large N. However, the growth of a much larger

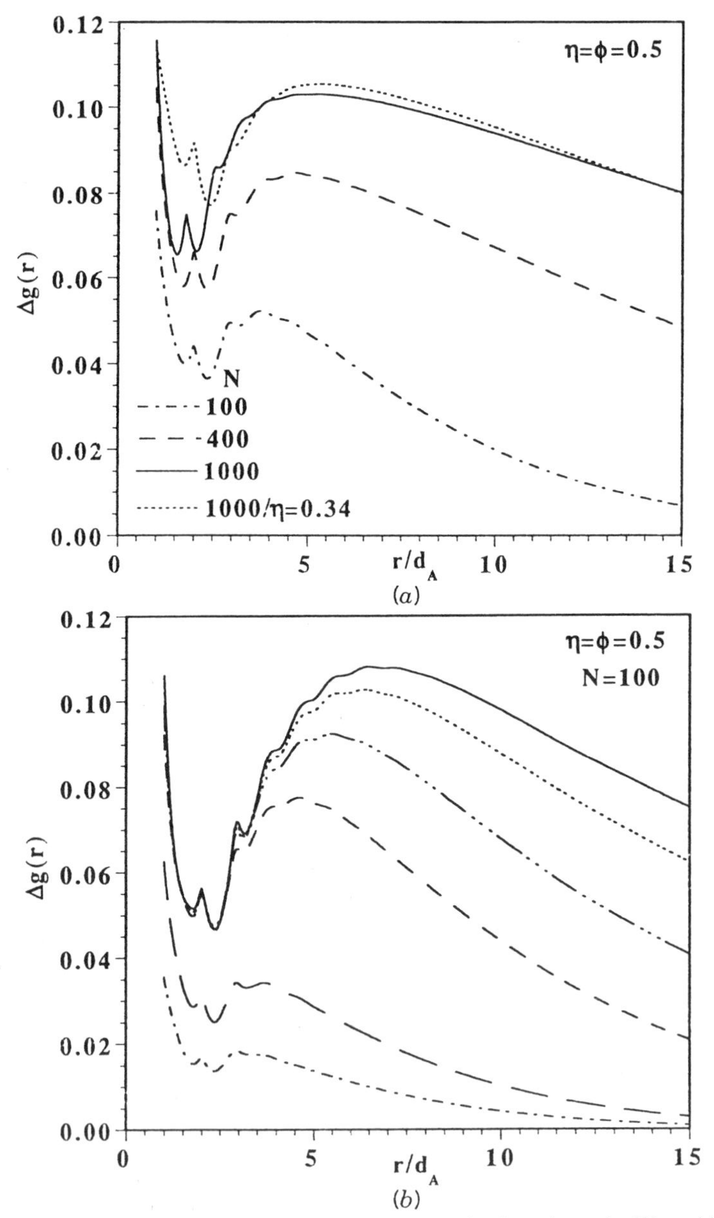

Figure 18. Intermolecular pairing function in the equimolar athermal stiffness blend.[100] Except as explicitly noted all curves are for the melt like packing fraction of 0.5. (a) Results for fixed aspect ratio asymmetry of $\gamma = \Gamma_B/\Gamma_A = 1.49$ and various values of N. (b) Dependence on aspect ratio asymmetry for fixed $N = 100$. From top to bottom the curves correspond to $\gamma = 2.319$ (spinodal boundary), 2.199, 1.979, 1.734, 1.343, and 1.219.

amplitude and longer spatial range component is dramatic. It coincides with increasingly large long-wavelength concentration fluctuations and ultimately spinodal phase separation. This N-dependent correlation feature is associated with the chain connectivity and inpenetrability on intermediate "correlation hole" length scales.

The dependence of $\Delta g(r)$ on chain stiffness mismatch at fixed $N = 100$ is shown in Figure 18b. Again, local segregation increases with stiffness mismatch but tends to saturate. However, *non*local segregation continues to grow in both amplitude and range, ultimately resulting in phase separation. The latter behavior is particularly striking since at fixed N the size of the polymers, and hence the spatial range the correlation hole associated with meltlike density fluctuations, is nearly fixed. This plot also illustrates an intriguing empirical observation that the contact value $\Delta g(r = d)$ tends to acquire a common critical value at the spinodal phase boundary (roughly 0.1 in the cases of Fig. 18), which is nearly independent of how the phase transition is driven (increasing N at fixed $\gamma = \Gamma_B/\Gamma_A$, or increasing γ at fixed N).[100]

The detailed nature of single-phase blend correlations, effective interaction parameters, and the entropy-driven phase separation phenomenon have been found to be sensitive to system-specific factors such as composition, differences in A and B site diameter, density, and local architectural details.[100] However, the basic conclusions summarized above seem qualitatively general.

A limited number of numerical studies have been carried out using the free energy charging approach to blend thermodynamics instead of the compressibility route described above.[100] In the miscible region, quantitative differences in the excess free energy of mixing, or interaction parameter, have been found, but these differences are generally no more than a factor of 2 or 3. More work is required to establish the severity of the thermodynamic inconsistency problem for this athermal stiffness blend system. Another caveat is that it is well known that PRISM does not properly incorporate "nematiclike" orientational correlations in the one phase fluid.[8,39,40] This should represent a technical limitation to describing liquid–liquid phase separation, which becomes more severe as the chain aspect ratios increase. Description of isotropic-nematic phase separation is not possible. However, for the experimental applications to conformationally flexible polyolefins and polydienes of current interest, strong nematic correlations seem improbable.

D. Analytic Gaussian Thread Model

PRISM theory for the athermal stiffness asymmetric blend model described in Section IV.C can be analytically solved[29,30] in the idealized

Gaussian thread limit for $N \rightarrow \infty$. For many physical problems (e.g., polymer solutions and melts, liquid–vapor equilibria, and thermal polymer blends and block copolymers), the Gaussian thread model has been shown to be reliable *in the sense* that it is qualitatively consistent with many aspects of the behavior predicted by numerical PRISM for more realistic semiflexible, nonzero thickness chain models. However, there are classes of physical problems where this is not the case. The athermal stiffness blend in certain regions of parameter space is one case, both in the bulk[100] and near surfaces.[110] Nevertheless, even for this problem the thread model does correctly capture certain aspects, and when it does fail this provides considerable insight into the key factors that control the behavior of real polymer systems.[100]

Employing the simplified Gaussian forms of Eq. (3.1), and enforcing the three pointlike core conditions within the PY closure approximation $[g_{MM'}(r=0)=0]$, results in three nonlinear transcendental equations for the three direct correlation parameters $C_{MM'}$. In the long-chain limit, the Gaussian chain structure factors take on a perfectly self-similar form: $\hat{\omega}_M(k) = 12(k\sigma_M)^{-2}$, and no large k crossover to locally rigid behavior, $\hat{\omega}(k) \propto (k\xi_p)^{-1}$, occurs. This mathematical feature allows an exact analytic solution that has a scaling form[29,30,100]

$$C_{BB} = \gamma^4 C_{AA} \qquad C_{AB} = \gamma^2 C_{AA} \tag{4.6}$$

$$\sigma_A^{-3} C_{AA}(\phi) = \frac{-\pi\eta\Gamma^3}{18}[\phi + \gamma^2(1-\phi)] \tag{4.7}$$

where $\Gamma = \sigma_A/d$ and $\gamma = \sigma_B/\sigma_A$ is the "structural or conformational" asymmetry variable. The perfect scaling relations among the blend composition-dependent $C_{MM'}$ are a consequence of the self-similar intrachain correlations and the PY closure treatment of the (pointlike) excluded volume constraints.

Equation (4.6) predicts that the repulsive effective potential between segments of like species, $-k_B T C_{MM}$, is larger in the blend for the conformationally stiffer chain than the more flexible one. This result is in accord with physical intuition and is mathematically required to satisfy the local hard-core (point) exclusion constraint. Moreover, the repulsive pseudopotential between the more flexible (stiffer) species is predicted to increase (decrease) upon transfer from a one-component melt to the blend environment. This trend is intuitively sensible, consistent with the signs of the species-dependent effective chi parameters derived elsewhere,[100] and agrees with numerical SFC PRISM calculations and Monte Carlo simulations of the partial excess free energies of mixing and effective chi parameters.[97]

The thread model is predicted to be miscible under all conditions due to the perfect scaling relations of Eq. (4.6). Although this complete miscibility conclusion is in agreement with numerical SFC PRISM for low aspect ratio chains (those best described by a Gaussian model), it misses the entropy-driven phase separation phenomenon found numerically for sufficiently stiff polymers.[100] This point emphasizes the limitations of the thread model. It ignores the consequences of local chain rigidity on packing, which appears to be central to the athermal phase segregation phenomenon. Another example of the limitations of the Gaussian thread model is its prediction[29,30] of interchain random mixing or conformal solution behavior, corresponding to $g_{MM'}(r) = g_{eff}(r; \phi)$. Again, this is a consequence of the (assumed) perfectly self-similar Gaussian single polymer structure and the $d \rightarrow 0$ idealization. In reality, there are always differences in the $g_{MM'}(r)$ functions due to local, nonuniversal breaking of the self-similar chain correlations.

Finally, field theoretic approaches[99] have recently predicted athermal phase separation driven by nonlocal-entropic considerations for *incompressible blends of Gaussian thread polymers*. This prediction is at odds with PRISM theory in the thread limit. However, for the effective chi parameter PRISM theory has been shown[100] to be equivalent to the field theory *if* the free energy route is employed in conjunction with the extremely simple RPA closure (not PY). The RPA closure, $C_{MM'}(r) = -\beta u_{MM'}(r)$ *for all r*, is known to be very poor for repulsive force systems and violates the hard-core impenetrability condition. Thus, the field-theoretic prediction has been suggested to be a consequence of the combined use of a long-wavelength incompressibility approximation in conjunction with a RPA closure.[100]

V. THERMAL EFFECTS IN POLYMER BLENDS: PERTURBATION APPROACH

In reality, polymer mixtures are not athermal and attractive interactions can play a crucial role in determining their miscibility and thermodynamic properties.[1,93–95] This fact is evident in the mean-field Flory–Huggins theory where phase separation and the interaction parameter are entirely of an enthalpic origin. For a binary blend of A and B single-site chains composed of N repeat units, the free energy of mixing per segment is given in the *incompressible* Flory theory as[1]

$$\beta\,\Delta F_{\min} = \frac{\phi_A \ln \phi_A}{N_A} + \frac{\phi_B \ln \phi_B}{N_B} + \chi_0 \phi_A \phi_B \tag{5.1a}$$

where $\phi_A + \phi_B = 1$, and the interaction, or chi, parameter in its off-lattice version is of an extremely simple, purely energetic mean-field form

$$\chi_0 = \frac{-\rho}{2k_B T} \int d\mathbf{r}[v_{AA}(r) + v_{BB}(r) - 2v_{AB}(r)] \tag{5.1b}$$

Here $v_{MM'}(r)$ are the (generally attractive) tail potentials between species M and M′. The spinodal instability corresponds to a vanishing of the second compositional derivative of the free energy of mixing yielding the condition

$$2\chi_{0,s} N\phi(1-\phi) = 1 \tag{5.2a}$$

where $\phi \equiv \phi_A$ and $N_A = N_B = N$. This implies a liquid–liquid spinodal phase separation temperature of

$$k_B T_s = N \frac{\rho\phi(1-\phi)}{2} \int d\mathbf{r}[2v_{AB}(r) - v_{AA}(r) - v_{BB}(r)] \tag{5.2b}$$

The critical composition is $\phi_A = \phi_B = \frac{1}{2}$ by symmetry.

The two prime predictions of Flory theory are as follows: (i) $T_s \propto N$ due to the nearly complete loss of ideal entropy of mixing due to chain connectivity constraints. Hence, it is correctly predicted to be generally very difficult to create a miscible polymer blend.[93–95] (ii) Immiscibility is promoted as the A and B monomers become more *chemically* distinct as quantified by their intermolecular tail potentials (e.g., London dispersion interactions).

Equation (5.1) includes only the ideal, combinatorial entropy of mixing and the simplest conceivable "regular solution" type estimate of the enthalpy of mixing based on completely random mixing of monomers: $g_{MM'}(r) = 1$ in the liquid state language; χ_0 is referred to as the "bare" chi parameter since it ignores all aspects of polymer architecture and interchain nonrandom correlations. For these reasons, the model blend for which Eq. (5.1) is thought to be most appropriate for is an *interaction and structurally symmetric* polymer mixture. The latter is defined such that the *only difference* between A and B chains is a $v_{AB}(r)$ tail potential, which favors phase separation at low temperatures. The closest real system to this idealized mixture is an *isotopic blend*, where the A and B chains are hydrogenated and deuterated versions of the same polymer.[111] The symmetric model has played a central role in theoretical and simulation studies due to its great simplicity from a chemical viewpoint.[112]

In real systems, nonrandom mixing effects, potentially caused by local polymer architecture and interchain forces, can have profound consequences on how intermolecular attractive potentials influence miscibility. Such nonideal effects can lead to large corrections, of both excess entropic and enthalpic origin, to the mean-field Flory–Huggins theory. As discussed in Section IV, for flexible chain blends of prime experimental interest the excess entropic contribution seems very small. Thus, attractive interactions, or enthalpy of mixing effects, are expected to often play a dominant role in determining blend miscibility. In this section we examine these enthalpic effects within the context of thermodynamic pertubation theory for atomistic, semiflexible, and Gaussian thread models. In addition, the validity of a Hildebrand-like molecular solubility parameter approach based on pure component properties is examined.

A. Thermodynamic Perturbation Theory

For *dense* nonpolar polymers the intersite interactions are of the van der Waals type, and one anticipates that the attractive branch of the potential may exert little influence on interchain packing. Although obviously true if the tail potentials are weak relative to $k_B T$, such "repulsive force screening"[5,11] may also be operative in polymer mixtures for several reasons discussed below. Although there will undoubtably be errors made by such a simplification, it represents a conceptually and computationally convenient starting point. In such a thermodynamic perturbation, or high-temperature approximation (HTA), approach the polymer liquid structure is *assumed* to be determined solely by an appropriately constructed repulsive reference system.

The starting point is the reduced Helmholtz free energy of the *blend* in the standard "charging parameters" form[8]

$$\frac{\beta F}{V} = \frac{\beta F_0}{V} + \frac{1}{2} \sum_{\alpha\gamma} \rho_\alpha \rho_\gamma \int_0^1 d\lambda \int \beta v_{\alpha\gamma}(r) g^{\lambda}_{\alpha\gamma}(r)\, d\mathbf{r} \tag{5.3}$$

where the attractive branch of the potential $\lambda v_{\alpha\gamma}(r)$ is gradually turned on as the charging parameter λ varies from 0 to 1; F_0 is the free energy of the corresponding athermal, or reference repulsive force, system and may depend on temperature *implicitly* through the density, single-chain conformation, and/or effective hard core diameter. Equation (5.3) ignores *single-chain intramolecular* contributions associated with the attractive branch of the potentials. This generally represents an additional approximation, *but* within the context of the conformational ideality simplification such contributions would only contribute terms *linear* in polymer concentration and blend density. Such terms then *cancel out* in the free energy of mixing relevant to blend mixing thermodynamics. Since all PRISM work on blend thermodynamics to date has employed this conformational ideality assumption, Eq. (5.3) is appropriate. Future work based on the self-consistent formulation of PRISM discussed in Section VIII needs to be done in order to investigate the corrections to blend thermodynamics due to *nonideal* conformational effects (e.g., changes in polymer structure on going from the melt to blend, or mixture composition-dependent conformational changes).

Within the HTA scheme, the liquid structure of the mixture is approximated by the structure of the athermal system, that is, $g^{\lambda}_{\alpha\gamma}(r) \approx g^{0}_{\alpha\gamma}(r)$. Thus to first order, Eq. (5.3) can be approximated as

$$\frac{\beta F}{V} \approx \frac{\beta F_0}{V} + \frac{1}{2} \sum_{\alpha\gamma} \rho_\alpha \rho_\gamma \int \beta v_{\alpha\gamma}(r) g^{0}_{\alpha\gamma}(r)\, d\mathbf{r} \tag{5.4}$$

Such a HTA might be expected to be particularly accurate for polymers since the critical temperature grows without bound as N increases. Thus, the literal perturbative condition that $\beta v_{\alpha\gamma}(r) \ll 1$ might be expected to hold in the one-phase region for long chains. Although this argument is sound *in principle*, in practice the experimentally accessible temperatures are restricted to $T = 200\text{–}500$ K so such a weak coupling condition will not necessarily be valid for laboratory blends *for which* the demixing transition is measurable.

At constant pressure P, the Gibbs free energy of mixing ΔG_{mix} of the blend relative to the pure components can be expressed as

$$\Delta G_{\text{mix}} = \Delta F_{\text{mix}} + P\, \Delta V_{\text{mix}} \tag{5.5}$$

where ΔV_{mix} is the volume change of mixing. In first-order perturbation theory, the Helmholtz free energy of the reference athermal system is entirely due to the entropy of mixing, that is, $\Delta F^{0}_{\text{mix}} = -T\, \Delta S_{\text{mix}}$. Thus, from a knowledge of the structure of the athermal reference blend, one can calculate the free energy of mixing and phase behavior of the general blend. Any theory based on Eqs. (5.4) and (5.5) is expected to yield classical critical exponents.

Incompressible Flory mean-field theory[1] is recovered from Eqs. (5.4) and (5.5) if one assumes the following: (i) no excess volume of mixing; (ii) a blend composition-independent total packing fraction; (iii) the athermal reference system is an ideal solution, that is, zero excess entropy of mixing; and (iv) literal random mixing, corresponding to $g_{MM'}(r) = 1$.

In the PRISM studies carried out to date, simplification (i) has been invoked. An effective interaction, or chi parameter, can be defined in the usual manner as the second derivative of the *excess* free energy of mixing:

$$\begin{aligned}\chi &\equiv -\frac{1}{2}\frac{\partial^2}{\partial \phi^2}\Delta F_{\text{mix}}^{\text{exc}} \\ &= -\frac{1}{2}\frac{\partial^2}{\partial \phi^2}[-T\,\Delta S_{\text{mix}}^{\text{exc}} + \Delta H_{\text{mix}}] \\ &\equiv \chi_a + \chi_{\text{B}}\end{aligned} \tag{5.6}$$

where the third line defines excess entropic and enthalpic interaction parameters. Spinodal phase boundaries are determinable from this quantity.

B. Phase Behavior of Atomistic Models

Rajasekaran and co-workers[32] applied Eq. (5.4) to the PE/i-PP blend using the 10 radial distribution functions determined from the athermal mixture as discussed in Section IV.B. For this case Eq. (5.5) can be written in the form

$$\frac{\beta\,\Delta G_{\text{mix}}}{V} = -\frac{\Delta S_{\text{mix}}}{k_{\text{B}}V} + \left(\frac{\eta}{v}\right)\chi_H \phi(1-\phi) + \frac{\beta P\,\Delta V_{\text{mix}}}{V} \tag{5.7}$$

with an enthalpic interaction parameter χ_H defined implicitly in terms of the heat of mixing. Alternative definitions of an effective chi parameter, such as Eq. (5.6) are equally valid since *as a matter of principle* a single interaction parameter cannot completely characterize the nonideal aspects of a compressible binary mixture. For purposes of computing the spinodal phase boundary, the precise definition of a homogeneous phase interaction parameter is *irrelevant*. For the PE/i-PP blend, χ_H can be written in the form

$$\chi_H(\phi) = -\frac{1}{2}\left(\frac{\eta}{v}\right)[\Delta H_{\text{PE}} + \Delta H_{\text{PP}} - 2\,\Delta H_{\text{PE/PP}}] \tag{5.8a}$$

with the various contributions to the heat of mixing taking the form

$$\Delta H_{\text{PE}} = (1-\phi)^{-1}\int \beta v_{\text{DD}}[g_{\text{DD}}^{\text{melt}} - \phi g_{\text{DD}}^{0}]\,d\mathbf{r} \tag{5.8b}$$

$$\Delta H_{\mathrm{PP}} = \frac{1}{9\phi} \sum_{a,b=A}^{c} \int \beta v_{ab}[g_{ab}^{\mathrm{melt}} - (1-\phi)g_{ab}^{0}]\, d\mathbf{r} \tag{5.8c}$$

$$\Delta H_{\mathrm{PE/PP}} = \frac{1}{3} \sum_{a=A}^{c} \int \beta v_{aD} g_{aD}^{0}\, d\mathbf{r} \tag{5.8d}$$

where the arguments of the relevant functions have been suppressed. In the literal random mixing limit, in which all the radial distribution functions are unity, Eqs. (5.8) reduce to the continuum analog of the Flory–Huggins "bare" chi parameter:

$$\chi_0 = \frac{16\pi\beta\varepsilon\sigma^3}{9}\left(\frac{\eta}{\nu}\right)\left(\frac{1+\lambda_1+\lambda_2}{3} - 1\right)^2 \tag{5.9}$$

where the λ's are ratios of attractive well depth parameters as defined in Eq. (3.15), and σ and ε are the Lennard-Jones parameters for interactions between a pair of methylene sites.

Using Eq. (4.3) for the second derivative of the entropy of mixing together with Eqs. (5.8) for the heat of mixing permits the evaluation of the Gibbs free energy of mixing as a function of volume fraction ϕ of polyethylene. In their application to the PE/i-PP blend, Rajasekaran and co-workers[32] approximated the volume change of mixing as zero. This is equivalent to approximating the partial molar volumes in the mixture by the pure component molar volumes. It should be emphasized that making the assumption that $\Delta V_{\mathrm{mix}} = 0$ does *not* neglect the effect of density fluctuations (or equation-of-state effects), which are still present in the free energy of mixing through the composition dependence of the packing fraction $\eta(\phi)$. If the polarizability ratios are estimated from group additivity tables,[86] or from Jorgensen's potential functions[90] for alkanes, the heat of mixing for PE/i-PP is found to be positive. Furthermore the critical temperatures found from PRISM theory[32] are much higher than the corresponding Flory–Huggins estimates. This can be seen from the spinodal curves plotted in Figure 19 obtained using Small's estimate[86] for λ_1 and λ_2. The critical temperature for this mixture is predicted by PRISM theory to be approximately 11 times larger than from the corresponding mean-field Flory value. This is in qualitative accordance with experimental observations indicating a high degree of incompatibility between polyethylene and polypropylene. Thus, one concludes that local nonrandom packing effects, induced by local structural asymmetries in the monomeric structure of PE and i-PP, lead to gross (primarily enthalpic) destabilization of the blend.

In the case of polyolefin chains one expects the site polarizability

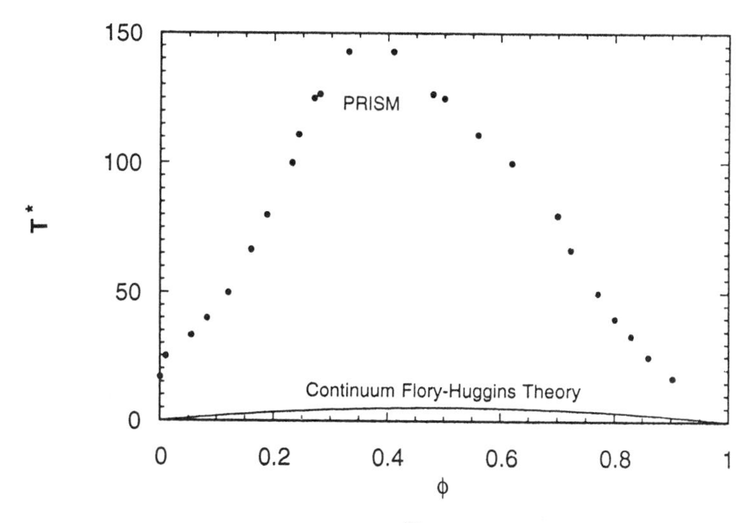

Figure 19. The spinodal curve calculated[32] using PRISM theory for a blend of polyethylene and isotactic polypropylene ($N_{PE} = N_{PP} = 200$). The ratios λ_1 and λ_2 were estimated from the group contribution tables of Small.[86] The ordinate is a reduced temperature defined as $T^* = k_B T/\varepsilon$ where ε is the united atom, Lennard-Jones parameter for a pair of methylene sites.

increases with the number of hydrogens present. For example, the polarizability of CH_3 is larger than for a CH group. Thus, in terms of Eq. (3.15) $\lambda_1 > 1$ and $\lambda_2 < 1$ for polyolefins. It is interesting to observe that if λ_1 and λ_2 are switched [keeping the bare Flory–Huggins chi parameter in Eq. (5.9) *unchanged*], then PRISM theory predicts a *negative* heat of mixing and a miscibility for this hypothetical PE/i-PP blend. Such an intriguing possibility of "compensation" of the demixing consequences of structural and interaction potential differences between species was first discovered by Singh and Schweizer[113] as described in the next section.

Finally, Honeycutt[114] has applied blend PRISM theory at an atomistic RIS model level to study the effect of tacticity (stereochemical differences) on the phase behavior of a commercially important binary polymer mixture. Tacticity is found to result in significant changes of the computed spinodal boundaries, which serves to again emphasize the importance of monomer structure and local packing on the free energy of mixing.

C. Miscibility of Semiflexible Chain Models

The thermodynamic behavior of binary blends of the semiflexible chain model ($l_b = d/2$), discussed in Sections III.A.2 and IV.C, have been thoroughly investigated numerically using PRISM theory within the HTA framework.[113,115] Calculations have focused on the experimentally relevant range of chain aspect ratios ($\Gamma = \sigma/d = 0.8$–1.4) for polyolefins, polydienes, and other flexible polymers. The mismatch of Lennard-Jones-*like* attractions ($r > d$)

$$v_{\mathrm{MM'}}(r) = \varepsilon_{\mathrm{MM'}}\left[\left(\frac{d}{r}\right)^{12} - 2\left(\frac{d}{r}\right)^{6}\right] \tag{5.10}$$

is characterized by a *monomer-averaged* well depth ratio parameter $\lambda = (\varepsilon_{\mathrm{BB}}/\varepsilon_{\mathrm{AA}})^{1/2}$ (typically in the range 0.9–1.1). In the spirit of constructing a minimalist model, the Berthelot geometric combining law is adopted for the AB cross term, and a meltlike reduced density of $\rho d^3 = 1.375$ is employed. The pure components are taken to have the identical reduced melt densities and zero volume change upon mixing is assumed. Experiments[91] on polyolefins support the latter behavior, and the consequences of the former simplifications have been discussed.[69,115]

A primary goal is to investigate the combined influences of conformational asymmetry (characterized by $\gamma = \Gamma_{\mathrm{B}}/\Gamma_{\mathrm{A}}$) and interaction or chemical asymmetry (characterized by λ) on blend miscibility as conveniently quantified by the effective chi parameter of Eq. (5.6). Investigating the validity of regular solution, or Hildebrand, approaches[85] is also of interest since it has been recently suggested[91] to work surprisingly well for hydrocarbon polymer alloys based on experimental studies of polyolefin blends. For simplicity, we focus here on equimolar mixtures ($\phi = \frac{1}{2}$), although the blend composition–dependence of the effective chi parameter is found to be very weak under the conditions of the calculations stated above.[115]

As argued in Section IV and elsewhere,[69,100,115] the excess entropic contribution to the chi parameter appears to be small in an absolute sense for experimentally relevant chain lengths, aspect ratios, and conformational asymmetries. Moreover, Singh and Schweizer[69,100,115] have estimated (based on the SFC model blend) that the enthalpic contribution to the chi parameter χ_{B}, is generally much larger (1–3 orders of magnitude) than the excess entropic contribution χ_a in Eq. (5.6). Thus, only the enthalpic contribution, defined as the second term in Eq. (5.6), is considered here. It is instructive to express χ_{B} in terms of two distinct contributions:

$$\chi_{\mathrm{B}} = \frac{-\beta\rho_m}{2}\int d\mathbf{r}\, v_{\mathrm{AA}}(r)[g_{\mathrm{AA}}(r) + \lambda^2 g_{\mathrm{BB}}(r) - 2g_{\mathrm{AB}}(r)] + \Delta\chi_{\mathrm{B}} \tag{5.11}$$

where the second term is implicitly defined. Incompressible Flory or random mixing theory is recovered if a literal mean-field approximation is invoked, that is, $g_{\mathrm{MM'}}(r) = 1$, resulting in a "bare" $\chi_0 = (10\pi/9) \times$

$\rho d^3 \beta \varepsilon_{AA}(\lambda - 1)^2$. The first term in Eq. (5.11) includes nonrandom packing corrections and arises from taking compositional derivatives only on the explicit factors $\rho_A \rho_B$ in Eq. (5.6). The second term describes enthalpic contributions associated with *composition-dependent changes in local (athermal) blend structure*. Although the latter are found to be small in absolute sense, they are significant in a polymer blend since miscibility is so difficult to achieve ($\chi N = 2$ at the critical point). The precise magnitude and direction of the local structural changes are *non*universal, but there is a general tendency for clustering of like species that increases as the mismatch of the A and B polymer aspect ratios grows. We note in passing that the χ_B of Eq. (5.6) appears to generally be in excellent agreement with the direct enthalpy of mixing definition implicit in Eq. (5.7), that is, $\chi_H = \Delta H_{mix}/\phi(1-\phi)$.

In Figure 20 representative results for the dependence of the thermal chi parameter χ_B on chain aspect ratios and the energetic asymmetry variable are presented.[113] Incompressible Flory theory would predict a

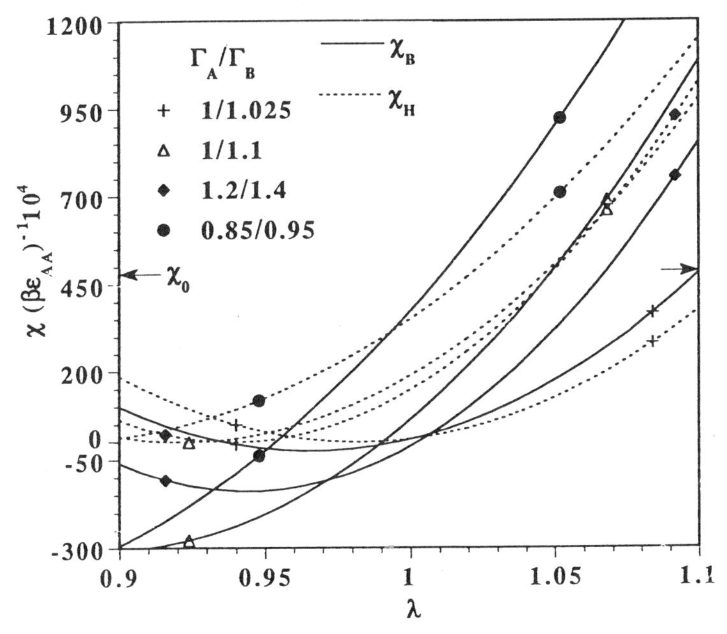

Figure 20. Blend (solid lines) and Hildebrand (dashed lines) effective chi parameters in units of the reduced thermal energy[113] computed within the HTA scheme. Results are shown as a function of the chemical asymmetry variable for four cases of structural (aspect ratio) mismatch. The symbols attached to the curves are guides to the eye for identifying the various cases. The horizontal arrows label the value of the bare interaction parameter for $\lambda = 0.9$ and 1.1.

parabola centered about $\lambda = 1$ and $\chi_B = 0$. (Also shown are predictions based on the melt solubility parameter approach, which are discussed in the next section.) The major predicted trends for χ_B, which all have major implications for understanding recent experiments on hydrocarbon blends that do not conform to simple mean-field theory,[91,98,116] are as follows:

1. χ_B is a nonmonotonic, *nonadditive* function of the conformational and energetic asymmetry variables. A minimum chi parameter, or "most miscible state," occurs for an optimal value of chemical asymmetry.

2. χ_B is relatively large (compared to critical value of $2/N = 0.002$ for present calculations) and positive (unfavorable to mixing) when the structural and energetic asymmetries "reinforce," that is, if $\gamma > 1$, then $\lambda > 1$. This situation corresponds to the common experimental situation for polyolefins where the higher aspect ratio polymer (less branched) is characterized by stronger (more attractive) van der Waals interactions. In the "reinforcement" regime the positive chi parameter increases strongly as γ increases, and decreases as the mean blend aspect ratio increases at fixed conformational asymmetry.

3. Negative values of χ_B can occur when the conformational and energetic asymmetries are sufficiently *large and if* the system is in a "compensation" regime defined as $\gamma > 1$ and $\lambda < 1$ or vice versa. From the perspective of Flory or regular solution theory, this is a surprising and unanticipated result for nonpolar polymer blends. The system behaves as if there were a specific AB attraction that favors AB contacts, although no such *bare* energetic driving force has been included in the calculation. This result suggests novel strategies for achieving miscibility of hydrocarbon polymers of significantly different monomer structures, and may provide a basis for understanding recent "anomalous" experimental observations of such facile mixing of special hydrocarbons.[91]

4. Complex non-mean-field T dependences of χ_B may occur ($\neq T^{-1}$) due to polymer-specific thermal variations of chain aspect ratios. Since $\beta\varepsilon$ is of order unity at typical experimental temperatures, the predicted absolute magnitudes of χ_B are consistent with experimental measurements for most hydrocarbon blends.[69,115]

The origin of the rich predictions for χ_B is the two terms in Eq. (5.11). The leading *exchange energy* type of contribution is dominant in the asymmetry reinforcement parameter regime. However, near or in the "compensation" regime, the second term in Eq. (5.11) can become negative and dominant over the first (positive) contribution. Thus, the physical origin of a $\chi_B < 0$ is the composition-dependent *changes* in blend packing.[113,115]

One might ask whether a useful analytic theory at the thread model level might be constructed based on the PRISM/HTA framework, the free energy route, and the results of Section IV.D. The answer is *no* because of the unrealistic random structural mixing prediction, $g_{MM'}(r) = g_{eff}(r; \phi)$, of the athermal thread blend model, which is inadequate for capturing the consequences of conformational asymmetry on the blend enthalpy of mixing.

The influence of several physical features on the predictions of the PRISM/HTA theory of blends and the simplified SFC and Berthelot models of polymer structure and interactions need to be investigated. These include (i) the effect of atomistic-level structure features such as explicit chain branching, (ii) mixing volume changes and composition-dependent blend packing fraction, (iii) nonideal conformational perturbations, and (iv) possible T-dependent modifications of local packing (beyond HTA).

Points (iii) and (iv) have begun to be addressed within the context of the simplest "symmetric" blend model as discussed elsewhere[117–120] and in Sections VIII and VI, respectively. Some aspects of point (ii) has been examined elsewhere.[69,115] Point (i) has been addressed for the PE/PP blend discussed in Section V.B above. Based on the proposed mappings of real structures to the effective SFC model,[52] and the preaveraging of the site–site Lennard-Jones well depth parameters, we have found[115] that the atomistic and coarse-grained theories predict thermal chi parameters that differ by at most a factor of 2. The generality of this encouraging level of agreement is unknown at present, and we certainly expect there will be systems for which the coarse-grained SFC approach will be poor. However, systematic and detailed applications of the minimalist SFC model approach have been recently carried out for over 50 polyolefin blends and compared against experiments.[115] The agreement between the complicated non-mean-field experimental behavior and the coarse-grained PRISM/HTA theory is very good, and we believe establishes it as a generally reliable zeroth-order basis for understanding and predicting miscibility of polyolefin alloys. The cases not well treated primarily involve *particular* homopolymer/homopolymer blends with small positive chi parameters.[115] For these cases subtle packing effects associated with the nonspherical monomer structure may become dominant.

D. Molecular Solubility Parameter Theory

A second part of the study of Singh and Schweizer[113,115] described in Section V.C was to investigate the validity of a solubility parameter approach to polymer blend miscibility. Solubility parameter, or Hildebrand, theory[85] is potentially extremely useful from a practical viewpoint since it expresses (by assumption) blend properties *solely* in terms of the pure one-component liquid properties. From a fundamental theoretical perspective, it is an interesting question to ask how much of the mixture thermodynamic (and structural) behavior is "foretold" in the pure melt systems?

The *additional* approximations required to obtain a solubility theory from Eq. (5.11) are well known.[69,85] The diagonal $M = M'$ contributions to χ_B are computed using the *pure one-component* melt radial distribution functions $g_{MM}(r) \approx g_{M,\text{melt}}(r)$. The AB cross term is approximated by a geometric combining law involving the pure component diagonal terms

$$-\int d\mathbf{r}\, v_{AB}(r) g_{AB}(r) \approx \sqrt{\int -d\mathbf{r}\, v_{AA}(r) g_{AA,\text{melt}}(r)} \sqrt{\int -d\mathbf{r}\, v_{BB}(r) g_{BB,\text{melt}}(r)} \tag{5.12}$$

Note that such an approximation is more severe, and does not have the same theoretical basis, as the well depth parameter Berthelot analog of $\varepsilon_{AB} = (\varepsilon_{BB}\varepsilon_{AA})^{1/2}$. Under the above conditions, the second term in Eq. (5.11) *vanishes,* and the first term reduces to Hildebrand form, denoted here as χ_h, and given by

$$\chi_h \equiv \frac{\beta}{2}(\delta_B - \delta_A)^2 \tag{5.13a}$$

$$\delta_M^2 \equiv -\rho_m \int d\mathbf{r}\, v_{MM}(r) g_{M,\text{melt}}(r) \tag{5.13b}$$

In Eq. (5.13b) pure component melt cohesive energy parameters have been defined. Note that χ_h is rigorously blend composition-independent and can never be negative. However, subtle, *nonadditive* competitions between conformational and energetic asymmetries are captured since *local* chain architecture enters via its influence on radial distributions.[52,69]

Representative results based on Eq. (5.12) are shown in Figure 20. In the asymmetry "reinforcement" regime (characteristic of most polyolefin blends), there is remarkable agreement between χ_h and χ_B, both with regard to the general shape of the curves and the absolute magnitudes. The level of agreement between χ_h and χ_B tends to improve as the aspect ratio mismatch decreases and/or the mean aspect ratio increases. However, significant differences occur as the "compensation" regime is approached and/or entered, and χ_h can never be less than zero. The "negative" deviation case ($\chi_B < \chi_h$) are generally predicted to be more dramatic.

Clearly, both experimental studies[91,116] and our theoretical results suggest the solubility parameter approach is very valuable, at least for hydrocarbon polymer blends. This has motivated the derivation and application by Schweizer and Singh of an *analytic* molecular solubility parameter PRISM theory at the level of the Gaussian thread model.[69] The thermal chi parameter is given by Eq. (5.13) where the melt cohesive energy parameter of species M, δ_M, are given by Eq. (3.17). The analytic thread predictions for the dependence of χ_h on blend packing fraction, aspect ratios, conformational and energetic asymmetry parameters and spatial range of the attractive branch of the potential are all in *qualitative* agreement with numerically derived trends based on the more realistic SFC model (although quantitative differences occur as expected).[69] The reason for this agreement is the *qualitative* adequacy of the thread model for pure component melt packing and cohesive energy as discussed in Section III.C. The simple analytic form is thus very valuable for both physical understanding of numerically derived trends and qualitative prediction of the consequences of changing system parameters.

VI. BEYOND THERMODYNAMIC PERTURBATION THEORY: MOLECULAR CLOSURE APPROXIMATIONS

The ideas of van der Waals, and the elegant modern statistical mechanical formulations, have clearly established that for *high-density one-component* fluids of atoms or molecules interacting via a harsh repulsion plus slowly varying attractive tail(s), the fluid *structure* is dominated by the nearly hard-core repulsions.[5,11] The effect of attractive interactions on packing is "screened," and their important thermodynamic consequences can be treated within a perturbative or HTA scheme. At moderate and low fluid densities, or in multicomponent fluids, this great simplification is far less tenable, and attractive potentials can significantly modify intermolecular packing correlations.[5,11] The formulation of microscopic integral equation theories and closure approximations that accurately capture the consequences of *both* repulsive and attractive forces under all situations has proven to be a difficult problem even for simple atomic fluids.[5,11] Many approaches exist, but the concept of a "reference" repulsive force fluid treated by one closure scheme (e.g., PY) and a second different closure approximation to treat slowly varying forces (e.g., HNC) is common and quite successful.[5] The "best" closure scheme often depends on system-specific details such as temperature, fluid density, and the strength and spatial form of the tail potentials. Even for atomic fluids, most work along these lines has focused on the variable density *one-component* case, and simple fluid mixtures have received far less attention.[121,122] Theoretical work on molecular liquid mixtures within the interaction site model framework is also sparse.[123] Although some progress has been made recently within the integral equation framework on the question of a proper treatment of the critical divergence (exponents) problem,[18] this aspect has only minor effects on phase diagrams, effective interaction parameters, free energy of mixing, etc., and decreasingly so for polymer mixtures as N increases and mean-field critical behavior becomes increasingly more relevant.[2,112]

Thus, the development of suitable closure approximations to simultaneously treat *both* attractive and repulsive forces in macromolecular fluids and blends has a relatively weak theoretical foundation upon which to build. Moreover, for polymer mixtures phase separation and long wavelength (radius-of-gyration scale) structure and scattering are known to be very molecule weight dependent,[2] a feature with no analog in simple fluids. Considerable difficulties were encountered in our early attempts to use the PRISM approach to treat such thermal macromolecular mixtures based on atomiclike site–site closure approximations.[28,30,31] This motivated the development of novel approximation schemes by Yethiraj and Schweizer,[118,119] referred to as *molecular closures*, which explicitly account for polymer connectivity at the level of *both* the generalized Ornstein–Zernike-like equations and the closure approximation. In this section we discuss the current situation for the above problems using the binary polymer mixture as the prototypical example.

A. Atomic Versus Molecular Closures

The most straightforward approach to treat the effect of the tail potentials $v_{MM'}(r)$ is to employ the site–site analog of closures developed for atoms. Based on the standard site–site MSA closure[5,8]

$$C_{\alpha\gamma}(r) = -\beta v_{\alpha\gamma}(r) \qquad r > d_{\alpha\gamma} \tag{6.1}$$

the predictions of PRISM theory for idealized *symmetric* binary mixture of homopolymers [see discussion below Eq. (5.2)] were obtained by Curro and Schweizer numerically for various single-chain models and analytically within the Gaussian thread idealization.[28,30,124] The PRISM–MSA predictions resulted in *qualitative N-dependent errors* for long wavelength collective fluctuations [$\hat{S}_{MM'}(k)$ for small $kR_g < 1$] as well as for the critical and spinodal temperatures *based on the compressibility route to the thermodynamics*. In particular, the large N-dependence of the effective chi parameter [which controls the amplitude of $\hat{S}_{MM'}(k=0)$ for the symmetric blend] and spinodal temperature are given by

$$\chi_{\text{eff}} = \frac{\rho}{2}[\hat{C}_{\text{AA}}(0) + \hat{C}_{\text{BB}}(0) - 2\hat{C}_{\text{AB}}(0)] \propto \chi_0 N^{-1/2} \tag{6.2}$$

$$T_s \propto \rho d^3 |\varepsilon_{\text{AA}}| \phi(1-\phi)\sqrt{N} \tag{6.3}$$

where χ_0 is the "bare" result defined in Eq. (5.1b), which ignores all aspects of polymer connectivity and intermolecular correlations. Thus, PRISM–MSA theory predicts a massive N-dependent renormalization of the chi parameter, and radical modification of the mean-field[1] scaling law of $T_c \propto \mathrm{N}$.

Surprisingly, the Flory N-scaling behavior[1] had never been verified experimentally at the time the PRISM–MSA prediction was obtained in 1988. Subsequently, both specially designed experiments by Bates et al. on model isotopic polymer blends,[125] and lattice Monte Carlo simulations by Deutch and Binder,[126] clearly showed the PRISM-MSA N-scaling predictions were *incorrect* and the mean-field Flory scaling was essentially exact. PRISM theory based on alternative site–site closures (PY, HNC, MS; see Section II) *and the compressibility route* were shown to disagree *even worse* with the mean-field N-scaling law than the MSA approximation.[118] Thus, the unsettling conclusion is that PRISM theory with any standard atomiclike closure to treat attractions makes qualitative errors for the N-dependence of long-wavelength fluctuations and compressibility route spinodal phase boundaries.[118,119]

As shown by Chandler,[127] for binary symmetric model mixtures the erroneous phase boundary prediction problem can be avoided by employing the free energy route to the thermodynamics. From Eq. (5.3), one sees that the long-wavelength errors in interchain pair correlations due to the tail potentials are *cutoff* for fluid systems interacting via relatively short-range tail potentials. In the large N limit the HTA becomes exact and simple Flory theory is recovered (with quantitative corrections due to local density correlations). Thus, one conclusion that can be drawn is that atomic closures for attractions yield results that are "thermodynamically inconsistent" in a *qualitative* manner for long-wavelength concentration fluctuation processes.[127] However, questions concerning the long-wavelength structure, scattering, and physical clustering driven by thermal interactions *cannot* be addressed by invoking a HTA or the free energy route. Moreover, from a basic theoretical perspective it would be very satisfying to construct new closure approximations that are thermodynamically consistent at least in a *qualitative* N-scaling sense. Of course, there would remain the unavoidable *quantitative* inconsistencies, but such a development would place PRISM theory on the same footing as simple liquid-state theory.[5]

The development of new *molecular closure* schemes was guided by analysis of the nature of the failure of the MSA closure. In particular, the analytic predictions derived by Schweizer and Curro[124] for the renormalized chi parameter and critical temperature of a binary symmetric blend of *linear polymeric fractals* of mass fractal dimension d_f embedded in a spatial dimension D are especially revealing. The key aspect of the mass fractal model is the scaling relation or growth law between polymer size and degree of polymerization: $R_g \propto N_f^{1/d} \sigma$. The non-mean-field scaling, or chi-parameter renormalization, was shown to be directly correlated with the average number of "close contacts" between a pair of polymer fractals in D space dimensions: $N^2/R_g^D \propto N^{2-(D/d_f)}$. If the polymer and/or space is sufficiently "open" ($d_f < D/2$), then the number of contacts between a pair of macromolecules is intensive (N-independent), and the predicted N-dependent renormalization of the effective chi parameter disappears. Moreover, the predicted renormalization remains qualitatively unchanged with varying temperature, and thus applies *even at very high temperatures in the weak coupling, small concentration fluctuation limit*. These features strongly suggest the origin of the N-dependent renormalization is fundamentally linked to the *two-polymer* problem and not directly with correlations at low temperature.[119,124]

With this motivation, a family of new molecular closures was formulated by Yethiraj and Schweizer[118,119] *at the level of two macromolecules, not two interaction sites*. Even at the two-macromolecule level, the presence of a large number of interaction sites implies there are many *indirect* correlation pathways (mediated by chain connectivity) between a pair of tagged sites on different macromolecules [the leading $\underset{\sim}{\Omega} * \underset{\sim}{C} * \underset{\sim}{\Omega}$ terms in the PRISM equation (2.2)]. Such *indirect* correlation pathways are ignored in the standard atomic site–site closure approximations that retain *only the direct interaction* between the pair of tagged sites. One might anticipate that the importance of such indirect pathways increases with N, and their neglect by atomiclike closures is the origin of the N-dependent errors of such schemes.

PRISM with the molecular closures has been applied both numerically[118,120] and analytically[119,120,128] to several different macromolecular mixture problems. All the (limited) studies to date support the conclusion that this molecular closure approach does properly describe the N dependence of long-wavelength concentration fluctuations and spinodal phase boundaries. As true of simple RISM, these new closures are not really "derived," but their heuristic construction is motivated by a combination of intuition and empirical experience with liquid-state theory. Their ultimate validity and usefulness can be established only by comparison of the theoretical predictions with simulations, experiments, and known exact limiting results.

The strategy for explicitly formulating the molecular closures was guided by three considerations.[118,119] (1) Use of the commonly employed "reference" approach.[5] The successful site–site PY closure is retained to describe the repulsive force reference fluid but a molecular closure scheme is adopted to describe the attractive, slowly varying forces. (2) The approximation scheme is required to provide an exact description of the structural consequences of the tail potentials at the *two-molecule* level in the weak coupling limit $[\beta v_{MM'}(r) \ll 1]$. (3) Use of an appropriate site–site approximation for the *direct* attractive interaction contribution motivated by experience in simple fluids.[5,8]

Taken as a whole, the ideas discussed led Yethiraj and Schweizer to propose the following *reference molecular closure* approximations for site interaction potentials consisting of a hard core plus tail[118,119]

$$\underset{\sim}{\Omega}{}^{*}\underset{\sim}{C}{}^{*}\underset{\sim}{\Omega}(r) = \underset{\sim}{\Omega}{}^{*}\underset{\sim}{C}{}^{(0)*}\underset{\sim}{\Omega}(r) + \underset{\sim}{\Omega}{}^{*}\,\Delta\underset{\sim}{C}{}^{*}\underset{\sim}{\Omega}(r) \qquad r > d_{\alpha\gamma} \tag{6.4}$$

where the symbol * represents spatial convolution integrals. From the PRISM equation (2.2b) one sees that the quantity being approximated on the left-hand side of this equation is the *two-molecule component* of the interchain pair correlation $h_{MM'}(r)$. The convolution structure implies that the full site–site direct correlation function inside and outside the hard core are coupled, which is not true for the standard atomic closures. Explicit results require specification of the attractive potential contribution to the direct correlation function $\Delta C_{MM'}$. For nonpolar fluids three schemes have been proposed and investigated based on analogy with the MSA and linearized PY closures of simple fluids[119]

$$\Delta C_{MM'}(r) = -\beta v_{MM'}(r)\Theta_{MM'}(r) \tag{6.5}$$

$$\Theta_{MM'}(r) = 1 \quad \text{(R-MMSA)} \qquad g^{0}_{MM'}(r) \quad \text{(R-MPY/HTA)} \qquad g_{MM'}(r) \quad \text{(LR-MPY)} \tag{6.6}$$

where the tail potential is nonzero only *outside* the hard-core diameter. In the "full" R-MPY closure $\Theta_{MM'}(r) \equiv [1 - \exp(\beta v_{MM'}(r))]g_{MM'}(r)$. Within the two-molecule framework, the R-MMSA is "mean-field-like" in the sense that *only* the bare tail potential enters. On the other hand, the linearized R-MPY approximation can be thought of as correcting the bare attractions by weighting their importance by the full interchain pair correlation function. Thus a self-consistent "feedback" mechanism between fluid packing and the effective potentials $[C_{\mathrm{MM'}}(r)]$ enters in a manner akin to a partial enthalpy quantity. The HTA version of the linearized R-MPY does not have such a feedback aspect and corrects the bare potential only via the packing correlations of the reference repulsive force fluid. The PRISM equations, with molecular closures plus the exact hard-core impenetrability constraint, can be numerically solved using the Picard algorithm.[5]

Analytic solutions are also possible based on the idealized Gaussian thread model since the molecular closures simplify dramatically. Because the hard-core diameter is shrunk to zero, Eq. (6.4) applies for all r, thereby allowing "cancellation" of the convolution integrals and all factors of ω. Hence, the thread analogs of Eqs. (6.5) and (6.6) become[119]

$$C_{\mathrm{MM'}}(r) = C^{0}_{MM'}\delta(\mathbf{r}) - \beta v_{\mathrm{MM'}}(r)\Theta_{\mathrm{MM'}}(r) \tag{6.7}$$

Note that the R-MMSA and R-MPY/HTA approximations for the direct correlation functions are now "deterministic," that is, uncoupled from the determination of the *full* pair correlation functions $g_{\mathrm{MM'}}(r)$. However, for the most complex R-MPY closure the direct correlation functions are still self-consistently linked to the *full* radial distribution functions and tail potentials.

Although essentially all studies to date using PRISM and the molecular closures have involved macromolecules, it is conceivable such closures may be of value even for small or intermediate-sized flexible and/or rigid molecules. A careful documentation of the accuracy of the new molecular closures as a broad function of thermodynamic state and molecular fluid type remains an important future direction. In addition, recent interesting alternative approaches to liquid theory for polymer mixtures with attractions have been developed within the general PRISM framework by Melenkevitz and Curro[129] based on the optimized RPA(ORPA) approach, and Donley et. al.[41] based on density functional theory and also from a field-theoretic perspective by Chandler.[127] Application of these approaches to treat the effect of attractive interactions on fluid structure and phase transitions remains to be worked out.

B. Fluctuation Phenomena in Symmetric Blends

As described briefly in Section V, the "structurally and interaction symmetric" binary blend is an idealized model in which the A and B homopolymer (single site) chains are identical in every respect except they interact via a potential $v_{\mathrm{AB}}(r)$, which disfavors mixing. By symmetry, the critical concentration is equimolar, $\phi_c = 0.5$. Many "symmetric" models can be constructed depending on the choice of single-chain model and form of the intermolecular tail potentials $v_{MM'}(r)$. In the athermal limit, this symmetric blend reduces to a homopolymer fluid. Although the symmetric model is not of direct experimental relevance, it has been the prime subject of the large majority of computer simulations and field-theoretic studies of thermal polymer alloys.[112] It is believed to contain the generic aspects of demixing and concentration fluctuations in polymer blends of central interest to polymer physicists. Such a symmetric model has also been investigated in the simple atomic liquids community.[121] One origin of the mathematical simplicity of this model is the near decoupling of density and concentration fluctuations due to the assumed symmetries.

Based on lattice model simulations (for N typically in the range 16–256), Binder and co-workers have discovered a host of non-mean-field, or fluctuation, phenomena in the symmetric blend model, most of which appear to be nonuniversal.[126,130] In particular, in contrast to Flory–Huggins theory, an effective, or renormalized, chi parameter that deviates from the mean-field behavior [see Eq. (5.2)] in many ways was observed:

$$\frac{\chi_{\text{eff}}}{\chi_0} = F(T, \rho, \phi, N, \{v_{MM'}(r)\}) \tag{6.8}$$

where F is a complicated function(al) that also depends on polymer intramolecular structure. This *renormalization ratio* is generally less than unity (fluctuation stabilized). The fluctuation effects discovered via simulations, though obtained for lattice models, provide stringent semiquantitative tests of the ability of PRISM theory with the new molecular closures to describe density and concentration fluctuation processes in polymer blends.

In this section some of the recent results obtained for the symmetric blend based on PRISM with the molecular closures are summarized.[118–120] In the numerical studies a tangent SFC chain model with zero bending energy is employed (which is the closest continuum chain model used in the lattice simulations). For simplicity, the single-chain structure factor is taken to be *ideal* (blend composition and density independent), and *temperature independent*. The tail potentials are taken to be of a Yukawa form

$$v_{MM'}(r) = \varepsilon_{MM'} \frac{d}{r} \exp\left[\frac{-(r-d)}{ad}\right] \qquad r \geq d \tag{6.9}$$

where d is the site hard-core diameter, a is the spatial range, and $\varepsilon_{MM'}$ is the energy parameter. The effective chi parameter is computed from the first equality in Eq. (6.2), which is adequate for the symmetric blend model but not in general due to nonlinear compressibility effects.[128,131]

1. *Numerical Results*

A very extensive numerical study has been recently carried out by Singh and co-workers.[120] Except for the critical divergence Ising exponent, excellent agreement is found between PRISM theory *based on the R-MPY closure* and simulation[126,130] for *all* the fluctuation effects mentioned above. Some of the fluctuation effects are found to be intrinsic, that is, survive in the long chain limit, while others are finite size effects that arise from chain-connectivity-induced coupled local and long-wavelength concentration fluctuations. *However,* due to the multiple sources of the fluctuation effects, even asymptotic finite size effects can *appear intrinsic* over extended ranges of N as observed in the relatively small N simulation studies.

All the fluctuation effects can be understood in simple terms by examining the enthalpy of mixing and local interchain pair correlations.[120] The key physical process is thermally driven local interchain rearrangements corresponding to clustering of like species. These fluctuation processes are driven largely by the enthalpic desire to reduce unfavorable AB contacts, which becomes increasingly important for smaller N, lower blend density, shorter tail potential range, and/or more equimolar concentrations.

Here we give a couple of representative examples for both thermodynamic and structural properties.[120] *At fixed temperature*, PRISM/R-MPY calculations of $\chi_{\mathrm{eff}}(\phi)$ generally show a parabolic dependence on blend composition with a minimum at $\phi = 0.5$ in qualitiative accord with simulation. This behavior is due to the fact that at fixed T the blend is closer to the demixing transition the closer the blend composition is to equimolar. Thus, fluctuation stabilization effects become weaker as the pure one-component fluid state is approached leading to the observed parabolic-like behavior. An example of the amplitude of this effect as a function of N is shown in Figure 21 for a fluid packing fraction [$\eta = 0.2$ (concentrated solution)] representative of the lattice simulations. Results are plotted for various temperatures close to the critical temperature (corresponding to $\chi_{\mathrm{eff}} N = 2$), spatial range of tail potentials, and thermodynamic route. Most results are for an *S ordering* of tail potentials corresponding to $\varepsilon_{\mathrm{AA}} = \varepsilon_{\mathrm{BB}} = 0$ and $\varepsilon_{\mathrm{AB}} > 0$, except for the one case labeled "MS" which corresponds to a "Most symmetric" choice of tail potentials: $\varepsilon_{\mathrm{AA}} = \varepsilon_{\mathrm{BB}} = -\varepsilon_{\mathrm{AB}} = \varepsilon < 0$ also studied by simulation.[126] Note the reduced amplitude for the MS potential model case, and the near N independence of the amplitude for N of the order of 10^2 and smaller,

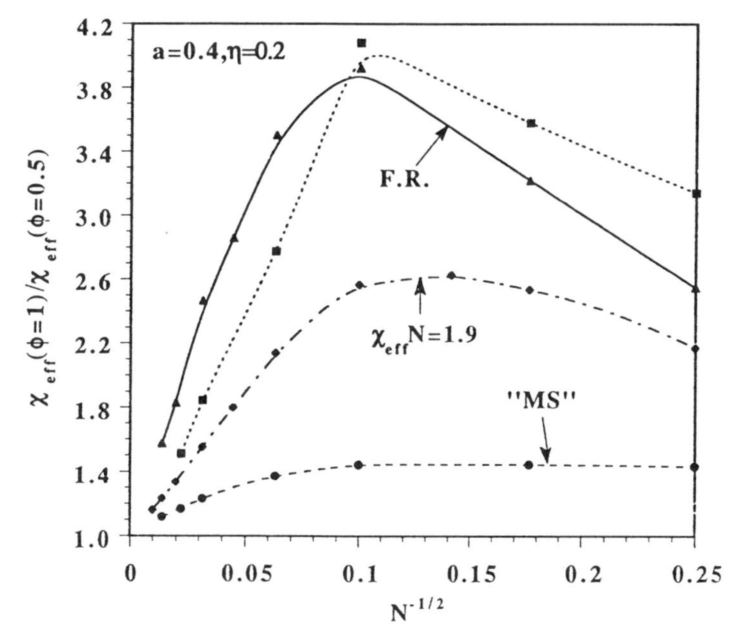

Figure 21. PRISM/R-MPY predictions[120] for the amplitude of the composition dependence of the effective interaction parameter of an equimolar, symmetric binary mixture as a function of chain degree of polymerization. The spatial range parameter of the Yukawa tail potential is $a = 0.4$. Except as noted, all the curves are based on the compressibility route to the thermodynamics (F.R. denotes free energy route), are at the critical point ($\chi_{\mathrm{eff}}N = 2$), and the S-ordering choice of the tail potentials.

features in agreement with the lattice simulations.[126] However, in the large N limit the chi parameter becomes concentration independent since the relevant temperature scale diverges ($T_c \propto N$), and the tail potentials cannot then induce structural rearrangements and physical clustering. At meltlike packing fractions of $\eta = 0.5$ (not shown here),[120] the blend is far less compressible and the fluctuation effects as embodied in $\chi_{\mathrm{eff}}(\phi)$ are suppressed to the level of a 20–30% effect for N of the order of 100 and less.

The predicted critical temperature relative to the mean-field Flory value is shown in Figure 22. Note the significant reduction of T_c in all cases due to thermally induced changes in blend structure that reduce (enhance) the number of close AB (AA and BB) contacts. However, the N dependence of this fluctuation effect and approach to the long-chain limiting behavior is highly nonuniversal. Simple physical arguments have been given to explain such trends.[120] The results of Figure 22 agree with the limited simulation studies[126,130] available for the effect of tail

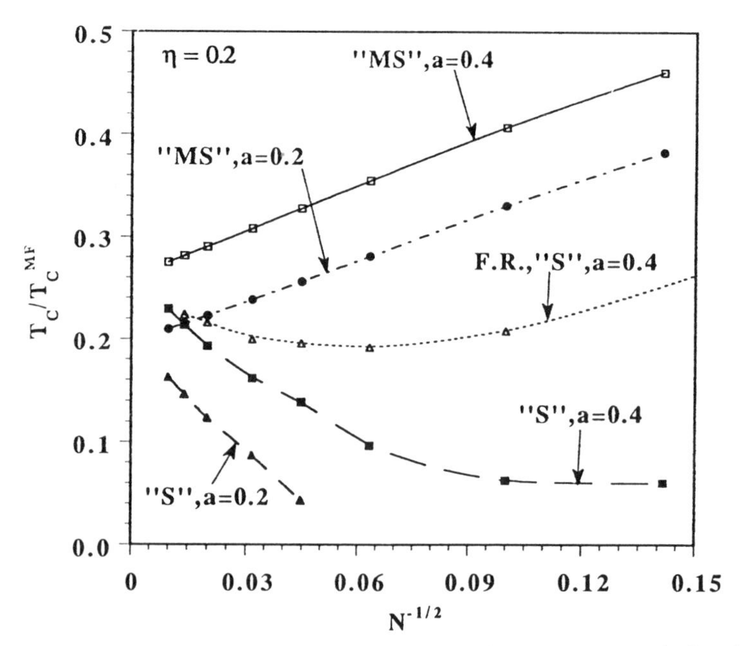

Figure 22. Concentrated solution symmetric model critical temperature (reduced by its mean-field value) as a function of N as predicted by PRISM/R-MPY theory.[120] Results for two choices of tail potential orderings and spatial range are shown. Except as noted, all curves are for the compressibility route to the thermodynamics. Smooth interpolative curves through the theoretical points are a guide to the eye.

potential ordering, spatial range, and N on the renormalization of the critical temperature. At higher meltlike packing fractions, trends similar to Figure 22 are obtained (with some subtle changes), but the overall scale of the normalization effects is much smaller, for example, 0.8–1.1, consistent with the less compressible nature of the melt state.[120]

An example of the changes in blend pair correlations upon cooling from the athermal (pure melt) limit to the critical temperature is shown in Figure 23 for the two values of N and a highly asymmetric blend composition. Strong deviations of the $g_{MM'}(r)$ from the athermal melt behavior are seen especially for the smaller N case. Moreover, the clustering of the minority species (A) is more dramatic than the majority (B) species. For the $N = 2000$ case, the majority species pair correlations are barely discernable from the athermal melt behavior, consistent with the approach to the HTA limit. Similar trends are seen at higher meltlike packing fractions but again are less dramatic due to the less compressible nature of the fluid.[120] Continuum molecular dynamics simulations are underway to test the influence of attractive tail potentials on blend

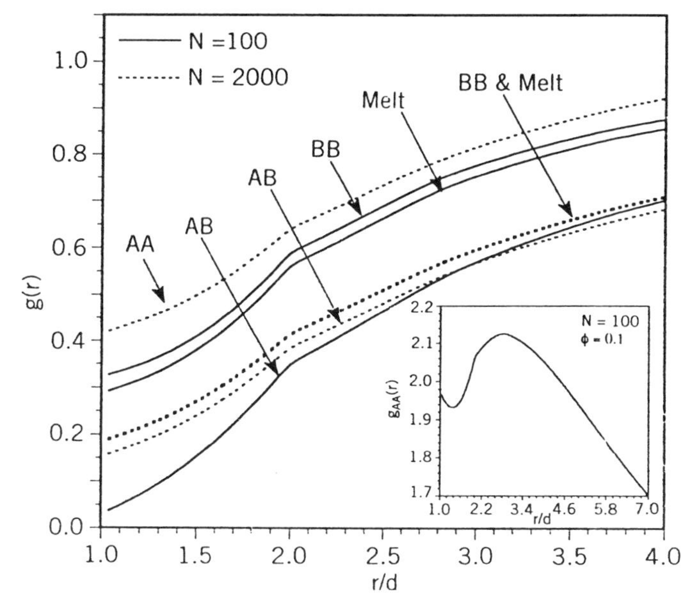

Figure 23. PRISM/R-MPY predictions for the local intermolecular blend structure of a 10/90 mixture of A and B chains.[120] The corresponding athermal limit (homopolymer melt) curves are also plotted. Results are shown *at the $\phi = 0.5$ critical temperature* for two choices of degree of polymerization, the S-ordering of the tail potential ($a = 0.4$) and a concentrated solution packing fraction of $\eta = 0.2$. The inset shows $g_{AA}(r)$ for $N = 100$, which is off scale on the main plot.

structure and the molecular closure approximations.[132] Very recent work by Gromov and de Pablo[132] has shown for the symmetric blend model that PRISM with the R-MPY closure is in excellent agreement with continuous space simulations for the structure, mixing thermodynamic properties, and the coexistence curve.

Finally, for PRISM/R-MPY theory the thermodynamic consistency between the free energy and compressibility route calculations of the chi parameter and spinodal phase boundaries has been shown to be remarkably good.[120] Moreover, in the long-chain limit the predicted chi parameter and phase boundary appear to be *exactly equivalent*, which is a unique circumstance for liquid-state theories. However, this is not a general feature of PRISM with the molecular closures but rather derives from the fact that in the long-chain limit the critical temperature becomes arbitrarily high, the HTA is rigorous, and thus the *symmetric blend* reference system reduces to a *composition-independent* homopolymer melt.

2. *Analytic Predictions for Gaussian Threads*

For the symmetric Gaussian thread blend interacting via the Yukawa tail potentials of Eq. (3.16), a nearly complete analytic treatment can be carried out for all three molecular closures of Eq. (6.7) within the S potential ordering.[119,120] Remarkably, all the analytically derived trends are consistent with numerical studies based on the compressibility route to the thermodynamics. Results based on the free energy route have also been obtained.[120]

Based on the simplest R-MMSA closure the effective chi parameter is given by[119]

$$\chi_{\rm eff} = \beta\rho \int d\mathbf{r}[v_{\rm AA}(r) - v_{\rm AB}(r)] = \chi_0 \tag{6.10}$$

Thus, in the thread limit PRISM with the R-MMSA closure reduces precisely to mean-field Flory theory! The corresponding spinodal temperature is given by Eq. (5.2). The inverse osmotic compressibility, or concentration fluctuation scattering intensity, is nearly exactly given by the RPA form[2]

$$\hat{S}^{-1}(0) = \frac{1}{N\phi} + \frac{1}{N(1-\phi)} - 2\chi_0 \tag{6.11}$$

that is, $\hat{S}^{-1}(0)$ goes to zero linearly with inverse temperature.

The R-MPY/HTA closure includes the athermal density fluctuation correction (ϕ-independent) due to the $g_0(r) \neq 1$ factor in Eq. (6.5). The predicted chi parameter, denoted $\chi_{\rm HTA}$, is given by[119]

$$\frac{\chi_{\rm HTA}}{\chi_0} = 1 - \frac{3}{\pi a \rho \sigma^2}\left[\frac{1}{1 + a/\xi_c} - \frac{1}{1 + a/\xi_\rho}\right]$$

$$\rightarrow \frac{a}{a + \xi_\rho} = \frac{T_{s,\rm HTA}}{T_{s,\rm FH}} \quad \text{as } N \rightarrow \infty \tag{6.12}$$

where ξ_ρ and ξ_c are defined in Eq. (3.4), and $T_{s,\rm FH}$ is the mean-field Flory–Huggins (or thread PRISM/R-MMSA) spinodal temperature. The inverse osmotic compressibility still displays the standard T^{-1} law of Eq. (6.11), only the absolute magnitude of the effective chi parameter is modified.

Within the compressibility route to the thermodynamics, the above two closures yield predictions that miss entirely the blend composition-dependent fluctuation corrections obtained numerically in the preceding subsection based on the R-MPY closure. Application of the linearized version of the latter (LR-MPY of Eq. (6.6)) for the thread model results in a nonlinear, self-consistent equation for χ_{eff}, which does capture these fluctuation effects. Straightforward analysis yields the nonlinear transcendental equation[120]

$$\frac{\chi_{\text{eff}}}{\chi_0} = 1 + \frac{3(\xi_c^{-2} - \xi_\rho^{-2})}{\pi\rho\sigma^3\Theta}$$
$$\Theta \equiv (\xi_+^{-1} + \xi_-^{-1})(1 + a/\xi_+)(1 + a/\xi_-) \tag{6.13}$$

where a density-fluctuation-like length scale ξ_+ and concentration fluctuation length scale ξ_- are given by

$$\xi_+^{-2} = \xi_\rho^{-2}[1 + 2\phi(1-\phi)\chi_{\text{eff}} NR^{-2}]$$
$$\xi_-^{-2} = \xi_\rho^{-2}[1 - 2\phi(1-\phi)\chi_{\text{eff}} N] \qquad R = \xi_\rho/\xi_c \tag{6.14}$$

Note that the above two length scales are coupled via χ_{eff} and the nonlinear self-consistent Eq. (6.13). Thus, the blend density and concentration fluctuations are coupled in a manner that becomes stronger for lower density (more compressible fluid), smaller chain lengths, and/or thermodynamic states closer to the spinodal phase boundary.

For long chains and moderate densities, Eq. (6.13) predicts the amplitude of the composite dependence of the chi parameter to be[120]

$$\frac{\chi_{\text{eff}}(\phi = 1)}{\chi_{\text{eff}}(\phi = \frac{1}{2})} = 1 + \frac{\xi_\rho}{\xi_c}\left(1 + \frac{\xi_\rho}{\xi_c}\right)[1 - \sqrt{1 - \chi_{\text{eff}} N/2}] \tag{6.15}$$

This simple equation explains all the numerical results for $\chi_{\text{eff}}(\phi)$ as a function of N, a, and ρ. It is clear that the composition dependence of the chi parameter is a finite size effect that is inversely proportional to the correlation hole length scale ξ_c, which is enhanced as the critical point is approached. With increasing density, the screening length ξ_ρ decreases, which reduces the composition dependence of χ_{eff}. The tail potential range a competes with the density screening length.

For high densities and/or large N, the R-MPY closure can be shown to yield[120]

$$\frac{T_{c,\text{MPY}}}{T_{c,\text{HTA}}} \cong \left[1 + \frac{\xi_\rho}{\xi_c}\left(1 + \frac{\xi_\rho}{\xi_c}\right)\right]^{-1} \tag{6.16}$$

which quantifies the effect of concentration fluctuations on blend stabilization. The general spinodal temperature, T_s, follows from Eq. (6.13) by setting $2\chi_{eff}N\phi(1-\phi)=1$. The resulting expressions show that the *shape* of the predicted spinodal envelope remains of the mean-field Flory *form*, that is, the net renormalization *at the spinodal* is ϕ independent.

As discussed elsewhere,[120] a consequence of the concentration fluctuation process captured by the linearized or full R-MPY approximation is a *finite size nonlinear* region in the $\hat{S}^{-1}(k=0)$ versus T^{-1} scattering curve near the spinodal temperature corresponding to a stabilization relative to the mean-field/RPA behavior. Analytic expressions have also been derived for the interchain pair correlations, $g_{MM'}(r)$. Physical clustering, as characterized by $\Delta g(r)$ of Eq. (4.1), depends explicitly on all three length scales, liquid density, and (implicitly) spatial range of the tail potential. Its amplitude approaches zero as $N^{-1/2}$ for large N, that is, as the ratio of microscopic length scales to the macromolecular size. The chain length asymmetric case, $N_A \neq N_B$, has also been studied[120] and distinctive modifications of the fluctuation effects are predicted that can be tested by computer simulation.

In the long-chain limit, all concentration fluctuation effects on the chi parameter disappear and a density correlation corrected version of mean-field theory is obtained since the R-MPY closure reduces to its HTA version. Thus, *perfect* thermodynamic consistency between the free energy and compressibility route predictions are obtained in the $N\rightarrow\infty$ asymptotic limit for the idealized *symmetric* blend system.[120]

C. Conformational and Interaction Asymmetric Blends

The idealized symmetric blend model is not representative of the behavior of most polymer alloys due to the artificial symmetries invoked.[128] Predictions of spinodal phase boundaries of binary blends of conformationally and interaction potential asymmetric Gaussian thread chains have been worked out by Schweizer[128] within the R-MMSA and R-MPY/HTA closures and the compressibility route to the thermodynamics. Explicit analytic results can be derived for the species-dependent direct correlation functions $C_{MM'}$, effective chi parameter, small-angle partial collective scattering functions, and spinodal temperature for arbitrary choices of the Yukawa tail potentials. Here we discuss only the spinodal boundary for the simplest Berthelot model of the $v_{MM'}(r)$ tail potentials discussed in Section V. For simplicity, the A and B polymers are taken to have the same degree of polymerization N.

The spinodal temperature corresponding to liquid–liquid phase separation follows from the condition $\hat{S}^{-1}_{MM'}(0)=0$ of Eq. (4.5). The analytic result is[128]

$$k_B T_s = \frac{\rho|\tilde{H}_{AA}|}{\phi + \gamma^4(1-\phi)} \left[N\phi(1-\phi)(\gamma^2 - \lambda)^2 + \left\{ \frac{\phi + \lambda^2(1-\phi)}{-\rho C^0_{AA}(\phi)} \right\} \right] \tag{6.17}$$

where the attractive energy scale variable $\tilde{H}_{AA}$ is given by

$$|\tilde{H}_{AA}| \equiv \begin{cases} |\hat{v}_{AA}(0)| & \text{R-MMSA} \\ |\hat{v}_{AA}(0)| \dfrac{a}{a + \xi_{eff}} & \text{R-MPY/HTA} \end{cases} \tag{6.18}$$

In the above equations, $\xi_{eff} = 36\sigma_A/[-\pi\sigma_A^{-3} C^0_{AA}]$ is a ϕ-dependent effective density screening length in the reference athermal blend, $\gamma = \sigma_B/\sigma_A$ quantifies the conformational asymmetry, $\lambda^2 = \varepsilon_{BB}/\varepsilon_{AA}$ quantifies the interaction asymmetry, and C^0_{AA} is the integrated strength of the AA direct correlations in the athermal thread blend [see Eq. (4.7)]. The leading term displays the classic N scaling associated with macromolecular concentration fluctuations. It predicts a strong, nonadditive connection between miscibility and conformational and energetic asymmetries. Note that even if there is no "bare" energetic driving force for phase separation ($\lambda = 1$; $\chi_0 = 0$), strong immiscibility can occur due to the conformational differences that influence the direct correlation functions, and hence local packing and enthalpy of mixing. This striking non-mean-field prediction has been very recently verified by Kumar and Weinhold (preprint, 1996) using off-lattice Monte Carlo simulation. The possibility of *reinforcement* or *cancellation* of the conformational and energetic asymmetries is also predicted. This behavior was also found numerically based on thermodynamic perturbation theory[113,115] (see Section V.C), and analytically within the molecular solubility parameter approach[69] of Section V.D. However, subtle differences between the compressibility and (numerical) free energy routes for this and other aspects do occur due to the thermodynamic inconsistency problem with integral equation theory. For example, the emergence of a *negative* chi parameter in the asymmetry *compensation* regime (see Fig. 20) is not predicted. The shape of the predicted spinodal envelope, and location of the critical composition, is not of the classic symmetric inverted-parabola form due to both stiffness asymmetry and explicit compressibility corrections. The application of Eq. (6.17) to qualitatively interpret a range of experiments on polyolefin and polydiene alloys has been given by Schweizer.[128]

The second term in Eq. (6.17) can be viewed as a *compressibility contribution*, which is N independent and vanishes in the hypothetical incompressible limit where $-C^0_{AA} \rightarrow \infty$. For long chains it is generally expected to be a minor correction.

We also note that in principle, Eq. (6.17) is really a transcendental equation for the spinodal temperature if the polymer density and/or statistical segments length depend on temperature as generally true for real systems. This aspect can lead to apparent N dependences of the spinodal temperature, which do not follow the classic Flory scaling relation of $T_s \propto N$. A liquid–gas-type transition is also predicted,[128] but since it is driven by density, not concentration, fluctuations the relevant temperature scale is independent of N, and thus is expected to be well below T_s for macromolecules.

Numerical PRISM studies based on finite thickness SFC models of the stiffness and interaction binary blend, and molecular closures and compressibility route, have not yet been widely pursued. Preliminary blend calculations by David and Schweizer[133] using the R-MPY/HTA closure do find trends qualitatively consistent with the analytic thread predictions. These studies also suggest that for chains in the range of $N = 500$–5000 that temperature-induced changes in local packing in the homogeneous phase blend are quite weak, thereby providing some support for the use of the perturbative HTA approach to thermodynamics discussed in Section V.

D. Other Physical Problems and Systems

There are many other physical systems and fundamental questions involving the role of attractive forces in macromolecular fluids that have begun to be addressed.

1. *The effect of attractions on the structure of dense one-component polymer melts*. According to the van der Waals ideas, attractions should have very little effect. Surprisingly, we are unaware of simulations that have probed this question, although they are now in progress.[132] Recent PRISM studies by Butler and Schweizer[134] using atomic and molecular closures have been carried out. Repulsive force "screening" of the effects of attractions on structure is recovered for many, but not all, closure approximations.

2. *The effect of attractions on solvent quality (good, theta, poor), and low temperature polymer–solvent phase separation*. Some tentative analytic work has been done on the latter problem by Schweizer and Yethiraj based on the molecular closures.[119] Non-mean-field dependences of the critical polymer volume fraction on N have been found.

3. *Liquid–vapor equilibrium of chain molecule fluids.* Both analytic and numerical work has been recently done by Schweizer and coworkers.[119,134] The compressibility route predictions of PRISM for this problem are *extremely sensitive* to closure approximation since the relevant fluid densities are very low and large-scale density fluctuations are present.[134] The atomiclike MSA closure leads to qualitatively incorrect results as does the R-MMSA closure. However, the R-MPY/HTA approximation appears to be in excellent accord with the computer simulation studies of n-alkanes[135a] and model chain polymers,[135b] including a critical density that decreases weakly with N and a critical temperature that increases approximately logarithmically with N.

4. *Physical clustering, long-wavelength concentration fluctuations, and microphase separation of self-assembling block copolymers and surfactant-like chain molecules.* These systems are discussed in the next section.

5. *Heating-induced phase separation (the so-called lower critical solution temperature case) often occurs in nonaqeous polymer solutions and blends.*[93,94,101] The physical origin of this phenomenon, and the degree of universality of mechanism, remains largely a mystery. *One* possible mechanism for chemically complex monomers is a competition between dispersive forces, which favor low-temperature demixing, and "specific," more rapidly varying attractive forces (e.g., hydrogen bonding or charge transfer) between A and B monomers, which favor low-temperature miscibility. In such a situation, thermally induced structure changes in the blends may be very important and could lead to a lower critical solution temperature (LCST). Such a mechanism is presently under study using PRISM theory and the molecule closures. An alternative mechanism based on nonadditive hard-core volumes has been explored by Honeycutt using PRISM theory.[114]

VII. SELF-ASSEMBLING BLOCK COPOLYMERS

Block copolymers are molecules composed of two or more distinct monomers chemically bonded in the same chain.[95,136] We consider the simplest case where there are two types of elementary units A and B. These units are arranged into bonded linear sequences, or *blocks*, of variable length that are then repeated a variable number of times. To date, only one-component fluids composed of "periodic" block copolymers where the A and B block lengths are unique have been studied based on PRISM theory. However, random or statistical copolymers where there is quenched chemical or sequence disorder associated with the polymerization process are also of great interest.[137,138]

For simplicity, we focus on the simplest linear *di*block copolymer, which consists of one chain of N_A units connected at a *junction* to one chain of N_B units forming a polymer of $N = N_A + N_B$ units. The copolymer composition is denoted by a variable $f = N_A/N$. Simple diblock copolymers are macromolecular versions of classic surfactant molecules composed of chemically distinct *head* and *tail* portions.

Within the PRISM approach, one can formally view a block copolymer as a special case of the multisite systems discussed in Section III.B. However, the block structure results in novel physical phenomena, not displayed by simple polymer melts, which is important both scientifically and from a materials engineering viewpoint.[95,136] The most prominent feature is the ability of such a fluid to spontaneously self-assemble into *microdomains* of variable purity, spatial symmetry, and *N*-dependent size. Ultimately a first-order microphase separation transition, or weak crystallization, into an ordered superlattice-type structure can occur where the characteristic domain size or lattice constant is typically 50–500 Å. Such microscopic segregation represents a dramatic structural reorganization of the fluid and can be driven by changing temperature, increasing degree of polymerization, or increasing copolymer density in solution.

To treat in a general manner the self-assembling block copolymer fluid using PRISM theory requires developing appropriate macromolecular closures for describing the structural consequences of both repulsive (hard core) and attractive intermonomer interactions. From a theoretical polymer physics perspective, fundamental questions include (i) the existence (or nonexistence) of a critical point or spinodal instability on a finite length scale (microphase separation), (ii) possible breakdown of the conformational ideality approximation as physical clustering of like species and microdomain formation emerges, and (iii) the proper treatment, and physical origin, of fluctuation processes that stabilize a disordered, but highly correlated, liquid phase against microphase separation.

In this section we summarize progress made over the past few years by David and Schweizer,[33,133,139,140] including both the basic theoretical modifications of PRISM theory that have been proposed and examples of specific predictions and their relationship to experiments,[136,141,142] phenomenological field-theoretic studies,[143–145] and computer simulations.[146–148] The application of the new molecular closures that describe structure and thermodynamics in a qualitatively correct manner are emphasized, along with addressing points (i) and (iii) listed above. Point (ii) requires a fully "self-consistent" treatment, which is briefly addressed in Section VIII. Here, we retain the conformational ideality approximation. Recent self-consistent block copolymer PRISM studies[140] suggest corrections to the ideality assumption are a small, second-order effect for *macromolecular* systems in disordered fluid states. Description of first-order microphase transitions into ordered structures of various symmetries have been addressed within phenomenological field-theoretic approaches.[143–145,149] However, the treatment of this aspect *within* a liquid-state framework remains a largely unsolved problem, although promising density functional schemes are presently being developed by Nath et al.[150] Such density functional approaches necessarily require the homogeneous phase correlations as input.

As was true for the homopolymer melt and blend problems, mathematical tractability requires greatly reducing the number of coupled integral equations of order N^2. At the simplest level for the block copolymer case, both chain end and AB junction effects are preaveraged.[33] That is, the site–site direct correlation functions are assumed to depend only on species type, not location within the block copolymer molecule. For the single-site models of present interest, only three distinct direct and intermolecular site–site pair correlation functions are required for an AB copolymer. The PRISM equations are again given by Eq. (2.2), with an effective Ω matrix given by[138]

$$\begin{aligned} \hat{\Omega}_{\mathrm{MM}'}(k) &= \rho_{\mathrm{M}} N_{\mathrm{M}}^{-1} \sum_{\alpha,\gamma \in \mathrm{M}} \hat{\omega}_{\alpha\gamma}(k) \\ \hat{\Omega}_{\mathrm{AB}}(k) &= \rho_m (N_{\mathrm{A}} + N_{\mathrm{B}})^{-1} \sum_{\alpha \in \mathrm{A}} \sum_{\gamma \in \mathrm{B}} \hat{\omega}_{\alpha\gamma}(k) \end{aligned} \tag{7.1}$$

where ρ_m is the total site number density. Various fluid densities and copolymer chain models have been studied (Gaussian thread, freely jointed, semiflexible).[33,133,139,140] The theory is general, but the numerical examples discussed here are for diblock *melts* ($\eta = 0.5$).

A. Athermal Limit

For the athermal diblock copolymer fluid the intersite potential is given by a hard-core potential between sites M and M′ with associated hard-sphere diameter $d_{\mathrm{MM}'}$, and the site–site PY closure of Eq. (2.5b) has been employed. Here, we focus on the $l_b/d = 0.5$ overlapping site SFC model for the linear chain blocks, which has been constructed to mimic the packing of real polymers as discussed in Section III.A. For simplicity, the hard-core diameters of the A and B sites are taken to be equal. Results for a series of four, $f = \frac{1}{2}$ (compositionally symmetric), $N = 500$ copolymers melts, which have a "common" A block with an aspect ratio of unity, are considered.[139] The relevant single-chain parameters are

TABLE II
Aspect Ratios, $\Gamma_M = \sigma_M/d$, of the Blocks That Compose the $l/d = 0.5$ Semiflexible Copolymer Systems Studied (labeled as A–D)[a]

Copolymer	Γ_A	Γ_B	$\gamma = \Gamma_B/\Gamma_A$	$R_{g,A}$	$R_{g,B}$	$D = 2\pi/k^*$
A	1	$\frac{4}{5}$	$\frac{4}{5}$	6.38	5.13	26.3
B	1	1	1	6.38	6.38	29.2
C	1	$\frac{5}{4}$	$\frac{5}{4}$	6.38	7.92	32.8
D	1	$\frac{3}{2}$	$\frac{3}{2}$	6.38	9.43	36.6

[a] Also listed in units of the site diameter are the block radii of gyration and microdomain size, D, for the $f = 0.5$, $N = 500$ athermal copolymer melts studied in Figs. 24 and 25.

listed in Table II, and the aspect ratios chosen are representative of the range appropriate for flexible polymers of experimental interest.[52,139] Conformational asymmetry is characterized by the variable $\gamma = \Gamma_B/\Gamma_A$.

Typical results[139] for the *common* block partial structure factors, $\hat{S}_{AA}(k)$, are shown in Figure 24. Only the low-angle regime is plotted to emphasize the prominent "correlation hole" maximum at $k = k^*$. Such a peak is a general consequence of the diblock architecture and is often characterized by a microdomain length scale of $D = 2\pi/k^*$, which is

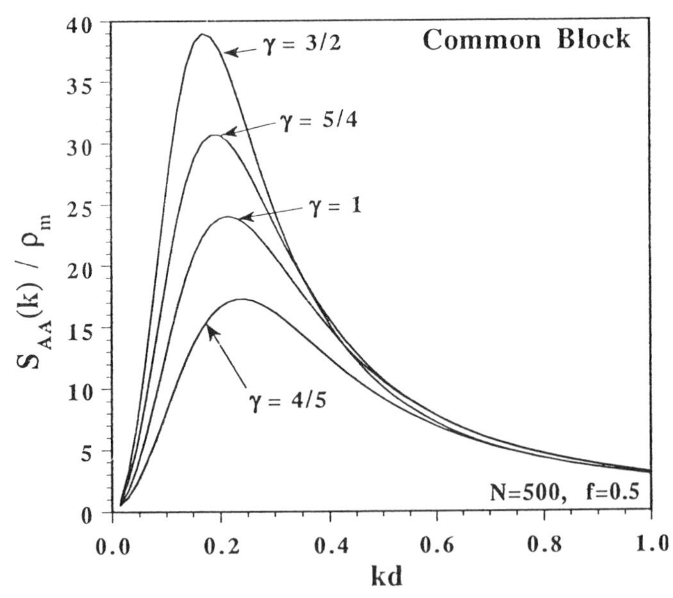

Figure 24. Predictions of PRISM theory with the PY closure[139] for the low wave vector common block collective structure factor for the athermal copolymer models listed in Table II. Note that the common block structure factor intensifies monotonically as the overall copolymer stiffness stiffness increases due to increasing B-block aspect ratio.

roughly equal to *three times* the overall copolymer radius of gyration. The curve labeled $\gamma = 1$ represents a homopolymer melt in which the two halves of the molecule have been artifically labeled. For this case the A and B monomers are obviously *randomly mixed* by construction. Thus, Figure 24 shows that if the A block is attached to a stiffer B block (higher aspect ratio), then collective fluctuations of the concentration of A monomers are enhanced, while if the B monomer is more flexible, the A collective fluctuations are suppressed *relative to* the random mixing behavior. In all cases, the changes relative to the random mixing situation are small, and no indications of incipient microphase separation are apparent in this athermal limit. Predictions for the wide-angle scattering regime are also obtainable, and subtle modifications of the local fluctuation regime relative to the homopolymer melt situation are found.[139]

The analog of Figure 24 for the variable block have also been obtained (not shown here).[139] Significant differences are found relative to the common block behavior. In particular, the peak partial scattering intensity is always found to be largest for the more flexible block. This nonequivalence of *partial compressibilities* on a length scale $2\pi/k^*$ is a result of the microscopic differences in packing (or direct correlations) of the A and B monomers. Such behavior is not captured by phenomenological field-theoretic approaches,[143–145] which describe long-wavelength concentration fluctuations by enforcing a literal *incompressibility* condition that introduces an artificial symmetry corresponding to $\hat{S}_{AA}(k) = \hat{S}_{BB}(k) = -\hat{S}_{AB}(k)$.

The common block site–site pair correlations on the *local* scale are shown in Figure 25. Relative to the homopolymer behavior, the *local* packing of A segments is enhanced (reduced) when the variable B block is stiffer (more flexible) than the A chain. This trend is physically sensible and corresponds to an "induced" change in common block packing due to the diblock connectivity. However, beyond separations of two or three site diameters in the longer distance scale "correlation hole" regime this trend *reverses*, for example, the packing of A segments is enhanced when the B block is more flexible than A. This behavior is largely a consequence of the fact that as the B block becomes more flexible, the overall diblock copolymer size decreases and the spatial range of the correlation hole is reduced, and thus $g_{AA}(r)$ approaches its random value of unity more quickly. Locally, packing trends for the *variable* block, $g_{BB}(r)$, are weakly perturbed[139] from the corresponding homopolymer melt behavior shown in Figure 2.

For a diblock copolymer fluid approaching a microphase separation transition, one expects long-distance oscillations in the $g(r)$'s to develop with a period given by the microdomain size $D = 2\pi/k^*$ (see Table II).

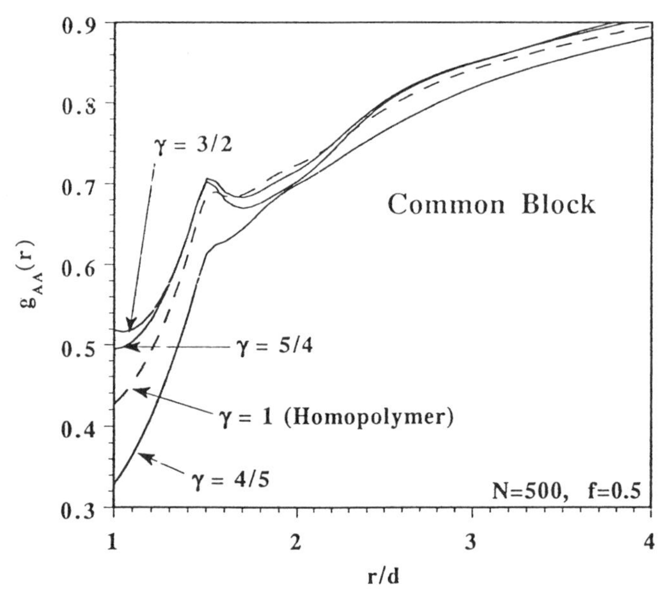

Figure 25. Common block intermolecular radial distribution function for the same athermal conditions stated in Figure 24. Note the monotonic increase of local packing as the B-block stiffness is increased.

For the athermal diblock melts such oscillations are found to be extremely low amplitude. In addition, the difference correlation function $\Delta g(r)$ as defined in Eq. (4.1) has also been studied, which is a spatially resolved measure of physical clustering into microdomains. It has been found to be extremely low amplitude (typically, 0.02–0.04 for the cases considered), indicative of very little segregation of the A and B segments.[139] These structural features emphasize the high degree of miscibility predicted by PRISM theory for conformationally asymmetric diblocks of experimentally relevant N values in the purely athermal limit. Thus, to properly describe microphase separation and physical clustering of block copolymer fluids, it is necessary to include thermal effects and attractive interactions.[33,133,138]

B. Thermally Driven Assembly

Consider the same model system as above, but where there are attractive tail potentials, $v_{MM'}(r)$, between sites of type M and M' on different copolymers. In the numerical examples presented later, the shifted Lennard-Jones attraction of Eq. (5.10) is appended to the SFC diblock model of Section VII.A. For simplicity, the Berthelot potential model is again adopted: $\varepsilon_{AB} = (\varepsilon_{AA}\varepsilon_{BB})^{1/2}$, and thus the chemical or interaction asymmetry between the A and B monomers is contained in the energy ratio parameter $\varepsilon_{BB}/\varepsilon_{AA} \equiv \lambda^2$. The temperature is nondimensionalized in units of ε_{AA}. Within the simplest, incompressible mean-field theory based on random mixing, the relevant "bare" energetic quantity for controlling thermodynamic incompatibility, and hence microphase segregation, is the Flory chi parameter of Eq. (5.1b).

As was found for the analogous blend, the application of site–site atomiclike closure (e.g., MSA) to the diblock copolymer fluid problem predicts a qualitatively incorrect N dependence of the long-wavelength thermal concentrations fluctuations.[33] Thus, use of the molecular closure approximations[118,119] are necessary. For the AB diblock copolymer fluid of present interest, there are three closure equations of the form of Eq. (6.5). It is important to note that the closure approximations utilized are identical in form for diblock copolymers and homopolymer blends, even though the physical processes involved (microphase separation versus liquid–liquid macrophase separation) are very different. The matrix structure of the molecular closures implies the direct correlation functions outside the hard core are not explicitly given by the closure approximation.[33] Within the interaction site model approach, it is well known that in such cases the site–site direct correlation functions acquire a low amplitude, very small k, divergent part as a necessary consequence of the defining generalized Ornstein–Zernike-like equations.[151] For the block copolymer problems this peculiar mathematical feature has been shown to be essentially irrelevant in that it has no discernable effect on the physically relevant theoretical predictions obtained numerically.[33]

The basic qualitative features of the numerical PRISM predictions for nonzero hard-core diameter models are similar for all the closure approximations, although significant differences can occur for certain properties.[33] We defer discussion of the latter point to Section VII.C and here present numerical results[133] based on the linearized R-MPY/HTA closure defined in Eqs. (6.4) and (6.5). Qualitative comparisons[33,133] with the idealized *structurally and interaction symmetric* diblock model lattice (continuum) simulations of Binder and co-workers[146] (Grest[148]) show good agreement with the PRISM predictions.

Consider first the $f = \frac{1}{2}$ compositionally symmetric diblock models of Table II and the simplest situation of no (bare) chemical symmetry, that is, $\lambda = 1$. For this case mean-field theory predicts that thermal energetic effects vanish, $\chi_0 = 0$. On the other hand, within liquid-state theory since the A and B monomers form chains of different aspect ratios, nonrandom correlations exist that result in a correlated contribution to the fluid enthalpy. As the diblock melt is cooled, further structural reorganization, *driven by the attractive tail potentials*, occurs in order to reduce energetically unfavorable contacts.

Explicit examples of thermal PRISM predictions[133] are shown in Figures 26–29. The growth of the peak in the long-wavelength partial structure factors upon cooling is shown in Figure 26. There are several important features: (1) The peak intensity grows very strongly with cooling but does not diverge. Initially, the behavior is mean-field-like, that is, a nearly linear decrease of the inverse peak intensity with inverse temperature. However, at very low temperatures a new "fluctuation" regime is entered characterized by nonlinearity of the plot and the avoidance of a true spinodal divergence. (2) As was true in the high-temperature (athermal) limit, the more flexible species exhibits larger microdomain scale fluctuations. This feature is interpretable in several ways.[133] For example, since the more flexible block packs more poorly in the liquid, it can more easily undergo large-scale concentration fluctuations since the A-rich domains are more compressible. (3) An *apparent* spinodal can be defined by extrapolation of the peak scattering curves in the vicinity of the nonlinear regime. The deduced temperature is

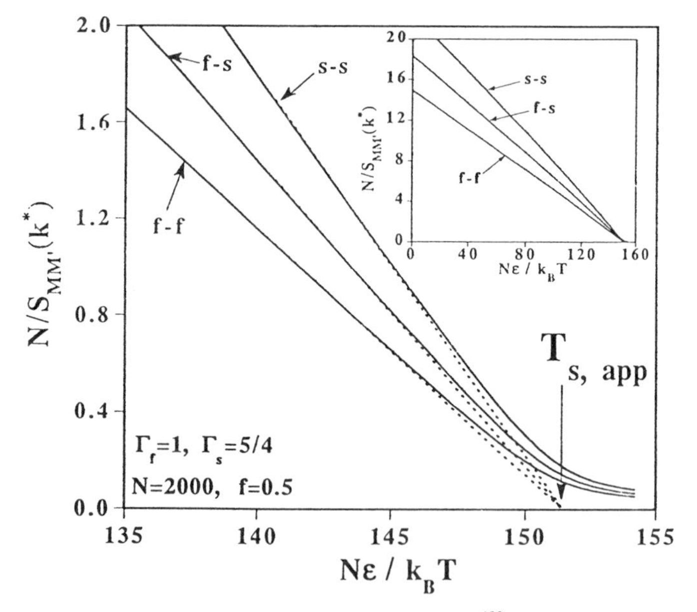

Figure 26. Reciprocal partial peak scattering intensities[133] as a function of dimensionless inverse temperature for an asymmetric diblock melt (model C) with $N = 2000$ and $f = 0.5$. High-temperature linear (mean-field) extrapolations, shown as dotted lines, converge to a unique apparent spinodal temperature.

essentially identical for all three partial scattering functions consistent with an extrapolated microphase spinodal interpretation.

The example of Figure 26 is for a single choice of N, f, and conformational asymmetry $\gamma = \Gamma_B/\Gamma_A$. Systematic numerical calculations[33,133] and the thread model analytic analysis[33] (see Section VII.C) reveal that (i) the extrapolated spinodal temperature grows strongly with N (roughly linearly), (ii) the temperature interval over which the nonlinear behavior of the inverse peak scattering intensity versus T^{-1} decreases as the chains get larger, and (iii) at fixed N and f the apparent (extrapolated) microphase spinodal transition temperature increases strongly as the conformational asymmetry increases. All these trends are in qualitative accord with experiments on polyolefin diblock melts,[136,141,142] and point (ii) implies that the fluctuation stabilization of the disordered phase is a finite size (N) process.

An example of point (iii) is shown in Figure 27 for two values of the chemical asymmetry variable λ. This figure illustrates that conformational mismatch strongly destabilizes the homogeneous diblock phase. This effect arises from block-dependent packing differences that result in a

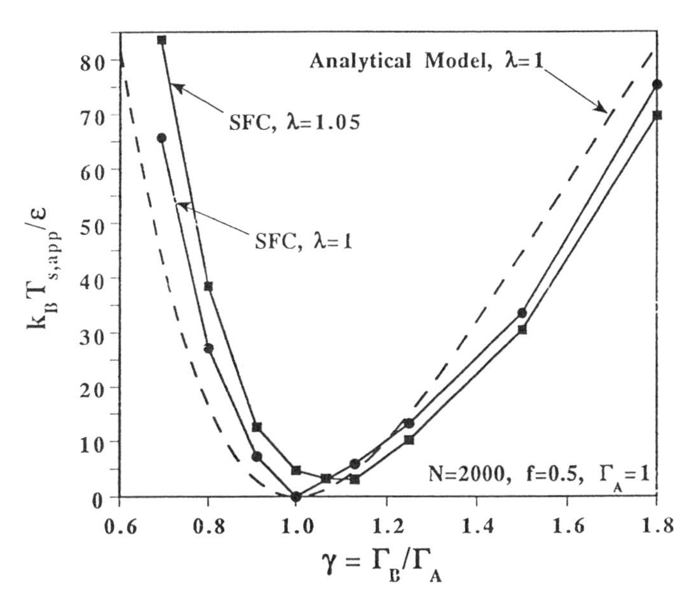

Figure 27. Apparent spinodal phase diagram for $N = 2000$, $f = \frac{1}{2}$ stiffness asymmetric diblock copolymer melt.[133] Here $\Gamma_A = 1$, while Γ_B was varied. Numerical results (SFC) are shown for two choices of the bare attractive energy asymmetry variable λ. The dashed curve shows the corresponding analytical thread predictions discussed in Section VII.C.

correlated contribution to fluid enthalpy. In addition, Figure 27 clearly shows that the bare energetic and conformational asymmetries do not influence the spinodal temperature in an additive manner. The asymmetry factors, which individually tend to destabilize the randomly mixed state, can either "reinforce" or tend to "compensate" each other thereby giving rise to novel mechanisms of copolymer miscibility and immiscibility. The curve labeled "analytic" arises from PRISM theory for the Gaussian thread model as discussed in Section VII.C.

Species-dependent interchain pair correlations at low temperatures are shown in Figure 28 and significant changes from the high T athermal behavior are found (see, e.g., Fig. 25). As expected physically, on a spatially local and global scale as the temperature is lowered there is an enhancement (reduction) of like (unlike) contacts due to emerging microdomains and AB interfaces. The flexible block reorganizes more upon cooling since its microdomains are more compressible. This is dramatically illustrated by the crossing of the stiff–stiff and flexible–flexible pair correlations with increasing site separation. On macromolecular length scales, strong oscillations develop indicative of phase separation on the microdomain length scale D. Consistent with the long

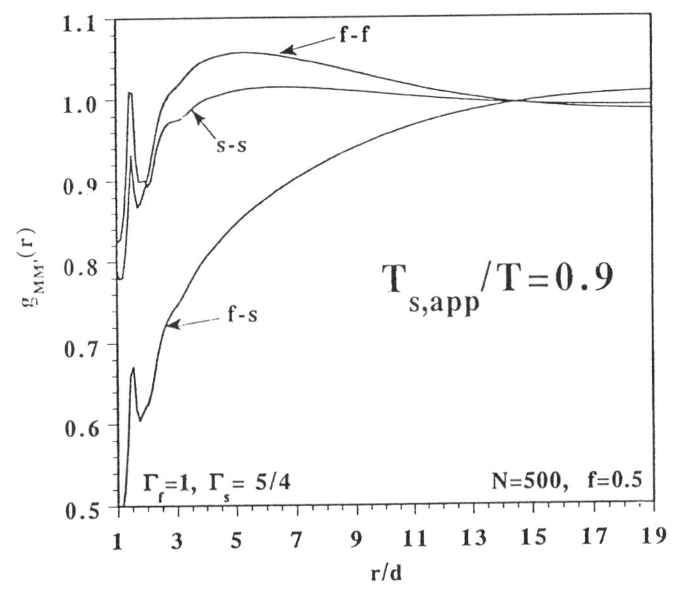

Figure 28. Low-temperature radial distribution functions[133] for model C melt with $N = 500$ and $f = 0.5$.

wavelength scattering peaks of Figure 24, the flexible block displays *larger* amplitude long-wavelength fluctuations.

The influence of block aspect ratios and overall degree of polymerization on the physical clustering *at a fixed* melt packing fraction, fixed ratio of aspect ratios, and fixed distance from the (extrapolated) spinodal temperature is displayed in the form of $\Delta g(r)$ in Figure 29 [defined here to be *one-half* the quantity given in Eq. (4.1)]. There are two key trends.[133] (i) Segregation is enhanced when the B block is more flexible or equivalently when the *average* diblock copolymer aspect ratio is smaller. This is again due to the enhanced compressibility of the lower mean aspect ratio fluid and emphasizes that the absolute value of both block aspect ratios, not just their relative value or difference, are important. (ii) Segregation or physical clustering monotonically decreases with increasing N. This trend clearly establishes that the physical process

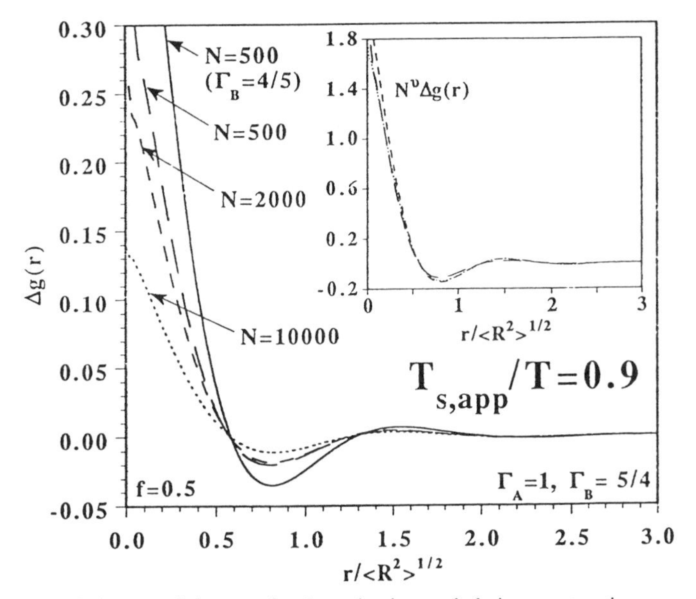

Figure 29. Influence of degree of polymerization and chain aspect ratio asymmetry on the pairing function,[133] defined here as $\Delta g(r) = 0.5[g_{AA}(r) + g_{BB}(r)] - g_{AB}(r)$, at a relatively low-temperature.[132] Results are shown at a fixed (low) temperature relative to the apparent microphase spinodal point for model C and for several values of overall degree of polymerization. One case for model A ($\Gamma_B = \frac{4}{5}$) is shown to demonstrate the effect of overall copolymer backbone stiffness. The inset shows a reduced or scaled plot for the three different N cases of model C. Optimum collapse of the different curves occurs when the exponent $\nu \cong 0.275$.

responsible for generating concentration fluctuation phenomena is of a "finite size" nature, which would appear to vanish in the hypothetical $N \rightarrow \infty$ limit. The possibility of simple N scaling of this fluctuation process is explored in the inset of Figure 29, which shows that, surprisingly, a near superposition of different N cases is obtained *over all length scales*. This observation suggests a tight correlation, and common origin, for the local and global thermally induced structural changes in the diblock copolymer liquid. The apparent scaling exponent, ν, is roughly 0.275. The interpretation of this value is postponed to Section VII.C.

Another interesting *spatially resolved* measure of segregation and microdomain formation is a "length-scale-dependent composition" of species M′ in a spherical volume of radius R surrounding a tagged monomer of type M defined as[133,139]

$$\Phi_{\mathrm{MM}'}(R) \equiv \frac{\rho_{\mathrm{M}'} \int_0^R r^2 g_{\mathrm{MM}'}(r)\, dr}{\sum_{\mathrm{M}''} \rho_{\mathrm{M}''} \int_0^R r^2 g_{\mathrm{MM}''}(r)\, dr} \tag{7.2}$$

In either the large R limit or the random mixing limit of $g_{\mathrm{MM}'}(r) = g(r)$ for all M and M′, these functions reduce to the stoichiometric values defined by the copolymer composition. Indirect experimental measurements of such local compositions have recently been attempted (with the help of a model) by several methods that probe equilibrium and/or dynamic structure on length scales varying from a few monomer diameters (light scattering, nuclear magnetic resonance (NMR); relevant to the glass transition), to the block size (dielectrics for polar polymers), up to the microdomain length scale (small-angle neutron scattering). This definition is constructed to count only segments on different chains surrounding a tagged monomer. Such a definition would seem the most relevant one for describing length-scale-dependent *inter*chain friction in condensed phases. Inclusion of the self, intramolecular contribution is trivial.

An example[133] of the predictions of PRISM for these effective compositions is shown in Figure 30 for $f = \frac{1}{2}$ and a range of length scales R. At high temperatures, deviations from the random mixing, stoichiometric value of $\frac{1}{2}$ is small, consistent with the discussion of the athermal limit in Section VII.A. A rapid growth occurs as the extrapolated spinodal temperature is approached with increasingly pure domains emerging on short length scales. The flexible block reorganizes more than the stiff block due to its enhanced compressibility. As expected, the segregation decreases as the correlation volume examined increases. On the microdomain length scale, $D \cong 33d$, probed by SANS there is roughly

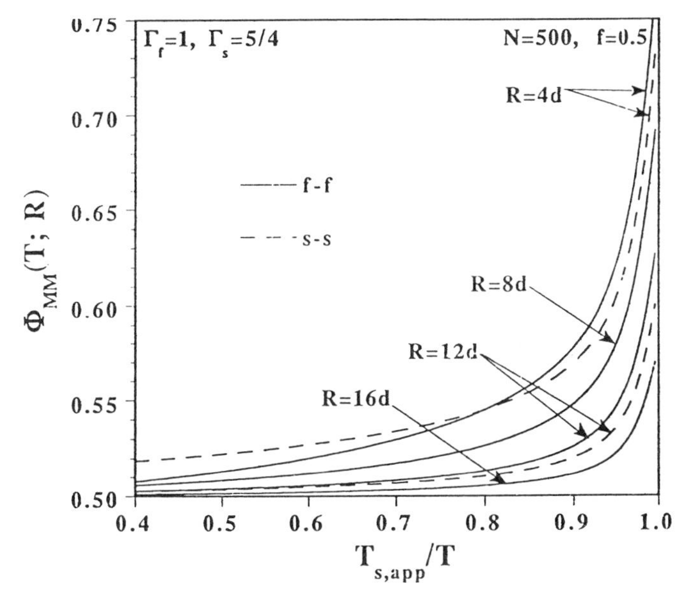

Figure 30. Diagonal values (flexible-flexible and stiff-stiff) of the effective composition[133] as a function of reduced inverse temperature for various choices of the correlation radius R and $f = \frac{1}{2}$ for model C. The size of the flexible block is $R_{g,f} = 6.4d$, the overall diblock size is $R_g = 10.2d$, the microdomain length scale is $D = 32.8d$.

a few percent enrichment. Consistent with the results of Figure 30, such segregation does decrease with increasing N (at fixed $T/T_{s,\text{app}}$), but does so *very slowly*. For experimentally relevant values of diblock copolymer $N < 1000$ strong fluctuation effects are always predicted.

Results (not shown) for the very compositionally asymmetric case of $f = \frac{1}{10}$ have also been obtained.[133] As opposed to the $f = \frac{1}{2}$ case, which is known to organize into lamellar or "sheetlike" domains, the $f = \frac{1}{10}$ case generally acquires a body-center-cubic-type structure at low T corresponding to spherical domains of the minority components dispersed in the continuous majority phase. Thus, very weak perturbations of the effective composition from its bulk value is found for the majority domain, and very strong deviations emerge for the minority domain (e.g., $\Phi = 0.2 - 0.4$ for $T = 0.9T_{s,\text{app}}$). We note that PRISM theory does not predict microdomain symmetries of ordered phase but rather the pair structure of the highly fluctuating precursor or supercooled fluid.

Detailed and successful applications of the PRISM theory of diblock polyolefin melts to interpret the strong experimental correlation[98] of the microphase ordering temperature with polyolefin monomer structure (e.g., chain stiffness and monomer branching) have been presented.[132]

Finally, we make three general comments about PRISM theory in the low-temperature limit where convergence of the numerical algorthim becomes increasingly difficult.[33,133] (1) At very low temperatures, defined as $T \ll T_{s,\mathrm{app}}$, the cross term $g_{\mathrm{AB}}(r)$ can attain unphysical negative values for small r. This is perhaps not surprising since nothing in the integral equation approach explicitly forbids such behavior, and at low temperature the true physical system does try to mimimize contact between the A and B segments (narrow interface limit). (2) There are indications that at very low temperatures PRISM predicts the emergence of a highly correlated, but globally disordered, fluid phase that attains a *ground-state structure*. By the latter term we mean the peak scattering intensity, indicative of the purity and spatial coherence of microdomains, and the interchain pair correlations, approach limiting low T values. Analytic support[33] for this assertion is given in Section VII.C. As a purely speculative remark, for $f \cong \frac{1}{2}$ such a diblock copolymer fluid phase may have the form of a quenched, spinodally decomposed fluid (perhaps of random bicontinuous morphology). Such an idea has been previously advocated based on experimental studies by Bates, Fredrickson, and co-workers.[136,141] (3) At very low temperatures one expects[136] the diblock copolymer will "stretch" in order to further reduce AB contacts, thereby violating the assumed "Flory ideality" conformational behavior. PRISM theory has been generalized to treat this aspect, although the amount of stretching appears small for long chains not too far below the extrapolated spinodal temperature.[140]

C. Analytic Predictions in the Gaussian Thread Limit

In the Gaussian thread limit analytic results have been derived for copolymer fluids using the molecular closures.[33,128,140] The analytic results provide insights to several key questions and behaviors that emerge from the numerical PRISM studies. These include: (1) the role of nonzero monomer hard-core diameter, density fluctuations, and concentration fluctuations on diblock liquid-phase behavior and structure; (2) relationship between phenomenological field-theoretic approaches[143–145] and the molecular closure-based versions of PRISM theory; and (3) the influence of molecular weight, composition, solution density, and chemical and conformational asymmetries of the blocks on copolymer microphase separation temperatures.

Most field-theoretic work,[143–145] and all computer simulation studies to date,[146–148] have focused on the idealized "structurally and interaction symmetric" AB copolymer model. As was true for the blend (see Section VI.B), in the high-temperature athermal limit this model diblock reduces to a homopolymer fluid. This oversimplified model has the symmetry-related virtues of near decoupling of density and concentration fluctuations and allows the cleanest study of the polymer physics aspects of points (1) and (2) listed above. Thus we consider it first.

1. *Idealized Symmetric Model*

For simplicity, we consider the S-ordering tail potential choice: $v_{\text{MM}} = 0$ and $v_{\text{AB}} > 0$. The R-MMSA and linearized R-MPY/HTA expressions for the direct correlations functions and effective chi parameter are identical to the blend cases and hence are "deterministic." Thus, at low temperature the reduced repulsive tail potential, $\beta v_{\text{AB}}(r)$, can become arbitrarily large leading to a microphase separation spinodal instability defined as $\hat{S}_{\text{MM}'}(k^*) \rightarrow \infty$. The latter condition is precisely given by[33]

$$\hat{\Lambda}(k^*) = 0 = 1 - \rho_{\text{A}}\hat{\omega}_{\text{AA}}(k^*)\hat{C}_{\text{AA}}(k^*) - \rho_{\text{B}}\hat{\omega}_{\text{BB}}(k^*)\hat{C}_{\text{BB}}(k^*)$$
$$- 2\rho_{\text{m}}\hat{\omega}_{\text{AB}}(k^*)\hat{C}_{\text{AB}}(k^*) + \rho_{\text{A}}\rho_{\text{B}}\,\delta\hat{\omega}(k^*)\,\delta\hat{C}(k^*) \tag{7.3}$$

$$\delta\hat{\omega} \equiv \hat{\omega}_{\text{AA}}\hat{\omega}_{\text{BB}} - f^{-1}(1 - f^{-1})\hat{\omega}_{\text{AB}}^2 \qquad \delta\hat{C} \equiv \hat{C}_{\text{AA}}\hat{C}_{\text{BB}} - \hat{C}_{\text{AB}}^2 \tag{7.4}$$

where $\rho_{\text{m}}\hat{\omega}_{\text{MM}'}(k^*) = \hat{\Omega}_{\text{MM}'}(k^*)$ from Eq. (7.1), and k^* is implicitly defined as the wave vector for which the diveregence first occurs. One can easily show[33] that for this symmetric thread diblock model at high meltlike densities, PRISM plus the R-MMSA closure reduces precisely to the incompressible mean-field theory of homogeneous block copolymers derived by Leibler[143] with $\chi_{\text{eff}} = \chi_0$. The R-MPY/HTA closure leads to a nearly identical theory,[33] but the chi parameter is corrected for local density fluctuations via the athermal pair correlation function in a manner essentially identical to the analogous blend result of Eqs. (6.12).

The reduction of thread PRISM with the R-MMSA closure for the idealized fully symmetric block copolymer problem to the well-known incompressible RPA approach[143] is reassuring. However, in contrast with the blend case, for copolymers that tend to *micro*phase separate on a finite length scale, the existence of critical or spinodal instabilities is expected to be an artifact of the crude statistical mechanical approximations. That is, finite N "fluctuation effects" are expected to destroy all such spinodal divergences and result in only first-order phase transitions in block copolymers [i.e., Eq. (7.3) is never satisfied]. Indeed, when PRISM theory is numerically implemented for finite thickness chain models using the R-MMSA or R-MPY/HTA closures spinodal divergences do not occur.[33] Thus, one learns that even within the simpler molecular closures, the finite hard-core excluded volume constraint results in a "fluctuation effect" that destroys the mean-field divergences.

Thus, taking the thread limit results in an "ideal" phase transition that allows a precise microphase separation transition temperature to be defined which is relevant to extrapolated spinodals that can be measured experimentally and via simulation. This aspect is exploited in the next subsection.

Although the nonzero hard-core diameter constraint itself is sufficient to destroy spinodal and critical instabilities, one suspects this may not be the dominant mechanism for fluctuation stabilization of the low-temperature copolymer fluid.[33] The enthalpic feedback mechanism contained in the R-MPY closure, which as discussed in Section VI.B correctly describes the concentration fluctuation effects in the macroscopically phase separating symmetric *blend*,[120] will also result in an arrest of the divergent concentration fluctuations on the microdomain length scale. Although these two sources of fluctuation stabilization are not rigorously separable, comparison[33] of numerical PRISM studies with the simulations of Binder and Fried[146] suggest the enthalpic feedback mechanism is dominant, at least for the idealized symmetric diblock model.

The linearized R-MPY version of the thread molecular closure condition of Eq. (6.7) can be shown to result in a nonlinear, self-consistent integral equation for the effective chi parameter[33] [or equivalently the concentration fluctuation part of the collective structure factor $\hat{S}(k)$]

$$\hat{\chi}_{\text{eff}}(k) = \hat{\chi}_{\text{HTA}}(k) - \frac{1}{4\pi^3 f(1-f)\rho_m} \int d\mathbf{k}'\, \hat{\chi}_0(|\mathbf{k}-\mathbf{k}'|)\hat{\chi}_{\text{eff}}(k')\hat{F}^{-1}(k')\hat{S}(k') \tag{7.5}$$

where χ_{HTA} is the R-MPY/HTA closure prediction, $\hat{S}(k)$ is of the incompressible random phase approximation (IRPA) *form*

$$[\rho_m \hat{S}(k)]^{-1} = \hat{F}(k) - 2\hat{\chi}_{\text{eff}}(k) \tag{7.6}$$

and $\hat{F}(k)$ is defined as

$$\hat{F}(k) \equiv \frac{f\hat{\omega}_{\text{AA}}(k) + (1-f)\hat{\omega}_{\text{BB}}(k) + 2\hat{\omega}_{\text{AB}}(k)}{f(1-f)\,\delta\hat{\omega}(k)} \tag{7.7}$$

Analytic expressions for F are available for Gaussian chain models.[143] The self-consistent fluctuation correction arises from the (finite size) coupling of local, thermally driven changes of $g_{\text{AB}}(r)$ with the microdomain scale concentration fluctuations. Such a coupling is mediated by the chain and block connectivity constraints, and vanishes if the HTA becomes exact, that is, $g_{\text{AB}}(r) - g_0(r) \to 0$. This condition can be achieved in many ways: $T \to \infty$ (or $\chi_{\text{eff}} \to 0$), $f \to 0$ or 1 (homopolymer limit), $\rho \to \infty$ (incompressible limit), $N \to \infty$, or the "infinitely weak, infinitely long range" Kac tail potential limit ($a \to \infty$).

Equations (7.5)–(7.7) have been solved numerically for diblock copolymers,[33] but their mathematical form alone guarantees there is no $k^* \neq 0$ divergences. The general form of Eq. (7.5) is similar to that derived based on the incompressible phenomenological field-theoretic treatment of fluctuations of Brazovskii (a self-consistent harmonic or Hartree approximation) as applied to diblock copolymer problem by Fredrickson and Helfand.[144] However, the physical origin of the feedback stabilization effect in the field theory is coupling to nonlinear (quartic and cubic) purely entropic single-chain corrections to the Landau free energy expansion, and thus is entirely distinct from the PRISM/R-MPY correlated enthalpic mechanism.[33] These differences have many consequences, including the very different conditions under which fluctuation effects are "turned off" in the two theories. Recent diagrammatic field-theoretic analysis by Stepanow[145] has argued that a non-RPA-based approach leads to results very different than the Brazovskii analysis. Stepanow's main results appear to be qualitatively consistent with the PRISM predictions.

Further analytic progress can be made by approximating the form of the structure factor in Eq. (7.5) as a delta function, that is, a dominant wave vector approximation. The resulting self-consistent equation can be written as a cubic equation for the square root of the peak intensity variable[33]

$$\hat{S}^{3/2}(k^*) + \frac{N}{\Theta}\left(\frac{\chi_s - \chi_{\text{HTA}}}{\chi_0}\right)\hat{S}(k^*) - \frac{\hat{S}^{1/2}(k^*)}{2\chi_s} - \frac{N}{2\Theta\chi_0} = 0 \qquad (7.8)$$

where χ_s is the critical (but unattainable) value required for a spinodal divergence in Eq. (7.6) (e.g., $\chi_s = 10.495/N$ for $f = \frac{1}{2}$). The factor Θ is known and approaches an N-independent value for long chains. It is given by

$$\Theta = \frac{x^*}{\alpha f(1-f)(\xi_\rho^{-1} - \xi_c^{-1})} \qquad (7.8a)$$

where $\alpha^2 = (\frac{1}{3})x^*[\partial^2 F/\partial x^2]_{x=x^*}\sigma^2$, which is tabulated elsewhere,[143,144] $x = (kR_g)^2$, and σ is the statistical segment length.

Analytic solution of Eq. (7.8) can be obtained in three special regimes.[33] Here we focus solely on the N-scaling behavior of the maximum concentration fluctuation amplitude, that is, $\hat{S}(k^*) \propto N^\nu$. (i) The high-temperature, or low N, "mean-field" limit of $\chi_0 N \ll (\chi_0 N)_{s,\text{HTA}}$ where the latter is the spinodal value within the PRISM/R-MPY/HTA theory. Here $\nu = 1$, $\chi_{\text{eff}} \propto \hat{S}^{-1}(k^*) \propto T^{-1}$. (ii) A second case is at the spinodal predicted by the R-MPY/HTA level theory where one can show that $\hat{S}(k^*) \propto N^{4/3}$. The predicted N-scaling exponent agrees with the Brazovski field-theoretic approach[144] *at the order–disorder microphase transition or its extrapolated apparent spinodal*. The amplitude of $\Delta g(r)$ is predicted to scale as $N^{-1/3}$. The latter scaling law is close to the behavior found numerically in Figure 29 for a finite thickness chain model ($d \neq 0$) and R-MPY/HTA closure. (iii) In the very low temperature or high N limit, $\chi_0 N \gg (\chi_0 N)_{s,\text{HTA}}$, Eq. (7.8) predicts a limiting "ground-state" behavior given by $\hat{S}(k^*) \propto N^2$. This prediction is very different than the corresponding field-theoretic result[144] for the (metastable) disordered fluid phase of $\hat{S}(k^*) \propto N^2(\chi_0 N)^2$, which displays a different N scaling and continues to grow without bound as the temperature is lowered. This striking difference again highlights the fundamentally different origin of the predicted "fluctuation" stabilization effect: correlated enthalpic in PRISM theory and entropic in the field theory. In practice, one expects a first-order phase transition to occur in the vicinity of the extrapolated spinodal, so the above low-temperature results are appropriate for a (metastable) supercooled fluid phase.

Examples[33] of the predictions of Eq. (7.8) for $f = \frac{1}{2}$ symmetric diblocks are given in Figure 31. The high-temperature linear part of the curves is well described by the R-MPY/HTA theory. At lower temperatures, concentration fluctuations become manifest and nonlinear behavior emerges. With increasing chain length, the fluctuation regime becomes narrower. These features are in accord with numerical PRISM studies[33,133] and experimental SANS measurements.[136,141] The inset of Figure 31 shows the growth of the effective N-scaling exponent over the entire temperature regime. The exponent crosses over from the mean-field value to the strong coupling behavior in a rather small interval of temperature (or N).

The results in Figure 31 are for $f = \frac{1}{2}$, a meltlike fluid density, a particular choice of spatial range of the tail potential, and variation of $\chi_0 N$ was achieved by cooling. Very similar results[33] are found a constant T by increasing N, and for asymmetric diblock compositions $f \neq \frac{1}{2}$. However, at fixed temperature (but not fixed distance from the extrapolated spinodal), the effective chi parameter increases as the diblock composition becomes purer. This is similar to the blend situation where $\chi_{\text{eff}}(\phi)$ is roughly parabolic and concave-upward, but there are distinct differences in the shape of the block copolymer $\chi_{\text{eff}}(f)$, and significantly larger amplitudes are predicted since the relevant phase separation length scale is finite.[33] The characteristic microdomain length scale, $D = 2\pi/k^*$,

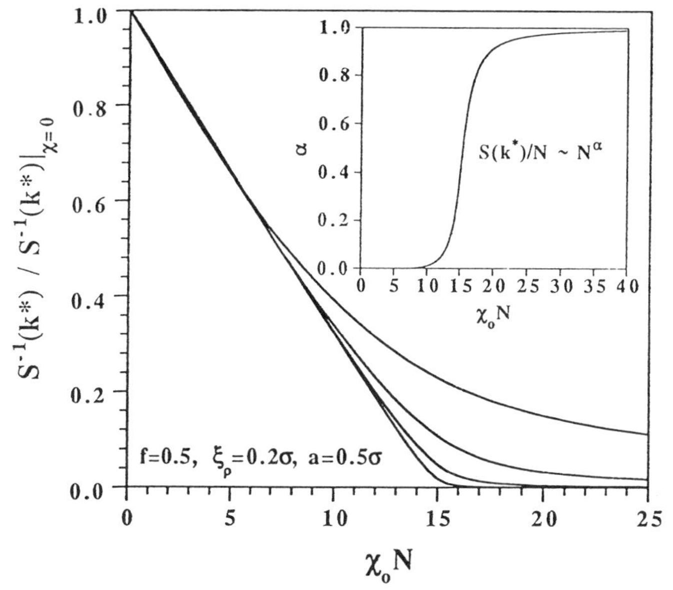

Figure 31. PRISM plus linearized R-MPY closure predictions[33] for the normalized inverse peak scattering intensity of the $f = \frac{1}{2}$ symmetric Gaussian thread diblock copolymer model. Results (top to bottom) for $N = 20$, 200, 2000, and 20,000 are shown at fixed melt density and a Yukawa tail potential range parameter of $a = 0.5$. Here, the bare driving force for microphase separation, $\chi_0 N \propto NT^{-1}$, is varied by changing temperature. The inset shows the apparent exponent that describes the scaling relation between peak scattering intensity and N as a function of inverse temperature (as extracted from the three largest chain lengths).

is predicted to increase upon cooling even under the conformational ideality condition of fixed block copolymer chain dimensions. Although this is a finite size fluctuation effect, it emphasizes the physical fact that k^* is affected by both single-chain size and collective correlations.[146]

As the overall fluid density is decreased (corresponding to diblock copolymers in a nonselective solvent), the fluctuation regime of the peak scattering intensity curve becomes broader and begins farther away from the extrapolated spinodal temperature.[33] Physically, this is because the lower density solution is more compressible, and hence thermally driven clustering and segregation of like monomers is more easily achieved. A qualitatively equivalent explanation is that the separation between the microdomain scale and the density correlation length scale is smaller at lower copolymer densities, and thus the finite size coupling mechanism is more efficient. Similarly, if the spatial range of the attractive tail potential, a, is *reduced*, then thermally driven local clustering of like segments is *enhanced* and the fluctuation stabilization mechanism on the microdomain scale is *enhanced*.[33]

2. Role of Block Asymmetries

In analogy with the conformationally and interaction asymmetric thread blend analysis discussed in Section VI.C, Schweizer has derived analytic results for the microphase spinodal based on the R-MMSA and R-MPY/HTA closures.[128] The focus here is not on the rather universal fluctuation stabilization phenomenon discussed above but the influence of system-specific block stiffness and attractive interaction differences on the location of the (spinodal) microphase separation temperature.

One extra approximation must be invoked for the block copolymer case relative to the blend. The reference athermal system is not a mixture of A and B chains, but a connected block copolymer of A and B segments. Analytic solution of the thread PRISM equations has not been achieved for this case. Thus, as a technical approximation the reference hard-core direct correlations functions have been approximated by their blend values given in Section IV.D. Such an approximation should be excellent for large N and the diblock architecture, but will deteriorate in accuracy for multiblock architectures.

The predicted spinodal temperature for the Berthelot λ tail potential model is given by[128]

$$\frac{k_B T_s}{\rho_m |\tilde{H}_{AA}|} = f(1-f)N(\gamma^2 - \lambda)^2 \tilde{F}(k^*, f, \gamma) + \frac{\tilde{G}(k^*, f, \gamma)}{(-\rho C^0_{AA})} \tag{7.9}$$

where the symbols have the same meaning as for blends (see Section VI.C). The functions F and G are explicitly given elsewhere.[128] These functions are independent of N in the long-chain limit; for example, for the stiffness symmetric ($\gamma = 1$), $f = \frac{1}{2}$ diblock case $F = 2/10.495$. The ordering wave vector k^* is determined (numerically) by maximizing the right-hand side of Eq. (7.9).

For diblock copolymers, all the general trends predicted by Eq. (7.9) are qualitatively the same as the blend case discussed in Section VI.C. A fit[133] of an equation of the form of Eq. (7.9) to the numerically determined apparent spinodal temperatures for the finite thickness SFC model of Section VII.B is shown in Figure 27. The simple analytic form qualitatively captures the numerical PRISM results, although quantitative deviations occur as expected.

Since *thread* PRISM theory at the R-MMSA or R-MPY/HTA closure level predicts a spinodal instability, its description of the disordered phase can be combined with field-theoretic Landau expansion and Brazovskii methods to construct an analytic theory for the first-order microphase separation transition. Such a hybrid PRISM/field theory approach has been developed by Tang and Schweizer[152] and includes both global fluctuation and local correlation and asymmetry effects.

VIII. SOLVATION POTENTIALS AND SELF-CONSISTENT PRISM

All the theoretical work described so far has assumed *conformational ideality*. That is, the intramolecular pair correlations are presumed to be independent of fluid density (and composition in an alloy) and can be computed based on a chain model that only accounts for "short-range" interactions between monomers close in chemical sequence. This assumption can fail spectacularly in dilute "good" solution where the effective intrachain monomer–monomer interaction is repulsive in a second virial coefficient sense.[1–4] For such good solvent conditions, the polymer mass/size relationship no longer obeys the ideal random walk scaling law $R \propto N^{1/2}$, but follows the *self-avoiding walk* (SAW) law $R \propto N^{\nu}$ with $\nu \cong \frac{3}{5}$ corresponding to the "swollen" coil behavior. As the dilute solution is concentrated by increasing a dimensionless measure of monomer concentration (e.g., ρ), the polymers begin to interpenetrate, and the excluded volume swelling effect is progressively screened. The precise manner this occurs is predicted by scaling theories[2] to be of a power law form *under large N, semidilute* solution conditions, $R \propto \rho^{-1/8}$, for $\rho^* \ll \rho \ll 1$. Here, $\rho = 1$ corresponds to the neat melt, and in good solvents $\rho^* \propto N^{-4/5}$, which represents the *semidilute overlap* density when different chains just begin to touch. The semidilute regime is characterized by strong interchain overlap conditions, but still small *overall* concentration of polymer. At high melt concentrations the polymer behaves as an ideal random walk, and it is widely believed (but not proven) that chain dimensions "saturate," that is, become ρ independent.[1–3]

Phenomenological scaling theories, based on analogies with critical phenomena, have been developed to qualitatively describe semidilute solutions in the asymptotic long-chain limit ($N \to \infty$).[2] Self-consistent field-theoretic approaches have also been constructed by Edwards and co-workers to describe the physical behavior summarized above.[3] However, such theories are based on the most idealized Gaussian thread chain model, and *integrable delta function* two- and three-body psuedopotentials between monomers. The latter can loosely be identified as describing effective monomer–monomer interactions in solution at the second and third virial coefficient level; in practice they are treated as empirical parameters. Neither scaling nor field-theoretic approaches are appropriate for dense solutions and melts.

As Edwards has recently emphasized,[153] a truly microscopic theory of phenomena such as discussed above would be very valuable since it would provide not only quantitative system-specific information but also could establish the range of validity of phenomenological scaling ansatzes. In this section we summarize recent progress toward this goal in the framework of PRISM and liquid-state theory. We note that Chandler and Pratt carried out pioneering work in this area by developing a statistical mechanical framework for the self-consistent calculation of a single molecule conformation within the RISM formalism.[8,12]

The effective single macromolecule potential surface, $U(\underset{\sim}{R})$, consists of three physically distinct contributions[43]

$$U(\underset{\sim}{R}) = U_0 + U_E + W(\underset{\sim}{R}) \tag{8.1}$$

where $\underset{\sim}{R}$ denotes a complete set of coordinates required to specify the configuration of a polymer molecule composed of N sites. The first term, U_0, describes the "bare" local intramolecular interactions that specify chemical bonding constraints (e.g., fixed bond length and bond angle) and local chain flexibility (e.g., dihedral angle rotations in an atomistic model or bending energy in a SFC model). All the "ideal" models discussed in Section III contain *only* this type of energy term. The second contribution, U_E, describes nonbonded "long-range" (in chemical sequence) excluded volume-type intramolecular interactions that are taken to be *pair decomposable*. Under good solvent conditions, U_E favors chain swelling to reduce *intra*molecular repulsive contacts. The third term, $W(\underset{\sim}{R})$, describes the "solvation free energy," that is, the reversible work required to achieve a configuration $\underset{\sim}{R}$ in the condensed phase due to *inter*molecular solute–solvent potentials only. In principle, this solvation potential is an N-body function. However, mathematical tractability would seem to require reduction to an (effective) pair-decomposable form, that is,

$$W(\underset{\sim}{R}) = \sum_{\alpha<\gamma=1}^{N} W_{\alpha\gamma}(|\mathbf{r}_\alpha - \mathbf{r}_\gamma|) \tag{8.2}$$

All work to date has employed this simplification.

Pratt, Chandler, and others have developed and applied approximate *solvation potentials* for flexible n-alkane fluids such as butane and decane and other relatively small molecules.[2,154,155] The original approaches invoked a "superposition" approximation,[12] which in its most naive form corresponds to assuming pair decomposability of $W(\underset{\sim}{R})$ *and* calculation of the effective pair potential based solely on consideration of the two sites.

A more accurate solvation potential approach was then developed by Chandler and co-workers[23,24] in the context of a quantum electron in dilute (classical) solution that led to new and more accurate self-consistent approximations for the effective pair potential in $W(\tilde{R})$.[155] This work represents the starting point for our self-consistent PRISM-based studies of polymer fluids and the development of novel solvation potential approximations required for an adequate qualitative treatment of some macromolecular systems.

A. Solvation Potential Theories

Chandler and co-workers[23,24] have proposed a self-consistent pair interaction, called the *Gaussian fluctuation* (GF) potential, of the form

$$\beta W_{\alpha\gamma}(r) = -\sum_{\lambda\delta} \int d\mathbf{r}' \int d\mathbf{r}'' C_{\alpha\lambda}(|\mathbf{r} - \mathbf{r}'|) S_{\lambda\delta}(|\mathbf{r}' - \mathbf{r}''|) C_{\delta\gamma}(r'') \quad (8.3)$$

where $S_{\alpha\gamma}(r)$ is the total density–density correlation function between sites α and γ separated by a distance r. For molecules composed of symmetry equivalent sites, Eq. (8.3) simplifies in Fourier space to

$$-\beta \hat{W}(k) = -\hat{C}^2(k)\hat{S}(k) \quad (8.4)$$

This approximate form can be derived many ways: from renormalized second-order perturbation theory,[23] Gaussian (linear response) density field theory,[23] or via density functional expansions.[44,156] Pictorially, the medium-induced potential between a pair of tagged sites is determined by coupling to the surroundings via an effective potential (direct correlation function), which is mediated by density fluctuations of the condensed phase. The integrated strength of the medium-induced potential is

$$\hat{W}(0) = -k_B T \hat{C}^2(0)\hat{S}(0) = -\frac{1}{\rho^2 \kappa_T} \quad (8.5)$$

where the second equality follows from use of the PRISM Eq. (2.2). Equation (8.5) shows $W(r)$ is (on average) attractive, favoring compressed polymer conformations, and hence will tend to cancel or "screen" the expansive consequences of the bare intramolecular repulsions. Thus, the effective attraction is expected to become stronger as fluid density increases, chain length decreases, and/or polymer backbone stiffness increases, since all these changes enhance intermolecular packing and reduce bulk compressibility. Moreover, $W(r)$ becomes increasingly structured (e.g., oscillatory) as the above changes are made.[24,43]

The basic feature of a self-consistent mean pair approximation such as

Eqs. (8.2) and (8.3) is that the effective interaction between two sites on the polymer depends on their instantaneous position, and only on the entire macromolecule conformation in an *average* (implicit) sense via the direct and collective pair correlations. This simplification does *not* preclude describing situations of broken conformational symmetry, such as polymer collapse, solvated electron localization, or spatially inhomogeneous conformational characteristics such as occur in star polymers (see Section IX).[23,24,46]

Grayce and Schweizer,[46] based on graph-theoretic and hueristic arguments, suggested a modified form for the solvation potential ("PY style") that is in the spirit of the Percus–Yevick closure

$$\beta W^{\mathrm{PY}}_{\alpha\gamma}(r) = -\ln\left[1 + \sum_{\lambda,\delta} \int d\mathbf{r}' \int d\mathbf{r}'' C_{\alpha\lambda}(|\mathbf{r}-\mathbf{r}'|) S_{\lambda\delta}(|\mathbf{r}'-\mathbf{r}''|) C_{\delta\gamma}(r'')\right] \tag{8.6}$$

Equation (8.3), by contrast, was argued to be in the HNC closure spirit (HNC-style solvation potential). On general grounds, one expects the PY-style solvation potential to be *weaker* (e.g., less compressive) than its HNC-style analog (see Fig. 32). However, the two solvation potentials do become equal in the limit of weak interactions, $\beta W(r) \ll 1$. This limit may occur for many reasons such as high temperature, low fluid density, and/or large intersite separations. Moreover, the theoretical *model* adopted (e.g., thread, SFC, RIS) and/or the statistical mechanical approximations invoked may result in such a weak coupling regime for differing regimes of thermodynamic state.

For simple fluids,[5] the PY and HNC closure approximations are useful in different contexts, and this is probably also true of the corresponding solvation potentials. Based on analogies with atomic and small-molecule liquids, and homopolymer fluids, one might expect the PY potential is better for bare interactions, which are spatially short ranged and repulsive.[46] This is the situation for the polymer/good solvent system of present interest. However, one might expect that the HNC-style potential is useful for longer-range, more slowly varying potentials such as Coulombic or possibly Lennard-Jones attractive tails. The role of macromolecular architecture (or precise single-polymer model) in such qualitative ideas may be subtle and is not well understood at present.

B. Self-Consistent Solution of Single Macromolecule Problem

Implementation of self-consistent PRISM theory requires addressing the difficult technical question of how to iteratively solve the effective N-body

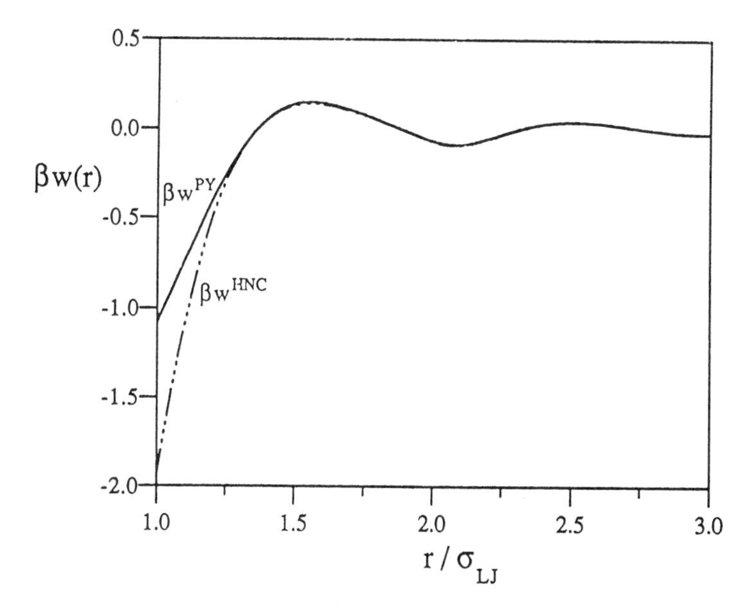

Figure 32. Typical high-density ($\rho\sigma_{LJ}^3 = \beta\varepsilon_{LJ} = 1$) medium-induced potential between monomers of a repulsive Lennard-Jones $N = 100$ chain as predicted by self-consistent PRISM theory based on the HNC-style and PY-style solvation potentials and the variational generating functional method.[47]

single-polymer problem with "long-range" intramolecular interactions, *in conjunction* with PRISM theory for interchain pair correlations and hence the solvation potential itself. For the homopolymer/good solvent problem, successful field-theoretic approximate approaches have been developed.[2–4] However, these methods *explicitly* rely on the use of an *integrable*, delta-function psuedopotential description of segment–segment interactions within an *effective* Gaussian thread framework. The use of more realistic models with local chemical constraints and a finite chain thickness (hard-core constraint) requires the development of different approaches.

Over the past several years, we and our collaborators have constructed and applied several approaches[43–47] that vary greatly in both computational convenience and level of statistical mechanical approximation. These various approaches often have distinct (and often limited) regimes of applicability and level of accuracy. Here we sketch the essential physical features and statistical mechanical approximations of the different numerical approaches.

1. *Single-Chain Monte Carlo Simulation*

The most rigorous *and* computationally demanding approach to solving the statistical mechanics of an effective Hamiltonian given by Eqs. (8.1) and (8.2) is to perform a *single-chain* Monte Carlo simulation in the framework of a self-consistent PRISM calculation. This approach was first explored by Melenkevitz et al.[44] using a standard "kink-jump" algorithm and subsequently by Grayce and co-workers[47] using the more accurate "pivot algorithm".[157]

Although straightforward in principle, and simpler than a many-chain simulation, there are several practical difficulties associated with proper sampling and equilibration. For example, at high fluid densities the solvation pair potential is strong and oscillatory (see Fig. 32), which makes sampling and equilibration difficult. Early work using the kink-jump method was found to fail at such high densities,[44] and the pivot algorithm with its "long-range moves" was required for proper statistical sampling.[47]

2. *Free Energy Variational Approaches*

A simpler, analytic approach is to employ the standard idea of replacing the real system with a computationally tractable "reference system" (polymer chain model in our case), the parameters of which are variationally optimized using an approximate single-chain free energy expression. Two such schemes have been explored[45,47] that differ in *both* the choice of reference chain model and the form of the approximate free energy.

Earlier work by Melenkevitz et al.[45] was based on the standard Gibbs–Bogoliubov inequality for the single-chain free energy:

$$\beta F \leq \beta F^0 + \sum_{\alpha<\gamma=1}^{N} \int d\mathbf{r}\beta\, \Delta\phi_{\alpha\gamma}(r)\omega^0_{\alpha\gamma}(r) \tag{8.7}$$

where $\omega^0_{\alpha\gamma}(r)$ is the *reference system* intrachain pair correlation function matrix, and $\Delta\phi_{\alpha\gamma}(r)$ is the *difference* in single-chain effective interaction pair potential between the real and reference systems.

In numerical applications, the "solvent" is treated as a vacuum, that is, it enters solely in determining the polymer concentration or packing fraction η. Thus, the real two-component polymer/solvent mixture is abstracted to a one-component polymer fluid of variable density. For the homopolymer/good solvent system, the reference system is chosen to be an *ideal* SFC (as computed with the discrete Koyama approximation)[55] with an effective, or "renormalized," bending energy ε_r. The repulsive interactions between sites separated by two bonds $[\omega^0_{\alpha,\alpha+2}(r)]$ is also included in the reference system and treated *exactly*. Thus, the effective bending energy describes both the *bare* bending efficiency of the real chain, plus the combined consequences of the bare intramolecular repulsions and medium-induced potential. Since the explicit effect of intrachain excluded volume is absent in the ideal SFC model, it is introduced *approximately* by defining an effective bond length, l_{eff}, determined such that in the dilute solution limit SAW scaling for $\langle R^2 \rangle$ is recovered. This approach is in the spirit of field-theoretic studies[3] and is capable of treating the entire polymer concentration regime in good solvents. Values of the density-independent quantity l_{eff} as a function of N are determined from Monte Carlo simulation of a *single self-avoiding* chain.[45] Explicit forms of Eq. (8.7) for this model and variational theory are given elsewhere.[45] It is important to mention that the bare (singular) intrachain repulsions are *omitted* in the evaluation of $\Delta\phi_{\alpha\gamma}(r)$ in Eq. (8.7).

More recently, a second free energy variational approach has been developed by Grayce and co-workers.[47] An approximate single-chain free energy is constructed in the spirit of a functional expansion about an ideal reference state with an explicit accounting of intrachain excluded volume. The latter cannot be naively treated via perturbation theory since it is singular and nonintegrable for the finite hard-core diameter models of interest. Rather, a virial-like treatment of the nonbonded intrachain repulsions and medium-induced potentials is employed by carrying out an expansion about an ideal reference system through second order in the appropriate Mayer f bonds.

In applications, the reference system is again chosen as an ideal SFC with renormalized bending energy ε_r and next nearest neighbor pair correlations, $\omega^0_{\alpha,\alpha+2}(r)$, exactly computed. The resulting approximate single-chain free energy is[47]

$$\beta F(\varepsilon_r, \varepsilon_{\text{eq}}) \approx \beta F^0(\varepsilon_r) + \beta \int d\mathbf{r} \Bigg\{ \sum_{\alpha=1} \omega^0_{\alpha,\alpha+2}(r; \varepsilon_r)\, \Delta\phi_{\alpha,\alpha+2}(r) - \sum_{\alpha+2<\gamma} \omega^0_{\alpha,\gamma}(r; \varepsilon_r)[-1 + \exp(-\beta\, \Delta\phi_{\alpha\gamma}(r))] \Bigg\} \tag{8.8}$$

where $\Delta\phi_{\alpha\gamma}(r) = u(r) + W(r; \varepsilon_{\text{eq}})$ for $|\alpha - \gamma| \geq 3$ (where u is the bare intrapolymer "excluded volume" potential), and $\Delta\phi_{\alpha\gamma}(r)$ also includes the difference between the bare and renormalized bending potential for $|\alpha - \gamma| = 2$. Here, ε_{eq} and ε_r must be determined self-consistently by free energy minimization and a recursive generating functional procedure.[47]

Under certain conditions it is possible to bypass this process by treating W not as an external field but as a functional of the intrachain correlations, and therefore directly minimizing F. That is, $\varepsilon_{\text{eq}} = \varepsilon_r$ is enforced at the start of the self-consistent iteration procedure and $F(\varepsilon_r)$ is mimimized. Such a "direct" method[47] is generally much faster to numerically implement than the full recursive procedure. As a matter of principle, it does not yield the same predictions as the recursive approach, but in applications studied to date *based* on the PY-style solvation potential the results are quite similar (see Section VIII.C).

Both the accuracy of variational approaches and their range of applicability are expected to depend strongly on the physical problem, thermodynamic state conditions, and the choice of reference system and form of the approximate free energy. The approach of Melenkevitz et al.[45] is relatively crude but is expected to be qualitatively useful over the entire range of polymer concentration and degree of polymerization. The approach of Grayce et al.[47] is more general with regards to describing the chemical structure of macromolecules, but by construction is not capable of recovering the SAW dilute solution behavior due to the perturbative (in an f-bond sense) treatment of intrachain excluded volume interactions. Thus, this approach is expected to be most appropriate for dense solutions and melts and hence is complementary to the heavily coarse-grained scaling and field-theoretic approaches.[2–4]

3. *Optimized Perturbation Theory*

The first self-consistent PRISM studies by Schweizer et al.[43] considered only the HNC-style solvation potential and were based on an *optimized perturbative*, not variational, determination of the ideal reference system effective bending energy. The starting point is a simple functional expansion of the true single-chain free energy about an ideal reference system[43]

$$\beta F = \beta F^0 + \sum_{\alpha>\gamma}^{N} \int d\mathbf{r} \left.\frac{\delta \beta F}{\delta f_{\alpha\gamma}(r)}\right]_0 [f_{\alpha\gamma}(r) - f^0_{\alpha\gamma}(r)] + \cdots \tag{8.9a}$$

where the full and reference Mayer f-bonds are

$$\begin{aligned} f_{\alpha\gamma}(r) &\equiv \Theta_+(r-d)\exp[-\beta(v_b(\varepsilon_0)+W)] - 1 \qquad |\alpha-\gamma| = 2 \\ &\equiv \Theta_+(r-d)\exp(-\beta W) - 1 \qquad |\alpha-\gamma| \geq 3 \\ f^0_{\alpha\gamma}(r) &\equiv \Theta_+(r-d)\exp[-\beta v_b(\varepsilon_r)] - 1 \qquad |\alpha-\gamma| = 2 \\ &\equiv 0 \qquad |\alpha-\gamma| \geq 3 \end{aligned} \tag{8.9b}$$

where ε_0 is the bare SFC bending energy that quantifies the potential v_b, and ε_r is the corresponding renormalized value of the reference system. The required functional derivative (evaluated in the reference system) is easily determined. The reference system effective bending energy is chosen by requiring the first correction to $F - F^0$ vanishes. This scheme is analogous to well-known "blip-function" theories of Chandler and co-workers[5,11] for determining a hard-core diameter of soft repulsive force simple fluids. Such a simple approach is *not* adequate for *low* polymer densities and does not recover SAW scaling in the dilute limit. It is believed to be most adequate for describing "small" nonideal conformational corrections, for example, nearly meltlike conditions. Moreover, this nonvariational approach can only be implemented for ideal reference systems characterized by only *one* parameter, and it has been shown to *not* properly describe the *rigid rod* limit where all condensed phase modifications of chain conformation must vanish.[43]

C. Theory/Simulation Comparisons for Homopolymer Good Solutions

The best test of self-consistent PRISM theory and the different solvation potential approximations is via comparison of its predictions against exact computer simulation studies of the *same model*. The drawback is that present computer power limits such comparisons to short and intermediate length chains (N less than roughly 200). Many detailed comparisons have been carried out at all levels of approximation discussed in Section VIII.B. Here we give a few examples along with summarizing remarks. The reader is referred to the original studies for details and a complete discussion.

Two similar freely jointed (zero bare bending energy) models of a homopolymer/solvent system have been studied by many-chain simulation. Yethiraj and Hall[60] investigated a purely hard-core, tangent bead model for $N = 20$–100 and monomer packing fractions of $\eta = 0.1 - 0.35$ (concentrated solution). A shifted repulsive Lennard-Jones model of nearly tangent chains has been studied by Kremer and Grest ($N = 20$–200 at fixed $\eta \cong 0.45$) and by Gao and Weiner[158] ($N = 16$, $\eta = 0.1$–0.47). The intramolecular structure factor $\hat{\omega}(k)$ and chain-averaged $g(r)$ were monitored, along with average chain dimensions R and R_g. Here we focus on R_g, which for dense solutions is also a direct measure of the chain persistence length or stiffness.

We begin with the most "rigorous" version of self-consistent PRISM based on a Monte Carlo evaluation of the effective single-chain problem. Theoretical predictions of Grayce and co-workers[47] are compared with many-chain simulation results for the mean-square end-to-end distance of the hard-core chain model as a function of polymer packing fraction in

Figure 33 for $N = 20$ and 100. (The corresponding *estimates* of the dilute–semidilute crossover points are $\eta^* \cong 0.06$ and 0.015, respectively). *Over* the many-chain simulation range of density studied, the PRISM/Monte Carlo results based on the pivot algorithm and full self-consistent evaluation of $\hat{\omega}(k)$ are in good (but not perfect) agreement with the simplified self-consistent PRISM/Monte Carlo approach of Melenkevitz et al.[45] based on the kink-jump algorithm. Moreover, the theoretical predictions based on both the HNC-style and PY-style solvation potential are in *qualitative* agreement with the exact results. Motivated by blob scaling arguments for *semidilute* solutions,[2] the density dependence of $\langle R^2 \rangle$ was fit to a power law form: $\langle R^2 \rangle \propto \eta^{-\alpha}$ for $0.1 < \eta < 0.35$. Over this very limited range, the power law form is adequate and an exponent in agreement (to within statistical error) with the many-chain simulation value[60] of 0.25 ± 0.1 was found. The latter value is consistent with the long-chain blob scaling prediction, but the agreement may be accidental since the chains studied are rather short and the density regime very limited.

Quantitatively, the calculations based on the HNC-style solvation potential appear to be superior *for the density range studied by the full many-chain simulations.* This is misleading, however, since (as seen in Fig. 33) the more compressive HNC-style solvation potential leads to a very strong, rapidly varying reduction of chain dimensions at high concentrations. This behavior conflicts with basic Flory arguments[1–3] and the limited experimental data available suggesting that chain dimensions tend to saturate, or at least become very slowly varying in dense melts.[159,160] Examination of typical chain configurations in the PRISM Monte Carlo calculation based on the HNC-style solvation potential reveals a form of local collapse, or condensation, of closely bound sites along the chain.[44,47]

Further comparisons are given in Table III for relatively dense solutions and two values of N. Note that the polymer size is very different than the ideal freely jointed, ideal Koyama with local swelling, or the self-avoiding walk behavior. The accuracy and predicted trends of the HNC-style and PY-style approaches are relatively (but not completely) insensitive to N over the range of 20–100 for the choice of purely hard-core interactions. Summarizing, it appears the HNC-style solvation potential predicts too strong a compressive solvation force at high densities and is inadequate in this regime. Based on all the studies to date,[44,47] the PY-style solvation potential coupled with PRISM/Monte Carlo seems qualitatively sensible under all conditions, and typically makes errors of 10–20% in the prediction of the *absolute* magnitude of R^2 and R_g^2. *Relative* trends seem to be predicted significantly more accurately.

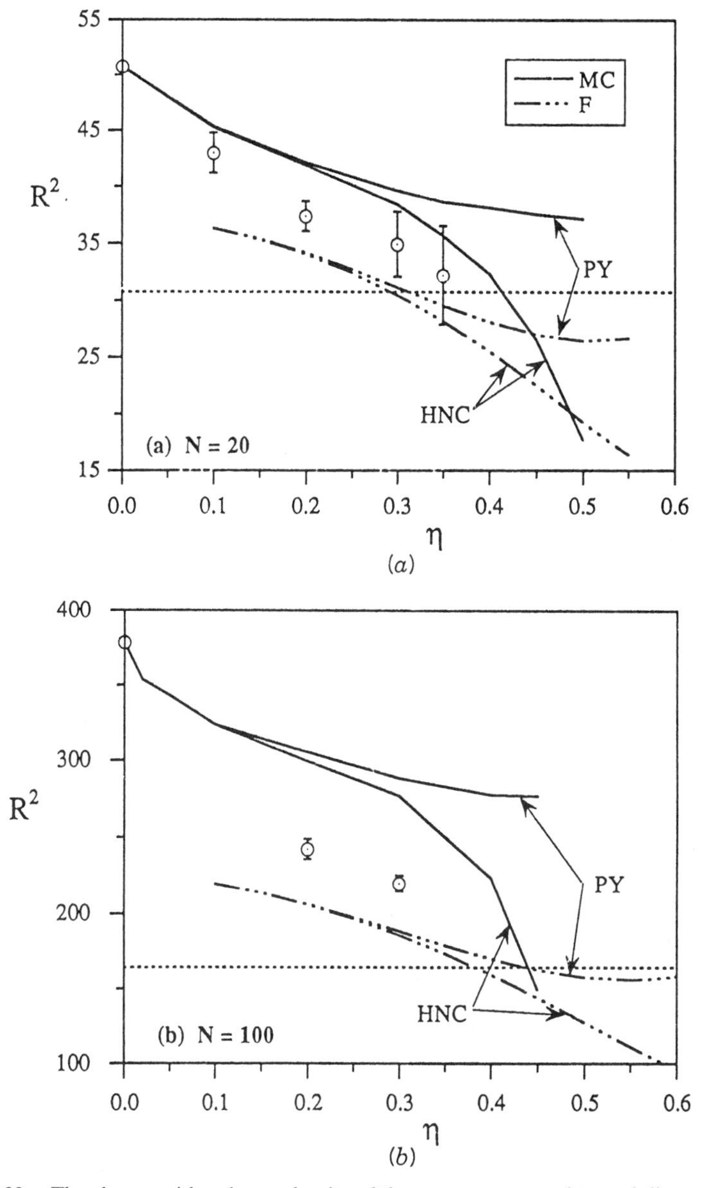

Figure 33. The change with polymer density of the mean-square end-to-end distance of hard-core chains of length (a) $N = 20$ and (b) $N = 100$. The data points are exact many-chain simulation results[60] and the solid (dash-dot) lines are the self-consistent PRISM/Monte Carlo (free energy generating functional) predictions using the two solvation potentials.[47] The dashed horizontal line is the value of R^2 for an ideal freely joined chain *with* a minimum next nearest neighbor bending angle of 60°, which mimics the local hard-core repulsion.

TABLE III
Mean-Squared End-to-End distances, R^2 [a]

	$\eta = 0.2, N = 100$		$\eta = 0.3, N = 100$		$\eta = 0.35, N = 20$	
Method	R^2	% error	R^2	% error	R^2	% error
Simulation	242.5	±2.7	220.1	±2.1	32.2	±13.3
MC (HNC)	299.6	23.5	276.7	25.7	35.7	10.7
MC (PY)	305.4	26.0	288.3	31.0	38.6	19.9
F (HNC)	206.1	−15.0	185.7	−15.6	28.2	−12.5
F (PY)	206.5	−14.9	188.4	−14.4	29.6	−8.2
F (HNC)[b]	221.2	−8.8	201.3	−8.5	30.4	−5.6
F (PY)[b]	220.8	−8.9	200.6	−8.8	30.7	−4.8
SAW[c]	348.5	43.7	348.5	58.3	50.8	57.6
FJC[d]	99.0	−59.2	99.0	−55.0	19.0	−41.1
SFC[e]	164.1	−32.3	164.1	−25.4	30.8	−4.5

[a] In units of the hard-core diameter of $N = 20$ and 100 linear hard-core chains at several packing fractions η. The many-chain simulation results of Yethiraj and Hall[60] are listed along with theoretical predictions based on various approximate implementations of the self-consistent PRISM scheme described in detail in the text.[47] MC refers to the single-chain Monte Carlo method and F refers to the variational generating functional method. The HNC or PY in parentheses refers to the style of solvation potential approximation employed. Percent error of the theoretical predictions relative to the simulation values are also listed. For the simulation data the percentage error gives the statistical margin of error.

[b] Using a direct minimization of F as described in the text and Appendix of Ref. 47.

[c] Value for a self-avoiding random walk.

[d] Value for an ideal freely jointed chain.

[e] Value for the SFC model with local intrachain repulsion between next nearest neighbors included.

The predictions of the variational free energy method [see Eq. (8.8)] of Grayce et al.[47] are also listed in Table III. Results are shown for both the rigorous generating functional approach and the more approximate, but computationally simpler, "direct" minimization scheme. Reasonable agreement between the two technical implementations are found. Generally, the generating functional method predicts consistently smaller polymer sizes than found based on the Monte Carlo method. This is true for both the HNC and PY-style potentials, and presumably reflects nonideality effects lost by the truncation of the virial expansion in Eq. (8.8) and by the assumption of ideal trial conformations invoked by the generating functional approach. For the shifted repulsive Lennard-Jones models (not shown), the PY-style approach is much more accurate than the HNC style, and significantly more accurate (typical errors of 5–8%) in an absolute sense than for the hard-core models (typical errors of 14–19%).

The gross qualitative trend of the HNC-style potential predicting much too small, "collapsed" conformations at high density is again found based on the approximate variational/generating functional approach of Grayce et al.[47] A more traditional polymer science demonstration of the origin of this fact is the trend with polymer density of the *monomeric* second virial coefficient, defined in normalized dimensionless form as

$$v_0 \equiv -\frac{3}{4\pi\sigma^3} \int d\mathbf{r}(\exp\{-\beta[u(r) + W(r)]\} - 1) \tag{8.10}$$

where $u(r)$ and $W(r)$ are the bare intrachain and solvation pair potentials, respectively, between nonbonded sites. Traditionally, v_0 is taken as a measure of "solvent quality," with positive values indicating a good solvent and coil swelling, negative values indicating a poor solvent and a collapsed conformation, and a zero value defining an ideal, "theta" state.[1–4] It has been shown based on the generating functional approach that $v_0 > 0$ for the PY-style potential, and approaches zero gradually at high melt densities consistent with the Flory concept of the melt as a theta solvent.[47] However, for the HNC-style potential v_0 changes from positive to negative around $\eta \cong 0.4$–0.45, consistent with the observation of overly compressed polymer dimensions.

Comparisons between theory and exact many-chain simulations have also been carried for the structural properties $g(r)$ and $\hat{\omega}(k)$ by Grayce and co-workers[47] based on the PY-style potential and the variational generating functional method. An example is given in Figure 34 for $N = 100$ and the medium-density case of $\eta = 0.3$. The general shape of $\hat{\omega}(k)$ is well predicted by the effective SFC ideal model, although the "plateau" at intermediate wave vectors is too high since the theory underpredicts chain dimensions (see Table III) and hence local stiffness. Other comparisons suggests improved agreement of both $g(r)$ and $\hat{\omega}(k)$ (see, e.g., Fig. 3) at higher, meltlike densities where the chains are conformationally more ideal, and where the differences between self-consistent and non-self-consistent PRISM predictions become increasingly small.

Based on all the above results, and many others not shown, it has been demonstrated that the generating functional method (with PY-style solvation potential) is more accurate at high densities than the single-chain Monte Carlo/PRISM method when compared against the exact many-chain simulation results. This suggests the additional approximations employed by the generating functional method compensate for other errors in the PRISM-based theory. Thus, the generating functional method at high density seems complementary to the PRISM/Monte

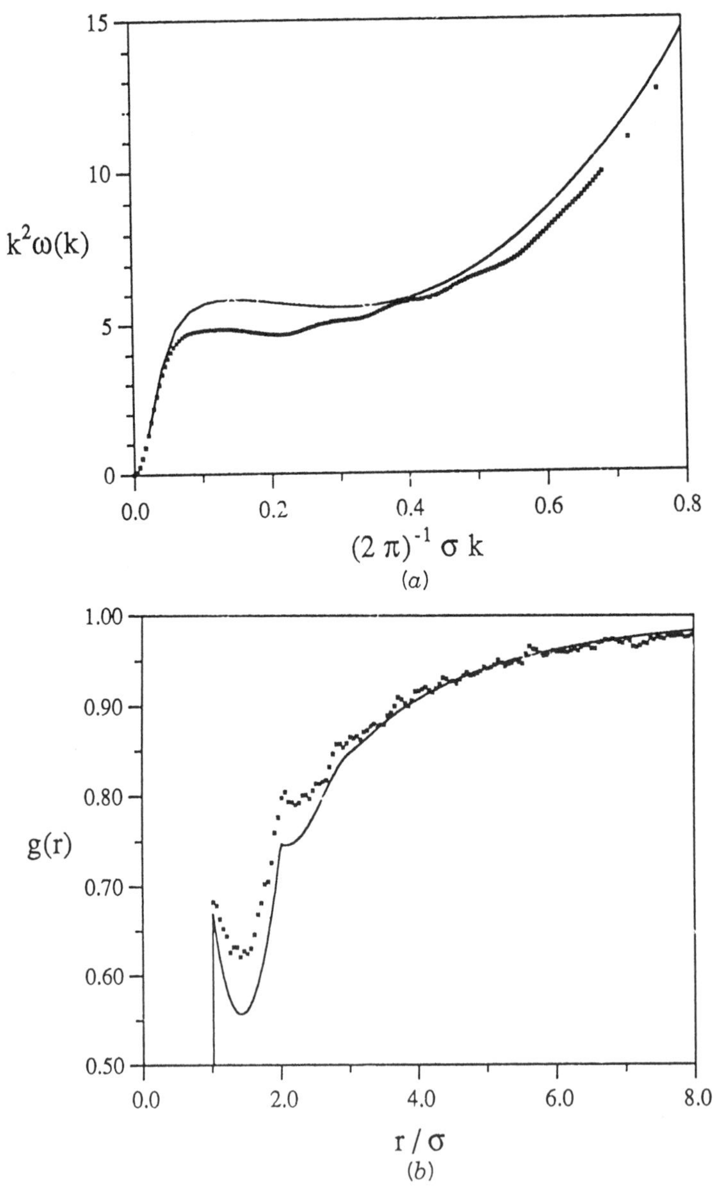

Figure 34. Self-consistent PRISM structural predictions for (a) average intramolecular structure factor plotted in Kratky form, and (b) site–site intermolecular radial distribution function for $N = 100$ hard-core chains at a concentrated solution packing fraction of 0.3. The points are the many chain simulation results[60] and the lines are the PRISM results based on the PY-style solvation potential and the simplified version of the variational generating functional method of Grayce et al. discussed in the appendix of Ref. 47.

Carlo method since the computational demands of the latter become very heavy at high densities and/or high N.

Application and generalization of the self-consistent PRISM theory to flexible trimer fluids, and detailed comparison with many molecule simulations, has also been performed by both Grayce and dePablo[42] and Yethiraj.[161]

Finally, we briefly summarize the first application of self-consistent PRISM theory by Schweizer and co-workers[43] based on the HNC-style potential and a simple optimized perturbation ("blip function") approach. Detailed comparisons of the predicted chain dimensions and $g(r)$ for the $N = 50$, 100 and 150 repulsive Lennard-Jones fluid model of concentrated solutions studied by Kremer and Grest[51] have been carried out. R^2 was predicted to be 3–7% *too small*, and in all cases deviations from ideal behavior (SFC with next nearest neighbor 1–3 repulsions explicitly accounted for) was extremely weak. The self-consistently determined $g(r)$ was in excellent agreement with simulations except near the first peak where the theory underestimated the maximum by roughly 10% as a consequence of the underpredicted chain dimensions and hence local stiffness. A key finding here is that at very high meltlike densities, all but the most local aspects of intrachain repulsion (and concommitant swelling) are screened out consistent with the basic tenets of the Flory ideality ansatz. Thus, as expected, the need for a fully self-consistent theory becomes far less important under high-density conditions. However, the fact that such "reasonable" behavior is found based on the HNC-style potential is again an indication of the subtle role played by the approximate single-chain theory, choice of reference system, and solvation potential in determining the theoretical predictions for the screening problem.

D. Numerical and Analytic Model Calculations

Besides benchmark comparisons with exact simulation results, model calculations have been performed to numerically explore additional issues.

1. The large N behavior, inaccessible to many-chain simulation (and very difficult for PRISM/Monte Carlo), but relevant to experiments and field-theoretic and scaling predictions, has been studied numerically based mainly on the HNC-style/variational approach of Melenkevitz and co-workers.[45] For fixed large N of the order of 10^3, a power law scaling behavior of $\langle R^2 \rangle$ with density has been found for intermediate (semidilute) solution densities in rough accord with phenomenological scaling predictions.[2] The question of global screening of intrachain excluded

volume interactions, as quantified in the effective exponent ν in the relationship $\langle R^2 \rangle \propto N^{\nu}$, has also been studied. Over a range of N values less than or equal to 2000, Melenkevitz et al.[45] find $\nu \cong 0.56$ for $\eta = 0.4$ (intermediate between the ideal 0.5 and SAW 0.6 values), $\nu \cong 0.51$ for $\eta = 0.54$ (meltlike), and $\nu \cong 0.50$ for higher liquid packing fractions in agreement with the Flory ideality hypothesis. The latter result has also been derived based on the Edwards psuedopotential field-theoretic method,[3] but is a highly nontrivial achievement for a microscopic theory based on *non*integrable, singular hard-core interactions between monomers. Thus, a truly microscopic basis for the ideality and semidilute scaling ideas has been established, and the nature of the corrections and limitations of these simple concepts can be elucidated.

2. The subtle question of "incomplete" screening of the excluded volume swelling at very high densities has been studied by several PRISM-based approaches.[43,45,47] At issue is whether chain dimensions approach a density-independent value, and whether it is truly "ideal," in the high packing fraction limit. Experiments are unclear on this point,[160] and PRISM studies yield differing conclusions depending on solvation potential choice and approximate free energy based scheme to evaluate the effective single-chain problem. Benchmark simulations that address this question would be of great value.

3. The subtle question of whether the *relative* changes in polymer dimensions as a function of solution density become N independent in the semidilute and concentrated regime has been considered by Grayce et al.[47] within the PY-style/generating function framework.

4. The influence of variable bare chain stiffness (aspect ratio) on the predicted nonideal conformational corrections. Predictions of the renormalized persistence length, relative to its bare value, have been obtained by Grayce and Schweizer[162] based on the PY-style solvation potential and the variational generating functional method, and also by Schweizer and co-workers[43] based on the optimized perturbation approach and the HNC-style potential. At high densities the predicted renormalizations are rather modest (typically $\pm 10\%$ or less). The most interesting feature is the nonmonotonic dependence of the renormalization ratio or bare persistence length, indicating a crossover from the condensed phase effects favoring chain compression (relative to the ideal limit) to a type of "induced rigidity." The latter is a subtle consequence of the emergence of strong solvation shells in the liquid for stiff polymers at high density. However, we do not overaly emphasize this feature for two reasons: (i) For high aspect ratio chains nematic/orientational correlations are expected to become increasingly important, a physical

feature not properly described by RISM-based approaches.[8,38–40] (ii) The theories[43,47] do not properly recover the "trivial" rod limit where the renormalization ratio should approach unity. Many-chain simulations would be particularly helpful in guiding the further development of the self-consistent approach for such semiflexible polymer systems.

Finally, we point out that *analytic* results have been derived and discussed based on the HNC-style solvation potential in the idealized Gaussian thread limit.[43] The variational free energy theory of Melenkevitz et al.[45] has also been worked out within the analytic thread model framework. Interesting connections and differences between the thread PRISM theory, the field-theoretic approaches[3] of Edwards, Muthukumar, and others, and blob scaling arguments[2] have been established. The work of Melenkevitz et al.[45] is novel in the sense that it combines the liquid-state PRISM approach to effective interactions (direct correlations, solvation potentials) with field-theoretic schemes for solving the effective single-chain problem in a manner that is qualitatively correct for all density. The construction of "hybrid" liquid-state/field-theoretic approaches, applicable for large N macromolecular systems, is an attractive direction for future development.

E. Other Applications

There are many other physical problems and macromolecular systems for which the self-consistent PRISM approach should be useful. The following represents an incomplete list of problems for which preliminary work has been done or which appear to be attackable based on the present state of the art.

1. *Theta and Poor Solvents*. A key question here is how to generalize the intermolecular closure approximation and the intramolecular solvation potential to simultaneously treat the competing repulsive and attractive bare forces. The problem of low-temperature polymer collapse in dilute solution is a classic problem in this area.

2. *Polymer Alloys*. Perturbation of meltlike conformation upon transfer to a multicomponent environment is not understood. The influence of proximity to phase boundaries, coupled density and concentration fluctuations, and mixture composition on both single-chain dimensions and miscibility are problems that have begun to be addressed within the PRISM formalism for the simple "symmetric" blend model by Singh and Schweizer[163] and symmetric diblock copolymer model by David and Schweizer,[140] and other more coarse-grained field-theoretic approaches.[164] Comparisons with the few available simulations[126,130,146–148] have also

been performed. However, the experimentally relevant conformationally and interaction asymmetric alloy cases[140] remains to be carefully considered.

3. *Atomistic Models*. Self-consistent conformational calculations at the atomistic level have not been studied although a tractable scheme for RIS models has been proposed.[43] One might expect much less perturbation of single-chain structure at an atomistic level where there are constant bond lengths, bond angles and so forth. However, rotational isomers generally differ in energy only of order $k_B T$, so a priori it is not clear what happens for specific polymer systems.

4. *Constrained Polymers*. The conformation of polymers constrained in various ways, for example, grafted to a flat surface ("brush"), adsorbed on a spherical colloidal particle, or tethered to a central branch point as in *star* polymers.[165] All such problems involve potentially large "nonideal" conformational effects and also introduce additional complications associated with site inequivalence within the PRISM formalism. Progress for star polymers is briefly described in the next section.

IX. STAR-BRANCHED POLYMER FLUIDS

Star polymers represent an interesting and important class of macromolecules of nonlinear global architecture. They consist of f linear chains or arms (with N_a monomers per arm) connected at a central branch structure, as schematically shown in Figure 35. This architecture is characterized by a spatially inhomogeneous single-molecule density that decreases as one moves away from the star center. Thus, relative to linear chains, a new and stronger form of *site nonequivalency* is present that results in spatially nonuniform screening of the intramolecular excluded volume interactions. In particular, the reduced exposure of the central "core" regions of the star to other polymers suggest less screening and hence more "swollen" conformations near the star center. However, very far from the branch point the star is still a low-density fractal object, and hence is expected to display conformational behavior similar to the linear chain case.

The above physical features imply that a fully self-consistent treatment of intramolecular and intermolecular pair correlations is more important for star polymers than linear chains, and the concept of "ideality" is expected to be of much less utility even at high melt densities. The treatment of star polymers within a self-consistent PRISM formalism has been very recently pursued by Grayce and Schweizer.[166,167] Here we give a brief description of some of the essential theoretical modifications

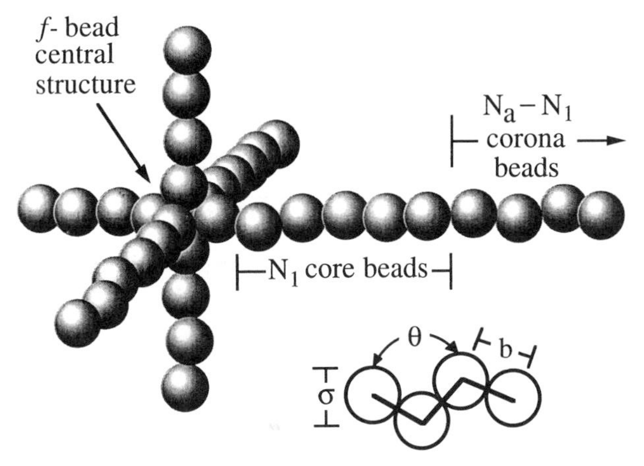

Figure 35. The star-branched tangent SFC model of an "$f \times N_a$" star, where N_a is the number of sites per arm and the arm number $f = 6$ in the example shown. The total number of sites comprising the star is $f(N_a + 1)$. The number of "core" sites is denoted N_1 and is determined variationally as described in the text.

required to treat stars, and a few conformational and structural results that emphasize some of the distinctive new phenomena characteristic of the star architecture. We note that the star polymer fluid is a model for other physical systems such as spherical micellar fluids and sterically stabilized colloids.[165]

A. Basic Model and Theory

The basic theory of star polymer fluids developed by Grayce and Schweizer is general in its ability to treat polymer models of variable chemical detail.[166] For simplicity, we discuss the theory in the context of the tangent, semiflexible chain model. As true for most of the results discussed in Section VIII, the bare bending energy is set equal to zero, and pure hard-core interactions (athermal or good solvent conditions) are employed in numerical studies carried out so far.

A sketch of the model star polymer is shown in Figure 35. Site equivalence is clearly broken by the presence of the central branch point, and in principle one expects that the local stiffness of the arms varies continuously as one proceeds from the star core to its outer "corona" region. The development of a technically tractable theory requires some simplification, or "preaveraging," of this complete site *in*equivalence. As indicated in Figure 35, a *three-region* scheme has been adopted motivated by both conceptual simplicity and by analogy with coarse-grained polymer physics type approaches.[168,169] The model star has a rigid branch point structure consisting of f sites (site type 0), a *core* region of fN_1 sites (site type 1), and an outer arm or corona region (site type 2, $N_a - N_1$ sites per arm).

For the *three site* model there are six independent intramolecular partial structure factors, radial distribution functions, and solvation pair potentials that must be determined self-consistently. PRISM theory has been implemented at the level of the variational generating functional theory of Grayce et al.[47] as described in Section VIII.B.2 using the PY-style solvation potential. However, two new features arise for star polymer. (i) Two distinct effective bending energies, ε_1 and ε_2, associated with the core and corona regions, respectively, are required. (ii) The number of sites that comprise the core region is not a priori known, but is treated as a variational parameter in the free energy minimization process. Thus, there are three variational parameters to be determined self-consistently.[166]

B. Conformation and Liquid Structure

Detailed numerical predictions have been obtained for average conformational properties such as the mean-square end-to-end distance of the core region, the overall star radius of gyration, the number of sites in the core, the effective persistence lengths of the core and corona regions, and the single-star structure factor $\hat{\omega}(k)$. The influence of variables such as number of arms ($f = 4$–12), arm degree of polymerization and fluid packing fraction on these properties has been established.[166] In a rough qualitative sense, the chain segments that comprise the corona region behave similarly to the analogous linear chain case. However, the core region is always found to be strongly swollen relative to the "ideal" state even at melt densities and results in a transfer of monomer density outward from the central region of the star to the coronal region. As a consequence, there is an increase of overall star dimensions, and distinctive changes in the star collective structure factor and intramolecular radial distribution function occur, relative to *either* the analogous linear chain case *or* the ideal Gaussian star model behavior.[166]

The predicted overall star dimensions as a function of (concentrated) fluid packing fraction and arm number is shown in Figure 36 for a macromolecule of modest size ($N_a = 100$). A simple power law form, $R_g^2 \propto f^{0.38}\eta^{-0.3}$, fits the numerical results very well. Curiously, the predicted density scaling exponent is rather close to the scaling theory[2] exponent of 0.25 *for long chains in semidilute solution*, although the significance of this near agreement is unclear. Note that no density-independent chain dimensions are attained, in conflict with common assumptions of ideal behavior at high meltlike densities.

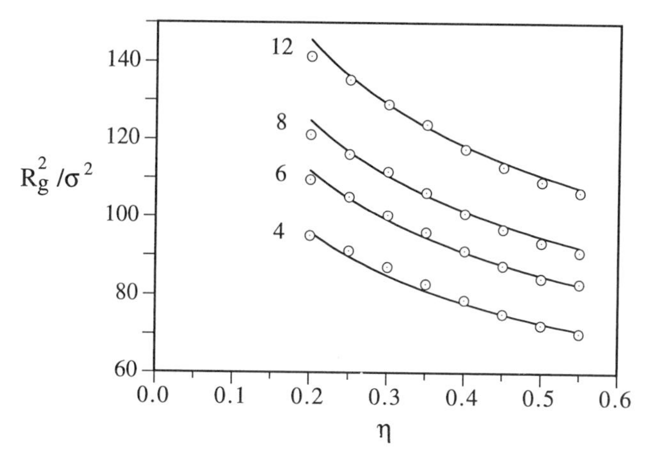

Figure 36. Mean-squared radius-of-gyration of $f \times 100$ stars as a function of monomer packing fraction and the indicated number of arms.[166] The curves are an empirical power law fit given by $35f^{0.38}\eta^{-0.30}$.

Both the core swelling, and the fraction of arm sites in the core grow as the fluid density decreases, the number of arms increases, and/or the arm molecular weight increases.[166] The latter trend is rather surprising since the core size appears to (weakly) increase without bound as arm degree of polymerization increases according to the power law $\langle R_1^2 \rangle \propto N_a^{1/3}$. Convincing physical interpretation of this intriguing behavior is at present lacking.

The fundamental origin of all the conformational trends discussed above is the strong reduction of the solvation potential, and hence screening effect, in the core region relative to the corona region. An example of the predicted self-consistent solvation pair potentials $W_{ij}(r)$ is given in Figure 37.

The predicted nonideal conformational effects can be probed by SANS experiments, and theoretical/experimental comparisons are given elsewhere.[166] A detailed physical picture of the origin of the nonideal conformational behavior in terms of the thermodynamic forces a star experiences has been constructed. Comparison of the self-consistent PRISM theory results with phenomenological scaling, and other coarse-grained polymer physics approaches have also been presented, and distinctive qualitative and quantitative differences have been identified.[166]

The self-consistently determined intermolecular pair correlations, $g_{ij}(r)$, and collective partial structure factors, $\hat{S}_{ij}(k)$, also display several unique physical features due to the star-branched architecture.[167] An example for the pair correlations is shown in Figure 38 for a meltlike

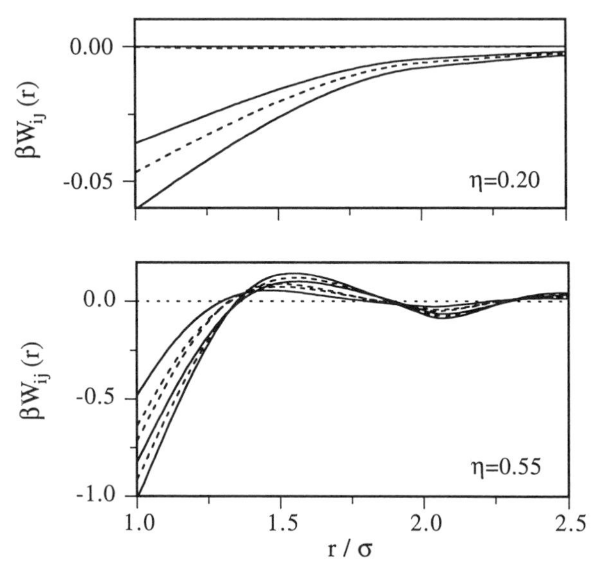

Figure 37. The six-pair components W_{ij} ($\{i,j\} = 0, 1, 2$) of the self-consistently determined solvation potential W acting on a 8×100 star for two values of fluid packing fraction. The three "diagonal" ("off-diagonal") components W_{ii} (W_{ij}, $j \neq i$) are drawn with solid (dashed) lines. The strengths of the components at contact ($r = \sigma$) are, from weakest (least negative) to strongest (most negative), $\{0,0\}$, $\{0,1\}$, $\{0,2\}$, $\{1,1\}$, $\{1,2\}$, and $\{2,2\}$, respectively. Only the three strongest (W_{11}, W_{12}, W_{22}) can be discerned on the lower density plot.

density. The corona region is quite similar to the analogous linear tangent SFC[55] case: a local solvation shell regime followed by a power law correlation hole region out to intersite separations of order R_g. However, the central branch structure and core region pair correlations show reduced local ordering (as expected), plus two new structural features: (i) a broad intermediate region where $h(r)$ grows in a nearly linear fashion with increasing site–site separation and (ii) a weak oscillation in the vicinity of the global chain dimension separation. The latter feature is hard to see on the scale of Figure 38, but is clearly visible on an expanded scale. Close examination reveals the characteristic oscillation wavelength scales as $N_a^{1/3}$, a distance corresponding to the mean separation of cores on different stars within the correlation volume containing of order $N_a^{1/2}$ interpenetrating star molecules.

The macromolecular scale solvation shell feature is clearly seen in the partial and total structure factors in Fourier space. An example is given in

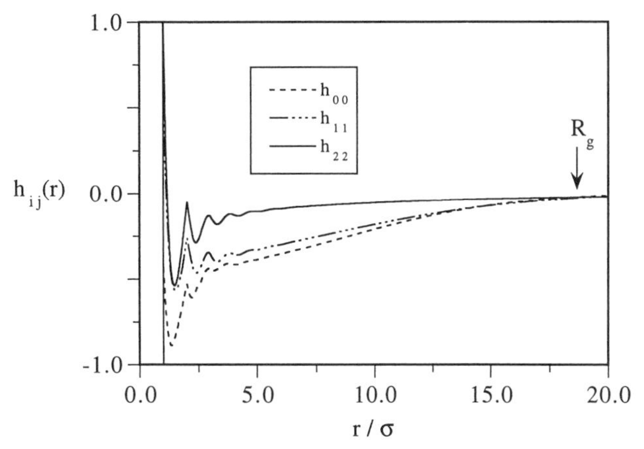

Figure 38. Diagonal components of the self-consistently determined intermolecular site–site pair correlation functions, $h_{ij}(r) = g_{ij}(r) - 1$ for a dense melt ($\eta = 0.55$) of 8×400 star polymers. The overall star radius-of-gyration is indicated.[167]

Figure 39. The very small and high k regions are weakly dependent on arm number and are nearly identical to the analogous linear chain case. However, the low-angle broad maximum at $k^* \propto N_a^{-1/3}$ is a unique star feature that becomes more intense, and shifts to slightly smaller wave vector, with increasing arm number. This feature implies there is a type of macroscale colloidal ordering in star polymer melts, which is also predicted to occur under semidilute and concentrated solution conditions.[167]

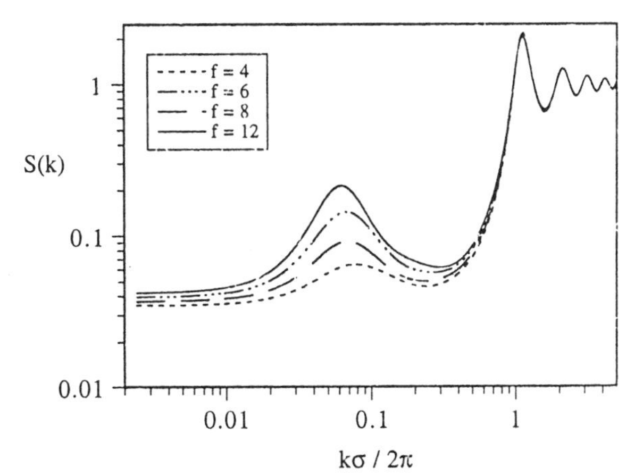

Figure 39. Total collective structure factor of a dense melt of $f \times 100$ stars. Note the logarithmic scales.

The signature of macroscale colloidal ordering in semidilute star polymer solutions has been found experimentally using SANS,[170] and these measurements are in good agreement with the PRISM predictions. Moreover, phenomenological scaling-type arguments have been advanced by Witten and Pincus[169] that also predict such a low-angle scattering maximum at $k^* \propto R_g^{-1} \propto N_a^{-3/5}$ *but only under semidilute* ($\rho \cong \rho^*$), *long arm, good solvent* conditions. In the latter situation, the stars do not interpenetrate appreciably and hence behave roughly as fuzzy spheres. However, PRISM theory predicts the colloidal ordering feature is a generic consequence of the star architecture and persists in dense solutions up to the melt state. As a cautionary remark, the latter conclusion is sensitive to small quantitative theoretical errors since the low angle peak in $S(k)$ emerges as a small difference between partial structure factors which are large and of opposite sign. Appropriate experiments to test the melt PRISM predictions have apparently not yet been carried out.

X. OTHER INTEGRAL EQUATION APPROACHES

There has been an explosive growth, particularly over the past couple of years, in developing continuum space integral equation theories for chain molecule and polymer fluids that are *not* based on the RISM formalism. There are at least three distinct classes of new theories. Our goal is to briefly summarize the various approaches and cite relevant references. A more detailed description of the technical aspects of the various theories, and the distinctive similarities and differences between them, will be the subject of a future review.[9] We also point out that much progress has recently been made in developing *statistical thermodynamic* theories of sophisticated polymer *lattice* models by Freed and co-workers,[171] and continuum polymer fluids by Hall and coworkers.[79] Description of these advances are beyond the scope of this chapter.

The first and earliest class of new approaches to polymer fluids is the *associating fluid model* pioneered by Wertheim.[172,173] In the associating fluid model one considers a liquid of particles interacting with a potential given by

$$u(r_{12}) = u_{\mathrm{R}}(r) + u_{\mathrm{AB}}(r_{\mathrm{AB}}^{12}) + u_{\mathrm{BA}}(r_{\mathrm{BA}}^{21}) \tag{10.1}$$

where $u_{\mathrm{R}}(r)$ is a hard-core interaction between the centers of particle 1 and particle 2. Each particle is taken to have two attractive sites or "glue spots" A and B located randomly on its surface. Sites A and B interact with an attractive potential $u_{\mathrm{AB}}(r_{\mathrm{AB}}^{12})$ where r_{AB}^{12} is the distance between site A on particle 1 and site B on particle 2. In practice the range of the attractive interactions is taken to zero such that $f_{AB}(r_{AB}^{12}) = \exp[-\beta u_{AB}] - 1 \propto \delta(r - \sigma)$. Because of steric restrictions, an attractive site cannot form more than one bond with a site on another particle. Thus the potential in Eq. (10.1) leads to a *polydisperse* mixture of freely jointed, tangent hard-sphere chains. A somewhat different model for asssociating particles, but in the same spirit, was suggested by Smith and Nezbeda.[174]

Wertheim[172] developed a novel graph-theoretic expansion for the grand canonical partition function of an associating fluid. Unlike the conventional Mayer cluster expansion, which involves only the overall density, the Wertheim expansion involves the four singlet densities ρ_0, ρ_A, ρ_B, and ρ_{AB} where the subscript denotes the bonded sites. Using first-order thermodynamic perturbation theory (TPT1) about the hard-sphere fluid, Wertheim found an analytical expression, via a compressibility route, for the equation of state of the polydisperse chain liquid

$$\frac{\beta P}{\rho_m} = \frac{1+\eta+\eta^2-\eta^3}{(1-\eta)^3} - \left(1-\frac{1}{\nu}\right)\left(1-\frac{\eta}{2-\eta}+\frac{3}{1-\eta}\right) \tag{10.2}$$

where ν is the average chain length in the mixture. Equation (10.2) employs the well-known Carnahan–Starling equation of state[5] for the hard-sphere reference system as the first term on the right-hand side. The TPT1 equation of state is in remarkably good agreement with Monte Carlo simulations of Dickman and Hall[175] for monodisperse chain liquids of 4, 8 and 16 units. Somewhat improved results were found by carrying out the perturbation expansion to second order.[173] These ideas were extended to chain mixtures by Chapman and co-workers.[176] Most recently, Kalyuzhnyi and Cummings[177] obtained an analytic solution of the general multicomponent version of Wertheim's model based on the Baxter factorization method.

Wertheim has also formulated an integral equation for the structure of chain molecule liquids using the associating fluid model.[173] This formulation involves an Ornstein–Zernike-like equation that incorporates bonding between attractive sites; various closures such as the Percus–Yevick and HNC closures can be used. Analytical solutions[178,179] to the integral equations are possible for the PY closure for hard-sphere chains. Chang and Sandler[179] found good agreement between the Wertheim theory and Monte Carlo calculations for the intermolecular radial distribution function for short-chain liquids of $N = 4$ at high packing fraction. As the chains become longer ($N = 8$ and 16), the theory tends to overestimate $g(r)$.

Chiew[180,181] has also developed an integral equation approach to describe chain molecule liquids in the spirit of the associating particle theory. Chiew views the collection of particles as a multicomponent system in which the particles interact with species-dependent attractive interactions to generate a mixture of chain molecules. The multicomponent Ornstein–Zernike equation is then solved (with the PY closure), subject to a chain connectivity constraint, for the structure of the liquid. Chain connectivity is maintained at the level of the correlation functions (rather than through the potentials as in Wertheim theory), by requiring that $g_{i,i+1}(r) \propto \delta(r-\sigma)$ between adjacent particles making up the chain backbone. Analytical solutions have been obtained by Chiew. The equation of state from the Chiew theory,[180] via the compressibility route, has a similar form to that of TPT1 in Eq. (10.2).

$$\frac{\beta P}{\rho_m} = \frac{1+\eta+\eta^2-\eta^3}{(1-\eta)^3} - \left(1-\frac{1}{\nu}\right)\frac{(1+\eta/2)}{(1-\eta)^2} \tag{10.3}$$

As in the Wertheim equation of state, Chiew also made use of the Carnahan–Starling expression for the pressure of the corresponding hard-sphere liquid. Similar agreement between TPT1 and the Chiew theory was found with Monte Carlo simulations of Dickman and Hall[175] for the pressure.

The contact value of the intermolecular radial distribution function from the Chiew theory is identical[179] to the Wertheim prediction although differences occur for larger values of r. Comparisons have been made between the Chiew theory and simulations of Yethiraj and co-workers[182] ($N = 4$, 8), and the simulations of Chang and Sandler[179] ($N = 4$–16). Both the Wertheim and Chiew integral equations predict radial distribution functions somewhat higher than seen from simulations. As in the case of PRISM theory, better agreement with simulation is found as the density increases. To our knowledge, no direct comparisons between PRISM theory and the integral equations of Wertheim and Chiew have been made as yet. Based on agreement with the Chang and Sandler[179] simulations for $N = 4$–16, and the Grest and Kremer simulations in Figure 3 for $N = 50$–150, we tentatively conclude that PRISM theory is more accurate for the structure of long-chain molecule athermal liquids at high densities. On the other hand, the Wertheim and Chiew theories give significantly better predictions for the hard-core fluid equation of state than the PRISM based results.

Wertheim's predictions for athermal binary mixtures have been tested against simulation by Chang and Sandler.[183] Prausnitz and co-workers[184] have developed modified versions of Chiew's equation of state for athermal mixtures to treat hard-core copolymer mixtures. The thermodynamic consequences of attractive interactions are treated in the HTA spirit. Phase diagram and miscibility map predictions have been made.

An important distinction between PRISM theory and the associating fluid approaches is that the latter treats intramolecular and intermolecular correlations on the same footing. It is not obvious, however, that the PY closure is equally valid for both intramolecular and intermolecular correlations. Many years ago PY calculations on a single athermal chain polymer (dilute, good solvent case) by Curro et al.[185] suggested that the PY closure leads to unphysical results. PRISM theory, by contrast, treats correlations between sites on the same chain and between sites on different chains separately. In Section VIII we discussed the self-consistent calculation of intramolecular correlations with PRISM theory. For realistic polymer chain models at liquidlike densities, the excluded volume along the chain appears to be effectively screened out for interaction sites separated by few backbone bonds. This allows one to determine intramolecular correlations for realistic polymer models from a separate single-chain calculation, which can then serve as input to PRISM theory to calculate interchain packing effects.

A second, entirely different class of new polymer integral equation theories have been developed by Lipson and co-workers,[186–188] Eu and Gan,[189–192] and Attard[193] based on the site–site version of the Born–Green–Yvon (BGY) equation.[5] The earliest work in this direction was apparently by Whittington and Dunfield,[194] but they addressed only a special aspect of the isolated polymer problem (dilute solution). The central quantity in the BGY approaches is the formally exact expressions that relate two and three (or more) intramolecular and intermolecular distribution functions. The generalized site–site Ornstein–Zernike equations and direct correlation functions do not enter. In the BGY schemes the closure approximation(s) enter as approximate relations between the two- and three-body distribution functions supplemented with exact normalization and asymptotic conditions. In the recent BGY work of Taylor and Lipson[188] a four-point distribution function also enters.

A polymeric complication of all BGY approaches is the need for several different types of three-point correlation functions. For the purely *inter*molecular distribution function the standard Kirkwood superposition approximation[5,8] is invoked, that is, a real space product of the three corresponding intermolecular pair functions. However, for the "intramolecular triplet" distribution functions (involving one, two, or three sites on the same polymer), there are many alternative schemes invoked by the different authors.[186–193] These can be viewed as different closure approximations. Eu and Gan[189–192] have focused on analyzing the Kirkwood hierarchy based on Kirkwood-like and Markov-like approximations for the intramolecular and intermolecular distribution functions. The technical treatment of site inequivalency, or the "too many equations problem," also varies among the different workers.

Intramolecular correlations are handled in different approximate manners in the various BGY approaches. Taylor and Lipson[188] treat pair correlations on the same chain as input to the theory in a manner similar to PRISM theory. In contrast, the formulations of Eu and Gan,[189–192] and also Attard,[193] yield closed integral equations for *both* the intra- and intermolecular pair distribution functions. Thus, in a sense the intra- and intermolecular pair correlations are treated on an equal footing, and a "self-consistent" integral equation theory is naturally obtained. Eu and Gan have recently presented a comparison[192] between their BGY approach and self-consistent and non-self-consistent PRISM theory, in both general conceptual terms and within the context of numerical predictions for specific model hard-core systems. For the jointed hard-core chain model studies, the theory of Eu and Gan appears quantitatively superior to PRISM predictions, particularly for the equation of state.[192]

All the BGY approaches[186–193] have yielded promising results for the structure and equation of state of hard core, jointed chain solutions and melts of modest degrees of polymerization. Eu and Gan have also successfully treated dilute homopolymer,[190] and AB copolymer,[191] solutions with soft-core interactions under both good and theta (ideal) solvent conditions. A noteworthy aspect of the approach of Lipson and Andrews is its ability to also be formulated and implemented for simple lattice polymer models.[186,187]

A third class of new polymer integral equation theories have been proposed by Kierlik and Rosinberg.[195] Their work is an extension of a density functional theory of inhomogeneous polyatomic fluids to treat the homogeneous phase. The Wertheim thermodynamic pertubation theory of polymerization is employed in an essential manner. Applications to calculate the intermolecular structure of rather short homopolymer solutions and melts have been made. Good results are found for short chains at high densities, but the authors comment that their earlier theory appears to be unsuited for long chains at low to moderate (semidilute) densities.[195]

As a general comment on the recent polymer integral equation work, we note that applications to date have focused primarily on the structure (intra- and intermolecular) and equation of state (based on a virial or free energy route) of the simple hard core, tangent jointed chain model of polymer solutions and melts. How tractable and generalizable the various approaches are for treating semiflexible and/or atomistic models of macromolecular fluids is unclear for most theories. Little, or no, work has been published on the effect of attractive forces on fluid structure, collective scattering functions, compressibility route thermodynamic predictions, semiflexible or atomistic chain models, or nonlinear macromolecular architectures. The associating fluid approach[172–181] has been generalized and applied to simple models of athermal polymer mixtures and copolymers, although the primary published emphasis is on thermodynamic properties and not structure. We anticipate many extensions and applications of the emerging integral equation theories in the near future, and feel it is premature to judge the relative advantages and disadvantages of the many distinct approaches. The latter task will undoubtedly be very difficult to simply summarize.

Finally, we mention an interesting recent study by Chandler[126] that extended the Gaussian field-theoretic model of Li and Kardar[196] to treat atomic and polymeric fluids. Remarkably, the atomic PY and MSA theories were derived from a Gaussian field-theoretic formalism *without* explicit use of the Ornstein–Zernike relation or direct correlation function concept. In addition, based on an additional preaveraging approximation, analytic PRISM theory[30] was recovered for hard-core thread chain model fluids. Nonperturbative applications of this field-theoretic approach to polymer liquids where the chains have nonzero thickness and/or attractive forces requires numerical work that, to the best of our knowledge, has not yet been pursued.

XI. DISCUSSION AND FUTURE DIRECTIONS

This chapter has focused on describing progress made over the past eight years on developing microscopic liquid-state theories of the conformation, structure, thermodynamics, and phase transitions of macromolecular fluids within the context of interaction site models and the RISM integral equation method. Even within this rather restrictive theoretical framework, many developments were not discussed. These include the following:

1. Development and application by Glandt and co-workers of a theory of chain molecule fluids in porous media[197] that heavily uses PRISM methods.[198]
2. Development and application of molecular closures to treat attractive forces in polymer solutions and melts (including the liquid–vapor transition).[119,134]
3. PRISM theory of polymer/colloid mixtures in the dilute colloid limit by Yethiraj and co-workers.[199] Shaw and Thirumalai have constructed a field-theoretic-type description for a single long

polymer chain in a colloidal solution that also employs PRISM-derived information.[200]

4. An exact analytic solution by Fuchs of the PRISM equations using the Baxter factorization method for special Gaussian chain models of melts and mixtures.[201]
5. Theory of self-assembling multiblock copolymer fluids.[202]
6. Analytic and numerical self-consistent PRISM theories of polymer blends and diblock copolymers.[163]
7. PRISM theory, particularly in its analytic Gaussian thread and string versions, has also been extensively employed by Schweizer and co-workers as the equilibrium input to microscopic generalized Langevin and mode-coupling theories of the dynamics of macromolecular fluids.[54,203,204]

Another important direction has been to construct microscopic theories of inhomogeneous systems by combining PRISM theory of disordered, bulk-phase structure with novel macromolecular versions of thermodynamic density functional theory. Building on the modern atomic[205] and site–site polyatomic[206] density functional theories, McCoy and co-workers have recently pursued this approach for treating first-order phase transitions such as polymer crystallization[68] and block copolymer microphase separation.[150] Extensions to treat chain molecule and polymer fluids near surfaces and interfaces have also appeared.[80,207,208] Surface segregation of polymer blends can also be treated using PRISM plus density functional theory.[207] A computationally simpler, but more approximate, alternative for treating polymers near surfaces, confined fluids, and alloy surface segregation is to employ an entirely integral equation approach, wall-PRISM theory, as proposed by Yethiraj and Hall.[78] Detailed applications of this approach to treat density waves and solvation forces of confined alkane fluids as a function of fluid–surface interaction potential, and direct tests of the theory against computer simulations and experiments, have been carried out by Walley and co-workers.[209] Athermal blend surface segregation has also been treated with wall-PRISM theory by Yethiraj and co-workers.[110]

Although we believe much progress has been made based on the PRISM theory approach, there remain important basic theoretical issues that require continuing attention in the future. The most obvious is the question of closure approximation. Even for the purely repulsive or hard-core polymer fluid, improved closures are desirable. Results for diatomic and polymer fluids based on the "diagramatically proper" Chandler–Silbey–Ladanyi (CSL) formulation[39] of RISM theory have been obtained by Yethiraj[38] based on the site–site PY closure. Unfortunately, for chain molecules this approach does not represent an improvement (indeed it appears worse) over the simpler, diagramatically "improper" PRISM theory. However, recent work by Lue and Blankschtien[210] based on the CSL theory has yielded encouraging results for some nonpolymeric fluid systems, and further consideration of the CSL approach is desirable. A promising, but computationally intensive, new approach has been proposed by Donley and co-workers[41] based on a 2-chain (not 2-site), density functional and solvation potential perspective.

The question of the effects of attractive tail potentials on macromolecular structure, and the appropriate closure(s), require much further study. Advances in the closure question may also result in more thermodynamically consistent theories, which is important for questions such as the equation of state, excess properties of multicomponent systems, and construction of constant pressure (not volume) theories. Other basic theoretical issues relate to the potential importance of corrections to site preaveraging, and the development of tractable schemes to systematically compute such corrections. The issue of "self-consistency" of intramolecular and intermolecular correlations is rather unsettled from the point of view of both how important these effects are (the answer is problem-specific), and how to best go about constructing a computationally convenient theory to treat them.

For all basic questions mentioned, one expects there is not a single unique answer for all systems and conditions of interest. Moreover, there is always a trade-off between computational convenience and numerical accuracy. Due to the rapid progress in computing power and simulation algorithms, we believe that computer simulation will become increasingly important in guiding the development of liquid-state theories of polymers. As has been true for simple atomic and molecular fluids, carefully designed simulations can provide invaluable benchmark results to unambiguously test the statistical mechanical (not polymer model) aspects of integral equation approaches to macromolecules. This activity will also be important in establishing the relative merits of the many different emerging integral equation theories of polymer fluids,[172–195] and the validity of coarse-grained scaling and phenomenological field-theoretic approaches and their relationship to microscopic liquid-state theories.

Finally, there remain many physical problems and systems that have either not been addressed at all or are just beginning to be seriously attacked, from a liquid-state integral equation perspective. An incomplete list is as follows: (1) Charged polymers, polyelectrolytes, and ionomers (strong dipolar interactions). The appropriate closure approximations and treatment of self-consistency in the presence of both hard-core forces and long-range Coulombic interactions are major unsolved problems. (2) Short-range orientational correlations in isotropic fluids, and nematic and other liquid crystalline phases, of rigid and semiflexible polymers. (3) Incorporation of quenched disorder (beyond naive pre-averaging) associated with intramolecular features such as tacticity variations, molecular weight polydispersity, and sequence disorder in copolymers. (4) Polymer gels, rubbers, and associating fluids where strong intermolecular "attractive" interactions can result in networklike and/or fractal structures that can be either quenched in or thermoreversible. (5) Self-assembly of intermediate-sized molecules (e.g., surfactants) into supramolecular structures (e.g., micelles, microemulsions). (6) Ternary and more complex mixtures of homopolymers and copolymers where there is a competition between macrophase and microphase separation. A related phenomenon is the role of low concentration additives on phase stability.

All these problems represent challenges to the further development of a general microscopic liquid-state theory of macromolecular systems, but we are hopeful that significant progress can be made in the near future.

ACKNOWLEDGEMENTS

The work described in this chapter has been done with many excellent collaborators. We gratefully acknowledge the essential theoretical contributions of our former postdocs: J. Donley, C. J. Grayce, K. G. Honnell, J. D. McCoy, J. Melenkevitz, J. J. Rajasekaharan, C. Singh, and A. Yethiraj. One of us (K.S.S.) expresses his thanks to former and present graduate students: C. Butler, E. F. David, K. Kolbet, and K. P. Walley, and also Drs. A. Chatterjee, M. Fuchs, M. Guenza, G. Szamel, and H. Tang for helpful discussions. Fruitful collaborations in the area of theory/simulation comparisons with G. Grest, K. Kremer, S. K. Kumar, J. D. McCoy, S. Plimpton and A. Yethiraj are gratefully acknowledged. We thank J. D. Londano, A. Habenshuss, A. Narten, and G. Wignall of Oak Ridge National Laboratory for experiment/theory collaborations. One of us (J.G.C.) acknowledges ongoing collaborations with J. D. McCoy and J. D. Honeycutt. Discussions and/or correspondence with many other scientists are also appreciated with a special thanks extended to D. Chandler. We are most grateful to E. F. David, C. J. Grayce, K. G. Honnell, and C. Singh for help in preparing the figures.

The PRISM work described in this chapter has benefitted from financial support from several sources. At Sandia National Laboratories we acknowledge the U.S. Department of Energy under contract DE-AC047DP00789, CRADA #1078, and the Basic Energy Sciences/Division of the Materials Science Program. At the University of Illinois primary support is from the Division of Materials Sciences, U.S. Department of Energy in cooperation with the UIUC Materials Research Laboratory (via grant No. DEFG02-91ER45439) and Oak Ridge National Laboratory, and also from DOE via CRADA No. 1078. Primary support for the block copolymer work was from NSF via the UIUC-MRL grant NSF-DMR-89-20538. Use of the central computing facilities of the UIUC-MRL, the massively parallel computing facilities at Sandia, and BIOSYM software are gratefully acknowledged.

REFERENCES

1. P. J. Flory, *Principles of Polymer Chemistry*, Cornell University Press, Ithaca, 1953.
2. P.-G. deGennes, *Scaling Concepts in Polymer Physics*, Cornell University Press, Ithaca, 1979.
3. M. Doi and S. F. Edwards, *Theory of Polymer Dynamics*, Oxford Press, Oxford, 1986.
4. K. F. Freed, *Renormalization Group Theory of Macromolecules*, Wiley, New York, 1987.
5. J. P. Hansen and I. R. McDonald, *Theory of Simple Liquids*, 2nd ed., Academic Press, London, 1986.
6. K. S. Schweizer and J. G. Curro, *Phys. Rev. Lett.* **58**, 246 (1987). J. G. Curro and K. S. Schweizer, *Macromolecules* **20**, 1928 (1987). J. G. Curro and K. S. Schweizer, *J. Chem. Phys.* **87**, 1842 (1987).
7. D. Chandler and H. C. Andersen, *J. Chem. Phys.* **57**, 1930 (1972).
8. D. Chandler, in *Studies in Statistical Mechanics*, Vol. VIII, E. W. Montroll and J. L. Lebowitz, eds., North-Holland, Amsterdam, 1982, p. 274; L. J. Lowden and D. Chandler, *J. Chem. Phys.* 61, 5228 (1974); **59**, 6587 (1973); **62**, 4246 (1975).
9. K. S. Schweizer and J. G. Curro, *Ann. Rev. Phys. Chem.*, in preparation.
10. K. S. Schweizer and J. G. Curro, *Adv. Polym. Sci.* **116**, 321 (1994).
11. H. C. Andersen, D. Chandler, and J. D. Weeks, *Adv. Chem. Phys.* **34**, 105 (1976); D. Chandler, J. D. Weeks, and H. C. Andersen, *Science* **220**, 787 (1983); J. A. Barker and D. Henderson, *Rev. Mod. Phys.* **48**, 587 (1976).
12. D. Chandler and L. R. Pratt, *J. Chem. Phys.* **65**, 2925 (1976); L. R. Pratt and D. Chandler, *J. Chem. Phys.* **66**, 147 (1977); L. R. Pratt, C. S. Hsu, and D. Chandler, *J. Chem. Phys.* **68**, 4202 and 4213 (1978).
13. P. J. Flory, *Statistical Mechanics of Chain Modecules,* Wiley, New York, 1969.
14. J. G. Kirkwood, *J. Chem. Phys.* **3**, 300 (1935); J. Yvon, *Actualities Scientifiques et Industrielles*, Hermann and Cie, Paris, 1935; M. Born and M. Green, *Proc. Roy. Soc.* **A188**, 10 (1946).
15. J. K. Percus, in *Classical Fluids*, H. L. Frisch and J. L. Lebowitz, eds., Wiley, New York, 1964.
16. J. Percus and G. Yevick, *Phys. Rev.* **110**, 1 (1958); J. von Leeuwen, J. Groeneveld, and J. deBoer, *Physica* **25**, 792 (1959); G. Stell, *Physica* **29**, 517 (1963).
17. See, for example, F. J. Rogers and D. A. Young, *Phys. Rev. A* **30** 999 (1984); P. Ballone, G. Pastore, G. Galli, and D. Gazillo, *Mol. Phys.* **59**, 275 (1986); B. Bernu, J. P. Hansen, Y. Hiwatari, and G. Pastore, *Phys. Rev. A* **36**, 4891 (1987).
18. Q. Zhang and J. P. Badiali, *Phys. Rev. A* **45**, 8666 (1992); A. Parola and L. Reatto, *Phys. Rev. A* **44**, 6600 (1991).
19. See, for example, B. D'Aguanno, U. Genz, and R. Klein, *J. Phys. Condens. Matter* **2**, SA379 (1990) and references cited therein.
20. P. A. Monson and G. P. Morris, *Adv. Chem. Phys.* **77**, 451 (1990); P. Rossky, *Ann. Rev. Phys. Chem.* **36**, 321 (1985).
21. D. Chandler, C. S. Hsu, and W. B. Streett, *J. Chem. Phys.* **66**, 5231 (1977); S. I. Sandler and A. H. Narten, *Mol. Phys.* **32**, 1543 (1976); A. H. Narten, *J. Chem. Phys.* **67**, 2102 (1977); C. S. Hsu and D. Chandler, *Mol. Phys.* **36**, 215 (1978); *Mol. Phys.* **37**, 299 (1979).
22. L. R. Pratt and D. Chandler, *J. Chem. Phys.* **67**, 3683 (1977); *Methods Enzymol.* **127**, 48 (1986).

23. D. Chandler, Y. Singh, and D. M. Richardson, *J. Chem. Phys.* **81**, 1975 (1984); A. L. Nichols, D. Chandler, Y. Singh, and D. M. Richardson, *J. Chem. Phys.* **81**, 5109 (1984); D. Laria, D. Wu, and D. Chandler, *J. Chem. Phys.* **95**, 4444 (1991).
24. D. Chandler, in *Les Houches Lectures on "Liquids, Freezing and Glass Transition,"* J. P. Hansen, D. Levesque, and J. Zinn-Justin, eds., North-Holland, Amsterdam, 1991, p. 193.
25. K. S. Schweizer and J. G. Curro, *Macromolecules* **21**, 3070, 3082 (1988).
26. J. G. Curro, *Macromolecules* **27**, 4665 (1994); J. J. Rajasekaran and J. G. Curro, *J. Chem. Soc. Faraday Trans.* **91**, 2427 (1995).
27. K. S. Schweizer and J. G. Curro, *Phys. Rev. Lett.* **60**, 809 (1988).
28. J. G. Curro and K. S. Schweizer, *J. Chem. Phys.* **88**, 7242 (1988); K. S. Schweizer and J. G. Curro, *J. Chem. Phys.* **91**, 5059 (1989).
29. J. G. Curro and K. S. Schweizer, *Macromolecules* **23**, 1402 (1990).
30. K. S. Schweizer and J. G. Curro, *Chem. Phys.* **149**, 105 (1990).
31. J. G. Curro and K. S. Schweizer, *Macromolecules* **24**, 6736 (1991).
32. J. J. Rajasekaran, J. G. Curro, and J. D. Honeycutt, *Macromolecules* **28**, 6843 (1995).
33. E. F. David and K. S. Schweizer, *J. Chem. Phys.* **100**, 7767, 7784 (1994).
34. D. Chandler, *Chem. Phys. Lett.* **139**, 108 (1987).
35. L. Lue and D. Blanckschtein, *J. Phys. Chem.* **96**, 8582 (1991); J. R. Elliot and U. S. Kanetar, *Mol. Phys.* **71**, 871 and 883 (1990).
36. K. Kojima and K. Arakawa, *Bull. Chem. Soc. Jpn.* **51**, 1977 (1978); **53**, 1795 (1980).
37. G. A. Martynov and G. N. Sarkisov, *Mol. Phys.* **49**, 1495 (1983).
38. A. Yethiraj, *Mol. Phys.* **80**, 695 (1993).
39. D. Chandler, R. Silvey, and B. M. Ladanyi, *Mol. Phys.* **46**, 1335 (1982).
40. D. M. Richardson and D. Chandler, *J. Chem. Phys.* **80**, 4484 (1984).
41. J. P. Donley, J. G. Curro, and J. D. McCoy, *J. Chem. Phys.* **101**, 3205 (1994).
42. C. J. Grayce and J. J. dePablo, *J. Chem. Phys.* **101**, 6013 (1994).
43. K. S. Schweizer, K. G. Honnell, and J. G. Curro, *J. Chem. Phys.* **96**, 3211 (1992).
44. J. Melenkevitz, K. S. Schweizer, and J. G. Curro, *Macromolecules* **26**, 6190 (1993).
45. J. Melenkevitz, J. G. Curro, and K. S. Schweizer, *J. Chem. Phys.* **99**, 5571 (1993).
46. C. J. Grayce and K. S. Schweizer, *J. Chem. Phys.* **100**, 6846 (1994).
47. C. J. Grayce, A. Yethiraj, and K. S. Schweizer, *J. Chem. Phys.* **100**, 6857 (1994).
48. P. J. Flory, *J. Chem. Phys.* **17**, 203 (1949).
49. D. G. Ballard, J. Schelton, and G. D. Wignall, *Eur. Polym. J.* **9**, 965 (1973); J. P. Cotton, D. Decker, H. Benoit, B. Farnoux, J. Higgins, G. Jannick, R. Ober, and J. des Cloizeaux, *Macromolecules* **7**, 863 (1974).
50. J. G. Curro, *J. Chem. Phys.* **64**, 2496 (1976); *Macromolecules* **12**, 463 (1979); M. Vacatello, G. Avitabile, P. Corradini, and A. Tuzi, *J. Chem. Phys.* **73**, 543 (1980).
51. J. G. Curro, K. S. Schweizer, G. S. Grest, and K. Kremer, *J. Chem. Phys.* **91**, 1357 (1989); K. Kremer and G. S. Grest, *J. Chem. Phys.* **92**, 5057 (1990).
52. K. S. Schweizer, E. F. David, C. Singh, J. G. Curro, and J. J. Rajasekaran, *Macromolecules* **28**, 1528 (1995).
53. L. J. Fetters, D. J. Lohse, D. Richter, T. A. Witten, and A. Zirkel, *Macromolecules* **27**, 4639 (1994).
54. K. S. Schweizer and G. Szamel, *J. Chem. Phys.* **103**, 1934 (1995).
55. K. G. Honnell, J. G. Curro, and K. S. Schweizer, *Macromolecules* **23**, 3496 (1990).

56. R. Koyama, *J. Phys. Soc. Jpn.* **22**, 1029 (1973).
57. See, for example, *Atomistic Modeling of Physical Properties*, Adv. Polym. Sci., vol. 116, L. Monnerie and U. W. Suter, eds., Springer-Verlag, Berlin, 1994.
58. D. Theodorou and U. W. Suter, *Macromolecules* **18**, 1467 (1985).
59. R. Khare, M. E. Paulaitis, and S. R. Lustig, *Macromolecules* **26**, 7203 (1993).
60. A. Yethiraj, C. K. Hall, and K. G. Honnell, *J. Chem. Phys.* **93**, 4453 (1990); A. Yethiraj and C. K. Hall, *J. Chem. Phys.* **96**, 797 (1992).
61. A. Yethiraj and K. S. Schweizer, *J. Chem. Phys.* **97**, 1455 (1992).
62. A. Yethiraj, *J. Chem. Phys.* **101**, 9104 (1994).
63. J. D. McCoy, K. G. Honnell, J. G. Curro, K. S. Schweizer, and J. D. Honeycutt, *Macromolecules* **25**, 4905 (1992).
64. K. G. Honnell, J. D. McCoy, J. G. Curro, K. S. Schweizer, A. H. Narten, and A. Habenschuss, *J. Chem. Phys.* **94**, 4905 (1991).
65. K. G. Honnell, J. D. McCoy, J. G. Curro, K. S. Schweizer, A. H. Narten, and A. Habenschuss, *Bull. Am. Phys. Soc.* **36**(3), 481 (1991).
66. A. H. Narten, A. Habenschuss, K. G. Honnell, J. D. McCoy, J. G. Curro, and K. S. Schweizer, *J. Chem. Soc. Faraday Trans.* **88**, 1791 (1992).
67. A. Yethiraj, J. G. Curro, K. S. Schweizer, and J. D. McCoy, *J. Chem. Phys.* **98**, 1635 (1993); J. G. Curro, A. Yethiraj, K. S. Schweizer, J. D. McCoy, and K. G. Honnell, *Macromolecules* **26** 2655 (1993).
68. J. D. McCoy, K. G. Honnell, K. S. Schweizer, and J. G. Curro, *Chem. Phys. Lett.* 179, 374 (1991); *J. Chem. Phys.* **95**, 9348 (1991).
69. K. S. Schweizer and C. Singh, *Macromolecules* **28**, 2063 (1995).
70. A. Yethiraj, J. G. Curro, and J. J. Rajasekaran, *J. Chem. Phys.* **103**, 2229 (1995).
71. See, for example, G. R. Mitchell and A. H. Windle, *Polymer* **25**, 906 (1984).
72. J. D. Londano, A. Habenschuss, J. G. Curro and J. J. Rajasekaran, *J. Polym. Sci. B, Polym. Phys. Ed.*, submitted, 1996.
73. U. W. Suter and P. J. Flory, *Macromolecules* **8**, 765 (1975).
74. D. G. H. Ballard, P. Cheshire, G. W. Longman, and J. Schelton, *Polymer* **19**, 379 (1978).
75. J. G. Curro, J. J. Rajasekaran, A. H. Habenschuss, D. Londono, and J. D. Honeycutt, *Macromolecules*, in preparation.
76. J. K. Percus, *J. Stat. Phys.* **15**, 423 (1976).
77. R. Dickman and C. K. Hall, *J. Chem. Phys.* **89**, 3168 (1988).
78. A. Yethiraj and C. K. Hall, *J. Chem. Phys.* **95**, 3749 (1991).
79. R. Dickman and C. K. Hall, *J. Chem. Phys.* **85**, 4108 (1986); K. G. Honnell and C. K. Hall, *J. Chem. Phys.* **90**, 1841 (1989).
80. S. Sen, J. M. Cohen, J. D. McCoy, and J. G. Curro, *J. Chem. Phys.* **101**, 9010 (1994); S. Sen, J. D. McCoy, S. K. Nath, J. P. Donley, and J. G. Curro, *J. Chem. Phys.* **102**, 3431 (1995).
81. (a) A. Yethiraj, *J. Chem. Phys.* **102**, 6874 (1995). (b) L. R. Dodd and D. Theodorou, *Adv. Polym. Sci.* **116**, 249 (1994).
82. J. A. Barker and D. Henderson, *Ann. Phys. Chem.* **23**, 439 (1972).
83. O. Olabisi and R. Simha, *Macromolecules* **8**, 206 (1975).
84. A. Yethiraj and C. K. Hall, *J. Chem. Phys.* **95**, 1999 (1991).
85. J. S. Rowlinson and F. L. Swinton, *Liquids and Liquid Mixtures*, Butterworth

Scientific, London, 1982; J. Hildebrand and R. Scott, *The Solubility of Nonelectrolytes*, 3rd Ed., Reinhold, New York, 1949.

86. E. A. Grulke, in *Polymer Handbook*, 3rd Ed., J. Brandrup and E. H. Immergut, eds., Wiley, New York, 1989; P. A. Small, *J. App. Chem.* **3**, 71 (1953); D. W. Van Krevelen, *Fuel* **44**, 229 (1965); K. L. Hoy, *J. Paint Tech.* **42**, 76 (1970).
87. P. A. Rodgers, *J. App. Polym. Sci.* **48**, 1061 (1993).
88. M. M. Coleman, C. J. Serman, D. E. Bhagwager, and P. Painter, *Polymer* **31**, 1187 (1990).
89. D. J. Walsh, W. W. Graessley, S. Datta, D. J. Lohse, and L. J. Fetters, *Macromolecules* **25**, 5236 (1992).
90. W. L. Jorgensen, J. D. Madura, and C. J. Swenson, *J. Am. Chem. Soc.* **106**, 6638 (1984).
91. W. W. Graessley, R. Krishnamoorti, N. P. Balsara, L. J. Fetters, D. J. Lohse, D. Schulz, and J. Sissano, *Macromolecules* **27**, 2574, 3073, 3896 (1994) and references cited therein; R. Krishnamoorti, W. W. Graessley, L. J. Fetters, R. J. Garner, and D. J. Lohse, *Macromolecules* **28**, 1252, 1260 (1995).
92. D. J. Lohse, private communication, 1995.
93. K. Solc, ed., *Polymer Compatibility and Incompatibility*, MMI, Midland, Michigan, 1981; O. Olabisi, L. M. Robeson, and M. Shaw, *Polymer–Polymer Miscibility*, Academic Press, New York, 1979; R. Koningsveld, *Adv. Colloid Interface Sci.* **2**, 151, 1986; P. J. Flory, *Proc. Roy. Soc. A* **234**, 60 (1956).
94. I. C. Sanchez, *Ann. Rev. Mater. Sci.* **13**, 387 (1983).
95. F. S. Bates, *Science* **251**, 898 (1991).
96. C. S. Stevenson, J. G. Curro, J. D. McCoy, and S. J. Plimpton, *J. Chem. Phys.* **103**, 1208 (1995); C. S. Stevenson, J. D. McCoy, S. J. Plimpton, and J. G. Curro, *J. Chem. Phys.* **103**, 1200 (1995).
97. J. D. Weinhold, S. K. Kumar, C. Singh, and K. S. Schweizer, *J. Chem. Phys.* **103**, 9460 (1995).
98. F. S. Bates, M. F. Schulz, and J. Rosedale, *Macromolecules* **25**, 5547 (1992); M. D. Gehlsen and F. S. Bates, *Macromolecules* **27**, 3611 (1994).
99. F. S. Bates and G. H. Fredrickson, *Macromolecules* **27**, 1065 (1994); G. H. Fredrickson, A. J. Liu and F. S. Bates, *Macromolecules* **27**, 2503 (1994); G. H. Fredrickson and A. J. Liu, *J. Poly. Sci. Polym. Phys.* **33**, 1203 (1995).
100. C. Singh and K. S. Schweizer, *J. Chem. Phys.* **103**, 5814 (1995).
101. I. C. Sanchez and R. H. Lacombe, *J. Phys. Chem.* **80**, 2352 and 2368 (1976); D. J. Walsh and S. Rostami, *Adv. Polym. Sci.* **70**, 119 (1985).
102. J. G. Kirkwood and F. P. Buff, *J. Chem. Phys.* **19**, 774 (1951).
103. M. Dijkstra, D. Frenkel, and J. P. Hansen, *J. Chem. Phys.* **101**, 3179 (1994), and references cited therein.
104. E. J. Meijer and D. Frenkel, *J. Chem. Phys.* **100**, 6833 (1994).
105. P. Bolhius and D. Frenkel, *J. Chem. Phys.* **101**, 9869 (1994).
106. D. Frenkel and A. A. Louis, *Phys. Rev. Lett.* **68**, 3363 (1992).
107. J. Lebowitz and J. Rowlinson, *J. Chem. Phys.* **41**, 133 (1964).
108. T. Biben and J. P. Hansen, *Phys. Rev. Lett.* **66**, 2215 (1991).

109. L. Onsager, *Proc. New York Acad. Sci.* **51**, 627 (1949).

110. A. Yethiraj, S. Kumar, A. Hariharan, and K. S. Schweizer, *J. Chem. Phys.* **100**, 4691 (1994); S. Kumar, A. Yethiraj, K. S. Schweizer and F. Leermakers, *J. Chem. Phys.* **103**, 10332 (1995).

111. F. S. Bates, L. J. Fetters, and G. D. Wignall, *Macromolecules* **21**, 1086 (1988); *Phys. Rev. Lett.* **55**, 2425 (1985); J. D. Londano, A. H. Narten, G. D. Wignall, K. G. Honnell, E. T. Hseih, T. W. Johnson, and F. S. Bates, *Macromolecules* **27**, 2864 (1994).

112. For a recent review of polymer alloy theories see K. Binder, *Adv. Polym. Sci.* **112**, 181 (1994).

113. C. Singh and KJ. S. Schweizer, *Macromolecules* **28**, 8692 (1995).

114. J. D. Honeycutt, *Makromol. Chem. Macromol. Symp.* **65**, 49 (1993); *Macromolecules* **27**, 5377 (1994); *ACS Polym. Preprints* **33**, 529 (1992).

115. C. Singh and K. S. Schweizer, *Macromolecules*, to be submitted (1996).

116. J. Rhee and B. Crist, *Macromolecules* **24**, 5663 (1991); J. C. Nicholson, T. M. Finerman, and B. Crist, *Polymer* **31**, 2287 (1990).

117. C. Singh and K. S. Schweizer, in preparation.

118. A. Yethiraj and K. S. Schweizer, *J. Chem. Phys.* **97**, 5927 (1992); **98**, 9080 (1993).

119. K. S. Schweizer and A. Yethiraj, *J. Chem. Phys.* **98**, 9053 (1993).

120. C. Singh, K. S. Schweizer, and A. Yethiraj, *J. Chem. Phys.* **102**, 2187 (1995).

121. X. S. Chen and F. Forstmann, *J. Chem. Phys.* **97**, 3696 (1992); G. Malescio, *J. Chem. Phys.* **96**, 648 (1992), and references cited therein; E. Arrieta, C. Jedrzejek and K. N. Marsh, *J. Chem. Phys.* **95**, 6806 and 6838 (1991).

122. L. Belloni, *J. Chem. Phys.* **98**, 8080 (1993); D. Gazillo, *J. Chem. Phys.* **95**, 4563 (1991).

123. G. P. Morris and D. J. Isbister, *Mol. Phys.* **59**, 911 (1986); **52**, 57 (1984).

124. K. S. Schweizer and J. G. Curro, *J. Chem. Phys.* **94**, 3986 (1991).

125. M. P. Gehlsen, J. Rosedale, F. S. Bates, G. D. Wignall, L. Hansen, and K. Almdal, *Phys. Rev. Lett.* **68**, 2452 (1992).

126. H. P. Deutsch and K. Binder, *Europhys. Lett.* **17**, 697 (1992); *Macromolecules* **25**, 6214 (1992); *J. Phys.* **3**, 1049 (1993).

127. D. Chandler, *Phys. Rev. E* **48**, 2893 (1993).

128. K. S. Schweizer, *Macromolecules* **26**, 6033 and 6050 (1993).

129. J. Melenkevitz and J. G. Curro, *J. Chem. Phys.*, submitted, 1996.

130. A. Sariban and K. Binder, *Macromolecules* **21**, 711 (1988).

131. I. C. Sanchez, *Macromolecules* **24**, 908 (1991).

132. P. Tillman, J. D. McCoy, J. G. Curro and S. J. Plimpton, *J. Chem. Phys.*, submitted, 1996; D. Gromov and J. J. de Pablo, *J. Chem. Phys.* **103**, 8247 (1995); J. J. de Pablo, private communication, 1996.

133. E. F. David and K. S. Schweizer, *Macromolecules*, submitted, (1996); E. F. David, Ph.D. thesis, Univ. of Illinois, June, 1995; M. Guenza and K. S. Schweizer, in preparation.

134. C. Butler and K. S. Schweizer, *Mol. Phys.*, to be submitted.

135. (a) B. Smit, S. Karaborni, and J. I. Siepmann, *J. Chem. Phys.* **102**, 2126 (1995); J. I.

Siepmann, S. Karaborni, and B. Smit, *Nature* **365**, 330 (1993). (b) Y.-J. Sheng, A. Z. Panagiotopoulos, S. K. Kumar, and K. Szleifer, *Macromolecules* **27**, 400 (1994).

136. For a recent reviews see F. S. Bates, and G. H. Fredrickson, G. H. *Ann. Rev. Phys. Chem.* **41**, 525 (1990).
137. R. L. Scott, *J. Poly. Sci.* **9**, 423 (1952), A. C. Balacz, I. C. Sanchez, I. R. Epstein, F. E. Karacz, and W. J. McKnight, *Macromolecules* **18**, 2188 (1985); D. R. Paul and J. Barlow, *Polymer* **25**, 487 (1984); C. Zhikuan, S. Rouna, and F. E. Karacz, *Polymer* **25**, 6113 (1992).
138. G. H. Fredrickson, S. T. Milner, and L. Leibler, *Macromolecules* **25**, 634 (1992).
139. E. F. David and K. S. Schweizer, *Macromolecules* **28**, 3980 (1995).
140. E. F. David and K. S. Schweizer, *J. Chem. Soc. Faraday Trans.* **91**, 2411 (1995).
141. K. Almdal, F. S. Bates, and K. Mortenson, *J. Chem. Phys.* **96**, 9122 (1992); J. Rosedale, F. S. Bates, K. Almdal, K. Mortensen, and G. D. Wignall, *Macromolecules* **28**, 1429 (1995), and references cited therein.
142. T. Hashimoto, T. Ogawa, and C. D. Han, *J. Phys. Soc. Jpn.* **63**, 2202 (1994); N. Sakamoto and T. Hashimoto, *Macromolecules* **28**, 6825 (1995).
143. L. Leibler, *Macromolecules* **13**, 1602 (1980).
144. G. H. Fredrickson and E. Helfand, *J. Chem. Phys.* **87**, 697 (1987); S. Brazovskii, *Sov. Phys. JETP* **41**, 85 (1975).
145. S. Stepanow, *Macromolecules* **28**, 2833 (1995).
146. H. Fried and K. Binder, *J. Chem. Phys.* **94**, 8349 (1991); K. Binder and H. Fried, *Macromolecules* **26**, 6878 (1993).
147. A. Weyerberg and T. A. Vilgis, *Phys. Rev. E* **48**, 377 (1993).
148. G. Grest, private communication, 1995.
149. J. Melenkevitz and M. Muthukumar, *Macromolecules* **24**, 4199 (1991), and references cited therein.
150. S. K. Nath, J. D. McCoy, J. G. Curro, and R. Saunders, *J. Chem. Phys.*, submitted, 1996.
151. P. T. Cummings and G. Stell, *Mol. Phys.* **46**, 383 (1982).
152. H. Tang and K. S. Schweizer, unpublished.
153. S. F. Edwards, *Faraday Dis.* **98**, 1 (1994).
154. B. M. Pettitt, M. Karplus, and P. J. Rossky, *J. Phys. Chem.* **90**, 6335 (1986); T. Ichiye and D. Chandler, *J. Phys. Chem.* **92**, 5257 (1988); H. Yu, B. M. Pettitt, and M. Karplus, *J. Am. Chem. Soc.* **113**, 2425 (1991).
155. For a review see G. E. Marlow, J. S. Perkyns, and B. M. Pettitt, *Chem. Rev.* **93**, 2503 (1993).
156. Y. Singh, *J. Phys. A-Math Gen.* **20**, 3949 (1987).
157. M. Lal, *Mol. Phys.* **17**, 57 (1965).
158. J. Gao and J. Weiner, *J. Chem. Phys.* **91**, 3168 (1989).
159. G. D. Wignall, in *Encyclopedia of Polymer Science and Engineering*, 2nd ed., Wiley, New York, Vol. 12, p. 112, 1987.
160. M. Daoud, J. P. Cotton, B. Farnoux, G. Jannick, G. Sarma, H. Benoit, R. Duplessix, C. Picot, and P. G. deGennes, *Macromolecules* **8**, 8041 (1975); J. S. King, W. Boyer, G. D. Wignall, and R. Ullman, *Macromolecules* **18**, 709 (1985).

161. A. Yethiraj, *Mol. Phys.* **82**, 957 (1994).
162. C. J. Grayce and K. S. Schweizer, in preparation.
163. C. Singh and K. S. Schweizer, *J. Chem. Phys.*, to be submitted.
164. J. Barrat and G. Fredrickson, *J. Chem. Phys.* **95**, 1281 (1991); T. A. Vilgis and R. Borsali, *Macromolecules* **23**, 3171 (1990); H. Tang and K. F. Freed, *J. Chem. Phys.* **96**, 8621 (1992); Z. G. Wang, *Macromolecules* **28**, 570 (1995).
165. S. T. Milner, *Science* **251**, 905 (1991); D. H. Napper, *Polymeric Stabilization of Colloidal Dispersions*, Academic Press, New York, 1983; G. S. Grest and M. Murat, in *Monte Carlo and Molecular Dynamics Simulations in Polymer Science*, K. Binder, ed., Clarendon Press, Oxford, England, 1994.
166. C. J. Grayce and K. S. Schweizer, *Macromolecules* **28**, 7461 (1995).
167. C. J. Grayce and K. S. Schweizer, *Macromolecules*, to be submitted, 1996.
168. M. Daoud and J. P. Cotton, *J. Phys.* **43**, 531 (1982); A. T. Boothroyd and R. C. Ball, *Macromolecules* **23**, 1729 (1990).
169. T. A. Witten and P. Pincus, *Macromolecules* **19**, 2509 (1986).
170. L. Willner, O. Jucknische, D. Richter, L. J. Fetters, and J. S. Huang, *Europhys. Lett.* **19**, 297 (1992).
171. See, for example, K. F. Freed and J. Dudowicz, *Theor. Chim. Acta* **82**, 357 (1992); J. Dudowicz and K. F. Freed, *Macromolecules* **26**, 213 (1993) and **28**, 6625 (1995), and references cited therein.
172. M. S. Wertheim, *J. Stat. Phys.* **35**, 19, 35 (1984); **42**, 459, 477 (1986).
173. M. S. Wertheim, *J. Chem. Phys.* **87**, 7323 (1987); **85**, 2929 (1986); **88**, 1145 (1988).
174. W. R. Smith and I. Nezbeda, *J. Chem. Phys.* **81**, 3694 (1984).
175. R. Dickman and C. K. Hall, *J. Chem. Phys.* **89**, 3168 (1988).
176. W. G. Chapman, G. Jackson, and K. E. Gubbins, *Mol. Phys.* **65**, 1057 (1988).
177. Y. V. Kalyuzhnyi and P. T. Cummings, *J. Chem. Phys.* **103**, 3265 (1995).
178. M. S. Wertheim, *J. Chem. Phys.* **88**, 1214 (1988).
179. K. E. Chang and S. I. Sandler, *Chem. Phys.* **102**, 437 (1995).
180. Y. C. Chiew, *Mol. Phys.* **70**, 129 (1990); **73**, 359 (1991).
181. Y. C. Chiew, *J. Chem. Phys.* **93**, 5067 (1990).
182. A. Yethiraj, C. K. Hall, and K. G. Honnell, *J. Chem. Phys.* **93**, 4453 (1990).
183. J. Chang and S. I. Sandler, *J. Chem. Phys.* **103**, 3196 (1995).
184. T. Hino, Y. Song, and J. M. Prausnitz, *Macromolecules* **27**, 5681 (1994); Y. Song, S. M. Lambert, and J. M. Prausnitz, *Macromolecules* **27**, 441 (1994).
185. J. G. Curro, P. J. Blatz, and C. J. Pings, *J. Chem. Phys.* **50**, 2199 (1969).
186. J. E. G. Lipson, *Macromolecules* **24**, 1334 (1991); *J. Chem. Phys.* **96**, 1418 (1992).
187. J. E. G. Lipson and S. S. Andrews, *J. Chem. Phys.* **96**, 1426 (1992); H. M. Sevian, P. K. Brazhnik, and J. E. G. Lipson, *J. Chem. Phys.* **99**, 4110 (1993).
188. M. P. Taylor and J. E. G. Lipson, *J. Chem. Phys.* **102**, 2118 and 6272 (1995).
189. B. C. Eu and H. H. Gan, *J. Chem. Phys.* **99**, 4084 (1993).
190. H. H. Gan and B. C. Eu, *J. Chem. Phys.* **99**, 4103 (1993); **100**, 5922 (1994).
191. H. H. Gan and B. C. Eu, *J. Chem. Phys.* **102**, 2261 (1995).
192. H. H. Gan and B. C. Eu, *J. Chem. Phys.* **103**, 2140 (1995).
193. P. Attard, *J. Chem. Phys.* **102**, 5411 (1995).

194. S. G. Whittington and L. G. Dunfield, *J. Phys. A* **6**, 484 (1973).

195. E. Kierlik and M. L. Rosinberg, *J. Chem. Phys.* **97**, 9222 (1992); **99**, 3950 (1993); **100**, 1716 (1994).

196. H. Li and M. Kardar, *Phys. Rev. A* **46**, 6490 (1992).

197. W. G. Madden and E. D. Glandt, *J. Stat. Phys.* **51**, 537 (1988); L. A. Fanti, E. D. Glandt, and W. G. Madden, *J. Chem. Phys.* **93**, 5945 (1990); C. Vega, R. D. Kaminsky, and P. A. Monson, *J. Chem. Phys.* **99**, 3003 (1993); M. L. Rosinberg, G. Tarjus, and G. Stell, *J. Chem. Phys.* **100**, 5172 (1994); D. Chandler, *J. Phys. Condens. Mat.* **42**, F1 (1991).

198. D. M. Ford, A. P. Thompson, and E. D. Glandt, *J. Chem. Phys.* **103**, 1099 (1995); A. P. Thompson and E. D. Glandt, *J. Chem. Phys.* **99**, 8325 (1993).

199. A. Yethiraj, C. K. Hall, and R. Dickman, *J. Colloid Interface Sci.* **151**, 102 (1992).

200. M. R. Shaw and D. Thirumalai, *Phys. Rev. A* **44**, R4797 (1991).

201. M. Fuchs, in preparation.

202. K. Kolbet and K. S. Schweizer, *J. Chem. Phys*, to be submitted.

203. K. S. Schweizer, *J. Chem. Phys.* **91**, 5802 and 5822 (1989); *J. Non-Cryst. Sol.* **131–133**, 643 (1991); *Phys. Scrip.* **T49**, 99 (1993).

204. K. S. Schweizer and G. Szamel, *Phil. Mag. B* **71**, 783 (1995); *Trans. Theo. Stat. Phys.* **24**, 947 (1995); H. Tang and K. S. Schweizer, *J. Chem. Phys.* **103**, 6296 (1995).

205. T. V. Ramakrishnan and M. Yussouff, *Phys. Rev. B* **19**, 2775 (1979); A. D. J. Haymet and D. W. Oxtoby, *J. Chem. Phys.* **74**, 2559 (1981); B. B. Laird, J. D. McCoy, and A. Haymet, *J. Chem. Phys.* **87**, 5451 (1987).

206. D. Chandler, J. D. McCoy, and S. J. Singer, *J. Chem. Phys.* **85**, 5977 (1986); J. D. McCoy, S. J. Singer, and D. Chandler, *J. Chem. Phys.* **87**, 4953 (1987).

207. J. P. Donley, J. J. Rajasekaran, J. D. McCoy, and J. G. Curro, *J. Chem. Phys.* **103**, 5061 (1995).

208. A. Yethiraj and C. E. Woodard, *J. Chem. Phys.* **102**, 5499 (1995); **100**, 3181 (1994).

209. K. P. Walley, K. S. Schweizer, J. Peanasky, L. Cai, and S. Granick, *J. Chem. Phys.* **100**, 3361 (1994); K. P. Walley, K. S. Schweizer, and A. Yethiraj, *J. Chem. Phys.*, to be submitted.

210. L. Lue and D. Blankschtein, *J. Chem. Phys.* **102**, 5460 (1995), and references cited therein.

DIELECTRIC PROPERTIES OF LIQUID CRYSTALS UNDER HIGH PRESSURE

STANISŁAW URBAN

Institute of Physics, Jagellonian University, Reymonta 4, 30-059 Kraków, Poland

ALBERT WÜRFLINGER

Physical Chemistry II, Ruhr-University, D-44780 Bochum, Germany

CONTENTS

Advances in Chemical Physics, Volume XCVIII, Edited by I. Prigogine and Stuart A. Rice.
ISBN 0-471-16285-X

I. INTRODUCTION

A. Usefulness of High-Pressure Studies

It is well known from thermodynamics that the properties of a macroscopic system can be fully described, if the equation of state is available, that is, $V = f(p, T)$. However, in real physical experiments only one parameter is easily changeable: the temperature at atmospheric pressure. High-pressure (HP) investigations require much more advanced experimental techniques and therefore such studies are less popular.

Pressure is an important thermodynamic parameter that is often neglected in experimental studies. The application of high pressure is a useful mean in order to study the mechanism of phase transformations[1] and to vary the intermolecular distance that is indispensable for any discussion of the intermolecular potential.[2] Only in the frame of high-pressure experiments can the equation of state be established that allows one to distinguish between isochoric temperature changes and isothermal density changes. High-pressure phase transformations are important subjects for materials science, providing a chance to synthesize metastable phases with valuable properties. The most prominent example is the well-known diamond synthesis, the role of high pressure is also highlighted in various fields of solid-state physics and spectroscopy.[3] Pressure-induced amorphization and memory effects have been observed where the crystal is transformed to a disordered solid under pressure.[4] The pressure dependence of dynamical properties (e.g., rate constants in chemical conversions[5] or relaxation times[6]) yields the volume of activation in the frame of transition-state theories. This quantity gives useful information for the space needed along a reaction path or in reorientation processes.

The term *high pressure* depends strongly on the problem. It seems to us that its use is justified only if the applied pressure changes significantly the property under study. In the case of molecular crystals or liquid systems there are only relatively weak intermolecular interactions, therefore pressures from a few tenths MPa up to ~300 MPa are usually sufficient in order to induce considerable changes in the physical properties of the system. Although pressure and temperature are in principle equivalent thermodynamic parameters, they affect the molecular system differently: Temperature causes mainly an excitation of rotational and vibrational energy states, whereas pressure changes the intermolecular distance. Thus HP studies can yield quite new insight into the properties and behavior of various systems.

The pressure dependence of the dielectric constant has been reviewed for fluids and liquids,[7] ionic crystals,[8] semiconductors,[9] ferroelectrics,[1,10] vitreous materials,[11] ice phases,[12] polymeric systems,[13] and plastic crystals.[14] These studies are helpful to probe theories of dielectric polarization and intermolecular forces.

The employment of high pressure is also indispensable to elucidate the phase situation of liquid crystals. Their rich polymorphism is most conveniently studied by high-pressure differential thermal analysis (DTA).[15–18] Not only may the phase behavior of the substance studied change significantly under pressure (pressure-induced and pressure-limited phases may occur), but also several peculiarities have been detected in liquid crystal phases, such as reentrant and tricritical phenomena.[19–21] Also the properties of ferroelectric[22] and polymeric[23] liquid crystals are markedly altered under pressure.

B. Liquid Crystals (LCs)

1. *Mesomorphic Behavior*

Liquid crystals are mesophases whose properties lie intermediate between isotropic liquids and long-range ordered crystals. They were discovered at the end of the last century by Reinitzer,[24] who was puzzled by two different melting points of cholesteryl benzoate. It took time until it was accepted that in fact a new state of matter had been found. At the "second melting point" cholesteryl benzoate transforms to an optically clear and isotropic liquid (clearing temperature). Substances that exhibit such a phase behavior as a single-component system are called *thermotropic*. There are also mesomorphic phases that exist only in the presence of a solvent. Such *lyotropic* liquid crystals will not be considered here, although they are of great biological interest.[25]

Friedel[26] was the first who distinguished three main classes of liquid crystals, according to the different kind of orders in the mesophases: nematic, smectic, and cholesteric. From the point of view of the geometrical shape of molecules, we divide the thermotropic LCs into *calamitic phases* (when the molecules are rodlike), *sanidic phases* (when the molecules are bricklike), and the *discotic phases* (when the molecules are disklike).[27–29]

a. The Nematic Mesophase. This is a turbid phase of low viscosity. Usually it is limited by the isotropic liquid at high temperature and by solid or a smectic phase at low temperature. The positional order of the molecules is destroyed. However, there is a certain parallel alignment of the long molecular axes of rodlike molecules, or a parallel arrangement of the disks in discotic phases. The preferred orientation will be called *director* and denoted by **n**. The director **n** corresponds to an axis of uniaxial symmetry with no polarity which means that **n** and −**n** are equivalent, although the individual molecules may be polar. For a quantitative treatment of the nematic properties it is necessary to define an order parameter that describes the distribution of the long molecular axes. The first nontrivial order parameters are[30]

$$\langle P_2(\cos\theta)\rangle = \tfrac{1}{2}\langle 3\cos^2\theta - 1\rangle \equiv S \tag{1}$$

$$\langle P_4(\cos\theta)\rangle = \tfrac{1}{8}\langle 35\cos^4\theta - 30\cos^2\theta + 3\rangle \tag{2}$$

where θ is the angle between the molecular symmetry axes and the nematic director. The brackets $\langle\cdots\rangle$ denote averaging over the orientations of all molecules.

The second rank-order parameter S can be derived from measurements of the macroscopic tensor properties such as birefringence and diamagnetic susceptibility. It varies typically between 0.4 at the clearing temperature to 0.7 at $T_{NI} - T \approx 20\,\mathrm{K}$ in nematic phase. The fourth rank-order parameter $\langle P_4\rangle$ may play an important role for a subtle analysis of the orientational distribution function and can be determined using polarized Raman spectroscopy.[31]

Optically active molecules may form a chiral modification of the nematic phase: the *cholesteric phase*. The local direction of **n** is slightly rotated, when we move perpendicularly to **n** to the adjacent region. Thus a helical structure is superimposed with a pitch comparable to the wavelengths of the visible light resulting in peculiar optical properties. Because of its relationship to the nematic state, this phase is also called *chiral nematic* and denoted by N*. In some cases cholesteric liquid crystals with a relatively short pitch exhibit so-called *blue phases*.[32] Surprisingly, the blue phases behave optically isotropic, whereas usually specific textures are observed with a polarizing microscope for liquid crystals. The helices are supposed to be arranged in a cubic superstructure.

b. The Smectic Mesophase. This is a turbid, viscous state, where the molecules are arranged in layers. The molecules in each layer are again more or less parallel oriented. Discotic molecules can be stacked one upon another to form columns (columnar smectic mesophase). We mention two important types: smectic A and C, where the director is orthogonal and tilted with respect to the layer. They can be considered as two-dimensional liquids with only a short range order of the molecular centers within the layers. Smectic A is optically uniaxial, similar to the nematic phase. In the case of smectic C the projection of **n** on the layer plane is not zero, defining a **c** vector in the layer plane. Contrary to **n**, **c** and −**c** are not equivalent. Provided we exclude chiral molecules, the smectic C phase is biaxial. Chiral smectics C will be denoted by C*. In smectic C* the **c** vector is continuously rotated, when we move to the next plane. Thus a helical structure is obtained where the helix axis is parallel to the layer normal. The symmetry of the C* phase allows for ferroelectricity that is also observed in some other chiral smectics.[33] More details of ferroelectric liquid crystals will be presented in Section IV.C.

c. Polymeric LCs. The mesogenic groups (usually rodlike) may be attached to the polymer backbone in the main chain itself or as side groups (comblike polymers).[34] The phase behavior of such polymers depends strongly on the polymer backbone, the shape of mesogenic units, and the length of spacers. The liquid crystalline phases (nematic, smectic, or cholesteric) exist between the clearing temperature and the glass temperature T_g. The LC phase may be macroscopically aligned by cooling the material from the isotropic melt into the LC state in the presence of a directing electric or magnetic field. Since polar groups are frequently incorporated in the mesogenic units and in the main chain, the dielectric relaxation spectroscopy is an effective method for studying molecular dynamics and the alignment of the sample.

Nowadays a wealth of experimental and theoretical methods are available to investigate the mesomorphic behavior of liquid crystals. For a detailed description of the liquid crystalline state the reader is referred to textbooks,[35–41] review articles,[42–46] and to proceedings of recent conferences.[47–49] We mention only briefly examples of mesogenic compounds with some thermodynamic properties, especially for those which have been studied dielectrically at elevated pressures (see below).

2. *Some Mesogenic Compounds and Phase Transitions*

We mentioned already cholesteryl benzoate as one of the first discovered liquid crystals, although the chemical structure of cholesterol was unknown at the time.[50] From this compound the name cholesteric phase was derived. *p*-Azoxyanisole (PAA), which belongs to the *n*OAOB homologous series (cf. Table I), was one of the first synthesized liquid crystals with known chemical structure. It is a typical representative of a rodlike compound. The first six members of this series form the nematic phase,

TABLE I
Examples of Mesogenic Compounds

	Name	Formula
MBBA	*N*-(*p*-methoxybenzilidene)-*p*-butylaniline	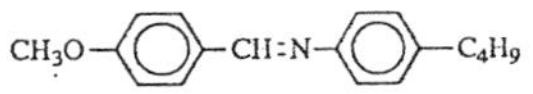 CH_3O–CH=N–C_4H_9
*n*OAOB	*p*,*p*′-di-akoxyazoxy-benzene	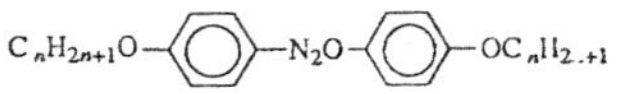 $C_nH_{2n+1}O$–N_2O–OC_nH_{2n+1}
*n*CB	4-*n*-alkyl-4′-cyanobiphenyl	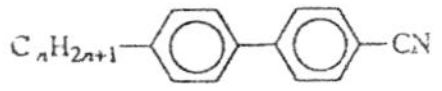 C_nH_{2n+1}–CN
*n*CHBT	4-(*trans*-4′-*n*-alkyl-cyclohexyl) isothiocyanatobenzene	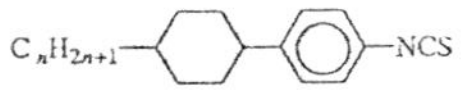 C_nH_{2n+1}–NCS
*n*PCH	4-*n*-alkyl-4′-cyclohexyl-cyano-phenyl	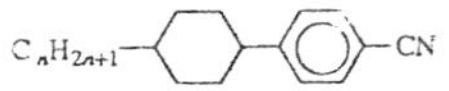 C_nH_{2n+1}–CN
*n*HCP	4-*n*-alkyl-4′-phenyl-cyano-cyclohexane	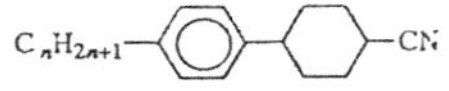 C_nH_{2n+1}–CN
*n*CCH	4-*n*-alkyl-4′-cyanobicyclohexane	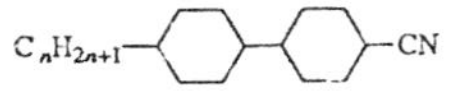 C_nH_{2n+1}–CN
*n*OCB	4-*n*-alkoxy-4′-cyanobiphenyl	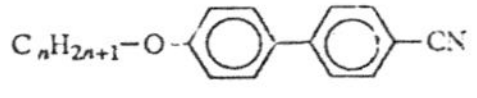 C_nH_{2n+1}–O–CN
DOBAMBC	*p*-decyloxybenzilidene-*p*′-amino-2-methyl-butyl-cinnamate	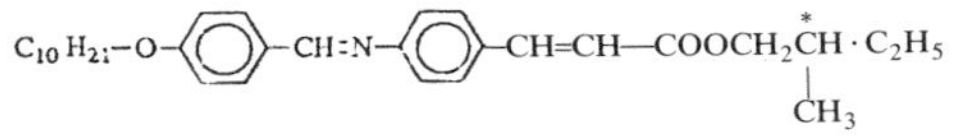 $C_{10}H_{21}$–O–CH=N–CH=CH–$COOCH_2\overset{*}{C}H(CH_3)\cdot C_2H_5$

but for $n \geq 7$ smectic phases are also observed. It is a common feature of rodlike liquid crystals that molecules with short end groups exhibit only nematic phases, whereas for longer homologues additional (or exclusively) smectic phases occur.

For PAA (1OAOB) the nematic phase exists between 118 and 135°C (at atmospheric pressure), which is inconvenient for practical applications. With the aim to find nematics with lower transition temperatures, Kelker et al.[51] synthesized some new liquid crystals, the best known example being *N*-(*p*-methoxybenzylidene)-*p*′-butylaniline (MBBA). At present more than 50,000 chemical compounds are known showing mesomorphic behavior.[52]

Wide scientific and practical interests are connected with the alkylcyanobiphenyls, *n*CB. For these compounds the dipole moment is parallel to the long molecular axis, which is advantageous for the interpretation of dielectric results (see below). It is interesting to investigate how the physical properties change, when we slightly alter the core or a specific group of the chemical compound. The series *n*PCH, *n*HCP, and *n*CCH are obtained by replacing one or both phenyls by the cyclohexyl group in *n*CB (see Table I).

In Table II we list some thermodynamic properties connected with phase transitions. The enthalpy changes for the NI transition are one order of magnitude smaller than the heat of melting, but still indicating a

TABLE II
Some Thermodynamic Properties of Selected Liquid Crystals at 1 atm[a]

		Melting		Smectic Phases		Nematic-Isotropic	
Substance	References	T (K)	ΔH (Jmol^{-1})	Type	T (K)	T_{NI} (K)	ΔH (Jmol^{-1})
PAA	57, 58	391	30340			408.6	733
MBBA	57, 59	295	15030			320	416
5CB	17, 60, 61	295.7	13390			308.2	390
6CB	38, 61	286	24267			301	293
7CB	38,61	301.7	25940			314.7	578
8CB	17, 60, 61	294.2	23430	$S_A \rightarrow N$	306.7	313.2	612
6CHBT	17, 62	285.7	26800			316.2	1600
5PCH	63–66	303	21350			328	960
6PCH	64	315				320	
7PCH	64–68	303		$N \rightarrow S_A$	290[b]	332	
8PCH	64	306				327	
5CCH	61, 64, 69	335	23900	$N \rightarrow S_B$	325[b]	358	
7CCH	61, 64, 69	344	31000			356	
8OCB	61	327	24686	$S_A \rightarrow N$	340	353	

[a] N, I, S_A, S_B stands for nematic, isotropic, smectic A, and smectic B phases.
[b] Monotropic transition.[56]

first-order transition. Transitions between smectic A and C (AC) or between smectic A–nematic (NA) are generally considered as second-order transitions, whereas smectic C–nematic (NC) is weakly first-order.[20] However, when approaching a NAC tricritical point (i.e., a multicritical point where the AC, NC, and NA phase boundaries meet) the latent heat of the NC transition disappears, i.e., the nature of the transition changes from first to second order.

Of special interest are the *n*CB and *n*PCH series, for which dielectric results under high pressure will be presented below. In the *n*OCB series 8OCB has attracted much attention because it exhibits the *reentrant* phase behavior.[53,54] This means the following unusual sequence of phase transitions on cooling under pressure:

$$\text{Isotropic} \rightarrow \text{nematic} \rightarrow \text{smectic A} \rightarrow \text{reentrant nematic}$$

There exists a maximum pressure beyond which the smectic A phase disappears (see Fig. 1d). The $p(T)$-phase boundary nematic–smectic A has an elliptic shape. Reentrant behavior has been found for many systems, and therefore several theoretical treatments have been considered to explain this phenomenon. The most successful is probably the frustrated spin-gas model, which also predicts the sensitivity to the number of carbon atoms in the alkyl chain of the molecule.[55]

It should be mentioned that for 8OCB the high-pressure smectic A and the reentrant nematic are only observed as supercooled, metastable phases that are not always found in a high-pressure experiment. Metastability has often been noticed in the study of liquid crystals.[17,38] The retarded onset of a phase transition is frequently accompanied by a so-called exothermic anomaly in DTA experiments, where the undercooled phase transforms on reheating. In some cases monotropic mesophases have been found, which are only observed on cooling, for example, smectic phases for 7PCH and 5CCH (cf. Table II).[56]

3. Phase Diagrams of Some LCs under Pressure

Figure 1 presents the pT-phase diagrams for some compounds listed in Table II. In general the transition temperatures have been determined with DTA and optical observations. The phase behaviors of the *n*-alkylcyanobiphenyls have been thoroughly studied by Shashidhar and Venkatesh,[70] who discuss a pronounced odd–even effect in the clearing temperature and in the slope dT_{NI}/dp. 8CB exhibits also a smectic A phase. Its smectic A–nematic transition was observed using a high-pressure optical microscope stage.[53] Literature data for the $T(p)$ dependencies of the phase boundaries of *n*CBs are partly discrepant.[70–72]

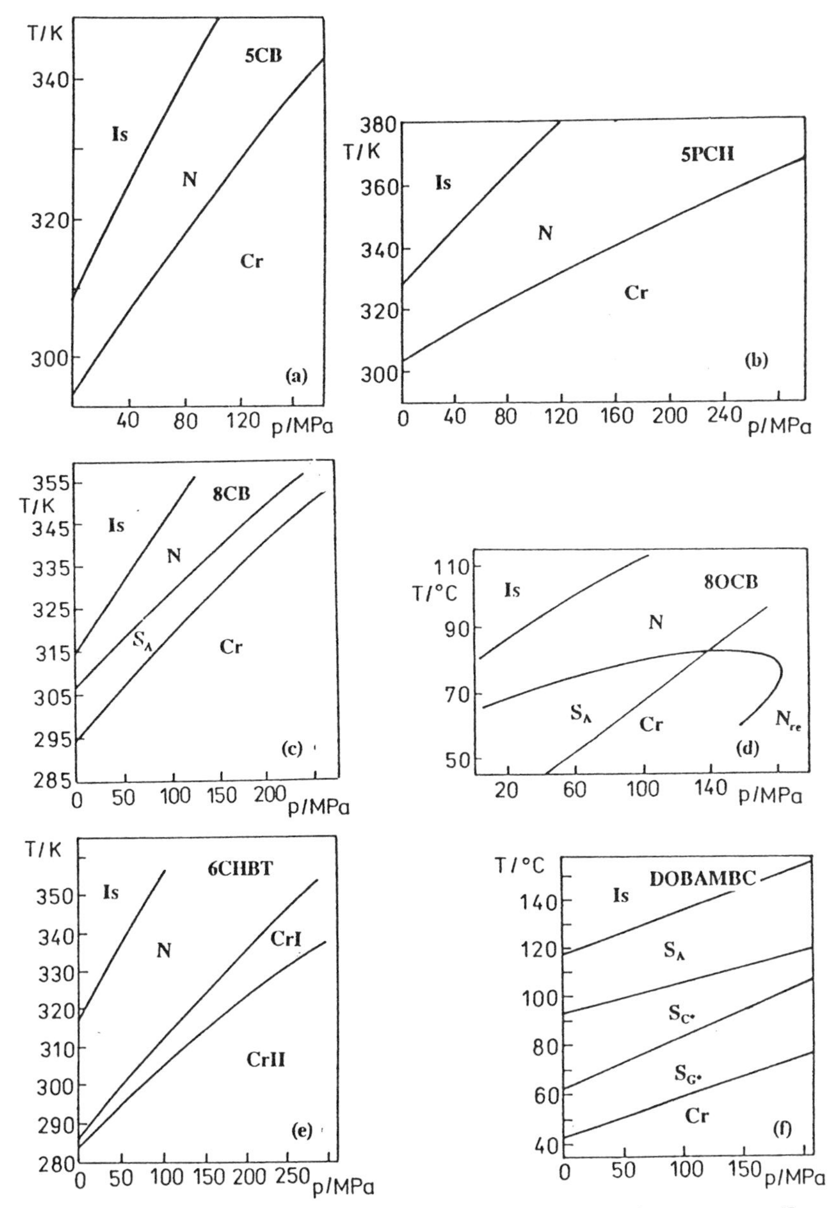

Figure 1. Temperature–pressure phase diagrams of selected LCs: (a) 5CB,[17] (b) 5PCH,[87] (c) 8CB,[17] (d) 8OCB,[53] (e) 6CHBT,[17] and DOBAMBC.[167] (Reprinted with kind permission from the authors and the editors.)

TABLE III
Transition Temperatures as Function of Pressure for Selected Liquid Crystals[a]

Substance	Transition	a	b	$c \times 10^4$	References
5CB	Cr→N	294.4	0.303	−1.739	17, 73–75
		295.7	0.264	−1.29	70
	N→I	308.4	0.419	−2.288	17, 73–75
		308.2	0.403	−2.64	70
6CB	Cr→N	286.2	0.264	−0.92	70
	N→I	301.2	0.337	−2.1	70
		301.7	0.372	−1.0	72
		301.4	0.411	−4.1	76, 83
7CB	Cr→N	301.7	0.246	−0.98	70
	N→I	314.7	0.479	−0.54	70
		315.3	0.339	−0.40	72
		316.2	0.364	−2.89	76, 84
8CB	Cr→S_A	294.5	0.250	−1.20	17, 75–77
	S_A→N	306.8	0.223	−0.71	53
	N→I	313.8	0.370	−2.40	17, 75–77, 85
6CHBT	Cr→N	286.5	0.270	−1.30	17, 75, 86
	N→I	316.7	0.420	−3.30	17, 75, 86
5PCH	Cr→N	303.1	0.246	−1.09	78, 87
	N→I	328.1	0.440	−1.11	78, 87
7PCH	Cr→N	303.5	0.311	−1.68	78, 88
	N→I	331.0	0.420	−2.86	78, 88
5CCH	Cr→N	338.0	0.330	−0.114	77
	N→I	359.5	0.536	−2.867	77
	N→S_B	325.1	0.355	+0.24	77

[a] $T\ (\mathrm{K}) = a + b\ (p/\mathrm{MPa}) + c\ (p/\mathrm{MPa})^2$.

Therefore we redetermined the transition lines for some selected liquid crystals in our laboratory.[73–78] The results (smoothed by polynomials) are listed in Table III together with some literature data. Further details of the polymorphism and crystal structures for some nPCHs and *n*CCHs can be found by Haase et al.,[67,79] Kuss,[66] and Schneider et al.[80] The phase behaviors are also noted in the presentations of the dielectric results.[81–89]

C. Remarks on the Theories of LCs

1. *Nematic Phase*

The first description of the nematic state was developed by Maier and Saupe.[90] They considered the molecules as simple rigid rods with no internal degrees of freedom. In the nematic state the rods are aligned on average parallel to the nematic director. The orientation-dependent potential energy of one molecule in the field of its neighbors is taken in the simple form

$$U(\theta) = -qP_2(\cos\theta) \tag{3}$$

where $P_2(\cos\theta)$ has the same meaning as in Eq. (1). Maier and Saupe (MS) assumed that the strength parameter q (the nematic potential) is proportional to the order parameter $S = \langle P_2(\cos\theta)\rangle$:

$$q = vS \tag{4}$$

The interaction coefficient v determines the energy scale of the potential and is temperature independent, however, it depends on the molar volume V.

It is interesting to note that the main assumptions of the MS theory $[U \sim P_2(\cos\theta),\ q \sim S]$, were satisfactorily justified by the model calculations with the use of the Gay–Berne potential.[91] Although the MS theory gives a qualitatively correct description of the main properties of the nematic phase, it cannot be used for a quantitative comparison with experiments.[92,93] That arises from the mean-field approximation and the neglect of other than the orientational components of the interaction energy. The discussion of the theoretical quantities with respect to the experimental ones presented by Urban et al.[82] for the nematic phase of 5CB sheds more light on this problem and generally support the above statement. Closer agreement with the experimental results have been obtained with theories in which some modifications in the form of the nematic potential were introduced.[94–97]

Humphries et al.[94] have considered in addition to the order parameter S the fourth rank-order parameter $\langle P_4\rangle$, which modifies the shape of the potential. They approximated the volume dependence of the interaction energy by

$$v = v_0 V^{-\gamma} \tag{5}$$

where γ is treated as a free parameter in fitting theory to experiment.[98–102] The coefficient v_0 is taken to be a constant, independent of pressure, volume, and temperature. Maier and Saupe assumed that $v \sim V^{-2}$, which means that only London dispersion forces are taken into account (corresponding to an r^{-6} dependence of the potential). McColl[103] introduced a thermodynamic coefficient Γ, defined as

$$\Gamma = -\left(\frac{\partial \ln T}{\partial \ln V}\right)_s \tag{6}$$

which is identical to γ for the potential above.[92] The γ values determined experimentally for different nematogens are usually larger than 2, indicating that repulsive forces must be taken into account.

Recently Tao et al.[96] extended the MS theory by adding to Eq. (3) the isotropic, density-dependent component of the molecular interactions $U_0(r)$ in the form of the Lennard-Jones potential $U_0(r) = 4\varepsilon_m[(\sigma/r)^{12} - (\sigma/r)^6]$. As a result they obtained a better agreement of the calculated and experimental quantities characterizing the nematic-isotropic transition, for example, volume change at T_{NI} and the values of dT_{NI}/dp. Chrzanowska and Sokalski[97] considered the case when the parameter σ in the Lennard-Jones potential is dependent on the orientation of molecules that allows one to predict properly for MBBA such properties as order parameters, elastic constants, and rotational viscosity coefficients.

To discuss the true contribution of particular components to the intermolecular interactions, it is necessary to find the volume dependence of the molecular field potential characterized by the exponent γ. For this the pressure–temperature studies of liquid crystals are indispensable. On the basis of high-pressure volumetric and nuclear magnetic resonance (NMR) studies, McColl and Shih[104] determined γ for PAA from the slope of the curves log T versus log V at constant S according to Eq. (6). Similar studies have been done for other members of the nOAOB series by Tranfield and Collings.[105] They found that γ decreases gradually from 4 for 1OAOB (PAA) to 1.9 for 6OAOB. The same procedure was applied by Urban et al.[82] for 5CB, using literature data on V and S from optical measurements,[71] yielding $\gamma = 5.3$. The MS theory predicts that

$$T_{NI}/v \sim T_{NI}V^{\gamma} = \text{const} \tag{7}$$

so that γ may be obtained from log T_{NI} versus log V plots. The result was $\gamma = 5.2$, in excellent agreement with the above value. This can be compared with findings of Emsley et al.[98] These authors found that Γ decreases steadily along the alkyl chain from 5.3 for the order parameter of the deuterons nearest to the aromatic core, to 4.2 for those from the methyl group. For 7PCH Emsley et al.[99] obtained $\gamma = 3.6$.

In a similar way Shirakawa et al.[106] determined γ for various liquid crystals belonging to the nCB and nPCH series. They found $\gamma = 5.24$ for 5PCH, which clearly reveals the inadequacy of the Maier–Saupe theory. The values γ decrease with increasing length of the alkyl chain that is contributed to the enlarged conformational freedom of the longer alkyl chain. It is interesting to note that the nPCH series yields smaller values for γ than the nCBs, which can be explained by the higher flexibility of the cyclohexyl core.

Collings and co-workers[107,108] calculated γ from

$$\gamma = \frac{\partial \ln(T/T_0)}{\partial \ln(V/V_0)} \tag{8}$$

where T_0 and V_0 represent a point on the nematic-isotropic coexistence curve. They obtained the following values for γ: 2.07 for 7OAOB, 3.20 for 8OCB, and about 3 for three 4,4′-di-*n*-alkylazoxybenzenes.

The comparison of γ obtained from high-pressure studies of different liquid crystals allows one to draw the following conclusions: (i) γ is usually larger than 2, indicating that besides dispersion repulsion forces also determine the intermolecular potential in the nematic phase. (ii) Different substances are characterized by different γ values. (iii) In a homologous series γ decreases with increasing chain length. (iv) γ decreases steadily along the alkyl chain. Thus, a balance between the long-range attractive and short-range repulsive interactions depends on the structure and flexibility of the molecule.

Another theoretical approach comes from Singh et al.[109] These authors analyzed the influence of short-range orientational order on the thermodynamic and orientational behavior of nematogens close to the N–I transition, considering both attractive and repulsive interactions. The potential parameters were chosen so as to reproduce quantitatively the clearing temperature of PAA. The estimated Γ values are considerably smaller than the experimental ones (ca. 2.3 against 4.0), while the slope dT/dp of the N–I transition line was one order of magnitude too high. Nevertheless the trends of pressure effect on the stability, ordering, and phase transition are in better agreement with experiments as compared to mean-field results.

Finally we note some recent molecular dynamics simulations with the use of atom–atom potential,[102,110–112] which can provide useful information concerning the relation between the detailed molecular composition, the phase behavior, and dynamical properties of the system. However, due to the large number of interaction sites in real molecular systems, the calculations are extremely time consuming, and therefore only a few liquid crystals have been studied in this way. Some simplifications are introduced to circumvent part of the difficulties.[112–114] In the approach by Cross and Fung[110,113] the phenyl rings in the LC molecules are considered as enlarged simple spheres, whereas other parts of the molecule are treated realistically. Inter- and intramolecular contributions arising from the alkyl chain to the conformational properties could be separated. In the model proposed by Yoshida and Toriumi,[114] the LC molecule is made up of linearly connected four spherical beads, each of which acts as a center of the Lennard-Jones interactions.

2. *Smectic A Phase*

Some variety of smectic A phases can be distinguished, if the ordering of molecules in layers is considered.[115–117] Generally two factors play an important role in the formation of different subtypes of smectic A phases: the dipolar effects and steric effects; and for rodlike molecules the dipolar features are most important.[116] The main property for discriminating the different subtypes is the ratio of the smectic layer spacing d to the molecular length l in its most extended form. The monolayer Sm A_1 can be regarded as the classical smectic A phase, which is characteristic for all nonpolar and many weakly polar compounds. Sm A_d is called the *partially bilayer smectic* and is formed by compounds with strongly polar terminal groups such as —CN, —CHO, and —NO_2.[115] As an example Figure 2 shows the arrangement of two neighbouring molecules for 8CB in the smectic A_d phase and for 5CB in the nematic phase.[118]

To describe the properties of Sm A phases, it is necessary to introduce—in addition to the order parameter S (defined for the nematic phase in Section I.B.)—a second-order parameter σ that accounts for the layer structure:

$$\sigma = \langle \cos 2\pi z/d \rangle \tag{9}$$

where z is the translational coordinate parallel to the director. On that basis McMillan[119] has extended the Maier–Saupe theory by assuming a coupling of the smectic order parameter σ to the orientational order parameter S. The intermolecular potential contained two model parameters, the strength of the potential and its effective range. McMillan's theory is in good accord with typical features of experimentally observed phase diagrams and thermodynamic quantities.[39,116,119] Then further refinements have been made to the theory,[120–123] which gave a better agreement with experiments. However, this mean-field theory cannot describe all properties of the smectic A phases, such as the various subtypes or the reentrant behavior.

Luckhurst and Simmonds[123] employed a new parametrization of the Gay–Berne potential that allowed more details of the liquid crystal behavior to be revealed. The main result was that the isotropic and nematic phases are dominated by short-range anisotropic forces, whereas the stability of the smectic A phase depends critically on the anisotropy of the attractive forces. The use of the Gay–Berne potential was criticized by others because it is too simple to find out how the particular molecular

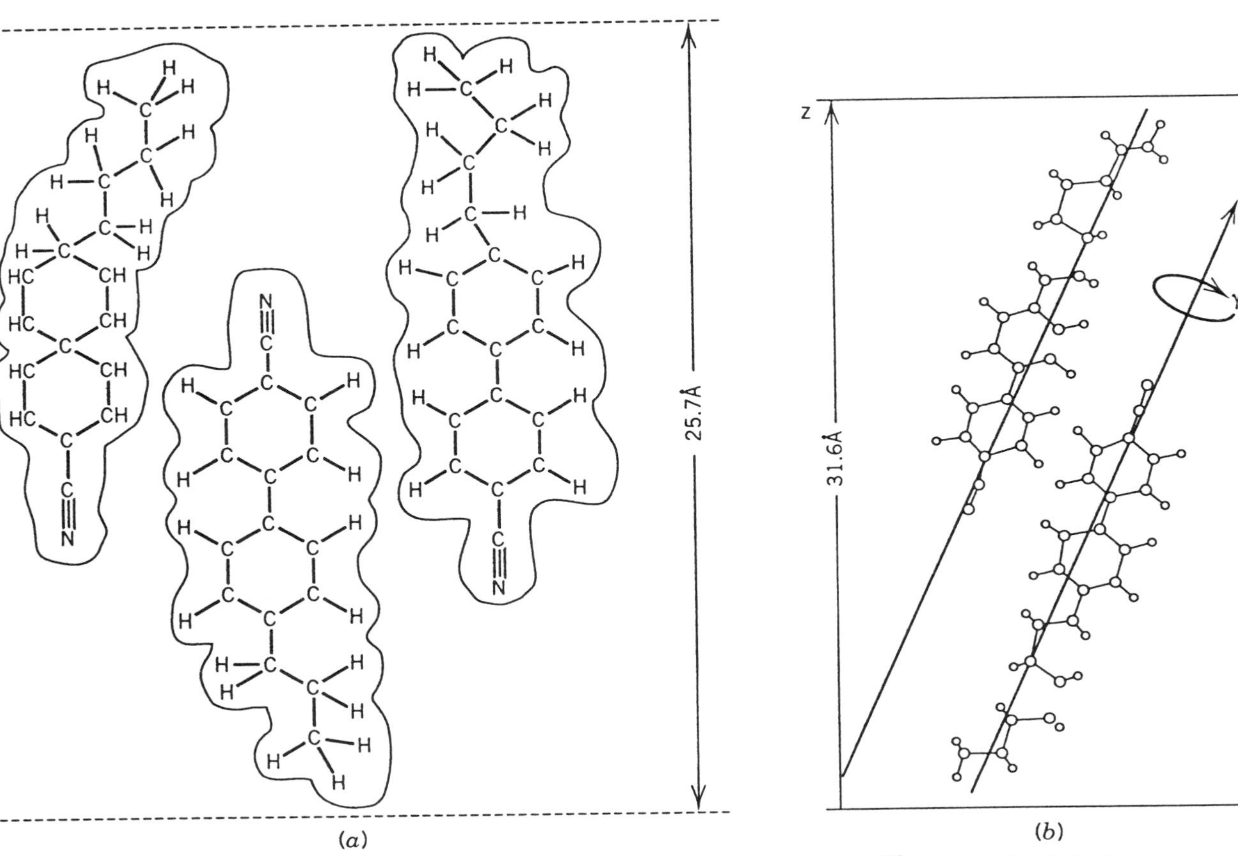

Figure 2. Schematic diagram of local structure in (a) 5CB and (b) 8CB.[118] (Reprinted with kind permission from the authors and the editor of *J. de Physique*.)

structure may affect the phase situation. Smectic phases have been identified in computer simulations of liquid crystal models with semiflexible molecules.[124,125] The well-known odd–even effects for the nematic properties could be recovered. Paolini et al.[126] employed a soft-core site–site potential and could reveal both nematic and smectic phases.

A number of models have been elaborated with the aim of explaining the reentrant phenomena in liquid crystals.[20,39,55,116,127] In the models the interactions between the molecules are described by means of long-range forces (dispersion, inductive, and dipole–dipole forces) as well as by short-range, mainly steric forces. The former forces are responsible for the creation of long-range order (smectic layering), whereas the latter ones tend to destabilize the smectic order. The competition between both types of forces may lead to the appearance of even multiple reentrant behavior being observed at decreasing temperature or increasing pressure.

II. DIELECTRIC PROPERTIES OF NEMATIC LIQUID CRYSTALS

A. Static Permittivity

The dielectric method is a powerful tool for studying the reorientational motions of dipolar molecules in condensed phases. The probing external electric field interacts with the dipole moment of a molecule that is embedded in a continuum consisting of the same molecules. Thus the molecule also interacts with its surrounding, and the net effect can be accounted for by introducing the local field, which is larger than the external one. The finding of a proper form of the local field is a crucial point of each theory of dielectrics. In the case of isotropic systems the Onsager approach is commonly accepted,[128–131] Maier and Meier[132] have used this approach to describe the dielectric properties of the nematic phase in spite of a strong anisotropy of molecular interaction in that phase. To calculate the orientational distribution function for the molecule, the Maier–Saupe potential was used. Therefore the obtained expressions for the dielectric permittivity components of the nematic phase may be treated as a rough estimation and cannot be used for quantitative analysis of the experimental results.

Due to the anisotropic features of liquid crystals we have to distinguish between different principal elements of the dielectric tensor. We consider uniaxial liquid crystalline phases, that is, nematic and smectic A, with the macroscopic z axis parallel to the director $\mathbf{n}$. In this case we have two principal elements: $\varepsilon_{\parallel} = \varepsilon_{zz}$, $\varepsilon_{\perp} = \varepsilon_{xx} = \varepsilon_{yy}$. Depending on whether the measuring field is kept parallel or perpendicular to the director $\mathbf{n}$, we obtain $\varepsilon_{\parallel}$ or $\varepsilon_{\perp}$, respectively. The dielectric anisotropy $\Delta\varepsilon = \varepsilon_{\parallel} - \varepsilon_{\perp}$ will certainly depend on the extent to which adjacent molecules are parallel aligned, that is, the dependence on the order parameter S will be particularly relevant.

Similar to the treatments for isotropic phases we have to compute the sum of the induced and orientation polarization.[37,38] Following Maier and Meier,[132,133] we consider a molecule with a permanent dipole moment μ that makes an angle β with the long molecular axis ζ. The anisotropy of the polarizability is accounted for by two principal elements α_l and α_t, along and transverse to ζ. The components of μ (in the molecule-fixed coordinate system) are: $\mu_l = \mu \cos\beta$, $\mu_t = \mu \sin\beta$. The components $\alpha_{\parallel}$, $\alpha_{\perp}$, $\mu_{\parallel}$, and $\mu_{\perp}$ (in the laboratory system) depend on the orientation of the molecular ζ axis, which makes an angle θ with the z axis, and thus are connected with the order parameter S:

$$\langle\alpha\rangle_{\parallel} = \tfrac{1}{3}[\alpha_l(2S+1) + \alpha_t(2-2S)] \tag{10}$$

$$\langle\alpha\rangle_{\perp} = \tfrac{1}{3}[\alpha_l(1-S) + \alpha_t(2+S)] \tag{11}$$

$$\langle\mu_{\parallel}^2\rangle = \tfrac{1}{3}[\mu_l^2(2S+1) + \mu_t^2(1-S)] = \tfrac{1}{3}\mu^2[1-(1-3\cos^2\beta)S] \tag{12}$$

$$\langle\mu_{\perp}^2\rangle = \tfrac{1}{3}[\mu_l^2(1-S) + \tfrac{1}{2}\mu_t^2(2+S)] = \tfrac{1}{3}\mu^2[1+\tfrac{1}{2}(1-3\cos^2\beta)S] \tag{13}$$

In order to derive equations for the dielectric components, the authors[132,133] used similar simplifications as in the Onsager theory.[128–131] The molecule is put in a spherical cavity with radius a, determined from the molar volume: $M/\rho = \frac{4}{3}\pi N_A a^3$ (M = molar mass, ρ = density, N_A = Avogadro number, $N \equiv N_A\rho/M$). In the cavity field factor $h = 3\bar{\varepsilon}/(2\bar{\varepsilon} + 1)$, and the reaction field factor $f = (\bar{\varepsilon} - 1)/[2\pi\varepsilon_0 a^3(2\bar{\varepsilon} + 1)]$, the anisotropy of the permittivity is ignored. In the expression $F = 1/(1 - f\alpha)$ an averaged value for the polarizability α is used. The resulting equations are,[37,38,132] where $\bar{\alpha} = \frac{1}{3}(\alpha_l + 2\alpha_t)$, $\Delta\alpha = \alpha_l - \alpha_t$:

$$\begin{aligned}\varepsilon_{s\parallel} - 1 &= (NhF/\varepsilon_0)[\langle\alpha\rangle_{\parallel} + F\langle\mu_{\parallel}^2\rangle/(kT)] \\ &= (NhF/\varepsilon_0)(\bar{\alpha} + \tfrac{2}{3}\,\Delta\alpha S + F\langle\mu^2\rangle/(3kT)[1-(1-3\cos^2\beta)S])\end{aligned} \tag{14}$$

$$\begin{aligned}\varepsilon_{s\perp} - 1 &= (NhF/\varepsilon_0)[\langle\alpha\rangle_{\perp} + F\langle\mu_{\perp}^2\rangle/(kT)] \\ &= (NhF/\varepsilon_0)(\bar{\alpha} - \tfrac{1}{3}\,\Delta\alpha S + F\langle\mu^2\rangle/(3kT)[1+\tfrac{1}{2}(1-3\cos^2\beta)S])\end{aligned} \tag{15}$$

$$\Delta\varepsilon = \varepsilon_{s\parallel} - \varepsilon_{s\perp} = (NhF/\varepsilon_0)[\Delta\alpha - F\langle\mu^2\rangle/(2kT)(1-3\cos^2\beta]S \tag{16}$$

where ε_s stands for the static permittivity and ε_0 is the permittivity of free space.

The last equation shows that the induced polarization is proportional to S, whereas the orientation polarization is governed by S/T. The experimental temperature dependence of $\Delta\varepsilon$ can roughly be explained with the Maier–Meier equations.

Using averaged quantities for ε and ε_∞: $\bar{\varepsilon} = \frac{1}{3}(\varepsilon_\parallel + 2\varepsilon_\perp)$, Onsager's equation for isotropic phase is obtained such that it can be used for estimation of permanent dipole moments of molecules ($\varepsilon_\infty \approx n^2$ being the high-frequency dielectric constant, n = refractive index):

$$\mu^2 = \frac{9kT\varepsilon_0 M(\overline{\varepsilon_s} - \overline{\varepsilon_\infty})(2\overline{\varepsilon_s} + \overline{\varepsilon_\infty})}{N_A \rho \overline{\varepsilon_s}(\overline{\varepsilon_\infty} + 2)^2} \tag{17}$$

There are many examples (e.g., for alkylcyanobiphenyls) where the calculated dipole moment is significantly smaller than expected from the polar substituents. This deviation can be formally accounted for in replacing μ^2 by

$$\mu_{\mathrm{eff}}^2 \equiv g_k \mu^2 \tag{18}$$

where the correlation factor g_k has been introduced analogously to the Kirkwood–Fröhlich theory.[129,130,134,135] In the case of the nCBs antiparallel correlations are dominant, leading to $g_k < 1$.

A more sophisticated treatment is given by Bordewijk and Böttcher,[136,137] who has taken into account the anisotropy of the nematic medium, but the derived expression for $g_{k,\parallel}$ and $g_{k,\perp}$ were obtained with other not very realistic assumptions.

Recently, Sharma[138] has proposed some extension of the Maier–Meier approach to the case of nematogens with antiparallel dipole–dipole correlations of the molecules. He treated a polar LC material as a mixture of unpaired molecules with a finite dipole moment μ and antiparallel pairs with zero dipole moment. The molecules interact with each other through a combination of the generalized Maier–Saupe pseudopotential for nematic mixtures and a reaction field energy term calculated from an extension of the Maier–Meier theory. Additionally, it was assumed that a dipole with dipole moment μ is embedded in a spherical cavity of dielectric permittivity n^2, which is surrounded by a medium of average dielectric permittivity $\bar{\varepsilon}$. In that case the expressions for the cavity field factor h and the reaction field factor f are given by $h = 3\bar{\varepsilon}/(2\bar{\varepsilon} + n^2)$, $f = (\bar{\varepsilon} - n^2)/[2\pi\varepsilon_0 a^3 n^2(2\bar{\varepsilon} + n^2)]$ and the left sides of Eqs. (14) and (15) should be replaced by $\varepsilon - n^2$. The calculations performed for 5CB and 7CB gave reasonable agreement with an overall behavior of the dielectric permittivity in the nematic phase. The mole fraction of the "associated" pairs appeared to be relatively small (~10%) and the association energy does not exceed $2kT$.

B. Dielectric Relaxation

1. Fundamentals

At sufficiently high frequencies the orientation polarization lags behind the measuring field, resulting in a decrease of the permittivity. Simultaneously, the system absorbs energy that appears as a "dielectric loss." The frequency-dependent dielectric constant is usually expressed as a complex permittivity: $\varepsilon^* = \varepsilon' - i\varepsilon''$. In the simplest case of a single relaxation time τ, the real and imaginary part of ε^* can be described with the well-known Debye equations:[129,131,136,139]

$$\varepsilon'(\omega) - \varepsilon_\infty = \frac{\varepsilon_s - \varepsilon_\infty}{1 + \omega^2\tau^2} \qquad \varepsilon''(\omega) = \frac{\varepsilon_s - \varepsilon_\infty}{1 + \omega^2\tau^2}\,\omega\tau \tag{19}$$

The dielectric relaxation time τ can be calculated from the frequency of maximum loss: $\tau = 1/\omega_{\max}$. A Cole–Cole plot of ε'' against ε' enables one to check easily deviations from a single Debye mechanism that can be accounted for by several extensions of the Debye theory. We use equations after Havriliak–Negami (which contains the Cole–Cole and Cole–Davidson equations as special cases) and Jonscher, preferentially for a precise evaluation of $\omega_{\max}$:

Havriliak–Negami:

$$\varepsilon^*(\omega) - \varepsilon_\infty = \frac{\varepsilon_s - \varepsilon_\infty}{[1 + (i\omega\tau_0)^{1-\alpha}]^\beta} \tag{20}$$

Jonscher:

$$\varepsilon''(\omega) = \frac{A}{(\omega/\omega_p)^{-m} + (\omega/\omega_p)^{1-n}} \tag{21}$$

The adjustable parameters ε_∞, α, β, τ_0 or A, m, n, ω_p are in general sufficient for an accurate fitting of the loss curves.[140] Moreover, they give useful information about the shape of the loss functions and distribution of relaxation times.

When the ionic conductivity σ_0 contributes to the absorption spectra, this effect has to be subtracted, where σ_0/ε_0 can be used as an adjustable parameter:

$$\varepsilon''_{\text{total}}(\omega) = \varepsilon''(\omega) + \sigma_0/\omega\varepsilon_0 \tag{22}$$

The temperature and pressure dependence of the dielectric relaxation time can be expressed in terms of activation parameters:[129,141,142]

Activation enthalpy:

$$\Delta^{\#}H = R\left[\frac{\partial \ln \tau}{\partial(1/T)}\right]_p = T\,\Delta^{\#}V\left(\frac{\partial p}{\partial T}\right)_\tau \tag{23}$$

Activation energy:

$$\Delta^{\#}U = R\left[\frac{\partial \ln \tau}{\partial(1/T)}\right]_V = \Delta^{\#}H - T\left(\frac{\partial p}{\partial T}\right)_V \Delta^{\#}V \tag{24}$$

Activation volume:

$$\Delta^{\#}V = RT\left(\frac{\partial \ln \tau}{\partial p}\right)_T \tag{25}$$

It is the aim of the present review to emphasize the pressure dependence of the activation quantities ($\Delta^{\#}V$ and $\partial\Delta^{\#}H/\partial p$), which is not obtainable in atmospheric pressure measurements.

In the case of nematic liquid crystals we have to take into account the anisotropic features, which results in significantly different reorientation processes parallel and orthogonal to the director **n**. Correspondingly, we have to distinguish between two relaxation times, $\tau_{\|}$ and $\tau_{\perp}$. The dipole components μ_l and μ_t relax through rotations about the short and long molecular axes, respectively. In measurements of $\varepsilon_{\|}$, when the field is parallel to **n**, rotations around the short molecular axis are strongly hindered by the more or less parallel aligned neighbors. Therefore this relaxation process will be shifted to low frequencies (l.f. mode). The molecule rotations about the long axes are weakly influenced by the nematic order, and therefore the corresponding relaxation process ($\varepsilon_{\perp}$) is observed at distinctly higher frequencies (h.f. mode). Due to the incomplete nematic order ($\theta > 0$, $S < 1$), there will also be a certain contribution from μ_l on $\varepsilon_{\perp}$ and μ_t on $\varepsilon_{\|}$. This can be understood by a reorientation of μ_l through an angle π over the latitude $\theta = \text{constant}$.[143] The influence of the nematic order on the interaction in the $\tau_{\|}$ process is accounted for by the nematic potential q (see below).

2. *Concept of the Retardation Factors*

According to the Maier–Saupe theory of the nematic phase, one can distinguish two energy minima, at $\theta = 0°$ and $\theta = 180°$, that correspond to molecular alignment with the nematic director. The energy barrier between them, the nematic potential q, is supposed to be proportional to the order parameter S. Thus, the molecular reorientation around the short axis is hindered by this potential, and this is why one observes a dramatic lowering of the relaxation rate when crossing the transition point from the isotropic phase ($S = 0$) to the nematic phase ($S \geq 0.3$ at T_{NI}). Then it seems to be reasonable to expect a straightforward relation between a degree of slowing down of the relaxation rate, characterized by the retardation factor $g_{\|} = \tau_{\|}/\tau_0$ (τ_0 being the relaxation time at $q = 0$) and the nematic potential q. It was first derived by Meier and Saupe[144] who considered the parallel relaxation process in the Maier–Saupe potential. They assumed that the perturbation of the angular distribution function by the probe electric field has the form of a cosine at all times after the field is switched off. In this manner they circumvented the need for solving the equation for the distribution function exactly. The obtained relation is the following:

$$g_{\|} = \frac{\tau_{\|}}{\tau_0} = \frac{e^{\sigma} - 1}{\sigma} \tag{26}$$

where the nematic potential barrier parameter $\sigma = q/RT$.

A detailed description of the molecular motions in the Maier–Saupe potential disturbed by the electric probe field has been done by Martin and co-workers.[145] These authors obtained the numerical solutions for the relaxation times $\tau_{\|}$ and $\tau_{\perp}$ and for polarization. They found that the relaxation process measured at $\mathbf{n}\|\mathbf{E}$ geometry is slowed down with respect to the Debye-type motion in the isotropic phase, whereas the second relaxation process connected with the molecular reorientations around the long axes ($\mathbf{n}\perp\mathbf{E}$ geometry) becomes faster in the presence of the nematic potential, that is, $g_{\|} > 1$ and $g_{\perp} < 1$.

Recently, Coffey et al.[146–151] developed a new approach to the problem of Brownian rotational motions of a single-axis rotator in a uniaxial potential. Using very sophisticated mathematical procedures, they were able to obtain the exact analytic solutions for the retardation factors $g_{\|}$ and $g_{\perp}$ in terms of σ. The following formula for the parallel retardation factor renders a close approximation to the exact solution for all σ

$$g_{\|} = \frac{\tau_{\|}}{\tau_0} = \frac{e^{\sigma} - 1}{\sigma}\left(\frac{2}{1 + 1/\sigma}\sqrt{\sigma/\pi} + 2^{-\sigma}\right)^{-1} \tag{27}$$

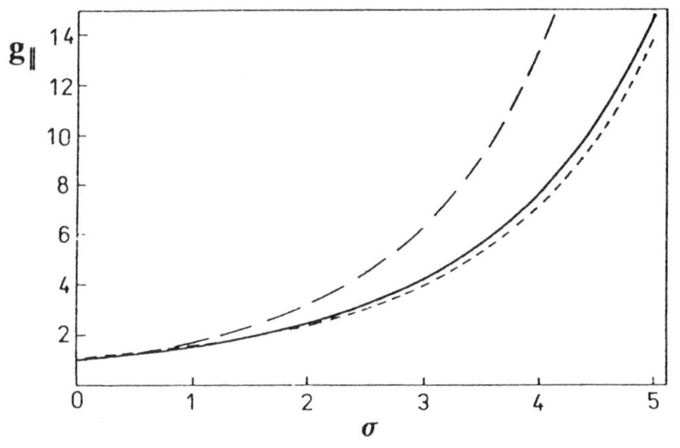

Figure 3. Retardation factor $g_\parallel = \tau_\parallel / \tau_0$ as function of the nematic potential barrier parameter $\sigma = q/RT$ according to Meier et al.,[144,145] Eq. (26) (broken line), and Coffey et al.,[146–150] Eq. (27) (dashed line) and exact solution (solid line). (Reprinted with kind permission from the authors and the editor of *Liq. Crystals*.)

In Figure 3 this formula is compared with the earlier formula of Meier and Saupe. It is clear that the Meier and Saupe formula considerably overestimates $g_\parallel$ for $\sigma > 1$, which causes the q values calculated with the use of Eq. (27) to be approximately 25% greater than those obtained by Eq. (26); see Refs. 88 and 152. Coffey et al.[149,151] found also the relations between the retardation factors $g_\parallel$ and $g_\perp$ and the order parameter S:

$$g_\parallel^S = \frac{2S + 1}{1 - S} \tag{28}$$

$$g_\perp^S = \frac{2 - 2S}{2 + S} \tag{29}$$

However, the expression for the longitudinal retardation factor does not provide a correct description of the behavior of $g_\parallel$ for $\sigma > 2$, whereas the one for $g_\perp$ is valid for all σ.[149,150] By comparison of the above equations one gets two general relationships between the retardation factors

$$g_\parallel = \frac{2 - g_\perp}{g_\perp} \qquad g_\perp = \frac{2}{g_\parallel + 1} \tag{30}$$

A discussion of the relationships (28)–(30) is given in Refs. 152 and 172.

III. EXPERIMENTAL METHODS

The measurement of the permittivity and the theoretical background is described in textbooks[129–131] and review articles.[14,20,46,153] In general the permittivity ε is determined from the increase of the capacitance of a capacitor filled with the substance under consideration. In the case of liquid crystals we have to take into consideration the principal elements of the dielectric tensor. At least two different principal values, $\varepsilon_{\parallel}$ and $\varepsilon_{\perp}$, have to be distinguished (e.g., for nematic and Sm A). For tilted smectic phases three main dielectric constants have to be measured.[153] The anisotropic permittivity of liquid crystals requires special designs for aligning the sample.[37] In order to measure the parallel ($\varepsilon_{\parallel}$) or perpendicular ($\varepsilon_{\perp}$) component, the electric field must be kept parallel or perpendicular to the director **n**. It is advantageous to align the sample with magnetic field that allows to measure $\varepsilon_{\parallel}$ and $\varepsilon_{\perp}$ with the same cell, yielding accurate values for the dielectric anisotropy $\Delta\varepsilon = \varepsilon_{\parallel} - \varepsilon_{\perp}$. At any rate the applied voltage should be low ($\approx$1 V) to avoid a hydrodynamic perturbation of the oriented sample.

Of course, three-terminal capacitors are most appropriate in order to avoid stray and lead capacitances. For measurements of isotropic liquids cylindrical capacitors are convenient. Parallel-plate capacitors are preferred for orienting the sample with a magnetic field. The cell is introduced between the poles of a magnet, providing the magnetic induction of the order of 1 T. Thus $\varepsilon_{\parallel}$ and $\varepsilon_{\perp}$ can be measured in the $\mathbf{E} \parallel \mathbf{B}$ and $\mathbf{E} \perp \mathbf{B}$ geometry, respectively. For studies at higher frequencies various techniques have been described for the investigation of the complex permittivity.[129,154–158] The alignment of the director can also be achieved in thin cells through interaction of the liquid crystals with the walls. The layer may be homeotropic or planar, depending on whether the director is perpendicular or parallel to the walls. This method can also be used in the time-domain spectroscopy (TDS).[154]

Dielectric studies on liquid crystals under high pressure are scarce, and only few high-pressure cells are described in the literature.[14,20,159,160] High-pressure experiments cause additional difficulties that are not encountered at normal pressure: The pressure transmitting medium must be separated from the sample. Dissolution of compressed gases, for example, in the substance under investigation can significantly change the measured properties.[161] Feedthroughs for the electrical leads to the capacitor require special sealings. The geometry of the capacitor and the connections may change under pressure that must be controlled by a careful calibration.

In the author's group dielectric cells were developed for the investigation of liquids and plastic crystals.[14,162] A scheme of the setup is presented in Figure 4. The autoclave A is pressurized with compressed argon that is created with a pump station up to a maximum pressure of 300 MPa. A heating and cooling jacket around the autoclave serves to adjust the temperature. The impedance is measured automatically with a Hewlett-Packard analyzer HP 4192, controlled by a personal computer.

The high-pressure vessel is shown in more detail in Figure 5. It is made of copper beryllium alloy and closed by a Bridgman piston with two bores for the electrical leads to the capacitor. Special care was necessary to design the sealings in order to avoid short circuits between the leads and the vessel.[162] In Figure 6 we present one example of a three-terminal dielectric cell.[74,163] The cylindrical capacitor is guarded by rings (g, x). It was shown recently that such cells can also be used to study the slow relaxation process observed in nematic phases.[81] A special moving piston, consisting of two parts (y, z) separated by an indium sealing (o), transmits the pressure on the substance, which can penetrate through l to the room (w) between the cylindrical electrodes (h, i). The compressed gas allows an accurate reading of the pressure. However, leakages cannot always be avoided, and the strong solubility of the gas then destroys the measurements. Therefore we established a new high-pressure equipment using compressed oil as a pressure transmitting medium.[77,88] Some changes in the dielectric cell and the lead connections were necessary in order to permit dielectric studies at somewhat higher temperatures up to 110°C. The new setup facilitates the measurements considerably. The frequency dependence of the permittivity was measured up to 13 MHz.

Very recently Urban[164] developed a high-pressure cell that enables one to measure both parallel and perpendicular components of the permittivity up to 70 MPa.

Sasabe and Ooizumi[165] measured the pressure dependence of the static permittivity of MBBA in order to determine the crystal–nematic transition line. They used a guarded electrode system that was pressurized hydrostatically.

Yasuda et al.[166] performed dielectric relaxation measurements on metastable solid MBBA, using a parallel-plate capacitor. Glass plates with strips of transparent tin-oxide-conducting coating served as electrodes. The cell was mounted in a copper beryllium pressure vessel, pressurized with liquid isopentane. A similar setup was used for the study of ferroelectric liquid crystals.[167] Ferroelectric liquid crystals have also been studied by Chandrasekhar and co-workers,[22] who used a sapphire cell setup in a high-pressure apparatus. The sample was sandwiched

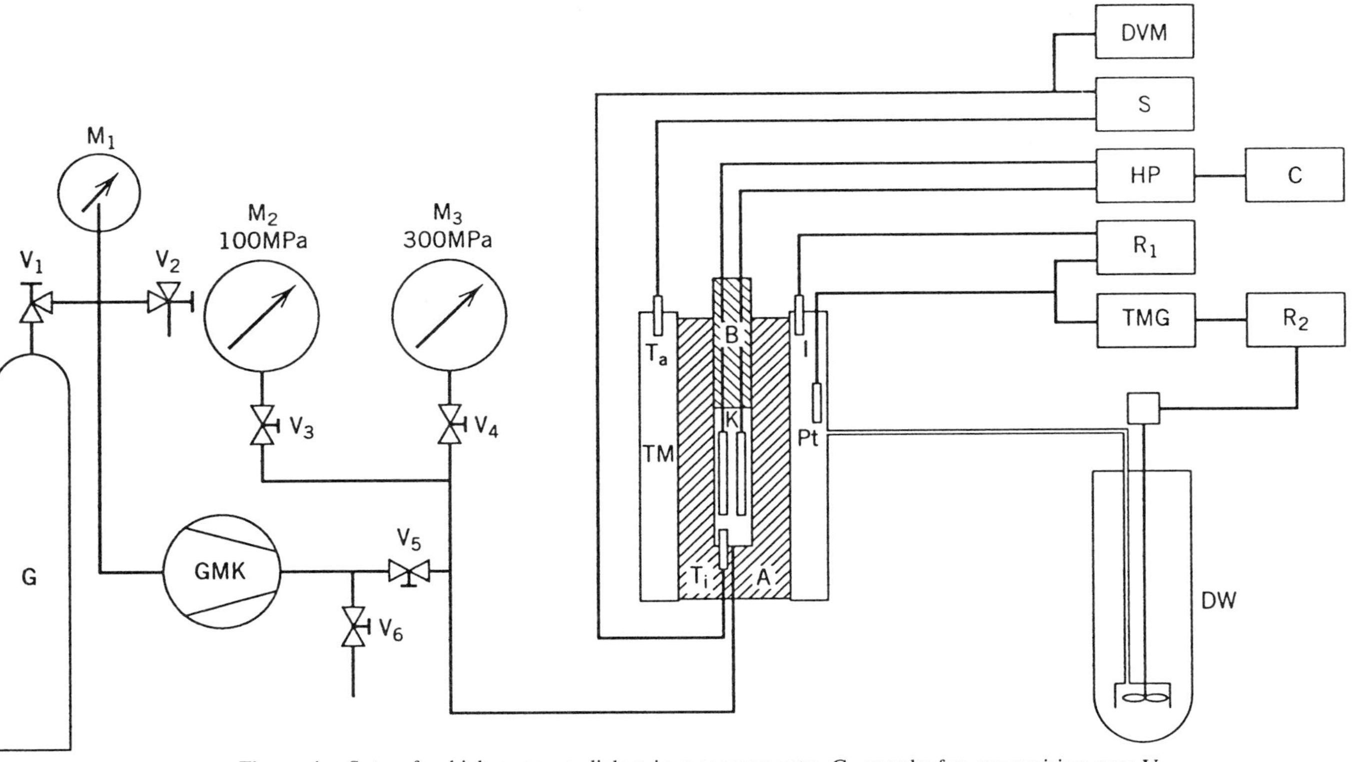

Figure 4. Setup for high-pressure dielectric measurements: G, supply for pressurising gas; V, valves; M, pressure gauge; GMK, gas compressor; TM, heating and cooling jacket; T_a, T_i, thermocouples; B, Bridgman piston; A, autoclave; K, capacitor; I, electrical connection to heating block; Pt, platinum resistance thermometer; DVM, digital voltmeter; S, recorder; HP, Hewlett-Packard impedance analyzer; C, computer; R_1, R_2, regulators; TMG, temperature registration; DW, dewar.

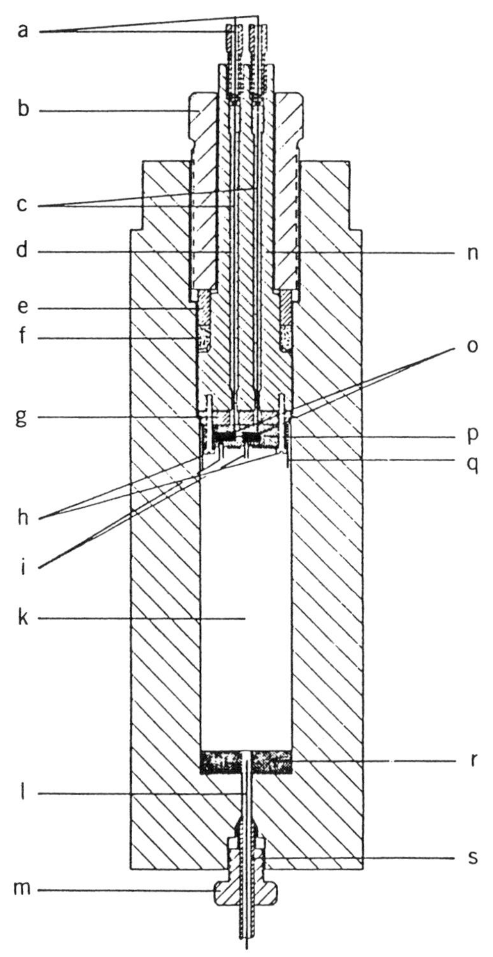

Figure 5. High-pressure autoclave: *a*, connections to the electrical leads; *b*, screw; *c*, electrical feedthrough; *d*, Teflon; *e*, pressure ring; *f*, sealing; *g*, stainless steel brace; *h*, screws; *i*, connections to the electrodes; *k*, location of the measuring cell; *l*, thermocouple; *m*, screw; *n*, Bridgman piston; *o*, part of the lead connections; *p*, *r*, Teflon spacers; *q*, clamp; *s*, connection to the pressurizing system.

between two steel cylinders, enclosed in an elastomer tube. The surface of the cylinders were coated with a polymer that was rubbed in such a way as to get a planar orientation of the sample.

W. Heinrich and B. Stoll[168] developed a high-pressure apparatus that was employed for the dielectric investigation of LC polyacrylates. The polymer specimens were pressed between stainless steel plates and

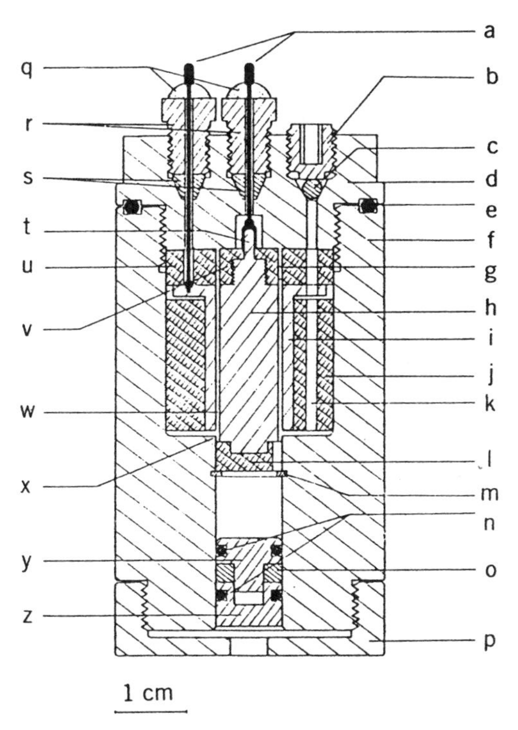

Figure 6. Dielectric measuring cell: *a*, electrical connections; *b*, screws; *c*, cone sealing; *d*, screw cover; *e*, o-rings; *f*, outer shielding of the capacitor; *g*, guard ring; *h*, inner electrode; *i*, outer electrode; *j*, spacer of Teflon; *k*, opening for substance inlet; *l*, clamp of Teflon; *m*, spring; *n*, o-rings; *o*, indium sealing; *p*, screwing with gas inlet; *q*, glue; *r*, screws; *s*, cone sealing; *t*, plug; *u*, Teflon ring; *v*, Teflon support; *w*, substance under study; *x*, shielding; *y*, *z*, parts of the piston that separates the substance from the pressurizing medium.

mounted in the high-pressure cell. So, the surface of the specimen had only little contact with the pressure transmitting silicon oil.

Carboni et al.[159] describe a specialized apparatus for the study of LCs at high pressure that include also electrooptic and electrical measurements.

IV. DIELECTRIC STUDIES AT ELEVATED PRESSURES

A. Isotropic Phase

The dielectric properties under high pressure of the isotropic phase of LC substances have been investigated to a much less extent than in the case of the nematic phase. This is due to a distinct shift of the relaxation spectra toward higher frequencies (see Fig. 14) and thus not accessible for usual impedance bridges. On the other hand, another experimental technique very well suited for that purpose, such as the time-domain spectroscopy (TDS) method, cannot be easily adopted to HP equipment.

Figure 7 presents the static permittivities $\varepsilon_{s,is}$ as function of pressure at constant temperatures within the isotropic phase of *n*CB substances measured in our lab. A small but noticeable decrease of the permittivity with the increase of pressure is observed. Thus, the effect is in opposition to that which can be expected from a change of the density of the samples. Analogous behavior has been observed for these substances at ambient pressure when the static permittivity was measured as a function of temperature starting from T_{NI}.[156,169–171] Namely, $\varepsilon_{s,is}(T)$ increases initially with temperature reaching a broad maximum at about T_{NI} + 10°C and then becomes a decreasing function of the temperature, as in a normal isotropic liquid. We observed such a behavior in the case of 7PCH.[88] This pretransitional effect was attributed to an increase of the number of dimers (with decreasing temperature) in a dynamic monomer–dimer equilibrium.[169,170] Additionally, some fluctuations of the nematic clusters embedded in the isotropic liquid near the transition temperature may also influence the static permittivity.[170]

In our HP studies of the isotropic phase of *n*CB and *n*PCH compounds the pressure was decreased from p_{NI} corresponding to the N–Is transition temperature at T = constant. According to the equation of state that corresponds to the increase in temperature at p = constant starting from T_{NI}. Thus, both observed changes of $\varepsilon_{s,is}$ have the same origin, that is, the breaking of the dipole–dipole associations of the cyano compounds

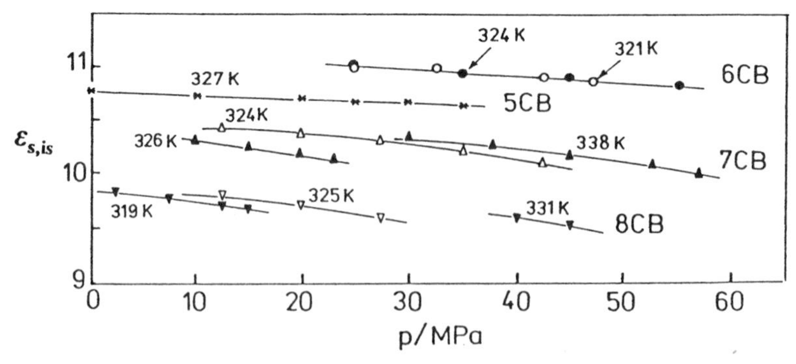

Figure 7. Static dielectric permittivity in the isotropic phase as a function of pressure for four members of *n*CB series.

caused by the temperature or pressure. However, due to the narrow pressure ranges of the isotropic phases attainable to our former experimental setup, a maximum of $\varepsilon_{s,is}(p)_T$ could not well be resolved.

In the case of 7PCH, however, the accessible range of the isotropic phase was broader, due to the establishment of a new high-pressure device that allows measurements at higher temperatures.[77,88] Figure 8 presents clearly maxima in $\varepsilon_{s,is}(p)_T$ for several constant temperatures. The effect is very weak, although well discernable. The positions of maxima are systematically shifted toward higher pressures when the temperature raises, but the distance from the transition line persists roughly constant: $p_{\rm NI} - p_{\rm max} \approx 40$ MPa.

As one can see in Figure 14 the relaxation spectra of the isotropic phase of all substances studied have maxima of losses above 100 MHz, so our HP setup can cover only a low-frequency part of the absorption bands. Therefore, to obtain the relaxation times vs. T or p we had to extrapolate the measured spectra to higher frequencies in order to find the critical frequency $f_c = 1/(2\pi\tau_{is})$. According to Parneix et al.[156] the Cole–Davidson skewed arcs should be used for that purpose. However, the recent measurements carried out by Gestblom and co-workers[152,154] with the use of the TDS method have shown that the spectra of the isotropic phase of 5CB[154] and 5PCH[152] could be well described by the Cole–Cole equation with a symmetric distribution of the relaxation times,

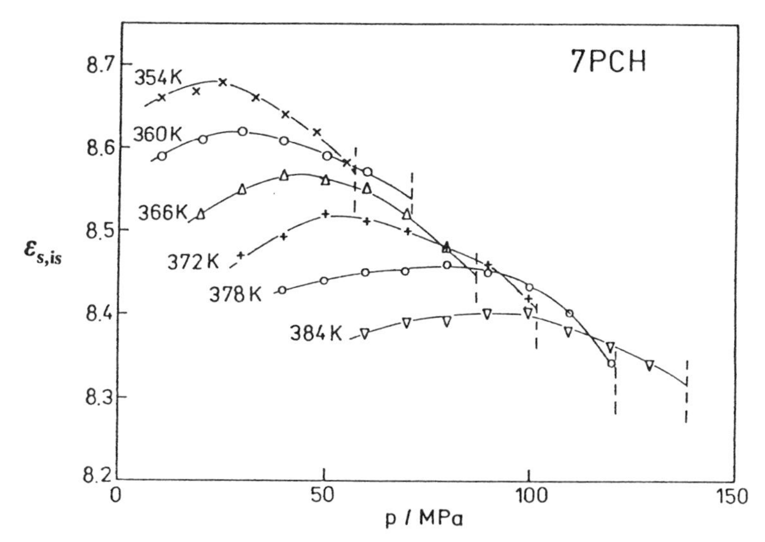

Figure 8. Static dielectric permittivity in the isotropic phase as a function of pressure for 7PCH.[88] Vertical lines mark the clearing points.

TABLE IV
Activation Volume $\Delta^{\#}V_{is}$ and Activation Enthalpy $\Delta^{\#}H_{is}$ for Different Substances in the Isotropic Phase

Substance	$\Delta^{\#}V_{is}$ (cm^3/mol)	$\Delta^{\#}H_{is}$ (kJ/mol)	References
5CB	30	30	81
6CB	33	36	83
7CB	36	46	84
8CB	47	41	85
8OCB	46	46	89
5PCH	34	33	87
7PCH	37	36	88
5CCH	32	37	77
6CHBT	35	32	86

whereas in the case of the isotropic phase of 7PCH the Cole–Davidson formula is more appropriate.[172] Fortunately, evaluation of the spectra does not influence the relaxation times markedly when τ_{is} is calculated from the f_c values.[140,172]

The isotropic phase of 5CB was studied with a special attention,[74,81] and the conclusions derived seem to be valid for all *n*CB substances: the activation volume $\Delta^{\#}V_{is}$ and activation enthalpy $\Delta^{\#}H_{is}$ can be treated as constant within the temperature and pressure ranges covered in the experiments. Table IV presents the obtained values of both quantities for particular substances studied. In case of *n*CB compounds one can observe some tendency of increasing activation quantities with the number of carbon atoms in the terminal group although the differences between the substances having similar geometrical shapes (5CB ↔ 5PCH ↔ 5CCH, 7CB ↔ 7PCH, 6CB ↔ 6CHBT, 8CB ↔ 8OCB) are rather small.

B. Nematic Phase

1. Static Permittivity

The temperature and pressure dependence of the parallel component of the static permittivity, $\varepsilon_{s\|}$, was studied for several *n*-alkylcyanobiphenyls (5CB,[74,81,164] 6CB,[83,164] 7CB,[84] and 8CB[85]), as well as for other similar compounds (6CHBT,[86] 5PCH,[76,77,87] 7PCH,[77,88] 8OCB,[89] and 5CCH[77]). However, the perpendicular component, $\varepsilon_{s\perp}$, and the dielectric anisotropy, $\Delta\varepsilon$, are known for 5CB and 6CB only.[164]

Figure 9 presents the static dielectric constants, $\varepsilon_{s\|}$ and $\varepsilon_{s\perp}$, as a function of pressure at two temperatures within the nematic phase of 5CB and 6CB. The average permittivity values, $\bar{\varepsilon} = (\varepsilon_{s\|} + 2\varepsilon_{s\perp})/3$, are practically independent of pressure and are the same at both temperatures.

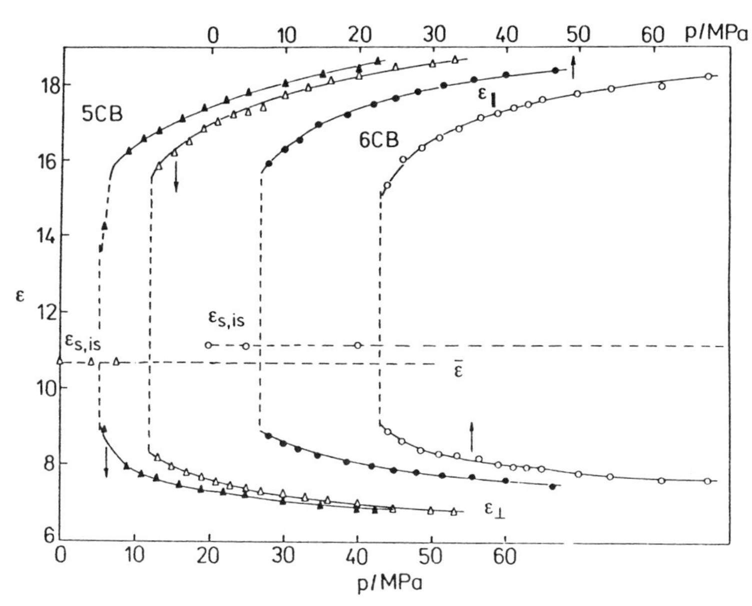

Figure 9. Pressure dependence of the principal static permittivities in the nematic ($\varepsilon_{\parallel}$, $\varepsilon_{\perp}$, $\bar{\varepsilon}$) and isotropic ($\varepsilon_{s,is}$) phase for 5CB (bottom scale, (▲) 310 K, (△) 313 K) and 6CB (top scale, (○) 304 K, (●) 310 K).[164] (Reprinted with kind permission from the editor of *Z. Naturforsch. A*.)

Moreover, they differ very slightly from the values $\varepsilon_{s,is}$ extrapolated from the isotropic phase. Characteristic steps between $\varepsilon_{s,is}$ and $\bar{\varepsilon}$, observed for all cyanobiphenyls at the I–N transition point at ambient pressure,[39,169–171] were practically not observed in the studies at $T =$ constant.[164] The dielectric anisotropy $\Delta\varepsilon$ presented in Figure 10 as a function of the reduced pressure, $p - p_{\mathrm{NI}}$, does not obey the Maier–Meier equation (16) well. The order parameter $S(p)$ normalized to the points close to the clearing temperature shows marked discrepancy as we go deeper into the nematic phase. The effect cannot be caused by the density change as the ratio $\Delta\varepsilon/\Delta\varepsilon^{\mathrm{MM}} \approx \Delta\varepsilon/\rho S F^2$, $\Delta\varepsilon^{\mathrm{MM}}$ being the anisotropy calculated with the use of Eq. (16), is not constant as a function of pressure (see Fig. 10). One can conclude therefore that the Maier–Meier equation (16) does not relate properly the permittivity anisotropy with the order parameter in the two cyanobiphenyls studied.[164]

The ratio $\Delta\varepsilon/\Delta\varepsilon^{\mathrm{MM}} \approx \mu_{\mathrm{eff}}^2/\mu^2 = g_k$ can give some information about a degree of dipole–dipole associations in the nematic phase of the substances under consideration. However, the obtained value of $g_k \leq 0.5$ seems to be unreliable in the light of the above discussion concerning the applicability of the Maier–Meier equations to real nematic phase prop-

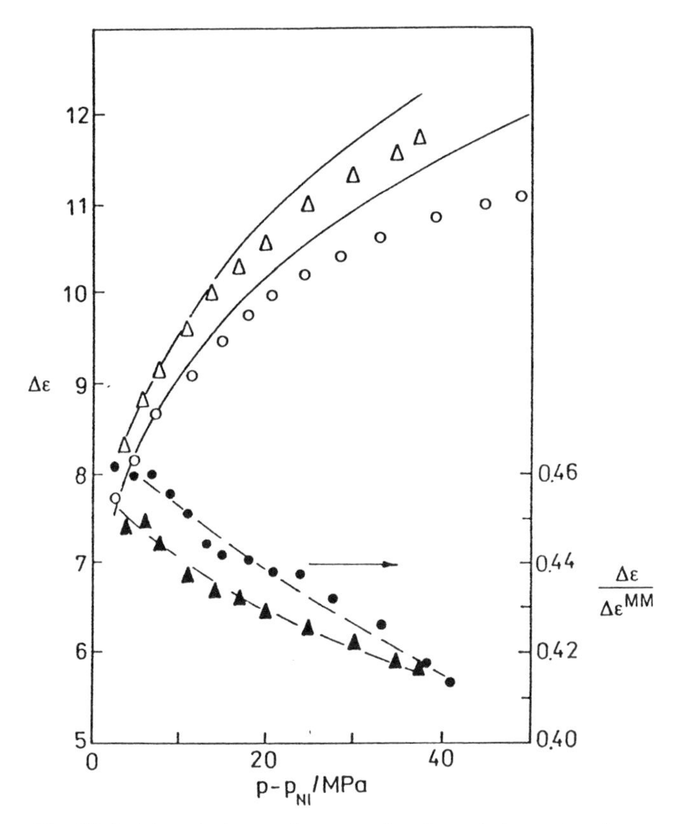

Figure 10. Dielectric anisotropy, $\Delta\varepsilon$, as a function of the reduced pressure in the nematic phase of (△) 5CB and (○) 6CB;[164] the lines correspond to the order parameter S normalized to the points close to p_{NI}. Full symbols [(●) 310 K, (▲) 313 K] concern the ratio of measured ($\Delta\varepsilon$) and calculated ($\Delta\varepsilon^{\mathrm{MM}}$) dielectric anisotropy for 5CB (right-hand scale); the lines are only guide for the eyes. (Reprinted with kind permission from the editor of *Z. Naturforsch. A.*)

erties. As mentioned in Section II.A, Sharma[138] has proposed the extension of this theory by including the electronic polarization to the cavity and reaction field factors. After doing that the Kirkwood g_k factor becomes more realistic (ca. 0.7).[164]

For other substances the parallel permittivity component, $\varepsilon_{s\|}$, is only known. Figure 11 shows a typical behavior of the static permittivity $\varepsilon_{s\|}$ in the nematic and isotropic phase of 7PCH at constant temperatures, while the pressure was successfully reduced from the points close to the nematic-solid transition line. Figure 12 presents the dependencies of $\varepsilon_{s\|}$ on the reduced pressure, $p - p_{\mathrm{NI}}$, for different substances studied. It is

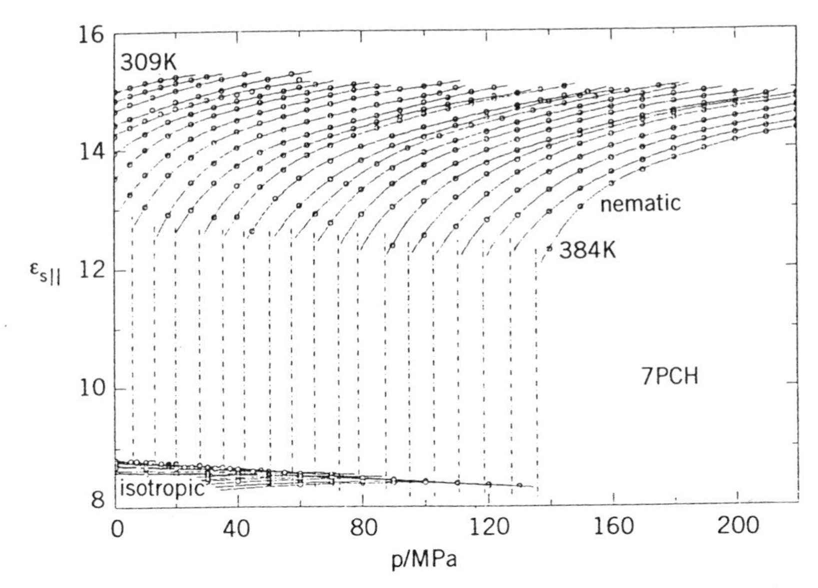

Figure 11. Pressure dependence of the static permittivity $\varepsilon_{s\|}$ in the nematic and isotropic phase of 7PCH; the temperatures for different isotherms vary in steps of 3 K.[88] The vertical lines mark the clearing points.

worthwhile to note that the absolute values of $\varepsilon_{s\|}$ obtained for the samples oriented by the electric field ($E \sim 300$ V/cm) are usually markedly greater than those measured in other labs when a B-field was applied.[83–85,87–89,172] Maybe the strengths of the magnetic fields used were not large enough. Two features of these behaviors should be noted: (i) the static permittivity decreases with increasing alkyl chain length and (ii) a stronger dependence of $\varepsilon_{s\|}(p)$ is observed close to the clearing temperature in the case of *n*CB compounds than for other ones. The first feature is probably connected with the conformation motions of the end group in the molecules, which become wider with longer chains. These motions smear out the position of the long molecular axis and make the molecule broader. The addition of CH_2 groups to the alkyl chain (e.g., 5PCH$\rightarrow$7PCH) produces looser molecular packing in the nematic phase.[173] It was found that the decrease of the permittivity between 5PCH and 7PCH (~15%) is not compensated by the change of the density and molar volume.[88] Figure 13 shows that the molar susceptibility $(\varepsilon_{s\|} - 1)M/\rho$ is appreciably reduced (to ~2%), but the difference remains in the whole range of pressures in the nematic phase of both compounds. The second feature can be connected with breaking of the dipole–dipole associations due to a change of monomer–dimer equilibrium caused by

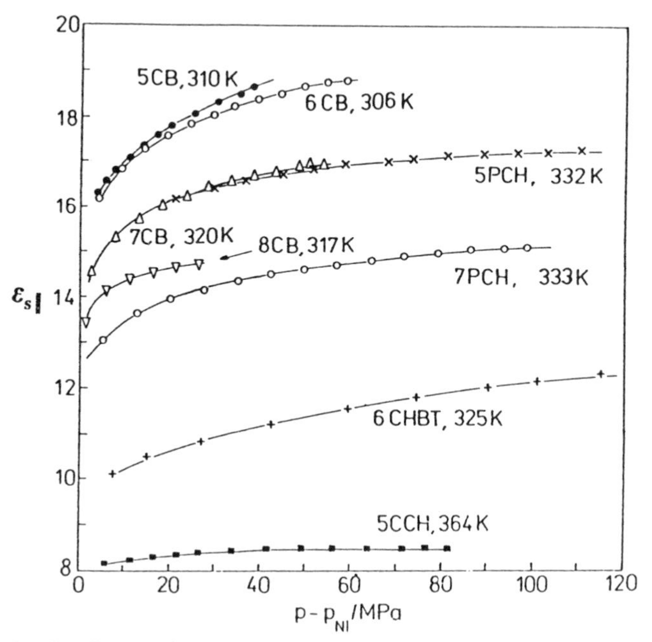

Figure 12. Parallel permittivity component, $\varepsilon_{s\parallel}$, as a function of the reduced pressure, $p - p_{\mathrm{NI}}$, for different substances in the nematic phase. The temperatures correspond to a few Kelvins above the clearing points at ambient pressure.

the increase of pressure.[81–89] However, a detailed analysis of these behaviors requires the knowledge of the order parameters in function of pressure, which is only exceptionally available.

2. Dielectric Relaxation

The alkylcyanobiphenyls are specially suited for the investigation of the dielectric relaxation because the two main modes of dipole relaxation are well separated. The hindered rotation about the short molecular axis is shifted to relatively low frequencies conveniently accessible in impedance bridge measurements. This relaxation process is coupled to the strength of the nematic potential q, which should be sensitive to a variation of the intermolecular distance. Hence a strong pressure dependence is expected for this relaxation process.

As an example in Figure 14 we show the real and imaginary parts of the complex permittivity for 6CB as a function of temperature at ambient pressure and as a function of pressure at constant temperature.[83] A similarity of the pictures is evident. In both cases the observed relaxation process can be described by a single Debye mechanism, which is easily

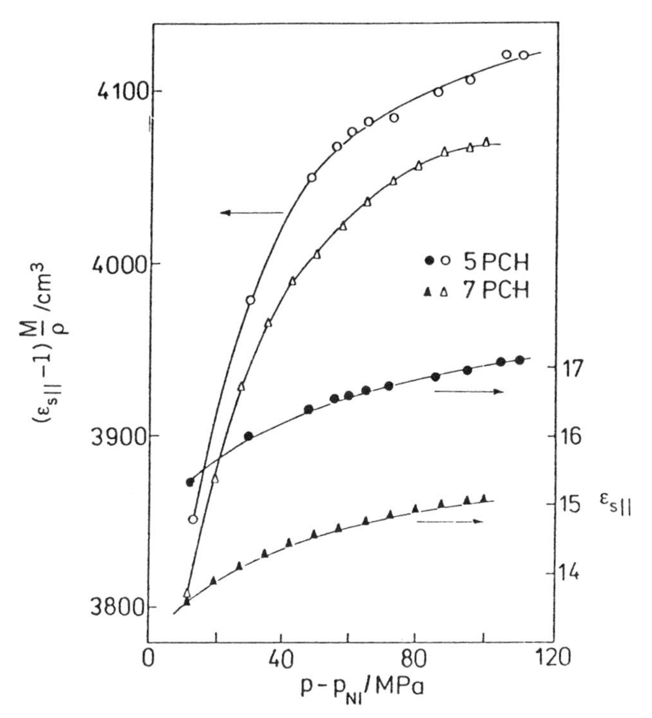

Figure 13. Pressure dependence of the parallel permittivity component, $\varepsilon_{s\|}$, and the molar susceptibility, $(\varepsilon_{s\|}-1)M/\rho$, for 5PCH and 7PCH in the nematic phase.

seen on the Cole–Cole plots (ε'' against ε'). This well-known behavior for the low-frequency relaxation process in the nematic phase can be checked more quantitatively in evaluating the shape parameters of the loss functions. Using the Jonscher equation, it was shown for 7CB that the parameters $m \approx 1 \pm 0.01$ and $1 > 1-n > 0.95$ are in fact very close to unity.[84]

The relaxation times $\tau_\|$ are strongly pressure dependent (see Fig. 15), and it is more convenient to present them in semilogarithmic form. In Figure 16 and 17 the logarithm of the dielectric relaxation time is plotted against pressure and the reciprocal temperature, respectively. Although the plots are not exactly straight lines (especially $\ln \tau_\|$ vs. p plots), we can determine the activation volume and activation enthalpy from the average slopes according to the equations presented in Section II.B. The activation energy can be derived from suitable cross plots of the isobaric relaxation time and isochors, or alternatively with Eq. (24), after

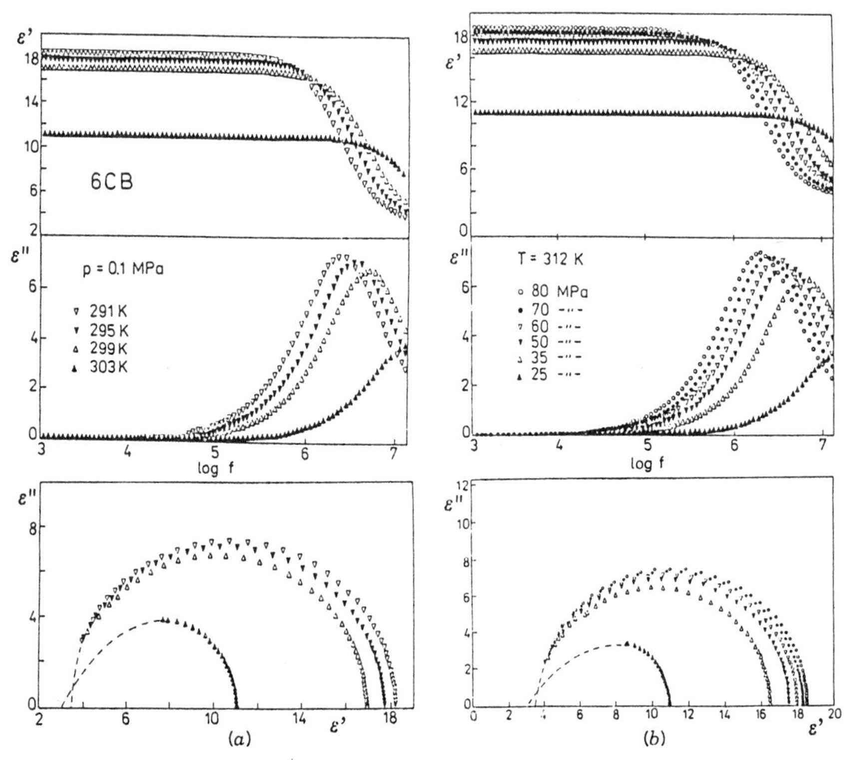

Figure 14. Dispersion and absorption spectra and Cole–Cole plots for 6CB: (a) at constant (ambient) pressure and different temperatures, and (b) at constant temperature and different pressures.[83] (Reprinted with kind permission from the editor of *Liq. Crystals*.)

evaluating $(\partial p/\partial T)_V = \alpha/\beta$ from pVT data. Both procedures yield the same result in the limits of experimental errors (see Table V).

The activation parameters are compared in Figures 18 and 19. The variation of the alkyl chain length in the alkylcyanobiphenyl homologues does not exhibit significant trends.[85] The most interesting result is the high activation volume that is an order of magnitude larger than those usually observed for reorientation processes in liquids and plastic crystals.[14] It amounts to about 25% of the molar volume of an LC substance. These high $\Delta^{\#}V_{\parallel}$ values reflect the strong sensitivity of the relaxation process against a change of the density of the phase. $\Delta^{\#}U_{\parallel}$ is appreciably smaller than $\Delta^{\#}H_{\parallel}$ due to the large activation volumes [cf. Eq. (24)]. The activation energy $\Delta^{\#}U_{\parallel}$ corresponds to the energy barrier at constant volume, which should be accessible from computer experiments. Indeed computer simulations have been performed for 5CB,[110,112] 5PCH,[174]

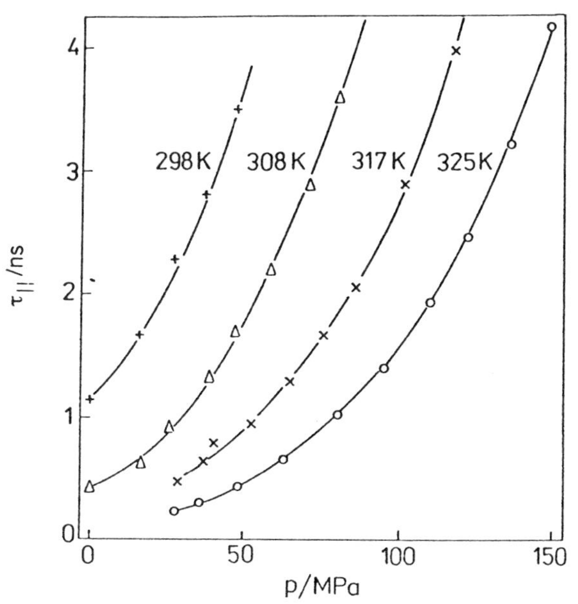

Figure 15. Pressure dependence of the relaxation time $\tau_{\|}$ at different temperatures in the nematic phase of 6CHBT.[86] (Reprinted with kind permission from the editor of *Liq. Crystals*.)

5CCH,[102] and *n*OCBs.[175] However, energy barriers for the reorientation processes in nematic phases have not been calculated so far.

In Figure 19 $\Delta^{\#}H_{\|}$ is plotted against pressure for several nematics. Surprisingly, $\Delta^{\#}H_{\|}$ decreases with increasing pressure and correspondingly we observed a negative density dependence for $\Delta^{\#}U_{\|}$. This behavior can be understood in the frame of a monomer–dimer equilibrium. The *n*CB molecules are supposed to form dimers that partly overlap[118] (see Fig. 2) and therefore require a relatively large volume. The specific arrangement is due to dipole–dipole correlations. Obviously, the average volume occupied by a dimer must be larger than the volume required by two monomers. Increasing pressure squeezes out the free volume and thus destroys the dimers. Because the monomers are assumed to reorient more freely, the activation enthalpy is reduced with increasing pressure. Once the dimers are destroyed, one should observe the "normal" pressure dependence of $\Delta^{\#}H_{\|}$, i.e., an increase with pressure. Probably the range of experimental pressure was not high enough to observe a minimum in $\Delta^{\#}H_{\|}(p)$. In this context we mention results for 6CHBT,[86] for which $\Delta^{\#}H_{\|}$ was practically independent of pressure (this substance does not show dipole–dipole correlations in the nematic phase[176]).

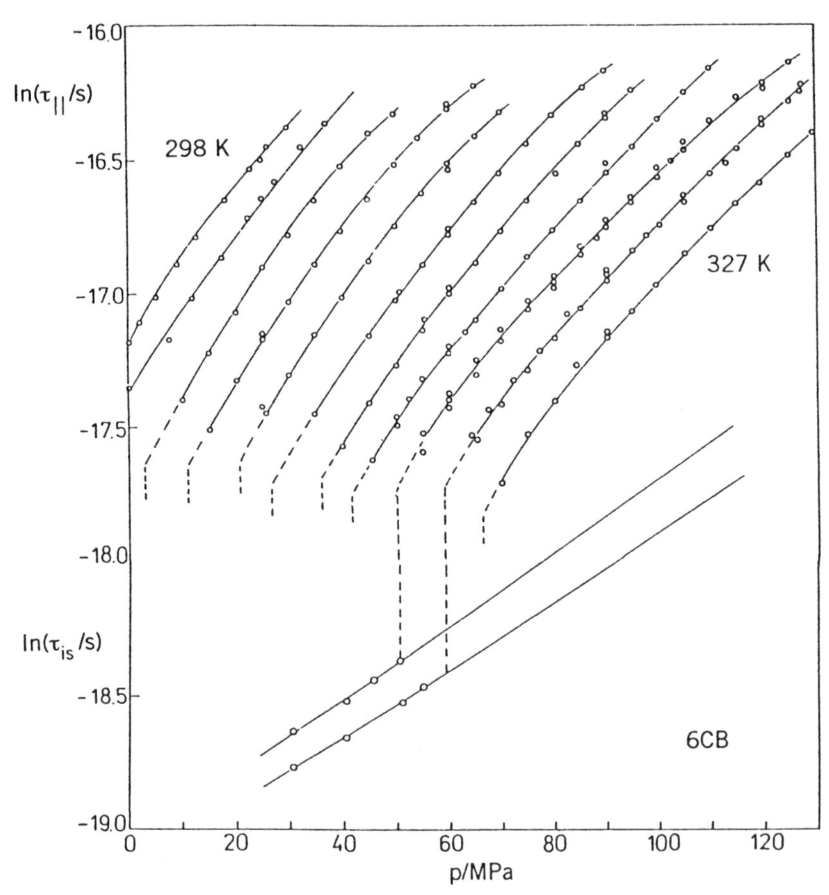

Figure 16. $\ln \tau_{\parallel}$ vs. p plots for the nematic and isotropic phase of 6CB. The temperatures for different isotherms vary in steps of 3 K.[83] (Reprinted with kind permission from the editor of *Liq. Crystals*.)

A comparison of 5CB with 5PCH[87] and 5CCH[77] and 7CB with 7PCH[88] enables us to analyze the influence of different molecular cores. Both the activation enthalpy and activation volume are considerably larger for *n*PCH and even more for 5CCH compounds (see Fig. 20). This could be connected with the higher flexibility of the cyclohexyl ring requiring more space for the reorientation. On the other hand the pressure dependence of $\Delta^{\#} H_{\parallel}$ is much less pronounced for *n*PCHs compared with *n*CBs. For the $\Delta^{\#} U_{\parallel}$ values the density dependence is even inverse (Fig. 21). Whereas $\Delta^{\#} U_{\parallel}$ decreases with increasing density for 5CB, we find an increase in the case of 5PCH and 7PCH, although for *n*PCHs the

TABLE V
Verification of the Relation Between Activation Parameters for 5CB and 5PCH

$$(\Delta^{\#} H_{\parallel})_p = (\Delta^{\#} U_{\parallel})_V + \frac{\alpha}{\beta} T(\Delta^{\#} V_{\parallel})_T$$

Substance	V_m (cm^3/mol)	T (K)	p (MPa)	$\alpha\ 10^4$ (K^{-1})	$\beta\ 10^4$ (MPa^{-1})	α/β (MPa/K)	$\Delta^{\#} U_{\parallel}$ (kJ/mol)	$\Delta^{\#} V_{\parallel}$ (cm^3/mol)	$\Delta^{\#} H_{\parallel}^{exp}$ (kJ/mol)	$\Delta^{\#} H_{\parallel}^{calc}$ (kJ/mol)
5CB	247	316	32	6.75	4.87	1.39	37.0	61.0	63.2	64.1
	245	327	64	8.00	4.94	1.62	28.6	58.5	58.0	60.0
	243	338	100	8.42	4.30	1.95	24.8	45.6	53.4	55.2
5PCH	265	323	40	7.09	5.47	1.30	38.4	71.4	70.4	68.7
	263	332	80	5.62	5.02	1.12	40.5	69.3	68.2	66.6
	261	344	100	5.23	4.77	1.10	40.9	69.3	66.9	67.3

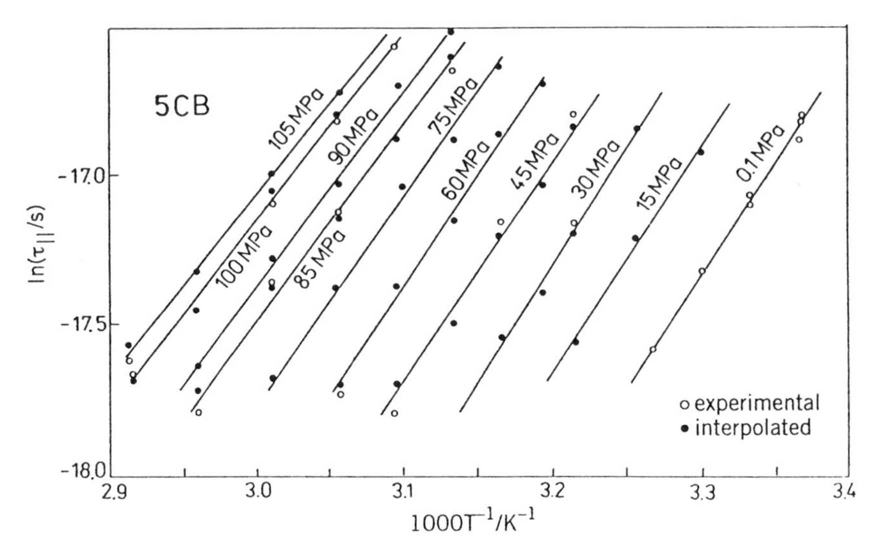

Figure 17. $\ln(\tau_{\|})_p$ vs. $1/T$ plots for the nematic phase of 5CB at different pressures.[81] (Reprinted with kind permission from the editor of *Phys. Rev. A.*)

formation of antiparallel dimers is generally accepted as well.[170,174,177] We assume that due to stronger dipole–dipole interactions in the nematic phase of 5PCH, the dimers cannot be so easily destroyed with pressure, and hence a weaker pressure dependence is observed. This conclusion is also supported by a similar pressure dependence found for the static permittivity (compare Fig. 12).

Interesting trends are observed in the case of 7PCH for which dielectric measurements were carried out in a distinctly broader range of temperatures and pressures owing to the new HP setup (see experimental part). Figure 20b shows that the activation volume decreases with increasing temperature like 7CB, but at higher temperatures (and at the same time at higher pressures—compare phase diagrams in Fig. 1) a plateau is reached. The activation enthalpy $\Delta^{\#}H_{\|}$ for 7PCH remains practically constant, whereas a distinct decrease was observed for *n*CB compounds (see Fig. 19). Combination of Eqs. (23) and (24) give the difference of two thermal pressure coefficients, $(\partial p/\partial T)_{\tau} - (\partial p/\partial T)_V = \Delta^{\#}U_{\|}/T\,\Delta^{\#}V_{\|}$, which is a measure for the isochoric activation energy. When the slopes $\partial p/\partial T$ were equal, then the relaxation time would be solely determined by steric conditions, and hence the activation energy $\Delta^{\#}U_{\|}$ would be zero. (A similar conclusion follows from the simplest free volume theory[23,142]). Figure 22 shows that both these slopes are evidently different, which yields distinctly nonzero $\Delta^{\#}U_{\|}$ values, so the $\tau_{\|}$ relaxa-

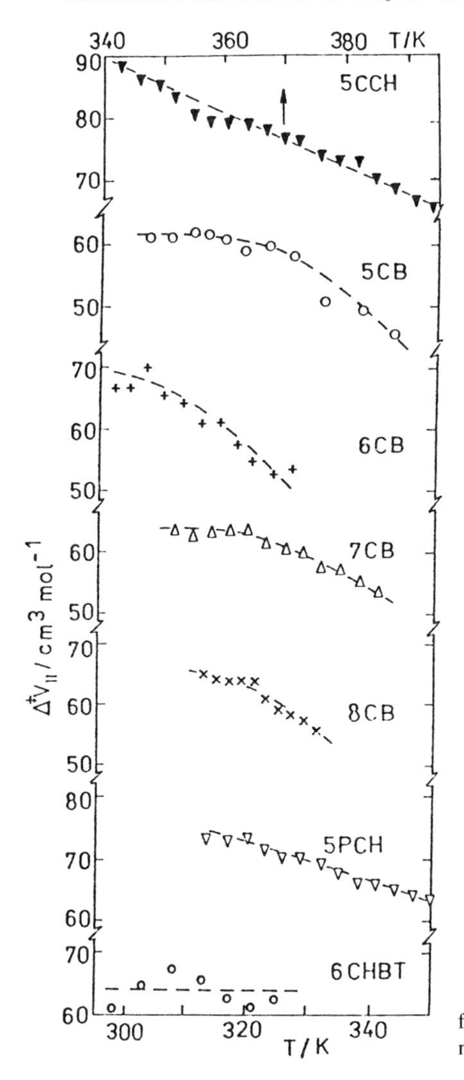

Figure 18. Activation volume $\Delta^{\#}V_{\|}$ as a function of T for different substances in the nematic phase.

tion process cannot be explained only by volume effects. It would be interesting to compare $\Delta^{\#}U_{\|}$ in more detail for liquid crystals, which do not form dimers, for example, 6CB with 6CHBT. However, p, V, T data are not available.

As outlined in Section II.B, Maier and Saupe have introduced the nematic potential q, which is a measure of the average strength of interaction between molecules treated as rigid bodies. However, the information about the values of the nematic potential is rather scarce. It can be determined from the deuterium NMR spectra[98,178–180] and from

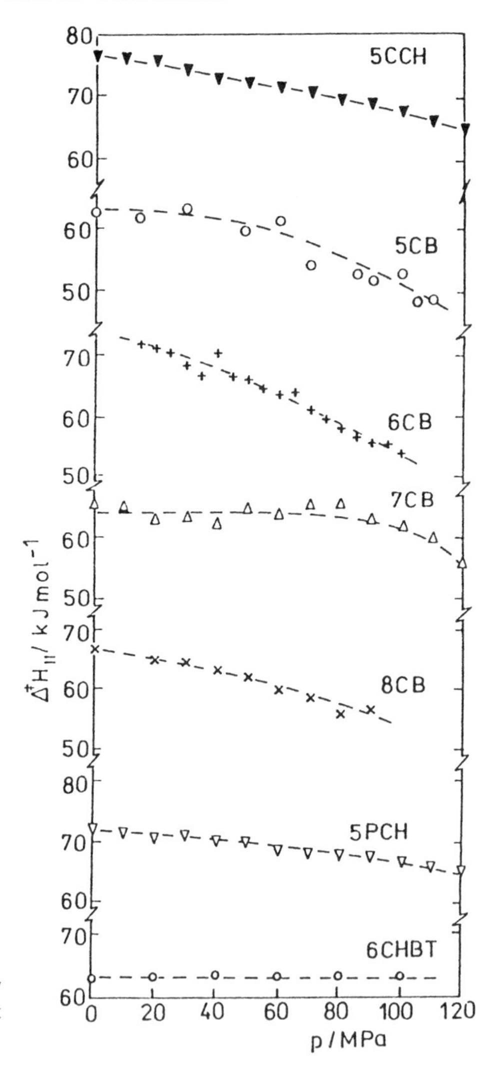

Figure 19. Activation enthalpy $\Delta^{\#}H_{\parallel}$ as a function of p for different substances in the nematic phase.

the dielectric relaxation studies using relations (26) and (27). In the latter case only compounds having dipole moments directed along the symmetry axis can be used. Only in such cases do the relaxation spectra measured in the isotropic phase give correct values of the relaxation times τ_0 characterizing the molecular reorientations about the short axes in the absence of the ordering potential. The compounds belonging to the *n*CB and *n*PCH homologous series seem to be good representatives in that respect. Therefore, using a "natural" extrapolation of $\ln[\tau_{is}(p)]_T$ from the isotropic to the nematic phase (see Fig. 16) one can obtain the values of

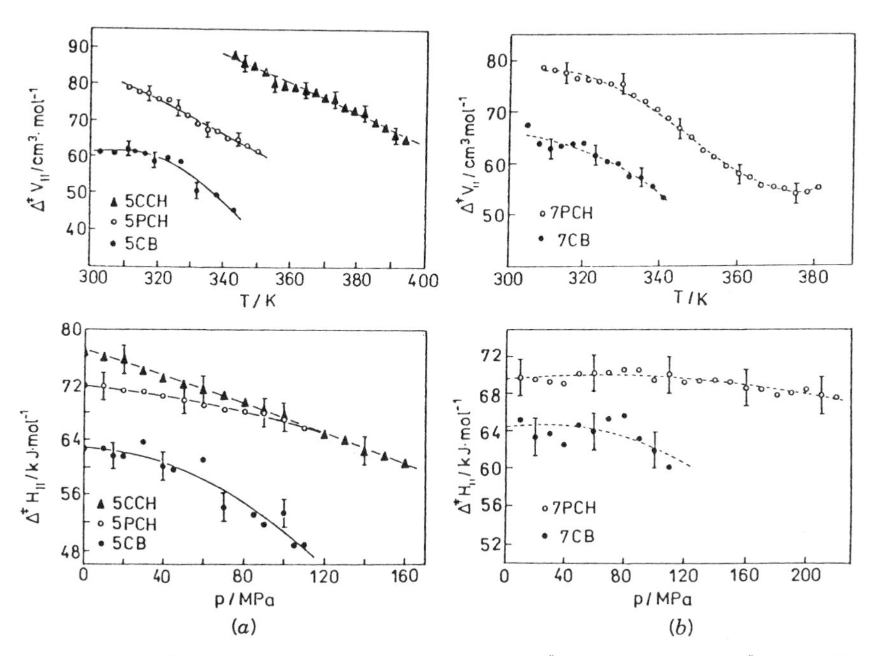

Figure 20. Comparison of the activation volumes $\Delta^{\#}V_{\parallel}$ and enthalpies $\Delta^{\#}H_{\parallel}$ for (a) 5CB, 5PCH,[87] and 5CCH[77] and (b) 7CB and 7PCH[88] in the nematic phase.

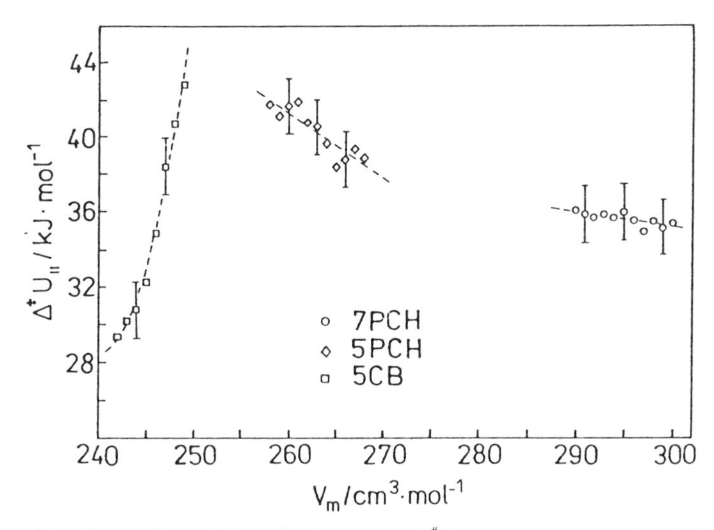

Figure 21. Comparison of the activation energy $\Delta^{\#}U_{\parallel}$ vs. molar volume V_m obtained for the nematic phase of 5CB,[81] 5PCH,[87] and 7PCH.[88]

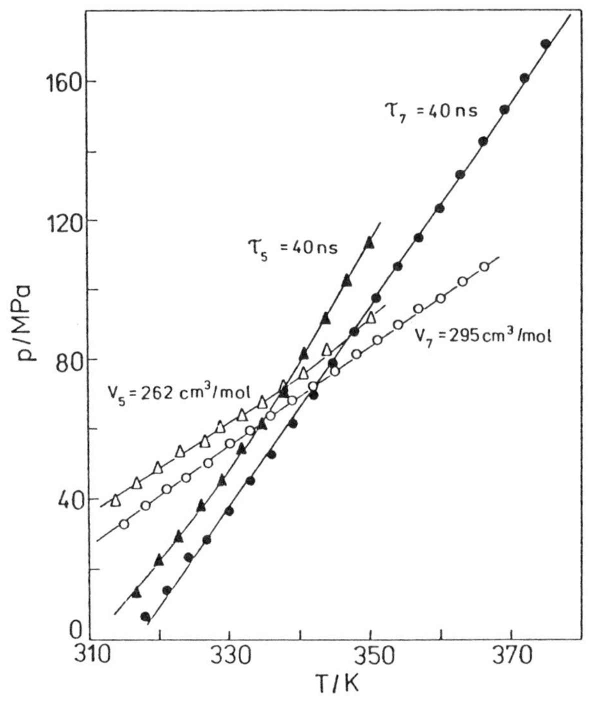

Figure 22. Comparison of $p(T)$ plots at $V=$ constant and at $\tau_{\|}=$ constant for 5PCH (τ_5, V_5) and 7PCH (τ_7, V_7).

$\tau_0(p)_T$ and then the $g_{\|}(p)_T$ factors. That procedure was applied to several substances studied.[82–84,87,88] Figure 23 (top) shows, as an example, the $g_{\|}$ factors for 5PCH calculated at different conditions. In the case of this substance the values of the retardation factor change from about 5 close to the N–I transition up to about 20 at the N–Cr transition. At $p=$ const and $T=$ const only minor differences are observed between values corresponding to different isotherms or isobars, and only the change of density (molar volume) involves considerable changes in $g_{\|}$ factors. For nCB compounds the $g_{\|}$ factors are significantly lower,[82–84] for example, for 6CB $g_{\|}$ changes from about 2 to about 4 within the nematic phase (Maier–Saupe theory predicts $g_{\|}=4$ at T_{NI}). Figure 24 presents the ratio $g_{\|}$ (experimental) to $g_{\|}^S$ (theoretical) calculated for 5CB at different conditions according to formula (28). It is clear from this figure that the relation predicted by formula (28) is not fulfilled; moreover, in the case of 6CB (and also 5CB) we have $g_{\|}^S > g_{\|}$, whereas for 7PCH it is $g_{\|}^S < g_{\|}$.[172] It should be noted, however, that the absolute values of the

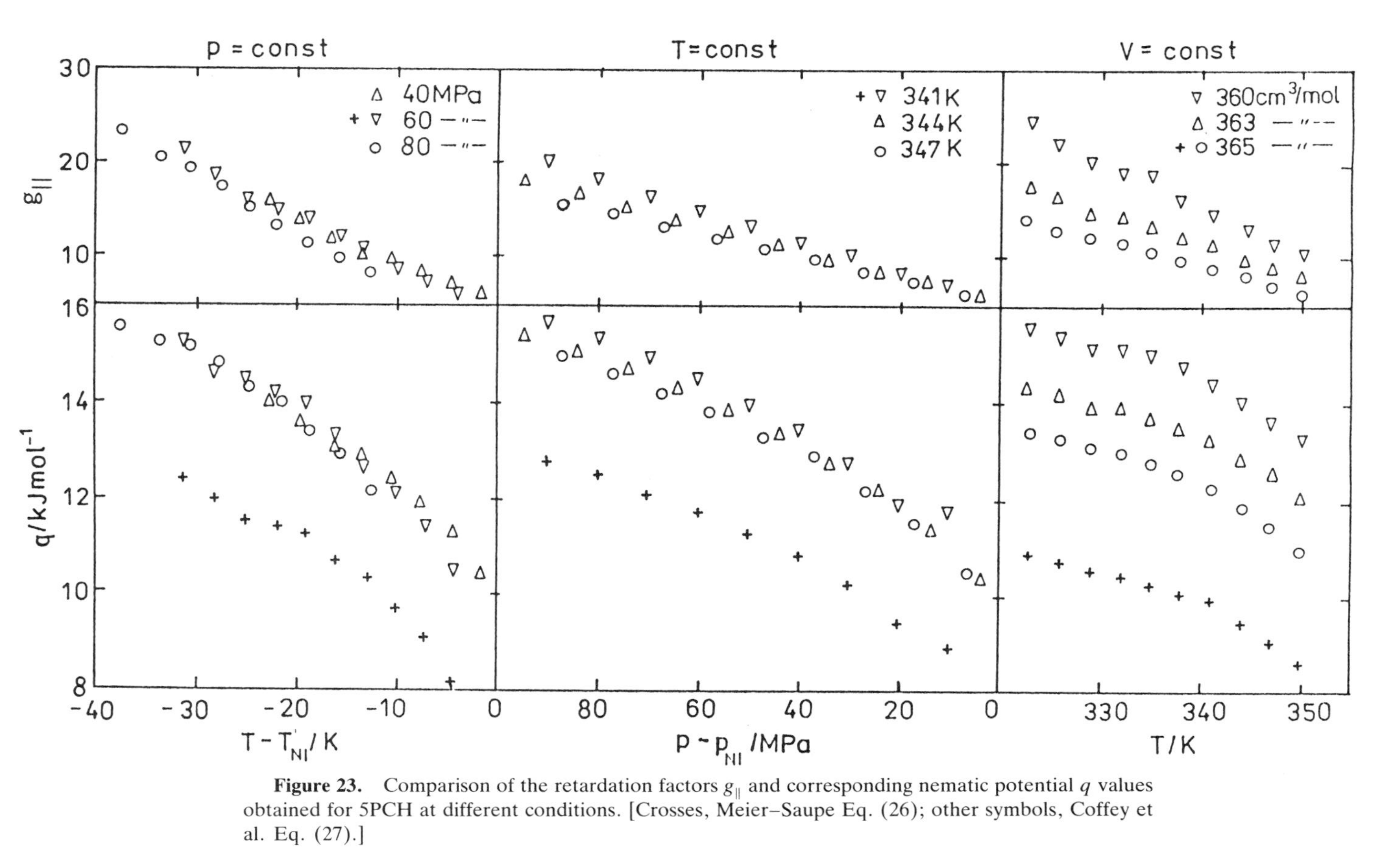

Figure 23. Comparison of the retardation factors $g_{\|}$ and corresponding nematic potential q values obtained for 5PCH at different conditions. [Crosses, Meier–Saupe Eq. (26); other symbols, Coffey et al. Eq. (27).]

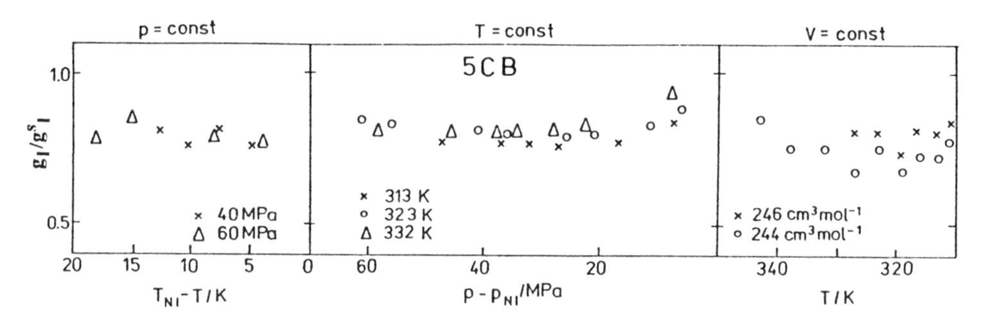

Figure 24. Ratio of experimental and theoretical $g_{\parallel}$ factors calculated for the nematic phase of 5CB at different conditions.

order parameters are hardly obtainable; this concerns especially the data calculated from the refractive index measurements.[172,181]

Having retardation factors, one can calculate the nematic potential q using equations derived originally by Meier and Saupe [Eq. (26)] or by Coffey and co-workers [Eq. (27), Fig. 3]. With the aid of Eq. (26) the nematic potential has been calculated as a function of temperature and pressure for various nematics.[82–84,87,88,152,172] For nCB compounds the q values vary between 3 kJ/mol near T_{NI} point and 6 kJ/mol near the melting temperature,[82–84] whereas in the case of nPCH substances they are higher[87,88,152,172] (ca. 8 and 16 kJ/mol, respectively). As pointed out in Section II.B.2, the relation between $g_{\parallel}$ and q derived by Coffey and co-workers seems to be more appropriate for characterizing the nematic phase. Figure 23 shows that the Coffey and co-workers formula gives systematically higher values (ca. 20–25%) of the nematic potential q than the Meier–Saupe equation, but both depend similarly on temperature and pressure.

The temperature and pressure dependencies of q yield important information about the validity of the assumptions of the theories describing the nematic state. In particular, having S values from independent experiments one can check the relation predicted by the Maier–Saupe theory (see Section I.C.1) that $q = vS$. However, the data on $S(T, p)$ are available for a few LC substances only.[71,72,98,99,104] Figure 25 presents the q versus S plots for three substances studied in our lab [the data on $S(T, p)$ were taken from Ref. 71 for 5CB, Ref. 72 for 6CB, and Ref. 99 for 7PCH]. In the case of two cyanobiphenyls some scatter of points obtained at different experimental conditions are observed, but essentially one can note a proportionality of both these quantities. For 7PCH, however, a nice proportionality was found for the results obtained at p = constant only,[88,172] whereas at V = constant it is completely failed[88]

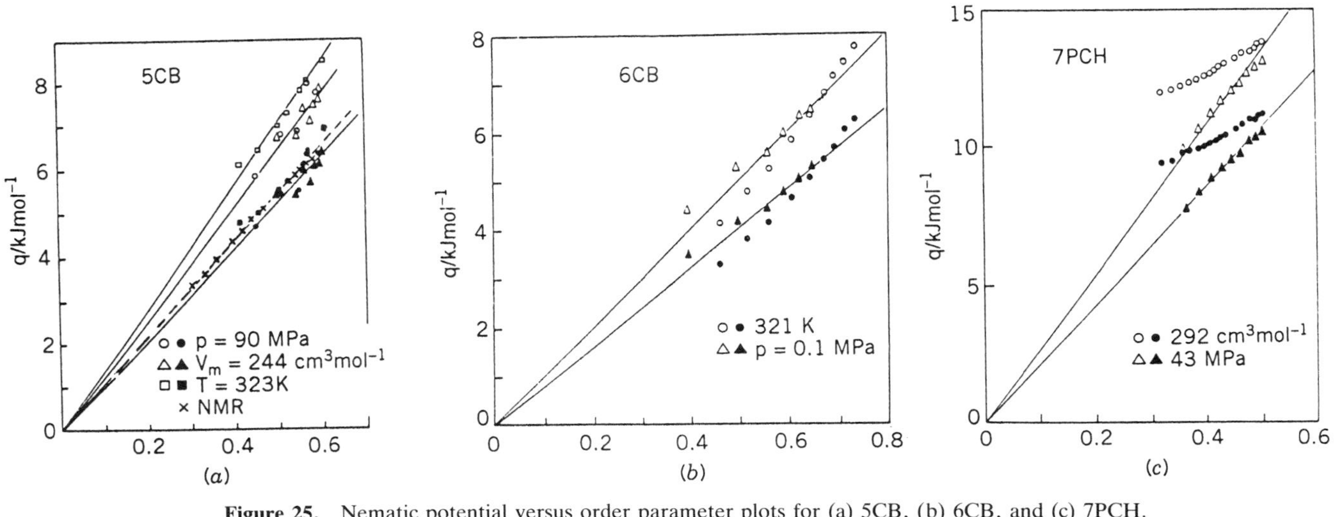

Figure 25. Nematic potential versus order parameter plots for (a) 5CB, (b) 6CB, and (c) 7PCH. [Solid symbols, Meier–Saupe Eq. (26), open symbols, Coffey et al. Eq. (27).] The NMR result in (a) comes from Ref. 178.

(the data on S were taken from Ref. 99 with a correction factor of 1.4 as described in Ref. 172). The calculated interaction coefficients $v = q/S$ are listed in Table VI together with other available data obtained with the aid of deuterium NMR experiments. One can note quite good consistency of the results coming from both experimental methods. This is easily seen in Figure 25a where the NMR data coincide nicely with the dielectric relaxation data obtained from the Meier–Saupe equation. Thus, the assumptions accepted on the way from the experimental relaxation times $\tau_{\|}$ and τ_{is} to the final interaction coefficient v (extrapolation procedure, theoretical mean-field approximation) seem to be justified.

The q values amount to only 15–20% of the activation enthalpy $\Delta^{\#}H_{\|}$.

TABLE VI
Mean-Field Interaction Coefficient v Determined from the Deuterium NMR Spectra (Aromatic Core) and the Dielectric Relaxation Studies for Some Members of Three Homologous Series, nCB, nPCH, and nOCB

	Interaction Coefficient v (kJ/mol)			
Substance	NMR	Diel. Relax.	Remarks	Refs.
5CB	10.9 ± 0.1		0.1 MPa	178
	11.5 ± 0.2		243 $cm^3 mol^{-1}$	98
	13.0 ± 0.1		238 $cm^3 mol^{-1}$	98
		11 ± 1	different conditions	82
6CB		8.1 ± 1[a]	p = const, T = const	
		(10.1 ± 1)[b]		
8CB	11.6		0.1 MPa	178
5PCH		17.4 ± 0.3	0.1 MPa	152
		(19.6 ± 0.3)		172
7PCH		14.4 ± 0.5	0.1 MPa	172
		(18.4 ± 0.5)		
		15.4 ± 0.6	43 MPa	
		(19.3 ± 0.6)		
3OCB	11.73		0.1 MPa	179
4OCB	11.73		0.1 MPa	179
5OCB		9.8 ± 0.2	0.1 MPa	228
		(12.4 ± 0.2)		
6OCB	11.73		0.1 MPa	179
	10.5			180
		10.5 ± 0.8	0.1 MPa	185
		(13.8 ± 1.2)		
7OCB		10.0 ± 0.4	0.1 MPa	185
		(12.5 ± 0.5)		
8OCB	12.4		0.1 MPa	179
		10 ± 1		185
		(13 ± 1)		

[a] Meier and Saupe formula (26).
[b] Coffey et al. formula (27).

According to de Jeu[37]

$$\Delta^{\#} H_{\parallel} = W_{\eta} + q \tag{31}$$

where W_{η} accounts for the viscosity effects. From the comparisons of Figures 19 and 23 and Table IV one can state that for 6CB the above relation is not well obeyed if it is assumed that $\Delta^{\#} H_{is} \approx W_{\eta}$. Moreover, the activation enthalpy $\Delta^{\#} H_{\parallel}$ and the nematic potential q behave in the opposite way with pressure (compare Figs. 19 and 23). This should mean that the activation energy W_{η} characterizing the viscosity in the nematic phase is a decreasing function of pressure! This unreasonable consequence may be avoided if a third component is added to Eq. (31).

$$\Delta^{\#} H_{\parallel} = W_{\eta} + q + W_{\mathrm{dd}} \tag{32}$$

where W_{dd} corresponds to the dipole–dipole association energy; it is estimated to be of the order of $(2 \div 3)kT$.[181,183] In this energy balance W_{dd} seems to be the only term that really decreases with increasing pressure due to the breaking of the dipole–dipole correlations between the molecules. A full microscopic understanding of the activation parameters is still lacking, however.

C. Smectic Phases

Only two LC substances with the smectic A_d phase[85,89] and one with smectic B phase[77] were studied under pressure dielectrically. In the case of the octylcyanobiphenyl (8CB) the temperature range of smectic phase shows minor increase with pressure, whereas for the second substance, octyloxycyanobiphenyl (8OCB), the smectic phase is limited; compare Figure 1. In the case of *trans*-4′-pentylbicyclohexyl-4-carbonitrile (5CCH) a metastable smectic B phase has been found at normal pressure.[64,69] Recently, we investigated the dielectric properties of this compound also under pressure.[77] For the smectic A phase the ratio of layer thickness to molecular length $d/l \approx 1.4$,[118] whereas the smectic B phase of 5CCH is close to a monolayer.[69] According to Figure 2 antiparallel dimers form the Sm A_d layers. Thus one could expect that this fact would influence the dielectric properties of the phase.

Figure 26 presents, as an example, the pressure dependence of the static permittivity and relaxation time for 8CB showing stepwise changes at the phase transitions.[85] The measurement runs were started from a well-oriented nematic phase, and the smectic phase was reached by slowly increasing the pressure at $T =$ constant or by decreasing the temperature at $p =$ constant. However, the permittivity always dropped to relatively

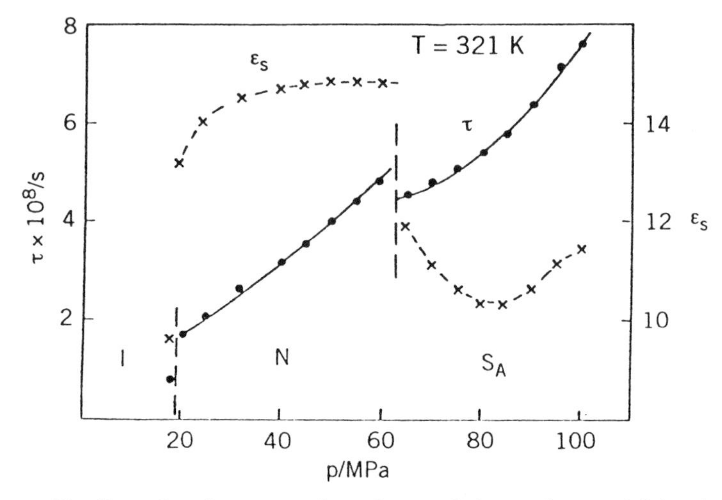

Figure 26. Example of pressure dependence of the static permittivity (×) and relaxation time (●) in the smectic, nematic, and isotropic phase of 8CB.[85] (Reprinted with kind permission from the editor of *Z. Naturforsch. A.*)

small values. In spite of the nonreproducibility of the values for ε' and ε'', the positions of maxima of losses on the frequency scale were very reproducible. That allowed us to calculate the relaxation times $\tau_{\|}$ also for the smectic phases of the studied LCs. Furthermore, Figure 26 shows the well-known feature of a continuous change of the relaxation time at the Sm A–nematic phase transition. In the case of 5CCH, however, a strong decrease of $\tau_{\|}$ by a factor of 210 is observed at the nematic–Sm B transition.[77] Similar strong steps have also been observed for other LCs with Sm B phases.[153]

Figure 27 presents the $\ln \tau_{\|}(p)$ dependencies obtained for 8CB. The activation enthalpies $\Delta^{\#} H_{\|}$ and activation volumes $\Delta^{\#} V_{\|}$ for the nematic and smectic phases are presented in Figure 28, both for 8CB and 8OCB. The activation enthalpies obtained for the smectic phases increase for 8CB but decrease for 8OCB with rising pressure, whereas it always decreases in the nematic phase. Also the plots for the activation volumes (Fig. 28b) exhibit similar trends. The opposite pressure dependencies observed for the nematic and smectic A phase of 8CB have been discussed in terms of a peculiar pressure influence on the molecular associations.[85] The existence of dimers in the smectic layers enlarges the free volume that facilitates the molecular reorientations; thus the activation enthalpy for the smectic phase is reduced, which has also been noted for other liquid crystals.[182]

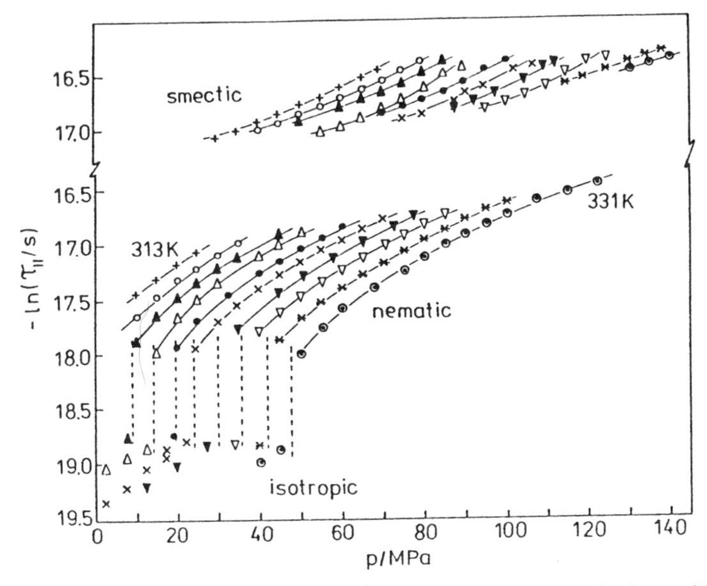

Figure 27. $\ln \tau_\parallel$ versus p plots for the smectic, nematic, and isotropic phase of 8CB at different temperatures; the temperatures for different isotherms vary in steps of 2 K.[85] (Reprinted with kind permission from the editor of *Z. Naturforsch. A*.)

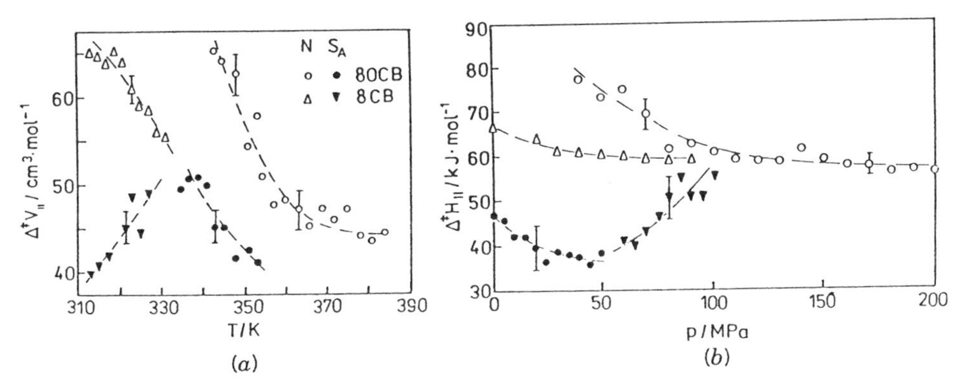

Figure 28. (a) Activation volume, $\Delta^{\#}V_\parallel$, as a function of temperature and (b) activation enthalpy, $\Delta^{\#}H_\parallel$, as a function of pressure for the nematic (open symbols) and smectic A (solid symbols) phase of 8CB[85] and 8OCB.[88]

If the pressure destroys the molecular associations, the excluded volume becomes smaller and the molecules will be more closely packed in the layers. Clearly, the layer thickness decreases with increasing pressure. both for 8CB[184] and 8OCB.[184,185a] However, this result does not explain the different pressure dependencies of $\Delta^{\#}H_{\|}$ in the smectic phases. In the case of 5CCH the activation enthalpy for the Sm B phase is practically independent of pressure.[77] It should be noted that the decrease of $\Delta^{\#}H_{\|}$ for the smectic phase of 8OCB concerns the low pressure region, whereas the increase of $\Delta^{\#}H_{\|}$ for 8CB was observed at higher pressures only. These pressure-limited phase behaviors of the smectic phases studied do not allow one to draw decisive conclusions at present. It would be desirable to study the $T(p)$ dependence of the dielectric relaxation in the smectic phase over a significantly larger pressure range that is still lacking.

D. Ferroelectric C* Phase

Ferroelectricity in LC materials was theoretically predicted by Meyer et al.[186] and found experimentally in DOBAMBC.[186,187] A proposal of practical use of various types of electro-optic devices with a high response speed (see, e.g., Ref. 188) has accelerated the synthesis of many ferroelectric liquid crystals (FLCs) and stimulated intensive theoretical[189–196] as well as experimental[189,194] investigations of pure substances or mixtures.[194–196]

Three main factors decide the appearance of the spontaneous polarization in an LC material: the molecule must possess a chiral group at a wing, its dipole moment must be deviated from the long molecular axis, and the substance must show the tilted smectic C phase (abbreviated as Sm C*). The substances under consideration have usually the following phase sequence:

$$\mathrm{Cr} \rightarrow \mathrm{Sm\ C}^* \rightarrow \mathrm{Sm\ A}^* \rightarrow \mathrm{N}^* \rightarrow \mathrm{Is}$$

In the higher temperature Sm A* phase of a chiral compound the molecular symmetry axes are normal to the smectic layers and the local point symmetry is $2/m$. This does not allow for spontaneous polarization to occur. In the Sm C* phase, on the other hand, the tilt of the long molecular axes with respect to the layer normal breaks the mirror plane symmetry and the spontaneous polarization is induced. Due to chirality of the molecules, the symmetry axis turns its direction on going from one smectic layer to another. This implies that the polarization shows a helicoidal order with a pitch distinctly larger than the layer thickness. The helix can be easily unwound by an external electric field normal to the helix. Some observation of the pressure dependence of the helix pitch is reported in Ref. 197.

However, the FLCs are improper ferroelectrics as the primary order parameter is the tilt angle and not the polarization.[189–196,198] Nevertheless, both these quantities are roughly proportional except the vicinity of the Sm C*–Sm A* phase transition,[189–196] which in most cases is of the second-order one.

Many relaxation processes influence the dielectric spectra of FLCs. Apart from the usual l.f. and h.f. modes characterizing the reorientations of molecules around their principal axes, the Sm C* phase shows at least two collective processes. One collective mode, the Goldstone mode (GM), is associated with the fluctuations of the azimuthal angle (the cone motion); it is observed in Sm C* phase at low frequencies and is not an activated process. The second mode, the soft mode (SM), is connected with the tilt fluctuations; its critical frequency falls in the kilohertz range, from ca. 50 to ca. 500 kHz. The soft mode shows a decrease of frequency in Sm A* phase on approaching the transition Sm A*–Sm C*, but it survives to the lower temperature phase. In special conditions (e.g., after applying an appropriate strength of the bias field[195]) yet another collective mode can be observed (domain mode).

All those modes can be observed via measurements of the perpendicular permittivity component ($\varepsilon_\perp$), so the influence of the l.f. molecular mode is eliminated, whereas the h.f. process can be observed at much higher frequencies than the previous ones. In most cases the collective modes are well described by a simple Debye-type relaxation model; however, they usually overlap, which makes the separation of particular components rather difficult.[195,196]

Main quantities characterizing the properties of FLCs (the values of the tilt angle θ, the spontaneous polarization P_s, the dielectric constant $\varepsilon_\perp^*$, the critical frequencies for particular collective modes) depend on intermolecular interactions caused by the chirality of molecules.[186,189–196,198] Thus the ferroelectricity of LCs must be sensitive to the intermolecular distance; it must then be pressure dependent. The pressure studies of FLCs have been undertaken in a few laboratories.[22,199–203]

Typical property of a ferroelectric system, the hysteresis loop, shows marked dependence on the pressure (Fig. 29). Observations of the hysteresis loops have shown that the spontaneous polarization P_s abruptly increases at the Sm A–Sm C* transition point when T decreases at $p = \text{const}$ (Fig. 30a) or when p increases at $T = \text{const}$[22,200,203] (Fig. 30b). That is accompanied by an increase of the tilt angle θ[200,201] (Fig. 30c). However, the coupling constant between P_s and θ is not purely

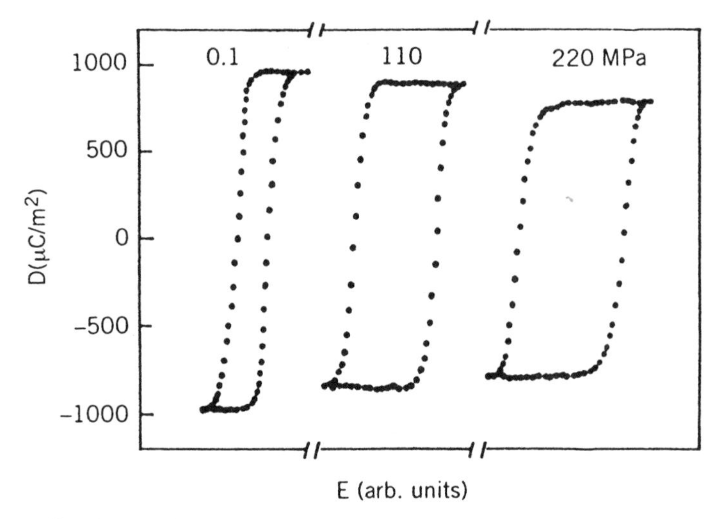

Figure 29. Hysteresis loops obtained for D8, [2*S*,3*S*]-4′-(2-chloro-3-methyl pentanoyloxy) phenyl *trans*-4″-*n*-octyloxy cinnamate, at $T_{NI} - 4$ K for different pressures.[203] (Reprinted with kind permission from the authors and the editor, Gordon and Breach Publishers, World Trade Center, 1000 Lausanne 30, Switzerland.)

linear,[22,200] which was predicted by the mean-field theories.[189–192] The magnitude of P_s as well as its rate of variation with temperature decreases with increasing pressure.[22,200]

The transverse dielectric constant $\varepsilon_\perp$ measured at low frequencies shows drastic change when the Sm A–Sm C* transition point is crossed

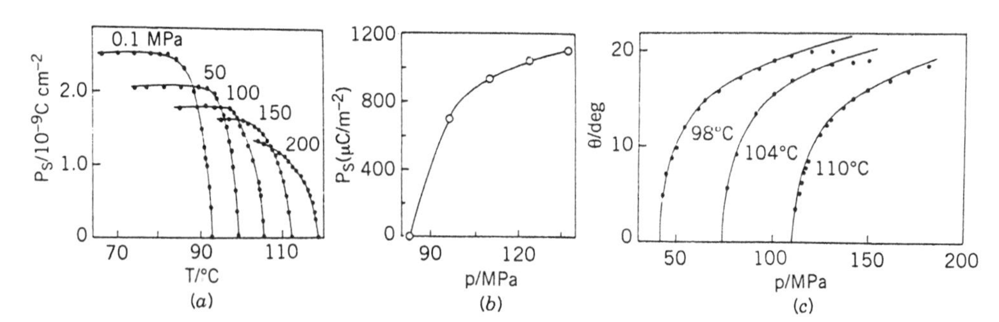

Figure 30. (a) Temperature dependence of the spontaneous polarization P_s of DOBAMBC at different pressures,[167] (b) Pressure dependence of P_s of D8 at $T = 87.7$°C.[203] (Reprinted with kind permission from the authors and the editor, Gordon and Breach Publishers, World Trade Center, 1000 Lausanne 30, Switzerland.) (c) Pressure dependence of the tilt angle θ at different temperatures in the Sm C* phase of DOBAMBC[167,199] [(a) and (c) reprinted with kind permission from the authors and the editor, Institute of Physics Publishing.]

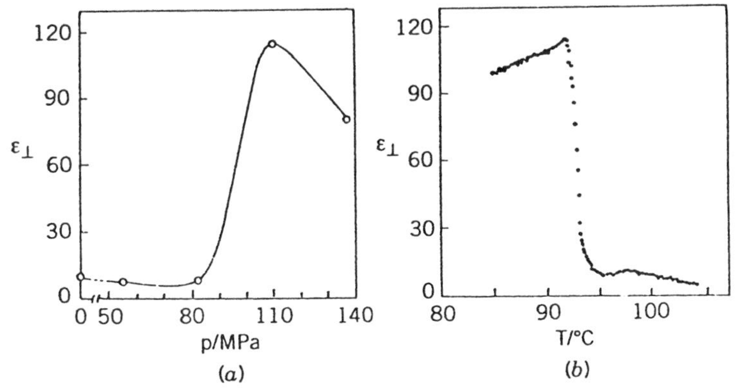

Figure 31. (a) Variation of $\varepsilon_\perp$ of D8 as a function of pressure at $T = 92°C$ and 50 Hz (the line is a guide to the eye). (b) Thermal variation of $\varepsilon_\perp$ at $p = 110$ MPa and $f = 50$ Hz.[22] (Reprinted with kind permission from the authors and the editor, World Scientific Pub. Co. Ptr. Ltd., Singapore.)

(Fig. 31).[22] Application of pressure drastically alters both the magnitude of $\varepsilon_\perp$ and its thermal and frequency variation; compare Figs. 32a, b, and c. At low frequency and low pressures the GM relaxation dominates (see Fig. 32a). With increasing pressure at constant frequency its contribution progressively decreases. The same can be said about the frequency dependence at a given pressure (cf., e.g., the values of $\varepsilon_\perp$ at 80 MPa in Figs. 32a, b, c). Because the Goldstone mode does not exist in the Sm A phase, the dielectric response of this phase allows for extraction of the information regarding the soft mode relaxation. It was found that the relaxation frequency of both these modes decreases on increasing the pressure.[22]

E. Polymeric Liquid Crystals

Polymeric liquid crystals have been extensively studied at ambient pressure because of their potential applications for electro-optical switching or storage effects.[204–206] However, high-pressure studies are very scarce.[23,207,208] At least five different relaxation processes are observed,[204–206,209] two of them above the glass transition (α and δ) and three below it (β, γ_1, γ_2).

The δ relaxation is observed in prominent form in some polyacrylates and polysiloxanes, if a strong dipole is attached to the mesogenic group parallel to its long axis. This relaxation is assigned to the rotation of the side group about the polymer backbone.[204,206–208,210,211,228] Its relaxation frequency falls in the kilohertz range, and the activation barriers hindering rotations are distinctly higher than for the low-frequency process in

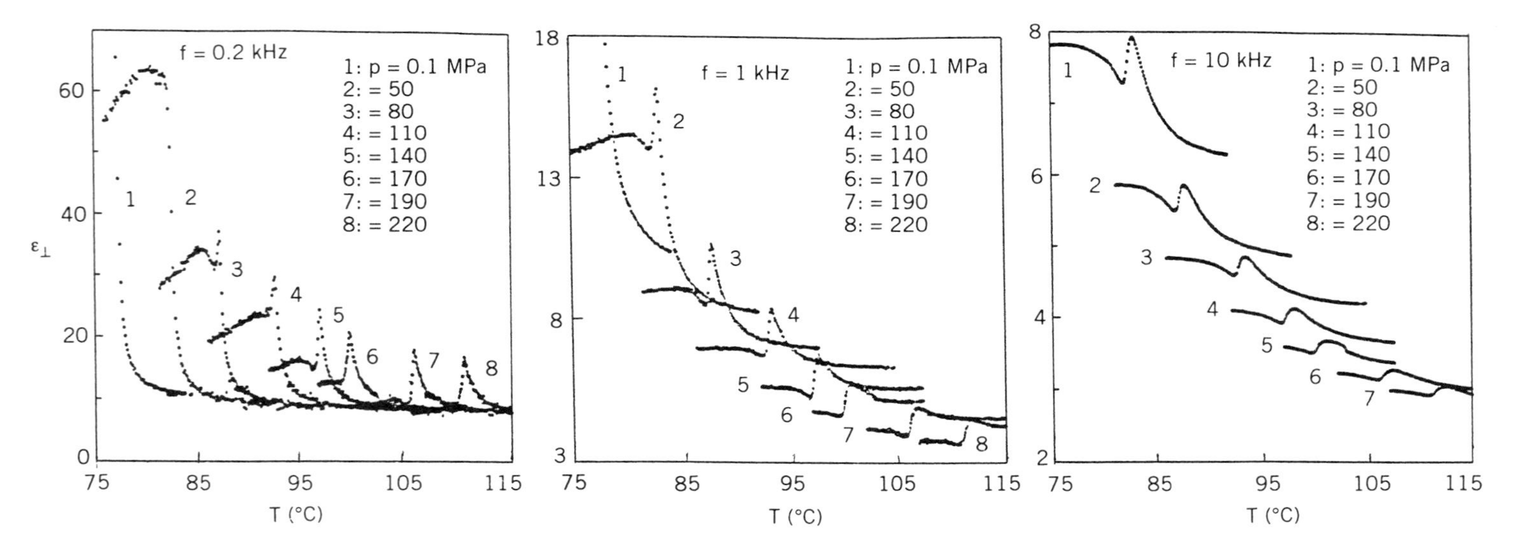

Figure 32. Temperature and pressure variation of $\varepsilon_{\perp}$ measured at different frequencies in the vicinity of the Sm A–Sm C* phase transition point of D8.[22] (Reprinted with kind permission from the authors and the editor, World Scientific Pub. Co. Ptr. Ltd., Singapore.)

the case of the low-molecular-weight mesogens.[206] The δ relaxation measured for parallel-oriented sample shows Debye behavior, although the proper permittivity values are hardly obtainable.

The α process is observed at higher frequencies (and correspondingly at temperatures below the δ relaxation). According to Attard,[210] it combines at least two other motions, the reorientations of the side group around the long axis and its stochastic precession around the director. Thus the α process must show a pronounced distribution. Both relaxation bands are seen when the samples are not aligned.[23,207,208] However, if the sample is cooled from the melt in the presence of a saturating ac electric field, a fully homeotropic alignment can be obtained.[23,210,211] In such a case the loss peak is dominated by the δ process. The β relaxation is connected with motions in the centre of the spacer, the γ relaxations with reorientations at the ends of the spacer group.[206,209] The pressure dependence of the γ relaxation was studied for non-liquid-crystalline polymer.[212]

Moura-Ramos and Williams[23,205] have studied LC siloxane polymer P/Si/8/CN in the pressure range 0.1–152 MPa, the temperature range 50–75°C, and the frequency range 10–10^5 Hz. The substance exhibits a smectic phase between 363 K (the clearing temperature) and 274 K (the glass transition temperature, T_g). Dielectric studies were performed with disk samples placed between metal electrodes of a three-terminal high-pressure cell. The results are presented as loss curves in terms of $G/\omega = \varepsilon'' C_a$, where G is the measured equivalent parallel conductivity of the sample, ε'' is the dielectric loss factor, and C_a the inter-electrode geometric capacitance.

Figure 33 shows typical spectra obtained at 70°C at different pressures. The sample was aligned homeotropically by cooling from the melt in the presence of an ac electric field. For this polymer sample the loss peak is dominated by the δ process. The α process appears only as a high-frequency shoulder. The curves have the same half-width of 1.46 being greater than 1.14 for a single relaxation time process as it happens in the case of low-molecular-weight smectic phases.[85] Thus, the attachment of the mesogenic group to a polymer chain means that its motions in the LC potential are strongly coupled to the main chain motions so that on decreasing the temperature toward glass temperature T_g (or increasing the pressure) the decrease in relaxation rate of the chain backbone also decreases the relaxation rate of the mesogenic group.[23]

As is seen in Figure 33 the peak height increases linearly with pressure (ca. 20%/kbar). Such a large effect could not be attributed to the increase of the order parameter nor to changes in sample dimensions and was interpreted as caused by a change in $\langle \mu^2 \rangle$ due to a change of

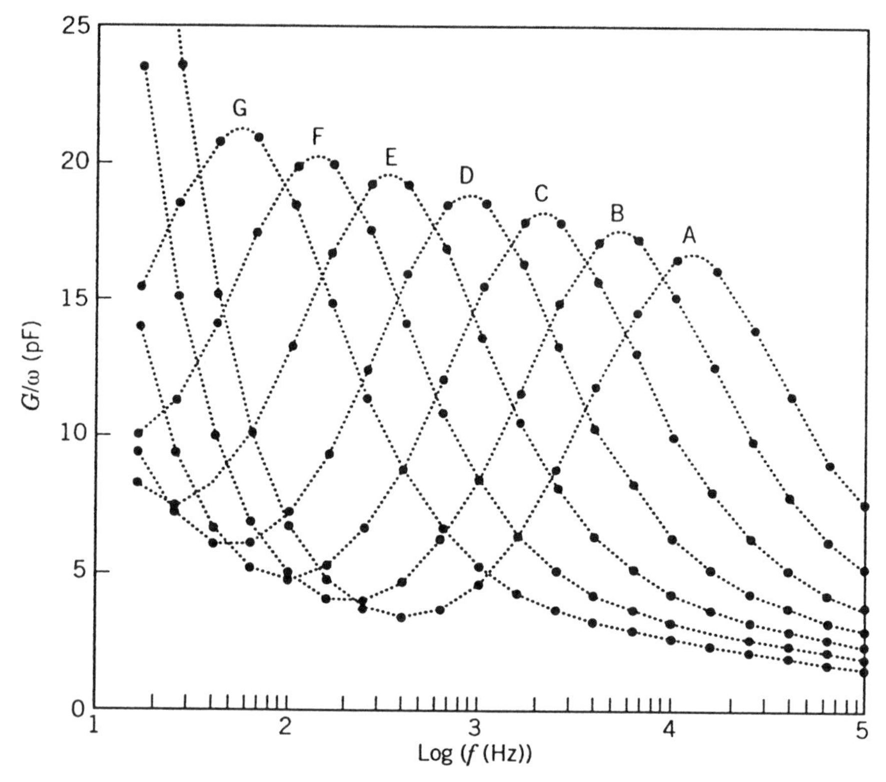

Figure 33. $\varepsilon'' C_a$ against log[f(Hz)] for homeotropically aligned sample at 70°C and different applied pressures.[23] Curves A–G correspond to 0.1, 25, 51, 76, 102, 128, and 152 MPa, respectively. (Reprinted from *Polymer* with kind permission from the authors and the editor, Butterworth–Heinemann journals, Elsevier Science Ltd, The Boulevard, Langford Lane, Kidington OX5 1GB, UK.)

conformation with pressure, which leads to an enhancement of the overall dipole moment.

The activation plots, presented in Figure 34 in the form of log f_m vs. T^{-1}, are markedly curved when approaching the apparent glass transition. The average slopes of the plots increase with pressure, indicating that the apparent activation energy ΔH_δ for the δ process increases with increasing pressure. At the same time the plots log f_m vs. p (Fig. 35) are downward at the higher pressures, and their average slopes decrease with rising temperature. That means that the activation volume ΔV_δ [cf. Eq. (25)] decreases with increasing temperature. Although pVT data are missing for the evaluation of the isochoric activation energy, the authors

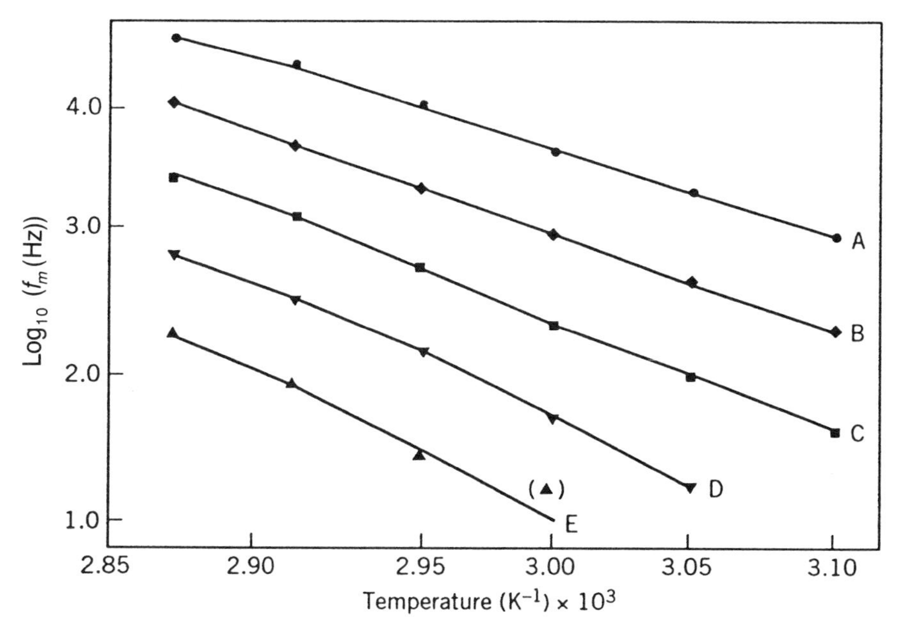

Figure 34. Log[f_m(Hz)] against reciprocal of temperature (K^{-1}) for the homeotropically aligned sample.[23] *A–E* correspond to 0.1, 34.5, 69, 104, and 138 MPa applied pressure, respectively. (Reprinted from *Polymer* with kind permission from the authors and the editor, Butterworth–Heinemann journals, Elsevier Science Ltd, The Boulevard, Langford Lane, Kidington OX5 1GB, UK.)

estimate the ratio $\Delta^{\#} U_{\delta} / \Delta^{\#} H_{\delta} \approx 0.7$–$0.8$, on using $(\partial p/\partial T)_v$ data for similar polymers.

It is interesting to note that an estimate of $(\partial p/\partial T)_\tau = \Delta^{\#} H/(T\,\Delta^{\#} V)$ gives constant slopes of 38.5 bar/K, which means that the changes in the activation enthalpies and activation volumes (multiplied by T) compensate for the δ process (the same was also observed for other LC polymers as well[207]). The authors present a detailed discussion in terms of current relaxation theories.[142–145] They emphasize the usefulness of high-pressure investigations that allows one to notice the fundamentally different mechanism for the α relaxation in amorphous polymers and the δ relaxation in LC polymers.

Another interesting high-pressure study on LC polymers was performed by McMullin et al.[208] They investigated an LC polymer sample that has been synthesized by Zentel et al.,[209,213] where 5% of the side groups have been substituted by an alkanol group. The polymer compound has a transition at 45°C from the glass phase to a smectic A phase

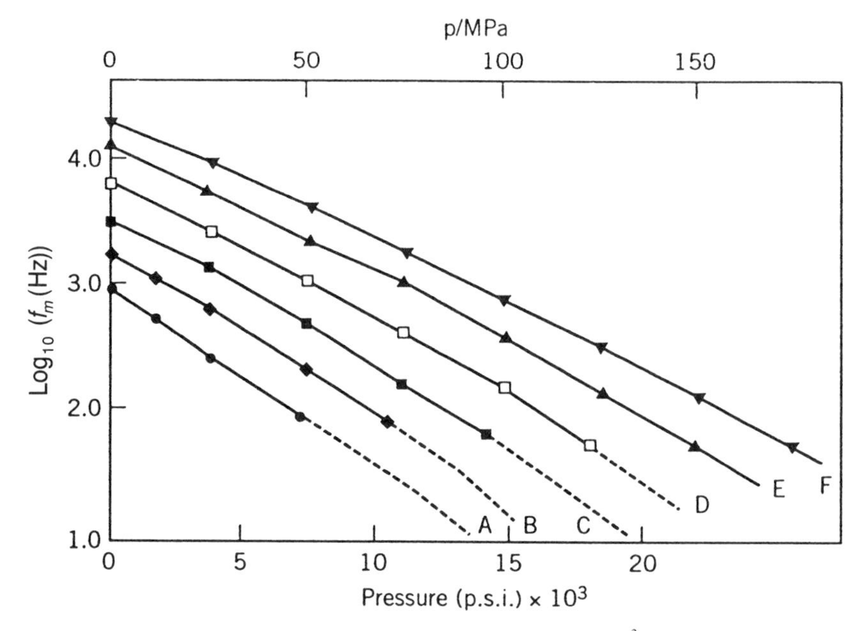

Figure 35. Log[f_m(Hz)] against pressure (1 psi = 6.895×10^{-3} MPa) for the homeotropically aligned sample.[23] *A–F* correspond to 50, 55, 60, 65, 70, and 75°C, respectively. (Reprinted from *Polymer* with kind permission from the authors and the editor, Butterworth–Heinemann journals, Elsevier Science Ltd, The Boulevard, Langford Lane, Kidington OX5 1GB, UK.)

that transforms at 102°C to the nematic state with a clearing temperature of 106°C. The measurements were carried out up to temperatures of 130°C and pressures of 300 MPa. The effective dipole moment calculated from the Onsager equation is largest for the planar-oriented sample and distinctly increases with rising pressure up to 150 MPa.

At lower temperatures the β process can be observed in the glass phase. The high-frequency permittivity above the β relaxation is estimated to be 3.48 that distinctly exceeds n^2, indicating probably the γ process.[209] The activation enthalpy for the β relaxation is 38 kJ/mol for nonaligned sample, compared with 54 kJ/mol for the homopolymer. The planar aligned sample yields 71 kJ/mol.

For the α process the halfwidths of the ε'' curves are of the order of 4 decades, which is distinctly larger than for the β process (2.6–3 decades). This big halfwidth implies a strong composition of the α-relaxation process. The activation enthalpies $\Delta^{\#}H_{\alpha}$ are 148 kJ/mol for the homeotropic and 236 kJ/mol for the planar sample. In the latter case the pressure dependence resulted in an activation volume of 100 cm^3/mol,

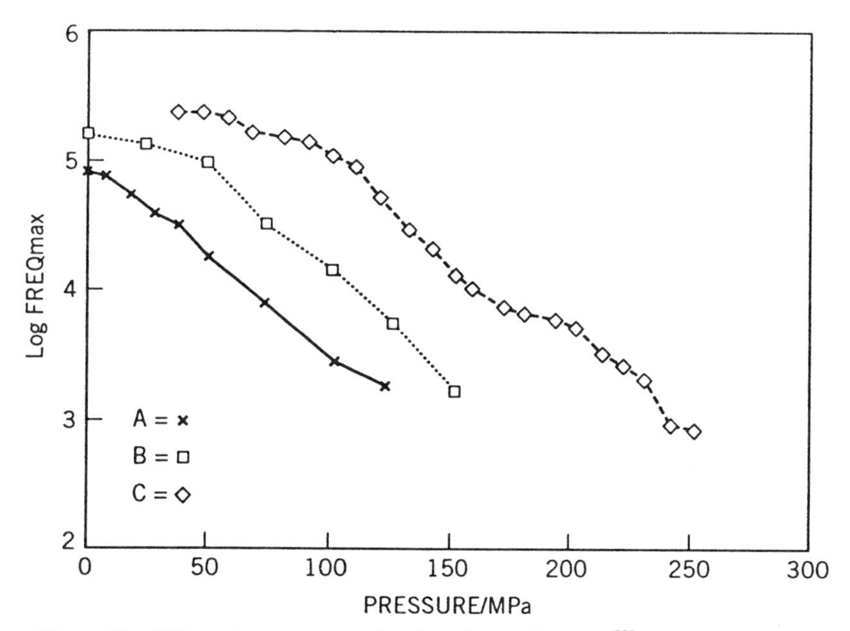

Figure 36. Effect of pressure on f_{max} for planar alignment.[208] Curves A, B and C correspond to temperatures of 90, 98, and 110°C, respectively. (Reprinted with kind permission from the authors and the editor of *Liq. Crystals*.)

slightly decreasing with increasing temperature, see Figures 36 and 37 (8 and 9 from Ref. 208). This corresponds to 25% of the repeating unit. It is interesting to note that activation volumes determined for the low-frequency relaxation in alkylcyanobiphenyls correspond also to about 25% of the molar volume (cf. Fig. 18). For a nonaligned sample the log f_m vs. p plot shows an increase of the slope with rising pressure, which means an increase of the activation volume with p. Also the halfwidth of the loss curve increases with p.

The β relaxation for the nonaligned sample yielded smaller activation volumes of 16.5–19.5 cm^3/mol. The presence of remarkable ionic conductivity made it difficult to discern the δ relaxation, and therefore activation parameters are not reported for this process. The origin of the weak δ process is probably different from the δ relaxation observed in other polymers. Heinrich and Stoll[168] and Gnoth[207] investigated two unaligned samples of polyacrylate, P/H/6/CN, and some mixtures with L7CN, respectively. They found for the α-process activation volumes ranging from 49% (80°C) to 31% (120°C) of the repeating molar volume. The low-frequency δ process behaved very similarly.

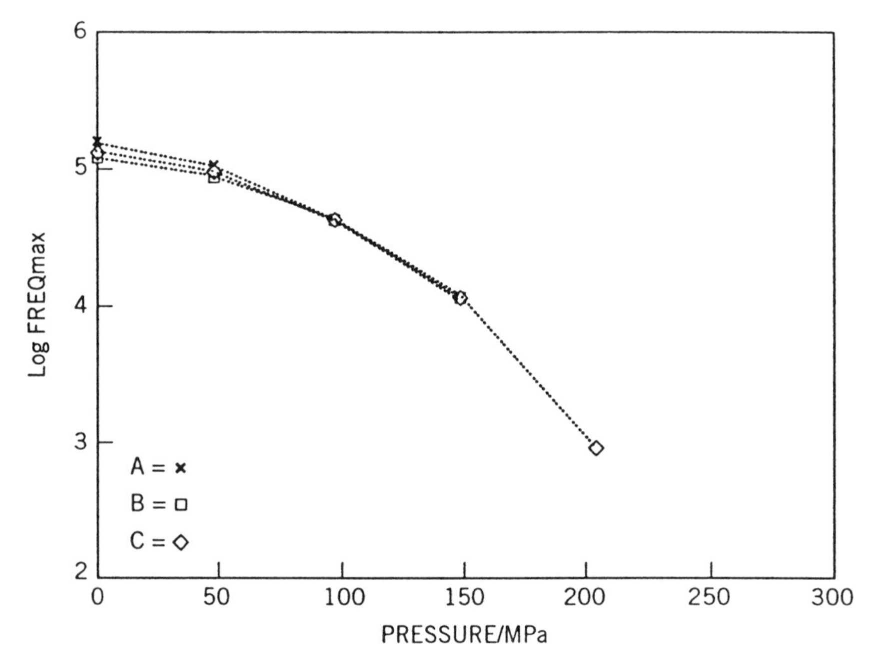

Figure 37. Effect of pressure on f_{max} for nonaligned sample.[208] Curves A, B and C correspond to temperatures of 100, 105 and 110°C, respectively. (Reprinted with kind permission from the authors and the editor of *Liq. Crystals*.)

F. Nonlinear Dielectric Effect Studies

The nonlinear dielectric effect (NDE) is an important tool for studying the properties of the isotropic-nematic transition.[131,214,215,229,230] The NDE is the measure of changes in electric permittivity due to the application of a strong electric field E: $\Delta\varepsilon^E = \varepsilon^E - \varepsilon$, where ε^E and ε denote the electric permittivity in the strong and weak electric fields, respectively. To a large extent the NDE is an analogue of the Kerr effect.

The NDE method cannot be applied to studies of the nematic phase because the strong electric field causes hydrodynamic flows that destroy the nematic order. This is not the case for the isotropic phase, if the conductivity is low enough.[131] It is well known that some nematiclike short-range order survives to the isotropic phase that influence many properties in the neighborhood of T_{NI}. Strong influence of the short-range orientational order on the phase transition properties is discussed in Ref. 109. Especially the Kerr effect, the Cotton–Mouton effect, the intensity of light scattering as well as the NDE are well suited for confirmation of the existence of strong orientational correlations between the molecules.[35,38,39,131,214,215] This enables one to determine certain molecular characteristics of the nematic phase in the isotropic phase. The behavior of these quantities may be described by the Landau–de Gennes model,[35] which predicts the same, mean-field type of temperature behavior for the parameters measured by the mentioned methods. In case of the NDE one has for $p = \text{const} = 1$ atm

$$\text{NDE} = \frac{\Delta\varepsilon^E}{E^2} \sim \frac{A}{(T - T^*)^{\psi}} \qquad T^* = T_{\text{NI}} - \Delta T \quad T > T_{\text{NI}} \tag{33}$$

where T^* is the temperature of the hypothetical, continuous phase transition, T_{NI} and ΔT are clearing temperature and the discontinuity of the isotropic-nematic transition, respectively, A is the amplitude, and ψ is the critical exponent.

The relation (33) was tested for a few LCs.[131,214,215] Figure 38 shows that it is fulfilled for MBBA and EBBA with the classical exponent $\psi = 1$. The ratio of the amplitudes corresponds roughly to the ratio of dielectric

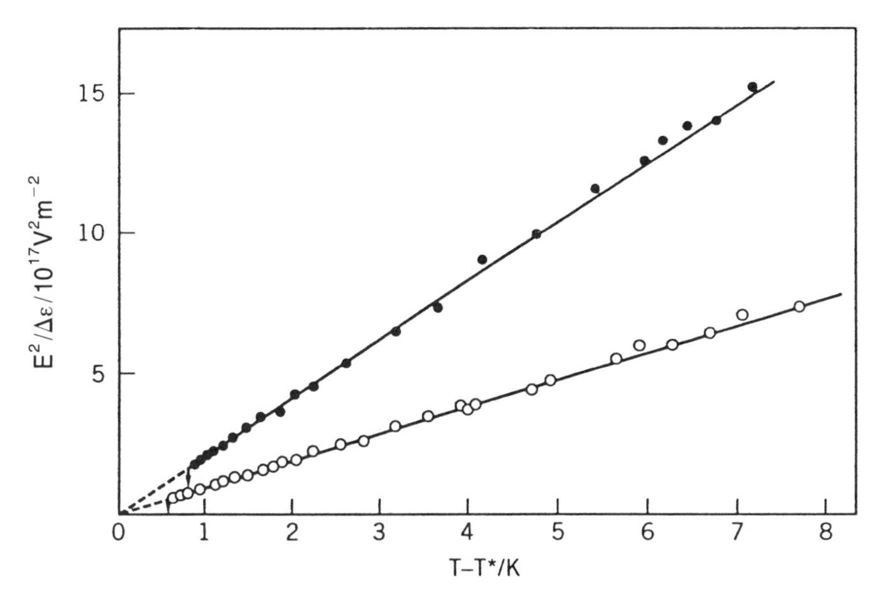

Figure 38. Inverse of the NDE versus temperature for MBBA (open circles) and EBBA (solid circles).[214] Arrows indicate the clearing point. (Reprinted with kind permission from the authors and the editor of *Liq. Crystals*.)

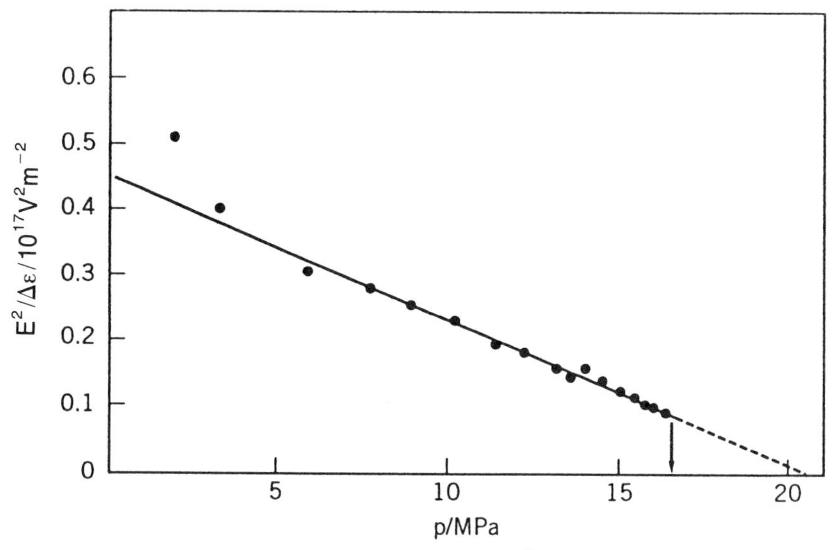

Figure 39. Isothermal dependence of the $(\mathrm{NDE})^{-1}$ versus hydrostatic pressure on approaching the clearing point for MBBA.[214] The arrow indicates the clearing point. Parameters: $T = 52°\mathrm{C}$, $p_{\mathrm{NI}} = 16.5\,\mathrm{MPa}$, $\Delta p = 5\,\mathrm{MPa}$. (Reprinted with kind permission from the authors and the editor of *Liq. Crystals*.)

anisotropies $\Delta\varepsilon$ multiplied by the ratio of the polarizabilities of both compounds, in accord with the theoretical prediction.[214]

The Landau–de Gennes model may also be considered for the isothermal pressure case. It leads to a relation analogous to (33)

$$\mathrm{NDE} = \frac{\Delta\varepsilon^E}{E^2} = \frac{A'}{(p - p^*)} \qquad p > p_{\mathrm{NI}},\ p^* = p_{\mathrm{NI}} - \Delta p,\ T = \mathrm{const} \quad (34)$$

with analogous meaning of the symbols.

In Figure 39 measurements are shown of the NDE versus pressure at constant temperature in the isotropic phase of MBBA. Clearly, the mean-field formula (34) is confirmed too.

The results emphasize the equivalence of the temperature and pressure variables,[216] which may be interpreted as a consequence of the postulate of isomorphism of critical phenomena.

V. CONCLUDING REMARKS

The presentation of the dielectric results of liquid crystals has shown that the application of high pressure elucidates more features of the intermolecular interaction and the dynamic behavior in the mesophases. When the equation of state and the pressure dependence of the order parameter S are known, current theories for the description of nematic LC phases can successfully be tested. In particular the different influences of temperature and volume can be separated and analyzed, if high-pressure volume data are available.[66,71,104–108,217–219] Moreover many liquid crystals exhibit peculiar high-pressure phase behaviors, such as reentrant and multicritical phenomena or pressure-induced phase transitions, metastable and pressure-limited phases, which are interesting subjects of their own.[19–22,53,54,80,185a,200,220–224] It is worthwhile to mention recent papers by Wolinski et al.,[225] who describe a fiberoptic high-pressure sensor with cholesteric liquid crystals.

We highlight some aspects of the high-pressure studies on the dielectric properties of liquid crystals reviewed in this chapter.

1. Pressure influences the dielectric permittivity of liquid crystalline phases considerably. For some LCs that form antiparallel dimers, the observation of a small maximum in $\varepsilon(T)$ at 1 atm above the clearing temperature is well established. In recent high-pressure investigations a similar effect was found, that is, a maximum in $\varepsilon(p)$ at $T = \text{const}$ in the isotropic phase in the neighbourhood of the NI transition.

2. The pressure dependence of the dielectric anisotropy cannot be solely explained by the change of the order parameter, as was predicted by the Maier–Meier equation. Moreover the variation of $\varepsilon_{\parallel}$ in a homologous series (e.g., 5PCH compared with 7PCH) is stronger than simple equations for the molar polarization predict. This is apparently due to the higher flexibility of the longer chain length that modifies the intermolecular interaction appreciably.

3. The longitudinal relaxation time $\tau_{\parallel}$ is strongly pressure dependent, yielding large activation volumes, $\Delta^{\#}V_{\parallel}$ that are an order of magnitude larger than those observed for other types of organic compounds. However, it should be noted that $\tau_{\parallel}(p)$ is not a purely exponential function, especially near the NI phase transition.

4. The activation enthalpy, $\Delta^{\#}H_{\parallel}$, and the activation energy, $\Delta^{\#}U_{\parallel}$, exhibit a peculiar pressure and density dependence, in particular for the nCB series that is opposite to the "normal" pressure dependence of the activation parameters. This behavior can be explained with specific dipole–dipole correlations between antiparallel associates, assuming that increasing pressure destroys voluminous dimers.

5. In spite of the variations with temperature and/or pressure of the activation parameters, the relation between them is in agreement with the predictions of the thermodynamics.

6. A "natural" extrapolation of $\ln \tau_{is}$ vs. p from the isotropic to the

nematic phase can serve for the calculation of the retardation factor $g_{\parallel}$ and then the nematic potential q as functions of T and p. That enables one to discuss the relation between q and S, thus to test the predictions of the mean-field theories.

7. The activation parameters found for the smectic A and B phases are significantly lower than those characterizing the nematic phase. This strange effect is similar to a behavior sometimes observed in plastic crystals, where the activation enthalpy for the reorientation in the ODIC phase is smaller, despite the higher density, than in the liquid phase. This was explained with the higher order in the solid state.[226]

8. Knowledge of the equation of state allows one to evaluate the exponent γ and thus to test theories for the intermolecular potential.

However, the number of liquid crystals that have been studied under pressure is very limited. In most cases neither the equation of state nor the pressure dependence of the order parameter is known. Only the mean-field theory of Maier and Saupe was extended to explain the dielectric properties of liquid crystalline phases. However, a recent approach by Photinos et al.[227] analyzed the nematic reentrance and phase stability based on the variational cluster method. The lack of a full theoretical description as well as insufficient experimental data should stimulate further high-pressure investigations in this field.

ACKNOWLEDGMENTS

Grants from the Deutsche Forschungsgemeinschaft (Wu 97/8-1 & 2) and the Fonds der Chemischen Industrie are gratefully acknowledged. The authors appreciate also financial support in the framework of cooperation between the Ruhr-Universität Bochum (Germany) and the Jagellonian University, Kraków (Poland). Moreover, the authors are indebted to many co-workers, especially to Dr. Thomas Brückert, who carried out most of the recent high-pressure studies on liquid crystals. We thank also Mrs. Galina Domosławska who prepared the majority of the presented drawings.

REFERENCES

1. K. Gesi, *Phase Transitions* **40**, 187 (1992); A. Gordon and S. Dorfman, *Phys. Rev.* **B50**, 13132 (1994).
2. G. A. Samara, *J. Phys. Chem.* **94**, 1127 (1990); *Physica B* **139–140**, 3 (1986).
3. S. C. Schmidt, J. W. Shaner, G. A. Samara, and M. Ross, eds., *High Pressure Science and Technology*, AIP Conference Proceedings 309 of the XIVth AIRAPT (held in Colorado, USA) New York, 1993.
4. J. S. Tse and D. D. Klug, *Phys. Rev. Lett.* **67**, 3559 (1991); *Science* **255**, 1559 (1992).
5. J. Jurczak, *High Press. Res.* **1**, 99 (1989); J. Troe, in *High Pressure Science and Technology*, p. 625, W. B. Holzapfel and P. G. Johannssen, eds., Conference Proceedings of the XIIth AIRAPT (held in Paderborn, Germany) Gordon & Breach,

New York, London, 1989, p. 625; U. Frey and A. E. Merbach, in *High Pressure Science and Technology*, W. B. Holzapfel and P. G. Johannssen, eds., Gordon & Breach, New York, London, 1989, p. 628.

6. E. Whalley, in *Advances in High Pressure Research*, Vol. 1, R. S. Bradley, ed., Academic Press, New York, 1966, p. 143.
7. E. U. Franck, *Pure & Appl. Chem.* **59**, 25 (1987); in *Organic Liquids*, E. Lippert and S. Bratos, eds., Wiley, New York, 1978, p. 181; W. G. S. Scaife, *J. Mol. Liq.* **49**, 155 (1991).
8. J. Shanker and S. Dixit, *Phys. Stat. Sol. A* **123**, 17 (1991).
9. G. A. Samara, *Phys. Rev. B* **27**, 3494 (1983).
10. G. A. Samara, *Ferroelectrics* **117**, 347 (1991).
11. R. N. Hampton, I. T. Collier, H. A. A. Sidek, and G. A. Saunders, *J. Non-Cryst. Solids* **110**, 213 (1989); I. T. Collier, R. N. Hampton, and G. A. Saunders, *High Temp. High Press.* **21**, 333 (1989).
12. E. Whalley, S. J. Jones, and L. W. Gold, *Physics and Chemistry of Ice*, Royal Soc. of Canada, Ottawa, 1973.
13. K. Matsushige, *Phase Trans.* **18**, 247 (1989).
14. A. Würflinger, *Int. Rev. Phys. Chem.* **12**, 89 (1993).
15. A. S. Reshamwala and R. Shashidhar, *J. Phys. E* **10**, 183 (1977).
16. C. Schmidt, M. Rittmeier-Kettner, H. Becker, J. Ellert, R. Krombach, and G. M. Schneider, *Themochim. Acta* **238**, 321 (1994).
17. M. Hartmann, M. Jenau, A. Würflinger, M. Godlewska, and S. Urban, *Z. Phys. Chem.* **177**, 195 (1992).
18. C. Rein and D. Demus, *Liq. Cryst.* **15**, 193 (1993); *Liq. Cryst.* **16**, 323 (1994).
19. S. Chandrasekhar and R. Shashidhar, in *Advances in Liquid Crystals*, Vol. 4, G. H. Brown, ed., Academic Press, London, 1979, p. 84.
20. R. Shashidhar, in *Phase Transitions in Liquid Crystals*, S. Martellucci and A. N. Chester, eds., Plenum Press, New York, London, 1992, p. 245.
21. A. K. Singh, ed., *High Pressure Science and Technology*, Proceedings of the XIIIth AIRAPT (held in Bangalore, India), 1991, pp. 514–531.
22. S. K. Prasad, S. M. Khened, S. Chandrasekhar, in *Modern Topics in Liquid Crystals*, A. Buka, ed., World Scientific, Singapore, 1993, p. 285; S. M. Khened, S. K. Prasad, V. N. Raja, S. Chandrasekhar, and B. Shivkumar, *Ferroelectrics* **121**, 307 (1991).
23. J. J. Moura-Ramos and G. Williams, *Polymer* **32**, 909 (1991).
24. F. Reinitzer, *Monatsh. Chem.* **9**, 421 (1888).
25. H. Ringsdorf, B. Schlarb, and J. Venzmer, *Angew. Chemie* **100**, 117 (1988); G. J. Vroege and H. N. W. Lekkerkorker, *Rep. Prog. Phys.* **55**, 1241 (1992).
26. G. Friedel, *Ann. Phys.* **18**, 273 (1922).
27. A. Adamczyk, in *Liquid and Solid State Crystals*, J. Zmija, ed., SPIE Proceedings, Vol. 1845, 1992, p. 2.
28. S. Chandrasekhar, *Mol. Cryst. Liq. Cryst.* **63**, 171 (1981).
29. C. Destrade, H. Foucher, H. Gasparoux, Nguyen Huu Tinh, A. M. Levelut, and J. Malthete, *Mol. Cryst. Liq. Cryst.* **106**, 121 (1984).
30. W. H. de Jeu, in *Phase Transitions in Liquid Crystals*, S. Martellucci and A. N. Chester, eds., Plenum Press, New York, London, 1992, p. 21.

31. S. Ye. Yakovenko, R. R. Ignatovich, S. Müller, and J. Pelzl, *Liq. Cryst.* **10**, 821 (1991); V. S. Ratchkevitch and S. Ye. Yakovenko, *Liq. Cryst.* **15**, 591 (1993).
32. H. Stegemeyer, Th. Blümel, K. Hiltrop, H. Onusseit, and F. Porsch, *Liq. Cryst.* **1**, 3 (1986).
33. A. M. Biradar, S. Wrobel, and W. Haase, *Phys. Rev. A* **39**, 2693 (1989); *Ferroelectrics* **99**, 149 (1989).
34. H. Finkelmann, *Angew. Chem.* **99**, 840 (1987).
35. P. G. de Gennes, *The Physics of Liquid Crystals*, Clarendon, Oxford, 1974.
36. G. R. Luckhurst and G. W. Gray, eds., *The Molecular Physics of Liquid Crystals*, Academic Press, New York, 1979.
37. W. H. de Jeu; *Physical Properties of Liquid Crystalline Materials*, Gordon & Breach, London, 1980.
38. G. Vertogen and W. H. de Jeu, *Thermotropic Liquid Crystals, Fundamentals*, Springer, Berlin, 1988.
39. S. Chandrasekhar, *Liquid Crystals*, 2nd ed., Cambridge University Press, Cambridge, 1992.
40. S. Martellucci and A. N. Chester, eds., *Phase Transitions in Liquid Crystals*, Plenum Press, New York, London, 1992.
41. A. Buka, ed., *Modern Topics in Liquid Crystals*, World Scientific, Singapore, 1993.
42. H. Kelker and R. Hatz, *Handbook of Liquid Crystals*, Verlag Chemie, Weinheim, 1980.
43. S. Chandrasekhar, *Mol. Cryst. Liq. Cryst.* **124**, 1 (1985).
44. J. Simon, J. J. André, and A. Skoulios, *New J. Chem.* **10**, 295 (1986); H. Gasparoux, F. Hardouin, C. Destrade, and H. T. Nguyen, *New J. Chem.* **16**, 295 (1992).
45. 100 Years Liquid Crystals, *Mol. Cryst. Liq. Cryst.* **165**, 1–572 (1988).
46. R. Steinsträßer and H. Krüger, *Flüssigkristalle*, Ullmanns Encyklop. Vol. 4 **11**, 657 (1976); R. Steinsträßer and L. Pohl, *Angew. Chem.* **85**, 706 (1973).
47. 15th International Liquid Crystal Conference, Budapest 1994, *Mol. Cryst. Liq. Cryst.* **260–265** (1995).
48. Bunsen-meeting, Leipzig 1993, *Neue Eigenschaften und Anwendungen von Flüssigkristallen*, Ber. Bunsenges. Phys. Chem. **97**(10), 1169–1410 (1993).
49. *Liquid and Solid State Crystals*, SPIE Proceedings, Vol. 2372, Washington, 1995.
50. D. Demus, *Mol. Cryst. Liq. Cryst.* **165**, 45 (1988).
51. H. Kelker, B. Scheurle, R. Hatz, and W. Bartsch, *Angew. Chem.* **82**, 984 (1970).
52. J. Thiem and V. Vill, (1992–1994) Landolt-Börnstein, New Series, Group IV, Macroscopic and technical properties of matter, Vol. 7, *Liquid Crystals*, Springer, Berlin-Heidelberg.
53. P. E. Cladis, R. K. Bogardus, W. B. Daniels, and G. N. Taylor, *Phys. Rev. Lett.* **39**, 720 (1977); P. E. Cladis, R. K. Bogardus, and D. Aadsen, *Phys. Rev. A* **18**, 2292 (1978).
54. J. Herrmann, H. D. Kleinhans, and G. M. Schneider, *J. Chim. Phys.* **80**, 111 (1983).
55. R. R. Netz and A. N. Berker, in *Phase Transitions in Liquid Crystals*, S. Martellucci and A. N. Chester, eds., Plenum Press, New York, London, 1992, p. 109.
56. D. Demus and H. Zaschke, *Flüssige Kristalle in Tabellen II*, VEB, Leipzig, 1984.

57. W. Spratte and G. M. Schneider, *Ber. Bunsenges. Phys. Chem.* **80**, 886 (1976).
58. L. C. Chow, *J. Phys. Chem.* **73**, 1127 (1969).
59. R. Chang, F. B. Jones, Jr., and J. J. Ratto, *Mol. Cryst. Liq. Cryst.* **33**, 13 (1976).
60. H. J. Coles and C. Strazielle, *Mol. Cryst. Liq. Cryst.* **55**, 237 (1979).
61. Product information, Merck, Germany.
62. R. Dąbrowski, J. Dziaduszek, and T. Szczuciński, *Mol. Cryst. Liq. Cryst. Lett.* **102**, 155 (1984); *Mol. Cryst. Liq. Cryst.* **124**, 241 (1985).
63. U. Finkenzeller, T. Geelhaar, G. Weber, and L. Pohl, *Liq. Cryst.* **5**, 313 (1989).
64. R. Eidenschink, D. Erdmann, J. Krause, and L. Pohl, *Angew. Chem.* **89**, 103 (1977); *Angew. Chem.* **90**, 133 (1978).
65. S. Sen, K. Kali, S. K. Roy, and S. B. Roy, *Mol. Cryst. Liq. Cryst.* **126**, 269 (1985).
66. E. Kuss, *Mol. Cryst. Liq. Cryst.* **76**, 199 (1981).
67. W. Haase and R. Pendzialek, *Mol. Cryst. Liq. Cryst.* **97**, 209 (1983).
68. K. Czupryński, *Thesis*, Military Technical University, No. 2301/95, Warszawa, Poland (1995).
69. G. J. Brownsey and A. J. Leadbetter, *J. Phys. Lett.* **42**, 135 (1981).
70. R. Shashidhar and G. Venkatesh, *J. Phys. Coll.* **40**, C3, 396 (1979).
71. R. G. Horn and T. E. Faber, *Proc. R. Soc. A* **368**, 199 (1979); R. G. Horn, *J. Phys. Paris* **39**, 105, 167 (1978).
72. G. P. Wallis and S. K. Roy, *J. Phys. Paris* **41**, 1165 (1980).
73. M. Hugo, *Diplom Thesis*, University of Bochum, Bochum, Germany (1991).
74. H. G. Kreul, *Doctoral Thesis*, University of Bochum, Bochum, Germany (1991).
75. M. Hartmann, *Doctoral Thesis*, University of Bochum, Bochum, Germany (1992).
76. T. Brückert, *Diplom Thesis*, University of Bochum, Bochum, Germany (1993).
77. T. Brückert, *Doctoral Thesis*, University of Bochum, Bochum, Germany (1996); T. Brückert and A. Würflinger, *Z. Naturf.* **51a**, 306 (1996).
78. D. Büsing, *Doctoral Thesis*, University of Bochum, in preparation.
79. W. Haase and H. Paulus, *Mol. Cryst. Liq. Cryst.* **100**, 111 (1983).
80. G. M. Schneider, A. Bartelt, J. Friedrich, H. Reisig, and A. Rothert, *Physica* **139 & 140B**, 616 (1986); A. Bartelt and G. M. Schneider: *Mol. Cryst. Liq. Cryst.* **173**, 75 (1989).
81. H. G. Kreul, S. Urban, and A. Würflinger, *Phys. Rev.* A **45**, 8624 (1992).
82. S. Urban, H. G. Kreul, and A. Würflinger, *Liq. Cryst.* **12**, 921 (1992).
83. S. Urban, T. Brückert, and A. Würflinger, *Liq. Cryst.* **15**, 919 (1993).
84. T. Brückert, A. Würflinger, and S. Urban, *Ber. Bunsenges. Phys. Chem.* **97**, 1209 (1993).
85. S. Urban, T. Brückert, and A. Würflinger, *Z. Naturforsch.* **49a**, 552 (1994).
86. S. Urban and A. Würflinger, *Liq. Cryst.* **12**, 931 (1992).
87. T. Brückert, D. Büsing, A. Würflinger, and S. Urban, *Mol. Cryst. Liq. Cryst.* **262**, 209 (1995).
88. T. Brückert, D. Büsing, A. Würflinger, and S. Urban, *Z. Naturforsch.* A **50**, 977
89. T. Brückert, A. Würflinger, and S. Urban, *Ber. Bunsenges. Phys. Chem.* **100**, 1133 (1996).
90. W. Maier and A. Saupe, *Z. Naturforsch. A* **14**, 982 (1959); **15**, 287 (1960).

91. A. P. J. Emerson, R. Hashim, and G. R. Luckhurst, *Mol. Phys.* **76**, 241 (1992).
92. G. R. Luckhurst, in *The Molecular Physics of Liquid Crystals*, G. R. Luckhurst and G. W. Gray, eds., Academic Press, New York, 1979, Chap. 4.
93. G. R. Luckhurst and C. Zanonni, *Nature (London)* **267**, 412 (1977).
94. R. L. Humphries, P. G. James, and G. R. Luckhurst, *JCS Faraday Trans.* 2 **68**, 1031 (1972).
95. R. M. Ernst, *Phys. Rev. B* **46**, 13679 (1992).
96. R. Tao, P. Sheng, and Z. F. Lin, *Phys. Rev. Lett.* **70**, 1271 (1993).
97. A. Chrzanowska and K. Sokalski, *Z. Naturforsch.* **47a**, 565 (1992).
98. J. W. Emsley, G. R. Luckhurst, and B. A. Timimi, *J. Phys. Paris* **48**, 473 (1987).
99. J. W. Emsley, G. R. Luckhurst, and S. W. Smith, *Mol. Phys.* **70**, 967 (1990).
100. J. W. Emsley, G. R. Luckhurst, W. E. Palke, and D. J. Tildesley, *Liq. Cryst.* **11**, 519 (1992).
101. B. Tjipo-Margo and G. T. Evans, *Mol. Phys.* **74**, 85 (1991).
102. M. R. Wilson and M.P. Allen, *Mol. Cryst. Liq. Cryst.* **198**, 465 (1991); *Liq. Cryst.* **12**, 157 (1992).
103. J. R. McColl, *Phys. Lett. A* **38**, 55 (1972).
104. J. R. McColl and S. C. Shih, *Phys. Rev. Lett.* **29**, 85 (1972).
105. R. V. Tranfield and P. J. Collings, *Phys. Rev. A* **25**, 2744 (1982).
106. H. Ichimura, T. Shirakawa, T. Tokuda, and T. Seimiya, *Bull. Chem. Soc. Jpn.* **56**, 2238 (1983); T. Shirakawa, T. Hayakawa, and T. Tokuda, *J. Phys. Chem.* **87**, 1406 (1983); T. Shirakawa, M. Arai, and T. Tokuda, *Mol. Cryst. Liq. Cryst.* **104**, 131 (1984).
107. C. S. Johnson and P. J. Collings, *J. Chem. Phys.* **79**, 4056 (1983).
108. W. M. Lampe and P. J. Collings, *Phys. Rev. A* **34**, 524 (1986).
109. S. Singh, T. K. Lahiri, and K. Singh, *Mol. Cryst. Liq. Cryst.* **225**, 361 (1993).
110. C. W. Cross and B. M. Fung, *J. Chem. Phys.* **101**, 6839 (1994).
111. I. Ono and S. Kondo, *Mol. Cryst. Liq. Cryst. Lett.* **8**, 69 (1991).
112. A. V. Komolkin, A. Laaksonen, and A. Maliniak, *J. Chem. Phys.* **101**, 4103 (1994).
113. C. W. Cross and B. M. Fung, *Mol. Cryst. Liq. Cryst.* **262**, 507 (1995).
114. M. Yoshida and H. Toriumi, *Mol. Cryst. Liq. Cryst.* **262**, 525 (1995).
115. R. Dąbrowski and K. Czupryński, in *Modern Topics in Liquid Crystals*, A. Buka, ed., World Scientific, Singapore, 1993, p. 125.
116. W. H. de Jeu, in *Phase Transitions in Liquid Crystals*, S. Martellucci and A. N. Chester, eds., Plenum Press, New York, London, 1992, Chaps. 1–3.
117. R. Shashidhar, in *Phase Transitions in Liquid Crystals*, S. Martellucci and A. N. Chester, eds., Plenum Press, New York, London, 1992, p. 227.
118. A. J. Leadbetter, R. M. Richardson, and C. N. Colling, *J. Phys. (Paris) Colloq.* **36**, C1 (1975); A. J. Leadbetter, J. C. Frost, J. P. Gaughan, G. W. Gray, and A. Mosley, *J. Phys. Paris* **40**, 375 (1979).
119. W. L. McMillan, *Phys. Rev. A* **4**, 1238 (1971); *Phys. Rev. A* **6**, 936 (1972).
120. M. R. Kuzma and D. W. Allender, *Phys. Rev. A* **25**, 2793 (1982).
121. J. Katriel and G. F. Kventsel, *Phys. Rev. A* **28**, 3037 (1983).

122. G. F. Kventsel, G. R. Luckhurst, and H. B. Zewdie, *Mol. Phys.* **56**, 589 (1985).
123. G. R. Luckhurst and P. S. J. Simmonds, *Mol. Phys.* **80**, 233 (1993).
124. K. Nicklas, P. Bopp, and J. Brickmann, *J. Chem. Phys.* **101**, 3157 (1994).
125. M. R. Wilson and M. P. Allen, *Mol. Phys.* **80**, 277 (1993).
126. G. V. Paolini, G. Ciccotti, and M. Ferrario, *Mol. Phys.* **80**, 297 (1993).
127. L. V. Mirantsev, *Mol. Cryst. Liq. Cryst.* **226**, 123 (1993).
128. L. Onsager, *J. Am. Chem. Soc.* **58**, 1486 (1936).
129. N. E. Hill, W. E. Vaughan, A. H. Price, and M. Davies, *Dielectric Properties and Molecular Behaviour*, T. M. Sugden, ed., van Nostrand, London, 1969.
130. C. J. F. Böttcher, *Theory of Electric Polarization*, Vol. I, Elsevier, Amsterdam, 1973.
131. A. Chełkowski, *Dielectric Physics*, PWN, Warszawa, 1980.
132. W. Maier and G. Meier, *Z. Naturforsch. A* **16**, 262, 1200 (1961).
133. G. Meier, in *Dielectric and Related Molecular Processes*, Vol. 2, Specialist Periodical Reports, The Chemical Society, London, 1975, p. 183.
134. J. G. Kirkwood, *J. Chem. Phys.* **7**, 911 (1939).
135. H. Fröhlich, *Theory of Dielectrics*, Clarendon Press, Oxford, 1949.
136. C. J. F. Böttcher and P. Bordewijk, *Theory of Electric Polarization*, Vol. II, Elsevier, Amsterdam, 1978.
137. P. Bordewijk, *Physica* **75**, 146 (1974).
138. S. R. Sharma, *Mol. Phys.* **78**, 733 (1993).
139. P. Debye, *Polar Molecules*, Chemical Catalog, New York, 1929.
140. A. Würflinger, *Ber. Bunsenges. Phys. Chem.* **95**, 1040 (1991).
141. N. Pingel, U. Poser, and A. Würflinger, *J. Chem. Soc. Faraday Trans. I* **80**, 3221 (1984).
142. G. Williams, *Trans. Faraday Soc.* **60**, 1548 (1964).
143. P. L. Nordio, G. Rigatti, and U. Segre, *Mol. Phys.* **25**, 129 (1973).
144. G. Meier and A. Saupe, *Mol. Cryst.* **1**, 515 (1966).
145. A. J. Martin, G. Meier, and A. Saupe, *Symp. Faraday Soc.* **5**, 119 (1971).
146. W. T. Coffey, Yu. P. Kalmykov, E. S. Massawe, and J. T. Waldron, *J. Chem. Phys.* **99**, 4011 (1993).
147. W. T. Coffey, Yu. P. Kalmykov, E. S. Massawe, *Adv. Chem. Phys.* **85**, 667 (1993).
148. W. T. Coffey, D. S. F. Crothers, and J. T. Waldron, *Physica A* **203**, 600 (1994).
149. W. T. Coffey, D. S. F. Crothers, Yu. P. Kalmykov, and J. T. Waldron, *Physica A* **213**, 551 (1995).
150. W. T. Coffey, Yu. P. Kalmykov, and E. S. Massawe, *Liq. Cryst.* **18**, 677 (1995).
151. W. T. Coffey and Yu. P. Kalmykov, *Liq. Cryst.* **14**, 1227 (1993).
152. B. Gestblom and S. Urban, *Z. Naturforsch.* **50a**, 595 (1995).
153. H. Kresse, *Adv. Liq. Cryst.* **6**, 109 (1983); *Fortschr. Phys.* **30**, 507 (1982).
154. B. Gestblom and S. Wróbel, *Liq. Cryst.* **18**, 31 (1995).
155. T. K. Bose, B. Campbell, S. Yagihara, and J. Thoen, *Phys. Rev. A* **36**, 5767 (1987).
156. J. P. Parneix, C. Legrand, and D. Decoster, *Mol. Cryst. Liq. Cryst.* **98**, 361 (1983); J. P. Parneix, C. Legrand, and S. Toutain, *IEEE Trans. Microwave Theory Tech.* **30**, 2015 (1982).

157. C. Druon and J. M. Wacrenier, *J. Phys. Paris* **38**, 47 (1977).
158. S. Wróbel, J. A. Janik, J. Mościcki, and S. Urban, *Acta Phys. Polon.* **A48**, 215 (1975).
159. C. Carboni, H. F. Gleeson, J. W. Goodby, and A. J. Slaney, *Liq. Cryst.* **14**, 1991 (1993).
160. W. G. Scaife and G. McMullin, *Measurement Sci. Techn.* **5**, 1576 (1994); *High Press. Res.* **13**, 77 (1994).
161. R. Krombach, *Doctoral Thesis*, University of Bochum, Bochum, Germany (1991).
162. A. Würflinger, *Habilitation Thesis*, University of Bochum (1981); A. Würflinger, *Ber. Bunsenges. Phys. Chem.* **82**, 1080 (1978); *Ber. Bunsenges. Phys. Chem.* **84**, 653 (1980).
163. U. Poser and A. Würflinger, *Ber. Bunsenges. Phys. Chem.* **92**, 765 (1988).
164. S. Urban, *Z. Naturforsch.* **50a**, 826 (1995).
165. H. Sasabe and K. Ooizumi, *Jpn. J. Appl. Phys.* **11**, 1751 (1972).
166. N. Yasuda, S. Fujimoto, S. Funado, and K. Tanaka, *J. Phys. D, Appl. Phys.* **17**, 1283 (1984).
167. N. Yasuda, S. Fujimoto, and S. Funado, *J. Phys. D, Appl. Phys.* **18**, 521 (1985); M. Ozaki, N. Yasuda, and K. Yoshino, *Jap. J. Appl. Phys.* **26**, L1927 (1987).
168. W. Heinrich and B. Stoll, *Colloid Polym. Sci.* **263**, 873, 895 (1985).
169. M. J. Bradshaw and E. P. Raynes, *Mol. Cryst. Liq. Cryst. Lett.* **72**, 73 (1981).
170. J. Thoen and G. Menu, *Mol. Cryst. Liq. Cryst.* **97**, 163 (1983).
171. G. Menu, Ph.D. Thesis, Catholic University of Leuven, Belgium (1988).
172. S. Urban, B. Gestblom, T. Brückert, and A. Würflinger, *Z. Naturforsch. A* **50**, 984 (1995).
173. V. V. Belyaev, M. F. Grebenkin, and V. F. Petrov, *Russian J. Phys. Chem.* **64**, 509 (1990).
174. G. Krömer, D. Paschek, and A. Geiger, *Ber Bunsenges. Phys. Chem.* **97**, 1188 (1993); S. Ye. Yakovenko, A. A. Minko, G. Krömer, and A. Geiger, *Liq. Cryst.* **17**, 127 (1994).
175. I. Ono and S. Kondo, *Bull. Chem. Soc. Jpn.* **66**, 633 (1993).
176. J. W. Baran, Z. Raszewski, R. Dąbrowski, J. Kędzierski, and J. Rutkowska, *Mol. Cryst. Liq. Cryst.* **123**, 237 (1985).
177. M. R. Wilson and D. A. Dunmur, *Liq. Cryst.* **5**, 987 (1989).
178. C. J. R. Counsell, J. W. Emsley, N. J. Heaton, and G. R. Luckhurst, *Mol. Phys.* **54**, 847 (1985).
179. C. J. R. Counsell, J. W. Emsley, G. R. Luckhurst, and H. S. Sachdev, *Mol. Phys.* **63**, 33 (1988).
180. R. Y. Dong and G. Ravindranath, *Liq. Cryst.* **17**, 47 (1994).
181. Z. Raszewski, J. Rutkowska, J. Kędzierski, J. Zieliński, P. Perkowski, W. Piecek, and J. Żmija, *Mol. Cryst. Liq. Cryst. Sci. Tec. A* **251**, 357 (1994).
182. S. Urban, E. Novotna, H. Kresse, and R. Dąbrowski, *Mol. Cryst. Liq. Cryst.* **262**, 257 (1995).
183. H. Schad and M. A. Osman, *J. Chem. Phys.* **75**, 880 (1981).
184. J. Przedmojski, J. Jędrzejewski, R. Dąbrowski, K. Czupryński, W. Tłaczała, and R. Wiśniewski, *Phase Trans.* **56**, 119 (1996).

185. S. Urban, B. Gestblom, H. Kresse, and R. Dąbrowski, *Z. Naturforsch. A* **51a**, 834 (1996).

185a. S. Chandrasekhar, R. Shashidhar, and K. V. Rao, *Advances in Liquid Crystal Research and Applications*, L. Bata, ed., Pergamon Press, Oxford, 1980, p. 123.

186. R. B. Meyer, L. Liebert, L. Strzelecki, and P. Keller, *J. Phys. (Paris)* **36**, L69 (1975).

187. B. I. Ostrovski, A. Z. Rabinovich, A. S. Sonin, and B. A. Srukov, *Sov. Phys.-JETP* **47**, 912 (1978).

188. M. Schadt, *Ber. Bunsenges. Phys. Chem.* **97**, 1213 (1993).

189. R. Blinc and B. Žekš, *Phys. Rev.* **18**, 740 (1978).

190. T. Carlsson, B. Žekš, A. Levstik, C. Filipič, I. Levstik, and R. Blinc, *Phys. Rev. A* **36**, 1484 (1987).

191. T. Carlsson, B. Žekš, C. Filipič, and A. Levstik, *Phys. Rev. A* **42**, 877 (1990).

192. B. Kutnjak-Urbanc and B. Žekš, *Liq. Cryst.* **18**, 483 (1995).

193. R. Blinc, in *Phase Transitions in Liquid Crystals*, S. Martellucci and A. N. Chester, eds., Plenum Press, New York, London, 1992, Chap. 22.

194. L. A. Beresnev, L. M. Blinov, M. A. Osipov, and S. A. Pikin, *Mol. Cryst. Liq. Cryst.* **158A**, 1 (1988).

195. M. Marzec, W. Haase, E. Jakob, M. Pfeiffer, S. Wróbel, and T. Geelhaar, *Liq. Cryst.* **14**, 1967 (1993).

196. S. Wróbel, A. M. Biradar, and W. Haase, *Ferroelectrics* **100**, 271 (1989).

197. M. Ozaki, T. Hatai, A. Tagawa, K. Nakao, H. Taniguchi, and K. Yoshono, *Jap. J. Appl. Phys.* **28** (Suppl. 28–2), 130 (1989).

198. R. B. Meyer, *Mol. Cryst. Liq. Cryst.* **40**, 33 (1977).

199. D. Guillon, J. Stamatoff, and P. E. Cladis, *J. Chem. Phys.* **76**, 2056 (1982).

200. G. G. Nair, S. K. Prasad, and S. Chandrasekhar, *Mol. Cryst. Liq. Cryst.* **263**, 311 (1995).

201. K. Yoshino, K. Nakao, M. Ozaki, R. Higuchi, N. Mikami, and T. Sakurai, *Liq. Cryst.* **5**, 1213 (1989).

202. M. Ozaki, K. Nakao, T. Hatai, and T. Yoshino, *Liq. Cryst.* **5**, 1219 (1989).

203. S. Krishna Prasad, S. M. Khened, and S. Chandrasekhar, *Ferroelectrics* **141**, 351 (1993).

204. H. Kresse, H. Stettin, E. Tennstedt, and S. Kostromin, *Mol. Cryst. Liq. Cryst.* **191**, 135 (1990).

205. G. S. Attard, K. Araki, J. J. Moura-Ramos, and G. Williams, *Liq. Cryst.* **3**, 861 (1988).

206. J. P. Parneix, R. Njeumo, C. Legrand, P. Le Barny, and J. C. Dubois, *Liq. Cryst.* **2**, 167 (1987).

207. M. Gnoth, Doctoral Thesis, University of Ulm, Germany, 1993.

208. G. McMullin, W. G. Scaife, and R. Zentel, *Liq. Cryst.* **18**, 529 (1995).

209. R. Zentel, G. R. Strobl, and H. Ringsdorf, *Macromolecules* **18**, 960 (1985).

210. G. S. Attard, *Mol. Phys.* **58**, 187 (1986).

211. A. Kozak, G. P. Simon, J. K. Mościcki, and G. Williams, *Mol. Cryst. Liq. Cryst.* **193**, 149 (1990).

212. H. W. Starkweather, Jr., P. Avakian, J. J. Fontanella, and M. C. Wintersgill, *Macromolecules* **25**, 7145 (1992).

213. R. Zentel and G. Reckert, *Makromolek. Chem.* **187**, 1915 (1986).

214. S. J. Rzoska and J. Zioło, *Liq. Cryst.* **17**, 629 (1994).

215. A. Drozd-Rzoska, S. J. Rzoska, M. Górny, and J. Zioło, *Mol. Cryst. Liq. Cryst.* **260**, 443 (1995).

216. M. A. Anisimov, *Critical Phenomena in Liquids and Liquid Crystals*, Gordon & Breach, London, 1994.

217. P. H. Keyes and W. B. Daniels, *J. Physique Coll.* **C3**, 380 (1979); R. Shashidhar, P. H. Keyes, and W. B. Daniels, *Mol. Cryst. Liq. Cryst. Lett.* **3**, 169 (1986).

218. R. A. Orwoll, V. J. Sulivan, and G. C. Campbell, *Mol. Cryst. Liq. Cryst.* **149**, 121 (1987).

219. S. Lakshminarayana, C. R. Prabhu, D. M. Potukuchi, N. V. S. Rao, V. G. K. M. Pisipati, and D. Saran, *Liq. Cryst.* **15**, 909 (1993).

220. A. Anakkar, J. M. Buisine, C. Alba-Simionesco, L. Ter Minassian, H. T. Nguyen, and C. Destrade, *J. Phys. III (Applied Physics, Materials Science, Fluids, Plasma and Instrumentation)* **2**, 1029 (1992); C. Legrand, N. Isaert, J. Hmine, J. M. Buisine, J. P. Parneix, H. T. Nguyen, and C. Destrade, *J. Phys. II France* **2**, 1545 (1992); A. Daoudi, A. Anakkar, and J. M. Buisine, *Thermochim. Acta* **245**, 219 (1994).

221. P. Pollmann, B. Wiege, and A. Rothert, *Liq. Cryst.* **3**, 225 (1988); P. Pollmann and K. Schulte, *Liq. Cryst.* **10**, 35 (1991); P. Pollmann and B. Wiege, *Liq. Cryst.* **6**, 657 (1989); E. Demikhov, J. Hollmann, and P. Pollmann, *Europhys. Lett.* **21**, 581 (1993).

222. A. N. Kalkura, R. Shashidhar, G. Venkatesh, D. Demus, and W. Weissflog, *Mol. Cryst. Liq. Cryst.* **84**, 275 (1982).

223. J. A. Janik, J. Mayer, S. Habryło, I. Natkaniec, W. Zając, J. M. Janik, and T. Stanek, *Phase Trans.* **37**, 239 (1992).

224. J. Rübesamen and G. M. Schneider, *Liq. Cryst.* **13**, 711 (1993).

225. T. R. Woliński, R. Dąbrowski, B. J. Bock, and S. Kłosowicz, *Thin Solid Films* **247**, 252 (1994); T. R. Woliński, R. Dąbrowski, A. Jamrozik, A. W. Domański, and B. J. Bock, *Proc. SPIE (USA)* **2372**, 393 (1994).

226. M. Davies, in *Organic Liquids*, Vol. II, A. D. Buckingham, E. Lippert, and S. Bratos, eds., Wiley, Chichester, 1978, p. 167.

227. A. G. Vanakaras and D. J. Photinos, *Mol. Phys.* **85**, 1089 (1995); A. F. Terzis and D. J. Photinos, *Mol. Phys.* **83**, 847 (1994).

228. F. J. Bormuth and W. Haase, *Liq. Cryst.* **3**, 881 (1988).

229. S. J. Rzoska, J. Zioło, and W. Pyżuk, *Chem. Phys. Lett.* **197**, 277 (1992).

230. A. Drozd-Rzoska, S. J. Rzoska, and J. Zioło, *Phase Trans.* **54**, 75 (1995).

ELECTRIC POLARIZATION OF POLAR TIME-DEPENDENT RIGID MATERIALS

S. HAVRILIAK, JR.

Rohm and Haas Research, Bristol Research Park, Bristol, Pennsylvania 19007

S. J. HAVRILIAK

Havriliak Software Development Co., Huntingdon Valley, Pennsylvania 19006

CONTENTS

Advances in Chemical Physics, Volume XCVIII, Edited by I. Prigogine and Stuart A. Rice.
ISBN 0-471-16285-X

I. INTRODUCTION

In this work we start with a limited but directed historical review of dielectric literature emphasizing the assumptions underlying dielectric theory. We outline the work of Debye[1–3] who developed the idea of point dipoles in neutral molecules. With the development of this idea he was able to explain why the dielectric constants of materials were greater than the square of the refractive index. Fixing these dipoles to spheres and assuming vicious drag forces to impede their motions lead to the now famous Debye relaxation process. With the development of this idea he was able to explain why relaxation were "anomalous" when compared to absorption curves at optical and infrared frequencies. These two contributions form a very profound and broad starting point in the field of dielectric relaxation theory and interpretation. For example, the point dipole concept was used by Onsager, Kirkwood, Frohlich, Scaife, Cole, and others as starting points for their theories,[4–8] to correct the inconsistencies of Debye's equilibrium theory with experimental results. The Debye process is used by all those who wish to discuss broad and skewed dielectric relaxation dispersions in terms of a distribution of Debye elements. Nearly all dielectric literature since his developments is based on these concepts, and seldom if ever are the underlying limitations inspected for their suitability.

Two limitations of Debye's original work form the basis of this work. First, associated with this polarization process there must be a distortion (strain) of the dielectric specimen because dipoles are fixed to unsymmetrical groups, and these become aligned in the electric field. The first question to be addressed is "What are the consequences of the strain energy associated with polarization process?" Second, the viscosity surrounding the sphere is assumed by Debye to be Newtonian. The second question to be addressed is "What are the consequences of a time-dependent (or complex) viscosity?" The present historical review is necessarily kept short and is selective for the purposes of illustrating the trend in this work. As a result many important contributions that are not directly in line with these questions are omitted.

II. HISTORICAL REVIEW

A. Debye Equilibrium Model

A convenient starting point for the analysis of dielectric effects is a sphere of the test material suspended in a vacuum to which an electric field is applied: see Figure 1. At equilibrium this electric field, **E**, induces a

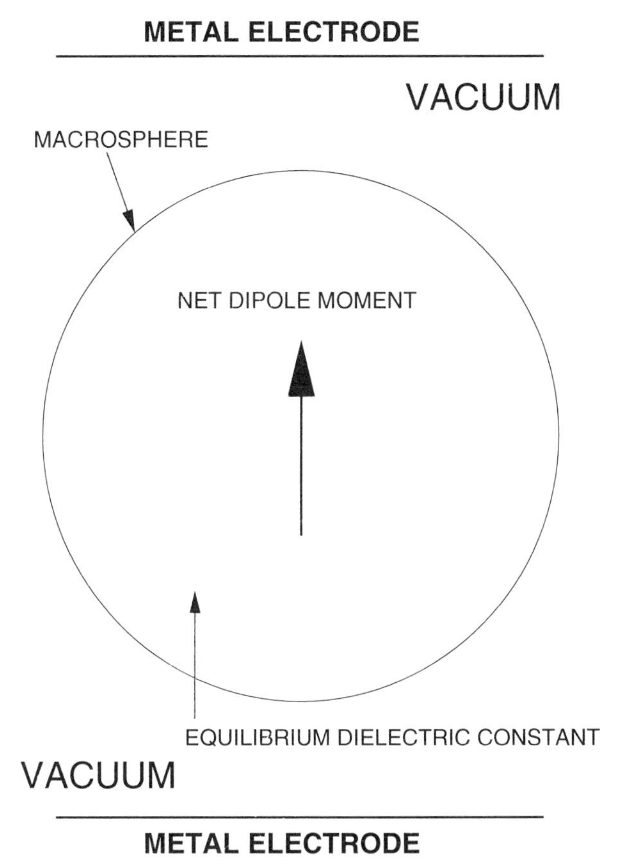

Figure 1. Starting point for most theoretical studies in a sphere of the material suspended in a vacuum to which an electric field **E** is applied to induce a moment **M**.

moment, **M**, in the sphere given by

$$\mathbf{M} = \frac{\varepsilon_0 - 1}{\varepsilon_0 + 2} \frac{3V}{4\pi} \mathbf{E} \tag{1}$$

In this expression ε_0 is the equilibrium dielectric constant while V is the volume of the sphere. The theoretical objective is to relate **M** to some molecular quantity. The earliest interpretation was in terms of molecular polarizability of the individual molecules in the sphere. This polarizability α_0 is due to the mobility of the electronic cloud that surrounds all atoms or molecules and is given by the Clausius–Mosotti equation:

$$\frac{\varepsilon_0 - 1}{\varepsilon_0 + 2} = \frac{4\, dN\pi}{3M} \alpha_0 \tag{2}$$

In this equation d is the density, N is Avogadro's number, and M is the molecular weight.

Dielectric measurements made in the audio-radio frequency range soon lead to a phenomenon called anomalous dispersion[9] because the frequency dependence of the real part of the complex dielectric constant did not behave at all like the frequency dependence of the real part of the refractive index in the region of significant absorption. Debye was the first to show that the polarization of this sphere is at least a two-step process. The first step is due to the nearly instantaneous ($\leq 10^{-13}$ s) displacements of electrons as represented by the Clausius–Mosotti equation. He then postulated that although these molecules are electrically neutral, the centers of positive and negative charges are not coincident thereby leading to a dipole moment. The dipole moment, generally represented in terms of Debye units, is the product of the separated charge and their separation. Idealization of a real dipole is referred to as a point dipole and corresponds to the condition of a very small charge separation.

B. Debye Dynamic Model

The Debye model proceeds by fixing a point dipole in the center of a microsphere itself centered in the macrosphere; see Figure 2. The result of his analysis, obtained by applying Maxwell–Boltzmann statistics is given by

$$\frac{\varepsilon_0 - 1}{\varepsilon_0 + 2} = \frac{4\, d\pi N}{3M} \left(\alpha_0 + \frac{\mu^2}{3kT} \right) \tag{3}$$

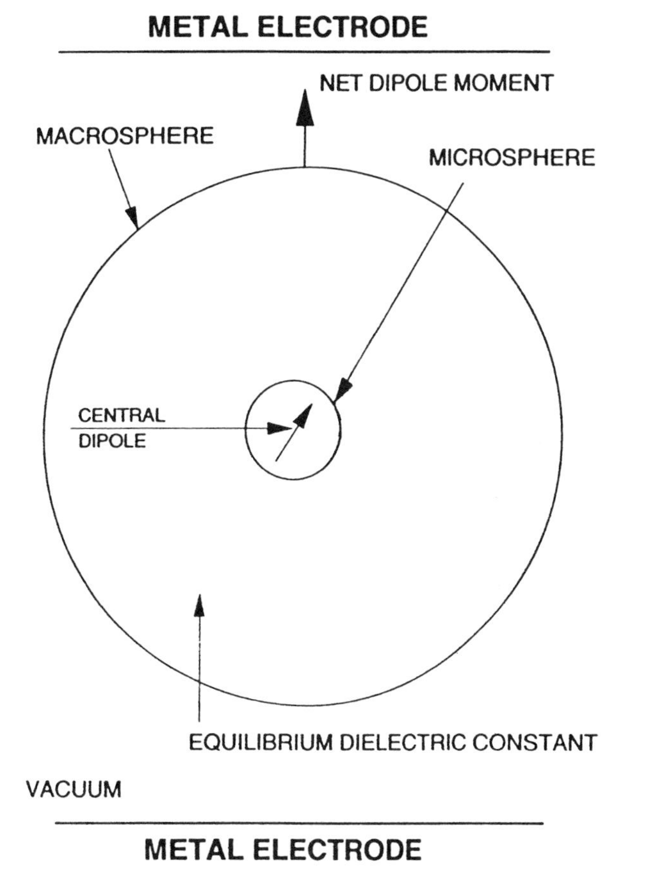

Figure 2. The geometry shown in Figure 1 is refined for dielectric analysis by fixing a microsphere concentric with the macrosphere and contains the elements of molecular structure.

In this expression k is the Boltzmann constant, T is temperature in Kelvin, and μ is the dipole moment in electrostatic unit (esu).

Debye's next contribution was to fix a point dipole on a microsphere of radius a and then assume the macrosphere to be a fluid with viscosity η. Under these conditions the relaxation time, τ, is given by

$$\tau = \frac{4\pi\eta a^3}{kT} \tag{4}$$

Debye's analysis yields for the complex dielectric constant $\varepsilon^*(\omega)$ at radian frequency ω, the following equation:

$$\frac{\varepsilon^*(\omega)-1}{\varepsilon^*(\omega)+2}=\frac{4\,d\pi N}{3M}\left(\alpha_0+\frac{\mu^2}{3kT}\right)\left(\frac{1}{1+i\omega\tau}\right) \tag{5}$$

In this equation $i=\sqrt{-1}$, and $\varepsilon^*(\omega)$ is the complex dielectric constant and is related to the real $\varepsilon'(\omega)$ and imaginary $\varepsilon''(\omega)$ components through the expression

$$\varepsilon^*(\omega)=\varepsilon'(\omega)-i\varepsilon''(\omega) \tag{6}$$

The real and imaginary components of Eq. (5) can be separated to yield

$$\varepsilon'(\omega)=\varepsilon_\infty+\frac{\varepsilon_0-\varepsilon_\infty}{1+X^2} \tag{7}$$

$$\varepsilon''(\omega)=+\frac{(\varepsilon-\varepsilon_\infty)X}{1+X^2} \tag{8}$$

with

$$X=\frac{\varepsilon_0+2}{\varepsilon_\infty+2}\,\omega\tau \tag{9}$$

An important result of Debye's model is that the relationship between the macroscopic, X, and molecular relaxation times, τ, is given by Eq. (9). For example, Eq. (9) predicts that materials with $\varepsilon_0\approx 100$ and $\varepsilon_\infty\approx 2$ the ratio of times is 25.5, with the molecular time being the faster one.

C. Criticisms of the Debye Model

There are several criticisms of Debye's results. First, let us consider Eq. (3) by assuming α_0 to be zero and defining a Curie temperature T_c to be given by

$$T_c=\frac{4\,d\pi N}{9M}\frac{\mu^2}{k} \tag{10}$$

The Debye Eq. (3) becomes

$$\frac{\varepsilon_0-1}{\varepsilon_0+2}=\frac{T_c}{T} \tag{11}$$

For reasonable choices of the parameters in Eq. (10), T_c can $\rightarrow T$; Eq.

(11) approaches 1; so that $\varepsilon_0 \rightarrow \infty$. In other words there is a temperature below which ε_0 is infinite, a prediction not consistent with observation.

D. Reaction Field Model

This criticism was solved by Onsager[10] for rigid dipoles and then Kirkwood[11] for polarizable dipoles who showed that point dipole(s) in the microsphere induce(s) a field (i.e., reaction field) in the macrosphere opposite that of the applied field. Onsager's result for nonpolarizable point dipoles is given by

$$\frac{\varepsilon_0 - 1}{\varepsilon_0 + 2} = \frac{\varepsilon_\infty - 1}{\varepsilon_\infty + 2} + \frac{3\varepsilon_0(\varepsilon_\infty + 2)}{(2\varepsilon_0 + \varepsilon_\infty)(\varepsilon_0 + 2)} \frac{4\,dN\pi\mu^2}{9kMT} \tag{12}$$

The quantity

$$\frac{3\varepsilon_0(\varepsilon_\infty + 2)}{(2\varepsilon_0 + \varepsilon_\infty)(\varepsilon_0 + 2)} \tag{13}$$

is due to the reaction field of the dipoles and significantly modifies the Debye result, that is, Eq. (3). For simplicity let us assume that $(\varepsilon_\infty - 1)/(\varepsilon_\infty + 2)$ can be ignored, that is, the case for modest to highly polar systems. Equation (12) becomes

$$\frac{\varepsilon_0 - 1}{\varepsilon_0 + 2} \approx \frac{3\varepsilon_0(\varepsilon_\infty + 2)}{(2\varepsilon_0 + \varepsilon_\infty)(\varepsilon_0 + 2)} \frac{4\,dN\pi\mu^2}{9kMT} \tag{14a}$$

or

$$\varepsilon_0 - 1 \approx \frac{3\varepsilon_0(\varepsilon_\infty + 2)}{(2\varepsilon_0 + \varepsilon_\infty)} \frac{4\,dN\pi\mu^2}{9kMT} \tag{14b}$$

It is important to recognize that there is a fortuitous cancellation of $\varepsilon_0 + 2$ on both sides of Eq. (14a) that yields the result in Eq. (14b). As a result of this cancellation, the equilibrium dielectric constant is now proportional to the square of the effective dipole moment and inversely proportional to temperature. Consequently, the Curie temperature occurs only as temperature approaches 0 K. The results are considered to be a significant advance over the Debye result.

E. Distribution of Relaxation Times

Another failure of the Debye theory is that the frequency dependence of the dielectric loss predicted by Eq. (8) is not what is observed experimentally. Experimental loss–log(frequency) curves are generally broader than those predicted by Eq. (8) in addition to being unsymmetrical. A plot of the loss, Eq. (8), against the real, Eq. (7), at the same frequency but with frequency as a running variable is a semicircle for the Debye process. Such Arrgand diagrams are generally referred to as Cole–Cole (C–C) plots and the semicircle is referred to as a Debye shape or single exponential process. Most materials do not exhibit such a simple shape. Rather, they are unsymmetrical, most often to the high-frequency side.

These experimental loss–log(frequency) or equivalently real–log(frequency) curves can be interpreted in terms of a distribution of Debye elements given by the following equation:

$$\frac{\varepsilon^*(\omega) - \varepsilon_\infty}{\varepsilon_0 - \varepsilon_\infty} = \int_{-\infty}^{\infty} F\left(\frac{\tau}{\tau_0}\right)\left(\frac{1}{1 + i\omega\tau_0}\right) d\ln\left(\frac{\tau}{\tau_0}\right) \tag{15}$$

In this equation the quantity $F(\tau/\tau_0)d\ln(\tau/\tau_0)$ is the distribution of relaxation times, while the quantity in brackets is the Debye relaxation function, that is, see Eq. (5). A particularly useful form for representing the complex dielectric constant dependence on frequency is given by[12,13] the following equation:

$$\frac{\varepsilon^*(\omega) - \varepsilon_\infty}{\varepsilon_0 - \varepsilon_\infty} = \frac{1}{[1 + (i\omega\tau_0)^\alpha]^\beta} \tag{16}$$

In this expression the parameters α, β represent the width and skewness of the relaxation process, respectively, and will be referred to as the H–N function. Since the left-hand side of Eq. (15) is given by the right-hand side of Eq. (16), a closed-form solution for the distribution times is possible. This solution is given by

$$F(y) = \left(\frac{1}{\pi}\right) y^{\alpha\beta}(\sin\beta\theta)(y^{2\alpha} + 2y^\alpha \cos\pi\alpha + 1)^{-\beta/2} \tag{17}$$

In this expression

$$y = \frac{\tau}{\tau_0} \quad \text{and} \quad \Theta = \arctan\left(\frac{\sin\pi\alpha}{y^\alpha \cos\pi\alpha}\right) \tag{18}$$

F. Time-Dependent Correlation Functions

An alternate interpretation of this broading-skewing of the relaxation process is given by Glarum[14] and later by Cole[15] who applied the irreversible-statistical-mechanics results of Kubo.[16] Once again the model chosen is a sphere suspended in a vacuum with a microsphere containing the elements of local structure centered in the macrosphere. Long-range interactions between the micro- and macrospheres is assumed to be given by the Onsager–Kirkwood reaction field. Glarum's result (assuming nonpolarizable dipole, i.e., Onsager's case) is given by

$$\frac{\varepsilon^*(\omega)-1}{\varepsilon^*(\omega)+2}=\frac{3[2\varepsilon_0+\varepsilon^*(\omega)]}{(2\varepsilon_0+1)[\varepsilon^*(\omega)+2]}\frac{4\pi N}{9kTV}(g\mu^2)I_\mu \tag{19}$$

In this expression $g\mu^2=\langle \boldsymbol{\mu}\cdot\mathbf{m}\rangle_0$ represents the equilibrium moment of the microsphere assuming all moments to be the same. The quantity I_μ is defined as the microscopic distribution of relaxation times given by

$$I_\mu=\int_{-\infty}^{\infty}F_\mu(\tau)\left(\frac{1}{1+i\omega\tau}\right)d\ln\tau \tag{20}$$

The microscopic distribution of relaxation times and the time-dependent correlation function $\Phi(t)$ is given by

$$\frac{-d\Phi(t)}{dt}=L[F(\alpha)] \tag{21}$$

One important result of Eq. (19) is that the quantity

$$\frac{3[2\varepsilon_0+\varepsilon^*(\omega)]}{(2\varepsilon_0+1)[\varepsilon^*(\omega)+2]} \tag{22}$$

is due to the complex reaction field due to dipoles in the microsphere. This term has a similar effect in the dynamic case as it does in the equilibrium case. In the dynamic case the quantity $\varepsilon^*(\omega)+2$ appears on both sides of Eq. (19) so that it is $\varepsilon^*(\omega)-1$ that is proportional to the square of the effective dipole moment. For the same reason $\varepsilon^*(\omega)-1$ is proportional to the molecular distribution function.

Another important result obtained from this analysis, as well as from Cole's analysis is that the relationship between the macroscopic and molecular relaxation times is given by the following equation:

$$\tau_0=\frac{3\varepsilon_0}{2\varepsilon_0+\varepsilon_\infty}\tau \tag{23}$$

This equation ranges from 1 to $\frac{3}{2}$ as ε_0 ranges from ε_∞ to ∞. This result is in stark contrast with Debye's result, that is, Eq. (9), which predicts a range of 1 to ∞ for the same ε_0 range. Cole and Glarum stated that they were "pleased" with this result, although the present authors do not know of any experimental evidence to support this contention.

G. Results from Fluctuation Theory

Scaife[17–19] proposed that the Cole–Cole plot method of analyzing dielectric data should be replaced by the use of the polarizability plot in which the imaginary coordinate of the complex polarizability, $\rho^*(\omega)$, defined by

$$\rho^*(\omega) = \frac{\varepsilon^*(\omega) - 1}{\varepsilon^*(\omega) + 2} \tag{24}$$

is plotted against its real component. The quantity, $\rho^*(\omega)$, is the complex polarizability of a dielectric sphere of unit radius suspended in a vacuum to which an electric field is applied while the quantity $\varepsilon^*(\omega)$ is the complex dielectric constant. Ordinarily $\varepsilon^*(\omega)$ is used to construct such plots followed by an analysis of the data. Scaife recommended $\rho^*(\omega)$ because in a sphere, long-range dipole–dipole coupling vanishes, and therefore $\rho^*(\omega)$ will be a good measure of the intrinsic properties of the substance. This method, according to Scaife, gives proper weighting to all polarization mechanisms and provides a ready means of comparing dielectric behavior of different substances. He applied this method to glycerol, an ethanol–water mixture, and to *n*-propanol. He found that deviations from expected behavior, which were interpreted in terms of a second relaxation process by others, were in fact an artifact due to the method of data representation. Another important result is that the ratio of microscopic (molecular) to macroscopic relaxation times is given by Eq. (9).

H. Models That Include a Time-Dependent Viscosity

Although the work of Glarum and Cole was criticized, we will not continue along these lines because for the present purposes they are relatively minor albeit elegant criticisms; see, for example, the work of Fatuzzo and Mason.[20] There are two other studies that need to be reviewed. The first one is by Gamant[21] who observed that the viscosity in Debye's model should be time dependent (see Fig. 3). By means of a series of heuristic arguments, he replaced η with a time-dependent one; that is, $\eta(t)$. We examined these equations[22] in some detail and could not obtain his results for the time dependence of $\varepsilon'(\omega)$.

The second work to be reviewed is that of DiMarzio and Bishop[23] (D–B) who introduced a time-dependent viscosity into the hydrodynamic equations of Debye's model and then solved the hydrodynamic equations

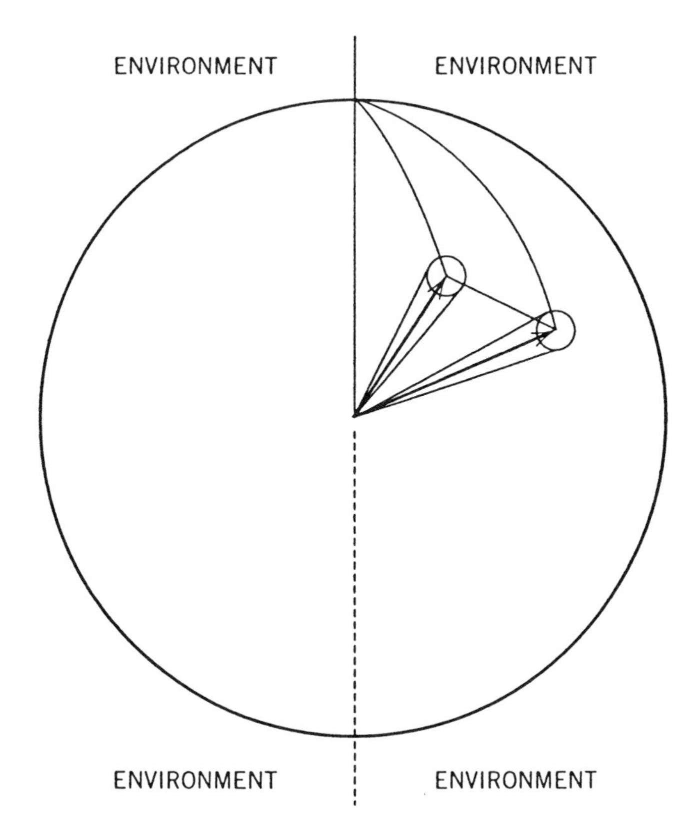

Figure 3. Dipole centered in a sphere (represented by the arrow) surrounded by an environment with properties that depend on the model. In the case of the Debye model, the environment has a viscosity η independent of time. In the DiMarzio–Bishop model the viscosity is a complex time-dependent viscosity $\eta^*(\omega = 1/t)$. In the Havriliak–Havriliak model the cavity is not spherical and the environment is taken to be represented by a complex tensile compliance $D(\omega = 1/t)$.

exactly. Their result is given by

$$\frac{\varepsilon^*(\omega) - \varepsilon_\infty}{\varepsilon_0 - \varepsilon_\infty} = \frac{1}{1 + KG^*(\omega)} \tag{25}$$

where

$$K = \frac{4\pi R^3}{kT}\left(\frac{\varepsilon_0 + 2}{\varepsilon_\infty + 2}\right) \tag{26}$$

In this equation R is the radius of the sphere and $G^*(\omega)$ is the "dynamic viscosity", which we will take to be given by the complex shear modulus. D–B concluded that this equation is either a one- or no-parameter equation, depending on whether or not R is measured or assumed. D–B considered the case of poly(n-octyl methacrylate), and we shall also consider the same case with some minor modifications in the treatment of data. These procedures permit us to calculate $\varepsilon^*(\omega)$ over an extensive frequency range and over a wide range of R. These values of $\varepsilon^*(\omega)$ were then represented in terms of the H–N function, that is, Eq. (16). In Figure 4 we have given complex plane plots using values for R similar to those used by D–B. The results in this figure are similar to those in Figure 2 of their work. Since we can use the H–N function to represent the calculated values of $\varepsilon^*(\omega)$, we can determine the influence of R on the H–N parameters. These results are given in Figure 5 for the α, β

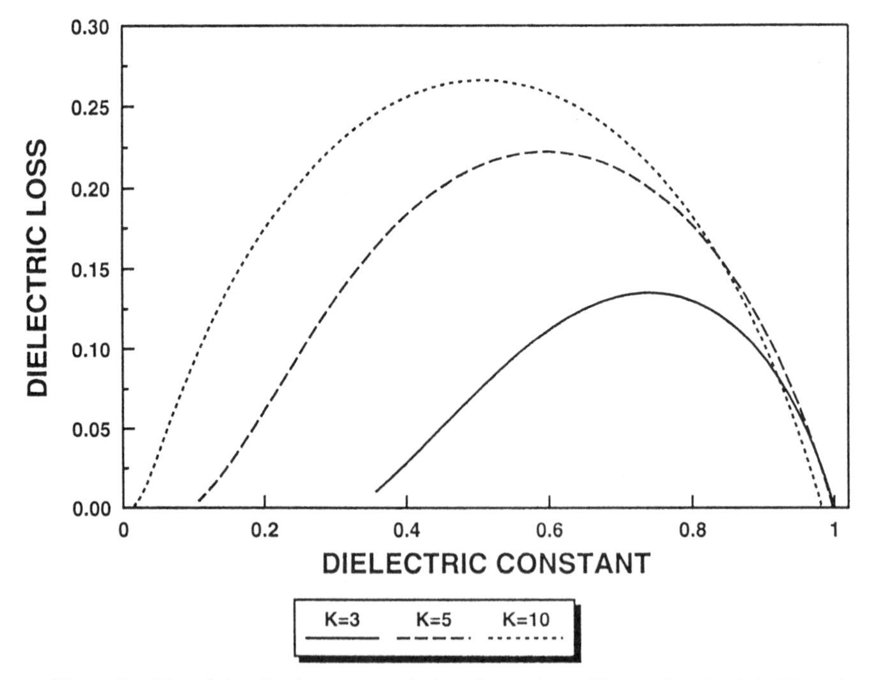

Figure 4. Plot of the elastic energy with time for various ellipse ratios. Included here is the electrostatic energy whose sign has been reversed. Complex plane plots calculated from the DiMarzio–Bishop model for various values of K (multiplied by 10^9) listed in the legend.

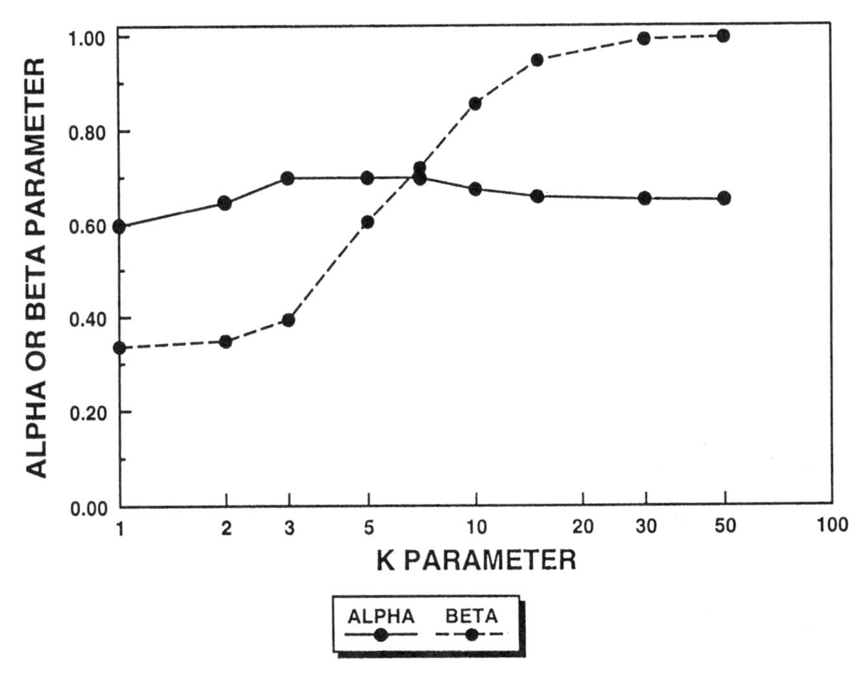

Figure 5. Plot of the α, β parameters as a function of the DiMarzio–Bishop K (multiplied by 10^9).

parameters, Figure 6 for the relaxation time and Figure 7 for ε_0, ε_∞, and $\Delta\varepsilon$. The parameter α is essentially constant while β varies from about 0.33 to 1.0 over the range of K's shown in Figure 5. Over the same range ε_0 and ε_∞ vary with K in such a way that both parameters range from 0 to 1 with their difference ($\Delta\varepsilon$) exhibiting a maximum. Finally in Figure 6 the relaxation time changes nearly 5 decades ($\log_{10}$) in time. In other words the effect of increasing K, hence radius of the sphere, is to decrease the relaxation frequency, decrease the skewness with keeping the breadth constant while $\Delta\varepsilon$ goes through a maximum. It is not clear why $\Delta\varepsilon$ goes through a maximum, and this point will be discussed in the next section. Much of these results are intuitive since the torque on $\boldsymbol{\mu}$ due to $\mathbf{E}$ is fixed, but the drag forces increase with surface area as R increases.

The important point of this discussion is that once the Debye process is modified to include a time-dependent local viscosity, realistic dielectric results are obtained for polymers. To repeat the D–B contention, Eq.

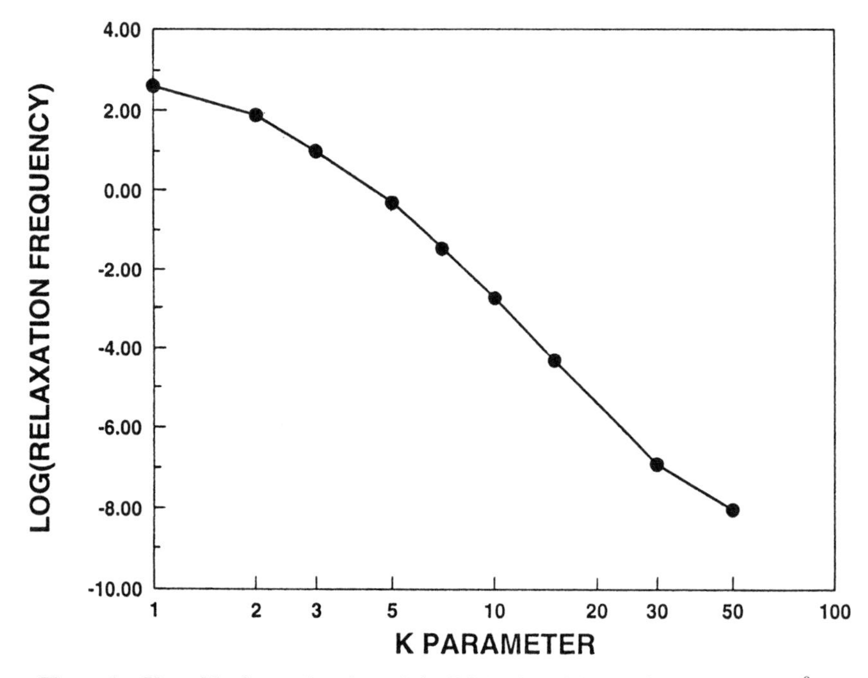

Figure 6. Plot of $\ln f_0$ as a function of the DiMarzio–Bishop K (multiplied by 10^9).

(25) is either a zero or a one-parameter model depending on our view of R. The net result is that the Debye relaxation process has been converted from a single exponential decay to a broad-skewed relaxation process.

III. STRAIN VS. POLARIZATION ENERGIES: NUMERICAL EXAMPLES

A. Equilibrium Case

The purpose of this section is to provide simple numerical examples that not only compare the magnitude of the elastic and electrostatic energies involved in the orientation process but also provide a simple example of what happens when the electric field is turned on. The electrostatic energy of a sphere[24] (U_e) suspended in a vacuum to which an electric field (E_0) is applied is given by (see Fig. 8)

$$U_e = -\left(\frac{\varepsilon_0 - 1}{\varepsilon_0 + 2}\right)\frac{3V}{4\pi}E_0^2 \qquad (27)$$

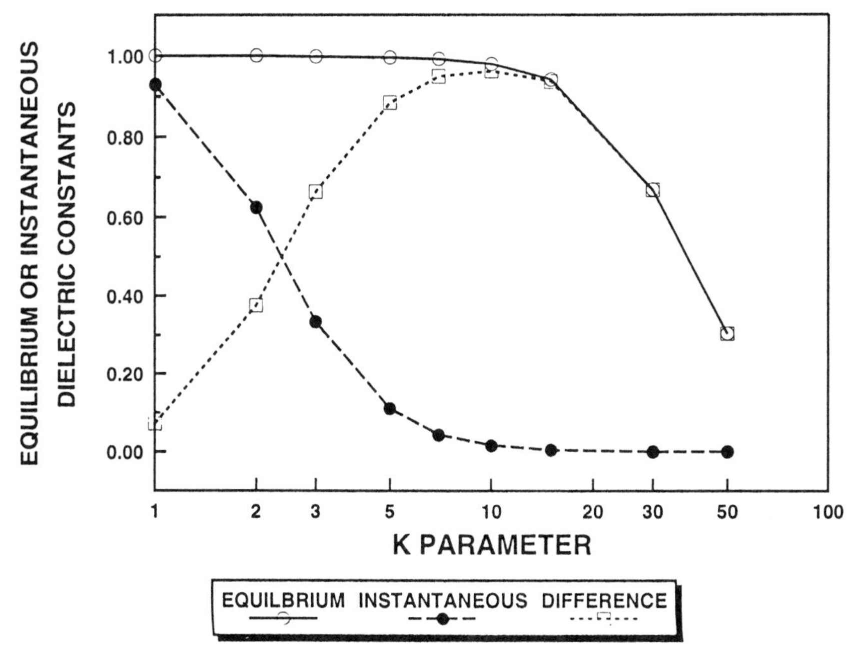

Figure 7. Plot of ε_0 and ε_∞ as a function of the DiMarzio–Bishop K (multiplied by 10^9).

For the case of $\varepsilon_0 = 8$, $V = 100\ \text{cm}^3$, and $E_0 = 300\ \text{V/cm}$ (1 esu), the electrostatic energy is -18.0 ergs. It is important to note that the energy is negative, that is, a reduction of the free energy has occurred.

Let us now consider the case of a polarized-distorted sphere such as the one in Figure 8. We would expect that associated with the polarization process there is a distortion process because dipoles are associated with some real molecular group that has a shape. In other words, when the dipoles orient in the electric field, there is an average projection in the field direction that is different from the average when the field is turned off. At this point it is necessary to postulate a molecular mechanism for the distorted sphere. Consider the sphere to consist of N ellipsoids with a major ($2a$) to minor (a) axis ratio of 2. The dipole moment $\mu_0 = 4D$ units, which was computed from Eq. (12), $\varepsilon_\infty = 1$, $\varepsilon_0 = 8$, and $T = \text{room}$ temperature is assumed to be in the direction of the major axis. Furthermore we assume a two-site model for the ellipsoids in such a way that the dipole moments of the sphere prior to the application of the sphere is zero. Consider a plane that contains 10^4 such sites in the z direction. The average projection of the ellipsoids in the z direction is $1.5a$. When the field is turned on (in the z direction), Frohlich[24] found

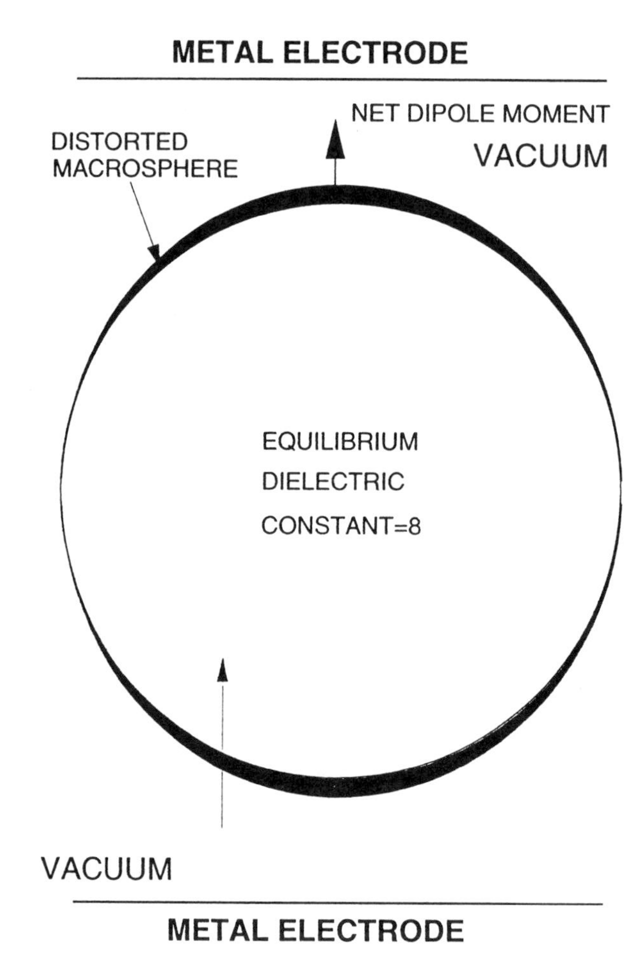

Figure 8. A polarized distorted sphere used as the basis for electrostatic or strain energy calculations.

that under these stated conditions (electric field and dipole moment at room temperature) the fraction of ellipsoids that orients is 10^{-4}. Computation of the average projection of the ellipsoids in the z direction when the field is turned on is $1.5001a$. The change in length/unit length due to the electric field, otherwise known as the strain (Γ) is 6.6×10^{-5}. For most polymers the tensile compliance is in the range of 10^{-6}–10^{-10} cm^2/

dyn. The strain energy under these conditions is given by[25]

$$U_s = \frac{1}{2}\Gamma\sigma \qquad V = \frac{\Gamma^2 V}{2D_0} \tag{28}$$

$$= 0.22 \text{ ergs} \quad \text{for } D_0 = 10^{-6} \text{ cm}^2/\text{dyn} \tag{28a}$$

$$= 2200 \text{ ergs} \quad \text{for } D_0 = 10^{-10} \text{ cm}^2/\text{dyn} \tag{28b}$$

In other words when the asymmetry of the orienting segment is as low as 2 (one might expect considerably higher values in the case of polymers) and for the range of compliance's that are usually observed in polymers, the strain energy can range from an insignificant fraction of the electrostatic energy to a level that dominates the entire process in polymers. The total energy of the system can never be positive for if it were, the polarization process would cease to exist because the driving force would vanish.

B. Time-Dependent Case

In the case of polymers the compliance depends on time (frequency). It is of interest to calculate the strain energy for Eq. (28) using the data for poly(*n*-octyl methacrylate). The time dependence of the tensile compliance data for this polymer is discussed in Reference 52. The results of this calculation, for different minor/major axis ratios is given in Figure 9. In addition, the negative (−) electrostatic energy is also given in that plot. The results clearly show that in the region of the α relaxation process, the elastic energy changes from a very large value to a very small value relative to the electrostatic energy, which does not change at all.

These simple numerical exercises illustrate the point that the total Hamiltonian of the system in the presence of the electric field must not only include the well-known electrostatic energy term but also the hitherto ignored strain energy term. The small strains that are encountered in the polarization process also justify the use of the electrostatic equations of the sphere. It is also quite clear that for very unsymmetrical species where equilibrium compliance approaches infinity, such as the case of bulky polar liquids, the strain energy term may also be zero and the Onsager–Kirkwood equations are once more applicable. It should be pointed out that the strain energy is also zero for finite values of D_0 but symmetrical species such as spherical dipoles.

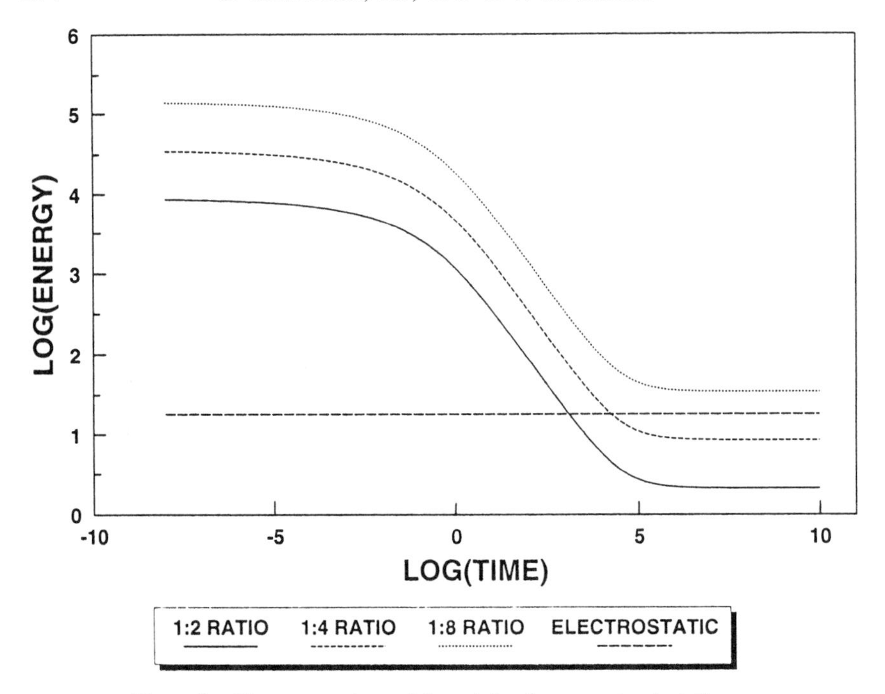

Figure 9. The geometric model used for the present calculations.

IV. EQUILIBRIUM POLARIZATION

A. General Case

The starting point for the present development is given in Figure 10. Let the total energy change (perturbation term in the Hamiltonian) caused by the action of the electric field on the sphere be given by

$$H'(q, E_0) = -\mathbf{M} \cdot \mathbf{E}_0 + U_s \tag{29}$$

To proceed further, we need to specify a microscopic mechanism for the deformation of the sphere. For the present purposes we choose a relatively simple mechanism that avoids all complications due to a variable and unknown Poisson ratio. We assume the material to be initially isotropic so that the deformation and the polarization are parallel to the applied field. Furthermore, the midsection of the deformed sphere, perpendicular to the direction of the field, will be assumed to be circular and with the original radius. Finally the deformation parallel to the field is proportional to the length of the original chord from the midplane of

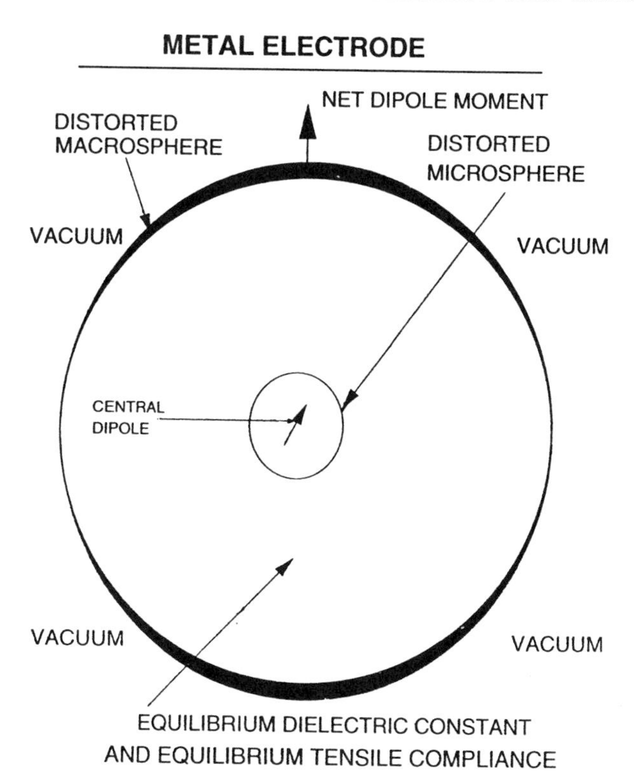

Figure 10. Plot of the elastic energy with time using the data for poly(*n*-octyl methacrylate) and various ellipse ratios. Included here is the electrostatic energy whose sign has been reversed.

the sphere. This model is one of pure tension, is uniformly strained in the direction of the electric field, and assumes the Poisson ratio to be zero.

The incremental energy (δU_s) for an incremental volume element (δV) of the sphere is given by $\Delta I_s/\Delta V_s = \sigma\Gamma/2$, where Γ is the strain of the sphere. Since the sphere is uniformly polarized, it is also uniformly strained and the strain energy will be given by $U_s = \sigma\Gamma V/2$. The Γ is an intensive property of the sphere, therefore we identify the strain of the sphere with the strain/unit. This quantity is also proportional to the total moment of the sphere per unit. In other words:

$$\Gamma = \frac{\xi M_z(q)}{N} = \frac{\sum \xi \mu_z}{N} \tag{30}$$

where $M_z(q)$ is the z component of the sphere moment. The total energy change of the sphere due to the presence of the field E_0 is, remembering that $\sigma = BE_0$,

$$H'(q, E_0) = -M_z(q)E_0 + \frac{\xi BV}{2N} M_z(q)E_0 \tag{31}$$

From the principles of statistical mechanics[26] we know that an experimental observable such as M is related to the statistical average **M** by

$$\frac{\varepsilon_0 - 1}{\varepsilon_0 + 2}\frac{3v}{4\pi} E_0 = \bar{M} = \int f(p, q, E_0)M_z(q)\, dp\, dg \tag{32}$$

where $f(p, q, E_0)$ is a normalized distribution function of the coordinates p, q in the presence of the field E_0. If we assume, as is customary to do, that $f(p, q, E_0) \cong \exp\{H(p, q, E_0)\}$, where $H(p, q, E_0)$ is the Hamiltonian of the system after application of the electric field and further assume that

$$H'(p, q, E_0) = H^0(p, q) + H(q, E_0) \tag{33}$$

where $H^0(p, q)$ is the unperturbed Hamiltonian of the system, then

$$f(p, q, E_0) = F_0(p, q)\left\{1 - \frac{M_z(q)E_0}{kT}\left(1 - \frac{\xi BV}{2N}\right)\right\} \tag{34}$$

and the average moment becomes

$$\bar{M} = \frac{\varepsilon_0 - 1}{\varepsilon_0 + 2}\frac{3V}{4\pi} = \frac{1}{3kT}\left\{1 - \frac{\xi BV}{2N}\right\}\langle \mathbf{M} \cdot \mathbf{M} \rangle_0 \tag{35}$$

In this expression $\mathbf{M} \cdot \mathbf{M}/3 = M_z(q)M_z(q)$ and the brackets $\langle \cdots \rangle_0$ mean an averaging with respect to the equilibrium distribution function $f_0(p, q)$. Equation (35) can be reduced one step further by noting that in an averaging procedure that will be used the units are assumd to be identical so that $\langle \mathbf{M} \cdot \mathbf{M} \rangle_0 = N\langle \boldsymbol{\mu} \cdot \mathbf{M} \rangle_0$. Equation (35) becomes

$$\frac{\varepsilon_0 - 1}{\varepsilon_0 + 2}\frac{3V}{4\pi} = \frac{N}{3kT}\left\{1 - \frac{\xi BV}{2N}\right\}\langle \boldsymbol{\mu} \cdot \mathbf{M} \rangle_0 \tag{36}$$

We now proceed to the evaluation of **B**. We know from Hooke's law[10] that the tensile stress is proportional to the average strain $\bar{\Gamma}$. Therefore,

$$\frac{\bar{\Gamma}}{D_0} = \sigma = BE_0 \quad \text{so that} \quad \bar{\Gamma} = BDE_0 \tag{37}$$

The average strain is given by

$$\bar{\Gamma} = BDE_0 = \int f_0(p, q)\left\{1 + \frac{\mathbf{M} \cdot \mathbf{E}}{kT}\left(1 - \frac{\xi BV}{2N}\right)\right\}\Gamma \, dp \, dq \tag{38}$$

Substituting Eq. (30) into (38) yields

$$BDE_0 = \left\{1 - \frac{\xi BV}{2n}\right\}\left\{\frac{\xi}{3NkT}\right\}E_0 \langle \mathbf{M} \cdot \mathbf{M} \rangle_0 \tag{39}$$

and finally

$$1 - \frac{\xi BV}{2N} = \left\{1 + \frac{\xi^2 V \langle \boldsymbol{\mu} \cdot \mathbf{M} \rangle_0}{6ND_0 kT}\right\}^{-1} \tag{40}$$

which in combination with Eq. (36) becomes

$$\frac{\varepsilon_0 - 1}{\varepsilon_0 + 2}\frac{3V}{4\pi} = \frac{N}{3kT}\left\{1 + \frac{\xi^2 V \langle \boldsymbol{\mu} \cdot \mathbf{M} \rangle_0}{6ND_0 kT}\right\}^{-1} \langle \boldsymbol{\mu} \cdot \mathbf{M} \rangle_0 \tag{41}$$

This is as far as the analysis can go without the simplifying assumptions needed to evaluate $\langle \boldsymbol{\mu} \cdot \mathbf{M} \rangle_0$.

B. Inconsequential Electrostatic Interactions

The assumption in theories based on point dipoles is that the nature of long-range coupling between the microsphere and the macrosphere is of electrostatic origin. This means that as the moment **m** in the sphere is rotated, the moment induced in the macrosphere, given by $-A(\varepsilon_0)\mathbf{m}$ also rotates (assuming the rotation to proceed infinitesimally slowly so that the system always remains in equilibrium). As the moment **m** in the microsphere is made to vanish, the induced moment in the microsphere $-A(\varepsilon_0)\mathbf{m}$ vanishes. Therefore in the limit of a nonpolar polymer there is no correlation between the two spheres. This assumption is probably acceptable for polar liquids but not for polymers where long-range correlations other than electrostatic one exist.

Let us now consider a microsphere containing a segment of a nonpolar polymer chain such as polystyrene, in the limit of very high holecular weight. As the segment in the microsphere chain rotates (or translates) slowly, a disturbance will be set up at any arbitrary distance away from the microsphere that is proportional to the magnitude of the rotation (or translation). Such disturbances are assumed to give rise to polymer orientation in the macrosphere. In other words a long-range nonelectrostatic correlation exists between the micro- and macrosphere that is based on polymer chain structure. These estimates are based on the stress birefringence measurements of Andrews.[5] In these experiments he was able to show that long-range correlations exist between the polymer chain and the phenyl side chains in polystyrene.

The problem is how to evaluate $\langle \boldsymbol{\mu} \cdot \mathbf{M} \rangle_0$. Ideally, this would come from the statistical theories of polymers. A review of the literature shows that such statistical theories are always molecular weight dependent, and these results are not applicable to the case of polar polymers in the bulk phase. Though it is not possible to develop a statistical theory, it is possible to consider Kirkwood's development (next section) as one limiting case and assume the other limiting case to consist of trivial (by comparison to steric factors) long-range electrostatic interactions.

We can proceed along lines similar to those of McCrum and coworkers as follows:

$$\langle \boldsymbol{\mu} \cdot \mathbf{M} \rangle_0 = \langle \boldsymbol{\mu}_i^{\mathrm{I}} \cdot \boldsymbol{M} \rangle_0 = \left\langle \boldsymbol{\mu}_i^{\mathrm{I}} \cdot \left\{ \boldsymbol{\mu}_i^{I} + \sum_{j \neq i} \mu_j^{\mathrm{I}} + \sum_j \boldsymbol{\mu}_j^{\mathrm{II}} \right\} \right\rangle_0 \tag{42}$$

In this expression $\boldsymbol{\mu}_i^{\mathrm{I}}$ is the ith unit on polymer molecule I and $\boldsymbol{\mu}_j^{\mathrm{I}}$ represents the remaining units on molecule I, and $\boldsymbol{\mu}_j^{\mathrm{II}}$ represents the units on all of the other polymer chains. Since all basic units are assumed to have the same dipole moment, we have

$$\langle \boldsymbol{\mu} \cdot \mathbf{M} \rangle_0 = \mu^2 \left\langle 1 + \sum_j \cos^{\mathrm{I}} \gamma_{ij} + \sum \cos^{\mathrm{II}} \gamma_{ij} \right\rangle_0 \tag{43}$$

or

$$\langle \boldsymbol{\mu} \cdot \mathbf{M} \rangle_0 = g_r \mu^2 \tag{44}$$

Combining this result with Eq. (41) we have

$$\frac{\varepsilon_0 - 1}{\varepsilon_0 + 2} \frac{3V}{4\pi} = \frac{N}{3kT} \left\{ 1 + \frac{\xi^2 V g_r \mu^2}{6 D_0 N k T} \right\}^{-1} g_r \mu^2 \tag{45}$$

$$= \frac{N}{3kT} \left\{ 1 + \frac{V g_r \gamma^2}{6 D_0 N k T} \right\}^{-1} g_r \mu^2 \tag{46}$$

remembering that $\xi \mu = \gamma$.

C. Onsager–Kirkwood Approximation

Though the quantity $\langle \boldsymbol{\mu} \cdot \mathbf{M} \rangle_0$ can be evaluated quite readily by a simple reference to Kirkwood,[2] we shall discuss the calculation at some length because its application to polymer systems is not done with certainty. Kirkwood's first assumption limits the elements of discrete local structure to a microsphere located inside the macroscopic one. Kirkwood's second assumption limits the correlation between the units of the microsphere and the remaining units to long-range dipole–dipole forces. The average correlation is given by an appropriate electrostatic calculation. Thus while the region is microscopically small, it may be quite large in terms of molecular dimensions. The average to be computed is $\langle \boldsymbol{\mu} \cdot \mathbf{M} \rangle_0$. This average may be performed by fixing the orientation of μ, averaging over all positions of $\boldsymbol{\mu}$. The last average is trivial for no terms in the Hamiltonian depend on the absolute position of $\boldsymbol{\mu}$.

To accomplish the first average let us denote a dipole within the small sphere by the suffix k and the one outside it by the suffix l. We can then average over all configurations of the $\boldsymbol{\mu}$ for a given configuration of the $\boldsymbol{\mu}_k$ and finally over all the $\boldsymbol{\mu}_l$. Thus

$$\langle \boldsymbol{\mu} \cdot \mathbf{M} \rangle_0 = \left\langle \boldsymbol{\mu}_i \cdot \sum \boldsymbol{\mu}_j \right\rangle \tag{47}$$

$$= \boldsymbol{\mu}_i \cdot \frac{\left\{ \int \sum \boldsymbol{\mu}_k e^{-H^0/kt} \, d\tau_{j \neq i} + \int d\tau_{j \neq i} \int \sum \mu_l \exp^{-H^0/kT} \, d\tau_l \right\}}{\int e^{-H^0/kT} \, d\tau_{j \neq i}} \tag{48}$$

The subscript i was added to $\boldsymbol{\mu}$ in order to keep track of the indexing in the sums and integrals. The moment induced in the outer sphere due to the presence of the moments in the microsphere is given by an electrostatic calculation and is equivalent to

$$\frac{\int \sum \boldsymbol{\mu}_l e^{-H^0/kT} \, d\tau_l}{\int e^{-H^0/kT} \, d\tau_l} \tag{49}$$

When this calculation is completed, it is found that the resulting moment equals

$$\frac{-2(\varepsilon_0^2 - 1)^2}{(2\varepsilon_0 + 1)(\varepsilon_0 + 2)} \sum \mu_k = -A(\varepsilon_0) \sum \mu_k \tag{50}$$

In order to perform this electrostatic calculation, Kirkwood had to invoke a double limiting process in that not only must the radius of the microsphere approach infinity but the ratio of the macrosphere to the radius of the microsphere must also approach infinity. This assumption clearly indicates that the microsphere, however large, is an insignificant portion of the macrosphere. Consequently, the second integral in Eq. (48) becomes

$$\int d\tau_{k \neq i} \int \sum \mu_l e^{-H^0/kT} \, d\tau_l = -A(\varepsilon_0) \int \sum \mu_k \, d\tau_{k \neq i} \int e^{-H^0/kT} \, d\tau_l \tag{51}$$

$$= -A(\varepsilon_0) \int \sum \mu_k e^{-H^0/kT} \, d\tau_{j \neq i} \tag{52}$$

and finally

$$\left\langle \boldsymbol{\mu}_i \cdot \sum \boldsymbol{\mu}_j \right\rangle_0 = (1 - A) \left\{ \frac{\boldsymbol{\mu}_i \cdot \int \sum \boldsymbol{\mu}_k e^{-H^0/kT} \, d\tau_{k \neq i}}{\int e^{-H^0/kT} \, d\tau_{j \neq i}} \right\} \tag{53}$$

If we let $\mathbf{m}$ equal the net moment of the microscopic region about $\boldsymbol{\mu}_i$, then

$$\left\langle \boldsymbol{\mu}_i \cdot \sum \boldsymbol{\mu}_j \right\rangle_0 = \frac{9\varepsilon_0}{(2\varepsilon_0 + 1)(\varepsilon_0 + 2)} \langle \boldsymbol{\mu} \cdot \mathbf{m} \rangle_0 \tag{54}$$

The evaluation of $\langle \boldsymbol{\mu} \cdot \Sigma \, \boldsymbol{\mu}_j \rangle_0$ is known as Kirkwood's electrostatic argument, its particular form here is due to Glarium,[28] though the same result can be obtained from somewhat different arguments; see, for example, Frohlich.[6] What is important is the recognition that there are two assumptions in Kirkwood's argument that may not apply to the case of polymers. Combining Eq. (54) with (41) leads to

$$\varepsilon_0 - 1 = \frac{4\pi N}{3kTV} \left\{ \frac{9\varepsilon_0}{2\varepsilon_0 + 1} \right\} \left\{ 1 + \frac{9\xi^2 V \varepsilon_0 \langle \boldsymbol{\mu} \cdot \mathbf{m} \rangle_0}{6ND_0 kT(2\varepsilon_0 + 1)(\varepsilon_0 + 2)} \right\}^{-1} \langle \boldsymbol{\mu} \cdot \mathbf{m} \rangle_0 \tag{55}$$

The quantity in the second brackets must, of course, be unitless. This is easily verified by noting that $\xi^2\langle \boldsymbol{\mu}\cdot\mathbf{m}\rangle_0 = \langle \gamma\, \Sigma\, \gamma\rangle_0$, which is unitless. Finally substitution of the values of V/D_0kT show that the entire fraction, hence the entire bracket, is unitless. A numerical check of the results is made by direct substitution into the square brackets and noting that its contribution is significant over the entire range of parameters studied in a previous section.

V. TIME-DEPENDENT POLARIZATION

A. General and Debye Equations

The time-dependent incremental strain energy, $\delta U_s(t)$, for an incremental volume element, δV, of the sphere is given by

$$\frac{\delta U_s(t)}{\delta V} = \frac{\sigma(t)\Gamma(q)}{2} \tag{56}$$

where $\Gamma(q)$ is the strain of the sphere. Since the strain of the sphere is uniform, the strain energy will be given by

$$U_s(t) = \tfrac{1}{2}\sigma(t)\Gamma(q)V \tag{57}$$

The strain $\Gamma(t)$ is an intensive property of the sphere, therefore, we identify the strain of the sphere with the strain/unit and is proportional to the total dipole moment of the sphere per unit. In other words,

$$\Gamma(q) = \sum \gamma/N = \zeta \sum \mu/N = \zeta M(q)/N \tag{58}$$

The strain energy of the sphere becomes

$$U_s(t) = \frac{\zeta BV}{2N} E(t)M_z(q) \tag{59}$$

In this expression, $M_z(q)$ is the z component of the moment and is assumed to be dependent on the coordinates q of the sphere. The perturbation term $H'(q, t)$ is given as the sum of the electrostatic and strain energy terms, that is,

$$H'(qt) = -M_z(q)E(t) + \frac{\zeta BV}{2N} E(t)M_z(q) \tag{60}$$

$$= -M_z(q)E(t)\left\{1 - \frac{\zeta BV}{2N}\right\} \tag{61}$$

In this expression $-M_z(q)E(t)$ is the electrostatic energy. The remaining arguments are similar to those of the previous section. The moment of the sphere suspended in the vacuum becomes

$$\frac{\varepsilon^*(\omega)-1}{\varepsilon^*(\omega)+2}=\frac{4\pi}{3kTV}\left\{1+\frac{\zeta^2V\langle M_z^2(0)\rangle_0 L[-\dot{\Phi}(t)]}{6kTN^2D(\omega)}\right\}^{-1} \tag{62}$$

In this expression L is the Laplace transform. In addition

$$L\{-\dot{\Phi}(t)\}=-\int_0^\infty e^{-i\omega t}\frac{d\Phi}{dt}\,dt \tag{63}$$

$$\Phi(t)=\frac{\langle M_z(0)M_z(t)\rangle_0}{\langle M_z^2(0)\rangle_0} \tag{64}$$

and

$$\langle M_z^2(0)\rangle_0=\frac{(\varepsilon_0+2)4\pi N^2}{(\varepsilon_0+1)9VkT}\left\{1-\frac{(\varepsilon_0-1)3V^2\zeta^2}{(\varepsilon_0+2)8\pi N^2D_0}\right\}^{-1} \tag{65}$$

This is as far as the analysis can go without the introduction of species schemes for evaluating the equilibrium averages. We shall, in this work, consider the case where the long-range interactions are electrostatic in origin.

B. Onsager–Kirkwood Approximation

This case assumes that the long-range interactions can be represented by electrostatic ones, that is, the Onsager–Kirkwood reaction field. The arguments of Fatuzzo and Mason leads to a relationship between the time-dependent moment of the macrosphere (Φ) and the time-dependent moment of the microsphere (ϕ), viz. Eq. (11).

$$L(-\dot{\Phi})=\frac{\varepsilon^*(\omega)(2\varepsilon_0+1)(\varepsilon_0+2)}{\varepsilon_0[2\varepsilon^*(\omega)+1][\varepsilon^*(\omega)+2]}\int(-\dot{\phi}) \tag{66}$$

If we apply these arguments to Eq. (7), the results are

$$\varepsilon^*(\omega)-1=\frac{4\pi N}{9kTV}\left\{1+\frac{\zeta^2VA'(\varepsilon^*)\langle\bar{\mu}\cdot\bar{m}\rangle_0L(-\phi)}{6kTND(\omega)}\right\}^{-1}$$
$$\times\frac{9\varepsilon^*(\omega)\langle\bar{\mu}\cdot\bar{m}\rangle_0}{2\varepsilon^*(\omega)+1}L(-\phi) \tag{67}$$

where

$$A'(\varepsilon^*) = \frac{9\varepsilon^*(\omega)}{[2\varepsilon^*(\omega)+1][\varepsilon^*(\omega)+2]} \tag{68}$$

VI. APPLICATION AND DISCUSSION OF THE RESULTS

A. Molecular Strain

The quantity γ, referred to as the molecule strain in previous sections, needs to be examined in greater detail. It is reasonable to assume[29] that if l_i represents the length of bond i and it makes an angle ϕ_i with the z axis, then the potential energy of the bond in a tensile field σ directed along the z axis is given by

$$BE_0 l_i \cos\phi_i = \sigma l_i \cos\phi_i = \boldsymbol{\sigma}\cdot\mathbf{l}_i \tag{69}$$

In other words we can associate a vector $\mathbf{l}_i$ with the bond, which behaves in much the same way as the vector representing the dipole moment $\boldsymbol{\mu}$. If $\mathbf{l}$ and $\boldsymbol{\mu}$ are specified, then a set of transformation equations can be constructed. The bond length $\mathbf{l}_i$ actually represents the difference between the overall length of two atoms and their diameter. In a similar way we can represent the strain as

$$\gamma = \frac{a-b}{b} \quad \text{or} \quad \boldsymbol{\gamma} = \frac{\mathbf{a}-\mathbf{b}}{b} \tag{70}$$

where a is the overall length of the bond and b is its diameter. The ratio in Eq. (70) is the change in length per unit length as the unit rotates about its axis. The vector $\mathbf{a}-\mathbf{b}$ is taken to be in the direction of the greater quantity a or b. The quantity γ refers to the strain of the molecule. What is needed is the strain at a particular point in the sphere. This will be proportional to the moment of the sphere per molecule.

$$\boldsymbol{\Gamma} = \frac{\xi\mathbf{M}}{N} = \frac{\xi\sum\boldsymbol{\mu}}{N} = \frac{\sum\boldsymbol{\gamma}}{N} \tag{71}$$

The direction of the dipole moment need not be limited to the direction of the major axis; it could lie in any direction. For these cases the sphere would be distorted but no longer in the direction of the electric field. In the special case of a dipole moment along the direction of the minor axis, the sphere would be an ellipse, but the major axis would be perpendicular to the direction of the applied field.

The interaction of the unit with its environment is best seen from a

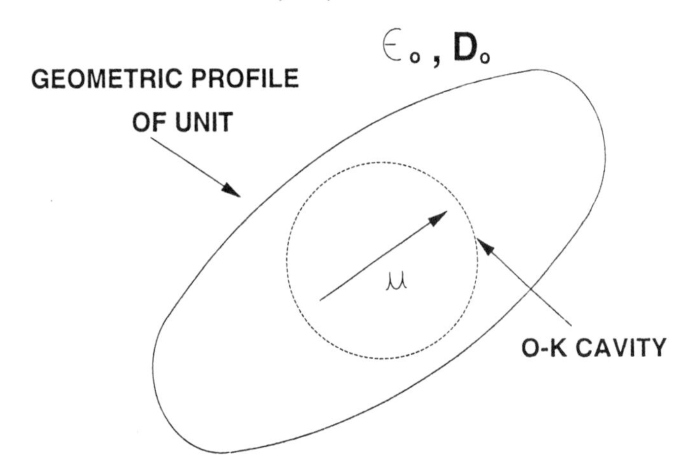

Figure 11. An elliptical unit is immersed in its own medium. The spherical cavity represented by the dashed line represents the Onsager cavity.

consideration of the Onsager model. Consider the following schematic representation given in Figure 11 of a polar species. The shape of the unit is an ellipse (solid line in Fig. 11), but the point dipole is considered to be centered in the sphere. The material constants are ε_0, D_0. In order for the dipole to orient, the ellipse must strain the environment surrounding the spherical cavity.

A more realistic model is the schematic representation of the one given in Figure 12. The portion depicted in Figure 12 could be part of a polymer molecule or of some large molecule. The dashed line represents the Onsager cavity with a dipole moment of magnitude μ located at its center. Surrounding the Onsager cavity is a medium of dielectric constant ε_0 and tensile compliance D_0. Rigidly connected to the sphere containing moment μ are two nonpolar spheres labeled A. When the field E_0 is turned on, the group μ tends to rotate. To do so in this model the sphere A must distort the medium giving rise to a strain energy term. To evaluate the strain the actual $\Delta l/l$ for the orienting species must be calculated.

B. Limiting Behavior

If we define $\{\cdots\}$ to be

$$\{\cdots\} = \left\{1 + \frac{\zeta^2 V \langle \boldsymbol{\mu} \cdot \mathbf{M} \rangle_0}{6ND_0kT}\right\}^{-1} \tag{72}$$

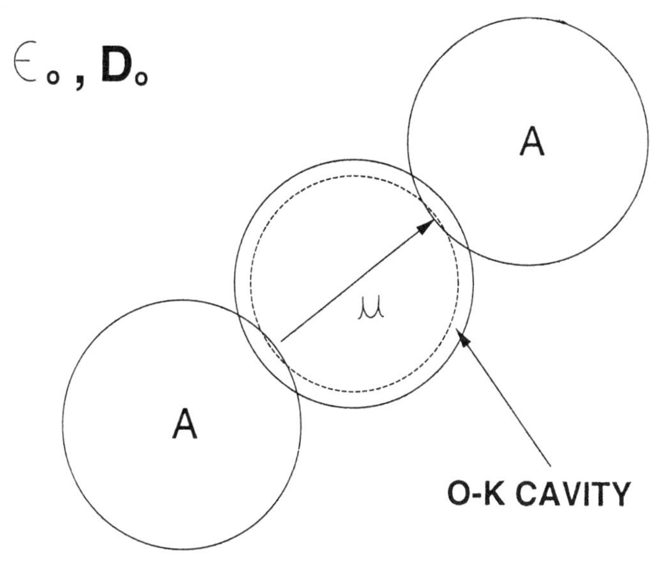

Figure 12. A spherical dipole in a cavity (dashed line) is rigidly connected to nonpolar sphere labeled A.

then the effects for various limiting values of ζ and D_0 on Eq. (46) can be evaluated.

Case I. Consider the case of symmetrical units, that is, $\zeta \to 0$ units and finite D_0 so that $\{\cdots\} \to 1$ and

$$\frac{\varepsilon_0 - 1}{\varepsilon_0 + 2} \frac{3V}{4\pi} \to \frac{N}{3kT} \langle \boldsymbol{\mu} \cdot \mathbf{M} \rangle_0 \tag{73}$$

Case II. Consider the case of highly unsymmetrical units or needle shaped units, that is, $\zeta \to \infty$, and finite D_0 so that $\{\cdots\} \to 0$ and

$$\varepsilon_0 \to 1 \tag{74}$$

Case III. Consider the case of a rigid environment, that is, $D_0 \to 0$, and $\zeta \neq 0$ so that $\{\cdots\} \to 0$ and

$$\varepsilon_0 \to 1 \tag{75}$$

Case IV. Consider the case of a flexible environment, that is, $D_0 \to \infty$, and $\zeta \neq 0$ so that $\{\cdots\} \to 1$ and

$$\frac{\varepsilon_0 - 1}{\varepsilon_0 + 2} \frac{3V}{4\pi} \to \frac{N}{3kT} \langle \boldsymbol{\mu} \cdot \mathbf{M} \rangle_0 \tag{76}$$

These limit results suggest that Eq. (46) is well behaved at the limits. For example, when $\{\cdots\} = 1$, Eq. (73) becomes the familiar polar liquid expression. On the other hand, when $\{\cdots\} = 0$, the equilibrium dielectric constant, $\varepsilon_0 = 1$, that is, that of a vacuum and the material behaves as if it were a nonpolar liquid. The results for the four cases just discussed in the previous section are the same and will not be reproduced here.

C. Curie Temperature

The second criticism of the Debye equation is that it predicts an electrical Curie temperature. Equation (46) does not lead to an electrical Curie temperature for certain choices of the parameters. This point can be seen from the following variation of Eq. (46).

$$\frac{\varepsilon_0 - 1}{\varepsilon_0 + 2}\frac{3V}{4\pi} = n\left\{\frac{2D_0N}{6D_0NkT + Vg_r\gamma^2}\right\}g_r\mu^2 \tag{77}$$

When $6D_0NkT < Vg_r\gamma^2$, Eq. (77) becomes

$$\frac{\varepsilon_0 - 1}{\varepsilon_0 + 2}\frac{3V}{4\pi} = \frac{2D_0N^2}{V}\frac{\mu^2}{\gamma^2} \tag{78}$$

In other words the polarizability and hence the dielectric constant are independent of temperature. Using the numerical constants in a previous section, i.e., Eqs. (28a) and (28b), the results are found to be quite reasonable and are given below:

For $D_0 = 10^{-6}$ dyn/cm^2: $$\frac{6D_0NkT}{Vg_r\gamma^2} = 2.49T$$

For $D_0 = 10^{-10}$ dyn/cm^2: $$\frac{6D_0NkT}{Vg_r\gamma^2} = 2.49 \times 10^{-4}T$$

In other words for reasonably rigid systems with $\gamma \sim 2$ the polarizability and hence ε_0 do not exhibit a Curie temperature.

D. Apparent Relaxation

Consider a case where the relaxation time of point dipoles in the system is much less than 10^{-7}. Then, in the experimental frequency range of 10^2–10^5 Hz, equilibrium dielectric constants will be measured. However, let us assume that γ^2/D_0 is such that $\varepsilon_0 - 1$ is reduced in magnitude. Furthermore, the equilibrium compliance D_0 is a function of time such that $D_0(t) = D_0\{1 - \exp(-t/\tau)\}$ where τ is the relaxation time somewhere in the experimental frequency range (10^2–10^5 Hz). This assumed relaxation process has a single relaxation time. The so-called equilibrium dielectric constant $\varepsilon_0(t)$ will have a value that is dependent on the frequency of measurement, simply because $D_0(t)$ is time dependent. This time-dependent dielectric constant may be given by redefining ε_0 to be $\varepsilon_0(t)$ in Eq. (46). This substitution yields the following equation:

$$\frac{\varepsilon_0(t)-1}{\varepsilon_0(t)+2}\frac{3V}{4\pi}=\frac{N}{3kT}\left\{1+\frac{\zeta^2 V g_r \mu^2}{6D_0(t)NkT}\right\}^{-1} g_r\mu^2 \tag{79}$$

Numerical estimates can be made with a few simplifying assumptions such as $\varepsilon_0(t)>1$ and $\varepsilon_0(t)D_1(t)\approx D(t)$ and finally scaling the results, that is, $\{\varepsilon_0(t)-1\}/\{\varepsilon_0-1\}$. In Figure 13 we have given the time dependence for a number of different values of $\gamma^2/D_0(t)$. In the first case, the value of $\gamma^2/D_0(t)$ was chosen such that part of the polarization process was partially "frozen out." In this case the relaxation times are nearly the same, the dielectric process is the faster one by a factor of about 1.5. However, in the other cases $\gamma^2/D_0(t)$ was chosen to be much greater.

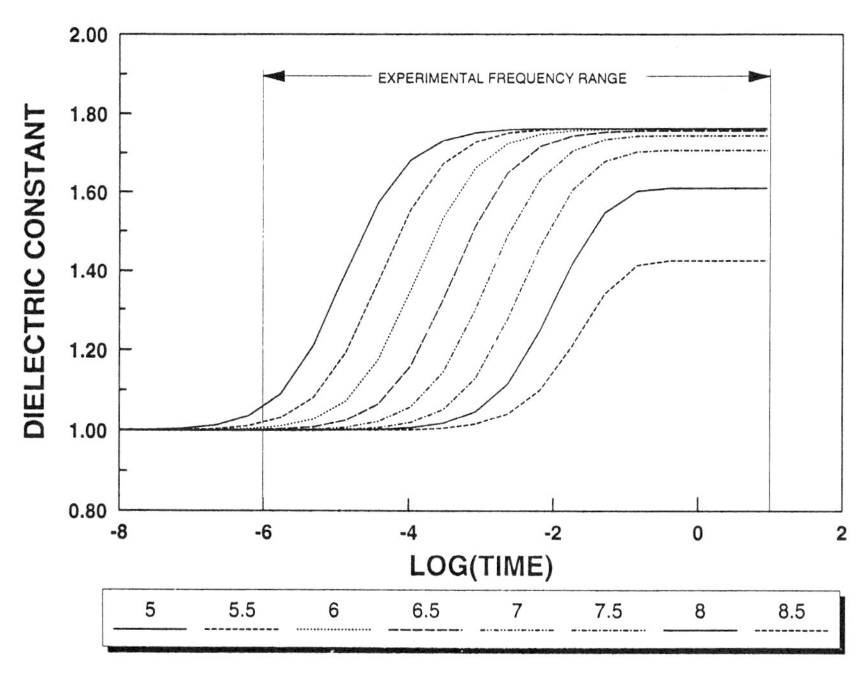

Figure 13. Plot of the time-dependent dielectric constant as a function of time for different values of the tensile compliance given in the legend.

Under these conditions, $\varepsilon_0(t)$ comes to its equilibrium value [when $\gamma^2/D_0(t)$ is such that it no longer freezes out the process] much sooner and its apparent relaxation time is much shorter. Clearly, from Figure 13, the relaxation time depends on the magnitude of D_0. This dependence can be seen by defining a relaxation time to be when $\varepsilon_0(t)$ reaches a certain value $\varepsilon_{0\tau}(t)$. This situation comes about when the quantity $D_0(t)$ reaches a specific value $D_{0\tau}(t)$ at which all other quantities have been assumed to be constant. For large values of D_0, this level will be reached in a short time so that $t/\tau < 1$. Making the substitution into the time dependence for $D_0(t)$ we have

$$D_0(t) = D_0 t/\tau \tag{80}$$

$$D_{0\tau}(t) = D_0 t_\tau/\tau \tag{81}$$

where t_τ implies a time that fixes $D_{0\tau}(t)$ for a given D_0. In other words the apparent dielectric relaxation time will be a linear function of D_0.

E. Multiple Dispersions

Many polymers exhibit multiple relaxation processes. For those polymers that are partially crystalline such as polyesters, the interpretation is quite simple; relaxation in the crystalline and noncrystalline domains have different dynamic parameters so that their respective relaxation processes occur at different experimental conditions. Acrylic polymers are not though to be partially crystalline, although there is some evidence to suggest that there may be a considerable amount of local structure. Not all acrylic polymers have two dispersions, those with short side chains exhibit two dispersions while those with longer side chains have only one, as the following results demonstrate.

A plot of the dielectric loss tangent with temperature for conventionally initiated polymethacrylates (PMA) taken from the data of Steck,[30] and Dyvik and Bartoe[31] is given in Figure 14. This series clearly demonstrates several important features. First, poly(methyl methacrylate) (PMMA) and poly(ethyl methacrylate) (PEMA) have two loss tangent maxima while the other polymethacrylates have only one. The relative magnitudes of the higher temperature loss peaks for these two polymers, that is, α peaks, are much smaller than their corresponding β peaks. The magnitude of the loss tangent for these two β peaks are similar to the α peaks of the other methacrylate polymers. The β peaks for PMMA and PEMA do not correlate with the glass temperature while the α peaks do correlate with that temperature. In other words, increasing the side chain length to about 3 or 4 carbon atoms appears to freeze out the β (glass

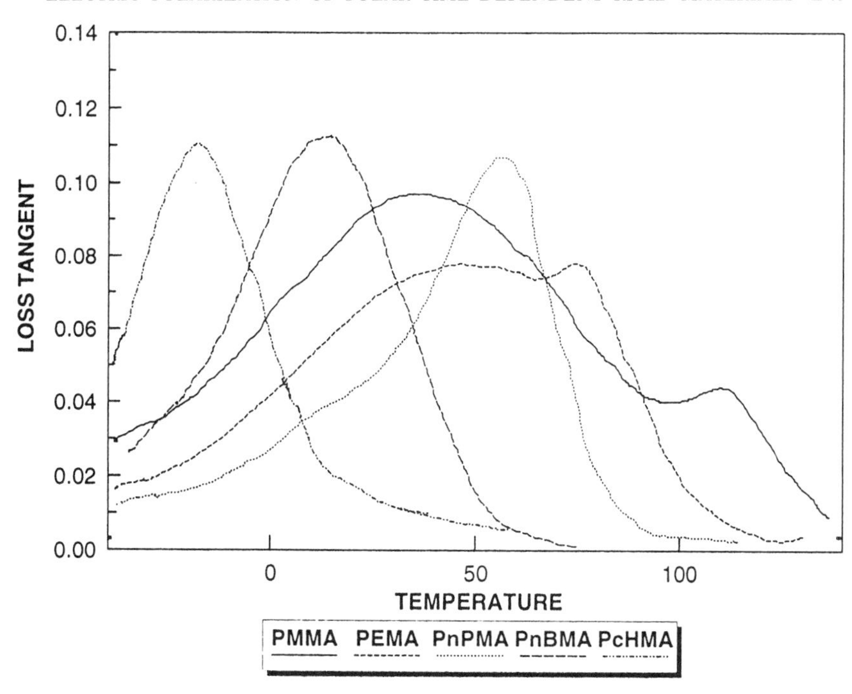

Figure 14. Plot of the loss tangent ($\times 100$) as a function of temperature for a number of methacrylate polymers. The polymers are listed in the legend.

phase) relaxation process. A complete set of equilibrium and instantaneous parameters for this acrylic series is given in Table I.

The dipole moments of the acrylic polymers just described lie in the planes of the ester groups, and nearly pointing in the direction of the carbonyl group.[32] The results from an infrared study and computer modeling of syndiotactic structures with increasing side chain lengths leads to the hypothetical structure shown in Figure 15.[33] Starting from the top of the figure, the first monomer unit is methyl methacrylate, the second, connected to the first in a syndiotactic placement is ethyl methacrylate, the third also connected in a syndiotactic placement is propyl methacrylate, and so on down the chain. Although this structure is hypothetical, it can be used to illustrate the following points. The labels in that figure identify the side chains and are placed near the carbonyl group. The carbonyl groups were positioned to be pointing in the same direction (toward the viewer) to demonstrate their proximity to the main chain. Dipoles pointing in the same direction are inconsistent with dipole moment calculations based on trial structures,[34] which suggest that they alternate toward and away from the viewer position. The side chains are

TABLE I
Instantaneous and Equilibrium Dielectric Constants for Some Acrylic Polymers

				β Process		α Process	
Polymer	η	η^2	$(1.07\eta)^2$	ε_∞	ε_0	ε_∞	ε_0
s-PMMA	1.47	2.16	2.47	2.50	4.6	4.6	5.0
PMMA	1.49	2.22	2.54				
PMA	1.48	2.19	2.51			4.11	7.42
PEMA	1.49	2.22	2.54	2.36	4.77		
PnPMA							
PnBMA	1.48	2.19	2.51	—	—	2.17	3.64
PiBMA	1.46	2.13	2.44	—	—	2.10	3.54
PAMA				—	—		
PcHMA	1.51	2.28	2.61	—	—	2.21	3.86
PnHMA	1.48	2.19	2.51	—	—	2.60	4.25
PnOMA				—	—	2.53	4.22
PnNMA				—	—	2.37	3.42

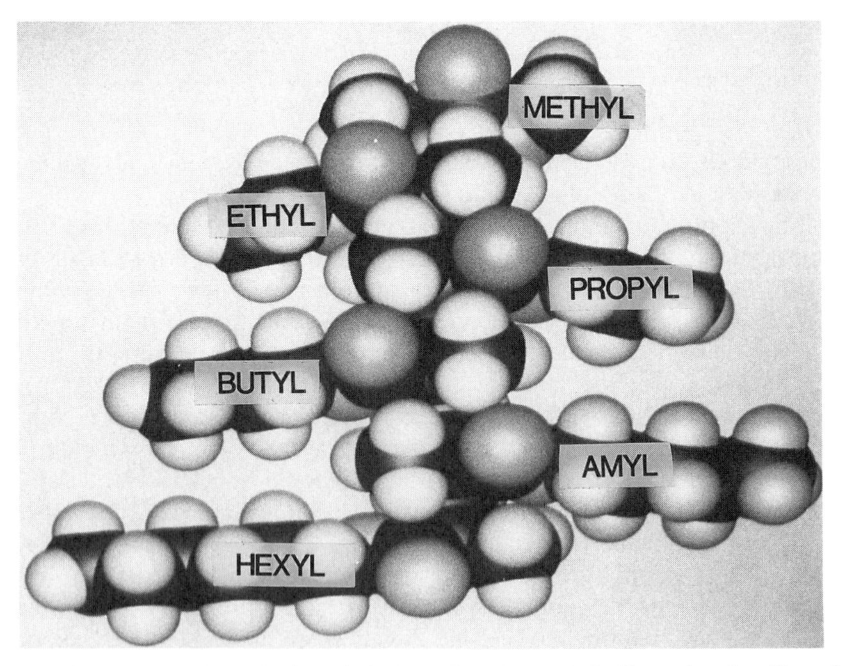

Figure 15. Top view of a hypothetical acrylic polymer chain illustrating the effect of polymer side chain length.

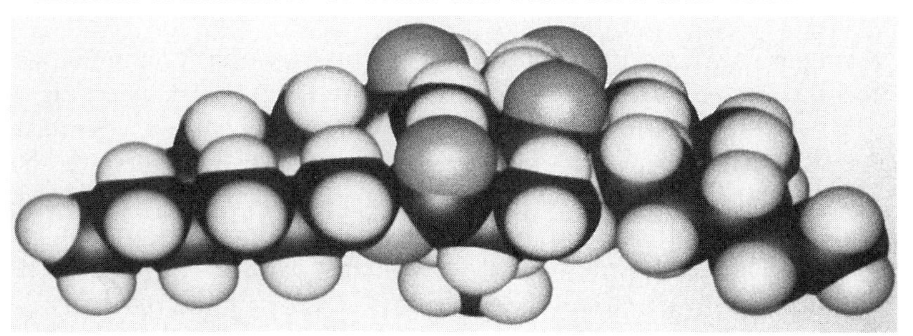

Figure 16. End view of the photograph in Figure 15.

in their extended conformation to emphasize their length. It is clear from this figure that increasing side chain length serves to separate main chains.

Figure 16 is an end view of the same structure shown in Figure 17. In this view the dipole moments would be pointing up. It is the dipole moment of several ester groups that interact with the applied electric field. Restricting that orientation at equilibrium is the equilibrium tensile compliance surrounding the unit.

Let us assume the following form for the time dependence of $D(t)$:

$$\frac{D(t)}{\gamma^2} = 1 \times 10^{-l} + \{1.0 \times 10^{-k} - 1 \times 10^{-l}\}[1 - \exp(t/\tau_0)] \qquad (82)$$

This expression is a simple exponential rise with limits at 0 time of 10^{-l} and at infinite time of 10^{-k}. Since the lower limit of tensile compliance for most polymers is about 10^{-12} cm^2/dyn for polymeric glasses at very

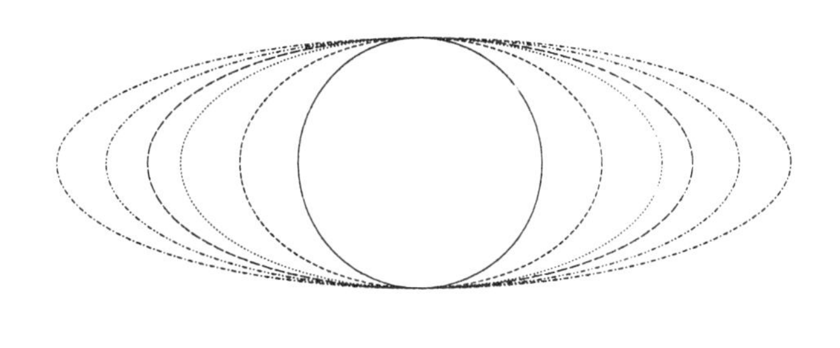

Figure 17. Elliptic cross section of acrylic polymers with increasing side chain length.

low temperatures and 10^{-5} cm^2/dyn for the rubbery transition region well above the glass transition temperature. For purposes of calculation let us set $l = -12$ and let k range from 5 to 8.5 in increments of 0.5. Since the quantity γ is a shape factor and is not known with certainty we shall assume that it is lumped in with $D(t)$. Presumably the range would nearly be zero for symmetrical units and perhaps 10–100 for highly unsymmetrical units. The results of this calculation are given in Figure 18. The relaxation time was adjusted to center the relaxation process in the experimental frequency range. At first, k shifts the relaxation process to longer times (lower frequencies), and finally decreases the magnitude of the relaxation process. The effects of l is observed by setting $k = 5$ and varying l from 11 to 14. The results of this calculation are given in Figure 19.

If we assume that the relaxation time, τ_0, is given by a typical Arrhenius rate plot:

$$\ln \tau_0 = \frac{\Delta E}{RT} + \ln(\Delta S) \tag{83}$$

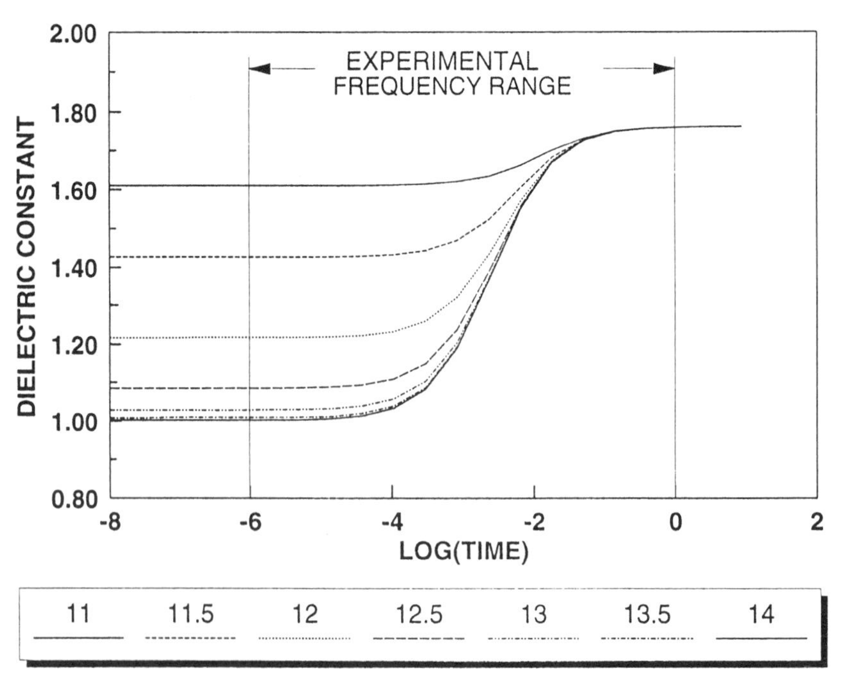

Figure 18. Dependence of the dielectric constant with time on various values of the instantaneous tensile compliance listed in the legend.

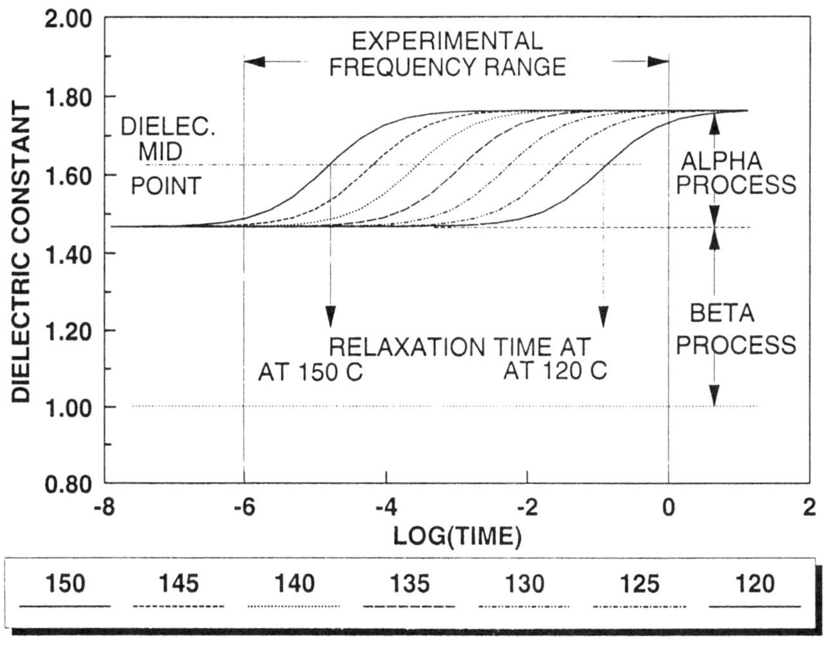

Figure 19. Dependence of the dielectric constant for fixed values of the instantaneous and equilibrium dielectric constants but for the relaxation times listed in the legend.

In this expression the activation energy ΔE is taken to be 100 kcal/mol, a number similar to the activation energy of s-PMMA $\ln(\Delta S)$ is set to -125, which places the rate plot in the experimental temperature and frequency range. The results of the calculation are given in Figure 20.

The α and β dispersions of s-PMMA can be discussed in terms of Figure 19. In the glass phase, γ^2/D_0 is such that the polymer segment orients with respect to the electric field that results in a value of $\varepsilon_0 = 4.6$; see Table I. At temperatures above the glass transition region the ratio γ^2/D_0 changes by several orders of magnitude not because γ^2 changes but because D_0 changes by 5 decades. The relaxation observed in Figure 19 labeled α process is entirely due to the assumed relaxation of D_0. In other words the polarization process is not complete until the strain energy term approaches zero. The activation energy of this dielectric α process is the same as is the activation energy of D_0 because it is the rate limiting step. In other words the rate of polarization (α process) due to softening of the matrix is determined by the matrix rate parameters even though the β process has a much faster rate.

The dielectric relaxation time can be defined as the time that the

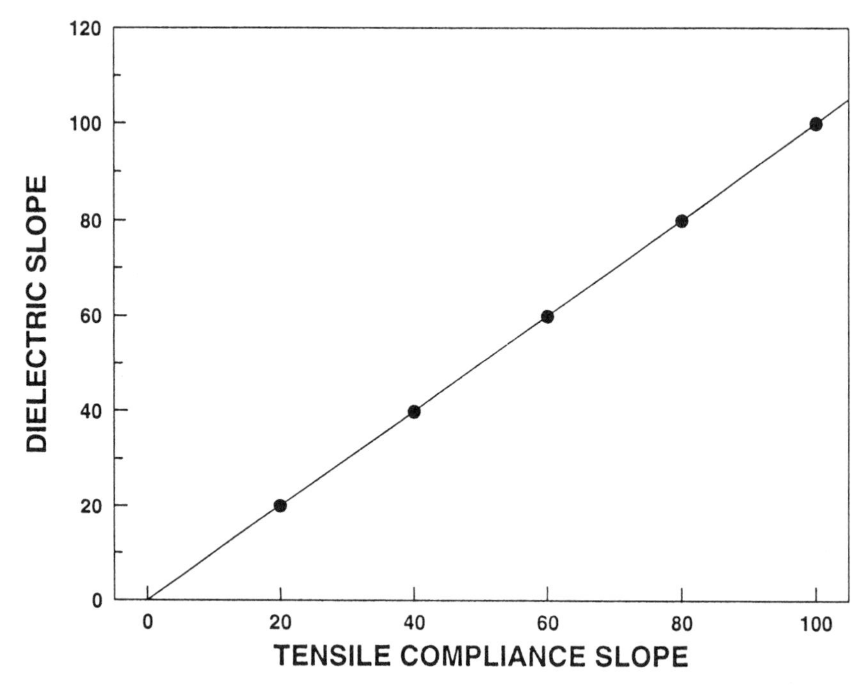

Figure 20. Dependence of the dielectric relaxation time on viscoelastic relaxation time.

dielectric constant (for the α process) reaches half its final value; see Figure 19. An activation energy can be calculated from a plot of these relaxation times with reciprocal temperature absolute. In Figure 20 we have plotted the dielectric activation energies for a range of assumed activation energies (20–100 in increments of 20) for the relaxation time in Eq. (83). $D(t)$'s were calculated using Eq. (83) and finally the relaxation time was calculated for $\varepsilon_0(t)$. The dielectric relaxation time was used to calculate the activation energy. As expected, the dielectric activation energy is determined by the activation for D_0.

Increasing the side chain length from methyl to ethyl, that is, from PMMA to PEMA alters some of the details for the α and β processes but the results, hence the interpretations, are similar. PMA is also similar to these two polymers, hence the interpretation is similar. No other acrylic polymer listed in Table I exhibits a β or glass phase process. This point is verified by the refractive indices (η) listed in Table I. Squaring η with or without some allowance for atomic polarization is similar if not the same as the ε_∞ for the other methacrylates. This agreement suggests that there are no higher frequency relaxation processes.

In as much as D_0's for acrylic polymers in the glass phase appear to be the same, it is the γ^2's that are different and consequently inhibit the orientation process. Figure 7 gives the reason of the dependence of γ^2 on side chain length. As the side chain increases in length, the cross section of the polymer becomes less cylindrical and more elliptical. At a side chain length of 4, that is, butyl, the ratio γ^2/D_0 in the glass phase inhibits the polarization process.

F. Broad Dielectric Dispersions

In the previous two sections we discussed the effects of representing a single-exponential time-dependent compliance on dielectric behavior. It is well known[35] that the time dependence of the tensile compliance of glassy materials such as polymers cannot be represented as a single exponential function. The decay can, however, be represented by a distribution of exponential functions. A considerable effort has been made in studying the origins and shapes of tensile compliance with time. In this work we simply asssume that an approximation to these distributions can take the form

$$\frac{D(t)}{\gamma^2} = 1 \times 10^{-(l)} + \{1.0 \times 10^{-(d)} - 1 \times 10^{-(l)}\}\{(1 - \exp[-(t/\tau_0)^c]\} \quad (84)$$

In this expression the last term in brackets is commonly known as the stretched exponential[36,37] and c is the stretched exponential constant which can range from 0 to 1.

For purposes of demonstrating the effect of c on $\varepsilon(t)$, let us set $l = 12$, $k = 8$, and let c range from 0.2 to 1.0 in Eqs. (79) and (84). The results for the time-dependent log(compliance) are shown in Figure 21. Since the quantity γ is a shape factor and is not known, we shall assume that it is lumped in with $D(t)$. Results of calculating $\varepsilon(t)$ for this range of tensile compliances from Eqs. (79) and (84) are given in Figure 22. These time-dependent dielectric constants can be transformed into the frequency domain, that is, complex dielectric constant, using the extended Scwarzl method.[38–43] A complex plane plot of the results are given in Figure 23 for the case of $l = 12$ and $d = 8$ and a range of c from 0.2 to 1.0. The plots exhibit a range of shapes when viewed in the complex plane. These complex plane plots can be represented by the H–N function using the techniques previously reported.

The H–N parameters and their confidence intervals for the four curves in Figure 23 (on page 258) are listed in Table II. A plot of the H–N parameter ε_0 as a function of c for different values of l are shown in Figure 24 (on page 259). The value for ε_0 is independent of c but depends on l. A plot of the H–N parameter α with c for various levels of l is given

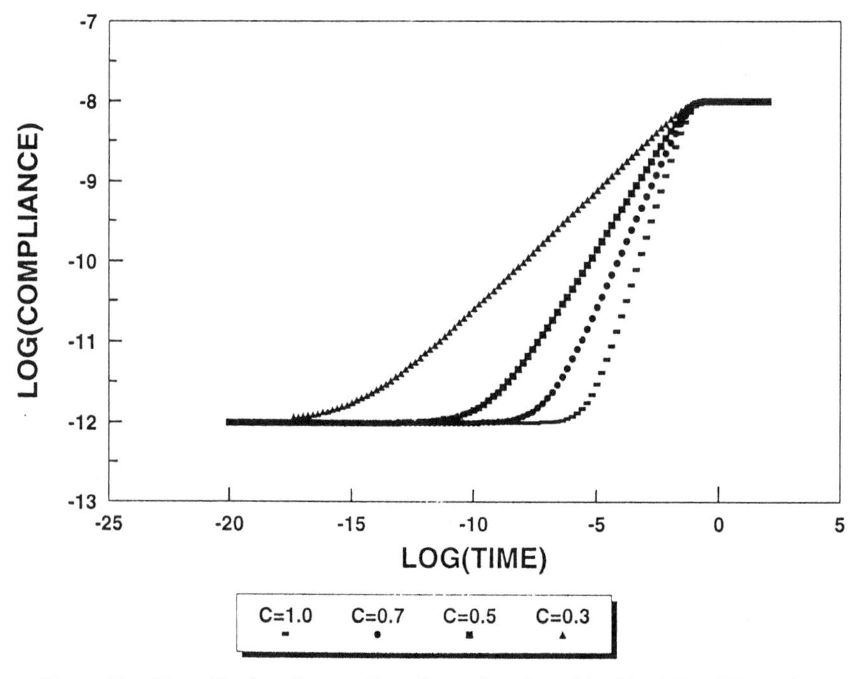

Figure 21. Plot of log(tensile compliance) as a function of log(time) for different levels of c and $l = 8$. Points indicate spacing and half the points used in the calculations.

in Figure 25 (on page 260). A plot of the H–N parameter β with c for various levels of l is given in Figure 26 (on page 261). A plot of H–N parameter $\ln(f_0)$ with c for various levels of l is given in Figure 27 (on page 262). Increasing the width of the tensile compliance curve, that is, decreasing c, has the effect of increasing the width of the dielectric relaxation process, that is, decreasing α. The magnitude of the change depends on l and at $l = 9.5$, there is no change and $\alpha = 1$. Increasing the width of the tensile compliance curve, that is, decreasing c, has the effect of decreasing β, that is, increasing the skewness of the dielectric relaxation process. Although the change depends on l, it does approach a limit as l approaches 9.5. There is no simple condition for $\beta = 1$. As in the case of β, $\ln(f_0)$ approaches a limit that is independent of c as l approaches 9.5.

A plot of α vs. β[44] for many materials have shown that the entire plane is populated, although not uniformly. Furthermore, this plot does not seem to sort out material by type. For example, such diverse materials as the β process in PMMA or PEMA are similar to those of cyclohexane, 124 Methyl 356 Chloro benzene, or even poly(trifluro monochloro ethylene). The list of materials for the case of $\beta = 1$ is even

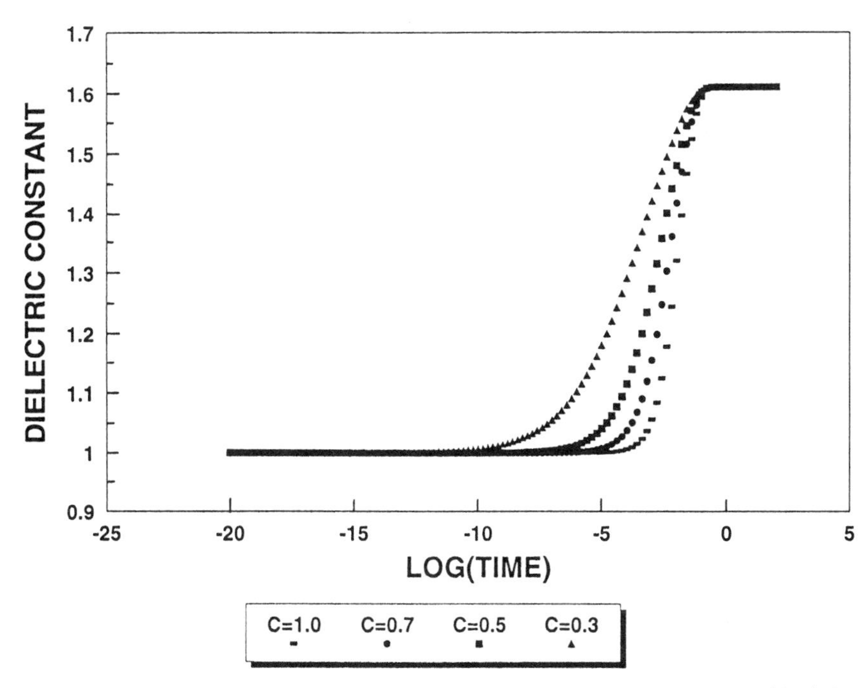

Figure 22. Plot of the time-dependent dielectric constant as a function of log(time) for different values of c and $l = 8$.

longer if one includes all of the original work of Cole et al.[45] In their work circular arcs were found to represent the dielectric relaxation process of such diverse materials as slate and halowax in addition to materials similar to the ones just cited. In other words there appears to be no clear delineation of the material type that is represented by the circular arc function. A plot of the values for these materials would be misleading because none of the researchers have tested the data to determine if two parameters were necessary. Values of $0.8 \leq \beta \leq 1.0$ when $\alpha \cong 0.7$ go undetected unless unbiased statistical techniques are used. A recent example of this problem is that of Naoki[46] who reported on the β dielectric relaxation process of poly(vinyl chloride) as a function of temperature and pressure. It has been demonstrated[47] using unbiased techniques that this process requires α and $\beta < 1$. Even so many other researchers continue to use the circular arc formalism even though it is probably not adequate. The important point is that a broad spectrum of materials are likely to fill the α vs. β plane[4] and without segregation according to material types.

Perhaps the most significant result of these calculations can be seen in a plot of α vs. β for different levels of c and l shown in Figure 28 (on

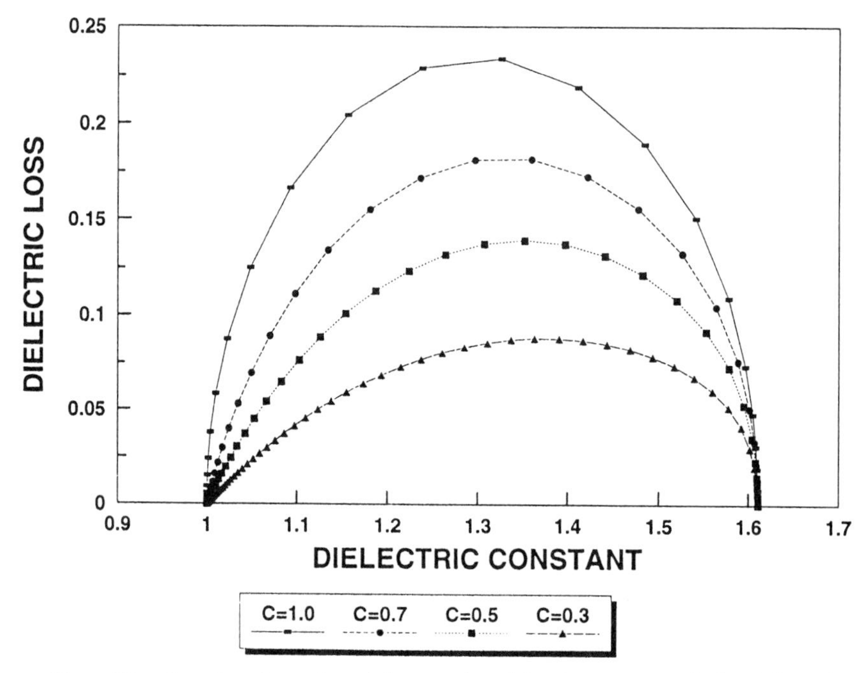

Figure 23. Complex plane plot of the complex dielectric constant calculated from the data in Figure 22.

TABLE II
Relaxation Function Parameters and Their Confidence Limits for Representing the Complex Dielectric Constant for Various Values of C and $L = 8$

Parameter	$C = 1$	$C = 0.7$	$C = 0.5$	$C = 0.3$
E_0	1.621	1.621	1.623	1.633
σ	0.001	0.001	0.001	0.001
ε_∞	1.000	1.000	1.000	1.000
σ	0.003	0.003	0.002	0.001
$\ln(f_0)$	6.19	6.35	6.6	7.7
σ	0.06	0.07	0.07	0.1
α	0.82	0.76	0.67	0.48
σ	0.01	0.01	0.01	0.01
β	1.09	0.77	0.60	0.50
σ	0.05	0.03	0.02	0.02

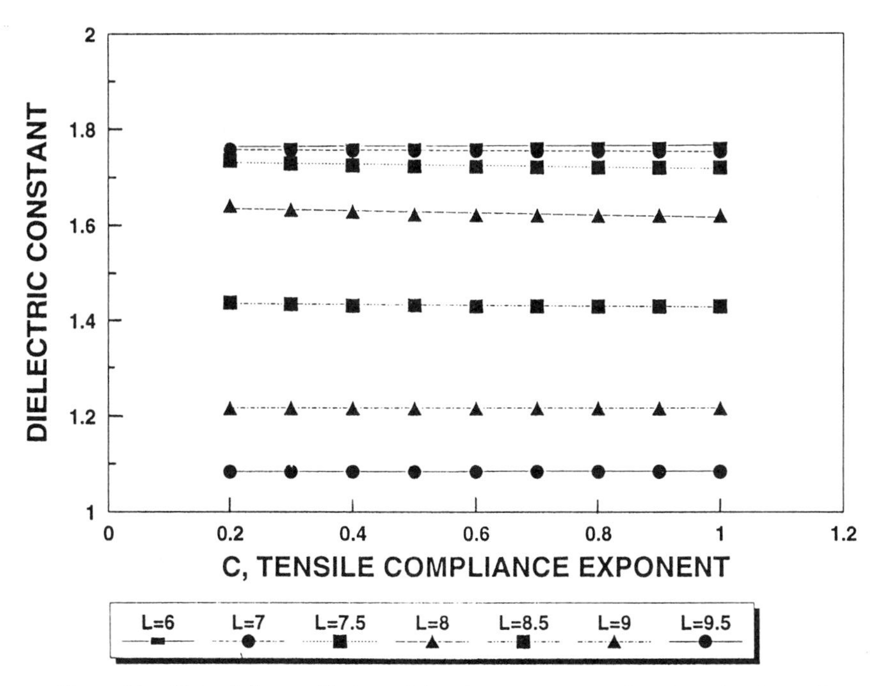

Figure 24. Plot of the equilibrium dielectric constant as a function of c for various levels of l.

page 263). In this plot the various c values have been connected at constant l. The different l values are indicated in the legend. A c of 1.0 is always in the upper right-hand corner of the graph for a given l, and the points shown are decreases in c by 0.1. The results shown in the figure are important because nearly the entire physically significant range of α and β is covered by the present range of c and l chosen for computations. Presumably if c was decreased to 0.1 and permitted to approach 0, the entire range, that is, $0 \leq \alpha \leq 1$ and $0 \leq \alpha \cdot \beta \leq 1$, would have been covered.

The well-known Cole–Cole plot ($0 \leq \alpha \leq 1.0$) forms a single line in this α, β plane; see Figure 29 (on page 264). The Kirkwood–Fuoss[48] as well as the Gaussian functions[49] are essentially coincident with the Cole–Cole function. The Cole–Davidson function $0 \leq \beta \leq 1.0$ is also a single line in this plane and forms one of the bounds that is perpendicular to the Cole–Cole line. The stretched exponential is represented by a hyperbolic-like line in the α, β plane and takes the form[50]

$$(\beta - 0.26 \pm 0.04)(1.092 \pm 0.007 - \alpha) = 0.058 \pm 0.006 \tag{85}$$

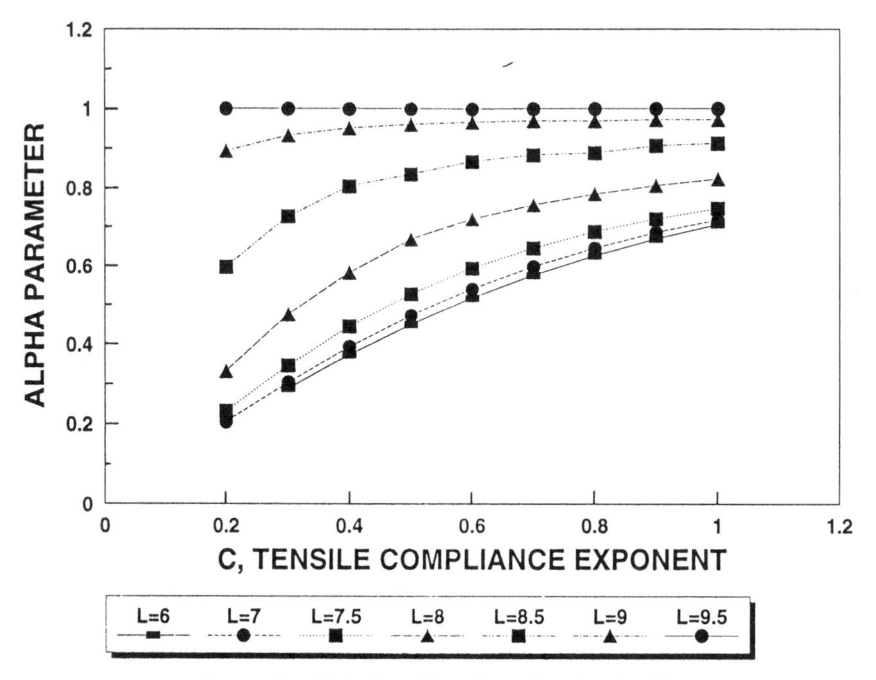

Figure 25. Plot of α with c for various levels of l.

Mansfield's model represents a range of α, β and accounts for about 16% of the total area.

The results in Figure 28 can be mapped into contour maps of constant l, or c as functions of α and β; see Figures 30 and 31 (on pages 265 and 266). These plots are important because they permit us to pick of the values of c and l for experimental values of α and β. In Figures 30 and 31 we have plotted the values of l and c for the Cole–Cole and Cole–Davidson functions, respectively. In addition there are a large number of theoretical correlation functions that form lines in this plane.[51]

The model proposed for dielectric relaxation assumes the inherent dielectric relaxation time to be much shorter than the time associated with the "softening up" of the environment. A simple distribution of exponential decays representing the softening of the environment accounts for a broad range of α, β pairs while changes in the asymmetry of the orienting unit completes the spreading out to all physically acceptable values of α and β. Hence two macroscopic parameters, that is, α and β have been related to two other parameters c and l. These two constants are closer to a molecular description of the system since one of them accounts for the macroscopic change in shape during the orientation

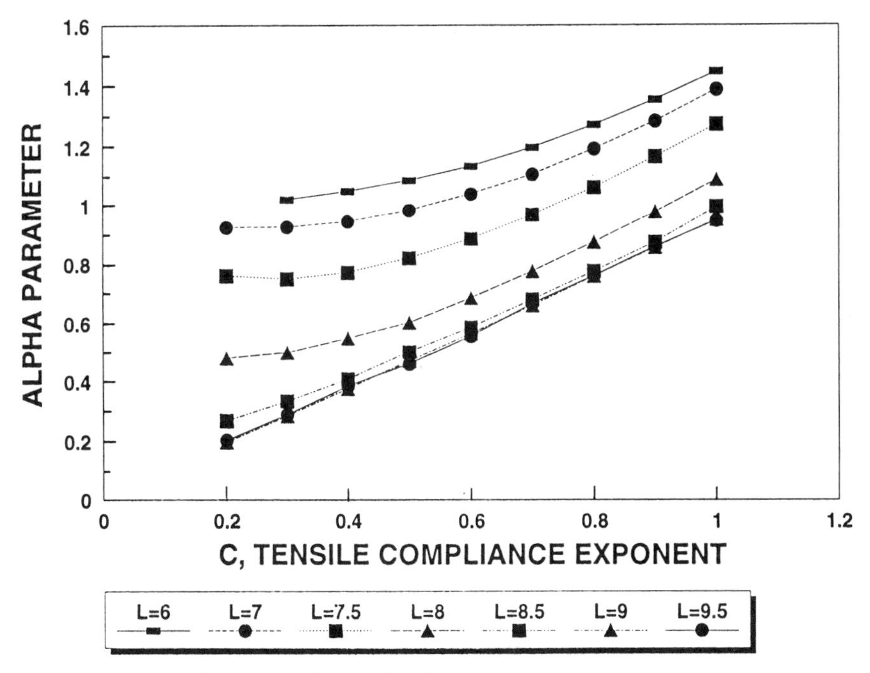

Figure 26. Plot of β with c for various levels of l.

process in terms of a molecular shape factor. A considerable effort has gone into understanding of the other parameter, c, in terms of vibrations along a polymer chain and the influence that the environment has on it. Equation (84) was chosen because of its simplicity. A more realistic choice would have been to use one of the functions used to represent polymer chain vibrations.

G. Comparison with the DiMarzio–Bishop Results[51]

Following the procedure adopted for evaluating the D–B model, we calculate $\varepsilon(t)$ for a given $D(t)$ using Eq. (79) and the viscoelastic data for poly(n-hexyl methacrylate). We use the frequency domain $J(\omega)$ data, see Reference 22 to calculate $D(t)$ in the time domain. Once the time-dependent dielectric constant is made, the frequency-dependent dielectric constant is calculated using the modified Schwarzl method.[52]

In Figure 32 (on page 267) we have given complex plane plots for different values of the shape parameters ξ, calculated from Eq. (79) and the time-dependent compliance for poly(n-hexyl methacrylate). The dependence of the H–N α, β parameters on the shape factor is given in

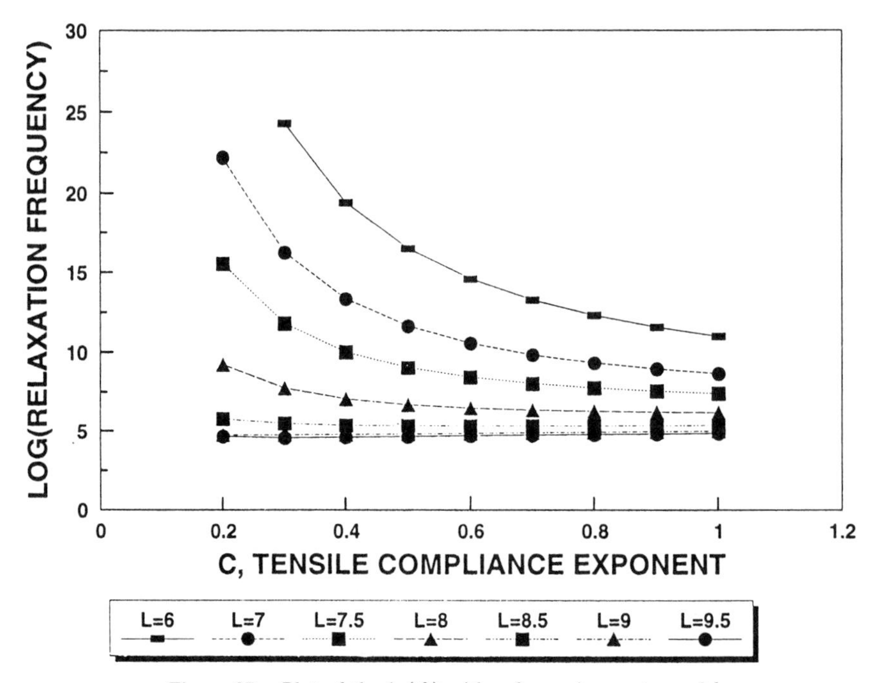

Figure 27. Plot of the $\ln(f_0)$ with c for various values of l.

Figure 33 (on page 268), while in Figure 34 (on page 269) the dependence of ε_0, ε_∞, and $\Delta\varepsilon$ on the shape factor is given. Finally in Figure 35 (on page 270) the dependence of $\ln f_0$ on the same parameter is given.

The results of the D–B and H–H models, as shown in a comparison of Figures 4–7 with Figures 32–35 are very similar. In both cases $\alpha \approx 0.6$ and approximately independent of either the K or the shape parameter ξ. The β parameter changes in a sigmoidal pattern from about 0.35 to 1.0 for the D–B model and from 0.35 to about the 0.7–0.8 range for the H–H model. In other words, both the D–B and the H–N models predict the shape of the dielectric relaxation process to have nearly the same dependence on sphere diameter or segment asymmetry ξ and both models have very different starting points.

In the case of DiMarzio and Bishop,[23] they solved the hydrodynamic equations for the Debye model and the non-Newtonian case exactly. The important result of their analysis is that the dielectric response is no longer a Debye type but depends explicitly on how the local viscosity depends on time. In other words the nature of the viscoelastic properties surrounding the sphere determines the shape of the dielectric relaxation process. This result is in marked contrast to the results of the model

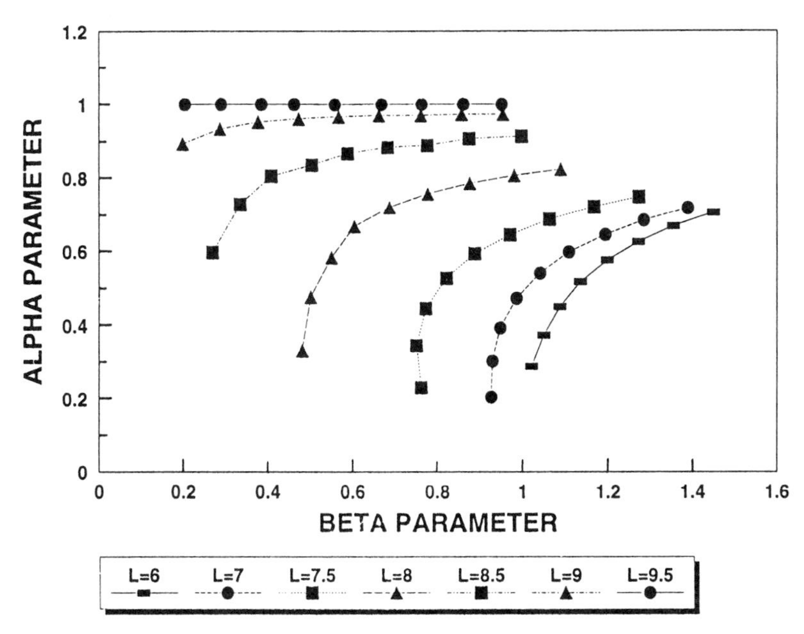

Figure 28. Plot of α against β for various levels of l indicated in the legend. In all cases $c = 1$ in the upper right-hand corner of the plot and decrease systematically to the left and down in increments of 0.1.

proposed by Cole[54] and extended by Shore and Zwanzig.[55] In these cases, the broadening of the dielectric relaxation process comes about because of time-dependent point dipole interactions. The shortcoming of the DiMarzio–Bishop result is based on the usual criticisms of the Debye model. In the view of the present authors, the criticisms of the Debye model is probably less serious than the point dipole assumptions that are customarily made.

The advantage of the H–H model is that the starting point is a more general formulation of the dielectric relaxation problem, that is, it is less specific than is the Debye model. Another advantage of this approach is that the relationship between strain and electrostatic energies is clearly incorporated into the model. This incorporation has the effect of approximating real molecules as point dipoles situated on bodies that have an arbitrary shape. Furthermore it is reasonable to assume that the relationship between dipole moment and shape factor is given by a tensor. In any case there is no reason to assume that the moment of the sphere and its distortion are collinear when the electric field is applied.

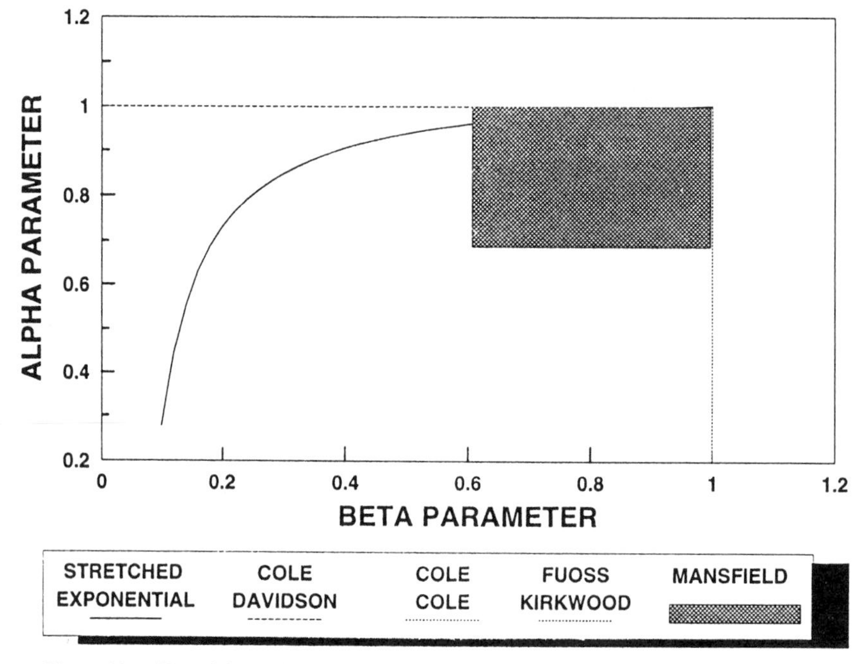

Figure 29. Plot of the various models for dielectric relaxation listed in the legend in the α, β plane.

For example, while the induced moment may be in the direction of the electric field, the distortion of the sphere may be perpendicular to the field.

The results of these comparisons show the advantages of incorporating the viscoelastic properties of the environment in predicting dielectric behavior. In this approach, the broadening of dielectric dispersions is due to the shape of the viscoelastic relaxation process. Equations (25) and (79) have at most only 1 adjustable parameter that is related to the volume or shape of the moving segment. This, of course, is a significant achievement, but the question now is why are viscoelastic dispersions shaped the way they are.

H. Comparison of Dielectric and Viscoelastic Dispersions

McCrum and co-workers[56] stated:

> In comparing dielectric and mechanical data it should be noted that the electric field **E** is analogous to the mechanical stress $\boldsymbol{\sigma}$ and that the dielectric displacement **D** is analogous to the mechanical strain γ. It follows

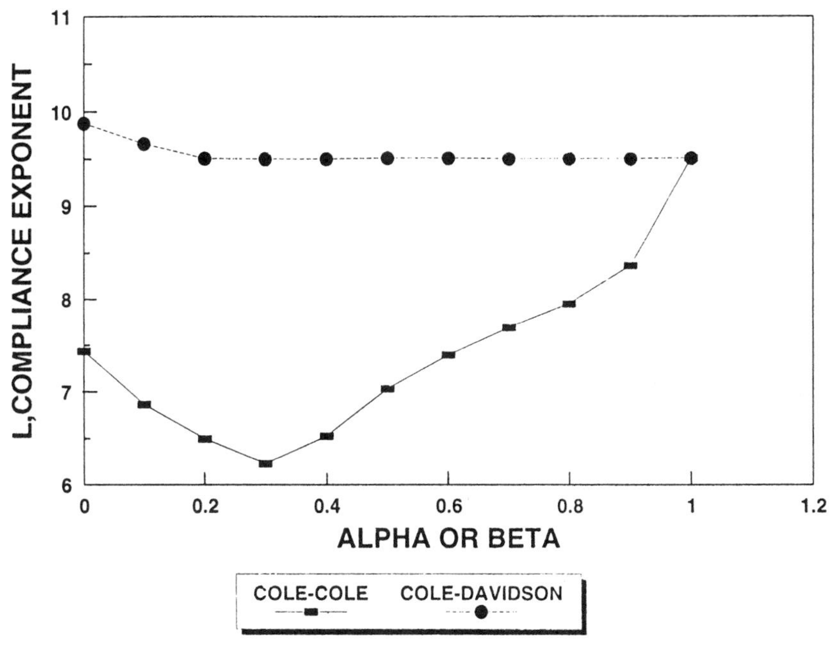

Figure 30. Plot of l as a function of α or β that meet the Cole–Cole or the Cole–Davidson condition.

that ε^* and τ_E are analogous to J^* and τ_σ respectively. Comparisons between dielectric and mechanical relaxation data might therefore be made, for example by noting the correlation which exists between the frequencies of maximum ε'' and J''.

An excellent example for such comparisons are the dielectric relaxation data of poly(n-hexyl methacrylate) by Strella and Chinai[57] and the viscoelastic relaxation data on the same polymer by Child and Ferry.[58] Complex plane plots for the dielectric process (normalized with respect to $\Delta\varepsilon = \varepsilon_0 - \varepsilon_\infty$) and the viscoelastic relaxation data are given in Figures 36 and 37 (on pages 271 and 272), respectively. There are several observations that can be made from these figures. First, the shapes are quite different for the two processes. Second, the frequency rages for the viscoelastic process is to the low side of the dielectric process and the temperatures are to the high side. An Arrhenius rate plot of the log(relaxation times) against temperature is given in Figure 38 for the $J^*(\omega)$ and the $\varepsilon^*(\omega)$ relaxation data. The $\varepsilon^*(\omega)$ data were treated using rigorous statistical techniques.[59] The two dashed lines represent 95% confidence intervals for the dielectric log(relaxation time). It was difficult

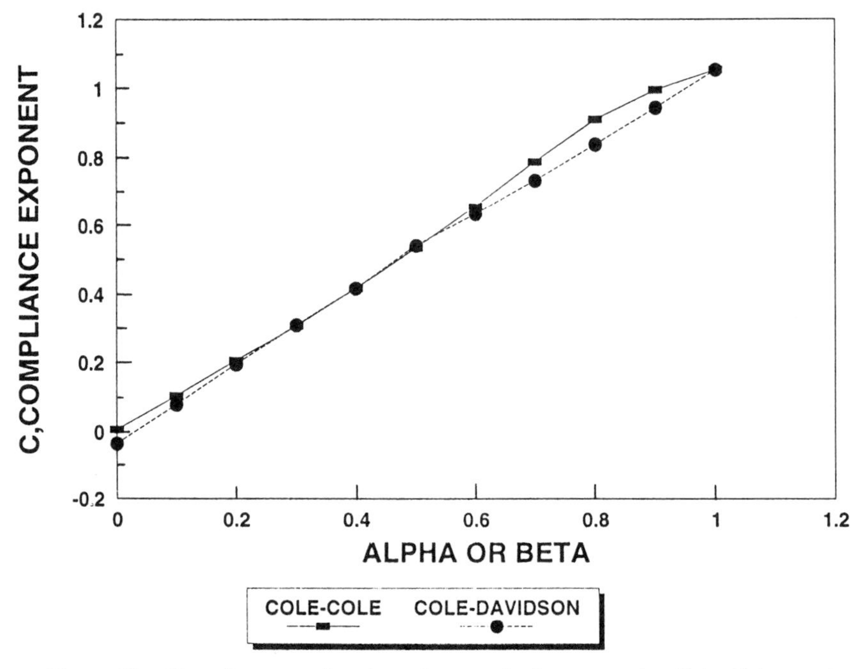

Figure 31. Plot of c as a function of α or β that meet the Cole–Cole or the Cole–Davidson condition.

to analyze the viscoelastic data using the same statistical techniques. For this reason we estimated the relaxation times by the procedure recommended by McCrum et al.,[56] that is, from the frequency of maximum loss. This estimation is plotted in Figure 38 (on page 273). Although one might expect small differences between the two methods of relaxation time estimates, that is, a factor of 2 or so, the results in Figure 38 show that there is a difference of about 4 orders of magnitude.

This observation is well known and in general there are two explanations. The first one is based on stating that a viscoelastic measurement is a macroscopic displacement while the dielectric measurement is a microscopic displacement. For this reason, several thousand microscopic displacements (dielectric) are required to make one macroscopic displacement (viscoelastic). Another method of comparison has been suggested and that is to compare the dielectric relaxation time to the viscoelastic relaxation time calculated form the modulus related to the compliance through $G^*(\omega) = 1/J^*(\omega)$. It is the view of the present authors that such comparisons are not properly made.

Since Scaife claims that $\rho^*(\omega)$ should be used for dielectric comparisons and the arguments given in Section IV.B support his contention,

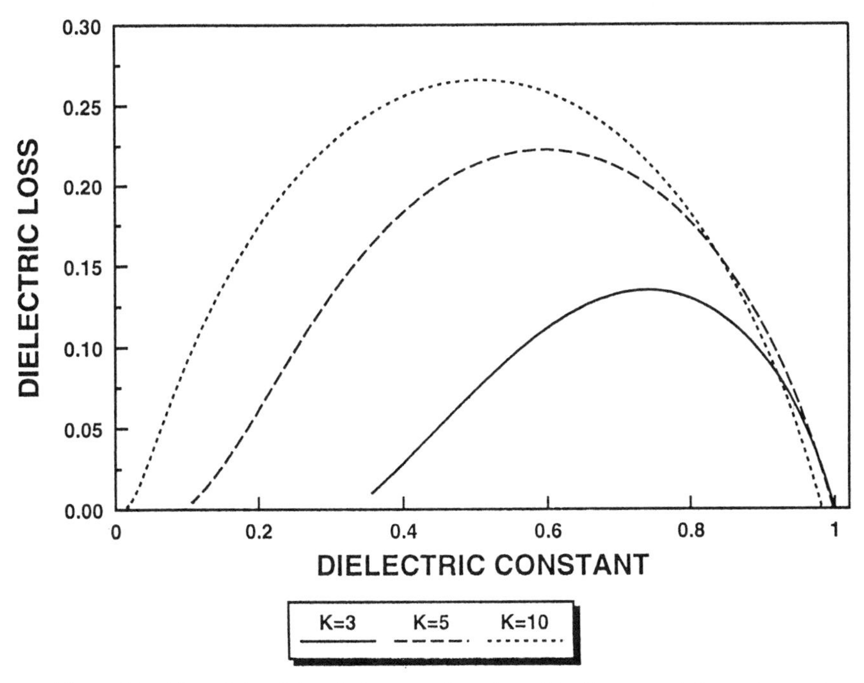

Figure 32. Plot of the dielectric loss as a function of log(frequency) for various values of the stretched exponential function parameter c for different values of l.

the question to be addressed is "What is the viscoelastic equivalent of the complex polarizability?" It is possible to construct a function analogous to the complex polarization of a sphere imbedded in a medium of ε_∞. This quantity is given by [complex distortion $\delta^*(\omega)$]

$$\delta^*(\omega) = \frac{J^*(\omega) - J_\infty}{J^*(\omega) + J_\infty} \tag{86}$$

In this equation J_∞ is the high-frequency limiting value of the real compliance and is the counterpart of $\varepsilon_\infty = 1$ in the corresponding electrostatic equations. The ratio of relaxation times by analogy with the dielectric case is given by

$$\tau_\delta = \frac{J_\infty}{J_0} \tau_J \tag{87}$$

In this expression J_∞ is the extrapolated value of $J'(\omega)$ to very high frequencies, while J_0 is the extrapolated value to very low frequencies.

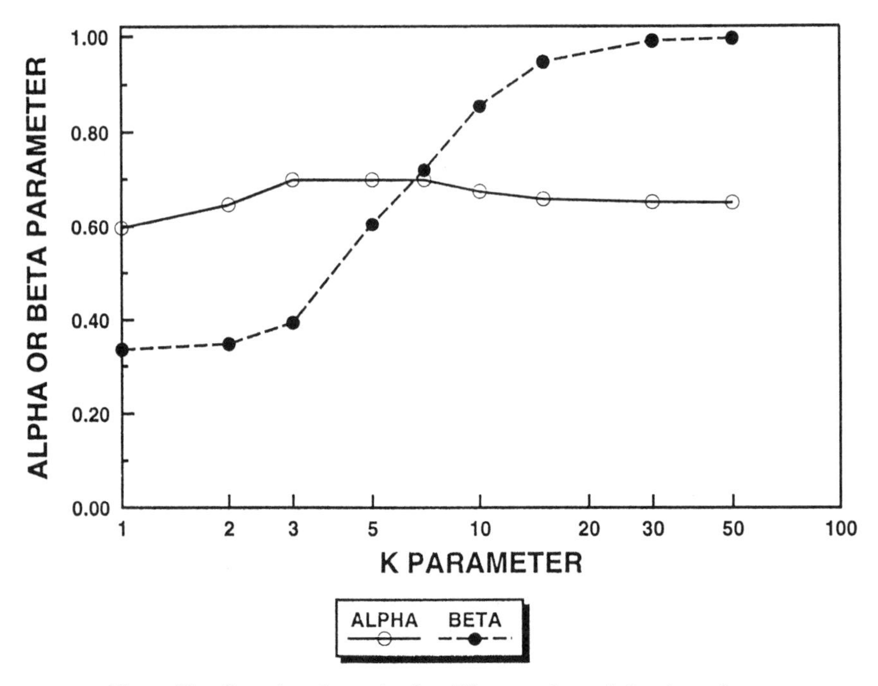

Figure 33. Complex plane plot for different values of the shape factor.

This extrapolation is readily accomplished in a complex plane plot and represents the condition of $J''(0) = 0$. Also in this expression τ_J is the relaxation time for the $J^*(\omega)$ data and τ_δ is the relaxation time for the $\delta^*(\omega)$ data. These relaxation times for $\delta^*(\omega)$ are plotted in Figure 38, and the agreement between the dielectric and viscoelastic relaxation times is now remarkable.

An equation similar to Eq. (86) can be derived rigorously from the basic theory of elasticity[60] by analogy to the model used to calculate $\rho^*(\omega)$;[61] see Figure 39 (on page 274). The viscoelastic analogue is a sphere suspended in a continuum of magnitude J_∞; the stress field is set by clamps. The result of an elastostatic analysis is given by Eq. (89).

$$\delta^*(\omega) = \frac{J^*(\omega) - J_\infty}{J^*(\omega) + \frac{2}{3}J_\infty} \tag{88}$$

The difference between Eq. (86) and (88) is slight.

The result of applying this equation to the viscoelastic α process of poly(n-hexyl methacrylate)[62,63] is given in Figure 40 (on page 275) for

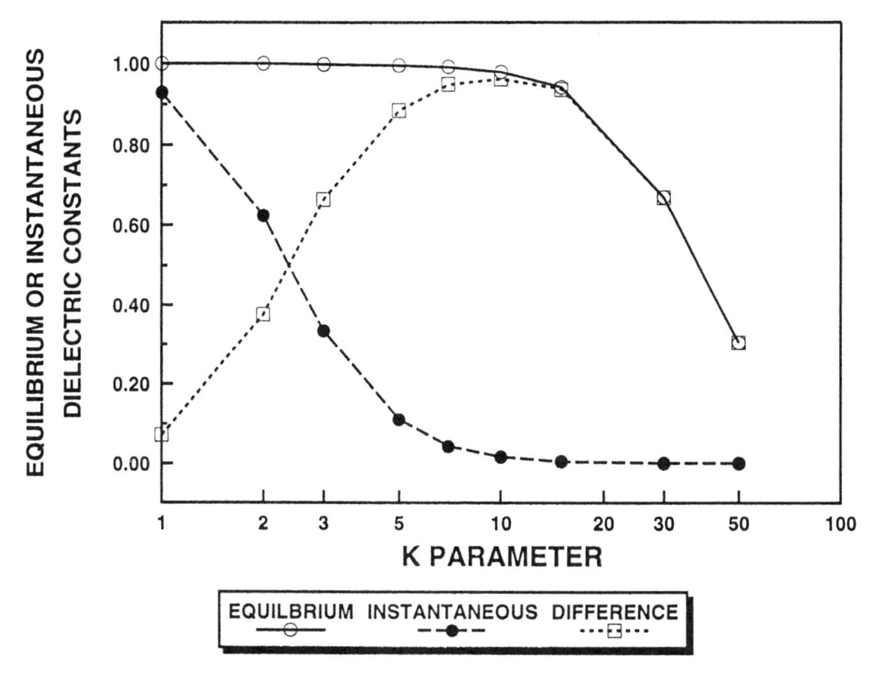

Figure 34. Dependence of the α and β parameters on the shape factor.

$\delta^*(\omega)$. The data in these figures can be analyzed using rigorous statistical techniques. Arrhenius rate plots for the relaxation times and their 95% confidence intervals for the dielectric and viscoelastic data are given in Figure 41 (on page 276). It is clear from these figures that the shapes of the complex plane plots are not only similar but that the rate plots overlap within their 95% confidence intervals.

Similar results[64] are obtained for poly(n-hexyl methacrylate), poly(n-butyl methacrylate), poly(methyl acrylate), and poly(vinyl acetate). In all cases the results follow the same pattern. The relaxation frequences for $J^*(\omega)$ data are 1000 to 10,000 times slower than they are for the corresponding dielectric ones. However, in the case of $\delta^*(\omega)$ and $\rho^*(\omega)$ comparisons, they are within the 95% confidence intervals. It should be pointed out that there is very little difference between the relaxation times estimated from $\rho^*(\omega)$ and $\varepsilon^*(\omega)$ data because the ratio $\varepsilon_0/\varepsilon_\infty \approx 3$–5.

A comparison of dielectric and viscoelastic relaxation α process has its own set of difficulties because the viscoelastic processes are small and the parameters J_0 and J_∞ highly temperature dependent. Nevertheless, the data can be analyzed and comparisons made. A comparison of the relaxation parameters for the α processes of poly(vinyl chloride),[65]

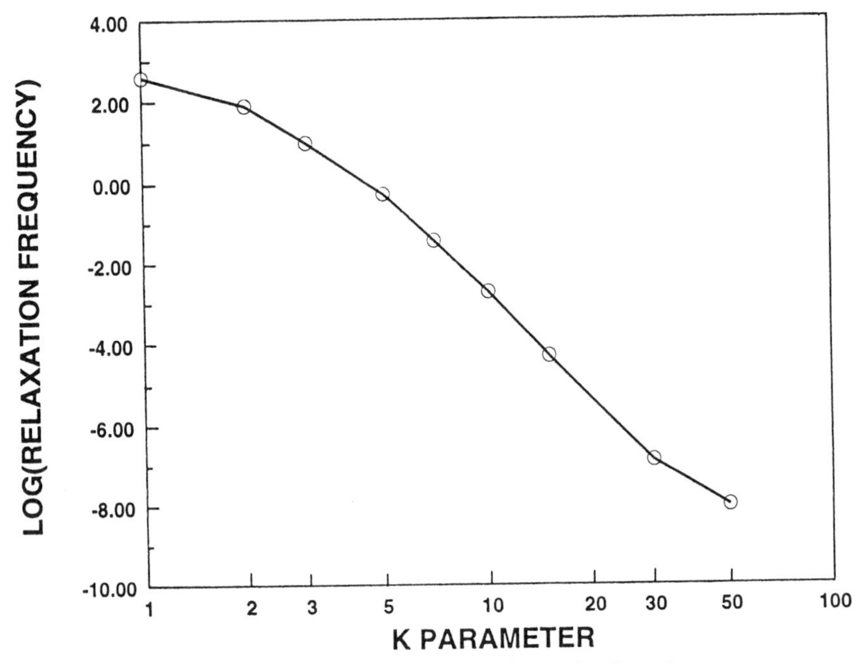

Figure 35. Dependence of ε_0, ε_∞, and $\Delta\varepsilon$ on the shape factor.

polycarbonate,[66,67] and poly(butelene theraphthalate)[68,69] show them to be the same within experimental error.

The results of this section show quite clearly that the application of Scaife's remarks to the comparison of dielectric and viscoelastic relaxation times changed their ratio (viscoelastic/dielectric) for the α process of polymers from several thousand to less than 2. It should be emphasized that since $\varepsilon_0/\varepsilon_\infty \approx 3$, a similar result would have been obtained from a comparison of $\varepsilon^*(\omega)$ and $\delta^*(\omega)$ data. Once again this method would be similar to comparing dissimilar quantities. In the case of small viscoelastic dispersions, that is, when $J_0/J_\infty \leq 3$, the relaxation times are also similar. This is important because it shows that scaling depends on the J_0/J_∞ ratio and not that viscoelastic responses are always several thousand times slower than the corresponding dielectric ones.

The results obtained from these dielectric-viscoelastic comparisons experimentally verify Scaife's remarks to be correct, that is, dielectric relaxation data should be analyzed in the form of $\rho^*(\omega)$ rather than $\varepsilon^*(\omega)$. A detailed analysis of the Glarum–Cole cavity model shows that their results, though mathematically elegant, simply do not represent the condition of real polar systems because there are many examples of

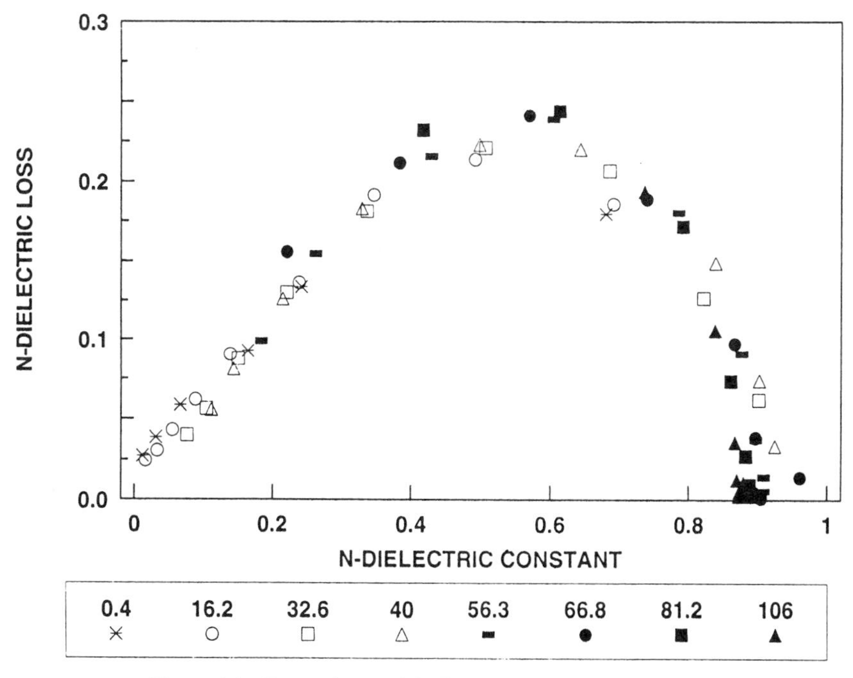

Figure 36. Dependence of ln f_0 on the shape parameter.

polar liquids that have relaxation shapes similar to those of polymers.[44] Consequently, even simple polar liquids probably contain long-range interactions during relaxation similar to those in polymers.

I. Small Molecules

Apart from an occasional reference to polymers, the equations developed in Sections IV and V are general and not necessarily limited to long-chain molecules. However, their application to small molecules is handicapped by the lack of information on D_0, though γ can usually be estimated reasonably well because of the preponderance of x-ray data on small molecules. Smyth[7] has reviewed, quite extensively, the dielectric properties of polar solids. In his work he attributed the low values of ε_0 to solidification, which usually fixes the molecule with such rigidity in the lattice that little or no orientation of the dipoles in an externally applied field is possible. Therefore the orientation polarization is zero, and the dielectric constant depends on the same factors as those in the nonpolar molecular solid. The dielectric constant temperature curves of these polar molecules show curves of great discontinuity at the melting point, for in

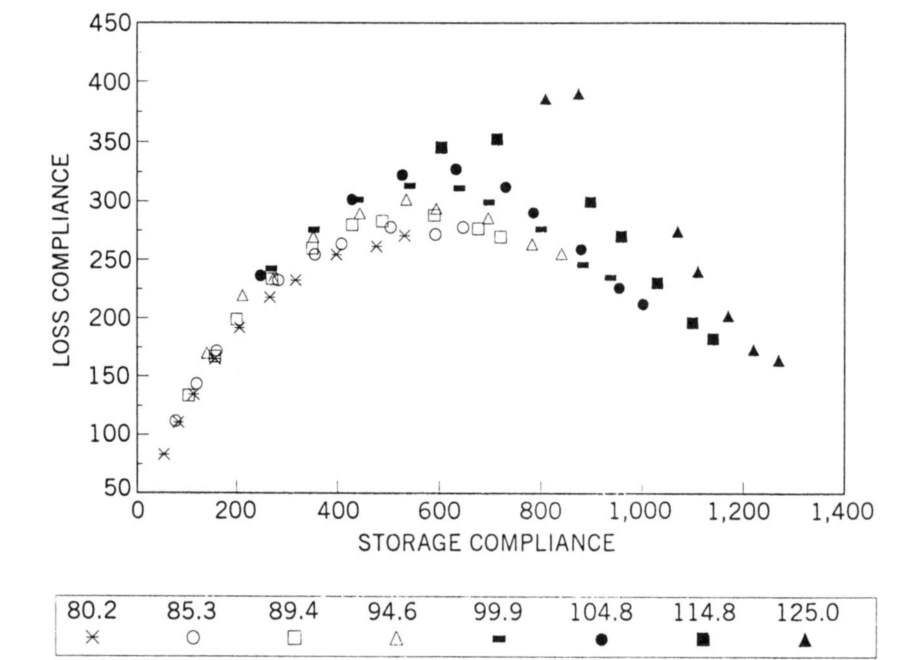

Figure 37. Complex plane plots of $\varepsilon^*(\omega)$ and $J^*(\omega)$ for poly(n-hexyl methacrylate).

the case of the nonpolar solids the changes in ε_0 at the melting point are small and are attributed to the slight changes in density.

Nitromethane, which has a large dipole moment and consequently makes a large contribution to the orientation polarization, affords a particularly striking example of this change at the melting point. For nitromethane the melting point is −29°C, the density 1.13, and the dielectric constant is 44 in the liquid phase while in the solid phase ε_0 is 3.5. If we assume the compliance to be very large, then $g\mu^2$ calculates to be 20.6. To compute the dielectric constant of the solid we need to know D_0, information that is not available. However, the tensile compliance of some salts such as lithium sulfate and rochell salt[70] are approximately 10^{-12} cm^2/dyn. We can assume D_0 to be the same because it is a very polar solid. The strain of the molecule can be estimated from the structure of the molecule. Nitromethane[71] is a Y-shaped molecule with a small fantail at one end due to the 2 hydrogen atoms. The dipole moment of the molecule lies along the axis, which bisects the ONO bond angle. Orientation polarization can take place in two ways, either by rotation in the plane of the molecule or by rotation about the axis perpendicular to the plane. The least restricted motion is by rotation in the plane of the

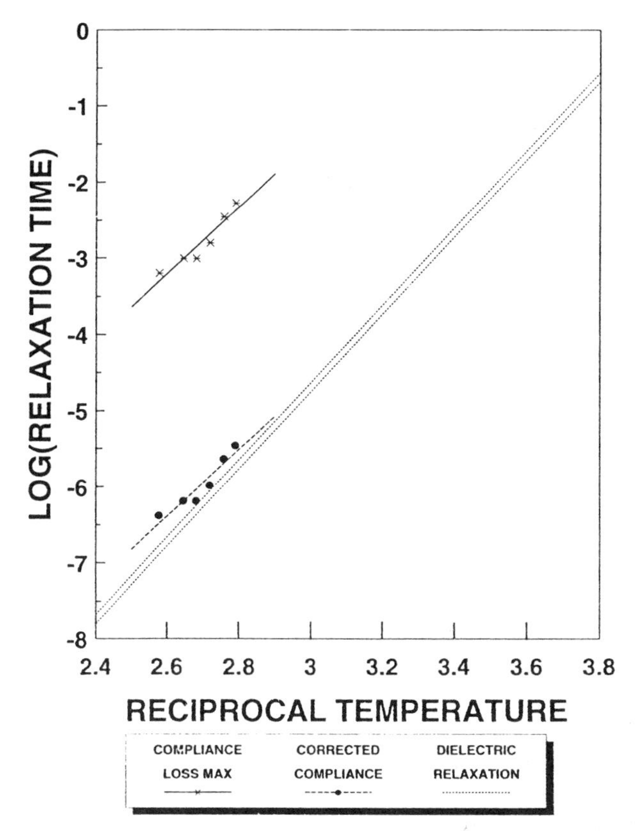

Figure 38. Plot of the log(relaxation times) for poly(*n*-hexyl methacrylate) determined from the data in Figure 37.

molecule. The cross-sectional shape of this molecule is approximately an ellipse with minor axis of 1.4 A and a major axis of 3.65 A. Therefore ε_0 calculates to be about 1.61 and 1×10^{-3} for the two directions. In other words the entire orientation process has been reduced in magnitude ("frozen out") to a very low and unobservable level. The actual value of 3.5 (experimental) contains contributions to ε_0, that is, ε_∞, which were specifically ignored in the present development.

Hydrogen iodide poses an equally interesting problem. Again, D_0 for this material is not known, but since it is a weakly polar solid, we shall assume D_0 to be about $10^{-11}\ \mathrm{cm^2/dyn}$, which is probably midway between salts and organic glasses. The molecule can be represented by an ellipse with a minor axis[72] of 1.4 A and a major axis of 1.6 A, so that γ calculates to be 0.12. The density of the liquid is taken to be $2.85\ \mathrm{g/cm^3}$,

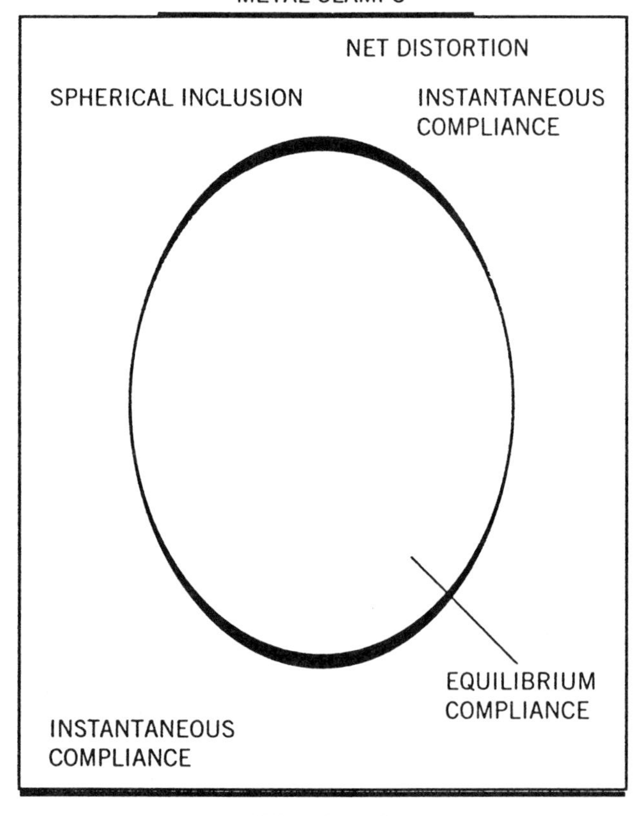

Figure 39. The viscoelastic model used to calculate the complex distortion.

the melting point −56°C and the dielectric constant of the liquid is 3.6 while that of the solid is 3.8. The slight increase of ε_0 upon solidification is attributed to an increase in the density. Under these conditions $g\mu^2$ computes to be 4.24, assuming D_0 of the liquid to be very large. Using this value for $g\mu^2$ and the stated values for γ and D_0, for the solid computes to be 2.7. In other words there may be a slight reduction in ε_0 but nothing significant. The failure to reduce the dielectric constant to a very low value is due to the nearly spherical shape of the molecule.

We could continue the discussion for the other small molecules described by Smyth with equally good results. However, in all cases the quantitative nature of the discussion is seriously limited because D_0 is unknown, though γ can be estimated with some certainty because x-ray

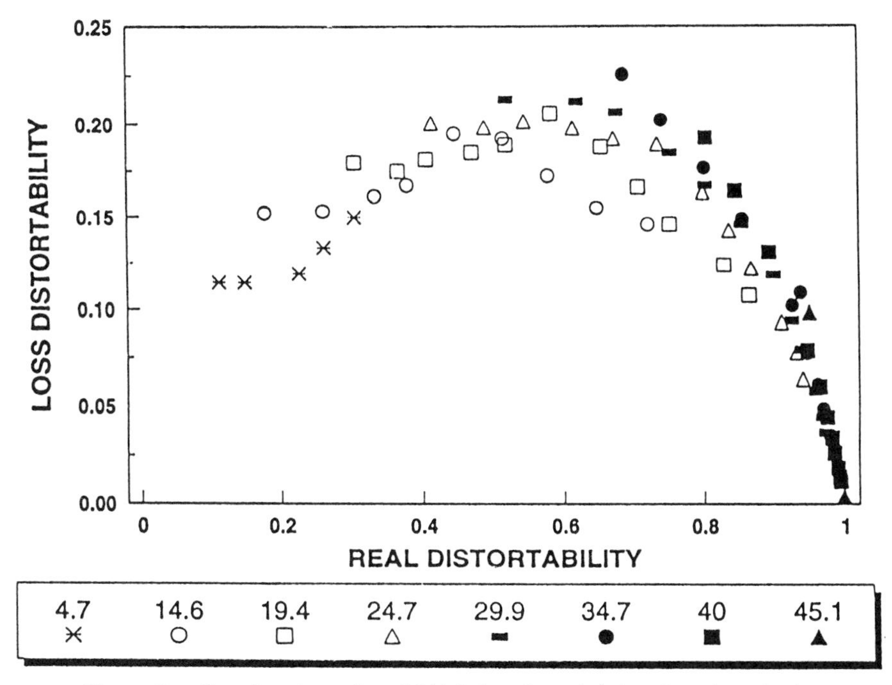

Figure 40. Complex plane plot of $\delta^*(\omega)$ data for poly(n-hexyl methacrylate).

crystallographic information is available. In the application of Eq. (46) to polymers, the reverse situation is usually the case. Compliances are well known but structural information is lacking, making it difficult to access γ.

J. Electromechanical Properties of Some Polymers

The basic assumption throughout this work is that an electric field induces a distortion of the test specimen, thereby giving rise to a strain energy that must be included in the perturbation term of the Hamiltonian. Let us now consider the following *gedanke* experiment based on the apparatus in Figure 42 (on page 277).[73] Starting from the bottom, we have a rigid optically flat support fabricated from glass or quartz. Sputtered onto this support is a gold electrode and filmed on this electrode by spinning from solution is a thin polymer specimen such as polycarbonate or PMMA. Deposited on this polymer film are two small gold disks that are electrically isolated from each other and are of optical quality to serve as electrode light reflectors. Above this compound electrode and shining a light on the two electrodes is an optical interferometer, which carries out the following functions. A colluminated laser light beam is split into two

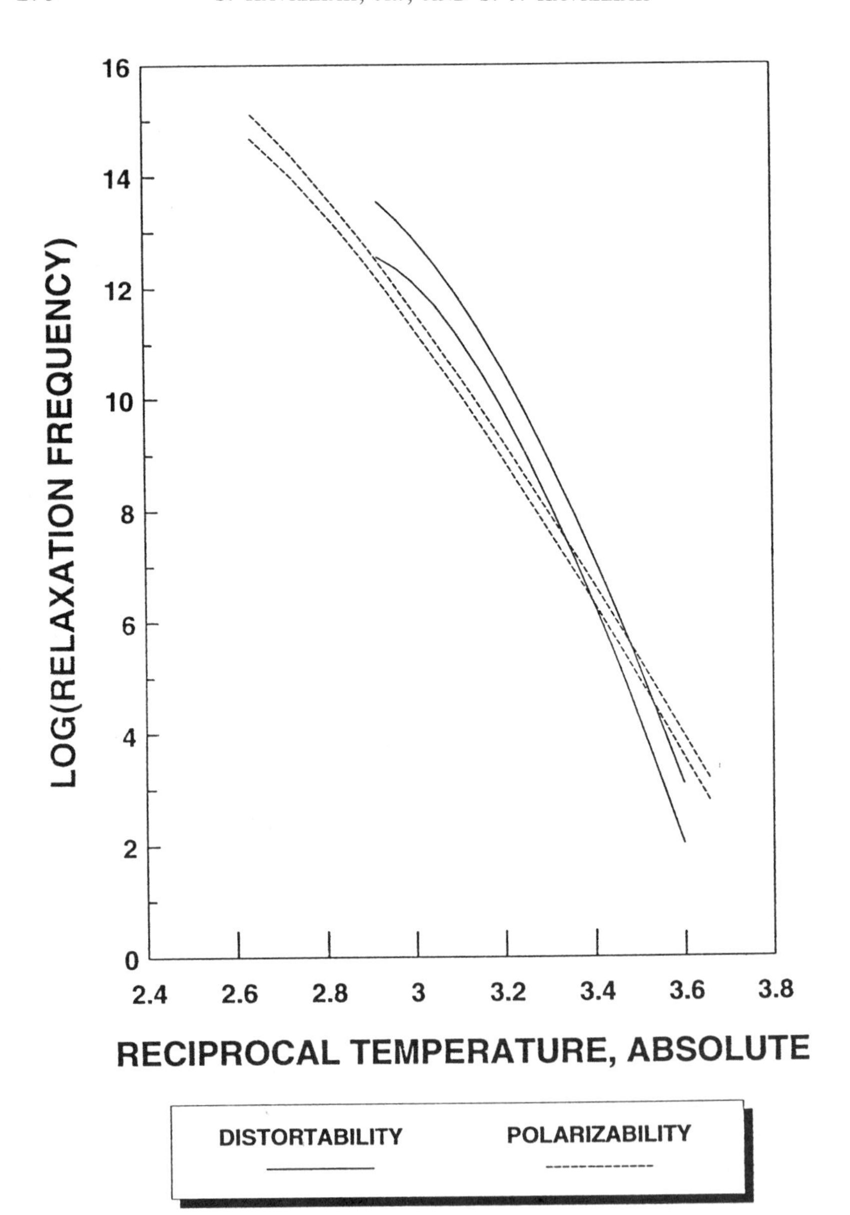

Figure 41. Plot of the log(relaxation times) for poly(n-hexyl methacrylate) determined from $\rho^*(\omega)$ data and $\delta^*(\omega)$ data.

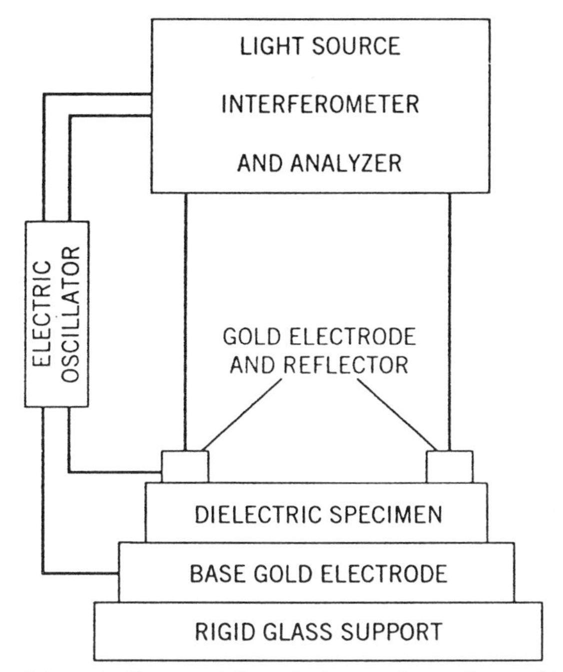

Figure 42. Schematic diagram of the *gedanke* experiment discussed in the text.

components and each is directed at one of the gold reflecting disks. When the light beams are reflected back to the interferometer an interference pattern is established in the analyzer. Any change in the optical path length, such as a thickness change in the polymer film thickness, will shift the pattern. A constant electric voltage is now applied across one of the top electrodes and the base electrode. If the electric field causes any dimensional changes in the thickness of the film, the interference pattern will shift and this shift is directly related to the optical path length change, hence a change in the film thickness. Since the film thickness is known, the strain (change in thickness/thickness) is measured. The constant electric voltage can be replaced by an oscillating electric voltage so that the strain can be measured as a function of oscilator frequency. Obviously, these measurements can be made as a function of temperature.

An elegant version of this simple device just described above has been constructed and reported by Winkelhahn and Neher[74] and is referred to as a Nomarski interferometer. This device uses a 2-μm film with a potential of 40 V/μm across the film, has a displacement sensitivity of 1 pm (picometer), and a frequency range of 100 Hz to 100 KHz. This device was used to measure the viscoelastic properties of PMMA. A plot

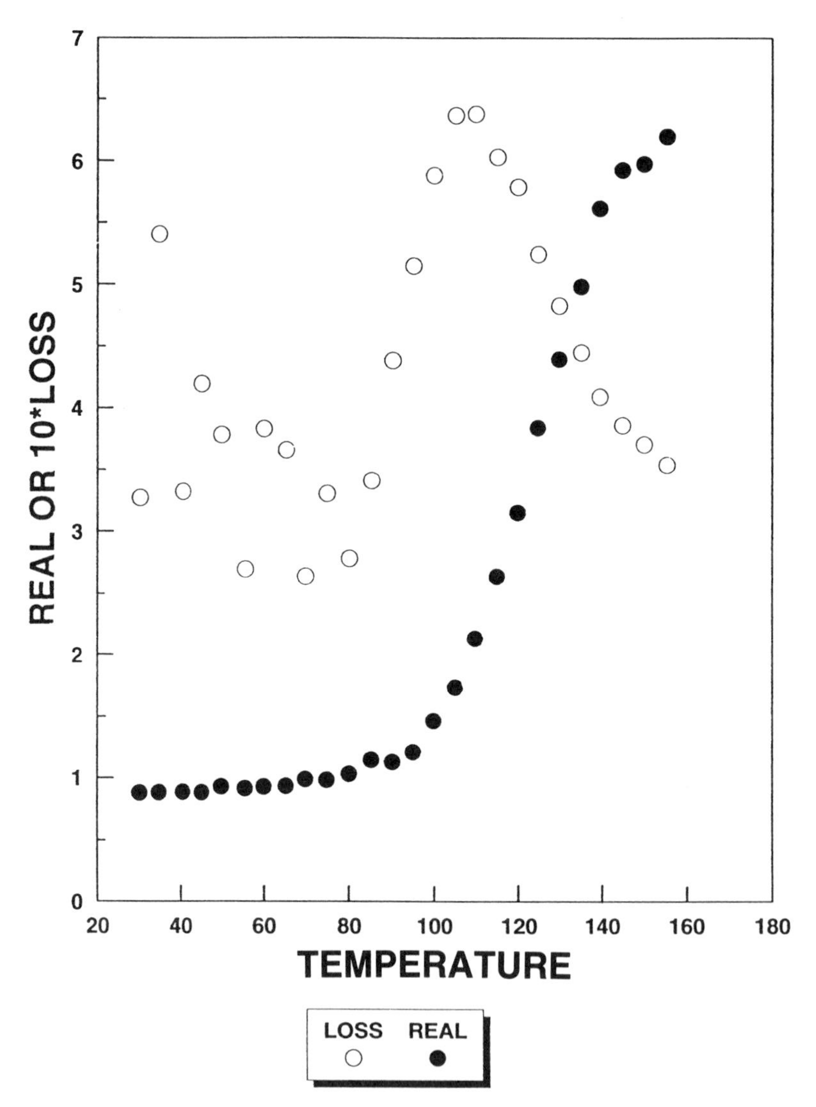

Figure 43. Plot of the real and 10* imaginary loss as a function of temperature.

of the real and (10 times the) imaginary parts of Youngs modulus calculated from the dimensional changes of the film and at a frequency of 999 Hz is given in Figure 43. There is obviously a dispersion in the glass transition region of PMMA nominally taken to be 100°C, since there is a loss maximum in these results at about 100°C. Below 80°C the data is

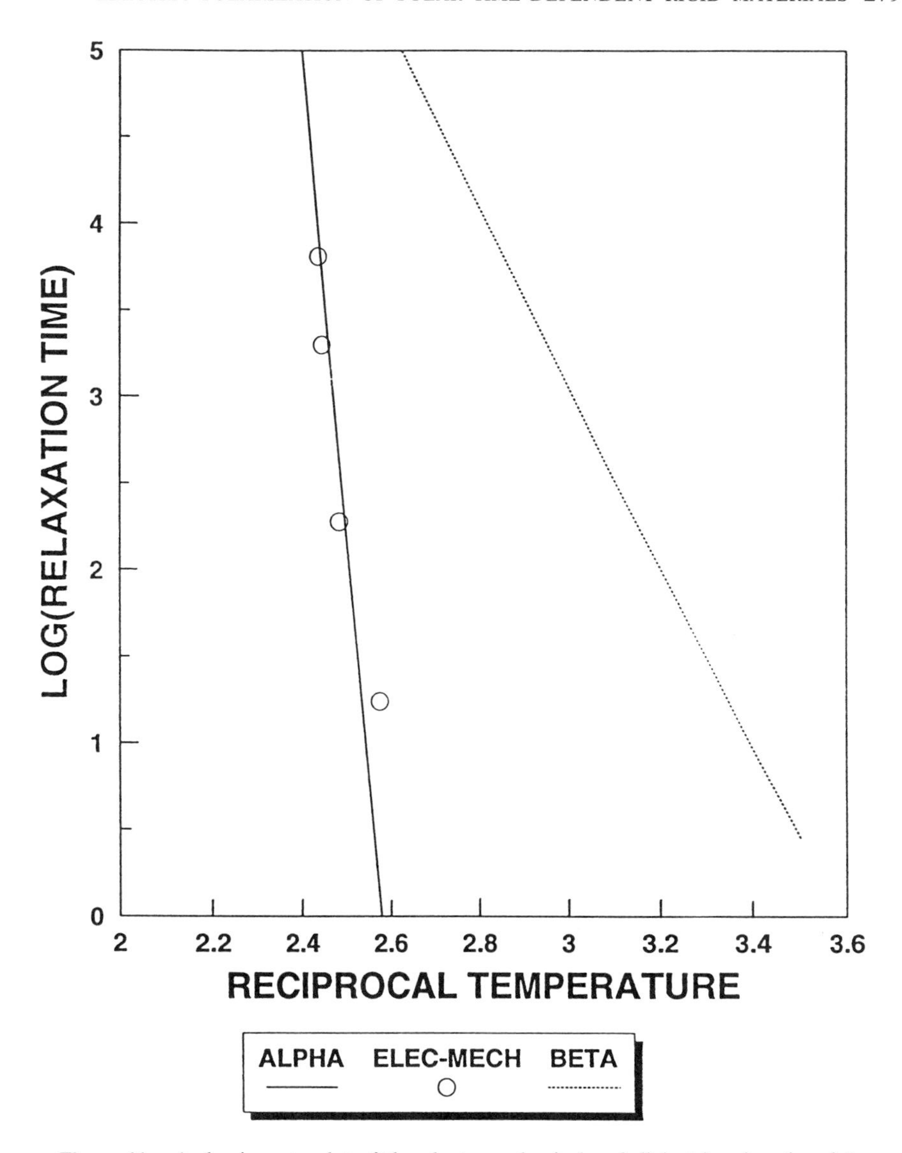

Figure 44. Arrhenius rate plot of the electromechanical and dielectric relaxation data for PMMA.

quite noisy and there may be another loss maximum which would be related to the α process of PMMA. In any case, performing these experiments at various frequencies and plotting the temperature at which the loss becomes a maximum yields the results shown in Figure 44. In addition, data taken from the literature[75] is also shown in Figure 44. The similarity of the electromechanical data with the dielectric α process is a further conformation that the two processes are similar.

These results clearly show that it is now possible to measure viscoelastic properties by means of dielectric measurements. The advantage here is that dielectric measurements can be made to very high frequencies, that is, at least 100 KHz relative to viscoelastic measurements, which can be made only up to 10–100 KHz.

VII. CONCLUSIONS

The issues mentioned in the Introduction were addressed. Some of the results were previously reported and before the elegant experimental work of Winkelhahn and Neher. These authors verified the basic assumption in this work, which is that the polarization process must be accompanied by a distortion of the dielectric test specimen, which gives rise to a strain energy term. The result of incorporating the strain energy into the perturbation term of the Hamiltonian at equilibrium is given by Eq. (46). These results were obtained neglecting the reaction field because the assumption was made that the long-range correlations cannot be represented by electrostatic forces.

The evidence presented in Section VI shows the importance of incorporating this strain energy term in the perturbation term of the Hamiltonian. In this work only the results of Eq. (46) were discussed for cases where the reaction field can represent long-range interactions. Although this equation was derived for the equilibrium case, the time dependence of the tensile (or shear) converts this into a time-dependent case. The agreement between the DiMarzio–Bishop model is also a verification of the basic assumption. Consequences of a fortuitous cancellation of terms in the denominator, that is, $(\varepsilon_0 + 2)$, due to the acceptance of the reaction field, was studied.

REFERENCES

1. P. Debye, *Phys. Z.* **13**, 97 (1912).
2. P. Debye, *Handbuch der Radiologie* **6**, 597 (1925).
3. P. Debye, *Polar Molecules*, Chemical Catalog, New York, 1929.

4. H. Frohlich, *Theory of Dielectrics*, Oxford University Press, New York, 1949.
5. J. W. Smith, *Electric Dipole Moments*, Butterworths Scientific, London, 1955.
6. J. W. Smith, *Dielectric Behavior and Structure*, McGraw-Hill, New York, 1955.
7. C. J. F. Bottcher, *Theory of Electric Polarization*, Elsevier Scientific, New York, 1973.
8. J. W. Smith, ibid., p. 21.
9. C. P. Smyth, *Dielectric Behavior and Structure*, McGraw-Hill, New York, 1955.
10. L. Onsager, *J. Am. Chem. Soc. I* **58**, 1486 (1936).
11. J. G. Kirkwood, *J. Chem. Phys.* **7**, 911 (1939).
12. S. Havriliak, Jr., and S. Negami, *J. Polym. Sci. Part C* **14**, 99 (1966).
13. S. Havriliak, Jr., and S. Negami, *Polymer* **8**, 161 (1967).
14. S. H. Glarium, *J. Chem. Phys.* **33**(5), 137 (1960).
15. R. H. Cole, *J. Chem. Phys.* **42**(2), 637 (1965).
16. R. Kubo, *J. Phys. Soc. Jpn.* **12**, 570 (1957).
17. B. K. P. Scaife, *Progress in Dielectrics*, London, Heywood, 1963, p. 143.
18. B. K. P. Scaife, *Proc. Phys. Soc.* **81**, 124 (1963).
19. B. K. P. Scaife, *Molecular Relaxation Processes*, Academic Press, New York, 1966, pp. 15–19.
20. E. Fatuzzo and P. R. Mason, *Proc. Phys. Soc.* **90**, 781 (1967).
21. A. Gemant, *Trans. Faraday Soc.* **31**, 1582 (1935).
22. S. Havriliak, Jr., and S. J. Havriliak, *J. Poly. Sci. Part B: Polym. Phys.* **33**, 2245 (1995).
23. E. A. DiMarzio and M. Bishop, *J. Chem. Phys.* **60**(10), 3802 (1974).
24. H. Frohlich, *Theory of Dielectrics*, Oxford University Press, London, 1949, p. 75.
25. S. Timosenko and J. N. Goodier, *Theory of Elasticity*, McGraw-Hill, New York, 1951.
26. R. C. Tolman, *The Principles of Statistical Mechanics*, Oxford University Press, London, 1938.
27. N. G. McCrum, B. E. Read, and G. Williams, *Anelastic and Dielectric Effects in Polymeric Solids*, Wiley, New York, 1967.
28. S. G. Glarum, Ph.D. Thesis, Brown University, Providence, RI (1960).
29. P. J. Flory, *Principles of Polymer Chemistry*, Cornell University Press, Ithaca, NY, 1953.
30. N. S. Steck, *S.P.E. Trans.* January, 34 (1964).
31. G. K. Dyvik, W. F. Bartoe, and N. S. Steck, *S.P.E. Trans.* April, 98 (1964).
32. S. Havriliak, Jr., and N. Roman, *Polymer* **7**, 387 (1966).
33. S. Havriliak, Jr., and S. J. Havriliak, *Polymer* **33**(5), 938 (1992).
34. S. Havriliak, Jr., *Polymer* **9**, 289 (1968).
35. J. D. Ferry, *Viscoelastic Properties of Polymers*, Wiley, New York, 1984.
36. R. Kohlarauch, *Pogg. Ann. Phys. Chem.* **91**, 179 (1854).
37. W. Williams and D. C. Watts, *Trans. Faraday Soc.* **66**, 80 (1970).
38. F. R. Schwarzl, and L. C. E. Struik, *Adv. Mol. Relax. Proc.* **1**, 210 (1967–1968).
39. F. R. Schwarzl, *Pure Appl. Chem. B.* **23**, 219 (1970).
40. F. R. Schwarzl, *Rheol. Acta* **8**, 6 (1969).
41. F. R. Schwarzl, *Rheol. Acta* **9**, 382 (1970).

42. F. R. Schwarzl, *Rheol. Acta* **10**, 166 (1971).
43. S. Havriliak, Jr., and S. J. Havriliak, *Polymer* **36**(14), 2675 (1995).
44. R. H. Cole and K. S. Cole, *Chem. Phys.* **10**, 98 (1942).
45. S. Havriliak, Jr., and S. J. Havriliak, *Dielectric and Mechanical Relaxation in Materials*, Chap. 14, Hanser, 1996.
46. M. J. Naoki, *J. Chem. Phys.* **918**, 5030 (1989).
47. S. Havriliak, Jr., and T. J. Shortridge, *Macromolecules* **23**, 648 (1990).
48. R. M. Fuoss and J. G. Kirkwood, *J. Am. Chem. Soc.* **63**, 385 (1941).
49. C. J. F. Bottcher and P. Bordewijk, *Theory of Electric Polarization, Vol. II, Dielectrics in Time-Dependent Fields*, Elsevier Scientific, Amsterdam, 1978.
50. F. Alverz, A. Alegria, and J. Colmenero, *J. Phys. Revs. B.* **44**(11), 7306 (1991).
51. S. Havriliak, Jr., and S. J. Havriliak, *J. Non-Crystalline Solids*, *172–174*, 297, 1994.
52. S. Havriliak, Jr., and S. J. Havriliak, *J. Poly. Sci.* **33**, 2245, 1995.
53. S. Havriliak, Jr., and S. J. Havriliak, *Polymer* **36**(14), 2675, 1995.
54. R. H. Cole, *J. Chem. Phys.* **42**(2), 637 (1965).
55. J. E. Shore and R. Zwanzig, *J. Chem. Phys.* **63**(12), 5445 (1975).
56. N. G. McCrum, B. E. Read and G. Williams, *Anelastic and Dielectric Effects in Polymeric Solids*, Wiley, London, 1967, p. 110.
57. S. Strella and S. Chinai, *J. Polym. Sci.* **31**, 45 (1958).
58. W. C. Child and J. D. Ferry, *J. Colloid Sci.* **13**, 389 (1958).
59. S. Havriliak, Jr., and S. J. Havriliak, *J. Mol. Liquids* **56**, 49 (1993).
60. S. Timoshenko and J. N. Goodier, *Theory of Elasticity*, McGraw-Hill, New York, 1951.
61. S. Havriliak, Jr., and S. Negami, *Brit. J. Appl. Phys. (J. Phys. D)* **2**(2), 1301 (1969).
62. S. Strella and S. N. Chinai, *J. Colloid Sci.* **31**, 24 (1959).
63. W. C. Child, W. Dannhauser, and J. D. Ferry, *J. Colloid. Sci.* **12**, 389 (1957).
64. S. Havriliak, Jr., and S. J. Havriliak, *J. Mol. Liquids* 1995, in press.
65. S. Havriliak, Jr., and T. J. Shortridge, *Macromolecules* **23**, 648 (1990).
66. S. Havriliak, Jr., and C. S. Pogonowski, *Macromolecules* **22**, 1466 (1989).
67. S. Matsuoka and S. Ishida, *J. Poly. Sci. Part C* (14), 247 (1966).
68. S. Havriliak, Jr., and C. A. Cruz, unpublished data.
69. W. P. Leung and C. L. Choy, *J. App. Poly. Sci.* **27**, 2693 (1982).
70. J. F. Nye, *Physical Properties of Crystals*, Oxford at the Clarendon Press, Oxford, 1957.
71. L. E. Sutton, Scientific Editor, *Tables of Interatomic Distances and Configuration in Molecules and Ions*, The Chemical Society, London, 1958.
72. F. P. Grigor'eva, T. M. Birshtein, and Yu. Ya. Gotlib, *Vysokomol. Soyed.* **A9**(3), 580 (1967).
73. The senior author (S. H., Jr.) wishes to express his gratitude to the staff members of the Max Planck Institute such as Prof. Fisher and Dr. Rickert for hosting his visit during April of 1995. In addition he wishes to thank Prof. Wegner and Dr. Neher for bringing their most recent work to his attention and in particular for supplying the information and graphs used in this section.
74. H.-J. Winhelhahn and D. Neher, *Proceedings SPIE Conference*, San Diego, 1994.
75. K. C. Rush, *J. Macromol. Sci.-Phys. B* **2**, 179 (1968).

MAGNETIC RELAXATION IN FINE-PARTICLE SYSTEMS

J. L. DORMANN

Laboratoire de Magnétisme et d'Optique de Versailles, CNRS-Université de Versailles–Saint Quentin, Bâtiment Fermat, 45 av. des Etats-Unis, 78035 Versailles, France

D. FIORANI

Istituto di Chimica dei Materiali, CNR, Area della Ricerca di Roma, CP10, 00016 Monteretondo Stazione, Italia

E. TRONC

Laboratoire de Chimie de la Matière Condensée (URA CNRS 1466), Université Pierre et Marie Curie, T54-E5, 4 Place Jussieu, 75252 Paris Cedex 05, France

CONTENTS

Advances in Chemical Physics, Volume XCVIII, Edited by I. Prigogine and Stuart A. Rice.
ISBN 0-471-16285-X

A. INTRODUCTION

Fine magnetic particles have generated continuous interest since the late 1940s as the study of their properties has revealed to be very challenging scientifically and technologically. The advancement of the understanding of the magnetic behavior of fine particles, since the pioneering work of Néel,[1] has been a very important contribution to the development of fundamental theories of magnetism and in modeling magnetic materials, as well as remarkable technological improvements, for example, in the information storage and data processing field, fostering the development of magnetorecording media with higher and higher density.

In the last few years the interest in fine magnetic particles[2,3] has grown enormously, with increasing attention devoted to the effect of nanoscale-size confinement on the physical properties and with regard to their potential in the nanoscale engineering of materials with very specific properties. Nanostructured materials, modulated on a length scale less than 100 nm, are known to exhibit properties different from, and often superior to, those of conventional materials that have phase or grain structures on a coarser size scale.[4]

Among the enhanced properties of nanostructured materials, the magnetic ones represent a very good example, with important technological implications, for example, enhanced remanence and giant coercivity in nanostructured permanent magnets. With decreasing particle size, an increasing fraction of atoms lies near or on the surface and interfacial regions (e.g., for grain sizes of 100, 10 and 5 nm, it corresponds to about 1–3%, 15–30%, and 30–60%, respectively), making the effect of the surface and interface electronic structure on the magnetic properties more and more important. As a matter of fact, the intrinsic magnetic properties of a material (e.g., spontaneous magnetization and magnetocrystalline anisotropy) are strongly influenced by the particle size. The more disordered atomic arrangement and the lower number of atomic neighbors on the surface with respect to the bulk are responsible for the decrease in the spontaneous magnetization of a ferromagnetic material with decreasing particle size. On the contrary, a nonmagnetic or antiferromagnetic material may acquire a net moment for low enough particle size. Moreover, the total anisotropy energy may increase with decreasing particle size, below a certain size, because of the growing surface anisotropy contribution.

Ultrafine magnetic particles are commonly present in different kinds of materials (e.g., rocks, living organisms, pigments, soils, ceramics, atmospheric aerosol, and corrosion products). For this reason, the study of superparamagnetic properties can be applied to different branches of science and technology,[2] allowing interesting data to be obtained. For example, in catalysis,[5] where the particle volume distribution can be determined and the effect of chemisorption and chemical reaction on surface electronic properties can be checked; in fine arts,[6] for the reconstruction of the production techniques of ancient ceramics and for authentification of paintings; in mineralogy[7] and in paleomagnetism,[8] for reading the geomagnetic record in rocks; in biology,[9] for a better insight in the structure and for a better understanding of the functions of iron storage proteins, like ferritine; in magnetorecording,[10] where the smallest suitable particle size can be determined, ultimately limited by superparamagnetic relaxation, allowing the maximum recording density to be reached.

The chapter is organized as follows: In Section A a general introduction to the scientific problem has been presented. In Section B the general properties of fine magnetic particles will be described and the basic concepts of superparamagnetism will be introduced. In Section C the different forms of anisotropy energy will be described in detail. In Section D the calculation of the relaxation time will be reported and the different proposed models will be described. In Section E the effects of interparticle interactions on the relaxation time will be discussed and their modeling will be reported. In Section F the experimental results obtained by different techniques will be reported and compared with the theoretical predictions. In Section G the quantum effects on the magnetic properties of fine particles (e.g., quantum tunneling of the magnetization) will be discussed and some experimental results will be reported. In Section H the properties of fine antiferromagnetic particles will be discussed. In Section I some conclusions will be drawn on the state of the art of the research on fine magnetic particles.

B. GENERAL PROPERTIES OF FINE PARTICLES: SUPERPARAMAGNETIC BEHAVIOR

B.1. Single-Domain Particles

It is well known that a magnetic body has a multidomain structure, that is, it is divided into uniformly magnetized regions (domains) separated by domain walls (Bloch walls) in order to minimize its magnetostatic energy. However, the energy to be minimized is the total energy, including, in addition to the magnetostatic term, the exchange and the anisotropic ones as well as the domain wall contribution. Therefore, it is the final balance of energies that determines the domain structure and shape. By reducing the dimension of the crystal, the size of the domains is also reduced and their structure may change, as well as the width and the structure of the walls. Due to the energy cost of the domain wall formation, the balance with the magnetostatic energy limits the subdivision in domains to a certain optimum domain size. As a matter of fact, there is a corresponding lower limit in the crystal size, below which a single-domain[11] structure does exist, since the energy increase due to the formation of domain walls is higher than the energy decrease obtained by dividing the single domain into smaller domains.

For typical magnetic materials the dimensional limit is in the range of 20–800 nm, depending on the spontaneous magnetization and on the anisotropy and exchange energies. For spherical crystals the characteristic radius is given by[12,13] $R_{sd} = 9E_\sigma/\mu_0 M_s^2$, where M_s is the saturated magnetization and E_σ is the total domain wall energy per unit area [$E_\sigma \cong 2(K/A)^{1/2}$, where K is the anisotropy energy constant and A is a parameter representing the exchange energy density]. Typical values[14] for

R_{sd} are about 15 nm for iron, 35 nm for Co, 100 nm for NdFeB, and 750 nm for $SmCo_5$.

The change from a multidomain to a single-domain structure is accompanied by a strong increase of the coercive field ($H_c \cong K/3M_s$ for uniaxial symmetry).

B.2. Superparamagnetic Particles

The anisotropy energy in a single-domain particle is proportional, in a first approximation, to the volume V. For uniaxial anisotropy the associated energy barrier, separating easy magnetization directions (i.e., the low-energy directions of the spin system) is $E_B = KV$. Thus, with decreasing particle size the anisotropy energy decreases, and for a grain size lower than a characteristic value, it may become so low as to be comparable to or lower than the thermal energy kT. This implies that the energy barrier for magnetization reversal may be overcome, and then the total magnetic moment of the particle can thermally fluctuate, like a single spin in a paramagnetic material. Thus the entire spin system may be rotated, the spins within the single-domain particles remaining magnetically coupled (ferromagnetically or antiferromagnetically). The magnetic behavior of an assembly of such ultrafine, independent magnetic particles is called *superparamagnetism*.[15–19]

The superparamagnetic behavior is exhibited by particles with dimensions in a defined range. If they are too small, almost all the atoms lie on the surface, leading to electronic and magnetic properties strongly modified with respect to the bulk ones, and the superparamagnetic model cannot be applied. This does not mean that no relaxation of the magnetic moment **m** occurs, but the laws governing it are expected to be different. It is difficult to state precisely a lower dimensional limit for superparamagnetic behavior, as it depends on several parameters. This should correspond, in our opinion, to a grain diameter of about 2 nm. As far as the upper limit is concerned, it is given in principle by the characteristic size for a single-domain particle, as long as the single-domain state and structure are effective (some uncertainties remain for some particular cases).[20–22] Actually the characteristic grain size of a magnetic material for superparamagnetic relaxation depends on the anisotropy constants and M_s values. As an example, for uniaxial anisotropy and $K = 5 \times 10^5$ erg/cm^3, for spherical particles this corresponds to a characteristic diameter $\phi_c \leq 20$ nm.

For fine magnetic particles the actual magnetic behavior depends on the value of the measuring time (τ_m) of the specific experimental technique with respect to the relaxation time (τ) associated with the overcoming of the energy barriers. As pointed out in the following sections, τ varies exponentially with the E_B/kT ratio. If $\tau_m \gg \tau$, the relaxation appears to be so fast that a time average of the magnetization orientation is observed in the experimental time window, and the assembly of particles behaves like a paramagnetic system (superparamagnetic state). On the contrary, if $\tau_m \ll \tau$, the relaxation appears so slow that quasi-static properties are observed (blocked state), like for magnetically ordered crystals, although strongly influenced by the particle surface structure. The blocking temperature T_B, separating the two states, is defined as the temperature at which $\tau_m = \tau$. Therefore T_B is not uniquely defined as well as ϕ_c, but it is related to the time scale of the experimental technique. As an example, for Fe_3O_4 ($K = 4.4\ 10^5$ erg/cm^3) at 290 K, the characteristic grain diameter for superparamagnetism, below which superparamagnetic relaxation and above which quasi-static properties are observed, is $\phi_c \cong 17$ nm for dc susceptibility measurements, while it is $\phi_c \cong 9$ nm for Mössbauer spectroscopy experiments, having a much shorter measuring time. The blocking temperature T_B for a magnetic particle increases with increasing size and for a given size increases with decreasing measuring time, and then the observation of a superparamagnetic or blocked state depends on the experimental technique (Fig. B.1). The highest value of T_B is represented by the Curie (or Néel) temperature, at which the transition from the superparamagnetic to the paramagnetic state occurs.

The techniques currently used to study the superparamagnetic relaxation are dc susceptibility (τ_m is not well defined, estimated to be around 100 s, but it depends on the type of magnetometer and on the measuring procedure), ac susceptiblity ($\tau_m = 10^2$–10^4 s for experiments at very low frequencies; $\tau_m = 10^{-1}$–10^{-5} s for classical experiments; $\tau_m = 10^{-5}$–10^{-8} s for measurements at very high frequencies, very difficult to realize, so far), Mössbauer spectroscopy (time window, 10^{-7}–10^{-9} s for ^{57}Fe), ferromagnetic resonance ($\tau_m = 10^{-9}$ s), and neutron diffraction (time window, 10^{-8}–10^{-12} s, depending on the type of experiments).

As pointed out in the next section, the magnetic anisotropy can often be considered uniaxial, with the total magnetic anisotropy given by

$$E(\theta) = E_B \sin^2\theta \tag{B.1}$$

where θ is the angle between the easy axis **zOz′** and the magnetization vector (Fig. B.2) and E_B ($=KV$) is the energy barrier (i.e., the maximum anisotropy energy value for $\theta = \pi/2$). In zero magnetic field, according to the Eq. (B.1), two symmetric minima exist for $\theta = 0$ and $\theta = \pi$ separated by an energy barrier equal to KV.

In the presence of a magnetic field H applied along the easy axis, the

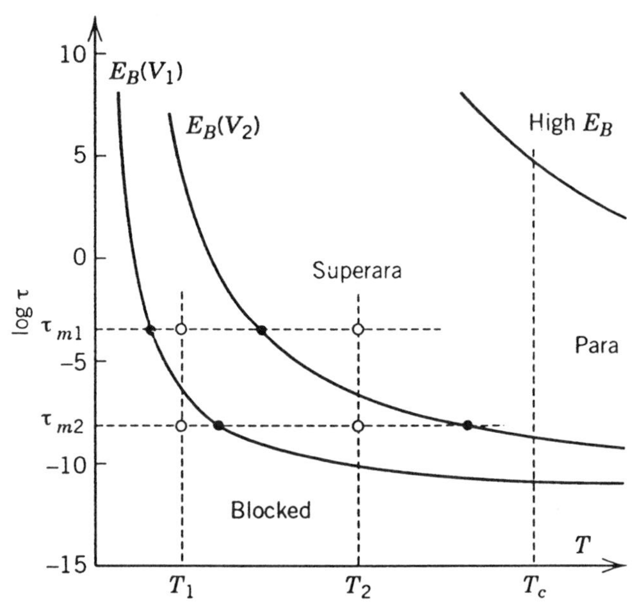

Figure B.1. Temperature dependence of the relaxation time τ for fine particles for three E_B values (full lines), T_B values (full circles) for two experiments with measuring time τ_{m1} and τ_{m2} and magnetic state seen by the measurements (marked by the open circles) at two given temperatures T_1 and T_2 according to E_B and τ_m values.

energy of a ferromagnetic particle becomes

$$E(\theta) = E_B \sin^2\theta - HVM_{\mathrm{nr}} \cos\theta \tag{B.2}$$

where M_{nr} is the nonrelaxing magnetization (see Section B.3). For $H < 2E_B/(VM_{\mathrm{nr}})$ $(=2K/M_{\mathrm{nr}})$ there are still two minima at $\theta = 0$ and $\theta = \pi$, but they are no longer equivalent, since the energy barrier between $\theta = 0$ and $\theta = \pi$ $[\Delta E = E_B(1 + HVM_{\mathrm{nr}}/2E_B)^2]$ is larger than that between $\theta = \pi$ and $\theta = 0$ $[\Delta E = E_B(1 - HVM_{\mathrm{nr}}/2E_B)^2]$. For $H > 2E_B/(VM_{\mathrm{nr}})$ there exists only one minimum, at $\theta = 0$. In Fig. B.3, $E(\theta)$ is reported for various values of $HVM_{\mathrm{nr}}/2E_B$.

At a thermodynamic equilibrium temperature T the probability $f(\theta)\,d\theta$ that the magnetization forms an angle θ with the easy direction is given by

$$f(\theta) = (1/Z)\exp(-E/kT)\sin\theta \tag{B.3}$$

where $Z = \int_0^\pi \exp(-E/kT)\sin\theta\, d\theta$. For $E \gg kT$ and $H = 0$, $f(\theta)$ is finite only in correspondence with the two minima (Fig. B.4). The probability

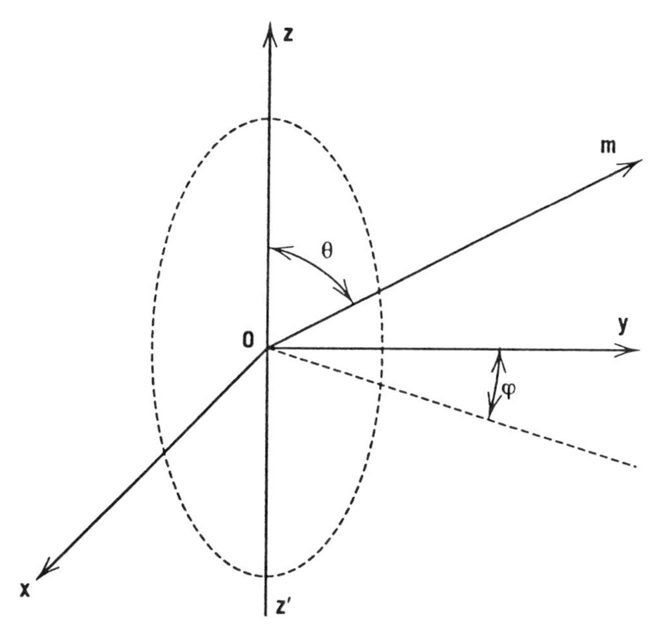

Figure B.2. Usual axis system for fine particles. The easy axis of the magnetization is along **zOz′**.

will be the same for $\theta = 0$ and $\theta = \pi$. The magnetization can then be considered fixed in one of two directions corresponding to the energy minima. This is the situation in large magnetically ordered crystals. If H is small the probability will be different for the two directions (higher for $\theta = 0$). If H is high so that $HVM_s \gg kT$, $f(\theta)$ is finite only for $\theta = 0$. For lower anisotropy energy values and/or H values, $f(\theta)$ becomes broader around the energy minima, and the magnetization can fluctuate around the easy directions. This corresponds to vibrations in the potential well, that is, transverse relaxation, implying an **m** component, leading to transverse susceptibility. Finally for $E \leq kT$ and $H < 2E_B/(VM_{\rm nr})$, $f(\theta)$ is finite for every θ value, that is, the magnetization has a significant probability of overcoming the energy barrier separating the two minima (superparamagnetic relaxation).

B.3. Complexity of Actual Fine-Particle Systems

Recently, new interesting phenomena, from both fundamental and technological points of view, exhibited by fine magnetic particles have attracted much attention, for example, macroscopic quantum tunneling of the magnetization,[23] magnetocaloric effect[24] well above liquid helium

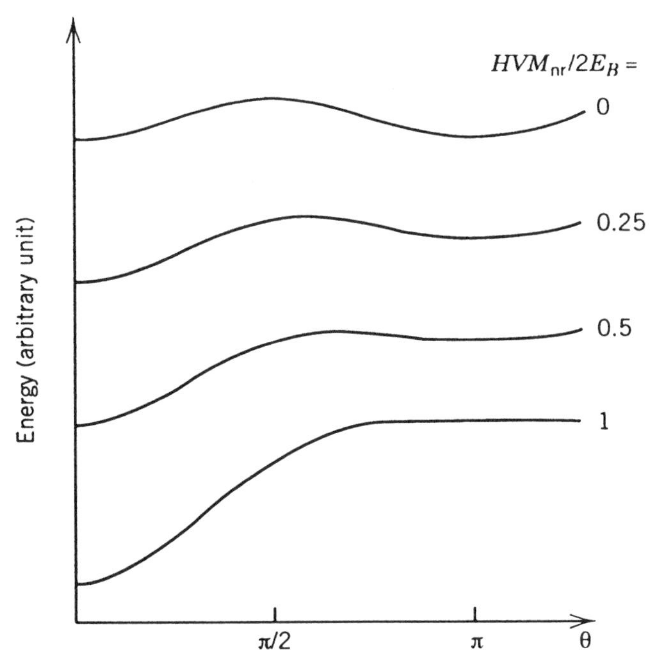

Figure B.3. Dependence of the total energy of an uniaxial particle as a function of the angle between the magnetization vector and the easy axis, with *H* applied along the easy axis.

temperature, and giant magnetoresistance.[25,26] These topics will be discussed in the next sections. The increasing technological prospects of fine magnetic particles make the need for a comprehensive theoretical description and modeling of their properties more and more important. Unfortunately, the actual situation in materials consisting of fine particles is very complex, and it is often necessary to account for the simultaneous presence of different factors.

First of all, in actual systems there is always a distribution of particle size, more or less broad. Moreover, different terms can contribute to the total anisotropy energy of a single-domain particle (see Section C), for example, magnetocrystalline, magnetostatic, shape, stress, and surface. The last one, which is closely related to the detailed chemical nature of surface and grain boundary, may become the dominant contribution to the anisotropy energy for particles smaller than about 4–5 nm, like for ultrathin films. Many studies have been devoted to surface magnetism[27–29] and to magnetism in films where the thickness becomes comparable with the range of exchange forces (for reviews see Refs. 30 and 31).

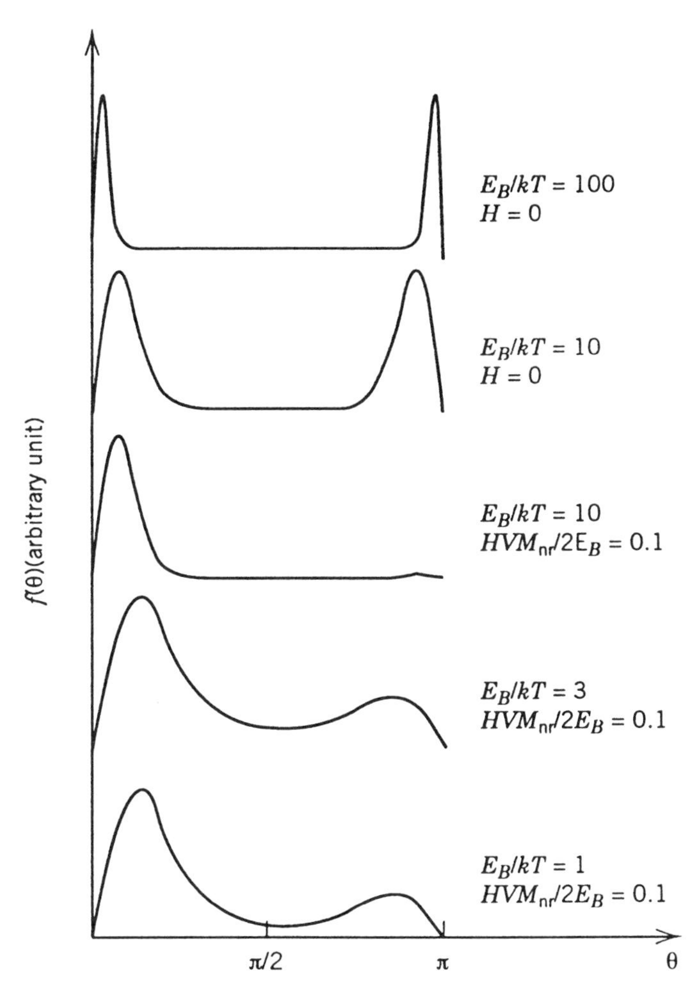

Figure B.4. Probability $f(\theta)$ of finding the magnetization at an angle θ from the easy axis, for an uniaxial particle, as a function of E_B/kT and of the applied field.

In this case, the broken symmetry in surfaces and interfaces results in a strong magnetic surface anisotropy (see Section C). Weakened exchange coupling, in combination with surface anisotropies, results for these virtually two-dimensional systems in the temperature dependence of the magnetic order, strongly different from that for three-dimensional bulk magnets. Moreover, the magnetic moment per spin is modified for a transition metal with respect to the bulk value, depending on the metal, the interface nature, the defects in the surface, and the depth of the layer.

This is effective in general for the two–three first layers. Likewise, magnetic perturbations are expected at the surface of fine particles. They are not negligible, becoming more and more important with decreasing particle size. For example, the two first layers represent about 40% of the volume for a spherical particle with a diameter of 4 nm. The possible effect of these perturbations on the relaxation time will be discussed in Section D.

Furthermore, surface disorders lead to a magnetic moment $|\mathbf{m}|$ value different from that expected from the saturation magnetization of the bulk material and the particle volume V. We define the nonrelaxing magnetization M_{nr} for fine particles as the ratio $|\mathbf{m}|/V$, in the absence of any relaxation effect. Thus, M_{nr} is measurable directly only if the relaxation is not appreciable. M_{nr} varies with temperature and depends on the magnetic field, in so far as the field modifies the magnetic irregularities occurring at the particle surface, and it could also depend on the interparticle interactions.

Experimental determinations of M_{nr} for fine particles show some inconsistencies. This is not surprising, as detailed studies on ultrathin films have shown that M_{nr} changes may depend on several parameters. Generally, M_{nr} is lowered with respect to the saturation magnetization value for the corresponding bulk material. The decrease should be mainly due to spin canting at the surface, that is, disorientations of spins directions, as evidenced by Mössbauer spectroscopy experiments.[32–34] A decrease of the magnetic moment per spin at the surface is also possible for metallic particles.[35] However, for the present it is difficult to predict such effects because of the lack of systematic studies on the different types of particles.

Another effect due to the disorder induced by the size confinement and by the surface contribution is the decrease of T_c with respect to the bulk ferromagnet value.

Finally, interparticle interactions (dipolar and exchange interactions too, if the particles are in contact or if superexchange is possible through a suitable medium) are almost always present, due to the range of the interactions and the difficulty in controlling the particle dispersal completely. The interactions can give an important contribution to the total anisotropy energy. This implies that in granular materials, where fine magnetic particles are dispersed in a nonmagnetic medium, the interparticle interactions and the specific particle arrangement need to be taken into account.

So far, most of the theoretical models did not account for such a complexity, often neglecting important terms, for example, the surface and interparticle interaction anisotropy energy. The development of theoretical models requires materials where the main parameters, for example particle size and shape, dispersion state, chemical state of the surface can be controlled independently. Synthesis methods supplying good systems for testing theoretical models actually need to be developed.

C. ANISOTROPIES IN FINE PARTICLES

In bulk materials, magnetocrystalline and magnetostatic energies are the main sources of anisotropy. In fine particles, other kinds of anisotropy can be of the same order of magnitude as these usual anisotropies. As the properties are stated by the relaxation time τ of the particle magnetic momert $\mathbf{m}$, τ being itself governed by the energy barrier E_B, it is important to know all the possible sources of anisotropies and their contribution to the total energy barrier. On the other hand, the calculation of τ has been performed precisely for the uniaxial symmetry, and approximative formulas exist for cubic symmetry and for the case where a field (unidirectional symmetry) is superimposed to uniaxial symmetry (see Section D).

We will discuss first the various anisotropies that can play a role in fine particles, not considering the effects of an applied field that have been included in the τ calculation (see Section D) and the effects of the interparticle interactions, treated in section E. Second, we will try to give some clues for resolving the complicated problem where either the magnetocrystalline anisotropy cannot be reduced to the first term K_1 or two anisotropies to be added have not the same symmetry. Reviews on the anisotropies encountered in fine particles can be found in Refs. 17 and 18.

C.1. Usual Anisotropy Energies

C.1.1. Magnetocrystalline Anisotropy

Magnetocrystalline energy can show various symmetries, but uniaxial and cubic forms cover the majority of cases. For uniaxial symmetry, the energy is given by

$$E_{\mathrm{cr}} = V(K_0 + K_1 \sin^2\theta + K_2 \sin^4\theta + \cdots) \tag{C.1}$$

while for cubic symmetry

$$E_{\mathrm{cr}} = V\left(K_0 + \frac{K_1}{4}(\sin^2 2\theta + \sin^4\theta \sin^2 2\varphi) + \frac{K_2}{16}\sin^2\theta \sin^2 2\theta \sin^2 2\varphi + \cdots\right) \tag{C.2}$$

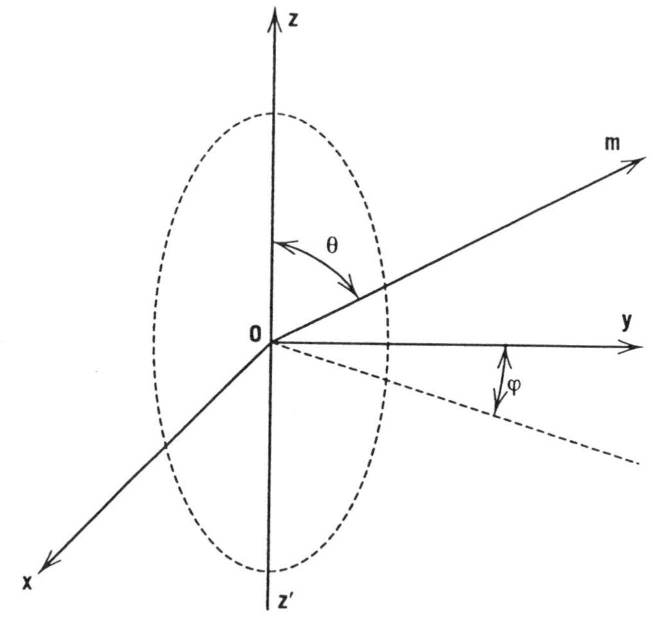

Figure C.1. Usual axis system for fine particles.

in the usual coordinate axes (Fig. C.1).

The τ calculations have been performed for these two symmetries by neglecting K_2 (Section D). But when K_2 is not negligible, the problem of the validity of the τ calculation and of the E_B value is raised. This situation corresponds to real cases. For example, numerous studies have been devoted to γ-Fe_2O_3 particles. For bulk γ-Fe_2O_3, K_1 is negative and the easy directions lie along [110]-type axes. This means that $K_2 > \frac{9}{4}|K_1|$ and therefore K_2 is not negligible with regard to K_1. We discuss this problem in Section C.3.

In uniaxial symmetry, a negative value for K_1 leads to the (110) easy plane ($\theta = \pi/2$). Then, **m** is able to rotate freely in this plane. However, there always exist other small anisotropies that cause energy barriers in this plane. Uniaxial τ formulation is not applicable and the effective E_B value is strongly lowered with regard to VK_1 as it is relative to the weak anisotropies in the (110) plane.

C.1.2. Magnetostatic Anisotropy

The second usual anisotropy comes from magnetostatic energy. As the particle is a single domain, this energy is related to the **m** components and can be expressed exactly only if the particle has an ellipsoidal shape. The determination of the demagnetizing field is a classical problem in magnetism. Here also, usual approximations can be utilized and the magnetostatic energy can be written as

$$(E_{\mathrm{ma}})_t = \frac{1}{2V}(N_x m_x^2 + N_y m_y^2 + N_z m_z^2) \tag{C.3}$$

where m_x, m_y, and m_z are the **m** components and N_x, N_y, and N_z are the demagnetizing factors relative to the particle shape (Fig. C.1) and V is the particle volume.

For an ellipsoid of revolution with the major axis along the [001] direction, the θ-dependent part of the energy is given by

$$E_{\mathrm{ma}} = (m^2/2V)(N_x - N_z)\sin^2\theta \tag{C.4}$$

with $m = |\mathbf{m}|$.

E_{ma} has the uniaxial symmetry and for a prolate ellipsoid the anisotropy constant is positive. For uniaxial magnetocrystalline energy and if the two [001] axes coincide, which is generally the case, the two constants add. For a sphere $E_{\mathrm{ma}} = 0$ and for an elongated particle of acicular type $N_x \simeq 2\pi$ and $N_z \sim 0$. In this case, the anisotropy constant is equal to $\pi m^2/V$. We note that the oblate ellipsoid shape does not seem usual, which generally prevents a negative value of the anisotropy constant.

C.2. Other Anisotropy Energies

C.2.1. Surface Anisotropy

The main origin of other anisotropy energies is related to surface effects. The first effect comes from the existence of the surface that represents a discontinuity for magnetic interactions. This leads to magnetostatic energy already discussed but also to a superficial energy, which for cubic symmetry has often the simple form,[36] writen by surface unit:

$$(E_{\mathrm{su}})_t = K_s \cos^2\theta'$$

where θ' is the angle between **m** and the perpendicular to the surface. K_s depends on the magnetostriction constant λ_{100} and λ_{111} and is of the order of magnitude of 0.1 to 1 erg/cm^2.[36] For an ellipsoid of revolution, the θ dependent part of the anisotropy energy is given in the usual axis system (Fig. C.1) by

$$E_{\mathrm{su}} = K_s F(e) S \sin^2\theta \tag{C.5}$$

with

$$F(e) = \frac{1}{2} \frac{(4 - 3/e^2)\arcsin e + (3/e^2 - 2)e\sqrt{1 - e^2}}{\arcsin e + e\sqrt{1 - e^2}} \tag{C.6}$$

where S is the particle surface, e is the ellipticity with $e^2 = 1 - b^2/a^2$ and $2a$ and $2b$ are the lengths of the major and minor axes, respectively. K_s can be positive or negative.

For weak e values, $F(e)$ reduces to

$$F(e) \simeq \tfrac{4}{15}e^2 + \tfrac{32}{315}e^4 \tag{C.7}$$

The Néel calculation is simple and it is not fully rigorous. However, it permits the description of the properties in a first approximation.

The effect of this anisotropy has been clearly evidenced for thin films where, for $K_s < 0$, perpendicular magnetization has been observed in many ultrathin film systems instead of parallel magnetization as expected from magnetostatic energy.[30,31] However, for a surface of low symmetry, a second term for the surface anisotropy occurs, which is written by surface unit as follows:

$$(E_{\text{su}})_t = K_{\text{sp}} \sin^2\theta' \cos^2\varphi'$$

where θ' and φ' are related to the magnetization in the axis system of the surface. The orders of magnitude of K_{sp} and K_s are similar.[30] For fine particles the **Oy** axis in the axis system of the surface has to be defined. One can consider that the angle between this axis and the tangent to the ellipsis is equal to ξ. In this case[37] we obtain by using the same method as in Néel calculation:

$$E_{\text{su}} = K_{\text{sp}} S[\tfrac{1}{2} + F(e)\cos^2\xi]\sin^2\theta \tag{C.8}$$

with $F(e)$ given by (C.6) or (C.7).

The second term in Eq. (C.8) has the same form as the Néel term (C.5), also leading to an additional anisotropy energy with uniaxial symmetry. However, it does not vanish for a perfect sphere unlike the Néel term. In fact, the second term only occurs for surfaces of low symmetry, but in our opinion, the imperfections in the surface lead to this low symmetry. Therefore, for any particle shape, surface anisotropy has to be taken into account, adding (or substracting according to the signs) to the magnetostatic E_{ma} and magnetocrystalline anisotropies.

This could lead to unexpected variations with V. For example, let us consider the variation of $E_{\text{ma}} + E_{\text{su}}$ vs. V, when $K_s < 0$ and $K_{\text{sp}} = 0$, and its effect on the relaxation time. For large V, E_{ma} dominates and [001] is the easy axis. Decreasing V, there is no contribution to the anisotropy for a certain volume V_c corresponding to a surface S_c such that

$$\frac{4}{15} e^2 |K_s| S_c = \frac{1}{2} \frac{m^2}{V_c} (N_x - N_z) \tag{C.9}$$

Below V_c, (110) becomes the easy plane.

Now, by taking into account the magnetocrystalline anisotropy, supposed with uniaxial symmetry, there is another limit volume such as

$$K_1 V = \frac{4}{15} e^2 |K_s| S - \frac{1}{2} \frac{m^2}{V} (N_x - N_z) \tag{C.10}$$

Generally, K_1 values are strong and the easy plane will be effective only for very small V. If the magnetocrystalline anisotropy is cubic with generally smaller K_1 values, the uniaxial anisotropy related to $E_{\text{ma}} + E_{\text{su}}$ dominates as we shall discuss in Section C.3.2. However, for V near V_c, the anisotropy becomes again cubic, and below V_c another change of regime occurs. This could be an explanation of the very unusual properties observed on Fe particles by Mössbauer spectroscopy[38] where by decreasing the particle size, the spectra evolve from magnetic toward superparamagnetic shape, afterward becoming again almost magnetic.

The effect of the surface[39–43] has been well demonstrated mainly by Mössbauer spectroscopy. Changes in the surface state induce variations of E_B and thus changes for the Mössbauer spectra.

C.2.2. Strain Anisotropy

The second effect due to the surface is related to strains. Because of magnetostriction, strains are effective in the **m** direction. But the corresponding energy is weak, and the θ dependence will be still weaker in cubic symmetry if $\lambda_{100} \sim \lambda_{111} = \lambda_s$. However, if exterior strains occur, the θ-dependent part of the magnetoelastic energy can be written:

$$E_{\text{st}} = -\tfrac{3}{2} \lambda_s \sigma S \cos^2\theta'' \tag{C.11}$$

where σ is the strain value by surface unit and θ'' the angle between **m** and the strain tensor axis.

Unfortunately, it is not possible to give a general formulation for accounting for strains because various cases occur depending on the sample and on the elaboration method. Nevertheless, if possible, an evaluation of strains and their effects will be useful.

C.3. Combination of Anisotropies

The τ calculation has been performed only in uniaxial and cubic symmetries by neglecting K_2. Then, there is a difficulty if K_2 cannot be neglected or if the anisotropies that act have not the same symmetry.

In this case, in our opinion, if the relaxation paths remain the same with regard to the symmetries where τ has been calculated, the τ formulation applied to the actual energy barrier is almost exact. Perhaps small modifications of the τ_0 factor (see Section D) occur. This is the case for uniaxial symmetry when the K_2 constant is taken into account. Then, $E_B = V(K_1 + K_2)$.

If now the relaxation paths are not too changed, the τ formulation using the actual E_B is still valid in first approximation. However the τ_0 factor is probably modified. For example, let us consider the asymptotic expressions of the eigenvalues for cubic symmetry (see Section D) when K_2 is neglected.

$$\lambda_1 \simeq \frac{E_B}{kT} \frac{16\sqrt{2}}{\pi} \exp\left[-\frac{E_B}{kT}\right] \tag{C.12}$$

with $E_B = K_1V/4$ for $K_1 > 0$ and

$$\lambda_1 \simeq \frac{E_B}{kT} \frac{8\sqrt{2}}{\pi} \exp\left[-\frac{E_B}{kT}\right] \tag{C.13}$$

with $E_B = |K_1|V/12$ for $K_1 < 0$.

For $K_1 > 0$, the easy axes are along [001]-type directions and the relaxation paths are through [011]-type axes while for $K_1 < 0$, the easy axes are along [111]-type directions and the relaxation paths are also through the [011]-type axes.

We can see from Eqs. (C.12) and (C.13) that the prefactor of the exponential for $K_1 > 0$ is two times the one corresponding to $K_1 < 0$.

Fortunately, the variation of the prefactor has an important role only if this variation is large because the τ value is mainly stated by E_B through the exponential. Nevertheless for an accurate adjustment, it will be more advisable to establish a precise formulation for τ.

We discuss below two cases, the cubic symmetry when K_2 is not negligible and the mixing of cubic and uniaxial symmetries. In fact, there are also other combinations that can occur like, for example, the addition of two uniaxial symmetries with different easy axes.[44] In this case, particular examinations will be needful.

TABLE I
Easy Axes, **m** Relaxation Paths, Estimation of the τ Calculation Validity and Energy Barriers for Cubic Symmetry When K_2 Is Not Negligible with Regard to K_1 (see text)

	Easy Axes	Paths through Axis	τ Calculation Validity	E_B/V
$K_1>0$				
$K_2>-2K_1$	[001]	[110]	Correct	$K_1/4$
$-3K_1<K_2<-2K_1$	[001]	D_2	Correct	$-\frac{K_1^2}{K_2^2}(K_1+K_2)$
$-9K_1<K_2<-3K_1$	[001]	D_1, [111], D_2	Approx.	$-\frac{K_1^2}{K_2^2}(K_1+K_2)$, $\left[-\frac{1}{K_2^2}\left(K_1+\frac{K_2}{3}\right)^3\right]$
$K_2<-9K_1$	[111]	D_1, [100], D_1	Approx.	$-\frac{1}{K_2^2}\left(K_1+\frac{K_2}{3}\right)^3$, $\left[-\frac{K_1^2}{K_2^2}(K_1+K_2)\right]$
$K_1<0$				
$K_2>(-\frac{9}{4})K_1$	[110]	D_2	Approx.	$-\frac{K_1}{4K_2^2}(2K_1+K_2)^2$
$-2K_1<K_2<(-\frac{9}{4})K_1$	[111]	D_1, [110], D_1	?	$-\frac{1}{K_2^2}\left(K_1+\frac{K_2}{3}\right)^3$, $-\frac{K_1}{4K_2^2}(2K_1+K_2)$
$K_2<-2K_1$	[111]	[110]	Correct	$-\frac{K_1}{12}-\frac{K_2}{27}$

C.3.1. Cublic Symmetry When K_2 Is Not Negligible

The results are summarized in Table I. In the table, D_1 and D_2 represent particular extremum directions for E_{cr}. However, these directions only exist in certain ranges of K_1 and K_2 values. For $0 \le \theta \le \pi/2$ and $0 \le \varphi \le \pi/2$ in the usual axis system (Fig. C.1), D_1 is given by (C.14), the other directions being deduced by symmetry.

$$\varphi = \frac{\pi}{4} \quad \text{and} \quad \theta = \arcsin\sqrt{-2\frac{K_1}{K_2}} \tag{C.14}$$

This extremum exists only if $0 \le -2K_1/K_2 \le 1$. For this extremum

$$E_{cr} = -V\left(\frac{K_1}{K_2}\right)^2 (K_1 + K_2) \tag{C.15}$$

Directions D_2 are defined by

$$\varphi = \frac{1}{2}\arccos\left|1 + \frac{2K_1}{K_1 + K_2}\right|$$
$$= \frac{\pi}{2} - \frac{1}{2}\arccos\left|1 + \frac{2K_1}{K_1 + K_2}\right| \tag{C.16}$$
$$\theta = \arccos\sqrt{-\frac{K_1}{K_2}}$$

This extremum exists for the same condition $0 \leq -2K_1/K_2 \leq 1$ and the E_{cr} value is the same (C.15).

Several axes are sometimes mentioned for the path (Table C.1). This means that **m** jumps over a main energy barrier and jumps also over a secondary energy barrier indicated between brackets. For $K_1 < 0$ and $-2K_1 < K_2 < (-\frac{9}{4})K_1$, the two energy barriers have the same order of magnitude, which leads to uncertainties in the τ formulation. But, in general, we think that the τ cubic formulation is usable with some modifications for the τ_0 prefactor.

It is interesting to define the order of size of E_B for γ-Fe_2O_3 because numerous studies have been devoted to fine particles of this material. For bulk, easy directions correspond to [110] axes with $K_1 \approx -4.7\ 10^4$ erg/cm^3. Therefore $K_1 < 0$ and $K_2 > -\frac{9}{4}K_1$, but no precise datum exists for K_2. From Table C.1, E_B is equal to $V|K_1|/4$ only for very high K_2 values. For $K_2 = -\frac{9}{4}K_1$, just at the limit, $E_B \approx 0.012V|K_1|/4$, when for $K_2 = 1.5(-\frac{9}{4}K_1)$ and $2(-\frac{9}{4}K_1)$, $E_B = 0.17V|K_1|/4$ and $0.31V|K_1|/4$, respectively. Therefore, except for high K_2 values, E_B is reduced with regard to the $V|K_1|/4$ value, which could be taken by assimilating the actual relaxation between [110] axes to the relaxation between [001] axes. Taking into account the low value for K_1, this means that the other anisotropies such as the magnetostatic anisotropy will dominate the magnetocrystalline one.

C.3.2. Mixing of Cubic and Uniaxial Symmetries

The problem of mixing of cubic and uniaxial symmetries is similar to the one discussed above. Either the relaxation paths are the same or similar to those resulting from one of the anisotropies, and then the corresponding formula can be applied with the actual energy barrier, or the contributions of the anisotropies to the total energy are similar and the τ prediction is difficult without particular calculations.

Let us consider the mixing of uniaxial anisotropy with a constant K_{1u}, first with simple cubic anisotropy with $K_1 > 0$ and $K_2 = 0$ and second with cubic anisotropy with $K_1 < 0$ and $K_2 > (-\frac{9}{4})K_1$, which leads to [110]-type easy axes.

C.3.2.1. Simple Cubic Symmetry ($K_1 > 0$, $K_2 = 0$) *and Uniaxial Symmetry*. In this case, the relaxation paths relative to uniaxial symmetry remain unchanged for $K_{1u} > K_1$, except that **m** now relaxes rather through [100] or [010] axes. The energy barrier is equal to $K_{1u}V$ and the uniaxial formulation is applicable.

For $K_1/2 < K_{1u} < K_1$, the maximum for $\theta = \pi/2$, $\varphi = 0$, or $\pi/2$ splits into a minimum and two maxima (Fig. C.2). There are a main and a secondary energy barriers. The effect of such a situation on the properties depends on the kind of measurements. For magnetization experienced along a direction, only the relaxation time τ relative to the main barrier is effective because τ being proportional to $\exp[E_B/kT]$, τ relative to the secondary barrier is much smaller than τ relative to the main barrier. In our opinion, the uniaxial formulation remains valid[45] with E_B given by

$$E_B = \frac{K_1 V}{4}\left(1 + \frac{K_{1u}}{K_1}\right)^2 \tag{C.17}$$

However, for measurements that experience all the τ, such as Mössbauer spectroscopy, there is a difficulty when the volumes are distributed. For example, supposing that the main E_B/V ratio is equal to three times the secondary E_B/V ratio, the relaxation time corresponding to the secondary barrier of a particle with a volume $3V$ will be the same as τ corresponding to the main barrier of particle with a volume V. This

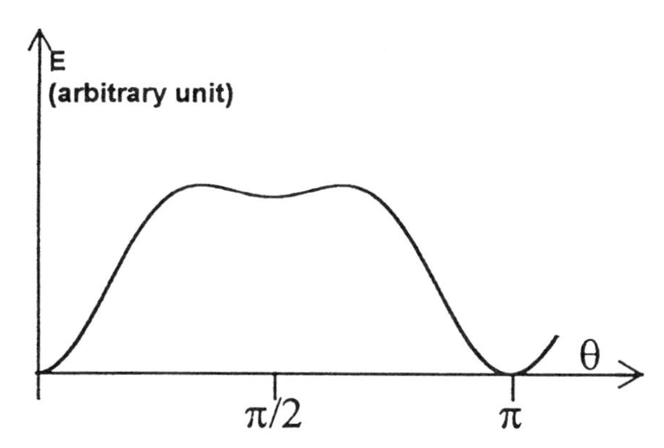

Figure C.2. θ-dependent part of the anisotropy energy for $\varphi = 0$ in the case of the sum of two components with the one having the uniaxial symmetry with a constant K_{1u} and the second having the cubic symmetry with a constant $K_1 > 0$ ($K_2 = 0$) and $K_{1u} > K_1$.

means that the results of the two kinds of experiments are not directly comparable.

For $0 < K_{1u}/K_1 < 0.5$, a new maximum occurs for $\varphi = \pi/4$ and

$$\theta = \arccos\sqrt{\frac{1}{3}\left(1 - 2\frac{K_{1u}}{K_1}\right)}$$

which leads to complex paths for $K_{1u}/K_1 \leq 0.3$–0.4. Then, the uniaxial formulation becomes doubtful.

For $K_{1u} < 0$, **m** relaxation occurs in the (110) plane with $E_B = VK_1/4$. In this case, we do not know what formula has to be applied. For $|K_{1u}|/K_1 \leq 0.3$–0.4, the same problem arises as for $K_{1u} > 0$, complex paths being present. In all cases the τ cubic formulation is valid only for very weak values of K_{1u}.

C.3.2.2. Cubic Symmetry with [110]*-Type Easy Axis and Uniaxial Symmetry*. For uniaxial anisotropy adding to cubic anisotropy with $K_1 < 0$ and $K_2 > (-\frac{9}{4})K_1$ such that [110]-type directions correspond to easy axes, the determination of the **m** relaxation paths is not easy because it depends on three parameters and for $\gamma\text{-}Fe_2O_3$, the K_2 value is unknown.

For positive K_{1u} values and $K_{1u} > |K_1|$, the relaxation paths are similar to those resulting from uniaxial symmetry though the relaxation occurs rather through the [110] axes in the (011) plane. In this case, the τ uniaxial formulation is applicable with $E_B = V(K_{1u} + K_1/4)$. For $K_{1u} < |K_1|$, the relaxation paths are complex. The easy directions lie along the $\varphi = 0$ and $\theta = (\frac{1}{2})\arccos(-K_{1u}/K_1)$-type directions, and there is transverse relaxation through D_3 axis defined by

$$\varphi = \frac{\pi}{4} \quad \text{and} \quad \cos^2\theta = \frac{1}{3K_2}\left[(3K_1 + 2K_2) \pm \sqrt{(3K_1 + K_2)^2 + 12K_{1u}K_2}\right] \tag{C.18}$$

and longitudinal relaxation through D_3 and the $\theta = \pi/2$, $\varphi = \pi/4$ axis.

The longitudinal E_B value, which is the correct parameter for magnetization measurements, is given by

$$E_B = \frac{VK_{1u}}{4}\left[2 - \frac{K_{1u}}{K_1}\right] \tag{C.19}$$

Due to the complicated paths, it is not clear that the τ uniaxial formulation is still valid. In our opinion it is correct in first approximation for $K_{1u}/|K_1| \geq 0.3$–0.4.

For measurements that experience all the relaxation times, those related to longitudinal as well as transverse relaxations are to be considered.

For negative K_{1u} values and $|K_{1u}| > |K_1|$, the relaxation paths are always in the (110) plane with easy directions along [110]-type axes in this plane and $E_B = |K_1|V/4$. For $|K_{1u}| < |K_1|$ and until $|K_{1u}|$ about 0.3–0.4$|K_1|$, the easy directions are the same but the relaxation paths are through D_4 axes defined by $\varphi = 0$ and $\theta = \arccos[(K_1 - K_{1u})/2K_1]$ and symmetrical axes and $E_B = (VK_{1u}/K_1)(K_{1u} - 2K_1)$. For lower $|K_{1u}|$ values the relaxation paths are complex. In all cases, it is difficult to define a τ formulation.

C.4. Conclusions

As demonstrated, the anisotropies for fine particles do not generally have a simple form, and this leads to uncertainties concerning the use of the known τ formulas. Fortunately, as we shall discuss in Section D, τ can be expressed for E_B/kT higher than about 3 by

$$\tau = \tau_0 \exp[E_B/kT] \qquad \text{(C.20)}$$

where τ_0 depends on various parameters and of the symmetries and the relaxation paths.

With such a formula, the main τ variation comes from E_B/kT through the exponential. From experimental data, it will be possible to determine E_B and τ_0, but it will be difficult to relate E_B to the anisotropies without a precise knowledge of their origins and their symmetries. The same problem will occur for τ_0, in addition to the uncertainties on the formulation.

Finally another problem comes from the measured phenomenon. The results of two experiments that experience the same phenomenon with the same process are evidently comparable. In the contrary case, precautions have to be taken.

D. RELAXATION TIME CALCULATION

The calculation of the relaxation time τ of the magnetic moment **m** of a particle is very important because the τ value states all the experimental results. Accurate formulas are needed but difficult to establish because an actual sample of fine particles does not correspond to any simple case, due to the various anisotropies, the interparticle interactions, the surface effects, and so forth. On the other hand, experiments do not always measure the same parameters and the formulas have to be adapted.

In this section, we state the problem and define the relaxation time τ in Section D.1. We describe Néel's model in Section D.2 and Brown's model in Section D.3. In this latter section, we discuss in some details the model hypotheses and their validity; we give the results of τ calculations and approximative formulas in the case of uniaxial symmetry with and without applied field and in the case of cubic symmetry. Some other cases are also discussed. Finally we give some comments in Section D.4.

D.1. Problem Statement

Let us consider a system of N particles with uniaxial symmetry. In the absence of a field, the anisotropy energy in the usual axis system (Fig. D.1) is equal to

$$E = KV \sin^2\theta \tag{D.1}$$

where K is the anisotropy constant and V the particle volume.

For $K > 0$, the easy axes are along **Oz** and **Oz**′ and the energies at the minima have the same value. Therefore, the two energy minima are

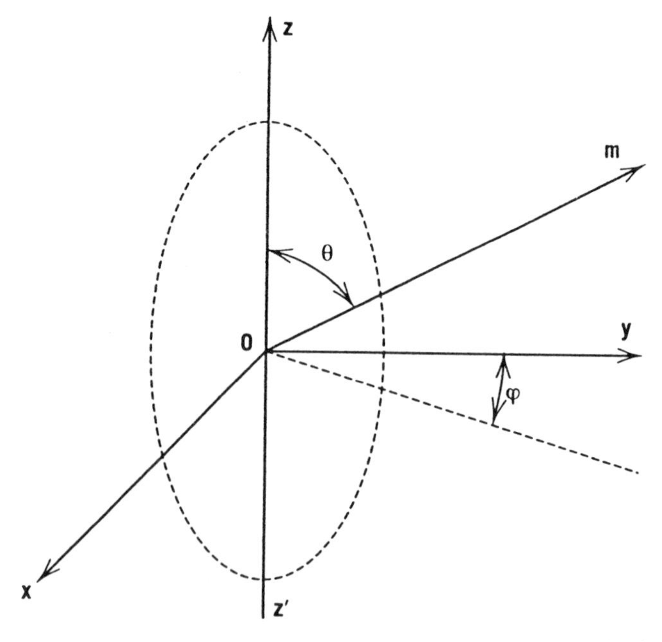

Figure D.1. Usual coordinate axis system. The easy axis of the magnetization is along **zOz**′.

equally populated when the thermodynamic equilibrium has been attained.

Now, let us consider that P particles have their magnetic moment $\mathbf{m}$ around $\mathbf{Oz}$ and $N - P$ particles around $\mathbf{Oz}'$ at a given time t. After a time dt, the numbers of particles that have $\mathbf{m}$ along $\mathbf{Oz}$ and $\mathbf{Oz}'$ are modified and the change is given by

$$dP = \frac{(N-P)}{2\tau} dt - \frac{P}{2\tau} dt \tag{D.2}$$

where $1/(2\tau)$ is the probability by time unit for a reversal of $\mathbf{m}$.

From Eq. (D.2), the magnetization along $\mathbf{Oz}$ can be written as

$$M = M_{nr} S_0 \exp(-t/\tau) \tag{D.3}$$

where M_{nr} is the value of the nonrelaxing magnetization of the particle (see Section D.3), S_0 is the initial value of $2P/N - 1$, and τ is identified with the relaxation time.

When a field H is applied along $\mathbf{Oz}$, E becomes

$$E = KV \sin^2\theta - HM_{nr} V \cos\theta \tag{D.4}$$

Now, the two energy values at the minima are not equal, and so two probabilities have to be defined: $1/\tau^+$ for the jump from the lowest minimum toward the upper minimum and $1/\tau^-$ for the opposite reversal (Fig. D.2).

Equation (D.2) becomes

$$dP = \frac{(N-P)}{\tau^-} dt - \frac{P}{\tau^+} dt \tag{D.5}$$

and the magnetization along $\mathbf{Oz}$ is now

$$M = M_{nr}[S_1 + (S_0 - S_1)\exp(-t/\tau)] \tag{D.6}$$

with

$$S_1 = \frac{\tau^+ - \tau^-}{\tau^+ + \tau^-} \tag{D.7}$$

and

$$\frac{1}{\tau} = \frac{1}{\tau^+} + \frac{1}{\tau^-} \tag{D.8}$$

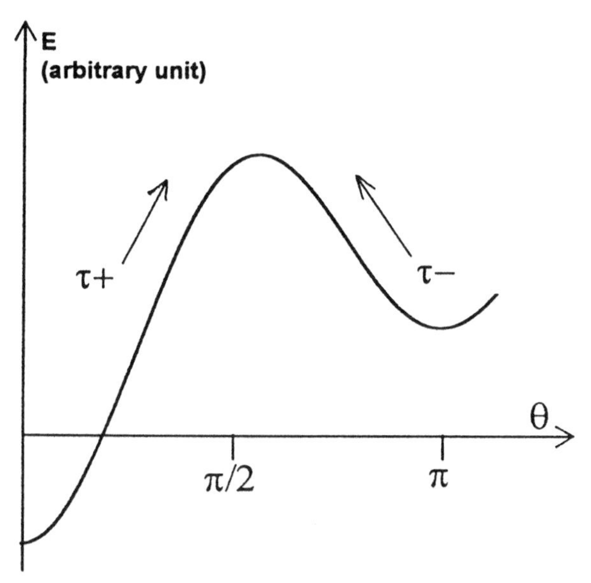

Figure D.2. θ dependence of the total energy when a field is applied along the easy axis. τ^+ and τ^- are the two relaxation times (see text).

The thermodynamic equilibrium is attained when $t \gg \tau$ or if $S_0 = S_1$. Now, the magnetization along **H** is not equal to zero. This means that, on average, a part of the magnetization is blocked in the lowest minimum, equal to S_1, the remaining part $1 - S_1$ relaxing with a relaxation time τ given by (D.8).

We note two points. First, the Eq. (D.8) gives for $H = 0$

$$\tau^+ = \tau^- = 2\tau$$

which is in agreement with the Eqs. (D.2) and (D.5). This means that the probability for **m** reversals without a field is $1/(2\tau)$, while with a field, it is $1/\tau^+$ and $1/\tau^-$ according to the minimum. These definitions have been made in order to retain the same formula [Eqs. (D.3) and (D.6)] for the decay of the magnetization. Second, without field, the probability is equal to 1 for $\Delta t = 2\tau$. Thus one may say that 2τ is the time for a direction change for **m** and to go forward and back, the time is 4τ.

D.2. Relaxation Time Calculation. Néel's Model

The first τ calculation was performed by Néel.[1] He supposed that the particle spins are rigidly coupled and that synchronous rotation of the spins occurs when **m** is reversed. He considered only the case of uniaxial symmetry and the calculation has been only performed for $E_B/kT \gg 1$, where E_B is the energy barrier, that is, when the spins are mainly along the easy directions. The equivalence of the spins to a gyroscopic system allows one to derive the following expression for the relaxation time:

$$\tau = \tau_{0N} \exp(\alpha) \tag{D.9}$$

with

$$\alpha = \frac{kV}{kT} \tag{D.10}$$

and

$$\tau_{0N} = \frac{1}{3} \sqrt{\frac{\pi}{2}} \frac{1}{\gamma_0} \frac{1}{|\lambda_s|} \sqrt{\frac{K}{G}} \frac{M_{nr}}{K} \frac{1}{\sqrt{\alpha}} \tag{D.11}$$

where γ_0 is the gyromagnetic ratio, λ_s is the longitudinal magnetostriction constant, and G is Young's modulus.

D.3. Relaxation Time Calculation. Brown's Model

Néel's calculation of τ was criticized by Brown[46–48] because the system is not explicitly treated as a gyromagnetic one and it supposes a discrete orientation approximation (**m** is essentially along the easy directions). He supposed that the magnetic moment orientations may be described by a Gilbert equation augmented by a random field term $h(t)$, which is assumed to be white noise:

$$\frac{d\mathbf{m}}{dt} = \gamma_0 \mathbf{m} \times \left[-\frac{\partial U}{\partial \mathbf{m}} - \eta \left(\frac{d\mathbf{m}}{dt} \right) + \mathbf{h}(t) \right] \tag{D.12}$$

where γ_0 is the gyromagnetic ratio, η the damping constant, and U the barrier potential. Now this is the Langevin equation of the process. With an applied field,

$$U = E - \mathbf{m} \cdot \mathbf{H}$$

The probability density of the orientations $W(\theta, \varphi, t)$ of **m** obeys a Fokker–Planck equation and can be written in the general form:

$$W = W_0 + \sum_{n=1}^{\infty} C_{in} U F_n(\theta, \varphi) \exp\left(\frac{-\lambda_n t}{\tau_D} \right) \tag{D.13}$$

with

$$\tau_D = \frac{V}{\eta k T}\left(\frac{1}{\gamma_0^2} + \eta^2 M_{\mathrm{nr}}^2\right) \quad \text{(D.14)}$$

where $F_n(\theta, \varphi)$ and λ_n/τ_D are, respectively, the eigenmodes and the eigenvalues of the Fokker–Planck equation, and C_{in} are coefficients that depend on the initial conditions.

For an axially symmetric potential, the Fokker–Planck equation becomes

$$\tau_D \frac{\partial w}{\partial t} = \frac{1}{\sin\theta}\frac{\partial}{\partial\theta}\left[\sin\theta\left(\frac{\partial W}{\partial\theta} + \frac{V}{kT}\frac{\partial U}{\partial\theta}W\right)\right] + \frac{1}{\sin\theta}\frac{\partial}{\partial\varphi}\left[\frac{V}{kT}\frac{1}{\eta\gamma_0 M_{\mathrm{nr}}}\frac{\partial U}{\partial\theta}W + \frac{1}{\sin\theta}\frac{\partial W}{\partial\varphi}\right] \quad \text{(D.15)}$$

A review devoted to an introduction and to detailed notes on aspects of the theory of magnetic relaxation, and which also treats the solid-state relaxation as well as the dynamic behavior of suspensions of single-domain ferromagnetic particles in fluids, can be found in Ref. 49.

D.3.1. Hypothesis Discussion

The use of Gilbert's equation and the meaning of the parameters used in it need some discussions. In fact, the Gilbert equation is concerned with spins and their influence on the determination of the relaxation time of the magnetic moment **m** set. Brown has overcome the difficulty of different directions of spins by supposing that the particle spins are parallel, with the same moment and that they are rigidly coupled. However, at the particle surface, disorders exist for the spin directions and variations of moments can be present (see Section B).

Our first question concerns the validity of the hypothesis of rigid coupling, necessary for synchronous rotation. Numerical experiments have shown the existence of new micromagnetic structures, such as the flower state, which leads to nonuniform modes for magnetization reversal in fine particles (see, e.g., Refs. 20 and 22). Furthermore, at the edge of the particles, the moments are not collinear with the core magnetization; and on decreasing the field, the magnetization reverses via fanning and (or) vortex formation.[50–52] Nevertheless, recent calculations[53] have considered very small square or circular particles with two kinds of anisotropies: (i) random bulk anisotropy, which is equivalent in the limit $J \rightarrow 0$ to the Stoner–Wohlfarth model[54] for a finite collection of random anisotropy noninteracting moments, and (ii) surface anisotropy only on the sites in the boundary of the particle. The results show that deviations from the uniform mode occur only for high anisotropy values with regard to the exchange coupling, K/J of an order of the unity or higher. Furthermore in this case, one may conjecture that for an assembly of particles, the properties do not differ greatly from those resulting from a uniform mode due to the distribution of particle shapes, which leads to a distribution of jumps in the hysteresis loop.

Recent measurements on an individual γ-Fe_2O_3 particle (prolate ellipsoid of 300 nm long and 65 nm wide)[55] have shown that for large angles (>30°) between the field and the particle axis, the switching field is in agreement with that predicted by the Stoner–Wohlfarth model (uniform mode), while for weak angle the switching is best represented by a curling mode. We note that in our opinion the concept of a single-domain state is questionable for such a particle size. Measurements of the switching probability $P(t)$ when a field is applied (several values around 1070 Oe have been used) have shown that the classical law $P(t) = 1 - \exp[-t/\tau)]$ (see Section D.1) is not obeyed. This questions the validity of the Néel–Brown model.[56] However, the measured properties correspond to a very particular case. Due to the large size, **m** relaxation does not occur without applied field. In order to observe thermal relaxation, the applied field has to be adjusted very precisely in order that the difference between the anisotropy energy and the Zeeman energy $\Sigma_j \mathbf{H} \cdot \mathbf{S}_j$ has the same order of magnitude as kT, in this case much smaller than each of the terms. Then, the disalignments of **S** in the particle surface due to its effect, including defects, are of fundamental importance because they give the value of the energy difference. Here the $P(t)$ calculation is complicated. Furthermore, it is not certain that after several cycles, the surface **S** state is the same because it is likely that for these spins multiple energy minima are present. Under low field, for fine particles that exhibit thermal **m** relaxation without field, the problem is completely different because the energy mainly comes from the anisotropic part. In our opinion, precautions must be taken for the interpretation of the results when the Zeeman and anisotropy energies are almost equal.

Very recent measurements[57] on single Co particle patterned by electron beam lithography with a thickness of 30 nm and an elliptic contour with axis between 50 and 150 nm have shown that only a particle of 80 × 50 nm size is single domain with a magnetization reversal in good agreement with the Stoner–Wohlfarth model, that is, the uniform mode. However, there is a distribution of switching fields, whose width increases with decreasing temperature, the temperature dependence of the mean switching field being in agreement for $T > 1$ K with the two-level model; $1 - P(t)$ is well described by a stretched exponential in agreement with the picture of multiple energy minima. Below 1 K, the properties are in favor of macroscopic quantum tunneling (see Section G).

In our opinion, in spite of spin disorders in the surface, the spins basically remain rigidly coupled mainly because the exchange force is strong with regard to the anisotropy, and the uniform mode is a good approximation, in agreement with Ref. 44. However, it is probable that some deviations occur for the surface spins during the **m** reversal, which perhaps lead to a slight modification of the τ_0 factor. Nevertheless, for very high values of the anisotropy or when a field is applied and when the Zeeman energy almost balances the anisotropy energy, the question remains open.

The second question concerns the meaning of the parameters in the Gilbert equation (D.12). The spin motion depends on γ_0, η, and the spin moment. Then, τ is also dependent on the same parameters. Nevertheless, the magnetic moment per spin is different for the surface and the core of particles, and it is probable that η is also different. However, due to the rigid coupling of the spins, τ is the same for all the particle spins. Therefore, the Gilbert equation must be considered as operating on a mean particle spin or in other words, on **m**. In this case, averaged parameters are to be used for η and the magnetization. For the former parameter, the problem is quite similar to those existing in bulk materials with some defects. We may expect that η values for particles will be larger than those observed for the corresponding bulk materials.[58] For the latter parameter, it is a little more complicated. The variation of the magnetic moment per spin at the surface is to be taken into account. However, the problem of the disorientations remains. With an applied field, the potential U depends on $\mathbf{M} \cdot \mathbf{H}$ and a useful parameter is the averaged magnetization along $\mathbf{H}$. Nevertheless, the kind of average to be performed on the Gilbert equation is not evident. In our opinion, the best way is to consider the averaged magnetization along the core spin direction, which is called the nonrelaxing magnetization M_{nr}.

This approach is supported by the results obtained on fine particles in which the bulk magnetic state is antiferromagnetic (see Section H). To be more specific, the magnetic state of these particles corresponds to an uncompensated antiferromagnetism, with a magnetic moment proportional to n^p, with $\frac{1}{3} \lesssim p \lesssim \frac{1}{2}$, n being the spin number of the particle. With our M_{nr} definition, M_{nr} is strongly lowered in comparison with ferromagnetic materials, which causes a large decrease of τ_0 (see below) in agreement with the observed results (see Section H).

Finally, it is useful to define a dimensionless constant η_r related to η[58]:

$$\eta_r = \eta \gamma_0 M_{\text{nr}}(0) \tag{D.16}$$

where $M_{nr}(0)$ is the value of the nonrelaxing magnetization at zero temperature. Changes of η_r with temperature are possible. In this case, the τ_D value is given by:

$$\tau_D = \frac{VM_{nr}(0)}{\gamma_0}\frac{1}{kT}\left[\frac{1}{\eta_r} + \eta_r\left(\frac{M_{nr}(T)}{M_{nr}(0)}\right)^2\right] \tag{D.17}$$

Brown[46–48] has suggested taking $\eta_r = 1$ and neglected the thermal variation of M_{nr}. At present, only a few significative results have been published. A value of η_r of the order of unity occurs[45,59] but lower values also occur.[59]

D.3.2. Calculation Without Applied Field in Uniaxial Symmetry

Equation (D.15) has been transformed into a Sturm–Liouville problem for the case of longitudinal relaxation.

Brown[48] assumed that the corresponding relaxation time is given by

$$\tau = \frac{\tau_D}{\lambda_1} \tag{D.18}$$

where λ_1 is the smallest nonvanishing eigenvalue of the Sturm–Liouville equation. This is true if $\lambda_1 \ll \lambda_k$, $k \geq 2$, since all the exponential functions $\exp(-\lambda_k t/\tau_D)$, $k \geq 2$, in (D.13) are small compared with $\exp(-\lambda_1 t/\tau_D)$ except in the very early steps of the approach to equilibrium. This is checked except for low α values.

Brown derived an approximate formula[47,48] for $\alpha \ll 1$, which has been refined later,[60] and one for $\alpha \gg 1$, which has been checked numerically.[60] For $\alpha \gg 1$, τ is given by

$$\tau = \tau_D \frac{\sqrt{\pi}}{4}\alpha^{-3/2}\exp(\alpha) \tag{D.19}$$

For $\alpha \ll 1$, τ is given by (D.18) with

$$\lambda_1 = 2(1 - \tfrac{3}{5}\alpha + \tfrac{48}{875}\alpha^2) \tag{D.20}$$

Recently, there has been a revival of interest in the problem of calculating τ for the purpose of obtaining a τ formula valid for any α value. In fact, this was important for experiments with short measuring time τ_m, for example, Mössbauer spectroscopy (see Section F.6), which allows one to detect relaxation times corresponding to α value in the range of 2–4.

First, Bessais et al.[58,61] have solved the Fokker–Planck equation by another method and have derived an expression for λ_1 valid for any α

value:

$$\lambda_1 = \left(1 + \frac{\alpha}{4}\right)^{5/2} \exp(-\alpha) \tag{D.21}$$

Later Aharoni[62] proposed a formula that seems a combination of the approximate Brown formulas for $\alpha \gg$ and $\alpha \ll 1$:

$$\lambda_1 = \left(\frac{2 + \frac{9}{5}\alpha + (4/\pi)^{1/3}\alpha^2}{2 + \alpha}\right)^{3/2} \exp(-\alpha) \tag{D.22}$$

Finally Coffey et al.[63,64] defined the longitudinal relaxation time τ as the correlation time, that is, the area under the curve of magnetization autocorrelation function. In this case

$$\frac{\tau}{\tau_D} = \frac{\sum_k A_k \lambda_k^{-1}}{\sum_k A_k} \tag{D.23}$$

where λ_k are the eigenvalues and A_k their amplitudes. They have shown that only λ_1 has to be considered in (D.23), τ being determined with very good accuracy. They have derived an exact analytical solution for τ and checked Brown's approximation for $\alpha \gg 1$. In fact, a better approximation is obtained by multiplying expression (D.19) by $1 + 1/\alpha$, yielding

$$\tau = \tau_D \frac{\sqrt{\pi}}{4} \alpha^{-3/2} \left(1 + \frac{1}{\alpha}\right) \exp(\alpha) \tag{D.24}$$

This formula has been compared to the exact solutions.[64] For $\alpha = 2$, 2.5, 3, and 3.5, the error is 41, 20, 9, and 3%, respectively. Therefore (D.24) can be used reasonably for $\alpha \gtrsim 3$.

Furthermore an approximate formula valid for any α has been derived[65]:

$$\tau = \tau_D \frac{1}{4\alpha} \frac{1}{\dfrac{\alpha}{\alpha + 1}\sqrt{\dfrac{\alpha}{\pi}} + 2^{-\alpha - 1}} [\exp(\alpha) - 1] \tag{D.25}$$

This formula has been also compared to the exact solutions and the other approximate formulas.[65] It is clear that (D.25) is the best formula with an error smaller than 8% (this value is obtained for $\alpha = 0.5$, the error decreasing as α increases). Nevertheless, in our opinion, if $\alpha > 3$, Eq. (D.24) is more convenient for application.

We note that in the notations of Coffey et al., λ_k differ by a factor of 2 from that used by Brown.

D.3.3. τ Calculation with Applied Field in Uniaxial Symmetry

The calculation was initially performed with the field direction along the easy axis because, in this case, the uniaxial symmetry is not broken and the calculations are not too complicated. However, for an assembly of particles, the easy axis is in a random position and the probability to find H_{app} parallel to easy axis is equal to zero.

For $\alpha \gg 1$, in the Néel model[1]:

$$\tau^{\pm} = 2\tau_{0\text{N}} \frac{1}{(1 \pm h)} \frac{1}{\sqrt{1 - h^2}} \exp[\alpha(1 \pm h)^2] \tag{D.26}$$

and in the Brown model[47,48]:

$$\tau^{\pm} = \frac{\sqrt{\pi}}{2} \tau_D \frac{\alpha^{-3/2}}{(1 \pm h)(1 - h)^2} \exp[\alpha(1 \pm h)^2] \tag{D.27}$$

with

$$h = \frac{M_{\text{nr}} H_{\text{app}}}{2K} \tag{D.28}$$

We note that in the two formulas, the only differences occur for the τ_0 factors, which has little importance because τ is mainly governed by the exponential factor.

This formula has been checked numerically.[66] It is a good approximation for $\alpha \gtrsim 3$ and low values of h depending on α, for example, for $h < 0.5$ and $\alpha \gtrsim 5$. However, as we will discuss below, Eq. (D.27) is valid only for low h values because (D.18) holds only for these values.

For $\alpha \ll 1$

$$\lambda_1 = \lambda_1^+ + \lambda_1^- = 2(1 - \tfrac{2}{5}\alpha + \tfrac{48}{875}\alpha^2 + \tfrac{2}{5}h^2\alpha^2) \tag{D.29}$$

This formula has been also checked numerically[66] and yields a good approximation for $\alpha \lesssim 1.5$.

Very recently, Coffey et al.[67] have calculated the different eigenvalues λ_k and their amplitudes A_k in the case of an applied field parallel to the easy direction. It is supposed that a weak constant applied field $\mathbf{\Delta H}$ is removed to the field $\mathbf{H}$ at $t = 0$. Then, the decay of the longitudinal component of the magnetization may be expressed as follows:

$$M_{\|} = \frac{|\mathbf{m}|^2 \, \Delta H}{VkT} \sum_k A_k \exp(-\lambda_k t/\tau_D) \tag{D.30}$$

where $|\mathbf{m}|$ is the modulus of the magnetic moment equal to $M_{nr}V$.

The longitudinal relaxation time τ is given by Eq. (D.23). If the smallest eigenvalue is only considered, τ reduces to

$$\tau = \tau_D / \lambda_1 \tag{D.31}$$

The results of the calculation performed for k values until 10 show that the longitudinal relaxation time has a behavior very different from $\tau_D \lambda_1^{-1}$ above certain critical values of the parameters α and h. For example, $\tau\lambda_1/\tau_D \approxeq 1$ for $\alpha \leq 2$ and $h \leq 0.1$, but for $h = 0.4$ and $\alpha = 20$, $\tau\lambda_1/\tau_D \approxeq 3$ 10^{-4}. This comes from the fact that the A_k values are much higher for $k = 4$ and 5 than for $k = 1$. For measurements performed over a long period, as for the relaxation of the thermoremanent magnetization, only the first mode, that is, the one corresponding to λ_1 operates because $\lambda_1 \ll \lambda_k$ $(k \neq 1)$, but for short measuring time as for frequency-dependent susceptibility, the mode $k = 5$ can dominate following the h values. This leads to the existence of high-frequency relaxation mode, in addition to that arising from the low-frequency mode associated with the $\mathbf{m}$ reversal. This results in a high-frequency loss, which for small h values displays itself as a shoulder in the conventional low-frequency peak for $h = 0$ and then predominates as h increases.

Calculations of the longitudinal relaxation time when the field is applied along a direction with an angle ψ with regard to the easy axis have also been performed very recently.[68] We recall that this case is of fundamental interest because in actual samples, the easy directions are in random position (except for very special samples). Indeed the calculation is very complicated as the axial symmetry is destroyed. Two assumptions have been made in order to simplify the calculations. First, the longitudinal relaxation time has been expressed using (D.31). However, the τ calculation for H_{app} parallel to the easy axis (see above) has shown that it is not valid for experiments with a short measuring time when h is high. Second, in the Fokker–Planck equation in spherical polar coordinates, the gyroscopic terms [those in $(\eta\gamma_0 M_{nr})^{-1}$] have been neglected. This is correct only if $(\eta\gamma_0 M_{nr})^{-1}$ is small, which is not the case. However, these terms would mainly influence the high-frequency behavior of the system and therefore can be neglected in first approximation.

The results show that a pronounced difference appears for $h > 0.1$ and

$\alpha > 0.5$ following ψ. For $\psi = \pi/2$, an approximate formula based on the Kramers transition state theory is given, valid for $h \geq 0.3$ and $\alpha \gg 10$, which is

$$\lambda_1\left(\frac{\pi}{2}\right) = 2\frac{\alpha}{\pi}\frac{(1-h)(1+h)^{1/2}}{h}\exp[-\alpha(1-h)^2] \tag{D.32}$$

Using a discrete orientation model, Pfeiffer[69] has calculated the relaxation time τ under applied field in any direction for the simplest case of the two-state model. Neglecting the variation of the preexponential factor τ_0 with the applied field, Pfeiffer writes:

$$\tau = \tau_0 \exp\left(\frac{\alpha g_1}{2}\right)\left[1 + \exp\left(-\frac{\alpha\,\Delta g}{2}\right)\right]^{-1} \tag{D.33}$$

where $KVg_1/2$ and $KVg_2/2$ are the barriers for the upper and lower minima, respectively, and $\Delta g = g_2 - g_1$.

This expression is only valid for sufficiently large α, that is, if **m** is mainly fixed in the minima. In addition the factor $[1 + \exp(-\alpha\,\Delta g/2)]^{-1}$, which varies between 0.5 and 1 may be neglected as the uncertainties on the τ_0 factor are probably of the same order of size.

Approximate expression is derived for g_1 [70]:

$$\frac{g_1}{2} = \left[1 - \frac{h}{h_c}\right]^{0.86+1.14h_c} \tag{D.34}$$

where h_c is the reduced field at which the lower barrier vanishes, given by[71]

$$(1-h_c^2)^3 - \tfrac{27}{4}h_c^4\sin^2 2\psi = 0 \tag{D.35}$$

or by[70]

$$h_c = [\cos^{2/3}\psi + \sin^{2/3}\psi]^{-3/2} \tag{D.36}$$

In fact, for actual samples with easy directions in random position and h not too small, the relaxation time is given, with a fair approximation, by its value for $\psi = \pi/2$.[71] On average the **m** population, which relaxes, is equal to (see Section D.1)

$$P = 1 - S_1 = \frac{2\tau^-}{\tau^+ + \tau^-} \tag{D.37}$$

S_1 being given by (D.7).

For ψ different from $\pi/2$, the potential wells are unequal and for h not too small $\tau^+ \gg \tau^-$ and $P \approxeq 0$. In addition, the probability to find H_{app} perpendicular to the easy direction is maximum. Then the effective longitudinal relaxation time is given by (D.32).

Usually, Brown's approximation for H_{app} parallel to the easy direction (D.27) is used (see Section F.2). For h not too small, $\tau = \tau^-$ from (D.8). In this case, one can compare $\tau(\psi = 0)$ and $\tau(\psi = \pi/2)$:

$$\frac{\tau(\pi/2)}{\tau(0)} = \frac{\lambda_1(0)}{\lambda_1(\pi/2)} = \sqrt{\pi}\sqrt{\alpha}h\,\frac{(1-h)^2}{\sqrt{1+h}}$$

For example, this ratio is equal to about 0.8 for $\alpha = 20$ and $h = 0.5$ but can strongly vary with α and h. Though the argument inside the exponential is the same in the two formulas, the error caused by the use of (D.27) is appreciable, depending on the values of h and α.

D.3.4. τ Calculation in Cubic Symmetry

In this case, the axial symmetry is also broken, and until now the eigenvalues λ_k of the Fokker–Planck equation have been obtained numerically. Approximations are also necessary for obtaining the asymptotical values ($\alpha \to 0$ and ∞).

For $K_1 > 0$ and $K_2 = 0$, six easy directions of [001] type exist. The three first λ_k have been calculated for $\alpha > 1$ and it has been shown that for $\alpha \gg 1$, the two first are enough for describing the **m** relaxation[72] with

$$\lambda_2 = 3\lambda_1/2 \tag{D.38}$$

For $\alpha \gg 1$, two approximate expressions have been derived, quite similar. The simplest can be written, valid for $\alpha \gtrsim 10$ as

$$\lambda_1 = \frac{8\sqrt{2}}{\pi}\,\alpha\,\frac{\eta_r \dfrac{M_{\text{nr}}(T)}{M_{\text{nr}}(0)}}{1 + \eta_r^2\left[\dfrac{M_{\text{nr}}(T)}{M_{\text{nr}}(0)}\right]^2}\exp\left(-\frac{\alpha}{4}\right) \tag{D.39}$$

with $\tau_1 = \tau_D/\lambda_1$.

For small α values, $\lambda_1 \approxeq 2$.[73] This value is a good approximation for $\alpha \lesssim 5$.

For $K_1 < 0$ and $K_2 = 0$, eight easy directions of [111] type occur. The four first λ_k have been calculated for $\alpha \geq 1$, and for $\alpha \gg 1$ the three first describe the **m** relaxation[72,74] with

$$\lambda_2 = 2\lambda_1 \quad \text{and} \quad \lambda_3 = 3\lambda_1 \tag{D.40}$$

The simplest approximate expression for λ_1 is

$$\lambda_1 = \frac{4\sqrt{2}}{3\pi}\alpha \frac{\eta_r \dfrac{M_{nr}(T)}{M_{nr}(0)}}{1 + \eta_r^2 \left[\dfrac{M_{nr}(T)}{M_{nr}(0)}\right]^2} \exp\left(-\frac{\alpha}{12}\right) \tag{D.41}$$

For small α values, $\lambda_1 \approxeq 2.$[73] The limit of validity of Eq. (D.41) is higher, $\alpha \gtrsim 30$, but this is also true for the low-energy approximation ($\lambda_1 \approxeq 2$) which can be used for $\alpha \lesssim 10$.

However, Eqs. (D.39) and (D.41) must be used with some caution for large α values due to the lack of numerical verification.

Two points have to be underlined. First, the high- and low-energy approximations do not overlap as it is the case for the first formulas for uniaxial anisotropy and a unique formula remains to be established. Second, the limits of validity of the formulas are related to α. If the limits are related to the argument inside the exponential, that is, $\alpha/4$ and $\alpha/12$, they are similar to those operating in uniaxial symmetry.

The existence for two ($K_1 > 0$) or three ($K_1 < 0$) relaxation times prevents the determination of an effective relaxation time valid for any experiment. Particular analysis for each technique will be required. For example, let us consider a particle including N spins with $K_1 > 0$. If at $t = 0$, all the spins are along the [001] axis ($\theta = 0$), at a given time after, the populations N_1, N_2, and N_3 along the $\theta = 0$, $\theta = \pi/2$, and $\varphi = 0$, $\pi/2$, π, $3\pi/2$, and $\theta = \pi$ axes, respectively, will be for $\alpha \gg 1$:

$$N_1 = \frac{N}{6}[1 + 3\exp(-t/\tau_1) + 2\exp(-t/\tau_2)]$$

$$N_2 = \frac{4N}{6}[1 - 3\exp(-t/\tau_2)]$$

$$N_3 = \frac{N}{6}[1 - 3\exp(-t/\tau_1) + 2\exp(-t/\tau_2)]$$

The magnetization M measured along the [001] axis is equal to

$$M = M_{nr}\frac{(N_1 - N_3)}{N} = M_{nr}\exp(-t/\tau_1)$$

Then, the magnetization depends only on the first relaxation time τ_1. On the contrary, for Mössbauer spectroscopy, which experiences any direction, the two relaxation times have to be considered.

Finally, we note that for $K_1 > 0$, λ_1 has been calculated when a field is applied for two α values.[75]

D.3.5. τ Calculations in Some Other Cases

D.3.5.1. Curling Mode. As we have discussed above in Section D.3.1, the curling mode for **m** rotation is brought into play for large particles, being energetically more favorable than the coherent mode. Several calculations for spheres of cubic symmetry have been published.[76–80] The main result concerns the τ variation with the particle diameter ϕ. For small particles, a coherent mode occurs, and the energy barrier (and therefore also τ) increases steadily up to a certain value of ϕ. When that ϕ is reached, reversal by curling becomes energetically easier than by coherent rotation, and the energy barrier (and therefore τ) decreases. A further increase of ϕ will increase the barrier for reversal by curling, and the increase in τ will be resumed.[79] In this calculation the anomalous decrease of τ with increasing ϕ was obtained as a general phenomenon, not just limited to cubic symmetry. However, the anomaly disappears if the anisotropy is either too large or too small, and it requires rather narrow range of anisotropy values for the effect to be observed.

This phenomenon could be a possible explanation[80] of the anomalies detected by Mössbauer spectroscopy on Fe particles where when increasing the particle size, a decrease followed by an increase of τ is observed.[38] In the same way, the change of rotation mode could be an explanation[81] of Weil's results for Ni particles.[80]

D.3.5.2. Elongated Particles. For elongated particles, nonuniform magnetization reversal occurs if the length of particles is higher than a domain-wall width. Calculation of the relaxation time τ^- out of the metastable state for **m** under an applied field has been performed.[82] A Fokker–Planck equation for the **m** dynamics based on the Kramers transition-state rate theory is constructed and the corresponding relaxation time is derived. This is applied to elongated particles with easy and hard axis anisotropies. For small length, uniform magnetization reversal occurs. A formula is given for τ^- that is identical to Brown's formula (D.27) except that in the preexponential factor $1 - h$ is replaced by $1 + h$. For large length, nonuniform magnetization reversal occurs and τ^- is numerically calculated. It is found that τ^- enormously decreases (several orders of magnitude) for the latter mode with regard to the former.

D.4. Some Discussions and Conclusion

The Néel and Brown models have been compared in some studies.[80,83] In fact, for uniaxial symmetry and $\alpha \gg 1$, the τ formulations are very similar [Eqs. (D.9) and (D.19) and (D.26) and (D.27)], the parameters inside the preexponential factor being only slightly different. However, Brown's model allows the τ calculation for any anisotropy energy and his approach constitutes a general formulation of the problem.

In the context of the Brownian model, other models have been developed for the interpretation of the results of particular experiments. For example, for Mössbauer spectroscopy where on the one hand transitions between nuclear spins are involved and on the other hand the measurement encompasses the whole τ set. These models will be briefly described in Section F.6. This is also the case for ferromagnetic resonance where models[84,85] have been developed from Raikher's calculations.[86] However the effective eigenvalue method, which allows one to derive the initial slope of the magnetization, is not valid when thermal activation process is present.[64] New calculations[87] have taken into account this problem. Theoretical results are summarized in Section F.7.

Finally, we want to underline that in our opinion, only one crucial problem remains to be resolved for the interpretation of numerous experiments, that is, the establishment of simple (approximate) analytical formulation for the relaxation times τ^+ and τ^- under applied field in any direction with regard to easy axis. This is the requisite condition for calculating the part of **m** blocked on average in the lowest minimum and the relaxation time τ of the other part, indispensable for modeling correctly experiments as zero-field-cooled and field-cooled magnetizations, thermoremanent magnetization. Pfeiffer's calculations[70] may be a way, but the discrete orientation model does not cover all the cases. This is not the case for Coffey's calculations,[68] though the field is assumed to be applied along a line of longitude. In addition, no analytical expression is available at the present for the latter model.

Two other interesting problems concern the derivation of approximate formulas for τ for cubic symmetry or more complex cases, and the study of the effect of the magnetic state at a surface due to the break of symmetry and the defects on the **m** reversal, therefore on τ. However, quasi-uniaxial symmetry is effective in the most cases (see Section C), and the corresponding τ formulation can be used. Nevertheless the τ calculation suggested in the former problem is important for experiments that determine τ for all barrier heights such as does Mössbauer spectroscopy. As discussed in the Section D.3.1, synchronous rotation is a good approximation and surface effects probably lead to only slight modifications of the preexponential factor. However, the effect can be of importance for experiments that determine τ for all barrier heights and when a field is applied in such manner that the Zeeman energy is practically the same as the anisotropy energy (see Section D.3.1).

E. INTERPARTICLE INTERACTIONS

E.1. Introduction

Magnetic interparticle interactions always exist inside fine-particle assemblies. They are more or less strong according to the volumic concentration C_v. Magnetic dipolar interactions are always present. If the matrix and the grains are metallic, RKKY interactions occur and depend on $1/d_{ij}^3$, where d_{ij} is the distance between particles, like dipolar interactions. When the matrix is insulating, superexchange interactions can exist according to the structure and the nature of the matrix and the bonding at the particle matrix interface. From insulating magnetic materials, we know that exchange interactions are short ranged, but if the bonding is favorable, superexchange interactions may extend until large distance. Other more complex effects are possible, as the one recently shown,[59] where it seems that the interactions modify the surface anisotropy. The determination of the interaction effects is complex because, on one hand, several causes can interplay and, on the other hand, the particle assembly shows, except for very exceptional samples, a disordered arrangement of particles with volume distribution and easy directions in random position. In addition, the thermal fluctuations of the particle magnetic moment **m** do not simplify the problem.

The interaction effects lead to a modification of the energy barrier E_B, which depends on the symmetry of the anisotropy of the single particle. We remark that if an **m** reversal does not change the anisotropy energy, which is the case for the usual symmetries, this reversal does not change neither the interaction energy of dipolar type. This does not mean that for a given particle the total energy is unchanged under the **m** reversal, but it is true for the particle assembly. Therefore any model accounting for the mean properties of the particle assembly must respect this symmetry condition. An E_B increase seems likely to occur, as discussed in detail in Section E.3.

Another possible effect of the interactions is a complete change of regime with the properties being no longer relevant to superparamagnetism and the **m** relaxation being no longer governed by E_B. Some studies[88–91] have considered the possibility of **m** dipolar order at low temperature, analogous to those predicted for spins.[92,93] A model leading to a transition from superparamagnetism toward **m** ferromagnetism has also been presented.[94,95] We discuss this point below.

E.2. Transition from Superparamagnetism Toward Collective State

For discussing the effects of magnetic interparticle interactions and the possible existence of a collective state at low temperature, let us consider that **m** correspond to superspins **S** with distance d_i between them. The **S** lattice is disordered, the **S** modulus is distributed, and the anisotropy is at random. The problem of interactions is then similar to those of spin interactions in bulk samples, but with another scale. Note that this is an ideal view, whereas the actual systems are much more complex, for example, $|\mathbf{m}|$ can vary with the interactions,[59,96,97] and the interaction effect may not be proportional to the particle volume (see Section E.3).

E.2.1. Magnetic State at Very Low Temperature

If the **S** fluctuations are not present or are very slow (large grains or low temperature), the **S** magnetic state results from the interactions and depends on the topology. If the exchange interactions are dominant, since the anisotropy directions are at random, the problem is similar to that resulting from random anisotropy, but with another scale. Several magnetic states have been predicted depending on the strength of the random anisotropy, the exchange interactions, and the applied field.[98,99] We can expect that **S** shows the same type of magnetic order. In particular, no long-range magnetic order should exist without applied field. The **S** phase is speromagnetic, which is very analogous to the spin glass phase, though the origin of disorders is not the same. An asperomagnetic phase could arise if a large field is applied and then removed, or if the particles are cooled under moderate field and the field is suppressed at low temperature. On the other hand, antiferromagnetic and ferromagnetic orders have been predicted for spins interacting via dipolar interactions in face centered cubic and body centered cubic structures, respectively.[92,93] Very recently, numerical simulation of hysteresis loops of fine-particle systems with random anisotropy directions and dipolar interactions have been made at zero temperature.[100] The results show that in the absence of anisotropy, a coercive field appears due to interactions. For high particle concentration, a short-range local order exists. Increasing the anisotropy, a crossover is observed between two regimes. For low anisotropy, the static properties are roughly the same as those observed in the absence of anisotropy. For high anisotropy a single particlelike behavior is observed, in agreement with the Stoner and Wohlfarth model.[54] However, no precise indication is given concerning the **S** state. It seems probable that an **S** disordered state arises, analogous to a spin glass phase or a spin glasslike state. We define this point below. We remark that the discussion does not apply to particles in perfect arrangement with the same volume and the same easy direction. An **S** ferromagnetic phase is expected in this case because the particles are indistinguishable. However, a crossover of the properties can be predicted as soon as disorders occur.

E.2.2. Temperature Dependence of the Properties

Consider now particles for which, without interaction, the blocking temperature T_B is smaller than the paramagnetic transition temperature T_C for any measuring time τ_m. Note that T_C can be lowered with regard to the bulk value due to the defects inherent to the fine-particle state. For a disordered arrangement, each particle possesses its own energy barrier E_B and its own T_B. When the interactions act, the energy barriers are no longer independent. The regime is still superparamagnetic if the **S** relaxation of a given particle is governed by its own E_B, of course, modified by the interactions. A collective state for **S** will be present if it is not possible to define E_B for each particle, but only the energy E_{ass} relative to the particle assembly. E_{ass} surely shows multiple minima in the phase space as for the spin glass phase.[101] A schematic diagram is shown in Fig. E.1 for a given measuring time τ_m. For a different τ_m, the T_B variation must be shifted upward (shorter τ_m) or downward (larger τ_m). If T_B increases with the interactions (continuous line), we shall observe, with decreasing temperature, paramagnetic, superparamagnetic, blocked, and collective states. In this hypothesis, the observation of a direct transition from paramagnetism toward blocked state is not possible

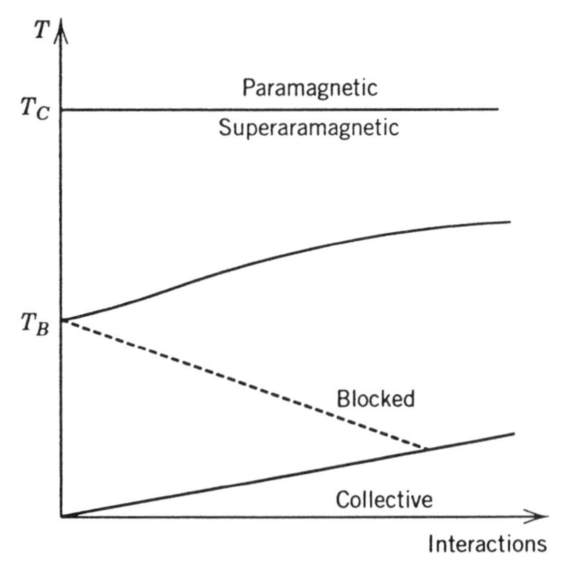

Figure E.1. Schematic phase diagram for fine particles with magnetic interparticle interactions. Case 1, the blocking temperature T_B increases with interactions (continuous line). Case 2, T_B decreases with interactions (dashed line) (see text).

because T_B cannot overstep T_C for which the interactions vanish (the interactions are proportional to the magnetization square). In the opposite case, T_B decreasing with increasing interactions (dashed line), we will observe the same sequence below a critical strength of the interactions. Above, the blocked state will be suppressed and a transition from superparamagnetism to a collective state will occur. We note that in this case, it becomes possible to observe a superparamagnetic state for particles for which this state does not occur in the absence of interactions ($T_B > T_C$). However, the same properties, that is, the disappearance of the blocked state, could be observed when T_B increases with the interactions, if the temperature T_{coll} of the appearance of the collective state increases more rapidly with the interactions. In this case, the line T_{coll} could in principle cross the line T_B (Fig. E.1) because the blocking process and the collective freezing are not governed by the same rules. In our opinion, this strong increase of T_{coll} with the interactions does not seem possible when the interactions are of the dipolar type (see Section E.2.5), but it could arise when the interactions are of a different type, for example, exchange or superexchange interactions. Finally, it is worth noting that the collective state cannot be ferromagnetic due to the disordered arrangement of particle magnetic moments.

However, a transition from superparamagnetic toward ferromagnetic state has been predicted when the interactions are of exchange type and the anisotropy is negligible with respect to the interactions.[94,95] This case is indeed very particular and, in our opinion, there is only a little chance that it occurs. The model actually appears to be oversimplified. The interactions among atoms at the particle interfaces are taken as equivalent to exchange interactions between the particle moments **m**; the particles are considered as identical and possible negative sign for the exchange constant is not foreseen. The taken hypothesis lead inevitably to a ferromagnetic **m** order. The model has been applied to Mössbauer results for fine particles of goethite,[95] which shows an antiferromagnetic state for bulk material (the exchange constant between the spins is negative). The observed properties may actually be due to a decrease of the Néel temperature T_N caused by defects[102] and no definitive proof of the model, like the existence of superparamagnetic or paramagnetic properties between T_N related to the bulk and the temperature of appearance of blocked (or antiferromagnetic) properties, have been supplied.

E.2.3. Properties of the Collective State

The sketching of the properties of the collective state is useful, but it is difficult to do from experimental data because there are very few results that can be retained. First, the experiments on fine-particle assemblies have generally been performed with one or two techniques, the usual case being zero-field-cooled magnetization M_{ZFC} or (and) Mössbauer spectroscopy. However, for the first experiment, τ_m is not well known (see Section F.2), and for the second there is difficulty in the determination of T_B due to the lack of an accepted model for lineshape in Mössbauer spectra under superparamagnetic relaxation (see Section F.6). In addition, the two τ_m values are very different ($\tau_m \simeq 10^{-8}$ s for Mössbauer spectroscopy and $\tau_m = 10^2$–10^3 s for M_{ZFC}). This leads to a large difference between the deduced T_B values, which does not allow the knowledge of the actual variation of the relaxation time τ with temperature. Second, there are very few data on the same set of particles with variable interparticle interactions, which would permit following the T_B variation with growing interactions, and finally there are also very few data on fine particles without (or negligible) interactions due to the difficulty in preparing such samples. Then some uncertainties remain on the validity of the Néel–Brown model.

Because of the **S**-disordered arrangement, the collective state properties must be of spin glass type. But are they similar to those of a spin glass phase or to those of a spin-glass-like state? At present there is a general agreement for considering that the transition toward a spin glass phase corresponds to a true thermodynamic transition with critical exponents and order parameter, the susceptibility pertinent for classical order being replaced by the nonlinear susceptibility χ_{nl}, which therefore diverges at the transition temperature. In addition, the variation of the freezing temperature T_f with τ_m is very small (Table II) and the low-temperature phase is nonergodic with a hierarchical and ultrametric

TABLE II
Values of the Criterion $\Delta T_f/(T_f \Delta \log \nu)$ Evaluated Near $\nu = 50$ Hz for Different Magnetic Disordered Systems (see text)

Systems	$\Delta T_f / T_f \Delta \log \nu$	References
CuMn 1–10%	0.005	103, 104
$CsNiFeF_6$	0.008	105
Mn aluminosilicate	0.013	104, 106
$CdCr_{1.7}In_{0.3}S_4$	0.015	107
$Eu_{0.4}Sr_{0.6}S$	0.021	104, 108
$Eu_{0.2}Sr_{0.8}S$	0.04	108
$ZnCr_{1.6}Ga_{0.4}O_4$	0.05	109
Interacting fine particles (theory)	0.05–0.13	—
$Fe–Al_2O_3$ particles	0.06	45
Noninteracting fine particles (theory)	0.10–0.13	—
$La_{0.994}Gd_{0.006}Al_2$	0.13	104

structure of the energy valleys in the phase space,[110,111] as revealed by the variation of the thermoremanent magnetization (TRM), which follows a stretched exponential and depends on the aging time[112] (for review studies on spin glass, see Refs. 113–115).

However, many bulk compounds with interaction disorders show properties similar to those of the spin glass phase: susceptibilty peak, whose maximum occurs at a temperature that depends on τ_m, but with a larger variation vs. τ_m, irreversibilities shown by zero-field-cooled and field-cooled magnetization (M_{FC}) experiments, similar variations of the TRM vs. time; but they do not show the specific properties cited above. All the properties have been interpreted as resulting from a magnetic disordered state called spin-glass-like but covering many different situations.[116–118] In several cases, the term cluster glass has been used, the properties being analyzed from the superparamagnetic model. This last attempt is reasonable because the properties of fine particles are also very similar. In fact, these compounds show in the studied range of temperature properties governed by thermally activated process and an absence of collective state. This phase could, however, be present at lower temperature. We roughly have two scenarios: a true thermodynamic transition toward a collective state, that is, the spin glass phase, and a progressive inhomogeneous freezing (or blocking) via a thermally activated process, which could lead at lower temperature to a collective state. This is exactly what is shown in Fig. E.1 for interacting particles.

E.2.4. Comparison with Experimental Data

Extensive studies have been performed on Fe particles embedded in alumina matrix[119] by means of numerous techniques (M_{ZFC} and M_{FC}, magnetization versus applied field and temperature, TRM, alternative susceptibility χ_{ac} (frequency ν in the range 210^{-3}–210^{4} Hz), Mössbauer spectroscopy, ferromagnetic resonance, neutron diffraction, and magnetooptic measurements). To our knowledge this is the only fine-particle system for which almost all magnetic techniques have been used. We shall comment on the data in Section F devoted to the experiments. Three samples with different mean particle volume ($V = 40$, 60, and $200\,\mathrm{nm}^3$) have been studied in greater detail. The volume distribution is relatively narrow, as seen by electron microscopy.[45,119] The ratio V/d^3 is similar for the three samples ($\simeq 0.17$), where d is the mean distance between two neighboring particles, and the interparticle interactions are strong. We focus for our purpose on the following results.

1. The T_B vs. τ_m variation has been obtained on a very large τ_m range. It is very well described with a superparamagnetic interaction model[45]

(see Section E.3). In the Table II, the ratio $\Delta T_B / T_B \, \Delta \log \nu$ deduced from χ_{ac} results is compared to similar data reported for some bulk magnetic disordered materials. This ratio, model independent, allows an evaluation of the variation of $T_f(T_B)$ vs. τ_m. The first four systems show a spin glass phase with thermodynamic transition. The ratio varies between 0.005 and 0.015. For the EuS–SrS system, the properties are not really clear and the existence of a transition is still debated for $Eu_{0.2}Sr_{0.8}S$. For $ZnCr_{1.6}Ga_{0.4}O_4$, clear evidence of a progressive freezing of clusters was found[120] and no critical behavior of χ_{nl} was observed.[121] $La_{0.994}Gd_{0.006}Al_2$ contains small noninteracting clusters, and the properties are relevant to superparamagnetism. Also included in the table are the ratio values evaluated for noninteracting particles from usual parameter values and for interacting particles from the model described below. It clearly appears that the Fe interacting particles cannot be classified among the spin glass compounds that show a thermodynamic transition. The ratio value is larger but not much different from the values corresponding to materials whose properties are explained in term of spin clusters. From the point of view of the T_B vs. τ_m variation, an inhomogeneous blocking process therefore appears to be more likely than **m** collective freezing for these strongly interacting particles. Discussions concerning the comparison of dynamical properties of spin glasses and fine particles as seen by χ_{ac} measurements and other techniques can be found in Refs. 122 and 123.

2. The thermal variation of the nonlinear terms of the magnetization $M(T)$ has been studied from field-cooled magnetization measurements.[124,125] The use of a simple expansion in terms of the field H:

$$M(T) = \chi_0(T)H - A_3(T)H^3 + A_5(T)H^5$$

gives a logarithmic divergence of the coefficient of the nonlinear term A_3 with a tendency to a saturation value close to T_B. This apparent divergence disappears if allowance is made for the thermal variation of χ_0 using the following expression in terms of $\chi_0 H$:

$$M = \chi_0 H - b_3(\chi_0 H)^3 + b_5(\chi_0 H)^5 \tag{E.1}$$

The need for introducing in the series expansion the actual thermal variation of χ_0 instead of a development in terms of $\mu H / k(T - \theta)$[126] comes from the fact that χ_0 does not obey a simple Curie–Weiss law near

T_B and that the value of θ is comparable to T_B. The new nonlinear terms b_3 and b_5 do not change significantly with the temperature. This is contrary to what is observed for the spin glass phase, where χ_0 is finite, but all the nonlinear terms appear to diverge at the transition.

3. Thermoremanent magnetization experiments[127] (see Section F.4) have shown that the waiting time has no detectable influence on the TRM decay in the studied temperature range and that this decay is described in a first approximation by a logarithmic law vs. time. The slope of the logarithmic decay is almost proportional to the temperature. This is exactly what is expected for the **m** relaxation of noninteracting particles with volume V distribution (we note that the proportionality to the temperature is in fact a kind of artefact when the V distribution is large, see Section F.4). A small anomaly in the slope is detected at low temperature, well below T_B related to $\tau_m = 10^{-2}$ s. This could be due to quantum effect (see Section G), but also to a change in the regime of TRM decay due to the appearance of a collective state. However, detailed experiments on interacting as well as noninteracting particles are needful for conclusion.

4. Inelastic neutron scattering[128–133] (see Section F.8) performed on two samples has evidenced two types of relaxation, the longitudinal and the transverse. The former corresponds to the usual relaxation between potential wells whereas the latter is related to vibrations inside the potential well. It is shown that several regimes occur for the transverse relaxation. Below a certain temperature T_{re} about two times smaller than T_B ($\tau_m = 10^{-2}$ s), a new regime appears where the interparticle interactions would destroy the local modes. However T_{re} depends on the volume and therefore cannot be a transition temperature toward a collective state. This regime could be the approach to such a state.

From these four experiments performed on particles strongly interacting via dipolar interactions, we can rule out the appearance of **m** freezing analogous to spin freezing in spin glass with thermodynamic transition, except perhaps at low temperature well below T_B. All the results can be explained from an inhomogeneous blocking process that obeys superparamagnetic laws with E_B modified by the interactions.

E.2.5. Temperature of Transition Toward a Collective State

What is the order of magnitude we can actually expect for a transition temperature toward a collective state? Calculation for three-dimensional cubic arrays of magnetic dipoles with identical magnetic moments μ that interact via dipolar interactions have shown that antiferromagnetic and ferromagnetic order occurs for bcc and fcc lattices, respectively. The transition temperature T_C is equal to[92,93]

$$T_C = \frac{a_0}{k} \frac{\mu_0}{4\pi} \frac{\mu^2}{d^3} \tag{E.2}$$

where a_0 is a numerical constant of the order of unity and d the distance between neighboring dipoles. This gives T_C of order of some mK. For fine ferromagnetic particles, μ^2/d^3 has to be multiplied by $NC_V/\sqrt{2}$,[90] where N is the number of particle spins and C_V the volumic concentration. But the particle assembly has a disordered arrangement, a volume distribution, and anisotropy axes in random orientation, and the prediction of the transition temperature value is difficult. For dipoles in a disordered arrangement, and with a μ distribution, Eq. (E.2) remains perhaps valid with $\langle \mu^2 \rangle$ replacing μ^2 and with an a_0 factor strongly lowered by supposing that the effect of disorders is similar to the effect of interaction disorders in bulk materials, which causes a strong decrease of the transition temperature. In addition, for fine particles, the effect of the anisotropies cannot be neglected. The order of magnitude of the anisotropy energy is generally not negligible with respect to the interaction energy (that depends on the samples: for example, for Fe particles embedded in Al_2O_3 [45] and γ-Fe_2O_3 particles in a polymer,[59] the interaction energy to anisotropy energy ratio is at low temperature about 4.6 and 2.6, respectively).

However, a recently published model[91] predicts a high transition temperature T_C toward a spin glass phase or a spin-glass-like state. This model neglects the anisotropy effect and seems based on a formula giving the transition temperature for spin glasses. But this formula is controversial, and it is not valid for concentrated spin glasses. Let us remember the difference between concentrated and "classical" spin glasses. For the latter, the interaction disorder is essentially due to the random dilution of the magnetic atoms, while for the former it is mainly due to competing interactions coming from neighbors. It is evident that the analogy with particle assemblies is possible only with concentrated spin glasses.

At present, we do not know any result for strongly interacting particles that gives a definite evidence of a transition from a superparamagnetic state toward a collective state. On the contrary, for γ-Fe_2O_3 particles in a polymer,[59,89,96,97,134–138] the properties of which will be discussed with some details below and in the Section F, TRM experiments[136,138] have shown that the relaxation rate is almost the same for samples showing negligible, weak, and strong interactions. For these samples, the particle set is identical, only the aggregation state and the distance between neighboring particles differ. This result definitely shows that in the investigated temperature range (the lowest temperature is well below T_B corresponding to χ_{ac} experiments; and for temperatures corresponding to T_B relative to Mössbauer spectroscopy, TRM is equal to zero) the regime is the same for all the samples and that no collective state appears.

E.3. Variation of the Blocking Temperature with Magnetic Interparticle Interactions

As we have underlined in the introduction of this section (E.1), the problem of interparticle interactions is complex because they can be of several kinds (dipolar, exchange, superexchange, and other more complicated effects),[88,89] and the arrangement of particles in the assembly is disordered with a volume distribution and easy directions in random orientation. In addition, the thermal fluctuations of the particle magnetic moment **m** do not simplify the question. In this section, we analyze the interaction effects for a disordered particle assembly when the temperature is sufficiently high for preventing the appearance of a collective state. We consider the case of dipolar interactions, but we think that the other kinds of interactions can be treated in the same way.

E.3.1. General Considerations for Dipolar Interactions

Let two particles with magnetic moments $\mathbf{m}_i$ and $\mathbf{m}_j$ be oriented along the unit vectors $\mathbf{u}_i$ and $\mathbf{u}_j$ respectively, with $|\mathbf{m}_i| = M_iV_i$ and $|\mathbf{m}_j| = M_jV_j$, where M is the magnetization for the considered temperature (M can depend on V) and V_i and V_j the particle volumes. The line joining the centers of the particles is oriented along the unit vector $\mathbf{r}_{ij}$ and the distance between the two centers is equal to d_{ij} (Fig. E.2).

The particle i sees a field $\mathbf{H}_{ij}$ due to the dipolar interaction with the

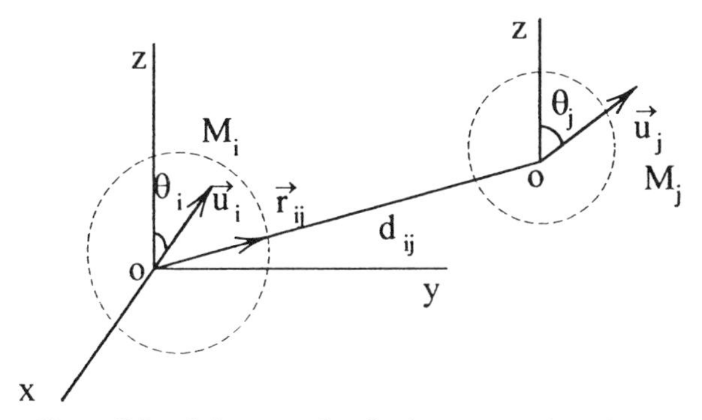

Figure E.2. Axis system for dipolar energy calculation.

particle j. The energy corresponding to the field is

$$E_{ij} = -\mathbf{H}_{ij} \cdot \mathbf{u}_i M_i V_i \tag{E.3}$$

with

$$\mathbf{H}_{ij} = (M_j V_j / d_{ij}^3)[3(\mathbf{u}_j \cdot \mathbf{r}_{ij})\mathbf{r}_{ij} - \mathbf{u}_j] \tag{E.4}$$

Equations (E.3) and (E.4) have been established by assuming **m** located in the particle center. Since the exchange interaction inside the particle is much stronger than the dipolar interactions due to the other particles, the directions of the spins remain parallel (or antiparallel), except at the surface and near the surface due to the broken symmetry at the surface and the defects. Then the interaction energy is given by Eq. (E.3), where $\mathbf{H}_{ij}$ is a mean interaction field. This mean field can be calculated numerically and compared to Eq. (E.4). For spherical particles the mean interaction field is equal to $\mathbf{H}_{ij}$ (E.4) multiplied by a coefficient significantly different from unity only in the case of very small particles in contact. For spheroidal particles, Eq. (E.4) is a good approximation. In the case of ellipsoidal particles relatively close to each other, this is not the case. The influence of disorders on the particle surface has also been evaluated, and it is negligible, except for very small particles.[45]

Therefore, the use of Eqs. (E.3) and (E.4) is restricted to some types of particles. For very small particles or elongated particles, like acicular particles, in close contact, modification of the interaction energy with respect to Eqs. (E.3) and (E.4) must be taken into account. This can be performed by using, for example, the method described in Ref. 139. On the other hand, it has been supposed that the exchange interactions inside the particles are much stronger than the dipolar interactions. However, if the intrinsic exchange interactions are balanced due to competing interactions inside the particles, the interactions may modify this balance and change the properties of the intrinsic magnetic state of the particle. This seems checked in $Fe_{64}Cr_{16}B_{20}$ amorphous particles embedded in an alumina matrix.[140] For bulk $Fe_{64}Cr_{16}B_{20}$, disordered magnetic state with reentrant properties at low temperature is observed.[141] We remark that it is still another case that has not been discussed in the above section. If in bulk state, the material shows disordered magnetic properties leading to a freezing temperature, what happens for fine particles? Is it possible to define T_B and what is the role of the interparticle interactions? In our opinion, that depends on the particle size with regard to the scale of freezing process, but this question remains for decoding.

E.3.2. Shtrikman and Wohlfarth Model

In a first attempt to model the interparticle interactions, Shtrikman and Wohlfarth[142] have proposed simple models for weak and strong interactions. For weak interactions, the authors consider a three-level system that leads to an additional energy barrier being added to the energy barrier of noninteracting particles, expressed as H_iMV, where H_i is an interacting field. Thermal averaging for a particle assembly is done by replacing H_i by $H_i \tanh(H_iMV/kT) \cong H_i^2MV/kT$, which leads to a Fulcher law for the relaxation time. For strong interactions, the authors introduce the concept of an effective volume that leads to a similar law. Later on, this model has been improved[143] by introducing the effects of the interactions due to distant neighbors. In our opinion, these phenomenological models are satisfying in a first approximation. However, they do not include a precise formulation for the parameters and therefore do not permit a quantitative comparison with the experiments.

E.3.3. Statistical Model

A microscopic calculation has been performed by Dormann et al.[45] for a disordered assembly of particles with volume distribution. In this case, the particle magnetic moments **m** relax with different relaxation times τ related to V. The probabilities to find $\mathbf{u}_i$ at (θ, φ) and $(\theta, \varphi + \pi)$ are assumed to be equal. In this case $\mathbf{H}_{ij}$ can be restricted to

$$\mathbf{H}_{ij} = (M_jV_j/d_{ij}^3)\mathbf{v}_{ij} \cos\theta_j \tag{E.5}$$

where $\mathbf{v}_{ij}$ is a vector depending on $\mathbf{r}_{ij}$ only. As $\mathbf{m}_j$ relaxes, $\cos\theta_j$ fluctuates, and the Boltzmann statistics can be used for calculating $\langle E_{ij} \rangle$. Then

$$\langle E_{ij} \rangle = -(M_iM_jV_iV_j/d_{ij}^3)\mathbf{v}_{ij} \cdot \mathbf{u}_i \mathscr{L}[(M_iM_jV_iV_j/d_{ij}^3)(\mathbf{v}_{ij} \cdot \mathbf{u}_i/kT)] \tag{E.6}$$

where $\mathscr{L}$ denotes the Langevin function.

The total interaction energy per particle i is then

$$E_i = -M_iV_i \sum_j (M_jV_j/d_{ij}^3)\mathbf{v}_{ij} \cdot \mathbf{u}_i \mathscr{L}[(M_iM_jV_iV_j/d_{ij}^3)(\mathbf{v}_{ij} \cdot \mathbf{u}_i/kT)] \tag{E.7}$$

The same symmetry condition for $\mathbf{u}_i$ as for $\mathbf{u}_j$ may be used, that is, the same probabilities to find $\mathbf{u}_i$ at (θ, φ) and $(\theta, \varphi + \pi)$. In this case

$$E_i = -M_iV_i|\cos\theta_i| \sum_j M_ja_{ij}\mathscr{L}[(M_iM_jV_ia_{ij}/kT)|\cos\theta_i|] \tag{E.8}$$

with $a_{ij} = (V_j/d_{ij}^3)(3\cos^2\xi_{ij} - 1)$, where ξ_{ij} is an angle parameter of $\mathbf{r}_{ij}$. The authors write:

$$E_i = -E_{Bi}\cos^2\theta_i \tag{E.9}$$

with

$$E_{Bi} = M_i V_i \sum_j M_j a_{ij} \mathscr{L}[(M_i M_j V_i a_{ij}/kT)] \tag{E.10}$$

A step not detailed in the initial study (there are also some typing errors corrected later) needs discussion. The energy given by Eq. (E.8) has the uniaxial symmetry and therefore satisfies the symmetry condition relative to dipolar interaction energy (see Section E.1). The barrier energy is given by (E.10). However, Eq. (E.9) is exact only if the argument inside $\mathscr{L}$ is small. That means that for a large argument value, the calculation of the relaxation time τ performed for uniaxial symmetry with $E = E_B\cos^2\theta$ is not exactly applicable. Nevertheless, we think that this introduces only small errors. In fact, the model is simple and, of course, cannot cover exactly very complicated phenomena. This is the reason why the authors have replaced a_{ij} outside the Langevin function by b_{ij} with b_{ij} near a_{ij}.

For the evaluation of the mean T_B, the mean energy barrier $(E_B)_{\text{int}}$ due to the interactions has been expressed by considering a regular arrangement of particles of mean volume V and mean magnetization M

$$(E_B)_{\text{int}} = M^2 V \sum_j b_j \mathscr{L}(M^2 V a_j/kT) \tag{E.11}$$

where now $a_j = V(3\cos^2\xi_j - 1)/d_j^3$ and b_j near a_j. d_j, a_j, and ξ_j correspond to a regular arrangement of particles.

E.3.4. Discussion of the Model

E.3.4.1. Average Problem. If the V distribution is narrow, the proposed method, that is, the replacing of the disordered arrangement by an averaged ordered compact arrangement, is correct. If the V distribution is large, the method remains a good approximation as long as mean properties are concerned such as, for example, for the mean T_B determination. However, the method could be questioned if the properties for a certain range of volume, more and less narrow, are needed to be known like, for example, for TRM experiments (see Section F.4). Nevertheless, for a disordered arrangement, the particles with a given V see all the possible arrangements, which justifies the use of a_j values relative to the corresponding regular arrangement. However, $(E_B)_{\text{int}}$ is mainly stated by the contribution of the first neighbors and it is unclear if a_1 relative to these neighbors has the same value if the origin particle is small or big. It seems more likely that for a small origin particle, d_1 is smaller than the mean value, leading to higher a_1 value. This means that the interactions narrow the E_B distribution, and therefore the relaxation time τ distribution. This trend seems checked for γ-Fe_2O_3 particles in a polymer[96] (see Section F.2). Two points are to be noted. First, the E_B distribution would be broadened with respect to the case without interaction if the interaction effects led to an E_B decrease. Second, the narrowing of the E_B distribution, which in the model increases when T decreases, is in agreement with an expected transition toward a collective state at low temperature governed, in a simple picture, by a main relaxation time.

Another question is related to the magnetization; M is temperature dependent and the effects on $(E_B)_{\text{int}}$ and τ will be discussed later. But M also depends on the volume due to the magnetic imperfections on the surface. If mean properties are concerned, an averaged $(\bar{M})^2$ value has to be considered. On the contrary, where the properties for a given V are needed, $\bar{M}M(V)$ is the good parameter where $\bar{M}$ is the volume averaged magnetization. Because M decreases with V, $(E_B)_{\text{int}}$ also decreases with V. This variation is opposite to that due to the a_1 variation. However, in our opinion, the effect is probably smaller than that due to the a_1 variation.

Now, let us try another approach in order to show that a model that is too simple could lead to an uncorrect conclusion. Let us consider a small particle P_1 with volume V_1 interacting with a large particle P_2 with volume V_2. With regard to the magnetic moment $\mathbf{m}(P_1)$ of P_1, $\mathbf{m}(P_2)$ appears static. It results in an interacting field acting on P_1, which appears also static at the scale of the P_1 relaxation. In this case, $\tau(P_1)$ decreases and a static magnetic moment appears. This moment creates a reaction field at the level of P_2 that modifies the direction of $\mathbf{m}(P_2)$, therefore modifies also the interacting field on P_1. If now the direction of $\mathbf{m}(P_2)$ is reversed, all the fields cited above are reversed. What is the final effect? It can be modelled, but if a third particle is introduced with volume intermediate between V_1 and V_2, it is easy to see that the model constructed for two particles is no longer valid. One can also understand that the use of a nonfluctuating unidirectional interacting field acting on the particle is not correct.

E.3.4.2. Effects of an Applied Field. If a field H_{app} is applied, two unequal minima appear for the anisotropy energy, leading to two relaxation times τ^- and τ^+ (see Section D). It results that a part of $\mathbf{m}$ is blocked on average in the deeper minimum and then it appears static, the other part relaxing with a τ value smaller than without field. We can separate the two parts.

For the "static" part, dipolar interactions occur. The effective field H_{eff} acting on a particle differs from H_{app}, depending on the volumic concentration C_v, the sample shape, the field direction and the magnitude of the static part; H_{eff} can be calculated through the Onsager model (for the method see Section F.2). However, this calculation has not been performed until now. Indeed, it is not simple because H_{eff} depends on the static part, which itself depends on H_{eff} through the τ^- and τ^+ values for which no well-established formulas exist at present (Section D).

For the "relaxing" part, dipolar interactions also act and the presented model can be used. However, it is not $|\mathbf{m}|$ that now acts, but only a part of $|\mathbf{m}|$. This means that $(E_B)_{\text{int}}$, which corresponds to the dynamic part of the interactions, decreases with respect to the case without H_{app}. Complete calculation of these effects remains to be done.

More complex effects are expected for TRM experiments. First, the calculation of M_{FC} before the cutting-off of the field has to be performed following the method described above. Second, after cutting-off of the field, the system is out of equilibrium. Then, the remaining magnetization creates an H_{eff} (which can add or substract to an eventual remaining H_{app}, depending on the experiment conditions) that decreases in modulus with the decay of the magnetization, modifying this decay. Here also, complete calculation is lacking.

E.3.4.3. Approximations of $(E_B)_{\text{int}}$ and Outcomes of the Model. Approximations of $(E_B)_{\text{int}}$ are useful in order to simplify the τ calculation. They can be done through the approximation of the Langevin function:

$$x \lesssim 1 \qquad \mathscr{L}(x) \sim x/3 \tag{E.12}$$

$$x \gtrsim 2 \qquad \mathscr{L}(x) \sim 1 - 1/x \tag{E.13}$$

If $M^2 V a_j / kT \lesssim 1$ for any j, which occurs at high temperature or for low C_v values (except at very low temperature), then

$$(E_B)_{\text{int}} = \left[\frac{(M^2 V)^2}{3kT} \right] \sum_j a_j b_j$$

For a close - packed arrangement, on supposing that $b_j / a_j = b_1 / a_1$, b_1 / a_1

being a constant value and with $a_1 = C_v/\sqrt{2}$

$$(E_B)_{\rm int} = 6.8(b_1/a_1)C_v^2(M^2V)^2/3kT$$

valid for $$M^2Va_1/kT = M^2VC_v/kT\sqrt{2} \lesssim 1 \tag{E.14}$$

This formula covers those resulting from the Shtrikman and Wohlfarth model (Section E.3.2), but here, the parameters are given. Results obtained for Fe particles in an Al_2O_3 matrix by means of neutron diffraction[129,130] are in agreement with Eq. (E.14) (see Section F.8).

A second approximation can be derived when Eq. (E.12) is fulfilled for the first neighbors while (E.13) is valid for the further neighbors. In this case

$$(E_B)_{\rm int} = -n_1\left(\frac{b_1}{a_1}\right)kT + M^2V\left(\frac{b_1}{a_1}\right)\left[n_1a_1 + \left(\frac{M^2V}{3kT}\right)\sum a_j^2\right] \tag{E.15}$$

where n_1 is the number of first neighbors and the sum is done over all the neighbors except the first.

For a close-packed arrangement, this relation reduces to

$$(E_B)_{\rm int} = -n_1\left(\frac{b_1}{a_1}\right)kT + M^2V\left(\frac{b_1}{a_1}\right)\left(\frac{C_v}{\sqrt{2}}\right)\left[n_1 + \frac{3}{2}\left(\frac{M^2VC_v}{3kT\sqrt{2}}\right)\right] \tag{E.16}$$

which is valid for

$$2 \lesssim \frac{M^2Va_1}{kT} = \frac{M^2VC_v}{kT\sqrt{2}} \lesssim 4 \tag{E.17}$$

Using the relaxation time τ formula valid for uniaxial symmetry and $E_B/kT \gtrsim 3$

$$\tau = \tau_0 \exp\left(\frac{E_B}{kT}\right) \tag{E.18}$$

where τ_0 depends on E_B, T, and other parameters (see Section D), and considering that the other anisotropy contributions have the uniaxial symmetry with a total constant K_t and by neglecting the effect of farther neighbors, one finds

$$\tau = \tau_0 \exp\left[\frac{-n_1b_1}{a_1}\right]\exp\left[\frac{K_tV + (n_1b_1/a_1)M^2Va_1}{kT}\right] \tag{E.19}$$

If the thermal variation of M is approximately given by

$$M(T) \cong M(0)[1 - a_t T] \tag{E.20}$$

and by keeping for $M(T)$ only the first-order term in T, then

$$\tau \simeq \tau_0 \exp\left\{ -\left(\frac{n_1 b_1}{a_1}\right)[1 + 2a_t M^2(0)Va_1/k] \right\} \times \exp\left\{ \frac{K_t V + (n_1 b_1/a_1)M^2(0)Va_1}{kT} \right\} \tag{E.21}$$

This law has the same form as (E.18) with an increased E_B and a decreased τ_0 value. In the limits given by (E.17) and by neglecting the weak change of τ_0 with T, this relation shows that the variation of $\log \tau$ vs. $1/T$ is linear and that the intercept with the $\log \tau$ axis depends mainly on $\log \tau_0$ and on the number n_1 of first neighbors. For a disordered arrangement of particles, $n_1 = 12$. This means that the intercept value is about $\log_{10}\tau = -17$ to -18 as $\log_{10}\tau_0$ lies about -10 to -11, almost independent of the sample parameters. We recall that b_1/a_1 is near to unity. The approximation given by Eq. (E.21) is well verified for Fe particles in an alumina matrix[45] (Fig. E.3) and for γ-Fe_2O_3 particles in a polymer[59] (Fig. E.4).

For precise adjustment, it is necessary to take into account the actual thermal variation of M, the weak variation of $\log \tau_0$ with T, and the effects of farther neighbors. For T outside the limits defined by (E.17), the full formula (E.11) has to be used.

We note that very recently this model has been checked for samples including particle chains, where $n_1 = 2$–3.[59]

For $M^2 Va_1/kT \gtrsim 4$, no simple approximation can be derived, the model becomes unprecise at low temperature, when τ becomes very slow. However, when $T \to 0$, $(E_B)_{\text{int}}$ remains finite when $\partial(E_B)_{\text{int}}/\partial T \to \infty$, which is in favor of a phase transition at zero temperature. On the other hand, it has been shown that a generalized Arrhenius law with an exponent about 2 simulates quite well the τ variation. This fact is also in favor of a transition at zero temperature.[115] Nevertheless, these features must be considered as a trend because of the uncertainty in the model at very low temperature. From this we can assert that if a collective state exists at zero temperature, the transition temperature is certainly very low.

E.3.5. Other Interparticle Interaction Models

Other attempts have been proposed to model the interparticle interactions. They correspond to an improvement of a model describing the

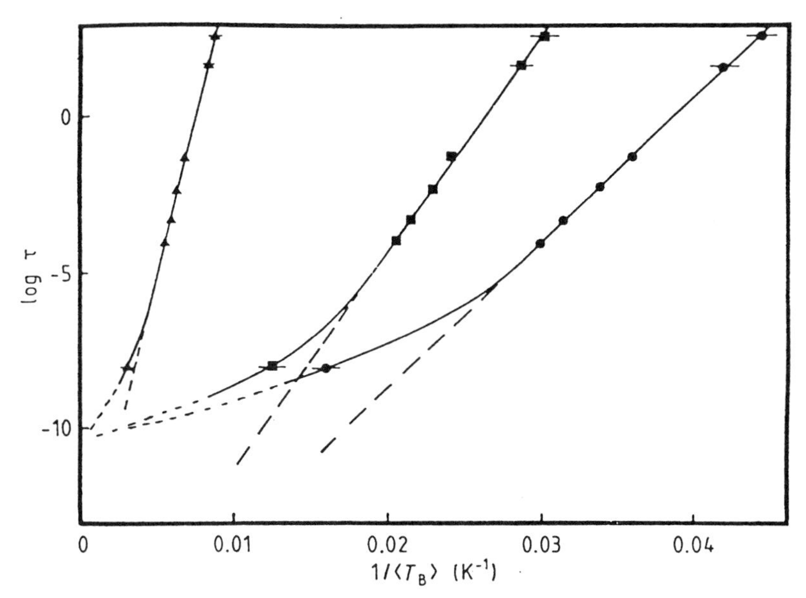

Figure E.3. Variation of the blocking temperature T_B in the classical plot of $\log_{10}\tau$ vs. $1/T_B$ for various samples of Fe particles embedded in an alumina matrix. (Reproduced with permission from Ref. 45.)

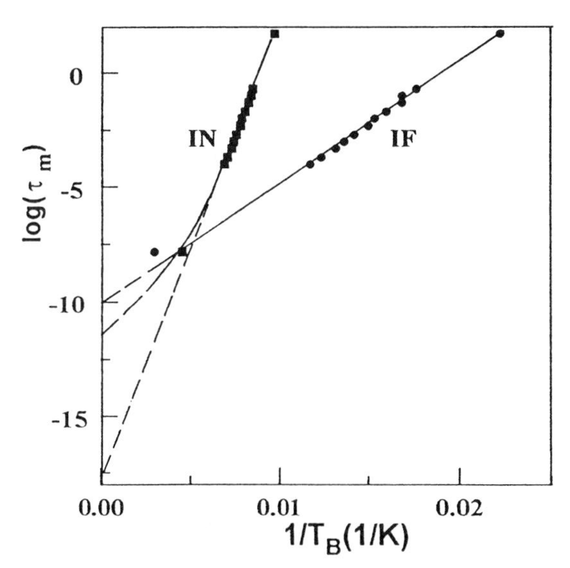

Figure E.4. Variation of the blocking temperature T_B in the classical plot of $\log_{10}\tau$ vs. $1/T_B$ for two samples of γ-Fe_2O_3 particles in a polymer. IN, interacting particles; IF, particles with very weak interactions.

properties resulting from a certain type of measurement, as, for example, for M_{ZFC} [144] (see Section F.2), or are more general, as, for example, the calculation of the time decay for a system of two identical interacting particles from the Fokker–Planck equation,[145] or the calculation of the energy for particle chains consisting of identical particles.[146] However, in all cases, the taken hypothesis are either too approximate for describing the real properties or too restrictive for generalization to actual samples. Recently, a new model has been proposed in the case of weak interactions[147] that leads to a decrease of T_B for growing interactions. This model is in contradiction with the other models, and in our opinion it is not justified from theoretical as well as from experimental points of view. The interparticle interactions act through a field that fluctuates. However, this fluctuation is treated independently of the **m** relaxation. This is not correct with regard to the problem symmetry (see Section E.1), which shows that these two fluctuations are not independent. The calculation also raises some objections. For example, the τ_0 variation with the interactions has been neglected. However, the calculated effect is of the same order of magnitude as this variation, because τ_0 is proportional to $E_B^{-3/2}$ (see Section D). The model has been used for explaining the results obtained by Mössbauer spectroscopy on γ-Fe_2O_3 particles in a polymer. From these experiments, the T_B relative to noninteracting particle samples are higher than T_B relative to interacting particle samples. However, an increase of T_B with the interactions for the two samples showing strong interparticle interactions is found. Experiments of χ_{ac} on a large frequency range show an increase of T_B with the interactions. In fact, all the results can be explained from the model presented in Section E.3.3, the discrepancy for Mössbauer T_B being mainly due to an effect induced by the interactions (variation of the damping constant and modification of the anisotropy energy form), and not to the interactions themselves.[59,137]

E.4. Some Conclusions

In our opinion, the model presented in Section E.3.3 allows a first explanation of all the results obtained with various techniques. However, the model is simple and cannot cover exactly the complex properties determined by the interparticle interactions. For other kinds of interactions, like RKKY interactions present in metallic samples or superexchange interactions which could occur in insulating samples, we think that the model is applicable after some adjustments, on condition that the temperature is sufficiently high for preventing the appearance of a collective state. In fact, the model is basically independent of the kind of interactions. For a disordered assembly of particles with volume distribution, the magnetic moments of particles always relax with different τ, and statistics are applicable for calculating the interaction energy. Progress in material sciences will allow the obtaining of more and more accurate results. Progress in modeling is also necessary.

F. EXPERIMENTAL PROPERTIES AND MODELING

F.1. General Considerations

Before discussing the experimental properties, it is useful to discuss the general framework in which the features will be analyzed. The dynamical aspect of the phenomenon implies that there exist several situations to be recognized for stating the method of analysis. This leads to a few basic questions. First, is there a change in the relaxation during the measurement? Second, the properties are measured during a certain time Δt. But, is the measurement repeated many times and then averaged or does the measurement follows the variation of the properties for successive Δt? The measuring time τ_m is defined as equal to Δt, but in the latter case the time elapsed between two successive Δt can be different to zero and a second characteristic time has to be defined. Third, what parameter is measured by the experiment, in what geometry? Is the measurement relative to one direction, any direction? Is it obtained by integrating over all directions? And finally, the classical question, does the measurement perturb the relaxation?

To answer these questions, it is necessary to know if the properties are measured at the thermodynamic equilibrium. Let us consider two unequal minima for the anisotropy energy. This leads to two relaxation times, τ^+ relative to the passage from the lowest minimum M^+ to the upper minimum M^- and τ^- relative to the reverse passage (see Section D). We can say that the thermodynamic equilibrium is reached if the populations of M^+ and M^-, which depend on the initial conditions, are proportional to τ^+ and τ^-, respectively. If the two minima are equivalent, $\tau^+ = \tau^-$ and the populations have to be equal. Of course, the equilibrium is always reached when there remains only one minimum due to the effect of an applied field.

Let us first consider the system at the equilibrium. If $\Delta t \gg \tau^+$, the increase of Δt does not change the results. For simulating the properties, statistics have to be applied to the measured parameter. This situation is typically achieved for magnetization under medium and high applied field, when only one minimum remains, and for the field-cooled and zero-field-cooled magnetizations when Δt is much larger than the τ^+ related to the appreciably biggest particle volume.

If now $\Delta t \ll \tau^-$, the results depend on the way the experiment measures the parameter and its possible variation. This variation is either at the τ^- scale (related, e.g., to the space orientation of the magnetic moment **m**) or at a time scale t_{sc} much smaller than Δt (due, e.g., to vibrations of **m** in the potential well). In the two cases, averaged values have to be considered, in the former case because the measurement is repeated numerous times and in the latter case as $t_{sc} \ll \Delta t$, but the averaging process is generally different. This situation corresponds to low temperature measurements, typically to a.c. susceptibility experiments where the magnetization is measured along a fixed direction, to Mössbauer spectroscopy where the hyperfine field is measured in all directions and to small-angle neutron-scattering experiments where the magnetization values are integrated over all directions.

For $\Delta t \approx \tau^+$ (or τ^-), complex phenomena appear. It is necessary in this case to do a particular analysis depending on the technique.

Now we consider the case where the equilibrium is not reached, that is, the relaxation regime varies with time. This is generally obtained by starting from an equilibrium state under field where the two minima are unequally populated and by changing the initial conditions, for example, by cutting off the field. A change in the magnetization value is detected only if the new set of relaxation times are not too slow with regard to the total measurement time.

The magnetization variation will depend on the initial state, its variation (with, e.g., the temperature) before the change, its variation during the time needed for changing the experimental conditions, and, finally, on the new conditions. An accurate analysis of all the process taking into account the kind of parameter measured by the experiment will be necessary for modeling the results. This situation is typically encountered in field-cooled and zero-field-cooled magnetization experiments and mainly for thermoremanent magnetization measurements.

In concluding this discussion, we want to emphasize that a precise analysis of the process is an indispensable preliminary to any interpretation of the results.

Regarding the interpretation of the data, an important point has to be emphasized. The formulation of the relaxation time τ includes the gyromagnetic ratio γ_0 (see Section D), which is generally expressed in angular frequency per Gauss. We must keep this unity in mind when we compare τ to the measuring time τ_m (see, e.g., the case of the ac susceptibility measurement).

Another difficulty concerns the choice of the models. On the one hand, the discrete orientation model, which leads to an exponential decay of the measured parameter with time [see Eq. (D.3)], does not exactly take into account the probability of the presence of **m** outside the energy minima. We believe that the resulting error is of second order in magnitude, but verification would be needful. On the other hand, the calculation from the Boltzmann statistics forgets the relaxation process and its gyromagnetic character. Then, this calculation is also approximate, with the order of magnitude of the errors depending on the cases.

In fact, the use of the probability densities of the **m** orientations resulting from the Fokker–Planck equation [Eq. (D.13)] would allow a correct calculation, but this is in general inextricable when it is applied for modeling an experiment. Hence, in practice there only remain the two first approaches, and one should not forget that these are mere approximations.

In Section F.2 we will discuss the properties related to the field-cooled and zero-field-cooled magnetizations at low field, the static susceptibility, and the determination of the effective field acting on a particle as related to the field applied to the particle assembly. In Section F.3, we will comment on the properties deduced from magnetization under moderate and high fields, and consider the determination of the non-relaxing magnetization. In Section F.4, we will discuss the remanence magnetizations, mainly the thermoremanence magnetization and make some comments on the coercive field. We will describe the features that result from ac susceptibility in Section F.5. In Section F.6, the results obtained by means of Mössbauer spectroscopy without and with applied field will be discussed. We note that this experiment has been widely used for the study of fine particles. In Section F.7, ferromagnetic resonance experiments will be discussed. We will discuss the results obtained by neutron diffraction in Section F.8. Finally, in Section F.9, we will briefly consider some other experiments.

F.2. Field-cooled and Zero-Field-cooled Magnetizations and Susceptibility in the Superparamagnetic State at Low Field

Field-cooled (FC) and zero-field-cooled (ZFC) magnetization experiments at low field are very useful for evidencing superparamagnetic properties. They are simple and point out the irreversible properties below a certain temperature, roughly the mean blocking temperature T_B related to the characteristic time of the experiment. However, quantitative features can be obtained only with a precise analysis of the phenomenon, which depends strongly on the experimental process and on the sample parameters, that is, the anisotropies, the volume distribution, the particle arrangement, the interaction effects, and so on. Much information can be derived from such an analysis concerning mainly the barrier energy distribution and the related volume distribution, the particle magnetic state, and the effect of the interparticle interactions from dynamical as well as static points of view. We will discuss the properties resulting from ZFC magnetization or χ_{dc} measurements in Section F.2.1, those corresponding to FC magnetization in Section F.2.2, and the effect of the interparticle interactions on the susceptibility in the superparamagnetic state and the meaning of the superparamagnetic temperature θ_{sp} in Section F.2.3.

F.2.1. Zero-Field-cooled Magnetization (M_{ZFC}) Experiments

Let us recall the usual process. The sample is cooled at once without applied field from a temperature where all the particles are in the superparamagnetic state until the lowest temperature T_{min}. Afterward a field is applied and the measurement is performed increasing the temperature. In fact, if the magnetization in the superparamagnetic state has to be studied, the above condition on the initial temperature must be fulfilled. But if only T_B is of interest, it is sufficient that the magnetic moments **m** of the particles are frozen in random orientation at T_{min}. This can easily be checked from the zero field magnetization value, which must be equal to zero.

An example of M_{ZFC} variations is shown in Fig. F.2.1 and concerns γ-Fe_2O_3 particles in a polymer with different degrees of dispersion.[96] The M_{ZFC} always follows the same variation. At very low temperature, when all the **m** are blocked, $M_{ZFC} = M_{nr}^2 H_{eff}/3K$, where M_{nr} represents the nonrelaxing magnetization, K the anisotropy constant, and H_{eff} the field seen by the particle. This expression corresponds to the magnetization of monodomains with uniaxial symmetry but does not take into account the existence of an eventual **m** order (see Section E.2). On increasing the temperature, M_{ZFC} increases and shows a maximum for a temperature T_{max} related to T_B. Afterward, M_{ZFC} decreases and from a certain temperature T_{bra}, M_{ZFC} shows thermodynamic equilibrium properties, called properties in the superparamagnetic state, when the **m** relaxation time for the appreciable largest particle is much smaller than the heating rate.

This variation has been observed for numerous systems of fine particles, for example: NiO antiferromagnetic particles,[148] Ni particles embedded in a SiO_2 matrix,[149] Fe particles in a SiO_2 matrix,[150,151] Fe_3O_4 colloids,[152] and Fe particles in a silica gel.[24]

In the superparamagnetic state, M_{ZFC} is given by the well-known formula, by neglecting the interparticle interactions, the volume distribution, and the anisotropies:

$$M_{ZFC} = M_{nr}\mathcal{L}\left[\frac{M_{nr}VH_{eff}}{kT}\right] \tag{F.2.1}$$

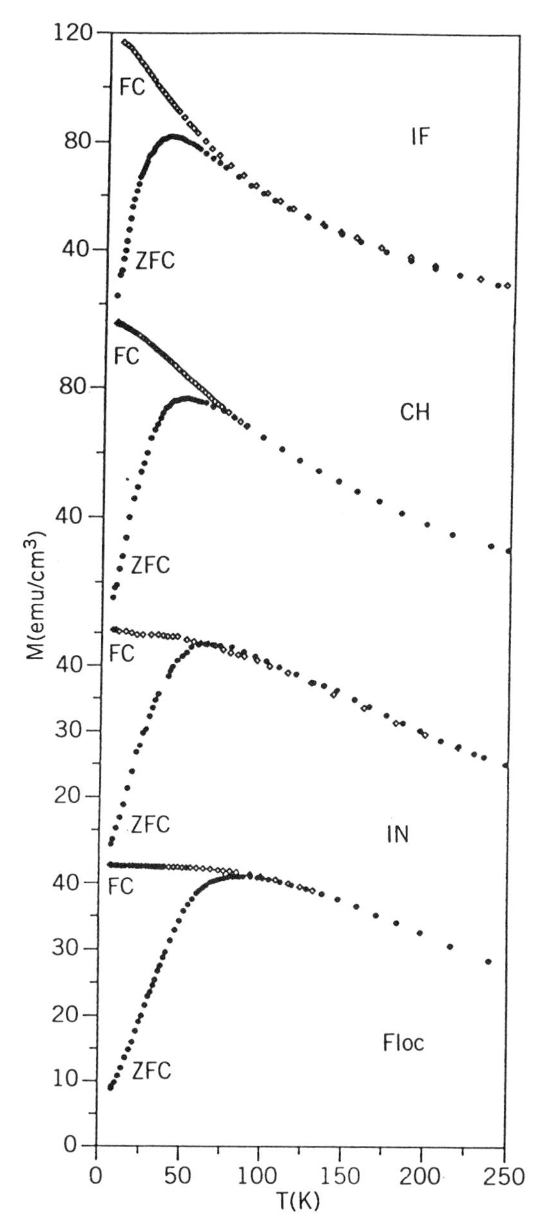

Figure F.2.1. Thermal variation of zero-field-cooled magnetizations of γ-Fe_2O_3 particles dispersed in a polymer. The averaged diameter corresponding to the mean volume is $\langle D \rangle = 7$ nm. Interparticle spacing of $5\langle D \rangle$ (IF), and $1.4\langle D \rangle$ (IN); chains (CH); strong aggregates (Floc).

where $\mathscr{L}$ is the Langevin function which reduces for $M_{nr}VH_{eff}/kT < 1$ to

$$M_{ZFC} = \frac{M_{nr}^2 V}{3kT} H_{eff} \tag{F.2.2}$$

However, this formula is a very rough approximation because almost all the actual parameters of the sample have been neglected.

For low H values, Chantrell et al.[153] have derived a precise expression valid for $M_{nr}VH_{eff}/kT \ll 1$:

$$M_{ZFC} = \frac{M_{nr}^2 V H_{eff}}{kT} \left[\cos^2\psi + \frac{1}{2}(1 - 3\cos^2\psi)\left(1 - \frac{I_2}{I_0}\right) \right] + \varepsilon(H_{eff}^3) \tag{F.2.3}$$

with
$$\frac{I_2}{I_0} = \frac{1}{\alpha}\left[-\frac{1}{2} + \frac{\exp(\alpha)}{I(\alpha)} \right]$$

and

$$\alpha = \frac{KV}{kT} \qquad I(\alpha) = \int_{-1}^{+1} \exp[\alpha x^2]\, dx$$

with ψ being the angle between $\mathbf{H}_{eff}$ and the easy direction.

With random easy axis Eq. (F.2.3) reduces to (F.2.2). The volume distribution can be easily introduced in (F.2.3). For random easy axis, we obtain

$$\langle M \rangle = \frac{\langle M_{nr}^2 V^2 \rangle}{\langle V \rangle} \frac{H_{eff}}{3kT} \tag{F.2.4}$$

Without interaction, that is, for a sample where the particles are very far from each other, H_{eff} is equal to the applied field H_{app}. In the usual case where the interparticle interactions are not negligible, H_{eff}, has to be expressed from H_{app}, which will be discussed in Section F.2.3.

We also note that the validity of Eq. (F.2.3) can be extended by calculating the terms in H_{eff}^3, H_{eff}^5 and so on (the even terms vanish). We note also that there are some typing errors in Ref. 153.

Recently, general expressions have been established for M_{ZFC} and the initial susceptibility.[144] However, some typing errors and mainly the utilization of unusual reduced variables complicate the understanding of the work. We agree with the general steps, but we think that the calculation is not valid for a narrow-volume distribution because the model of critical volume is too rough in this case. In addition, the treatment of the interactions is crude (see Section F.2.3). On the other hand, it seems that the authors have not weighted M_{ZFC} by V in the average calculation. We want to point out that all the experimental setup measure the magnetic moment (not the magnetization) and that the measured magnetization is

$$M_{\text{meas}} = \frac{\sum m}{\sum V} = \frac{\langle m(V) \rangle}{\langle V \rangle} = \frac{\langle VM(V) \rangle}{\langle V \rangle} \tag{F.2.5}$$

and not $M_{\text{meas}} = \langle M(V) \rangle$, which is often forgotten in studies.

Variations of the peak temperature $T_{\max}$ of the susceptibility have also been studied in relation to the particle concentration.[154] The interparticle interaction model presented in Section E.3.3 has been adapted in particular by including the volume distribution. However, the calculated change in $T_{\max}$ with concentration is smaller than that observed in experiments on Fe_3O_4 fine-particle system. This could be due to the difference between H_{app} and H_{eff} (see below) but could also be an intrinsic property. The same trend has been observed for γ-Fe_2O_3 fine particles[59] and is probably due to surface effects (see Section E.1).

The T_B variation with applied field have been calculated[71] from a simple model based on the fact that for $H_a = MH_{\text{eff}}/2K \gtrsim 0.05$, only the field orientations equal or near to $\psi = \pi/2$ contribute to the relaxing part of **m**. One error takes place in the statement of the model and concerns the energy barriers: only the two barriers for $\theta < \pi$ are of importance. Fortunately, the correction does not change the continuation of the calculation. In addition, some assumptions seem now too simple. However, a good agreement is obtained with experimental results (Fig. F.2.2). Other calculations have been performed later, taking into account the effects of the interparticle interactions.[155]

Finally, we want to emphasize that except for samples where the particles are very far from each other, a strong difference can appear between H_{app} and H_{eff}, depending on the geometry of the experiment. We show an example in Figure F.2.3 for Fe particles in an Al_2O_3 matrix.[89] The sample has a thick-film shape and H_{app} is parallel or perpendicular to the sample plane. Strong variations appear that lead to a noticeable difference for $T_{\max}$.

F.2.2. Field-cooled Magnetization (M_{FC}) Experiments

The process is the same as for M_{ZFC} experiments except that the cooling is done under H_{app} but here the condition related to the superparamagnetic state at the initial temperature is very important. If an appreciable

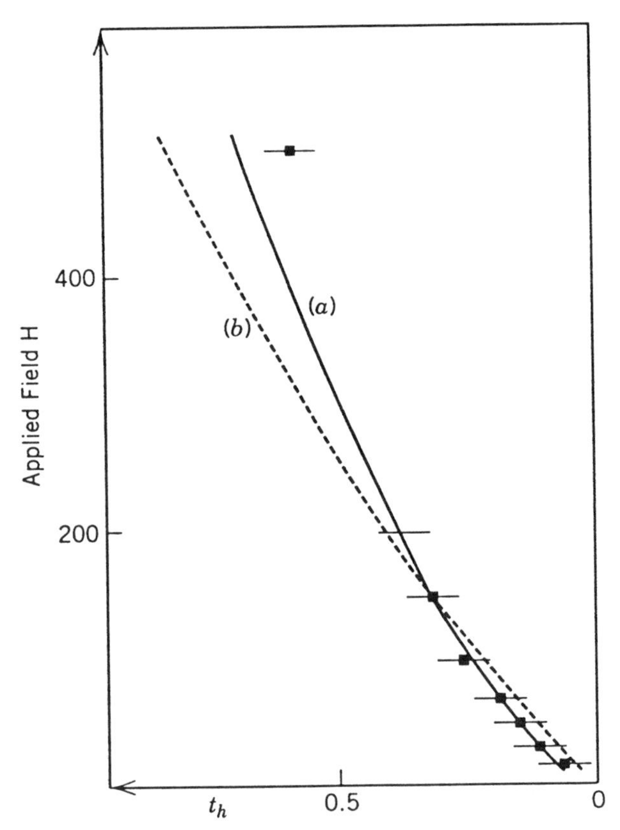

Figure F.2.2. Magnetic field versus reduced temperature $t_h = 1 - T_B(h)/T_B(0)$ for Fe particles in Al_2O_3. The dotted line (b) and the full line (a) show the best fit to the formula, without or with interacting field, respectively. (Reproduced with permission from Ref. 71.)

population of the largest particles is blocked at the initial temperature because their **m** relaxation times are still slow, the initial magnetic state of the sample is not well known, due to the fact that the magnetic state of the blocked particles depends on the thermal history of the sample. This alters the M_{FC} values by an unknown part that can vary with temperature and yields uncertain quantitative estimations.

Magnetization M_{FC} always follows the same variation (Fig. F.2.1). On decreasing the temperature, M_{FC} is merged with M_{ZFC} until T_{bra} (see previous section). Then M_{FC} continues to increase and from a certain temperature T_{sat}, M_{FC} remains constant. In the absence of an energy barrier (volume) distribution, the three temperatures T_{bra}, T_{max}, and T_{sat},

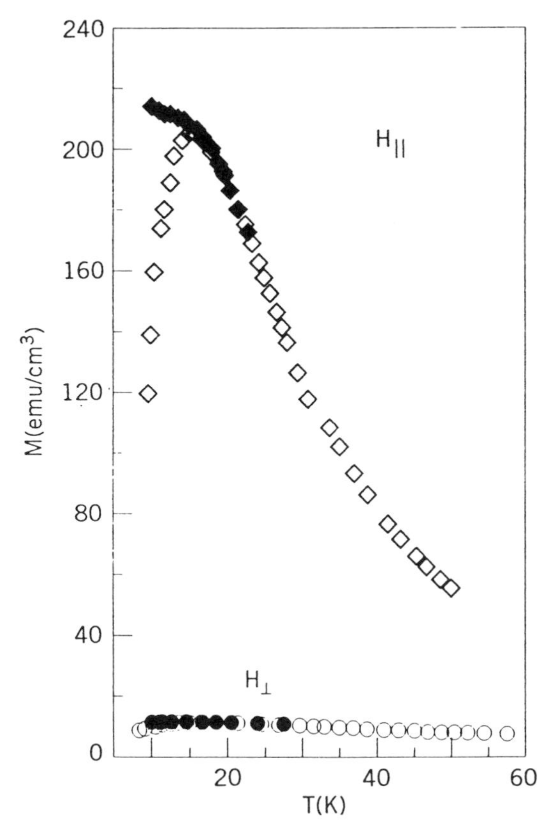

Figure F.2.3. Thermal variation of ZFC and FC magnetizations for Fe particles in Al_2O_3 for $H_{app} = 50$ Oe perpendicular and parallel to the plane of the thin sample.

are very near each other. Sometimes, M_{FC} shows a light maximum, which will be discussed later.

Chantrell and Wolhfarth[156] have developed a model that takes into account the cooling rate r_c. This model predicts that until a certain temperature $T_B(r_c)$ corresponding to the blocking temperature related to r_c, the system remains in thermal equilibrium and M_{FC} follows the superparamagnetic law, like M_{ZFC}. On passing through $T_B(r_c)$, the equilibrium magnetization is frozen in and below, M_{FC} remains constant-equal to $(M_{FC})_{max}$. An expression of $T_B(r_c)$ is given. We note that $T_B(r_c)$ is evaluated with respect to the T_B value for static measurement, which is roughly calculated and by supposing that the applied field is parallel to the easy direction. However, it is not too complicated to derive a more accurate expression from the model. Finally, the authors study the effect of a V distribution using a lognormal distribution. They show that the distribution leads to a smoothing of the abrupt variation predicted for single-volume particles.

For quantitative adjustment, the model needs some improvements, such as more accurate expressions for T_B and for the barrier under field with random easy axis. It is also necessary to take into account the interparticle interactions from dynamical as well as static points of view. We recall that H_{eff} can differ strongly from H_{app} (Fig. F.2.3). However, the general trends deduced from the model remain true:

1. Noticeable differences for M_{FC} are observed only if the r_c values are very different, by at least one order of magnitude.
2. For single-volume particles, M_{FC} remains constant below $T_B(r_c)$. That means that for an actual sample with a V distribution, M_{FC} will remain constant below $T_{\text{sat}} = T_B(r_c)$ relative to the appreciable smallest volume. It is interesting to point out that $(M_{\text{FC}})_{\text{max}}$ does not depend on V in first approximation, but only on r_c and on the field.
3. In the same way, $T_{\text{bra}} = T_B(r_c)$ relative to the appreciable largest volume.

We note that below $T_B(r_c)$, M_{FC} is always out of thermodynamic equilibrium and that M_{FC} variation with time will be observed if the observation time is sufficiently long.

Three last comments are worthwhile. First, the analysis of M_{FC} has been done on decreasing the temperature. In fact, the M_{FC} measurements are usually performed with increasing temperature. There is no appreciable difference if the cooling rate r_c and the heating rate r_h are not too different. But if r_h is much slower than r_c, which is quite possible as the measurements take place during the heating, $(M_{\text{FC}})_{\text{max}}(r_c)$ is smaller than the $(M_{\text{FC}})_{\text{max}}$ value obtained with a cooling rate equal to r_h. In this case a small maximum appears for M_{FC} (Fig. F.2.4).

Second, examination of Figure F.2.1 reveals that the M_{FC} increase is much more pronounced for particles without interaction than for particles with interactions and that T_{sat} is much lower for particles without interaction. As the volume distribution is the same for all the samples, a contradiction seems to appear except if we suppose that the energy barrrier distribution, in this case not related directly to the V distribution, is strongly narrowed for particles with interactions. In fact, we have shown in Section E.3.4 that the interactions could narrow the E_B distribution with regard to the V distribution. Other causes are possible but in our opinion, strong effects are not realistic. At present, it is

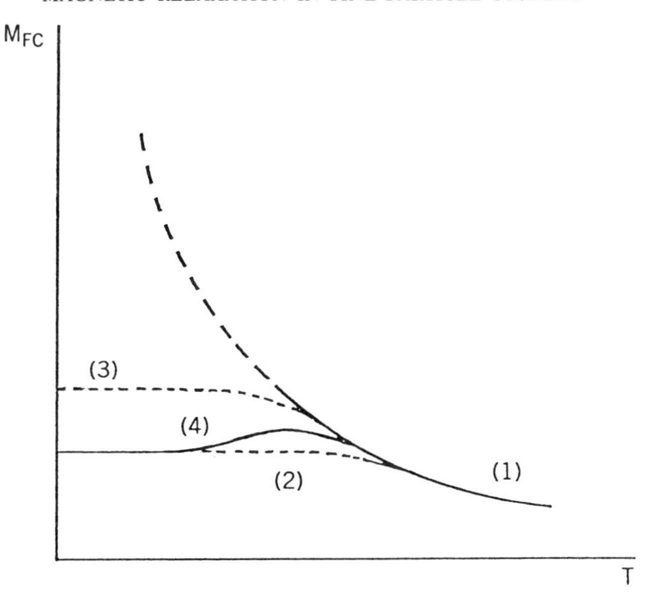

Figure F.2.4. Thermodynamic equilibrium M_{FC} [Eq. (F2.4)], curve (1). M_{FC} measured decreasing the temperature with cooling rates r_c and r_h, curves (2) and (3), respectively. M_{FC} cooled with r_c and measured increasing the temperature with heating rate r_h, curve (4).

difficult to conclude because M_{FC} has not yet been modeled taking into account the difference between H_{app} and H_{eff} due to the interparticle interactions, and the results shown in Figure F.2.1 have not been obtained with controlled parameters because the shape of the sample and the field direction were not well defined.

Finally, differential FC experiments proposed by Néel in his historical study[1] have never been performed, to our knowledge. The zero-field-cooled process is at once used until a given temperature, afterward the FC process is operated. This permits an exploration of the M_{nr} values vs. V, which can be very useful when M_{nr} varies with V.

F.2.3. Magnetization in the Superparamagnetic State

As discussed in the Section F.2.1, the component of the particle magnetization along a weak applied field can be written as follows[153]:

$$\langle M \rangle = \frac{C}{T} H_{eff} + \varepsilon(H_{eff}^3) \tag{F.2.6}$$

with

$$C = \frac{M_{nr}^2 V}{k}\left[\cos^2\psi' + \frac{1}{2}(1 - 3\cos^2\psi')\left(1 - \frac{I_2}{I_0}\right)\right] \tag{F.2.7}$$

where M_{nr} is the nonrelaxing magnetization of the particle, H_{eff} is the effective field acting on the particle, ψ' is the angle between $\mathbf{H}_{eff}$ and the easy direction (we consider uniaxial symmetry), and

$$I_n = \int_0^{\pi} \cos^n\theta \sin\theta \exp[-\alpha \sin^2\theta]\, d\theta \tag{F.2.8}$$

with $\alpha = KV/kT$.

Two problems have to be resolved. First, the determination of $\mathbf{H}_{eff}$ with regard to the field $\mathbf{H}_{app}$ applied on the sample, which contains the particle assembly; and second, the calculation of the mean magnetization $\langle \bar{M} \rangle$ of the particles along $\mathbf{H}_{app}$ by averaging on V and ψ', because a disordered arrangement of particles with a V distribution and easy directions at random is considered.

F.2.3.1. Problem Statement. If the volumic concentration C_v of particles in the sample is very low, $\mathbf{H}_{eff} \approx \mathbf{H}_{app}$, this relation is exact only when the sample contains one particle. For magnetic dipoles, two models have been proposed for the evaluation of H_{eff}, which account for the dipolar interactions between dipoles.

They define a spherical cavity containing one magnetic dipole and consider the continuum limit. The field H_{eff} acting on a dipole is the sum of the field H_{ext} acting on the cavity and a reaction field.

In the Lorentz model,[157,158] the medium is not polarizable. Then the reaction field is

$$H_{lor} = \frac{\mu - 1}{3} H_{ext} \tag{F.2.9}$$

where μ is the permeability of the medium referred to H_{ext}.

However, this hypothesis is a very rough approximation. Later, Onsager[158,159] has considered a polarizable medium. In this case, the reaction field is written as follows:

$$H_{ons} = \frac{\mu - 1}{2\mu + 1} H_{ext} \tag{F.2.10}$$

Finally H_{ext} can be expressed from H_{app} by considering the demagnetizing field H_d relative to the external shape of the sample. Then

$$\mathbf{H}_{ext} = \mathbf{H}_{app} + \mathbf{H}_d \tag{F.2.11}$$

Now, for application to fine-particle samples, it is possible to define in an analogous manner a spherical cavity containing one particle considered as a superdipole in such a manner that the volumic concentration of the particle in the cavity is equal to the C_v value valid for the sample.[89,90]

A first difficulty is related to the existence of anisotropies. Equations (F.2.9) and (F.2.10) have been established in the absence of anisotropy, which is correct for magnetic dipoles. But for fine particles, the time-averaged magnetization $\langle \mathbf{M} \rangle$ is not parallel to $\mathbf{H}_{eff}$ and Eqs. (F.2.6) and (F.2.7) give the component along $\mathbf{H}_{eff}$. On the other hand an averaged value of $\langle \mathbf{M} \rangle$ has to be considered. Due to the random orientation of the easy directions $\langle \bar{\mathbf{M}} \rangle$ is parallel to $\mathbf{H}_{ext}$, which leads to a constant value for $\mathbf{H}_{ext}$ by applying Eq. (F.2.11) with $\mathbf{H}_d = \mathbf{N}\langle \bar{\mathbf{M}} \rangle$ where $\mathbf{N}$ is the demagnetizing tensor.

Because of the μ definition,

$$\mu = 1 + 4\pi C_v \chi_{ext} \tag{F.2.12}$$

where χ_{ext} is the susceptibility of the particle related to H_{ext} and C_v the volumic concentration. If we admit that the two models consider only the component of the magnetization parallel to $\mathbf{H}_{ext}$ for the calculation of the reaction field, Eqs. (F.2.6) and (F.2.7) can be applied with $\psi' = \psi$, with ψ being the angle between $\mathbf{H}_{ext}$ and the easy direction.

F.2.3.2. Application of the Lorentz Model. In this model

$$H_{eff} = H_{ext} + \mathrm{Ni} C_v \langle M \rangle \tag{F.2.13}$$

with $\mathrm{Ni} = 4\pi/3$. Then,

$$\langle M \rangle = (H_{ext} + \mathrm{Ni} C_v \langle M \rangle) C/T \tag{F.2.14}$$

$$\text{and} \quad \langle M \rangle = \frac{C/T}{1 + \mathrm{Ni} C_v C/T} H_{ext} \tag{F.2.15}$$

Now, $\langle M \rangle$ has to be averaged on ψ. But, the Lorentz model leads to $\langle M \rangle$ divergence when $1 - \mathrm{Ni} C_v C/T = 0$. Therefore, the $\langle M \rangle$ averaged value can be determined only under conditions, that is, $1 - \mathrm{Ni} C_v (M_{nr}^2 V/2kT)(1 - I_2/I_0) \neq 0$ and if this value is positive, which is fulfilled when C_v is not too high, then $\mathrm{Ni} C_v (M_{nr}^2 V/kT) I_2/I_0 < 1$.

For this latter case, a complicated formula is obtained for $\langle \bar{M} \rangle$:

$$\langle \bar{M} \rangle = \frac{M_{\rm nr}^2 V}{kT} H_{\rm ext} \left[-\frac{1}{\beta'} + \frac{1}{2\beta'} \frac{1}{\left\{ \frac{\beta'}{2}\left(3\frac{I_2}{I_0} - 1\right)\left[1 - \frac{\beta'}{2}\left(1 - \frac{I_2}{I_0}\right)\right]\right\}^{1/2}} \right.$$
$$\left. \times \ln \frac{\left[1 - \frac{\beta'}{2}\left(1 - \frac{I_2}{I_0}\right)\right]^{1/2} + \left[\frac{\beta'}{2}\left(3\frac{I_2}{I_0} - 1\right)\right]^{1/2}}{\left[1 - \frac{\beta'}{2}\left(1 - \frac{I_2}{I_0}\right)\right]^{1/2} - \left[\frac{\beta'}{2}\left(3\frac{I_2}{I_0} - 1\right)\right]^{1/2}} \right] \qquad \text{(F.2.16)}$$

with $\beta' = \text{Ni}C_v(M_{\rm nr}^2 V/kT)$, and

$$\frac{I_2}{I_0} = \frac{1}{\alpha}\left[-\frac{1}{2} + \frac{\exp(\alpha)}{I(\alpha)} \right] \qquad \text{(F.2.17)}$$

with

$$I(\alpha) = \int_{-1}^{+1} \exp(\alpha x^2)\, \mathrm{d}x$$

For small β' value, Eq. (F.2.16) reduces to

$$\langle \bar{M} \rangle = \frac{M_{\rm nr}^2 V}{3kT} H_{\rm ext}\left[1 + \left(\frac{2}{5} - \frac{2}{5}\frac{I_2}{I_0} + \frac{3}{5}\left(\frac{I_2}{I_0}\right)^2 \right)\beta' + \cdots \right] \qquad \text{(F.2.18)}$$

From Eq. (F.2.17), $\frac{1}{3} < I_2/I_0 < 1$ for any α value, then the β' term varies between $\beta'/3$ ($\alpha = 0$) and $3\beta'/5$ ($\alpha = \infty$).

In fact, we can see that $\text{Ni}C_v\langle M \rangle$ is the component of the magnetizing field due to the spherical cavity. Now we can consider all the components of this field. In the axis system where the easy axis is along $\mathbf{O}_z$ and $\mathbf{H}_{\rm eff}$ in the zOy plane (Fig. F.2.5) with an angle ψ' with $\mathbf{O}_z$, $\langle \mathbf{M} \rangle$ is given by

$$\langle \mathbf{M} \rangle_x = 0$$

$$\langle \mathbf{M} \rangle_y = \frac{M_{\rm nr}^2 V H_{\rm eff}}{kT}\frac{1}{2}\sin\psi'\left(1 - \frac{I_2}{I_0}\right) = v_y H_{\rm eff} \sin\psi' \qquad \text{(F.2.19)}$$

$$\langle \mathbf{M} \rangle_z = \frac{M_{\rm nr}^2 V H_{\rm eff}}{kT}\cos\psi'\,\frac{I_2}{I_0} = v_z H_{\rm eff}\cos\psi'$$

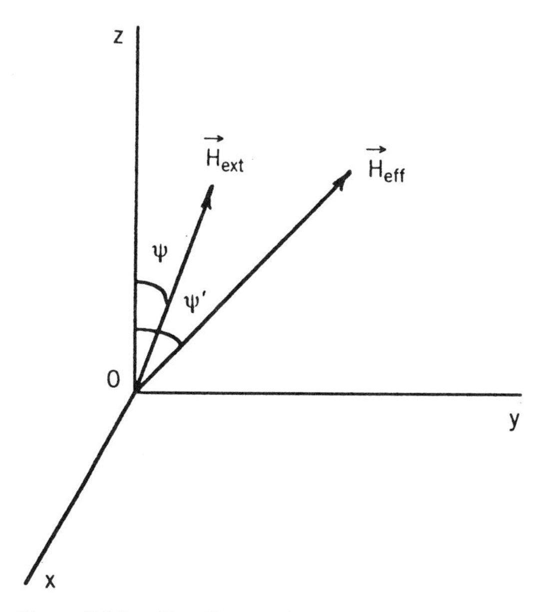

Figure F.2.5. Coordinate axis system for the particle.

As

$$\mathbf{H}_{\text{eff}} = \mathbf{H}_{\text{ext}} + \text{Ni}C_v\langle\mathbf{M}\rangle \tag{F.2.20}$$

$\mathbf{H}_{\text{ext}}$ is also in the zOy plane at an angle ψ with **Oz** (Fig. F.2.5). From Eq. (F.2.14), it is easy to calculate the $\langle\mathbf{M}\rangle$ component along $\mathbf{H}_{\text{eff}}$, which verifies (F.2.6) and (F.2.7).

Now, the $\langle\mathbf{M}\rangle$ component $\langle\mathbf{M}\rangle_{\|}$ along $\mathbf{H}_{\text{ext}}$ has to be calculated:

$$\langle\mathbf{M}\rangle_{\|} = \langle M\rangle_y \sin\psi' + \langle M\rangle_z \cos\psi' \tag{F.2.21}$$

From Eqs. (F.2.19) and (F.2.20)

$$\begin{aligned} \sin\psi' &= \frac{H_{\text{ext}}\sin\psi}{H_{\text{eff}} - \text{Ni}C_v H_{\text{eff}} v_y} \\ \cos\psi' &= \frac{H_{\text{ext}}\cos\psi}{H_{\text{eff}} - \text{Ni}C_v H_{\text{eff}} v_z} \end{aligned} \tag{F.2.22}$$

By reporting in Eq. (F.2.21) we find[160]:

$$\langle \mathbf{M} \rangle_{\|} = H_{\text{ext}} \left[\frac{v_y \sin^2 \psi}{1 - \text{Ni} C_v v_y} + \frac{v_z \cos^2 \psi}{1 - \text{Ni} C_v v_z} \right] \tag{F.2.23}$$

Because ψ is random, an averaged value $\langle \bar{M} \rangle_{\|}$ has to be considered:

$$\langle \bar{M} \rangle_{\|} = \frac{M_{\text{nr}}^2 V H_{\text{ext}}}{3kT} \left[\frac{\left(1 - \frac{I_2}{I_0}\right)}{1 - \frac{M_{\text{nr}}^2 V}{kT} C_v \text{Ni} \frac{1}{2} \left(1 - \frac{I_2}{I_0}\right)} + \frac{\frac{I_2}{I_0}}{1 - \frac{M_{\text{nr}}^2 V}{kT} C_v \text{Ni} \frac{I_2}{I_0}} \right] \tag{F.2.24}$$

We can check two points. If $C_v = 0$, we find for $\langle \bar{M} \rangle_{\|}$ the usual value $\langle \bar{M} \rangle_{\|} = M_{\text{nr}}^2 V H_{\text{ext}} / 3kT$. If $\alpha = 0$, $I_2 / I_0 = \frac{1}{3}$, and we have

$$\langle \bar{M} \rangle_{\|} = \frac{M_{\text{nr}}^2 V H_{\text{ext}}}{3kT} \frac{1}{1 - \frac{M_{\text{nr}}^2 V}{3kT} C_v \text{Ni}} \tag{F.2.25}$$

For this value of I_2 / I_0, from (F.2.7), $C = M_{\text{nr}}^2 V / 3kT$. Introducing this value in (F.2.15), we find, as expected, the same expression as (F.2.25).

We note that an average problem arises by considering only the component of the magnetization parallel to $\mathbf{H}_{\text{ext}}$ for the calculation of the reaction field, while this is not the case when considering all the components of this field. Nevertheless Eqs. (F.2.16) and (F.2.17) can be approximated by

$$\langle \bar{M} \rangle = \frac{M_{\text{nr}}^2 V H_{\text{ext}}}{3kT} \frac{1}{1 - \frac{M_{\text{nr}}^2 V}{3kT} C_v b \text{Ni}} \tag{F.2.26}$$

with $1 \leq b \lesssim 2$ depending mainly on α.

F.2.3.3. Application of the Onsager Model. For this model, it is not clear at present how to take into account the fact that $\langle \mathbf{M} \rangle$ is not parallel to $\mathbf{H}_{\text{eff}}$. We shall consider below only the $\langle \mathbf{M} \rangle$ component parallel to $\mathbf{H}_{\text{ext}}$

for the calculation of the reaction field, in this case along $\mathbf{H}_{\text{ext}}$. Then

$$H_{\text{eff}} = H_{\text{ext}} + \frac{\text{Ni}C_v\chi_{\text{ext}}}{2\text{Ni}C_v\chi_{\text{ext}} + 1} H_{\text{ext}}$$

It is possible to consider a cavity volume slightly smaller than that corresponding to C_v for the evaluation of the effect of the polarizability, in order to try a correction of the defectiveness of the continuum limit hypothesis.[90] Then

$$H_{\text{eff}} = H_{\text{ext}} + \frac{\text{Ni}C_v\chi_{\text{ext}}}{2\text{Ni}pC_v\chi_{\text{ext}} + 1} H_{\text{ext}} \tag{F.2.27}$$

With p near unity and χ_{ext} given by Eqs. (F.2.6) and F.2.7), we find

$$\chi_{\text{ext}} = \frac{C}{T} \frac{-1 + R + \left[(1+R)^2 - \dfrac{4R}{(2p+1)}\right]^{1/2}}{\dfrac{4p}{2p+1} R} \tag{F.2.28}$$

with $R = CC_v\text{Ni}(2p+1)/T$ and C given by Eq. (F.2.7).

The average on ψ of (F.2.28) leads to a very complicated formulation. Fortunately, Eq. (F.2.28) can be approximated in usual cases with an error smaller than 3% by

$$\chi_{\text{ext}} = \frac{C}{T}\left[1 + \frac{R/2p}{1+R}\right] \tag{F.2.29}$$

The average on ψ of Eq. (F.2.29) leads to

$$\begin{aligned}\bar{\chi}_{\text{ext}} = \frac{\langle \bar{M} \rangle}{H_{\text{ext}}} = \frac{M_{\text{nr}}^2 V}{3kT} \Bigg\{ & 1 + \frac{1}{2p} - \frac{1}{2pR'} \\ & \times \left(1 - \frac{1}{\left\{\frac{3}{2}R'\left(3\frac{I_2}{I_0} - 1\right)\left[1 + \frac{3}{2}R'\left(1 - \frac{I_2}{I_0}\right)\right]\right\}^{1/2}}\right) \\ & \times \operatorname{arctg}\left[\frac{\frac{3}{2}R'\left(3\frac{I_2}{I_0} - 1\right)}{1 + \frac{3}{2}R'\left(1 - \frac{I_2}{I_0}\right)}\right]^{1/2} \Bigg\}\end{aligned} \tag{F.2.30}$$

with

$$R' = \frac{M_{\mathrm{nr}}^2 V}{3kT} C_v \mathrm{Ni}(2p+1)$$

This formula can be approximated with an accuracy smaller than 2% by (F.2.31), except for small C_v values.

$$\bar{\chi}_{\mathrm{ext}} = \frac{M_{\mathrm{nr}}^2 V}{3kT}\left(1 + \frac{R'/2p}{1+R'}\right) \tag{F.2.31}$$

For low C_v values, the error coming from Eq. (F.2.31) can be about 10%. A better formula is

$$\bar{\chi}_{\mathrm{ext}} = \frac{M_{\mathrm{nr}}^2 V}{3kT}\left\{1 + \frac{M_{\mathrm{nr}}^2 V}{3kT}\mathrm{Ni}C_v\left[1 + \frac{1}{5}\left(3\frac{I_2}{I_0} - 1\right)^2\right]\right\} \tag{F.2.32}$$

F.2.3.4. H_{ext} vs. H_{app}. Let us consider the evaluation of $\mathbf{H}_{\mathrm{ext}}$ with regard to the applied field $\mathbf{H}_{\mathrm{app}}$. The usual difficulty concerning the sample shape also occurs, the demagnetizing field $\mathbf{H}_d$ being constant through the sample only if the shape is ellipsoidal. However, the usual approximations can be used and if C_v remains (approximately) constant till the sample limits:

$$\mathbf{H}_{\mathrm{ext}} = \mathbf{H}_{\mathrm{app}} - \mathbf{N}_e \langle \bar{\mathbf{M}} \rangle C_v$$

Here, $\mathbf{N}_e$ is the demagnetizing tensor relative to the external shape of the sample.

The angles (a, b) of the $\mathbf{H}_{\mathrm{app}}$ direction can be defined with respect to the ellipsoid axis (Fig. F.2.6). As $\langle \bar{\mathbf{M}} \rangle$ is parallel to $\mathbf{H}_{\mathrm{ext}}$, the susceptibility χ_{app} along $\mathbf{H}_{\mathrm{app}}$ is given by

$$\bar{\chi}_{\mathrm{app}} = \bar{\chi}_{\mathrm{ext}}\left(\frac{\sin^2 a \sin^2 b}{1 + N_{e_x} C_v \bar{\chi}_{\mathrm{ext}}} + \frac{\sin^2 a \cos^2 b}{1 + N_{e_y} C_v \bar{\chi}_{\mathrm{ext}}} + \frac{\cos^2 a}{1 + N_{e_z} C_v \bar{\chi}_{\mathrm{ext}}}\right) \tag{F.2.33}$$

For an ellipsoid of revolution and for $a = 0$ or $\pi/2$, we find the usual formula (also valid for a sphere)

$$\frac{1}{\bar{\chi}_{\mathrm{app}}} = \frac{1}{\bar{\chi}_{\mathrm{ext}}} + N_e C_v \tag{F.2.34}$$

where N_e is the demagnetizing factor for the $\mathbf{H}_{\mathrm{app}}$ direction. Equation (F.2.33) allows the χ_{app} determination for any $\mathbf{H}_{\mathrm{app}}$ direction. In par-

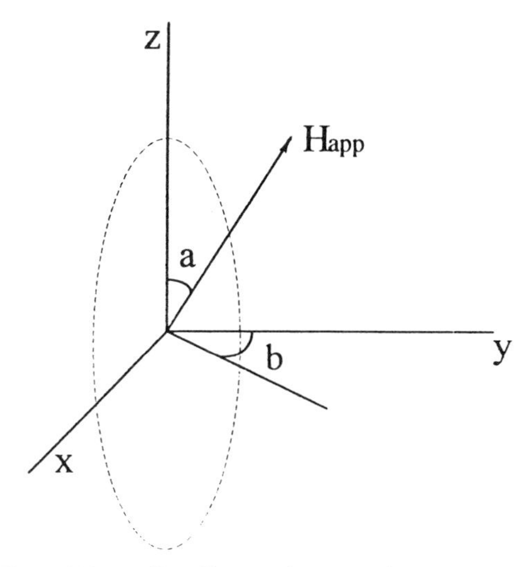

Figure F.2.6. Coordinate axis system for the sample.

ticular, it can be useful for the evaluation of χ_{app} where $\mathbf{H}_{\text{app}}$ is slightly disoriented with regard to $a = 0$ or $\pi/2$.

F.2.3.5. Final Formulas and Outcomes. We can now calculate $\bar{\chi}_{\text{app}}$ in the Onsager model. From the approximate formula (F.2.31), we find

$$\bar{\chi}_{\text{app}} = \frac{M_{\text{nr}}^2 V}{3kT} \frac{1}{1 + \dfrac{M_{\text{nr}}^2 V}{3kT} C_v \left(N_e - \dfrac{2p+1}{2p} \text{Ni} \right) + \dfrac{R'^2}{2p} \dfrac{1}{\dfrac{2p}{2p+1} + R'}} \tag{F.2.35}$$

For small C_v values, from (F.2.32)

$$\bar{\chi}_{\text{app}} = \frac{M_{\text{nr}}^2 V}{3k} \frac{1}{T + \dfrac{M_{\text{nr}}^2 V}{3k} C_v \left\{ N_e - \text{Ni} \left[1 + \dfrac{1}{5} \left(3 \dfrac{I_2}{I_0} - 1 \right)^2 \right] \right\}} \tag{F.2.36}$$

In the Lorentz model

$$\bar{\chi}_{\text{app}} = \frac{M_{\text{nr}}^2 V}{3k} \frac{1}{T + \dfrac{M_{\text{nr}}^2 V}{3k} C_v(N_e - \text{Ni})} \tag{F.2.37}$$

In all these calculations, we have not considered the presence of a volume distribution. However, for a precise adjustment, it has to be accounted for. In this case

$$\frac{1}{\langle \bar{\chi}_{\text{app}} \rangle_V} = \frac{1}{\langle \bar{\chi}_{\text{ext}} \rangle_V} + N_e C_v \tag{F.2.38}$$

and the volume average calculation is to be performed on Eqs. (F.2.31) or (F.2.32), by taking into account a possible M_{nr} dependence on V due to the surface magnetic disorders. At this step, it is important to recall that for the V average calculation, χ must be weighted by V. It is difficult to establish simple formulas for $\langle \bar{\chi}_{\text{app}} \rangle_V$ except for small C_v values where Eqs. (F.2.36) and (F.2.37) remain valid in a first approximation with $\langle M_{\text{nr}}^2 V^2 \rangle / \langle V \rangle$ replacing $M_{\text{nr}}^2 V$. We can see that $\langle \bar{\chi}_{\text{app}} \rangle_V$ depends mainly on $\langle M_{\text{nr}}^2 V^2 \rangle / \langle V \rangle$.

We examine now the outcomes of Eqs. (F.2.35)–(F.2.37). Equation (F.2.37) can be rewritten as

$$\frac{1}{\bar{\chi}_{\text{app}}} = \frac{3kT}{M_{\text{nr}}^2 V} + C_v(N_e - \text{Ni})$$

Or, because M_{nr} is temperature dependent,

$$\frac{1}{\bar{\chi}_{\text{app}}} = \frac{3k}{M_{\text{nr}}^2(0) V} \left[T \left(\frac{M_{\text{nr}}(0)}{M_{\text{nr}}(T)} \right)^2 \right] + C_v(N_e - \text{Ni}) \tag{F.2.39}$$

A straight line is observed for the thermal variation of $1/\bar{\chi}_{\text{app}}$ only if the variable $T' = T[M_{\text{nr}}(0)/M_{\text{nr}}(T)]^2$ is used. By using the usual writing of $\chi_{\text{app}} = C_{\text{sp}}/(T - \theta_{\text{sp}})$, we deduce,

$$\theta_{\text{sp}} = -C_v(N_c - \text{Ni}) \frac{M_{\text{nr}}^2(0) V}{3k} \tag{F.2.40}$$

In the Onsager model, a similar formula is obtained for small C_v values [we note that Eq. (F.2.32) is not valid for low T], but in the general case, a straight line is not obtained even if the T' variable is used. In Figure

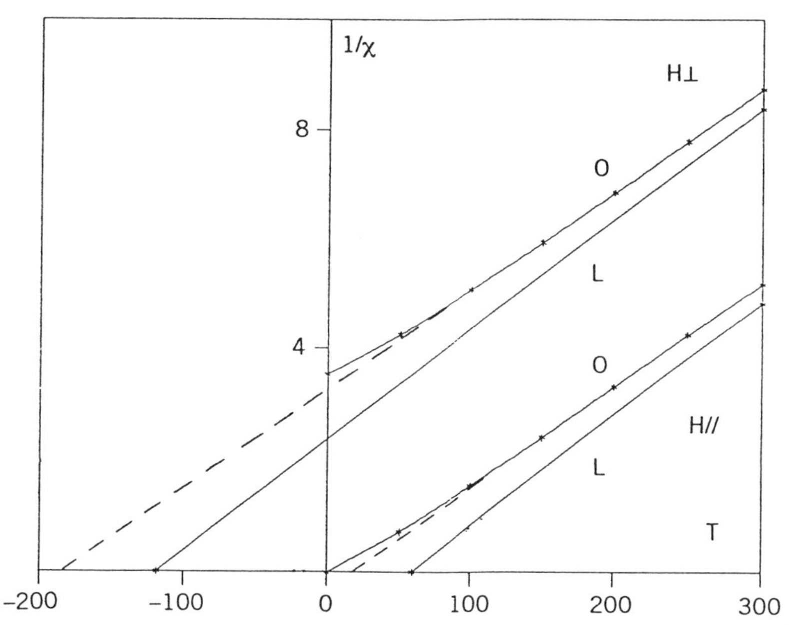

Figure F.2.7. $1/\chi$ variations in the Lorentz and Onsager models for demagnetizing factor $N = 0$ ($H_{\|}$) and 4π ($\hat{H}_{\perp}$) (see text). (Reproduced with permission from Ref. 90.)

F.2.7 the variation of $1/\bar{\chi}_{\text{app}}$ is represented for the two models for $N_e = 0$ and 4π, $M^2_{\text{nr}}V/3k = 50$, $C_v = 0.28$, and $p = 1$ for the Onsager model.

Two comments can be made. First, the Lorentz model leads to a true thermodynamical transition if $N_e < \text{Ni} = 4\pi/3$ at $T = \theta_{\text{sp}}$, which has never been observed. Indeed, when the models have been applied to paramagnetism, the question of the existence of such a transition was debated. No definitive conclusion was drawn because experimental evidences were lacking, mainly due to the difficulties in finding a material that was paramagnetic for $T < \theta_p$ and to very small χ values. Recently, it has been shown that the Lorentz model does not allow a good adjustment of the variation of χ vs. T for ferrocolloids.[161] From the results obtained (see below) with mastered parameters, that is, when the $\mathbf{H}_{\text{app}}$ direction is well defined with regard to the sample and when the demagnetizing field relative to the external shape is taken into account, the Lorentz model can also be ruled out for fine particles. This concludes the debate. Indeed, this result is not surprising as the Lorentz model neglects the medium polarizability, which is too rough an approximation.

Second, in the Onsager model, $1/\bar{\chi}_{\text{app}}$ shows a curvature (Fig. F.2.7). However, taking into account the temperature range where $\bar{\chi}_{\text{app}}$ at the thermodynamic equilibrium is measured, that is, T higher than T_B relative to the large volumes of the distribution with generally T_B smaller than about 100°C, this curvature is weak. For C_v not small, explicit formulation for C_{sp} cannot be derived and (F.2.38) has to be used.

F.2.3.6. Comparison with Experimental Results. The experimental results for which the measurement parameters are controlled, which permits only the derivation of the various parameters included in the formulas and a checking of the models, were until now very limited. They concern Fe particles in Al_2O_3 matrix[89,90] and γ-Fe_2O_3 particles[89,90,135] We show an example in Figure F.2.8 for γ-Fe_2O_3 particles. We can see from the figure that a weak curvature is observed for $1/\chi$ when $\mathbf{H}_{app}$ is parallel to the sample plane. It is also clear that the value of the intercept (the θ_{sp} value) of the quasi-linear part of $1/\chi$ with the temperature axis is strongly dependent of the thermal correction $M_{nr}^2(T)/M_{nr}^2(0)$. That raises the problem of the determination of this variable. In fact, it is only directly measurable from neutron diffraction experiments (see Section F.8). It can be deduced from magnetization under high-field experiment (see Section F.3) on the condition that M_{nr} does not vary much with H_{app}. If this is not fulfilled, only approximate values are obtained, which leads to difficulties for a precise determination of the intercept. We shall discuss the $M_{nr}^2(T)/M_{nr}^2(0)$ determination in Section F.3.

At low field, a good agreement is found for the difference $\theta_{sp\perp} - \theta_{sp\parallel}$, where $\theta_{sp\perp}$ and $\theta_{sp\parallel}$ correspond to the θ_{sp} values for the field perpendicular and parallel, respectively, to the sample plane, which means that

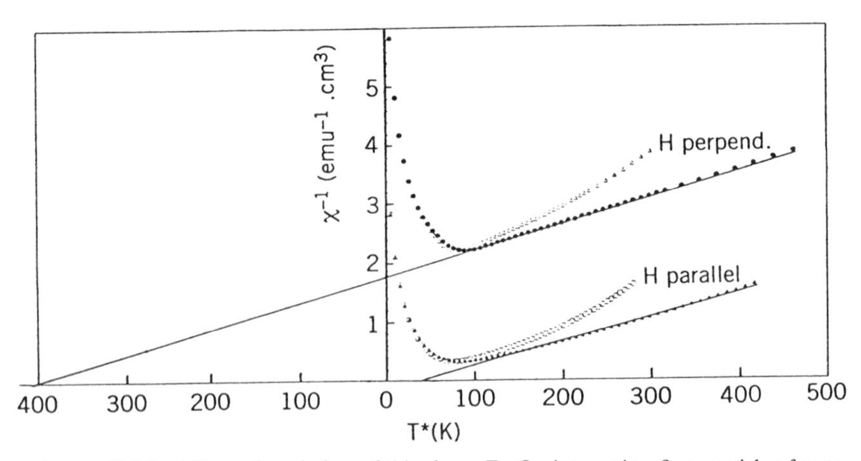

Figure F.2.8. Thermal variation of $1/\chi$ for γ-Fe_2O_3 interacting fine particles for an applied field $H = 5$ Oe perpendicular or parallel to the surface sample.

Eq. (F.2.34) is correct. A good agreement is also found for the two θ_{sp} values for high C_v values while for low C_v values some discrepancies occur. However, it is difficult at the present to know if these shifts are due to the lack of precision on the thermal correction or to other effects ignored in the model. We note that the calculation of Ref. 144 leads for $C_v = 0$ to a negative θ_{sp} value in particular cases.

Now, we examine some results obtained on other fine particles, but only qualitative trends can be deduced as the experiment parameters are not given. For Fe_3O_4 ferrofluids with a weak C_v value,[162] a negative θ_{sp} value is obtained. The experiment has perhaps been performed with H_{app} perpendicular, but $|\theta_{sp}|$ seems too high with regard to the above model. This result follows the same trend as those obtained for γ-Fe_2O_3 particles with weak C_v values (see above). For other Fe_3O_4 ferrofluids[152] with various C_v values, the authors indicate that θ_{sp} is negative and that $|\theta_{sp}|$ increases with C_v, which is in agreement with the model for H_{app} perpendicular. For Ni in SiO_2 matrix[149] and Fe in SiO_2 matrix[163] with medium and high C_v values, θ_{sp} is equal to $+20$ to $+40K$, in agreement with the model when H_{app} is parallel.

Finally, a last property has been evidenced very recently.[135] The magnetization in the superparamagnetic state has been studied vs. H_{app} on γ-Fe_2O_3 particles in a polymer. The effect of H_{app} is the same for samples with $C_v = 0.008$ and $C_v = 0.20$, independently of the H_{app} direction, which means that it does not depend on the interparticle interactions and is only related to the properties of a single particle. A small increase in C_{sp} vs. H_{app} is observed while a strong shift of θ_{sp} toward negative value (Fig. F.2.9) approximately proportional to H_{app} is observed. At the present no definite explanation is proposed.

F.2.4. Effect of the Interparticle Interactions and Conclusion

We have pointed out several times that the interparticle interactions considerably modify the properties. In accounting for the interactions, the following method can be used. Under field, a certain part of **m** is blocked in average, the other part relaxing. The model described in Section E.3.3 can be applied for the latter part, but it acts only on this part, thus reducing the interaction effects (see Section E.3.4). The former part has to be determined, taking into account the difference between H_{eff} and H_{app} as discussed above. The calculation is not simple because the two parts are interdependent and it remains to do.

In conclusion we want to underline that the magnetization in the superparamagnetic state depends strongly on different parameters such as the shape of the sample, the direction of H_{app}, the volumic concentration,

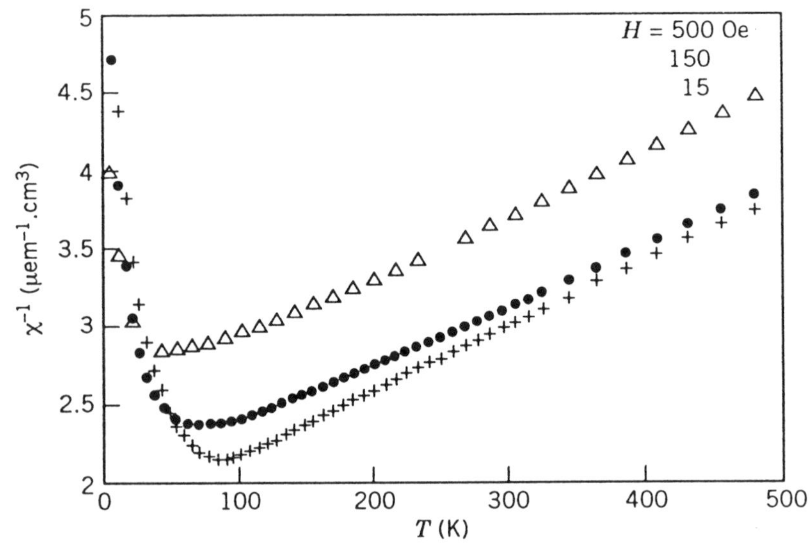

Figure F.2.9. Inverse susceptibility of γ-Fe_2O_3 fine particles for various applied fields parallel to the surface sample. (Reproduced with permission from Ref. 135.)

and the H_{app} value. These parameters also have a more and less strong influence on M_{ZFC} and M_{FC} values at low temperature. Any interpretation must take into account that these parameters have to be controlled.

F.3. Magnetization under Moderate and High Applied Field—Determination of the Nonrelaxing Magnetization

In this section we discuss the magnetization results under moderate as well as high applied field when the strength of the field leads to only one minimum for the anisotropy energy. In this case the longitudinal relaxation is suppressed, and only the transverse relaxation, due to vibrations of the magnetic moment **m** in the potential well, remains. The results are stated by the value of the nonrelaxing magnetization M_{nr} of the particle equal to $|\mathbf{m}|/V$, which depends on the temperature and the applied field. The magnetization M_{nr} is an important parameter, as on one hand, it is directly related to the magnetic state of the particle beside any relaxation phenomenon, and on the other hand all the results obtained from magnetic measurements will depend on its value.

F.3.1. Nonrelaxing Magnetization Determination

As we have discussed in Section B, the nonrelaxing magnetization M_{nr} differs from the saturation magnetization M_s of the bulk material because of the magnetic irregularities on the particle surface. The difference is not small and the M_{nr} vs. T variation can be very different from the M_s vs. T variation. All the experimental results depend on the M_{nr} value directly or through the preexponential factor τ_0 (see Section D). It is therefore very important to determine $M_{nr}(T)$ independently. However, this implies some difficulties. In fact, in principle any experiment whose results depend on M_{nr} could be used on the condition that the other parameters are known and a correct model is applied. However, generally some parameters need to be determined, and the model needs to be checked. In our opinion, three types of experiments allow the M_{nr} determination in particular conditions: small angle neutron scattering experiments (see Section F.8), magnetization measurements under moderate and high applied fields (the results will be discussed below), and field-cooled magnetization (M_{FC}) measurements under sufficiently high field.

For M_{FC} measurements, a sufficiently large field H_{app} has to be used in order to keep only one minimum for the anisotropy energy. For uniaxial symmetry $H_{app} > 2K_t/M_{nr}(H_{app}, T)$ (see Section D) where K_t is the anisotropy constant including all the anisotropy contributions (see Section C). Under field the result is the same as for magnetization measurements, which will be described below. If the field is removed below a certain temperature T_{cri} such that T_{cri} is smaller than the blocking temperature T_B of the smallest particles for which the population is appreciable, the relaxation of the remanent magnetization is very weak. In this case, $M_{nr}(T)$ can be deduced for $T < T_{cri}$ provided the anisotropy symmetry is known. This needs experiments at very low temperature if the V distribution includes an appreciable population of very small particles having low T_B. In any case, the determination of M_{nr} will be possible only in a limited temperature range. Moreover, the results cannot be directly related to M_{nr} if interparticle interactions lead to a collective magnetic state at low temperature (see Section E).

F.3.2. Magnetization under Moderate and High Applied Fields

For an applied field H_{app} higher than $2K_t/M_{nr}(H_{app}, T)$, in the case of uniaxial symmetry, only one minimum remains for the anisotropy energy, near the H_{app} direction. The longitudinal relaxation is suppressed and only vibrations in the potential well are possible.

A rough approximation of the resulting magnetization can be obtained by supposing the anisotropy negligible. In this case,

$$\langle M \rangle \simeq M_{\mathrm{nr}} \mathscr{L}\left(\frac{V M_{\mathrm{nr}} H_{\mathrm{app}}}{kT}\right) \tag{F.3.1}$$

where V is the particle volume and $\mathscr{L}$ is the Langevin function, which reduces for high-field H_{app} values to

$$\langle M \rangle \simeq M_{\mathrm{nr}} - \frac{kT}{V} \frac{1}{H_{\mathrm{app}}} \tag{F.3.2}$$

In fact M_{nr} depends on T, but also on H_{app} and V because spin disorders occur at the particle surface, which represent an important part of the particle due to its small size. $\langle M \rangle$ has to be averaged on V and the measured magnetization is given by

$$\langle \bar{M} \rangle \simeq \frac{\overline{M_{\mathrm{nr}} V}}{\bar{V}} - \frac{kT}{\bar{V}} \frac{1}{H_{\mathrm{app}}} \tag{F.3.3}$$

A better expression is obtained by taking into account the anisotropy and by applying the Boltzmann statistics.

In the usual axis system for uniaxial symmetry where ψ is the angle between $\mathbf{H}_{\mathrm{app}}$ and the easy axis, $\langle M \rangle$ is given by

$$\langle M \rangle = \frac{M_{\mathrm{nr}}}{Z} \int_0^{\pi} \int_0^{2\pi} \exp[-\alpha \sin^2\theta + \beta(\cos\theta \cos\psi + \sin\theta \sin\psi \cos\varphi)] \times (\cos\theta \cos\psi + \sin\theta \sin\psi \cos\varphi) \sin\theta \, d\theta \, d\varphi \tag{F.3.4}$$

with

$$z = \int_0^{\pi} \int_0^{2\pi} \exp[-\alpha \sin^2\theta + \beta(\cos\theta \cos\psi + \sin\theta \sin\psi \cos\varphi)] \sin\theta \, d\theta \, d\varphi$$

and

$$\alpha = \frac{K_t V}{kT} \qquad \beta = \frac{M_{\mathrm{nr}} V H_{\mathrm{app}}}{kT}$$

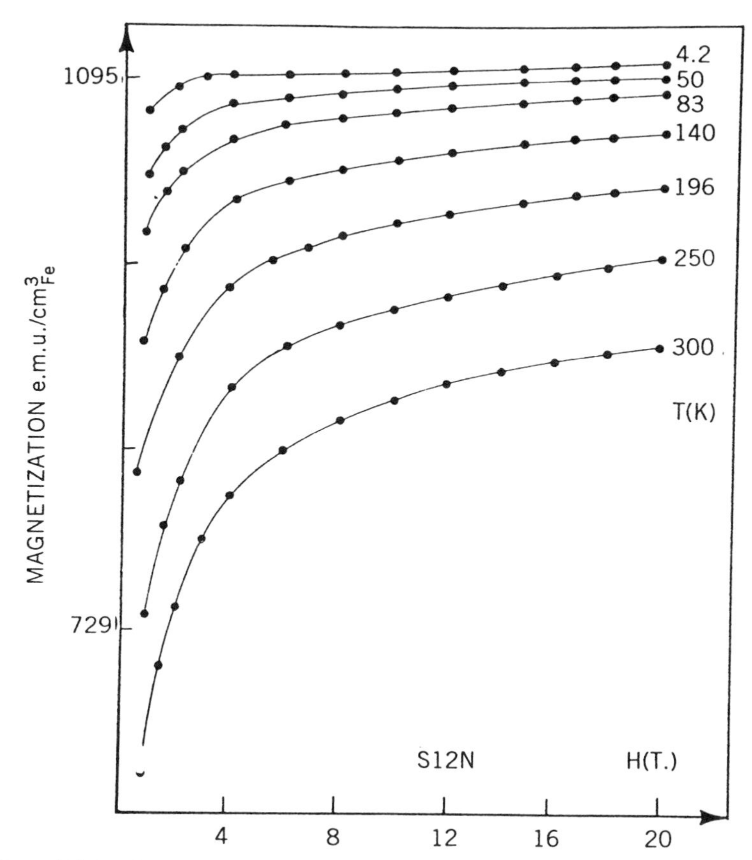

Figure F.3.1. Experimental and calculated magnetization values vs. applied field at various temperatures for Fe fine particles in Al_2O_3 matrix (Fe/Al_2O_3). (Reproduced with permission from Ref. 35.)

By series expansion and integration, the following is obtained[164]:

$$\frac{\langle M \rangle}{M_{nr}} = 1 - \frac{1}{\beta} - [\alpha(\sin^2\psi - 2\cos^2\psi) + 2\alpha^2 \sin^2\psi \cos^2\psi] \frac{1}{\beta^2}$$
$$- [-2\alpha(\sin^2\psi - 2\cos^2\psi) + 2\alpha^2(\sin^4\psi - 12\sin^2\psi\cos^2\psi$$
$$+ 2\cos^4\psi) + 8\alpha^3(\sin^2\psi - \cos^2\psi)\sin^2\psi\cos^2\psi] \frac{1}{\beta^3} \qquad \text{(F.3.5)}$$

High-order $1/H$ terms are indicated in Ref. 164.

For easy directions at random with regard to H_{app} direction, (F.3.5) reduces to

$$\frac{\langle M \rangle}{M_{nr}} = 1 - \frac{1}{\beta} - \frac{4}{15}\frac{\alpha^2}{\beta^2} + \frac{4}{3}\left(1 - \frac{4}{35}\alpha\right)\frac{\alpha^2}{\beta^3} \qquad \text{(F.3.6)}$$

Finally, by averaging on V, the following expression is obtained:

$$\langle \bar{M} \rangle = \frac{\overline{M_{nr}V}}{\bar{V}} - \frac{kT}{\bar{V}}\frac{1}{H_{app}} - \frac{4}{15}\left(\frac{\overline{K_t^2 V}}{M_{nr}}\right)\frac{1}{\bar{V}}\frac{1}{H_{app}^2} + \frac{4}{3}\left[\left(\frac{\overline{K_t^2}}{M_{nr}^2}\right)kT - \frac{4}{35}\left(\frac{\overline{K_t^3 V}}{M_{nr}^2}\right)\right]\frac{1}{\bar{V}}\frac{1}{H_{app}^3} + \cdots \qquad \text{(F.3.7)}$$

The two first terms of (F.3.7) cover the (F.3.3) equation, and the third term corresponds to the well-known effect of disalignment of **m** with regard to $\mathbf{H}_{app}$ due to the anisotropy. The formula has a more extended validity than (F.3.2), being applicable to lower fields. It also allows the K_t determination or at least a check of its value according to the accuracy.

For determining with good accuracy the parameters included in (F.3.7), magnetization results are necessary over an extended range of H_{app} values. Since the formula is valid for $H_{app} > 2K_t/M_{nr}$, corresponding to $H_{app} \approx 0.2$–$0.5T$ for the usual case, it is useful to perform measurements until about $5T$. In such a case, it is necessary to take into account the variation of M_{nr} vs. H_{app}. If it is not negligible and M_{nr} is kept constant, anomalies will be observed in the variation of the second term of (F.3.7). For weak M_{nr} vs. H_{app} variations, the following phenomenological relationship can be used:

$$M_{nr}(T, H_{app}) = M_{nr}(T, 0) + c_1(T)H_{app} + c_2(T)H_{app}^2 \qquad \text{(F.3.8)}$$

The result obtained on fine Fe particles embedded in an Al_2O_3 matrix[35] with $\bar{V}$ about $20\,\text{nm}^3$ are shown in Figure F.3.1 (on page 367). The data have been fitted with Eq. (F.3.7) excluding the $1/H_{app}^3$ term. The variation of M_{nr} vs. H_{app} has been estimated following Eq. (F.3.8). The variation of the second term of (F.3.7) is shown in Figure F.3.2 for the same set of particles labeled S12N and for two other sets labeled S16N and S14N with $\bar{V} \approx 100$ and $200\,\text{nm}^3$, respectively. One can see on Figure F.3.2 that some slight discrepancies occur at low temperature, probably due to the limits of the phenomenological (F.3.8) formula. The deduced $M_{nr}(T, 0)$ values normalized to bulk iron value are represented

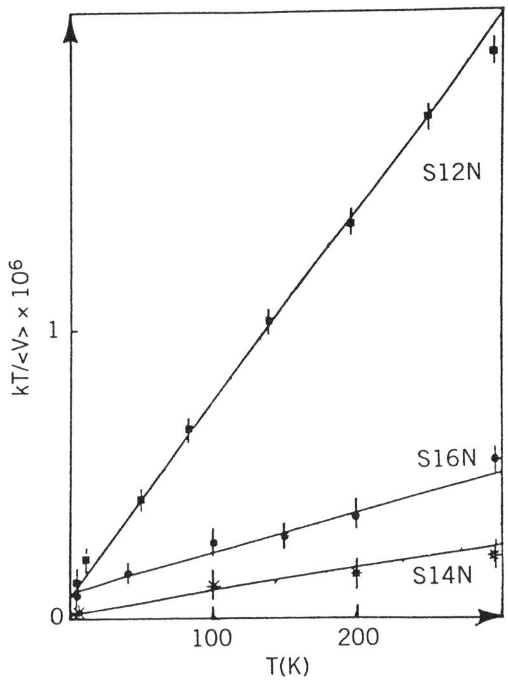

Figure F.3.2. Thermal variation of the $kT/\langle V \rangle$ term deduced from fitting procedure for three Fe/Al_2O_3 samples with different average particle volumes. (Reproduced with permission from Ref. 35.)

in the Figure F.3.3. $M_{nr}(0,0)$ is lower than the bulk value and decreases with $\bar{V}$. The thermal variation of $M_{nr}(T,0)$ is much more pronounced than for bulk, especially with decreasing $\bar{V}$. This is in agreement with what is expected for the influence of surface magnetic disorder (see Section B).

Nevertheless, Eq. (F.3.8) cannot be used for strong variations of M_{nr} vs. H_{app}, which seem effective for certain γ-Fe_2O_3 particles.[96,97] A term in $1/H_{app}$ would be reasonable in the expression of $M_{nr}(T, H_{app})$, but such a term mixes with the second term of (F.3.7). Therefore the $\bar{V}$ determination is not possible. The best way for determining $M_{nr}(T,0)$ is to fix $\bar{V}$, known from other measurements and to derive $M_{nr}(T,H)$ and $M_{nr}(T,0)$ by extrapolation. Of course, the accuracy on $M_{nr}(T,0)$ will depend on the accuracy on $\bar{V}$.

Finally, we note that Pfeiffer has calculated the variation of the magnetization vs. H_{app} for randomly oriented fine particles in the framework of a two-level model.[70] This calculation, valid in this model for

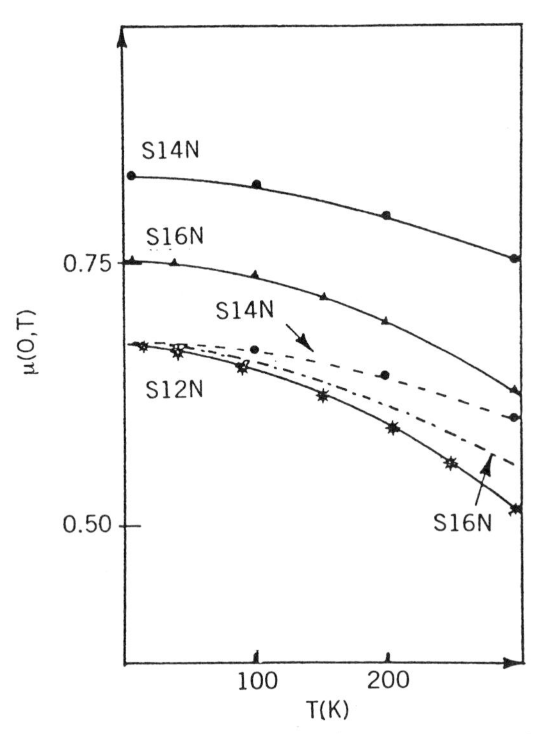

Figure F.3.3. Relative thermal variation of the magnetic moment $\mu(0, T) = \overline{M_{nr}V}$ for three Fe/Al_2O_3 samples (the same as in Fig. F.3.2) normalized to bulk iron (full lines) and for S16N and S14N samples normalized to S12N sample (dotted lines). (Reproduced with permission from Ref. 35.)

any H_{app} values, shows that noticeable deviations of the Langevin function occur as soon as α is not small.

F.4. Remanence Magnetization and Coercive Field

For particulate media used for magnetorecording, the most important parameters are the remanence magnetization (RM) and the switching field, that is, the field allowing the magnetization reversal, related to the coercive field H_c. For such technological reasons a large interest is currently devoted to the measurement of both RM and H_c. For particles, showing appreciable thermal relaxation of the magnetic moment **m**, the interest toward these measurements is mainly from a fundamental point of view.

In this section we will point out some aspects that will allow us to gain better insight into the **m** relaxation. We will first describe the different types of remanence curves; then we will discuss the expected behavior and the experimental results; and finally we will make some comments on coercive field measurements.

F.4.1. Remanence Magnetization Curves

In the case of dc applied field, three different primary remanent magnetization curves can be measured. They are thermoremanent magnetization (TRM), isothermal remanent magnetization (IRM), and dc demagnetization (DcD) curves.[165,166]

The remanence curves are measured as follows:

1. TRM: the field is applied at high enough temperature, that is, higher than the highest blocking temperature T_B corresponding to the biggest particles volume, whose population is appreciable. Then the sample is cooled down to the measuring temperature, at which the field is removed.
2. IRM: the sample is cooled down in zero field from high enough temperature, as for TRM, to the measuring temperature, at which the field is applied and subsequently removed. In this measurement it is necessary that before applying the field a thermodynamic equilibrium state, corresponding to zero magnetization, is actually reached.
3. DcD: the sample is cooled in zero field down to the measuring temperature, at which first the sample is saturated to a remanence $\text{IRM} = \text{IRM}(H = \infty)$, then a field is applied in the opposite direction, and finally it is removed. At a given temperature, in absence of interparticle interactions, DcD is given by

$$\text{DcD}(H) = \text{IRM}(\infty) - 2\text{IRM}(H) \quad \text{(F.4.1)}$$

The effect of interparticle interactions can be evaluated from $\Delta M(H)$,[166–168] defined by

$$\Delta M(H) = \text{DcD}(H) - [\text{IRM}(\infty) - 2\text{IRM}(H)] \quad \text{(F.4.2)}$$

$\Delta M(H)$ is a useful parameter for recording media because magnetic interactions have a close influence on the noise.

In an assembly of nonidentical particles each type of remanence arises because the moments of some particles have been taken over energy barriers, which they cannot overcome again (reversing their orientation) without external assistance from either an applied field or an increase of temperature. Thus the remanence is related to the distribution of energy barriers in the system under investigation:

$$\mathrm{RM} = M_{\mathrm{nr}} \int_{\Delta E_c}^{\infty} f(\Delta E) d\, \Delta E \tag{F.4.3}$$

where ΔE_c is a critical value of the energy barrier, related to the characteristic volume for superparamagnetic relaxation (Fig. F.4.1). At a given temperature and for a given previously applied magnetic field, only particles for which the energy barrier is higher than the critical one, that is, that are in the blocked regime, contribute to the measured remanence.

All the above reported remanences decrease with time. Other remanence curves can be obtained using ac fields, such as the anhysteretic remanence,[165] but they are essentially measured for magnetic recording media.

Recently some equations have been proposed,[169] that relate the ZFC magnetization and the field-cooled magnetization FC through the remanent magnetization values:

$$M_{\mathrm{ZFC}}(H, T, t) = M_{\mathrm{rev}}(H, T) + \mathrm{IRM}(H, T, t) \tag{F.4.4}$$

where M_{rev} represents the part of the magnetization that is time

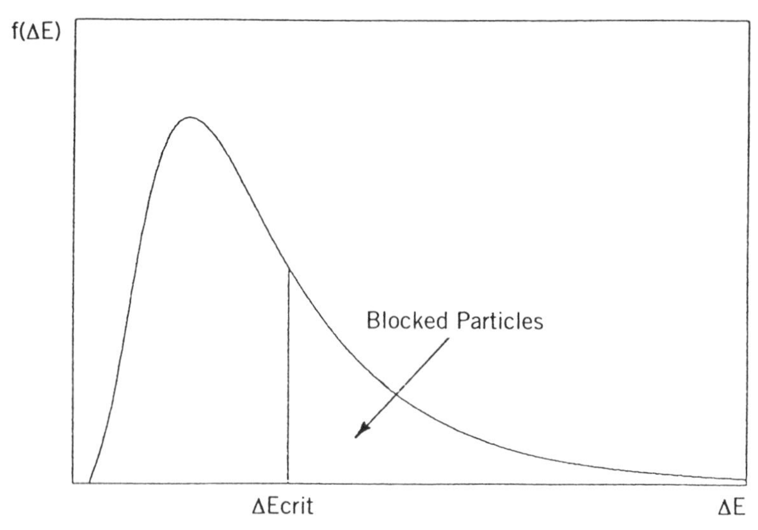

Figure F.4.1. Schematic representation of the energy barrier distribution showing the origin of the remanence. (Reproduced with permission from Ref. 166.)

independent and t is the time of the measurements.

$$M_{\mathrm{FC}}(H, T, t) = M_{\mathrm{ZFC}}(H, T, t) + \mathrm{TRM}(H, T, t) - \mathrm{IRM}(H, T, t) \quad \text{(F.4.5)}$$

Equation (F.4.5) was well checked for a system of fine magnetic particles,[168] but the verification needs a coherent process for the different experiments. In particular it is probable that the usual measuring process for M_{FC}, that is, by decreasing the temperature until the lowest one and then by measuring the magnetization increasing temperature, does not allow such a check. M_{FC} should be instead measured during the field cooling. It is also possible that the interparticle interactions cause small deviations from Eq. (F.4.5), like from Eq. (F.4.1).

F.4.2. Isothermal Remanence Magnetization (IRM) Studies

The studies on IRM have been mainly devoted to the measurement of the IRM values just after removing the field and to its dependence on the field.

IRM values have been predicted by models developed for a system of noninteracting fine particles with aligned easy axes,[170] randomly oriented easy axes,[171] and partially aligned easy axes,[172] with some degree of texture and taking into account the volume distribution, but neglecting the interparticle interactions. Another attempt[173] is based on the simulation method developed in Ref. 174. The cell under consideration consists of 1000 particles on both cubic and tetragonal lattices. Each spherical particle occupies a volume according to a Gaussian distribution, which essentially determines its switching field, and it is randomly assigned to a lattice point. The initial state for IRM is IRM = 0 and for DcD saturation remanence. Positive or negative dc field is applied. Each time the dc field is applied, the probability p_i that the magnetic moment reverses is calculated for a given particle. A random number x is generated, and, if $p_i > x$, the reversal is allowed. The local field is evaluated from applied field and interparticle interactions,[174] In our opinion, the assumptions for interparticle interactions are too simplified (see Section F.2), though refined calculations are used. However, it is difficult to know if the use of a more realistic formulation actually leads to important changes. IRM and DcD curves are derived from the simulation method and analyzed via the Henkel plot.

The time dependence of IRM has been modeled for particles with easy axes aligned parallel to a small field, applied to magnetize the system, and for a random distribution of easy axes. The treatment assumes noninteracting particles and takes into account the volume distribution $P(V)$. The field is applied during a certain time t' and the measurement is

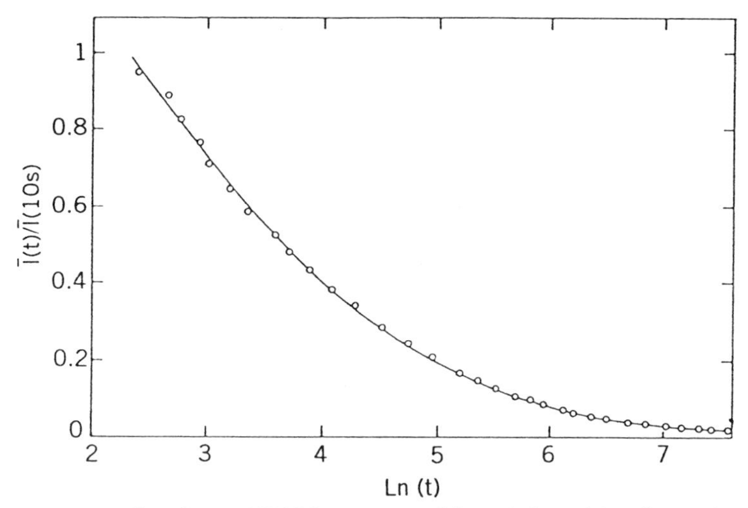

Figure F.4.2. Time decay of IRM for a system of fine cobalt particles: ○ experiment; – theory. (Reproduced with permission from Ref. 175.)

performed at a time t. An excellent agreement is obtained with results obtained for a system of fine cobalt particles (Fig. F.4.2).[175]

Let us briefly discuss an important step of the calculation. The decay of any magnetization (see Section D) was generally written following:

$$M = M_1 + (M_0 - M_1)\exp[-t/\tau(V)] \tag{F.4.6}$$

where M_1 corresponds to the **m** population blocked on the average in the lowest minima at the thermodynamic equilibrium. $M_1 = 0$ in absence of applied field. M_0 corresponds to the magnetization along the field direction at $t = 0$ and τ is the relaxation time depending on the volume V. We have considered the simplest case of uniaxial symmetry and the field along the easy axis.

Due to the volume distribution $P(V)$:

$$\bar{M} = \frac{\int_0^\infty [M_1 + (M_0 - M_1)\exp[-t/\tau(V)]VP(V)\,dV}{\int_0^\infty VP(V)\,dV} \tag{F.4.7}$$

Note that it is necessary to weight the M expression with V,[90] because magnetic moments proportional to V are measured.

For the second term of the integral, one can assume that $\exp[-t/$

$\tau(V)]=1$ and 0 for $t/\tau(V)<1$ and >1, respectively. Then a critical volume V_c is defined such that $\tau(V_c)=t$, and this second term reduces to

$$\frac{\int_{V_c}^{\infty}(M_0-M_1)VP(V)\,dV}{\int_0^{\infty}VP(V)\,dV} \tag{F.4.8}$$

This assumption considerably simplifies the calculation.

More refined approximations and numerical calculations[176,177] have shown that Eq. (F.4.8) is a good approximation as long as the volume distribution is not too narrow. The conditions are given in Ref. 176.

Other authors[178] have modeled the variation of DcD by considering that the easy axes are parallel to the applied field direction. Magnetic interparticle interactions have been introduced through an effective volume V_{eff}, which increases when T decreases. The process is the same as in Ref. 143. In our opinion, the V_{eff} approach is phenomenologically correct, although the exponential variation of V_{eff} vs. T is not really justified. Anyway, verifications are difficult, due to the existence of free parameters. Note that the authors indicate that the logarithmic slope S of the DcD decay reflects the shape of the distribution. This clearly comes from Eq. (F.4.7) where

$$S=\frac{d\bar{M}}{d\ln t}=\frac{\tau(V_c)}{\left(\frac{d\tau}{dV}\right)_{V=V_c}}\frac{[(M_0-M_1)VP(V)]_{V=V_c}}{\int_0^{\infty}VP(V)\,dV} \tag{F.4.9}$$

Several examples of DcD and IRM measurements on various particles can be found in the literature (see, e.g., Fe and Co particles).[179]

F.4.3. Thermoremanence Magnetization (TRM) Studies

F.4.3.1. Field Dependence. First, we will discuss the works devoted to the dependence of TRM values on the previously applied field, just after the removal of the field or a short time after (actually the calculations are valid for any time). A maximum in the TRM vs. H dependence, rather than a continuous increase to saturation (like for IRM), has been reported in the literature[169,180] in analogy with what is observed in spin glasses.[181,182] Two models have been presented, that take into account the TRM variation during the cutting-off of the field. In both cases a volume distribution is assumed, but the existence of a random distribution of easy axes and of interparticle interactions, usually present in real systems, is neglected.

In the first model,[183] the variation of the field value during the cutting-off is linear, but the model assumes that the system relaxes toward zero magnetization, rather than toward the appropriate value of equilibrium magnetization. The calculation leads to a maximum for TRM vs. field.

The second model[184] takes into account the correct values for the relaxed magnetization, and a maximum for TRM vs. field is found, becoming more pronounced with decreasing the width of energy barrier distribution, only if the variation of the field during the cutting off is nonlinear. On the other hand, some experimental results[185] showed a peak in TRM vs. H even when a constant time of field reduction was used.

Recently, two models have been developed accounting for the particle volume distribution and for the existence of interparticle interactions through a mean-field approach and Monte Carlo calculations.[186]

In the first approach, interaction effects are expressed by a field term (H_i) proportional to the magnetization of the system and acting in the same direction of the applied field ($H_t = H_a + N_d H_i$). A negative interaction field is assumed, leading to a decrease of magnetization. Interaction effects lower the saturation value of TRM, without giving rise to a peak, as with this formalism the interacting field at remanence saturates when the remanence is saturated.[186]

In the Monte Carlo approach, where the particles and their axes are generated randomly, the exact behavior of the interaction field is calculated. A peak in TRM is obtained, as a consequence of the interaction field (it is assumed to be negative and increasing with increasing field, but it does not saturate when remanence is saturated).[186]

F.4.3.2. Time Dependence. For a single particle an exponential decay of the remanent magnetization was predicted by Néel[1,187]:

$$\mathrm{TRM}(t) = \mathrm{TRM}(0)\exp(-t/\tau) \tag{F.4.10}$$

where the relaxation time follows an Arrhenius law [$\tau = \tau_0 \exp(E/kT)$], the magnetization reversal process occurring by a thermally activated mechanism. For an assembly of particles with volume distribution, implying a distribution of energy barriers, the time dependence comes out from the integration over the relaxation time distribution function:

$$\bar{M} = \frac{1}{Z} \int_0^\infty M(0) \exp[-t/\tau(V)] V P(V)\, dV \tag{F.4.11}$$

with

$$Z = \int_0^\infty V P(V)\, dV$$

The TRM relaxation data have been very often analyzed in terms of a logarithmic law[188–190]:

$$\mathrm{TRM}(t) = C - S \ln(t/t_0) \tag{F.4.12}$$

However, there are no physical reasons for using the above relationship. Moreover, the logarithmic dependence is observed only when the relaxation measurements are performed in a narrow time interval. In the few measurements extending over some decades of times, clear deviations from a straight line are observed in the plot of the TRM vs. $\ln(t)$,[127,191] depending on the type of energy barrier distribution, on the magnetic field value, and on the strength of interparticle interactions.

Time decay laws were derived for some specific relaxation time distribution functions. As an example for a gamma distribution function describing the relaxation time probability $P(\tau)$[20,44]:

$$P(\tau) = \frac{1}{\tau_0 \Gamma(p)} \left(\frac{\tau}{\tau_0}\right)^{p-1} \exp(-\tau/\tau_0) \tag{F.4.13}$$

The time dependence of TRM should be

$$\frac{M_r(t)}{M_r(0)} = \frac{2}{\Gamma(p)} \left(\frac{t}{\tau_0}\right)^{p/2} K_p\left[2\left(\frac{t}{\tau_0}\right)^{1/2}\right] \tag{F.4.14}$$

where K_p is the modified Bessel function of the third type.

This law allowed to fit satisfactorily relaxation data on a spin glass system,[192] but has not been applied to fine particles so far.

Khater et al.[193] calculated the time dependence of TRM for an assembly of noninteracting uniaxial particles with a Poisson volume distribution function $[P(V)\, dV = (4V/V_0)\exp(-2V/V_0)\, dV]$. Transforming the volume distribution in a distribution over relaxation times τ, described by an Arrhenius law $\tau = \tau_0 \exp(KV/kT)$, it follows that

$$P(\tau/\tau_0)d\ln(\tau/\tau_0) = \beta^2 \ln(\tau/\tau_0)\exp[-\beta \ln(\tau/\tau_0)]d\ln(\tau/\tau_0) \quad \text{(F.4.15)}$$

where $\beta = 2T/T_0$ with $T_0 = KV_0/k$.

For a time regime $1 \ll \tau/\tau_0 \ll \tau_m/\tau_0$, which can be experimentally probed, the thermoremanent magnetization can be expressed in a dimensionless form as follows:

$$\mathrm{TRM}(t) = \{[1 + \beta \ln(t/\tau_0) + \tfrac{1}{2}[\beta \ln(t/\tau_0)]^2 - (\beta/2)[1 + \beta \ln(t/\tau)]\}(t/\tau_0)^{-\beta} \quad \text{(F.4.16)}$$

The formula is the product of a polynomial in $\ln(t/\tau_0)$ and a power law $(t/\tau_0)^{-\beta}$. The detailed variation of TRM(t) depends on the values of T_0 and τ_0 characterizing the material, and it is expected to vary with temperature.

The results of numerical calculations of Eq. (F.4.16) are reported in Figure F.4.3 for $\beta = 0.08$, corresponding to low temperatures and for 0.20, corresponding to higher temperatures. At low temperatures TRM decays as $\ln t$, while at higher temperatures the decay deviates from the logarithmic law toward a more rapid decay. Intermediate β values determine a gradual transition from one type of decay to the other.

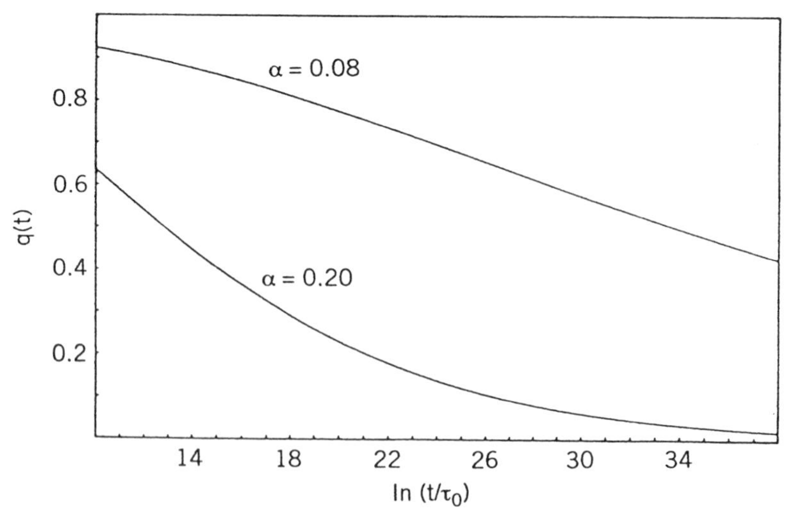

Figure F.4.3. Time decay of TRM[$q(t)$] for two different values of the β coefficient in Eq. (F.4.16). (Reproduced with permission from Ref. 193.)

The calculated remanent magnetization compares favorably with some experimental results reported on spin glasses,[182] described in terms of a clusters–fine particles model. However, as far as we know, the Khater model has not been applied to fine particles materials. Assumed distribution function is not realistic, as well as in most cases the absence of interparticle interactions.

Labarta et al.[176] and Iglesias et al.[194] proposed a scaling procedure allowing to extend the actual interval of relaxation measurements to much longer times. The experimental relaxation data on fine particles samples (e.g., Fe_3O_4 and FeC particles dispersed in a hydrocarbon oil) were found to scale with the variable $T \ln(t/\tau_0)$, as previously observed in spin glasses,[112,195] selecting an attempt frequency $1/\tau_0$ that brings all the curves onto one master curve. The same scaling behavior was found by Vincent et al.[196] on γ-Fe_2O_3 particles dispersed in a polymer. The $T \ln(t/\tau_0)$ scaling gives evidence that the reversal of the particles magnetic moments through the anisotropy barriers is governed by thermally activated dynamics. The scaling comes indeed from the fact that most of the magnetization change at temperature T and time t is due to crossing of barriers of order of $U_t = kT \ln(t/\tau_0)$. However, the scaling implies the absence of interparticle interactions and that the volume distribution is large. In this case, the scaling holds only for narrow temperature ranges. In Ref. 194 the experimental master curves were fitted to the theoretical one, taking the energy barrier distribution function to be a sum of two lognormal distribution functions (Fig. F.4.4).

Sanchez et al.[197] presented a computer simulation of the magnetic relaxation of an assembly of nonidentical, noninteracting, single-domain particles with uniaxial anisotropy. They analyzed the dependence of the magnetic viscosity $(dM/d \ln t)$ on the shape of the energy barrier distribution function, assumed to be lognormal in type.

Sampaio et al.[198] developed a model for interpreting magnetic viscosity $S(H, T)$ experiments at low temperature performed on small particles of Ba-ferrite. Their model, taking into account both particle size and switching field distribution, describes the experimental low-temperature dependence, $S(H, T) \propto T^{1/2}$, and predicts the observed scaling behavior on field and on temperature.

All the above-mentioned models do not account for the existence of interparticle interactions. Recently, Tronc et al.,[138] Fiorani et al.,[136] and Dormann and Fiorani[90] proposed a model for describing the time decay of TRM in a series of γ-Fe_2O_3 particles dispersed in a polymer, with different interparticle distances and then with different interactions strengths. The model accounts for the particle volume distribution as well as for the existence of interparticle interactions.

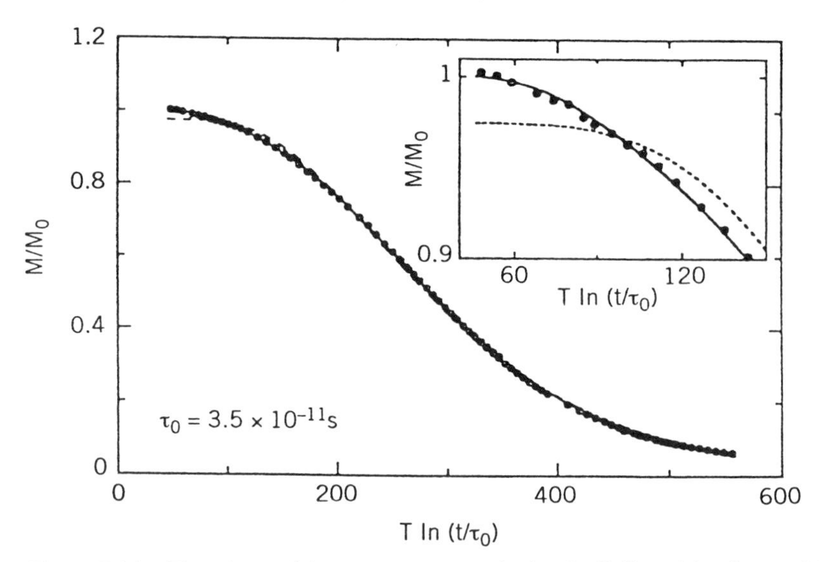

Figure F.4.4. Time decay of the remanent magnetization for FeC particles dispersed in a hydrocarbon oil in the temperature range 1.8–18 K. Dashed and solid lines correspond to the fits with one or two lognormal distributions. (Reproduced with permission from Ref. 194.)

Following Eq. (F.4.11), the time decay of TRM is given by

$$\mathrm{TRM}(t) = \frac{1}{Z}\int_0^\infty \Delta M_{\mathrm{FC}}(V)\exp(-t/\tau)VP(V)\,dV \tag{F.4.17}$$

with

$$Z = \int_0^\infty VP(V)\,dV$$

where ΔM_{FC} is the value of the magnetization immediately after switching off the field, at the end of the FC process.

Weighting the expression with V, because the effect of each particle is proportional to V, and with the usual approximation of $\exp(-t/\tau)$ by a step function (see above), the TRM(t) expression becomes

$$\mathrm{TRM}(t) = \frac{1}{Z}\int_{V_c}^\infty \Delta M_{\mathrm{FC}}(V)VP(V)\,dV \tag{F.4.18}$$

The relaxation rate is then given by

$$S = \frac{d\,\mathrm{TRM}(t)}{d\ln t} = -\frac{1}{Z}\left[\frac{kT}{K}\Delta M_{\mathrm{FC}}(V)VP(V)\right]_{V=V_c} \tag{F.4.19}$$

where $V_c = (kT/K)\ln(t/\tau_0)$ and K is the total anisotropy constant.

From Eq. (F.4.18) it comes out that TRM is scaled with $T\ln(t/\tau_0)$, provided that ΔM_{FC} and K remain constant. However, the ΔM_{FC} values depend on the cooling rate[156] and K can be considered constant only in absence of interparticle interactions.

The interparticle interactions lead to energy barrier modification following two processes (see Section E). The first is due to the part of the magnetic moment blocked on average in the lowest minimum and the second to the relaxing part. The blocked fraction creates a static interacting field that can be calculated from the Onsager model and that depends on the shape of the sample (see Section F.2). This field modifies the energy barrier and therefore K. The relaxing fraction of the magnetic moment also interacts, leading to an additional modification of the energy barrier, which can be calculated from the model presented in Section E (taking into account only this part). This leads to an increase of K, but clearly smaller than the one obtained when the entire **m** relaxes.

From Eq. (F.4.19), S is expected to be proportional to T, but this would imply a uniform volume distribution, which is not realistic. Equation (F.4.19) shows that in a first approximation S follows the distribution $V^2P(V)$.

The temperature dependence of the relaxation rate is reported in Figure F.4.5 for γ-Fe_2O_3 particles with different interparticle distances (IF, isolated-far particles; IN, isolated-near particles; FLOC, aggregates

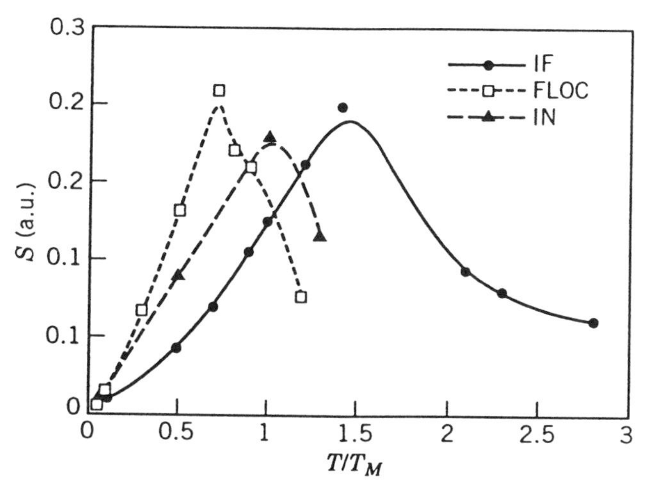

Figure F.4.5. Normalized magnetic viscosity $S = (1/M_0)\,dM/d\ln t$ vs. reduced temperature (T/T_M, where T_M is the temperature of the maximum of the dc low-field susceptibility) for γ-Fe_2O_3 particles dispersed in a polymer (see text for the meaning of IF, FLOC, IN). (Reproduced with permission from Ref. 136.)

of particles). A reduced temperature T/T_M is used, where T_M is the temperature of the low-field dc susceptibility maximum.[136] The results are in agreement with $V^2P(V)$ variation, as predicted by Eq. (F.4.18). However, the three curves do not superimpose, although the volume distribution $VP(V)$ is the same for the three samples. In fact, by scaling the temperature with T_M, we have taken into account the dynamical interactions as if they fully act. If now a reduced temperature T/T_{eff} is used, with T_{eff} varying slightly according to the samples (it is not the case for T_M), the three curves superimpose.[138] This is in agreement with the model described above. In addition, this latter scaling proves that the dynamics is of the same type of behavior for all dispersion states and excludes a low-temperature ordering between particle moments for the most aggregated particles.

The effect of aging on TRM was investigated on Fe-Al_2O_3 particles.[123,127] The "age" of the system is defined as the total time (t_a) elapsed since the sample was quenched in the blocked state. In Refs. 123 and 127 the sample was cooled down in a field to the measuring temperature; then the field was maintained applied for different times (waiting times: t_w) before removing it, and finally the time decay of TRM was recorded after each cooling sequence. The results show that M_{FC} (before removing the field) and TRM are sensitive to the previous magnetothermal history, their absolute values changing with t_w and with the cooling rate. The effect of the cooling rate on M_{FC} was theoretically predicted by Chantrell and Wohlfarth.[156] On the other hand the relaxation rate was found to be independent of t_w, unlike in spin glass systems where the dynamics slow down with increasing time spent in the low-temperature frozen phase ($t_a = t + t_w$).[112,199] In spin glasses the existence of a hierarchical organization of many energy valleys in the phase space accounts for the observed results, which were interpreted as a slow evolution of the system from a local energy minimum, in which it was trapped after the cooling down, toward the equilibrium probability distribution among the different valleys. This is not the case for an assembly of interacting fine particles with a volume distribution, although it is also characterized by a distribution of relaxation times and then by a distribution of energy minima. However, for noninteracting particles, the minima are independent. For interacting particles, the minima are not strictly independent but do not show a hierarchical organization like in spin glasses, except perhaps at very low temperature (see Section E).

Another relaxation experiment has been perfomed on a frozen ferrofluid with F_3O_4 fine particles.[200] In this case, the sample is cooled in zero field to the measurement temperature T_m, equilibrated for a waiting time t_w and then probed by applying a field of 3 Oe. The results show that the relaxation rate $S = dM/d \ln t$ depends on t_w for $T = 12.5$ K while for $T = 20$ K, corresponding to the χ_{dc} maximum, no variation is detected. In fact, at T_m before the application of the field, the particle system is not at the thermodynamic equililbrium due to the interparticle interactions. Decreasing the temperature, particles **m** progressively block according to their volume V. The blocked **m** create dipolar fields that modify the energy barriers of the neighboring particles and that change when the temperature is decreased, because new **m** are blocking. Of course, the average value of this field is equal to zero. Then, due to the finite value of the cooling rate, thermodynamic equilibrium is not exactly reached. The effect will be more especially pronounced as the blocked particle population is high. In addition, as in relaxation experiments, a narrow V range is experienced, the effect will also be more especially important as the particle population is large. These trends seem in agreement with the results cited above, but further experiments are necessary. We remark that the experiment described above remains to model.

F.4.4. Coercive Field

The coercive field H_c is very important in particulate media used for magnetorecording, as it is a measure of the switching field, that is, the field necessary for reversing the magnetization. At low temperature, where all the particle moments **m** are blocked. H_c is equal to the value expected for monodomains. In uniaxial symmetry $H_c = H_{c0} = (\frac{1}{2})H_k$ with $H_k = 2K/M_{nr}$, where K is the anisotropy constant corresponding to all contributions (see Section C) and M_{nr} is the nonrelaxing magnetization. At high temperature, when all **m** fluctuate with a relaxation time smaller than the measuring time, $H_c = 0$. For intermediate temperature, H_c can be evaluated from the formula[201]:

$$H_c = H_{c0}[(1 - \sqrt{(V_p/V)}] = H_{c0}[(1 - \sqrt{(T/T_B)}] \qquad \text{(F.4.20)}$$

where T_B is the blocking temperature of the particle corresponding to τ_m and V_p is the volume corresponding to T_B equal to the considered temperature. In the case of random distribution of easy axes[202,203] the following formula has been given:

$$H_c \cong H_{c0}(1 - V_p/V) \qquad \text{(F.4.21)}$$

Numerical calculations have been performed taking into account the V distribution according to a lognormal law and with easy axes in random positions.[204] This calculation shows that there is a strong dependence on standard deviation σ at large values of V/V_p. Note that the results for $\sigma = 0.4$ cover the formula (F.4.20) with a good approximation.

Finally, the calculation of the full hysteresis loop has been performed with aligned and randomly oriented easy axes.[69] The results show that H_c values for easy axes in random positions are similar to those obtained for the direction of the applied field at an angle of $\pi/4$ with respect to the easy axis. This leads to

$$H_c = H_{c0}\left[1 - \left(\frac{V_p}{V}\right)^{0.77}\right] \tag{F.4.22}$$

A more complicated formula is obtained in the case of a volume distribution.[69]

Several examples of H_c measurements can be found in the literature (see, e.g., studies on Fe–Co particles).[205,206]

F.4.5. Conclusions

The measurement of time decay of the remanent magnetization represents one of the most straightforward tools to investigate the dynamical behavior of fine particles and to study the magnetization reversal mechanisms. However, the interpretation of the experimental results is very difficult because of the complexity of actual fine particles systems (presence of size, shape, and interparticle distance distribution, random distribution of easy axes, existence of interparticle interactions, surface effects, etc.).

Remanence curves are very sensitive to the effects of interparticle interactions. Nevertheless, most of the theoretical models proposed so far do not take into account the existence of interparticle interactions, very often present in real samples, and their contribution to the total anisotropy energy. Therefore, great efforts are still necessary from the theoretical point of view, in order to interpret the remanence curves of actual assemblies of fine particles. As far as the study of the single-particle behavior is concerned, representing a necessary starting point for the comprehension of the behavior of complex systems, the recent development of micro SQUID susceptometers,[207] allowing to measure the magnetization of isolated particles, is expected to play an essential role for the knowledge of magnetization reversal mechanisms.

F.5. AC Susceptibility

Alternative susceptibility measurements at different frequencies ν (usually in the range 10–10^4 Hz, extensible down to 10^{-4} Hz and up to 10^8 Hz) represent a very useful tool for studying dynamical properties of magnetic nanoparticles, as they have the advantage of covering a large time window with the same technique. Another advantage of the ac technique is that the application of a small ac field, $H = H_0 \exp(i\omega t)$, where ω is the angular frequency ($= 2\pi\nu$), allows the initial susceptibility $\chi_i = (dM/dH)_{H\to 0}$ to be measured. This is determined by the spins able to follow the field variations. Moreover, as the strength of the used ac field is small, the barrier energy is very slightly modified (the variation is proportional to H^2, see Section D) and therefore the relaxation time τ value in the absence of applied field can be used with a good approximation.

In this section we will first discuss the models allowing the fit of the ac susceptibility (χ_{ac}) curves, then describe some results obtained on fine particles systems, and finally try to draw some conclusions.

F.5.1. Models

F.5.1.1. Gittleman Model. According to Gittleman[149] the susceptibility of an assembly of isolated single-domain particles, with a volume distribution, having their easy axes randomly oriented, is given by

$$\chi(T, \omega) = (1/Z) \int_0^\infty \chi_v(T, \omega) V f(V)\, dV \qquad \text{(F.5.1)}$$

with

$$Z = \int_0^\infty V f(V)\, dV$$

where $f(V)$ represents the volume distribution function, χ_v the volume susceptibility and $\chi_v V f(V)\, dV$ is the contribution to the total susceptibility, due to particles with volumes between V and $V + dV$. χ_v is calculated assuming that the ac field is a step function that turns on at time $t = 0$. We note that we have weighted $f(V)$ by V because a magnetic moment is measured.

The contribution of particles with volume V to the total magnetic moment is given by

$$m_v(t) = V H_0[\chi_0 - (\chi_0 - \chi_1) e^{-t/\tau}] \qquad \text{(F.5.2)}$$

where χ_0 is the superparamagnetic susceptibility (for $T > T_B$), corresponding to the thermodynamic equilibrium, given by

$$\chi_0 = M_{nr}^2 V / 3kT \qquad \text{(F.5.3)}$$

and χ_1 is the initial response of particle moments to both external field and anisotropy field in the so-called blocked state ($T < T_B$), given by

$$\chi_1 = (M_{\mathrm{nr}}^2/2K)\langle \sin^2\theta \rangle \tag{F.5.4}$$

where θ is the angle between the applied field and the easy magnetization direction, $\langle \sin^2\theta \rangle = \frac{2}{3}$, the average being made over all the particles.

The Fourier transform of Eq. (F.5.2) gives the complex susceptibility:

$$\chi = \frac{\chi_0 + i\omega\tau\chi_1}{1 + i\omega\tau} \tag{F.5.5}$$

The real part is given by

$$\chi' = \frac{\chi_0 + \omega^2\tau^2\chi_1}{1 + \omega^2\tau^2} \tag{F.5.6}$$

From Eq. (F.5.6) the application of an ac field predicts: (1) $\chi' = \chi_0$ when $\omega\tau \ll 1$ (at high temperature, where $KV \ll kT$, for uniaxial anisotropy: superparamagnetic regime, i.e., the particle moments relax via thermal fluctuations and may be reversed, even many times, during the measuring time τ_m). (2) $\chi' = \chi_1$ when $\omega\tau \gg 1$ (at low temperature, where $KV \gg kT$: blocked regime, i.e., the energy due to the magnetic field is not enough to reverse the particle moment during τ_m).

χ' can be calculated from Eq. (F.5.6) using the τ formula adapted to the anisotropy symmetry (see Section D). By considering a unique volume V, one can determine the temperature $T_{\max}$ of the χ_{ac} maximum. For example, for uniaxial symmetry with $\alpha = KV/kT \geq 3$:

$$\begin{aligned} \frac{KV}{kT_{\max}} &\cong -\ln(\omega\tau_0') + \frac{1}{2}\ln\left[\frac{\ln(\omega\tau_0')}{2\ln(\omega\tau_0') + 4}\right] \\ &\cong -\ln(\omega\tau_0') + \tfrac{1}{2}\ln(\tfrac{1}{2}) \end{aligned} \tag{F.5.7}$$

with

$$\tau_0' = \frac{\sqrt{\pi}}{4}\frac{VM_{\mathrm{nr}}(0)}{K\gamma_0}\left\{\frac{1}{\eta_r} + \eta_r\left[\frac{M_{\mathrm{nr}}(T)}{M_{\mathrm{nr}}(0)}\right]^2\right\}$$

At this step, it is important to emphasize that if for the gyromagnetic ratio the usual value $\gamma_0 \cong 2 \times 10^{-7}\,\mathrm{s}^{-1}\,\mathrm{G}^{-1}$ is used, given in angular frequency, ω has to be replaced by ν.

One can compare $T_{\max}$ to the blocking temperature T_B defined as the temperature at which $\tau = \tau_m$. For χ_{ac} measurements $\tau_m = 1/\nu$, taking into account the γ_0 value given above. Then

$$\frac{KV}{kT_B} \cong -\ln(\nu\tau_0') + (\tfrac{1}{2})\ln[-\ln(\nu\tau_0')] \qquad \text{(F.5.8)}$$

From this formula it comes out that $T_{\max}$ is slightly higher than T_B (about 10% in most cases).

For an assembly of particles with a size distribution, χ' can be calculated from (F.5.1) with χ_v' given by (F.5.6). However, the calculation is a little bit complicated, and we can use approximate formulas. In the case of broad V distribution, one can suppose that, at given temperature and for a certain measuring time, some particles will be in the superparamagnetic state and some of them in the blocked one. Thus the susceptibility is given by the sum of the two contributions:

$$\chi'(T,\nu) = \frac{M_{\mathrm{nr}}^2}{3kTZ}\int_0^{V_p(T,\nu)} V^2 f(V)\,dV + \frac{M_{\mathrm{nr}}^2}{3KZ}\int_{V_p(T,\nu)}^{\infty} Vf(V)\,dV \qquad \text{(F.5.9)}$$

where $V_p(T,\nu) = kT\ln(\nu/\tau_0)$ is the critical volume for superparamagnetic behavior and Z is given by (F.5.1); χ' is expected to show a maximum at a temperature $T_{\max}$ close to $\langle T_B\rangle$.

Actually Gittleman et al.[149] considered Eq. (F.5.6) in two separate parts. They fitted their data using Eq. (F.5.6) derived for the high-temperature limit and for the low-temperature limit separately.

At a temperature $T \gg T_B$ the susceptibility was calculated using $\chi' = (M_{\mathrm{nr}}^2/3kT)Vf(V)\,dV$, whereas at temperature $T \ll T_B$ they expressed the distribution function in terms of the power law, that is, $f(V) \propto V^n$ and the susceptibility was calculated using $\chi_1(1 + AT^{n+1})$, where A is a constant. Their high-temperature analysis gives information about the particle size distribution, while the low-temperature analysis gives an estimate of the effective anisotropy constant value.

Another simple approximation is to consider $\langle T_{\max}\rangle \cong \langle VT_{\max}(V)\rangle / \langle V\rangle$, as a magnetic moment is measured (see Section F.2). From Eq. (F.5.7):

$$T_{\max} \cong (KV/k)[-1/\ln(\nu\tau_0')]$$

Then

$$\langle T_{\max}\rangle \cong (\langle V^2\rangle K/\langle V\rangle k)[-1/\ln(\nu\tau_0')] \qquad \text{(F.5.10)}$$

Therefore, in the presence of a volume distribution, $T_{\max}$ can be identified in first approximation with T_B related to $\langle V^2\rangle/\langle V\rangle$.

The model has some limits, mainly as far as the calculation of χ in the blocked state ($\chi_1 = M_{nr}^2/3K$ for uniaxial symmetry) is concerned. The calculation does not account for vibrations of the magnetic moment **m** in the potential well, that is, for transverse relaxation. This leads to an abrupt variation of χ' close to $\langle T_B \rangle$ (for a single particle the variation should be steplike). Such a very rapid variation is never observed experimentally, neither for samples with a narrow volume distribution. In our opinion it is important to take into account the effect of transverse relaxation, which should smooth the χ' variation below T_B. Moreover, the volume distribution function assumed in the Gittleman calculation is not realistic. Finally, interactions between particles, almost always present in real systems, are neglected.

F.5.1.2 Other Models. Khater et al.[193] calculated the ac susceptibility for independent magnetic clusters (they can be treated as small magnetic particles) with a Poisson volume distribution. They give approximate equations for T_{max} and for T''_{max} (max of χ'') for a frequency range $\nu < 10^5$ Hz.

$$\frac{T_{max}}{T_0} = \frac{1}{2}\left[\left(1 + \frac{3}{x}\right)^{1/2} + \frac{1}{2x} - 1\right] \tag{F.5.11}$$

$$\frac{T_{max}}{T_0} = \frac{1}{2}\left[\left(\frac{1}{4} + \frac{3}{2x}\right)^{1/2} + \frac{1}{2x} - \frac{1}{2}\right] \tag{F.5.12}$$

with $T_0 = KV/k$ and $x = |\ln \omega\tau_0| - \ln \tan^{-1}(1/\omega\tau_0) \cong |\ln \omega\tau_0|$ for $\omega\tau_0 \ll 1$.
For $x \geq 2$

$$\frac{T_{max}}{T_0} = \frac{1}{|\ln(\omega\tau_0)|}\left[1 - \frac{9}{16}\frac{1}{x} + \frac{27}{32}\frac{1}{x^2} + \cdots\right] \tag{F.5.13}$$

The results do not differ substantially from the Gittleman equations. In the Khater model too, the distribution function is not realistic. Moreover, the Poisson distribution has the inconvenience of having only one parameter, which does not allow one to fix separately the width of the distribution, and its maximum.

Kumar and Dattagupta[208] studied the linear response of an assembly of noninteracting particles to a small oscillating field. An appropriate Fokker–Planck equation for the orientational distribution function of the particles was written down and solved approximately under the condition that $KV \gg kT$. Particle size distribution was not taken into account, but random orientation of easy axes with respect to the magnetic field is considered. This leads to modification of the χ_0 term, which becomes

$$\chi_0 = \frac{1}{3}\left(\frac{M_{\mathrm{nr}}^2 V}{3kT} + \frac{2M_{\mathrm{nr}}^2}{3K}\right) \tag{F.5.14}$$

Although no comparison for $T_{\max}$ has been made between the two models, some small differences are expected.

Recently Slade et al.[209] proposed a new model to determine the distribution of energy barriers E_B. It was shown that this distribution is given by

$$n(E) \approx \frac{1}{M_s^2 V[\ln(1/\tau_0 \bar{\omega})]} \frac{\partial[T\chi'(\omega T)]}{\partial T} \tag{F.5.15}$$

where $T = E_B/k \ln(1/\tau_0\omega)$.

As a consequence all plots of $n(E)$ vs. $T|\ln(\tau_0\omega)|$ for various ω should be superimposed. This was checked very well for a very large range of frequencies for small Co precipitates dispersed in an Ag matrix.[209]

However, the model needs some improvement in order to be generally applied to real systems. The interparticle interactions are neglected, as in other models. Moreover, the χ_1 contribution is neglected. This could be not too important, unless the transverse relaxation modifies χ' below T_B. Finally, the volume is taken as a constant, which is not realistic for fine particles, where the variation of E_B mainly comes from the volume variation.

F.5.1.3. T_B *vs.* $T_{\max}$. From the previous discussion it comes out clearly that $T_{\max}$ depends on the analytical forms of χ_1 and χ_0. In our opinion χ_1 is questionable because the effects of transverse relaxation have not been taken into account. A more accurate formula for χ_0, accounting in first approximation for interparticle interactions is $\chi_0 = M_{\mathrm{nr}}^2 V/3k(T - \theta_{\mathrm{sp}})$, where θ_{sp} depends on the volumic concentration, on the shape of the sample, and on other factors (see Section F.2). Of course, $\theta_{\mathrm{sp}} = 0$ in the absence of interparticle interactions.

It is clear that $T_{\max}$, and its relationship with $\langle T_B \rangle$, depends on the form of the volume distribution. Gittleman et al.[149] gave some simple relationships between $T_{\max}$ and $\langle T_B \rangle$ for different types of volume distribution functions, for example, $T_{\max} = AK\langle V \rangle/k|\ln(\omega\tau_0)| = AT_B$, where the constant A depends on the form of the size distribution and is equal to 2 and 1.8 for a rectangular and for a Poisson distribution, respectively.

Starting from the Gittleman model and approximating $[1 + \omega^2\tau_0^2$

$\exp(2KV/kT)]$ by a step function, we find

$$\frac{\chi'(T,\omega)}{\chi_1} \approx 1 + \frac{\int_0^{V_c} \left(\frac{KV}{kT} - 1\right) VP(V)\, dV}{\int_0^{\infty} VP(V)\, dV} \tag{F.5.16}$$

where $V_c = (kT/K)|\ln(\omega\tau_0)|$.

The maximum of $\chi'(T, \omega)$ corresponds to the maximum of the function G, given by

$$G = \frac{|\ln(\omega\tau_0)|}{V_c} \int_0^{V_c} V^2 P(V)\, dV - \int_0^{V_c} VP(V)\, dV \tag{F.5.17}$$

Therefore, for large $|\ln(\omega\tau_0)|$ the maximum of $\chi'(T, \omega)$ is obtained for a volume such that

$$\frac{1}{V_c} \int_0^{V_c} V^2 P(V)\, dV \tag{F.5.18}$$

is maximum.

This means that $T_{\max}$ can be related to a volume V_c that does not depend on ω, except when $|\ln \omega\tau_0|$ becomes small. In this case V_c can be determined from (F.5.17). Moreover, if the V distribution is known, V_c can be easily calculated from (F.5.18). Finally, we can define $T_{\max}$ as the blocking temperature related to the volume V_c. We remark that as long as $|\ln \omega\tau_0|$ is sufficently high, V_c does not vary with ω, and the results of two experiments with different ω are directly comparable. However, the Gittleman model is basically derived from the two-level model, like most of the models that we have presented and discussed. It is therefore, valid only if the probability P of finding the magnetization in an energy minimum is large. When ν increases, $T_{\max}$ increases and the χ data are shifted toward higher temperature. That leads to a decrease of P. As a consequence, the variation of V_c with ν resulting from Eq. (F.5.17) has perhaps no real meaning, the model becoming more imprecise by increasing ν.

A last point to discuss is the value of the measuring time. The problem is whether τ_m corresponds to $1/\omega = 1/2\pi\nu$ or to $1/\nu$. In fact, from the basic equation of the **m** relaxation (see Section D), the probability per time unit for **m** to overcome the barrier is equal to $\frac{1}{2\tau}$. In this case, we can roughly say that the time for **m** to go to from one minimum to another is equal to 2τ and to go forward and back is 4τ. In ac susceptibility measurements, the time for the field to go forward and back is $1/\nu$. Therefore τ is comparable to $1/4\nu$, more exactly $\frac{1}{2\pi\nu} = 1/\omega$. However, the choice between $\frac{1}{2\pi\nu}$ or $1/\nu$ depends on the unit of the gyromagnetic ratio γ_0, usually given in angular frequency, taken $= 2 \times 10^7\ \mathrm{s}^{-1}\ \mathrm{G}^{-1}$. In this case, $\tau_m = 1/\nu$, the factor 2π being in our opinion included in the γ_0 value. This is in agreement with most of the published results, where $\tau_m = 1/\nu$ has been considered. We note that for results obtained on Fe particles in an Al_2O_3 matrix[45] and γ-Fe_2O_3 particles in a polymer,[137] it is clear that the use of $1/\nu$ leads to a good agreement, unlike the case $\tau_m = \frac{1}{2\pi\nu}$.

F.5.2. Experimental Results

The temperature dependence of the ac susceptibility has been measured by many workers[45,162,180,209–215] at fixed as well as at variable frequency.

A typical behavior is reported in Figure F.5.1. At high temperature χ'

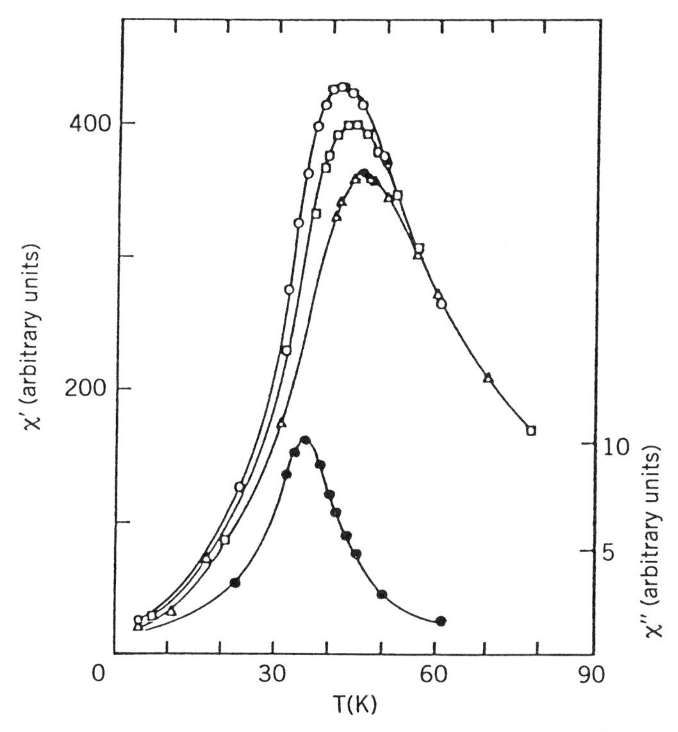

Figure F.5.1. AC susceptibility at different frequencies for a granular film consisting of iron grains dispersed in an alumina matrix (Fe-Al_2O_3). The out-of-phase component χ'' is reported for $\nu = 17$ Hz. (Reproduced with permission from Ref. 181.)

is frequency independent, indicating a situation of thermodynamic equilibrium, as expected in the temperature range where all the particles are in the superparamagnetic state. Just above T_{max} (it could also be well above, in the case of large distribution of energy barriers) and at $T \leq T_{max}$, χ' is frequency dependent, indicating a nonequilibrium situation, as expected in the blocked state. With increasing frequency T_{max} shifts to higher temperatures and χ' decreases.

The analysis of the frequency dependence of T_{max} allows one to check the models describing the temperature dependence of the relaxation time and to derive the related parameters, for example, the preexponential factor τ_0 and the energy barrier E_B.

Very few accurate determinations of τ_0, requiring measurements in a large frequency range, for independent magnetic particles are reported in the literature, mainly because of the difficulty of obtaining high dispersion in a suitable matrix.

Recently, Slade et al.[209] measured the ac susceptibility in a large frequency range (covering 8 decades) on alloy films of Co–Ag, having 5% at Co. By using their reported model, they derived the distribution of energy barriers and reported a value of $\tau_0 \cong 10^{-13}$ s. This value is too small for ferromagnetic particles, for which the theoretical predictions lead to a value between 10^{-9} and 10^{-11} s. The authors compare the τ_0 value to that obtained by Dickson et al.[210] on ferritine particles ($\cong 10^{-12}$ s), which are antiferromagnetic (they consist of a ferrihydrite core contained within a protein shell). For antiferromagnetic particles smaller τ_0 values are expected (10^{-12}–10^{-13} s) (see Section H).

A more physical τ_0 ($\simeq 10^{-10}$ s) value for ferromagnetic particles was obtained by Lazaro et al.[214,215] from measurements on iron particles embedded in a zeolite matrix, which, because of its microporous structure, allows the isolation of the magnetic grains. From a computational analysis of the experimental data, the authors also obtained information about the particle size distribution. However, the explored frequency range was too narrow ($\cong 10$–10^3 Hz).

A detailed analysis of the frequency dependence of T_{max} was reported for ferrimagnetic γ–Fe_2O_3 particles in a large frequency range ($2 \times 10^{-2} < \nu < 10^4$ Hz) by Dormann et al[59]. The frequency dependence of T_{max} was found to follow an Arrhenius law (for the first time clearly demonstrated for noninteracting particles in a large frequency range) with $\tau_0 \cong 10^{-10}$ s.

T_{max} was found to increase with the mean particle volume.[45,59,137,149,162] However, often the increase of the average size of particles also corresponds to an increase of particle concentration, due to the difficulty in controlling separately size and concentration, maintaining the same type of size distribution within a series of materials. This is why it is not easy to check the T_{max} vs. V dependence. For a controlled dispersion of γ-Fe_2O_3 particles T_{max} was found to be related, in first approximation, to $\langle V^2 \rangle / \langle V \rangle$ (see above).[59,137]

T_{max} was found to increase with particle concentration,[45,59,137,149,162] and its frequency dependence is expected to decrease. In most cases it is difficult to draw conclusions about the concentration dependence of T_{max}, since, as it was pointed out before, often the average particle size changes with concentration.

Dormann et al.[45,181] analyzed the frequency dependence of T_{max}, by measurements in a large frequency range ($2 \times 10^{-2} < \nu < 10^4$ Hz) and by Mössbauer spectroscopy, for granular films consisting of iron particles dispersed in an alumina matrix (Fe–Al_2O_3), and they compared it with that found in spin-glass systems. A simple Arrhenius law cannot describe the results with a physically meaning τ_0 value. A phenomenological Fulcher law ($\tau = \tau_0 \exp[E_a/K(T - T_0)]$, as well as the scaling laws describing spin-glass dynamics[45] were found to be inadequate, not allowing to fit the results with a unique set of parameters. The results were satisfactorily fitted by the model proposed by the authors,[45] which describes the effect of interparticle interaction on the relaxation time by means of a statistical calculation of the dipolar energy (see Section E). The model applies to an assembly of grains characterized by volume distribution, disordered arrangement, and direction of easy axes in random positions, as in most of real systems. From the fit (Fig. F.5.2) a value of $\tau_0 \cong 10^{-11}$ s was deduced, and the energy barrier values were derived for a series of samples with different iron content.

The Dormann model was further developed and applied to the analysis of ac measurements ($2 \times 10^{-2} < \nu < 10^4$ Hz) on γ–Fe_2O_3 particles in a polymer, with different degrees of dispersion.[59,137] The parameter values governing the relaxation time, deduced by applying the model, are consistent with the variation of particle diameter, interparticle distance, and number of first neighbors per particle within the series of samples.

Finally, a special mention of the experiments performed on ferrofluids[200,216–220] is needed. Fannin et al.[216–220] performed measurements of the complex susceptibility in the high-frequency range, up to 3 GHz,[216] developing the split-toroid technique. For magnetic fluids two relaxation mechanisms can occur, one by rotational Browning diffusion, dominating for large particles, characterized by a relaxation time $\tau_B = 3V_0^1\eta/kT$ (V_0^1 is the hydrodynamic volume, η is the viscosity)[221] and the other by superparamagnetic relaxation, dominating for smaller particles ($2 < \phi <$ 5 nm). As ferrofluids contain a distribution of particle sizes, both mechanisms will in general contribute, with an effective relaxation time[222]

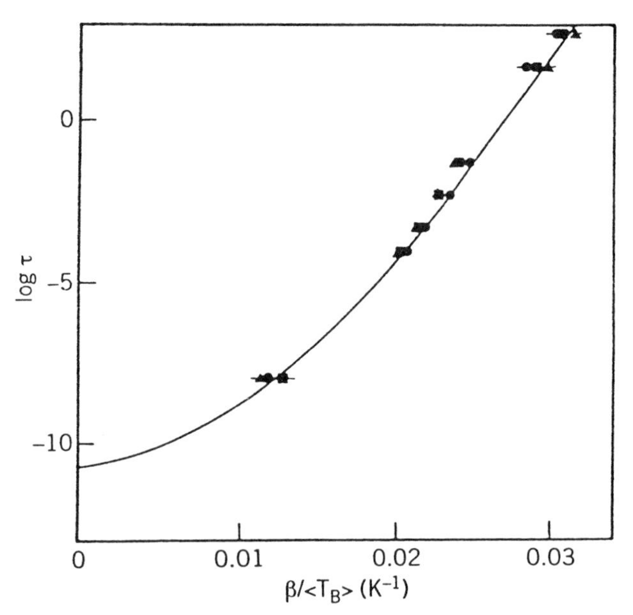

Figure F.5.2. Frequency variation of the blocking temperature (using scaling coefficent β) for various Fe-Al_2O_3 fine particle-samples. The full curve corresponds to the fit to the model of Dormann et al. (Reproduced with permission from Ref. 45.)

$\tau_{eff} = \tau\tau_B/(\tau + \tau_B)$. The dominant mechanism of a particle will be that with the shortest relaxation.

According to the Debye theory[221] of the complex susceptibility, χ' decreases with frequency, while χ'' has a maximum at a frequency ω_{max} $\tau_{eff} = 1$ (Fig. F.5.3). For particles relaxing by superparamagnetic mechanism the maximum of χ'' occurs for a frequency between 1 Mz and 100 MHz. For particles relaxing by the Brownian mechanism the frequency at which the maximum occurs is much lower. On manganese ferrite particles of mean diameter 9.4 nm in an hydrocarbon carrier Fannin et al.[216] observed the presence of a $\chi''(\omega)$ peak at 30 MHz, giving an indication of superparamagnetic relaxation, and a transition of $\chi'(\omega)$ to a negative value at 65 MHz, suggesting ferroresonance. In some ferrofluids two loss peaks were found, indicating the existence of two particle size distribution in the dispersion[217] (Fig. F.5.4).

F.5.3. Conclusions

The measurement of the ac susceptibility at variable frequency represents one of the most powerful tools for studying the dynamical properties of fine magnetic particles. The possibility of exploring a very large time

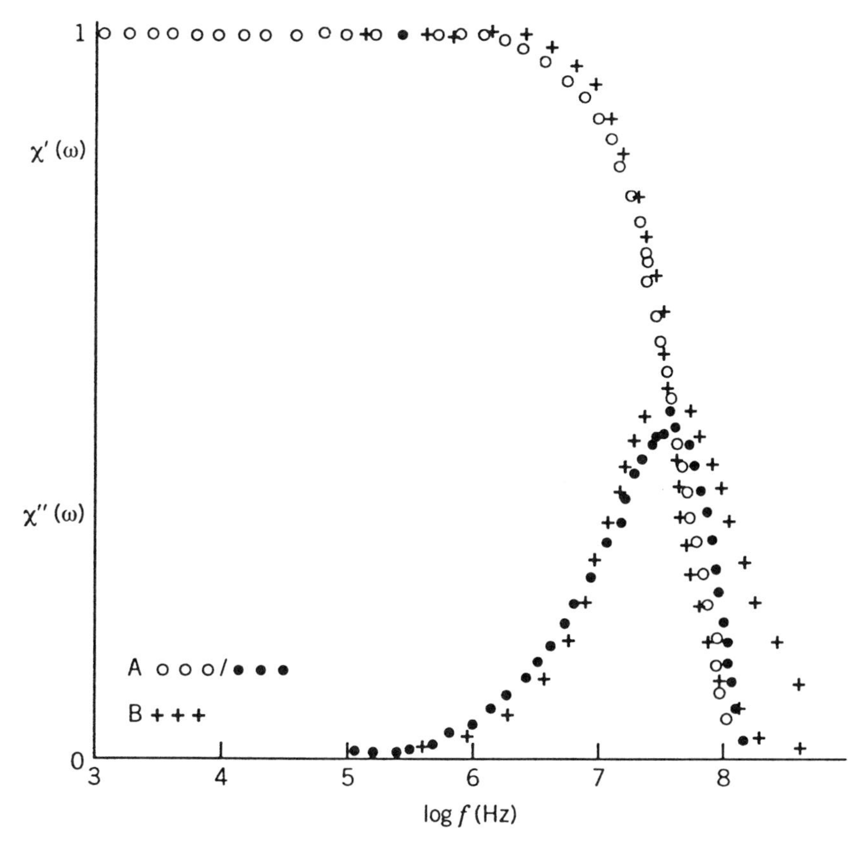

Figure F.5.3. (a) Normalized plot of $\chi'(\omega)$ vs. log f. (b) Debye profile for a ferrofluid consisting of a colloidal dispersion of magnetite ($\langle \phi \rangle = 3.05$ nm) in a hydrocarbon. (Reproduced with permission from Ref. 219.)

window, just by changing the frequency of the ac field, by using the same technique, makes this tool unique. This is very important, since the comparison between the results obtained by different techniques having different measuring time is not straightforward, especially when the volume distribution, and hence the relaxation time distribution, is not narrow.

The analysis of the frequency dependence of the susceptibility and of T_{max} allows one to check the models describing the temperature dependence of the relaxation time and to derive the related parameters, for example, the preexponential factor τ_0, the energy barrier, and the type and the width of its distribution. Unfortunately, in a very few cases the ac

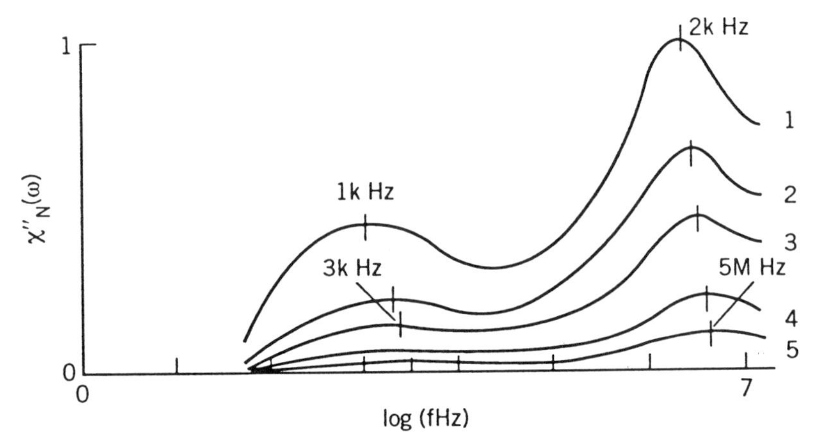

Figure F.5.4. $\chi''(\omega)$ vs. $\log f$ for some ferrofluids, consisting of magnetite in Isopar with different packing fraction. (Reproduced with permission from Ref. 217.)

measurements have been extended to a large enough frequency range. Further detailed analysis are needed to have a better insight on the fine-particles dynamics and in particular to make clear the relative role of different terms contributing to the total energy barrier for the magnetization reversal.

F.6. Mössbauer Spectroscopy

F.6.1. Introduction

Mössbauer spectroscopy is a technique utilizing γ-rays in the 10–100 keV range. It is based on the emission and resonant absorption of γ-rays in a recoil-free way in solids, that is, without any energy loss to the lattice. The extremely well-defined energy of these γ-rays makes it possible to study static as well as dynamic hyperfine interactions. Detailed chemical, structural, and magnetic information can be obtained about atoms on the surface or in the bulk of materials. In addition, in situ investigations, which are indispensable in case of particle structure or surface state changing with the surrounding conditions, can easily be achieved. Mössbauer spectroscopy is thus very well suited to studies of fine-particle systems as it has been shown in several reviews, for example, see Refs. 18, 32, 33, 223, and 224. The application of this technique is restricted to solids or frozen liquids containing a Mössbauer isotope. For practical reasons, ^{57}Fe is almost the only isotope easily utilizable for studies of magnetic materials. Fortunately, iron is present in most of the important magnetic materials. Magnetic compounds not containing iron can often be studied by doping the sample either with ^{57}Fe or the radioactive parent isotope, ^{57}Co.

Mössbauer spectroscopy sees the fluctuations of the magnetic moment of the particle, **m**, through the fluctuations of the magnetic hyperfine interaction. The phenomena are probed at the atomic level. This provides the technique with some distinctive advantages. First, the measurement is a local measurement. The result of the measurement, a spectrum and not a scalar quantity, arises from a superposition of local effects, not from the average. Second, the measurement can be achieved without applying a field. It is performed along any direction, not along one particular direction as in the case of magnetic measurements. And third, the characteristic time of the measurement, of the order of 10^{-8} s for ^{57}Fe, is intermediate between the measuring times accessible with usual ac susceptibility experiments (see Section F.5) and those relevant to neutron scattering experiments (see Section F.8). The timescale is thus rather short and yet long enough to allow observation of the blocked state to superparamagnetic state transition in a convenient temperature range in most cases. Moreover, the blocked state and the superparamagnetic state are easy to identify. This is illustrated in Figure F.6-1 for a ferritin sample,[210] with a six-line pattern for the blocked state and a doublet for the superparamagnetic state.

The nature of measurement, the Mössbauer process, has two major consequences. First, there is not a definite measuring time but a range of measuring times, which makes the blocking temperature difficult to define precisely. The relaxation spectrum of a given particle depends on the actual excursion of **m**, and the phenomena cannot be analyzed without the appropriate lineshape model. Second, Mössbauer atoms with different surroundings give distinct signals. Surface atoms, in particular, produce specific contributions that will be significant if the particle size is sufficiently small. Surface and superparamagnetic relaxation effects are generally difficult to disentangle from each other, particularly because of the lack of resolution produced by the distributions of particle size and shape. Therefore, surface-related phenomena are usually characterized in the limit of negligible relaxation effects whereas the effects of the relaxation are analyzed assuming that surface effects are negligible.

The contents of this section may be summarized as follows. An elementary introduction to Mössbauer spectroscopy and a brief description of the Mössbauer spectrum are given in Section F.6.2. For clarity, a digest of the basic concepts is reported in the Appendix.

Section F.6.3 is concerned with the investigation of static properties of fine particles with focus on the differences with respect to bulk material studies. Studies of surface magnetic properties are largely developed

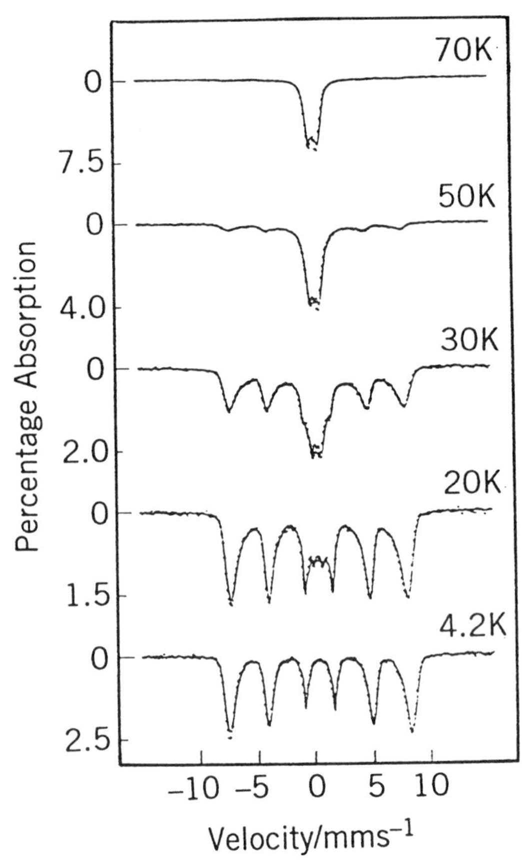

Figure F.6.1. Mössbauer spectra of a ferritin sample at various temperatures. (Reproduced with permission from Ref. 210.)

insofar as Mössbauer spectroscopy is, in this respect, quite a unique technique.

Section F.6.4 is devoted to superparamagnetic relaxation in zero applied field. Because the phenomena cannot be interpreted in details without a lineshape model, most representative models are surveyed. Their applicability to realistic situations is discussed. The exploitation of the magnetic splitting at low temperature and the determination of the blocking temperature are discussed. The difficulties and ambiguities in interpreting the Mössbauer spectra of actual samples are emphasized.

The influence of a large external magnetic field is considered in Section F.6.5 and that of a medium or weak field is discussed in Section F.6.6.

Finally, Section F.6.7 is devoted to concluding remarks.

F.6.2. The Mössbauer Spectrum

The Mössbauer effect is the recoil-free emission of a γ-photon by a radioactive nucleus and the subsequent resonant recoil-free reabsorption by another nucleus of the same type. In the process, the source nucleus goes from the excited state (nuclear spin I_e) down to the ground state (I_g), and the absorber nucleus in the ground state (I_g) is raised into its excited state (I_e).

The energy distribution of the emitted radiation and that of the resonant absorption cross section have identical lineshapes. Both are Lorentzian lines centered at the nuclear transition energy E_γ, with fullwidth at half height given by the uncertainty principle energy width of the nuclear excited state. This width, the natural linewidth, is defined by

$$\Gamma_N = \frac{\hbar}{\tau_N} \tag{F.6.1}$$

where τ_N is the mean lifetime (1.44 times the half life) of the excited state, and $\hbar$ is the Planck constant divided by 2π.

For the ^{57}Fe isotope, the transition of interest is $I_e = \frac{3}{2} \leftrightarrow I_g = \frac{1}{2}$ with $E_\gamma = 14.4$ keV and $\tau_N = 1.4 \times 10^{-7}$ s, thus $\Gamma_N = 4.6 \times 10^{-9}$ eV.

The very high definition of the γ-ray emitted in a recoil-free event allows the detection of the small energy variations associated with the hyperfine interactions. These energy variations are in the range of the Doppler shifts produced by small movements. Therefore, the energy of the transition is conveniently modulated by moving the source relative to the absorber (or vice versa). If v is the Doppler velocity, the energy is given by

$$E(v) = E_\gamma\left(1 + \frac{v}{c}\right) \tag{F.6.2}$$

where c is the velocity of light.

A typical Mössbauer experiment thus involves an oscillating radioactive source that contains a parent isotope (e.g., ^{57}Co for ^{57}Fe), a stationary absorber that is usually the sample, and a detector. The Mössbauer spectrum consists of a plot of γ-ray counts (relative absorption) as a function of the velocity of the source. In the source the radioactive isotope feeds the excited state of the Mössbauer isotope, which decays to the ground state. The energy of the recoil-free emitted radiation is Doppler modulated. Resonant absorption occurs when the energy of the γ-ray just matches the nuclear transition energy for a Mössbauer atom in the absorber. This is detected by the decreased transmission of the absorber. The shape of the line in a recorded absorption spectrum is given by the convolution of the emission and absorption lines. Therefore, the experimental line is a Lorentzian line of width $\Gamma = 2\Gamma_N$ (0.19 mm/s for ^{57}Fe). The actual linewidth is often larger as a result of inhomogeneity or thickness effects in the real samples.

The resonance absorption can also be detected in backscatter geometry by following the radiation emitted during decay of the excited absorbing nuclei. This geometry is especially useful for surface and catalysis studies.

A Mössbauer spectrum is characterized by its total absorption area, the number, position, relative intensity, and shape of the various absorption lines. These features result from the various interactions between the Mössbauer nuclei and their surroundings, as well as any motion of the Mössbauer atoms (see the Appendix).

The total absorption intensity depends on the concentration of the Mössbauer atoms in the absorber and their recoil-free fraction. The recoil-free fraction depends on the binding forces of the Mössbauer atom in the lattice (see Ap.1). The number and position of the absorption lines are determined by electric and magnetic electronic effects, the hyperfine interactions, which shift and split the nuclear levels. The allowed transitions between ground and excited substates are determined by the multipolarity of the nuclear transition. The $I_e = \frac{3}{2} \leftrightarrow I_g = \frac{1}{2}$ transition of ^{57}Fe is a magnetic dipole transition, and the allowed transitions obey the selection rules $\Delta m_I = 0, \pm 1$, where Δm_I is the variation of the spin quantum number between the excited substate and the ground substate. The electric and magnetic hyperfine interactions lead to the three basic parameters of a Mössbauer spectrum, namely the isomer shift (see Ap.2), the quadrupole splitting (see Ap.3), and the magnetic splitting (see Ap.4). The relative areas of the absorption lines are proportional to the probabilities of the corresponding transitions, and sensitive to polarization effects (see Ap.5). The basic lineshape is Lorentzian. Deviations from the Lorentzian lineshape are generally due to dynamical phenomena such as fluctuating hyperfine interactions or diffusion.

Time-dependent phenomena can influence the Mössbauer spectrum whenever they make the position of the Mössbauer nucleus or the properties of the nuclear environment and, hence, the hyperfine interactions change with time. Time-dependent effects can influence both the spectral lineshapes and the values of the Mössbauer hyperfine parameters. The nuclear transitions and the hyperfine interactions have characteristic times, and each type of relaxation phenomenon must be considered in the context of the appropriate time scale. In case of superparamagnetic relaxation, the magnetic hyperfine interaction fluctuates with time. The magnetic hyperfine field acting at a given Mössbauer nucleus is generally collinear with the magnetization vector, and the process can be considered in terms of a time dependence of the orientation of the hyperfine field.

Any motion of the Mössbauer nucleus can influence the spectrum in two ways, by affecting the absorption intensity of the spectrum itself, and also the linewidth, eventually the lineshape, as a result of a kind of additional Doppler motion.

F.6.3. Bulk and Surface Static Studies

Similar to bulk materials, through the recoilless fraction and static hyperfine interactions, Mössbauer spectroscopy can be sensitive to a wide range of phenomena relevant to fine-particle systems. The main differences with respect to bulk studies come from the specific properties of the atoms near the surface.

All hyperfine parameters of the atoms near the surface can be different from the hyperfine parameters of the atoms in the interior. In general, however, surface and bulk contributions are not clearly differentiated from each other. Because of the great variety of surface sites, the hyperfine parameters of the surface atoms, especially the hyperfine field, can be broadly distributed. These distributions combine with the effects of the distributions of particle size and shape, often making surface and volume effects difficult to distinguish from one another. Investigations based on the isotope selectivity of Mössbauer spectroscopy may then be particularly useful.

Using some representative examples, we shall focus on typical information that can be obtained in relation with the magnetic properties.

F.6.3.1. Recoil-Free Fraction. The viscosity of the fluctuations of the magnetization vector is influenced by spin–lattice interactions, in turn influenced by lattice vibrations. Studies of the recoil-free fraction, f, can in principle provide some information about the phenomena (see Ap.1).

Variations in lattice vibrations in fine particles with respect to the bulk may arise from (i) the reduced volume leading to lattice softening with resultant decrease of the Debye temperature, (ii) surface effects since the surface atoms are probably more weakly bound than the atoms in the interior, or (iii) changes in the lower and upper cut-off frequencies of the phonon spectrum.[225,226] The first two phenomena should decrease f while the latter could increase f. In general, one observes a recoil-free fraction in fine particle systems that is much smaller than that of bulk materials. However, most often this is not due to effects of the lattice vibrations but to the motion of the particle as a whole, which indeed drastically lowers the f factor.

A free nanoparticle is generally not massive enough to absorb all the recoil energy, therefore it recoils. The critical size for which the recoil energy gives rise to a line shift equal to the natural linewidth is 25 nm for iron oxide.[227] The Mössbauer effect is clearly observable for much smaller sizes. The recoil energy is actually shared by a large number of particles in case of agglomeration, or absorbed by the surrounding medium in case of a dispersion. The f factor is therefore strongly dependent on the sample fabrication, the particle preparation route and the sampling process via a wide range of parameters such as particle packing, adsorption phenomena, coupling between the particles and the matrix, or elastic properties of the suspending medium. This is illustrated in a number of studies, for instance for γ-Fe_2O_3,[226,228] Fe particles,[226,229–232] or Fe alloys.[233,234]

The reduction of the recoilless fraction is usually explained in terms of an oscillation of the particles. To characterize the phenomena within the particle, it is necessary to hinder the motion of the particles. For Fe particles embedded in a resin, Hayashi et al.[231,232] concluded that the metallic core of the particles is not softened as compared with the bulk, while the lattice vibration of the oxide layer is very soft. Van der Kraan,[235] investigating the surface properties of α-Fe_2O_3 by enriching the surface with ^{57}Fe, found nearly the same f fraction for the atoms at the surface and in the interior of 50-nm-sized particles. For 7 and 4 nm particles, the thermal variation of the f fraction appeared much faster for the atoms at the surface than for the atoms in the interior. As this effect increased with decreasing size, van der Kraan[235] suggested that the number of atomic layers behaving like the surface could depend on the size.

The dependence of the recoil-free fraction on particle size is complex since it can involve volume and surface effects. However, careful analyses should yield valuable information. Analyses of the f fraction can also reveal varying elastic properties of the matrix and be used to investigate dynamic effects including rheological properties of fine-particle systems.

F.6.3.2. Magnetic Hyperfine Field. The magnetic splitting of a Mössbauer spectrum yields the effective magnetic field at the Mössbauer atom (see Ap.4), and variations in the hyperfine field reflect variations in the magnetic and electronic properties. The thermal fluctuations of the magnetization vector near one easy direction lead to an apparent reduction of the magnetic splitting. This thermal effect (see Section F.6.4), must be negligible so that information about the hyperfine field can be deduced. Hence, the relevant experiments must be performed at low temperature or under a large applied field.

The effective magnetic field at a Mössbauer atom in a particle can be written as

$$\mathbf{H}_{\text{eff}} = \mathbf{H}_{\text{app}} + \mathbf{H}_D + \mathbf{H}_L + \mathbf{H}_{\text{int}} + \mathbf{H}_{\text{hf}} \tag{F.6.3}$$

where the contributions are in order of applied field, demagnetizing field, Lorentz field, dipole field due to the neighboring particles, and hyperfine field.

The main difference with respect to bulk studies for deducing the magnitude of the hyperfine field comes, in principle, from the demagnetizing field, and the main differences between the magnitude of the hyperfine field in fine particles and that in the corresponding bulk material generally arise from surface effects.

A. DEMAGNETIZING FIELD. In the absence of an applied field, the demagnetizing field in a bulk material is negligible because of the multidomain structure. It may be significant in the interior of a single-domain particle (see Section C.1.2). Therefore, the magnetic field acting at a Mössbauer nucleus can depend on the shape of the particle and the direction of magnetization, and the observed magnetic splitting can be larger or smaller than for the corresponding bulk material. In α-Fe, for instance, the hyperfine field (330 kOe at room temperature) is antiparallel to the magnetization. Hence, the demagnetizing field adds to the hyperfine field. The magnetic field acting at the nuclei in a spherical single-domain particle should therefore be larger than that of a multidomain particle, by about 7 kOe as deduced from the value of the magnetization.[229] This is well verified experimentally.[236] It is necessary to take into account the demagnetizing field for comparing the hyperfine field in fine particles with that in large particles where the domain structure appears. The effect of the demagnetizing field can be important especially in materials where the hyperfine field is rather small and the magnetization is large like in the case of α-Fe, it is usually much smaller in oxides.

B. SURFACE HYPERFINE FIELD. The demagnetizing field and the Lorentz field are not defined for the atoms at the surface. Calculations[237] of the magnetic dipole fields at atoms near the surface in fine particles and thin films of α-Fe show that only the first surface layer is perturbed, with variations of the order of 10 kOe depending on the position at the surface. Studies of thin and ultrathin metal films by Mössbauer spectroscopy (see, e.g., Refs. 31, 224, 238, and 239) show that one or two atomic layers are perturbed, as for the magnetization (see Section B.3). The surface hyperfine field may be larger or smaller than in the bulk at low temperature depending on the substrate and the material coating the surface, but generally the decrease with increasing temperature is faster.

For α-Fe particles,[33,39,240–243] at low temperature the hyperfine field of the atoms in the interior is similar to that of bulk α-Fe, and the surface hyperfine field is generally found larger, and often distributed. For instance an increase of 3% at 4.2 K was found for 2-nm particles in organic liquids.[242] The magnetization too was found to be 3% larger than for bulk iron. For 3.7-nm particles on a carbon support,[243] at 5 K the surface hyperfine fields appeared broadly distributed between 200 and 450 kOe, with an average value significantly larger than that of bulk iron, and with increasing temperature, the surface hyperfine field decreased faster than the values for the bulk. For 3.4-nm α-Fe particles in alumina,[244] we deduced surface hyperfine fields with two main components at 4.2 K, of ca. 380 and 290 kOe, which were assigned to the first and second surface layer, respectively, assuming a layer thickness of 0.23 nm. A normal value of the magnetic moment in the core and 50% reduction in the two surface layers explained the nonrelaxing magnetization that was 33% lower than that of bulk iron. In-field Mössbauer experiments produced spin canting effects (see Section F.6.3.3). By in situ studies of 2-nm particles on carbon supports,[39] it was found that chemisorption of oxygen results in the formation of a surface layer that is ferromagnetically coupled to the core of the particle, but with magnetic hyperfine fields similar to those found in thicker passivation layers, which have a disordered spin structure.

Metal nanoparticles are pyrophoric. To our knowledge, no ultra-high vacuum study has been reported to date, and reported studies are relative to particles passivated or embedded in some medium. Surface properties should then be dependent on the nature and structure of the interface.

Many ferrimagnetic oxides are chemically stable in air, therefore surface studies should be straightforward. However, to enhance the surface contribution coatings with ^{57}Fe or ^{57}Co are often applied. The hyperfine field at the surface of γ-Fe_2O_3 acicular particles, a few tenths of a micron in length, has been much investigated.[18,32,33,245–248] Particles coated either with ^{57}Fe[245,246] or radioactive ^{57}Co[247,248] were studied in absorption and emission Mössbauer experiments, respectively. A radioactive ^{57}Co nucleus decays by K capture to an excited state of $^{57}Fe^{2+}$. Before the emission of the γ-radiation, the ferrous ion has time to lose an electron, converting into the ferric state. All investigations showed that at low temperature the hyperfine field at the surface was similar to that of the bulk material but decreased more rapidly with increasing temperature. At 300 K for instance, the surface hyperfine field was found about 10% smaller than the bulk hyperfine field.

When a large particle made with natural iron is coated with ^{57}Fe or ^{57}Co, the Mössbauer spectrum contains a contribution arising from the ^{57}Fe nuclei in the underlying particle. Natural iron only contains 2.2% of the ^{57}Fe isotope, but because of the large volume-to-surface ratio the Mössbauer atoms in the interior are still in a significant proportion. A way round this problem is to use the ^{56}Fe isotope to prepare the particles and to coat them with ^{57}Fe. In this way the Mössbauer spectrum contains no contribution from the interior and the surface information can be clearly identified. The results so obtained by Shinjo et al.[249] for spherical α-Fe_2O_3 particles ca. 100 nm in diameter are shown in Figure F.6.2. It is clear (Fig. F.6.2a) that the hyperfine field of the surface sites at 300 K is smaller than in the bulk and distributed to a certain degree. The mean surface hyperfine field was found 4% smaller than the bulk value at 300 K, but very nearly the same at 4.2 K. The fact that the thermal decrease of the surface hyperfine field is less rapid for α-Fe_2O_3 than for γ-Fe_2O_3 was related to the difference in the Néel temperatures. As the data for the two materials can be scaled using reduced units (Fig. F.6.2b),

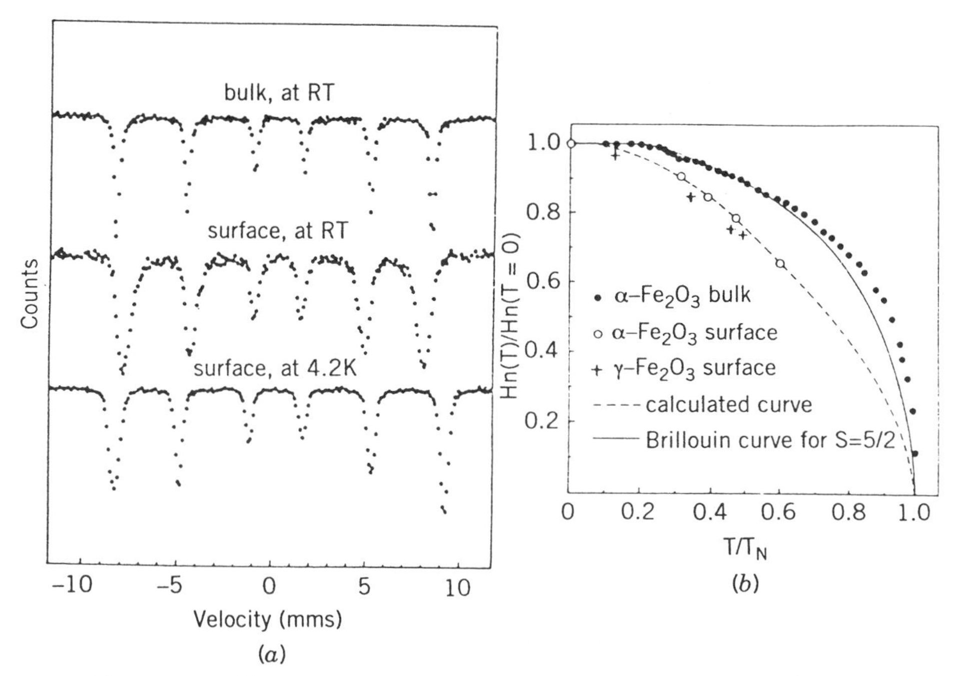

Figure F.6.2. (a) Mössbauer spectra of standard α-Fe_2O_3 at 300 K (top), ^{57}Fe-coated α-$^{56}Fe_2O_3$ at 300 K (middle), and 4.2 K (bottom); (b) temperature dependence of the hyperfine field in bulk α-Fe_2O_3, the surface of α-Fe_2O_3 and the surface of γ-Fe_2O_3. (Reproduced with permission from Ref. 249.)

it was suggested that the exchange interactions at the surface are reduced in a similar proportion (40%) with respect to the interactions in the bulk oxides. The results on ^{57}Fe-coated α-$^{56}Fe_2O_3$ also showed that the Morin transition takes place at the surface at the same temperature as in the bulk crystal. Similar investigations using the ^{56}Fe and ^{57}Fe isotopes were reported for α-FeOOH[250] and β-FeOOH,[251] but so far not for nanometer ferrimagnets. Such studies would be very worthwhile, especially regarding the influence of the particle size.

For quasi-non-interacting γ-Fe_2O_3 nanoparticles dispersed in a polymer (polyvinylic alcohol, or PVA),[252] with particle size decreasing from 10 to 4.6 nm, we found a slight decrease (2%) in the average hyperfine field deduced from the zero field Mössbauer spectra at 4.2 K and a large decrease (30%) for the nonrelaxing magnetization, $M_{nr}(0)$.[97] The fact that only the magnetization is strongly affected shows that the magnetic defects in the smallest particles are essentially due to incomplete spin alignment, and not to a large reduction of the magnetic moments. In zero field, the Mössbauer spectrum is only sensitive to the magnitude of the hyperfine field, not to its orientation. Orientational magnetic disorder is not detected, in contrast with magnetic measurements. By comparing the values of the hyperfine field and the magnetization with the bulk values, one can obtain information about the type of magnetic defects.

F.6.3.3. Spin Canting Effects. Mössbauer spectroscopy is appropriate for detecting a noncollinear magnetic structure. If an external field is applied parallel to the γ-ray beam, and if all the spins are collinear with this field, then the $\Delta m_I = 0$ transitions are forbidden (see Ap.5). The second and fifth lines of a six-line pattern will be absent. Conversely, nonzero second and fifth lines show that a noncollinearity is present in the sample.

A lack of spin alignment has been observed by Mössbauer spectroscopy experiments in applied fields up to 100 kOe for several nanometer ferrimagnetic oxides,[18,32,33,224] for instance, γ-Fe_2O_3[134,253–258] with adsorbed ^{57}Fe[245,246] or ^{57}Co,[247,248] Co-adsorbed γ-Fe_2O_3,[259,260] $CoFe_2O_4$,[261], $NiFe_2O_4$[262] coated with organic molecules,[263] $BaFe_{12}O_{19}$,[264] $Y_3Fe_5O_{12}$,[33] and CrO_2.[265] This is to be related to a reduction in the magnetization with respect to the bulk value due to the canting and a lack of saturation at low temperature in large applied fields which decrease the canting angle, as usually observed for ferrimagnetic oxide nanoparticles. A noncollinearity is less frequently observed in metallic particles. Complete or at least nearly complete alignment in fields of 40–50 kOe was reported for instance for α-Fe particles[39,243,266] and for FeNi and FeCo alloys,[233] but noncollinearity effects were also reported.[39,244]

Noncollinearity can result from incomplete alignment of all the spins as a result of large magnetic anisotropy, or from lack of alignment of a fraction of the spins mainly located in the surface due to the discontinuity (see Section B.3). The first assumption was recently suggested[260] for explaining Mössbauer spectroscopy results on Co-adsorbed γ-Fe_2O_3. It was clearly ruled out[257] by studies of the degree of spin alignment in frozen suspensions of γ-Fe_2O_3 nanoparticles as a function of the magnetic texture. The degree of spin alignment was found independent of the orientation of the easy directions in the samples when the applied field was larger than 7.5 kOe, and the observations were found all compatible with the concept of spin canting as suggested by Coey.[253]

A. γ-Fe_2O_3: A CHIEF EXAMPLE. γ-Fe_2O_3 has been the most studied nanomaterial by in-field Mössbauer spectroscopy. The influence of the applied field is illustrated in Figure F.6.3 for 6.8-nm γ-Fe_2O_3/PVA particles at 7 K.[267] In zero field the Mössbauer spectrum is a nearly symmetric six-line pattern. The slight asymmetry results from a slight difference in the hyperfine parameters of the A and B sites. For 0.1 T, apart from a slight broadening of the lines, no notable changes are observed as compared to zero applied field. For 0.5 T the intensity of the second and fifth lines is clearly reduced showing that alignment has mostly occurred, and only a slight decrease is observed when the field is further increased. The persisting second and fifth lines are the signature of a noncollinearity. The splitting of the first and sixth lines under 6 T shows that the spins in the A and B sites are mainly close to the field direction, parallel and antiparallel, respectively.

Coey,[253] who studied 6.5-nm γ-Fe_2O_3 particles, ascribed the noncollinear structure to random canting of surface spins. Studies of ^{57}Fe or ^{57}Co surface-coated acicular particles[247,248,254,255] supported the surface effect. This was recently contested for acicular particles,[245] and it was concluded that the canting probably resulted from structural defects. Vacancies on the A sites, for example, would lead to a local B-site spin canting as in ferrite with partial diamagnetic substitution.[268] The magnetic disorders near the surface too are analogous to the magnetic disorders encountered in substituted bulk ferrites and, from this point of view, the level of substitution to be considered in both A and B sites is indeed much larger near the surface of the particle than in the core. This along with the large values of the surface-to-volume ratio in nanoparticles strongly support the idea that the effects of the magnetic disorders in the surface should be prevailing. In our opinion, the problem of the surface magnetic properties should not be stated in the same terms for a nanoparticle as for an acicular particle of incomparably greater volume and generally made up of stacked, oriented crystallites.

Because of the complexity of the phenomena, it is evident that for

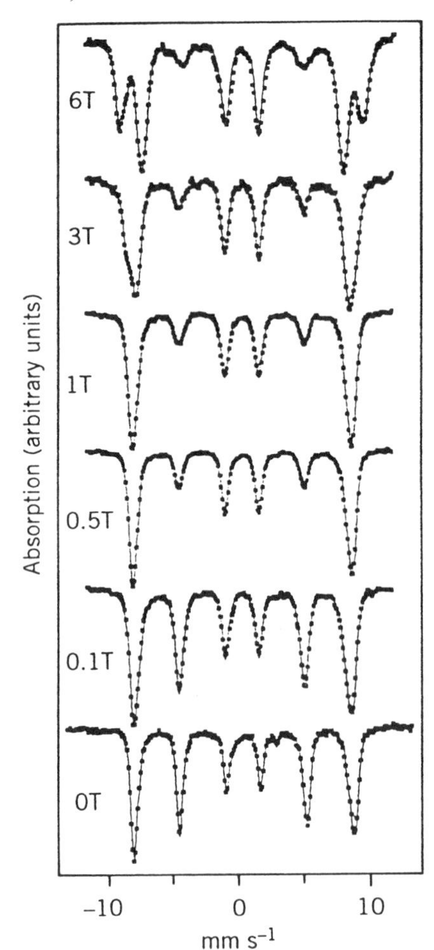

Figure F.6.3. Mössbauer spectra of 6.8-nm γ-Fe_2O_3 particles at 7 K in various applied magnetic fields parallel to the gamma-ray beam.[267]

characterizing individual properties of nanoparticles one must investigate systems where the individuals are well defined. The usual acicular γ-Fe_2O_3 particles are therefore not the best candidates in the present context. Nanoparticles obtained by coprecipitation routes, for instance, are more suitable. A particle is generally a single crystal. The particles can be dispersed in various media, and agglomerated particles are in a disordered arrangement except in very special conditions. For these particles, the degree of spin canting increases with decreasing particle size.[134,258] This is illustrated in Figure F.6.4 for γ-Fe_2O_3/PVA particles[134] with an average diameter varying between 10.1 and 2.7 nm (slightly different values are reported in Ref. 134 because of a different diameter average). A preliminary investigation of the influence of the interparticle

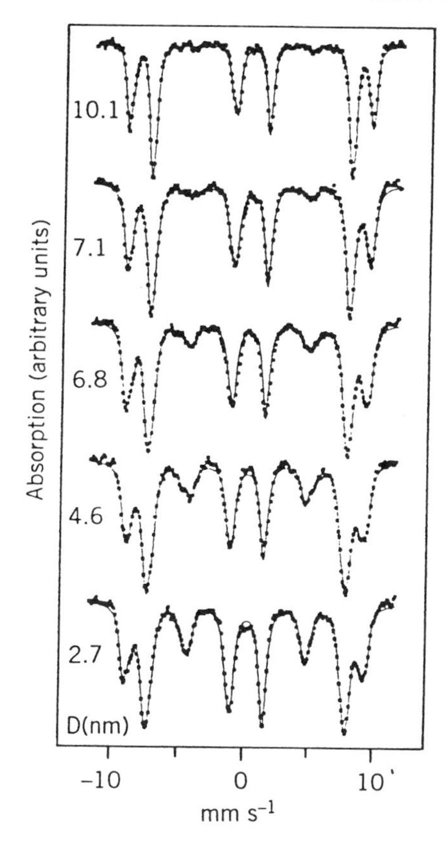

Figure F.6.4. Mössbauer spectra of γ-Fe_2O_3 particles at 7 K in a magnetic field of 6 T parallel to the gamma-ray beam. Influence of the average particle diameter (D). (Reproduced with permission from Refs. 134 and 267.)

interactions showed no major effect. All magnetic properties[59,97] as a function of the particle size and interparticle interactions suggest an important role of surface effects. These are, however, difficult to characterize precisely. Again, studies of fine particles made with ^{56}Fe and coated with ^{57}Fe would be very worthwhile.

The influence of the temperature on the recalcitrant spins was considered in a few investigations.[245,254,257] In each case the canting effect was found to decrease with increasing temperature, in small fields and in large fields. However, to our knowledge, no detailed study has been reported to date.

B. SPECTRUM EXPLOITATION. A noncollinearity of the spins in fine particles is generally easy to detect, but difficult to characterize precisely because of various distribution effects.

If for an atom, C, the hyperfine field vector forms an angle β with the direction of the applied field parallel to the γ-rays, the areas of the lines in the C pattern are in the ratio 3:2p:1:1:2p:3 with $p = 2\sin^2\beta/(1 + \cos^2\beta)$ [Eq. (Ap.12)], or equivalently, the relative area of one of the two and five lines is $A_{2,5} = \sin^2\beta/4$. In principle, both the canting angle and the fraction of canted moments can be deduced if at least another pair of lines of the C pattern is resolved. However, generally this is not the case due to the distributed parameters and other features that broaden the full spectrum, and neither the canting angle nor the canted fraction can be deduced from the line areas.

The effective field at a canted site is in a first approximation given by $H_{eff}(C) \sim H_{hf}(C) + H_{app}\cos\beta$. A distribution in canting angles results in a distribution in effective fields. This distribution effect will combine with the hyperfine field distribution due to volume and surface effects. If the canting occurs on several types of site, A and B in cubic spinels, for instance, the H_{eff} distributions may overlap. Usually, the experimental spectra exhibit broad lines (Figs. F.6.3 and F.6.4), and the analysis is ambiguous because of the mixed effects of the canting angle and hyperfine field distributions. De Bakker et al.[256] analyzed spectra of conglomerates of needlelike particles of γ-Fe_2O_3 by considering a simultaneous distribution in the canting angle and the hyperfine field, and found a linear correlation between β and H_{hf}. However, the spectra can probably be interpreted in several ways.

Because of the difficulties in making unequivocal analyses, usually the canting effect is only characterized from the relative area of the two and five lines. This allows one to deduce an average canting angle, or the thickness of the canted layer in a core–shell model with complete alignment of the spins in the core and random orientation at the periphery[253]. If D is the diameter of the particle and t the thickness of the canted layer, the fraction of canted spins is $q = 1 - (1 - 2t/D)^3$, with $q \sim 6t/D$ for $2t \ll D$. The relative area of the two or five line is then given by

$$A_{2,5} = \frac{q \int_0^{\pi} \sin^3\beta \, d\beta}{4 \int_0^{\pi} \sin\beta \, d\beta} = \frac{q}{6} \qquad \text{(F.6.4)}$$

The canted fraction q is related to the polarization factor p by $q = 3p/(2 + p)$. For a distribution of particle diameter, by assuming that t is independent of D_i and such that $2t \ll D_i$, one deduces $A_{2,5} = \Sigma_i n_i V_i q_i / 6\,\Sigma_i n_i V_i \sim t/D$ with $D = 6\langle V\rangle/\langle S\rangle$, where $\langle V\rangle$ is the mean volume and $\langle S\rangle$ the mean surface. This average diameter is suitable for pure surface effects. Note that the condition $2t \ll D_i$ may not hold for small particle sizes.

Random canting angles between 0 and $\pi/2$, or between π and $\pi/2$, lead to the same result as between 0 and π. The three situations may, however, be distinguished since they lead to two and five lines centered at different positions. These are given by the average effective fields, equal to H_{hf}, $H_{hf} + \frac{1}{2}H_{app}$, and $H_{hf} - \frac{1}{2}H_{app}$ for $(0, \pi)$, $(0, \pi/2)$, and $(\pi, \pi/2)$, respectively.

Studies of metallic surfaces at low temperature generally indicate perturbed properties over one or two atomic planes,[31,224] which yields a perturbed layer with a thickness of the order of 0.2 nm. For ferrimagnetic oxides, because of the presence of anionic planes in between the cationic planes, the perturbed layer should be somewhat thicker, possibly 1.5–2 times thicker depending on the structure. Consistently, with such an estimation, our preliminary results for γ-Fe_2O_3 particles (Fig. F.6.4) give t equal to ca. 0.35 nm for 2.7- to ca. 8-nm particles. For 10-nm particles, t appears unrealistically small. The t values deduced from reported data[253,256,258] vary between 0.3 and 0.8 nm. In particular, for γ-Fe_2O_3 particles prepared by coprecipitation with an average diameter of 6.5,[253] 7.5, and 9 nm,[258] t is ca. 0.5, 0.6, and 0.3 nm, respectively. The comparison of the various data suggests some influence of the preparative conditions. Systematic investigations would be very worthwhile.

F.6.3.4. Conclusion. Mössbauer experiments are appropriate for characterizing surface phenomena. The difficulties in obtaining clear information mainly come from the interplay of many distributed parameters in the materials. Surface phenomena are strongly dependent on the preparation and sampling techniques. It may therefore be difficult to establish some correlations between the results from different laboratories, eventually from one laboratory. To characterize the properties of the surface, it is indispensable to perform systematic investigations with, for instance, variation of the particle size for a similar surface state or variation of the surface state for a similar size. Utilization of the ^{56}Fe and ^{57}Fe isotopes is very promising. One should not, however, forget that the quality of the surface depends on the coating technique and that chemical and thermal treatments can modify the surface state significantly.

There is not much doubt that the magnetic properties of the atoms near the surface are different from the properties of the atoms in the interior. The thermal evolution of the hyperfine field, and probably that of the recoilless fraction are faster than for the bulk atoms. This may influence the way in which Mössbauer spectroscopy senses the superparamagnetic relaxation.

F.6.4. Superparamagnetic Relaxation in Zero Applied Field

F.6.4.1. Introduction. If the surface-related phenomena are negligible, the electronic spins in the particle will fluctuate in unison (see Section D.3.1), and the orientation of the hyperfine field will fluctuate like the orientation of the magnetic moment of the particle **m**. The Mössbauer nucleus senses the relaxation of **m** via the hyperfine interaction and thus the Mössbauer spectrum can, in principle, be influenced by all the trajectories of **m**.

The important characteristic times of the problem are the inverse of characteristic angular frequencies, which are related to characteristic energies via expressions of the type $E = \hbar\omega$. These characteristic times are (i) the mean life time of the Mössbauer excited state, τ_N, which determines the minimum width of the energy levels, $\Gamma_N = \tau_N^{-1}$, expressed in angular frequency; (ii) the time $\tau_L = \omega_L^{-1}$, where ω_L is the angular frequency of the Larmor precession of the magnetic moment of the nucleus in the excited state; and (iii) characteristic times determined by the relaxation process and related to the angular frequencies of the nuclear transitions. Note that the definition of the characteristic times and the relaxation times as the reciprocal of an angular frequency ensures consistency with the expression of the relaxation time as stated by Brown's equation[48] by using the usual value of the gyromagnetic ratio expressed in angular frequency ($\gamma_0 \sim 2 \times 10^7\,\mathrm{G}^{-1}\,\mathrm{s}^{-1}$ for iron) (see Section F.5.1).

The Zeeman splitting of the nuclear excited state (see Ap.4) is resolved if the spacing between the levels is larger than the width of the levels, that is, $\omega_L = |g_e|\mu_N H_{\mathrm{hf}}/\hbar > \Gamma_N$ or $\tau_L < \tau_N$. This condition is generally fulfilled for the ^{57}Fe isotope in magnetically ordered materials because of the large value of the hyperfine field. Typically, a hyperfine field $H_{\mathrm{hf}} = 500\,\mathrm{kOe}$ gives $\tau_L = 4 \times 10^{-9}\,\mathrm{s}$, whereas $\tau_N = 1.4 \times 10^{-7}\,\mathrm{s}$. In other words, there is always sufficient time for several complete Larmor precessions to take place before the nucleus decays, and τ_N is not the time scale determining the relaxation behavior.

τ_L can be thought of as the measuring time appropriate to the observation of a hyperfine interaction. Roughly speaking, if the relaxation time τ is such that $\tau \gg \tau_L$, the orientation of $\mathbf{H}_{\mathrm{hf}}$ hardly changes during the time of one Larmor precession. The nucleus practically senses the full interaction and produces a quasi-static six-line spectrum. If $\tau \ll \tau_L$, the orientation of $\mathbf{H}_{\mathrm{hf}}$ changes many times before the completion of one Larmor precession. The nucleus senses a zero time-averaged hyperfine field, and the spectrum resembles that of a paramagnet, with one or two lines depending on the quadrupole interaction.

The quadrupole interaction raises a first difficulty.[269] For a rigorous analysis, it is necessary to know the principal directions of the electric field gradient (EFG) tensor and the easy directions (see Ap.4). For the bulk material, if $2\varepsilon = eQV_{ZZ}/2 \ll \hbar\omega_L$, the accessible parameter is the quadrupole shift $\varepsilon_m = |E_Q| = \varepsilon(3\cos^2\zeta - 1 + \eta\sin^2\zeta\cos 2\xi)/2$ as given by Eq. (Ap.9). For fine particles, since **m** orientation changes are not instantaneous, the measured parameter is an average, $\langle \varepsilon_m \rangle$, which depends on the relevant range of **m** orientations. The quadrupole shift can therefore be different for the superparamagnetic state and the blocked state, and different from the bulk value. For instance, for an axial EFG tensor ($\eta = 0$) and uniaxial magnetic anisotropy with both symmetry axes coinciding, the quadrupole shift will be $\varepsilon_m = \varepsilon$ for the bulk material, and $\langle \varepsilon_m \rangle = \varepsilon/4$ for the superparamagnetic state assuming that all **m** orientations between the two easy directions are equiprobable.[17] In the blocked state, **m** fluctuates near one easy direction. Hence, the angles ζ and ξ are not fixed and, in general, it will be necessary to consider an average or a distribution of ε_m. In addition, if 2ε is not small compared to $\hbar\omega_L$, the problem becomes very complex and each case must be treated separately.

Because of the energy range of the nuclear transitions, perturbed spectra are produced over a significant range of relaxation times around τ_L, at least one order of magnitude above and below, but dependent on the static parameters of the bulk material and the relaxation process. The relaxation lineshape depends on the actual motion of the hyperfine field vector (**m**) and, in general, rather complicated lineshapes may be expected.

Due to the distribution of particle volume and shape, the experimental spectra consist of the sum of subspectra with different relaxation times. If only a small fraction of the particles have a relaxation time in the critical region, the relaxation lineshape effects will be weak and the Mössbauer spectra as a function of the temperature will be described as consisting of a bulklike component and a paramagnetic component in varying amounts. This is illustrated in Figure F.6.1 for a ferritin sample.[210] Figure F.6.5 shows a set of spectra for quasi-non-interacting γ-Fe_2O_3 particles. In this case, too, a superparamagnetic doublet grows at the expense of the magnetically split component as the temperature increases, but the six-line pattern also shows an increasing degree of asymmetrical line broadening, indicating significant relaxation effects. Relaxation effects are generally more important for nanometer ferro- and ferrimagnets than antiferromagnets. This is essentially an effect of the τ_0-factor in the expression of the relaxation time $\tau = \tau_0 \exp(KV/kT)$ where K is the effective anisotropy energy constant and V the volume of the particle.

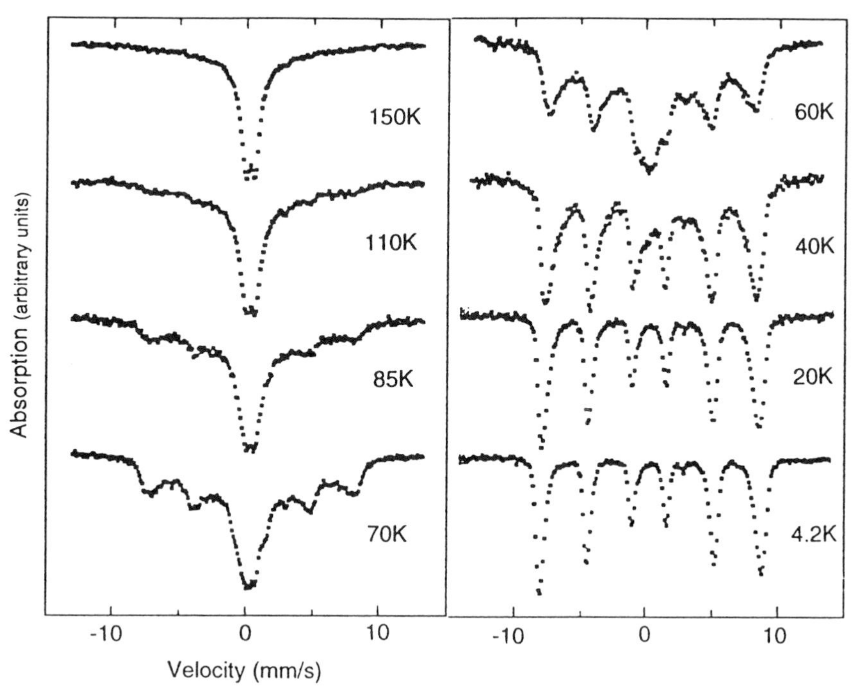

Figure F.6.5. Mössbauer spectra at various temperatures of quasi-non-aggregated 4.6-nm γ-Fe_2O_3 particles at low concentration in a polymer.[267]

The value of τ_0 determines the temperature (volume) interval required to span the Mössbauer time window. For ferromagnetic particles, τ_0 is normally of the order of 10^{-10}–10^{-11} s.[58,59,137] Because this is not very small compared to τ_L, the critical temperature (volume) interval is rather large and the population of particles with critical relaxation times can be significant even if the volume distribution is not narrow. In these conditions the problem of the relaxation lineshape can become crucial. This problem is generally much less critical for antiferromagnetic particles because τ_0 is smaller by approximately two orders of magnitude[58,210,270] (see Section H).

Numerical information about the **m** relaxation process cannot be deduced from the spectra without a model of the lineshape in the presence of superparamagnetic relaxation. The influence of related phenomena has been treated in many theoretical works; see, for example, Refs. 271–282. The rigorous analysis of the problem as stated by Brown's equation [Eq. (D.13)] is a complicated mathematical task even for the single uniaxial particle, and the lineshape has been determined in some special cases using simplifying assumptions. Because of the complexity of the phenomena and also lack of an amenable satisfactory lineshape model, one often simply determines an average blocking temperature or uses low-temperature data to deduce an effective anisotropy energy constant.

In the following, after a survey of most representative lineshape models and some considerations about analyzing the experimental spectra, the determination of the blocking temperature and the exploitation of low-temperature spectra will be discussed.

F.6.4.2. Lineshape Models. The Mössbauer lineshape can be influenced by all relaxation modes of the Fokker–Planck equation (see Section D.3). Because the relative importance of these modes depends on their population, it should be necessary to know both the eigenvalues of Brown's equation and the amplitudes of the associated modes. In fact, to determine the lineshape, it is necessary to connect the dynamics of the stochastic vector **m** given by Brown's equation with the quantum dynamics of the nuclear spin. This necessitates the use of superoperator Fokker–Planck equations[281] and, to our knowledge, the problem has not yet been completely solved.

A. THE STOCHASTIC RELAXATION MODEL. The most general theories of magnetic relaxation in Mössbauer spectroscopy involve stochastic models; see, for example, Ref. 283 for a review. A formalism using superoperators (Liouville operators) was introduced by Blume,[272] who presented a general solution for the lineshape of radiation emitted (absorbed) by a system whose Hamiltonian jumps at random as a function of time between a finite number of possible forms that do not necessarily commute with one another. The solution can be written down in a compact form using the superoperator formalism.

By using Afanas'ev's notations,[275] if the hyperfine field $\mathbf{H}_{\mathrm{hf}}(t) = \mathbf{n}(t)H_{\mathrm{hf}}$ takes a finite set of distinct orientations $\mathbf{n}_a$ $(a = 1, 2, \ldots, N)$, between which random transitions occur, the lineshape in the absorption spectrum is given by an expression of the type

$$\Phi_{\mathrm{abs}}(\omega) = -\mathrm{Im}\left\{(j\eta)\rho\hat{G}\left(\omega + \frac{i\Gamma}{2}\right)(j^{+}\eta^{*})u\right\} \qquad \text{(F.6.5)}$$

where Γ is the width of an excited nuclear level, η is the polarization vector of the incident γ-rays, j is the operator of the nuclear current responsible for the transitions between the ground $|m_g\rangle$ and excited $|m_e\rangle$ states of the nucleus, ρ is a row having the dimensionality N of the electronic spin system, and u is a unit column of the same dimensionality. The elements ρ_a are the relative probabilities of the electronic states $|a)$. $\hat{G}$ is a superoperator acting in the space of the $N(2I_e + 1)(2I_g + 1)$ functions $|m_g\rangle\langle m_e||a)$ and its form is

$$\hat{G}\left(\omega + \frac{i\Gamma}{2}\right) = \left(\omega - \hat{L}_{\text{hf}} - i\hat{R} + \frac{i\Gamma}{2}\right)^{-1} \tag{F.6.6}$$

The superoperator $\hat{L}_{\text{hf}}$ is the Liouville operator of the hyperfine interaction. It is diagonal in the electronic variables and is given by

$$\hat{L}_{\text{hf}}(a) = (A_e\mathbf{I}_e - A_g\mathbf{I}_g)\mathbf{n}_a \tag{F.6.7}$$

with $A_e = g_e\mu_N H_{\text{hf}}/\hbar$ and $A_g = g_g\mu_N H_{\text{hf}}/\hbar$, assuming $\varepsilon = 0$. The superoperator $\hat{R}$ describes the relaxation process. It is diagonal in the nuclear variables, and its elements are $R_{ab} = p_{ab}$ and $R_{aa} = -\Sigma_{b\neq a}\, p_{ab}$ where p_{ab} is the transition probability per time unit from state $|a)$ to state $|b)$.

These formulas completely determine the absorption spectrum. In the present case in its general form, the determination of the matrix elements of $\hat{G}$ is equivalent to solving the set of $(2I_e + 1)(2I_g + 1)$ differential equations of the Brown type.[284] The problem is complicated especially because the Hamiltonian does not commute with itself at different times. Because of the mathematical difficulties, approximate models have been used.

B. BASIC DISCRETE ORIENTATION MODELS. The simplest model is the symmetrical two-level model in which the vector $\mathbf{H}_{\text{hf}}$ hops between two opposite directions.[271] The lineshape has a simple analytic expression, which may be expressed[284] by

$$\Phi_{\text{abs}}(\omega) = -\text{Im} \sum_{j=1}^{3} \frac{\omega_1 c_j}{\omega_1\omega_2 - \omega_{\text{hf},j}^2} \tag{F.6.8}$$

with

$$\omega_1 = \omega + i\left(2p + \frac{\Gamma}{2}\right)$$

$$\omega_2 = \omega + \frac{i\Gamma}{2}$$

where $\omega_{\text{hf},1} = (-3A_e + A_g)/2$, $\omega_{\text{hf},2} = (-A_e + A_g)/2$, and $\omega_{\text{hf},3} = (A_e + A_g)/2$; c_j is the normalized intensity of the lines j and $7 - j$, and p is the probability of hopping per time unit. In Brown's model (see Section D.3.2), the relaxation mode given by the smallest nonvanishing eigenvalue, λ_1, of the Sturm–Liouville equation is the only significant mode.[64]

Hence, p^{-1} can be identified with $\tau_+ = \tau_- = 2\tau$, where τ_+ and τ_- are the relaxation times for the passage over the energy barrier in one direction and the reverse (see Section D.I), and τ is the relaxation time for **m** reversal (see Section D.3.2). Spectra calculated for different values of $p^{-1} = \tau_\pm = 2\tau$ are shown in Figure F.6.6. According to Ref. 273, a given

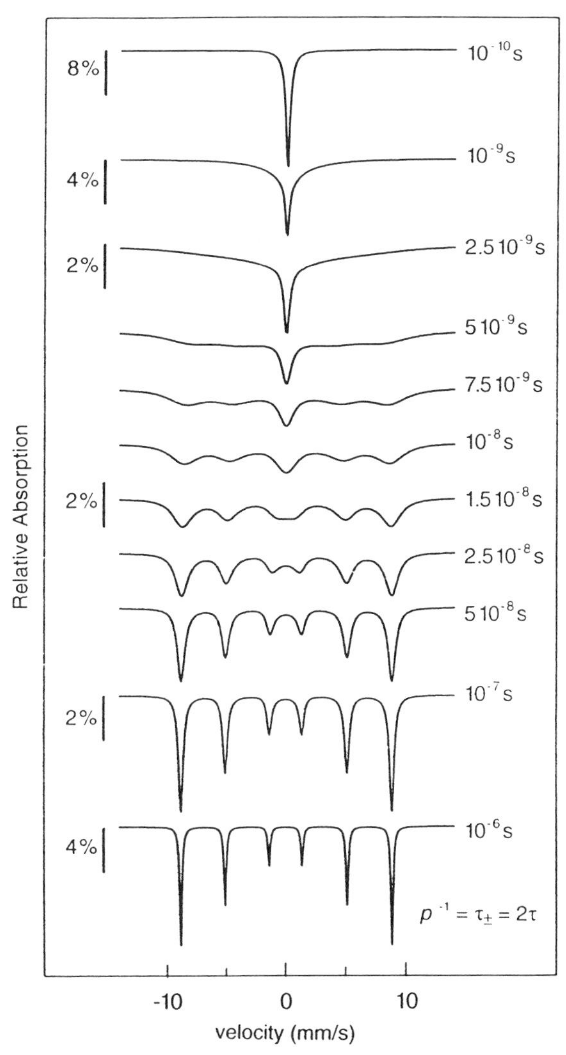

Figure F.6.6. Theoretical Mössbauer spectra calculated for various relaxation times in the discrete two-level relaxation model for uniaxial symmetry [Eq. (F.6.8)]. The linewidth is 0.2 mm/s and the hyperfine field is $55T$.

pair of lines, j and $7-j$, collapse at a relaxation time τ_j given by

$$2\tau_j = \tau_{\pm} = \frac{1}{\omega_{\mathrm{hf},j}} \tag{F.6.9}$$

The case of cubic symmetry has been treated by Afanas'ev and Onishchenko.[275] In case of [100] easy directions, the matrix of $\hat{G}$ is of rank 48, but the symmetry of the problem allows one to find an analytic expression given by

$$\Phi_{\mathrm{abs}}(\omega) = -\mathrm{Im}\,\frac{\varphi_0(\omega) + \varphi_1(\omega)}{6[1 - ip_a\varphi_0(\omega)]} \tag{F.6.10}$$

with

$$\varphi_0(\omega) = \sum_{j=1}^{3} \frac{\omega_1 c_j}{\omega_1\omega_2 - \omega^2_{\mathrm{hf},j}}$$

$$\varphi_1(\omega) = -16p_b(p_a - p_b)\prod_{j=1}^{3} \frac{1}{\omega_1\omega_2 - \omega^2_{\mathrm{hf},j}}$$

$$\omega_1 = \omega + i(4p_a + 2p_b + \tfrac{1}{2}\Gamma)$$

$$\omega_2 = \omega + i(6p_a + \tfrac{1}{2}\Gamma)$$

where p_a is the probability of transition per time unit between electronic states with change of direction of $\mathbf{H}_{\mathrm{hf}}$ by an angle $\pi/2$, and p_b is the probability of a transition with $\mathbf{H}_{\mathrm{hf}}$ reversal. If $p_b = 0$ or $p_b = p_a$ then $\varphi_1(\omega) = 0$. These two situations, no direct reversal or change of orientation by $\pi/2$ four times more probable than reversal, are nearly equivalent regarding the shape of the Mössbauer spectrum. A set of spectra calculated in the case $p_a = p_b$ is shown in Figure F.6.7. The comparison with the uniaxial case (Fig. F.6.6) shows that it may be essential to use the appropriate formulation for interpreting the effects of a size distribution. Such effects were investigated by Belozerskii et al. for uniaxial[276] and cubic symmetry.[277] In Brown's model, at least two ([100] easy axis) or three ([111] axis) modes are necessary to describe the relaxation of **m** (see Section D.3.4) The eigenvalues can be deduced from approximate formulas or by numerical calculation,[72–74] but the amplitudes of the eigenvalues are unknown (a formula analogous to (D.23) must be used). Hence, there is no appropriate formulation to date for the probabilities of transition relevant to Mössbauer spectroscopy.

The above models are discrete models. Then, **m** is constrained to lie along the easy directions and the jumps are supposed to be instantaneous.

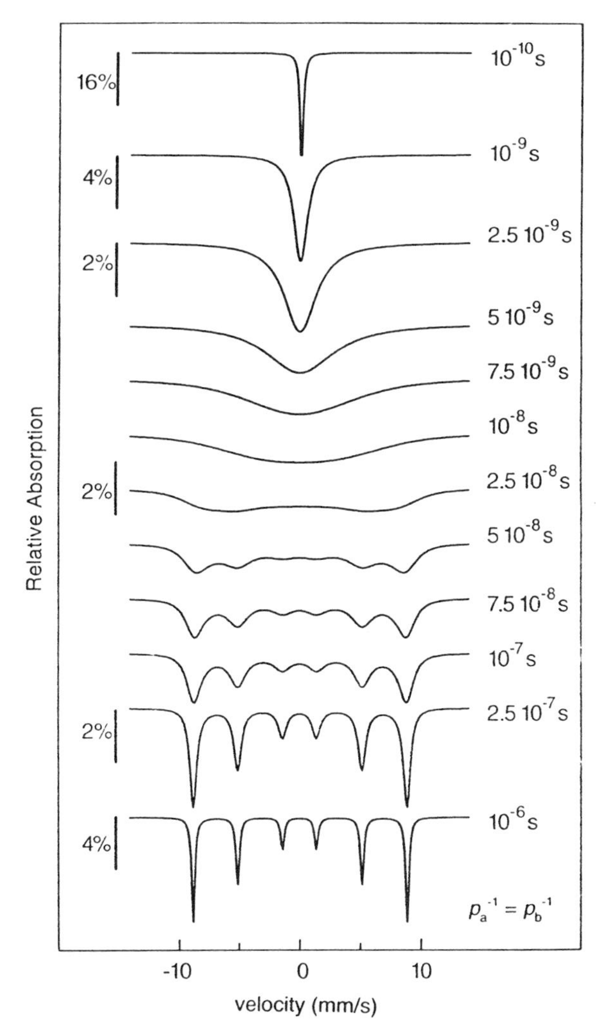

Figure F.6.7. Theoretical Mössbauer spectra calculated for various relaxation times in the discrete relaxation model for cubic symmetry with relaxation between the [100] directions [Eq. (F.6.10)]. The linewidth is 0.2 mm/s and the hyperfine field is $55T$.

The range of orientations between the easy directions is not taken into account and the fluctuations close to one easy direction are excluded.

C. DIFFUSIONAL MODELS. Jones and Srivastava[278] proposed a many-state model for uniaxial anisotropy in which all orientations of **m** with respect to the quantification axis are included in the calculation of the lineshape.

However, the absolute relaxation rate is a free parameter. Belozerskii et al.[279,280] proposed a model of discrete orientations allowing for the influence of precession of the magnetic moment around the easy direction. The lineshape has a relatively simple expression in the absence of relaxation, but the formulation becomes very complicated in case of precession and relaxation. Nevertheless, the lineshape can be computed for a given relaxation rate $p = 1/\tau$. Sedov[282] proposed a diffusion model for uniaxial anisotropy based on Brown's equation. The diffusion is assumed to proceed in the presence of regular precession that is fast compared to the nuclear Larmor frequencies. The lineshape is given by Eq. (F.6.6) where the superoperator $\hat{R}$ is given by a Sturm–Liouville equation similar to Eq. (D.15). Three temperature sets of spectra are presented in Ref. 282, as a function of a parameter somewhat equivalent to α/τ_D where $\alpha = KV/kT$ and τ_D is the time constant of diffusion given by $\tau_D = VM_{\text{nr}}(0)(\eta_r^{-1} + \eta_r)/\gamma_0 kT$ according to Eq. (D.17) neglecting the thermal variation of the magnetization.

By varying the rate of diffusion (precession), these models can provide a continuous connection between the two-level relaxation and a static hyperfine field distribution. The calculated spectra generally resemble the experimental spectra more than the spectra calculated using the discrete orientation models. In particular in the limit of slow diffusion (precession), they reproduce features that are frequently observed for strongly interacting particles (see Section F.6.4.3). However, such a situation is not in accordance with the rate of diffusion normally predicted by Brown's theory [Eq. (D.17)] with realistic values of the parameters and, to our knowledge, there is no other theory to date, except Néel's theory (see Section D.2), for predicting the rate of diffusion. In addition, the applicability of the above models is limited by the presence of free parameters, or by the mathematical complexity. The lineshape may be computed for given values of the parameters for one relaxation time, even a distribution of relaxation times, but there is much doubt about the success of the converse operation, that is, the determination of the parameters in the τ expression from the profile of the experimental spectra, because of various sources of ambiguity. It therefore seems difficult to apply these models, even to appreciate their validity.

D. LOW-TEMPERATURE LIMIT. In the low-temperature limit, the lineshape is essentially influenced by the fluctuations near one easy direction. These fluctuations are contained in the Fokker–Planck equation. Therefore, their amplitude and rate and, hence, the position and the shape of the spectral lines depend on the same parameters as τ, namely α and τ_D. The diffusional models[278–280,282] predict a strong asymmetrical broadening of the lines in the limit of slow diffusion (see comments above). As the diffusion rate increases, the line asymmetry reduces and the lines shift. An expression of the spectral line as a function of α and τ_D has been reported for small angular deviations.[282]

Mørup[18,94,285] considered the limiting case of fast diffusion, that is, much faster than the nuclear Larmor precession. For a particle with magnetic energy $E(\theta) = E_B \sin^2\theta$ and with $\mathbf{m}$ fluctuating about the easy direction at $\theta = 0$, the probability that $\mathbf{m}$ forms an angle between θ and $\theta + d\theta$ with the easy direction is given by

$$p(\theta)\, d\theta = \frac{\exp\{-E(\theta)/kT\} \sin\theta\, d\theta}{\int_0^{\pi/2} \exp\{-E(\theta)/kT\} \sin\theta\, d\theta} \tag{F.6.11}$$

The Mössbauer nucleus only senses the average projection of the hyperfine field ($\mathbf{m}$) onto the easy direction, that is,

$$H_{\text{obs}} = H_0 \overline{\cos\theta} \tag{F.6.12}$$

where H_0 is the hyperfine field in the absence of fluctuations and $\overline{\cos\theta}$ is given by

$$\overline{\cos\theta} = \int_0^{\pi/2} \cos\theta p(\theta)\, d\theta$$
$$= \frac{1 - \exp(\alpha)}{2\sqrt{\alpha} D(\sqrt{\alpha})} \tag{F.6.13}$$

where $D(\cdot)$ is the Dawson integral, and $\alpha = E_B/kT$.

For $\alpha \gtrsim 5$, $\overline{\cos\theta}$ is well approximated by[80]

$$\overline{\cos\theta} \sim 1 - \tfrac{1}{2}\alpha^{-1} - \tfrac{1}{2}\alpha^{-2} - \tfrac{5}{4}\alpha^{-3} - \tfrac{37}{8}\alpha^{-4} \tag{F.6.14}$$

The first-order approximation, satisfactory for $\alpha > 20$, results in an error not exceeding 5% for $\alpha \geq 5$.

For an arbitrary form of the magnetic energy, expressed by $E = E(u_x, u_y, u_z)$ as a function of the direction cosines of the magnetization vector, the average value of the hyperfine field is given by[94]

$$H_{\text{obs}} = H_0 \langle u_z \rangle \tag{F.6.15}$$

with the low-temperature approximation

$$H_{\text{obs}} \sim H_0 \left[1 - \tfrac{1}{2} kT \left\{ \left(\frac{\partial^2 E}{\partial u_x^2} \right)_0^{-1} + \left(\frac{\partial^2 E}{\partial u_y^2} \right)_0^{-1} \right\} \right] \tag{F.6.16}$$

This model can only be applied if the relaxation between the easy directions has a negligible influence on the magnetic splitting. As values of α larger than 5–10 normally result in relaxation times longer than 10^{-8} s, it is generally accepted that Eq. (F.6.12) can only be applied for values of α larger than 5–10.[94] The maximum reduction in the magnetic splitting solely due to the fluctuations near one easy direction is thus of the order of 5–15%. According to Eq. (F.6.12) H_{obs} depends on the particle size. Hence, a distribution of particle sizes will result in an apparent distribution of hyperfine fields. Because this can explain the asymmetrical line broadening often observed at low temperature, Mørup's model has been largely used to interpret low-temperature spectra (see Section F.6.4.5).

F.6.4.3. Modeling of Experimental Spectra. At present, the model combining the two-state relaxation model[271] [Eq. (F.6.8)] and Mørup's model for the low-temperature limit[94] [Eq. (F.6.12)] is the only amenable model for fitting experimental spectra, and also the only model for which the results can be compared with the results from other techniques. Such a model may be useful only if the observed spectra are little affected by relaxation phenomena, that is, when the temperature set of spectra can essentially be described in terms of two spectral components, a magnetic component and a superparamagnetic component, which coexist in varying amounts as a function of the temperature. This indeed implies that particles with a relaxation time in the Mössbauer window are in a relatively small amount at any temperature.

If the distribution in particle volumes is not known, one can determine a value of τ_0 and the distribution in the energy barriers.[286] If the volume distribution is known, one can determine the parameters in the expression of the relaxation time. This has mainly been used[18,269,287–289] with a simplified expression of the relaxation time, that is, with τ_0 either fixed, adjustable or given by Brown's approximation[48] [Eq. (D.19)], to determine the average anisotropy energy per volume unit, K. By using Coffey's approximation valid for all barrier heights[65] [Eqs. (D.25) and (D.17)] and fitting simultaneously variable temperature spectra, one can in principle determine the two parameters in the model, namely K and $q = M_{\text{nr}}(0)(\eta_r^{-1} + \eta_r)$, neglecting the thermal variation of the magnetization. For γ-Fe_2O_3 particles with varying mean size and negligible interparticle interactions,[252] we found that such a procedure could yield valuable information, but only from a semiquantitative point of view. The quality of the fits was generally satisfying, but K appeared significantly underestimated and q overestimated as compared to ac susceptibility data. This is due to the problem of the lineshape because of significant relaxation effects. For interacting particles, we did not take into account the temperature dependence of the interaction energy because of too many parameters and the fits were not as good as for the noninteracting particles. The results, however, suggested that the interactions decreased q, a feature clearly established by ac susceptibility experiments.[59,137]

The statistical model of interaction[45] (see Section E.3.3) is valid when most particles relax. At a given instant, the interaction energy may make the energy minima unequivalent, as in the case of a small static applied field, but the situation changes with time since the magnetic moments do not relax in unison because of the disordered arrangement of the particles and the volume distribution. In principle, a given particle can give two different Mössbauer spectra at two different times. Because of the duration of recording, the spectra corresponding to all possible configurations of the neighboring moments are contained in the experimental spectrum. Hence, in a first approximation the particle will be characterized by the sum spectrum (with somewhat broadened lines) corresponding to the average interaction energy.

The two-level relaxation model will be useless when relaxation phenomena are prevailing, as can be expected in the case of a sufficiently narrow distribution of relaxation times. Figure F.6.8 shows Mössbauer spectra of two samples of interacting Fe particles dispersed in alumina.[45,244,290] In the case of a broad distribution of particle sizes (Fig. F.6.8a), relaxation features are significant at intermediate temperatures, but one can clearly notice the superparamagnetic component growing at the expense of the blocked component as the temperature increases. Such a feature is not observed for the particles with a narrow distribution of sizes (Fig. F.6.8b). At 4.2 K the magnetic moment of the particle does not fluctuate, and the spectrum is characterized by a static distribution of hyperfine fields arising from surface effects (see Section F.6.3.2). As the temperature increases, it seems that there is a gradual evolution of the profile from the blocked component to the superparamagnetic component.

Figure F.6.9 shows Mössbauer spectra of floculated γ-Fe_2O_3 particles. As the temperature increases, the lines of the magnetic component present an increasing degree of asymmetrical broadening, and perturbed spectra without a visible superparamagnetic doublet prevail up to some

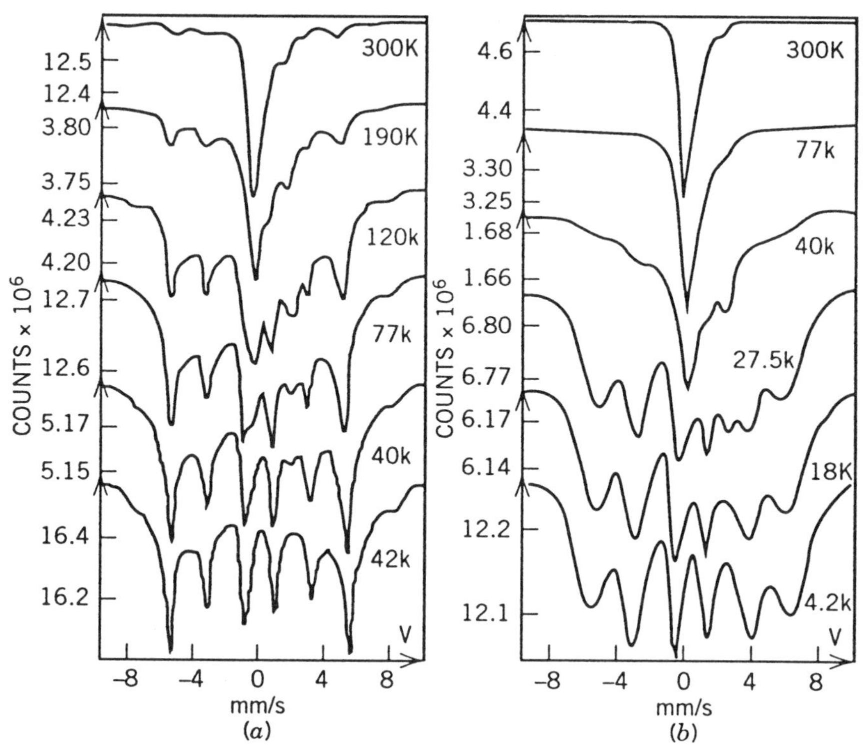

Figure F.6.8. Mössbauer spectra of α-Fe particles dispersed in an alumina matrix, with a rather broad (a) and a narrow (b) distribution of particle sizes.

temperature, after which there is a relatively sudden collapse of the magnetically split component. This is clearly different from the evolution observed for the same, quasi-non-interacting particles (Fig. F.6.5). The increased broadening of the outer lines as compared to the inner lines in the magnetically split component (Fig. F.6.9) cannot be reproduced by the two-level relaxation model, which predicts a collapse of the inner lines first (Fig. F.6.6). The diffusional models (see Section F.6.4.2) may reproduce the observed features, but in the limit of low diffusion rates, which cannot be correlated to the formulation of the relaxation time. The magnetic properties of these particles,[59,136,138] like those of the Fe/Al_2O_3 particles (Fig. F.6.8),[45,119,244] are governed by superparamagnetic relaxation.

Ferro- and ferrimagnetic particles most often give temperature sets of spectra that cannot be described in terms of a blocked component and a superparamagnetic component coexisting in varying amounts. Because

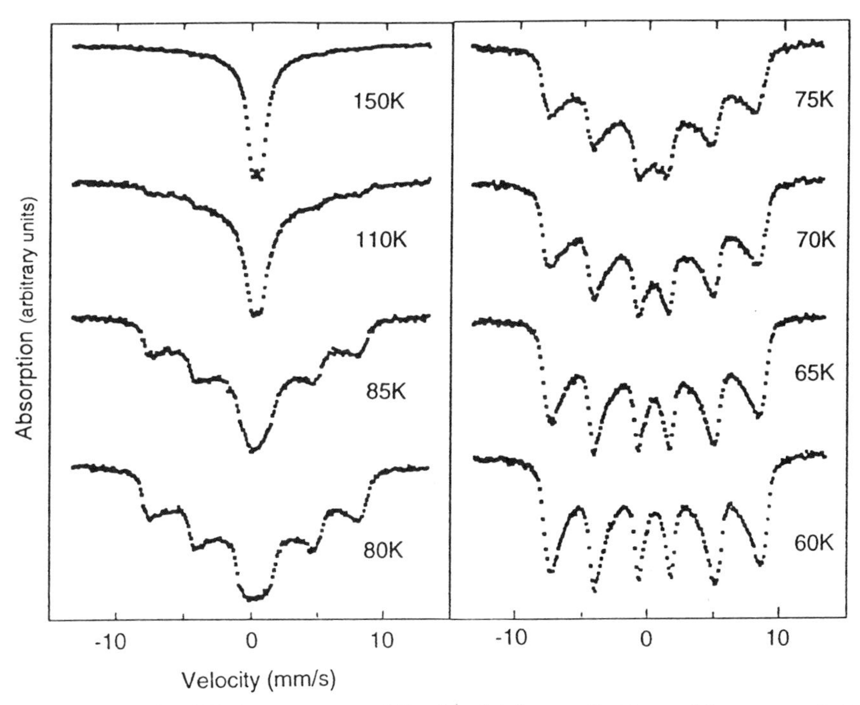

Figure F.6.9. Mössbauer spectra of floculated 4.6-nm γ-Fe_2O_3 particles measured at different temperatures.[267]

the particles are generally interacting, it is often inferred from the Mössbauer features that the properties are not relevant from superparamagnetic relaxation. However, due to the complexity of the phenomena, the analysis of the spectra is generally not unique, and achievable with dynamic and static models.[291] No sound analysis can be done without a previous, detailed characterization of the physical phenomena by other techniques. The effects of the distributions of particle volume and shape play a fundamental role in all measurements. It is our opinion that some of these Mössbauer spectra may be difficult to interpret because we do not know what is the actual distribution of relaxation times and because we lack a suitable model of the lineshape for superparamagnetic relaxation. It is also possible that the time dependence of the interactions introduce further complications.

F.6.4.4. Blocking Temperature. When the sextet and doublet (singlet) coexist, information concerning the distribution of relaxation times is contained in the temperature dependence of the relative areas of the two spectral components. This has been used to determine the distribution of particle volumes.[292] The procedure can be refined by considering an additional, perturbed component assigned to the particles with an intermediate relaxation time.[293]

Generally, the temperature dependence of the blocked and unblocked fractions is simply used to measure the median blocking temperature, T_B, that is, in general, the temperature at which magnetically split and unsplit components represent 50% each of the spectral area. The relevant average volume is the median volume, as a particle of volume V gives a spectrum whose area is proportional to V. The measuring time, τ_m, is difficult to determine precisely because it depends on the relaxation process and cannot be assessed without an assumption on the lineshape. Measuring time τ_m is generally set at 2.5×10^{-9} s or $\tau_{\pm} = 5 \times 10^{-9}$ s,[18] which approximately corresponds to the τ value for which a magnetic splitting is resolved in a two-level relaxation spectrum with a hyperfine field of 550 kOe (Fig. F.6.6).

In general, the experimental spectra show broadened lines. The value of T_B depends on the procedure used for the determination, which also determines the value of τ_m, whence possible difficulties in comparing the values of T_B for different samples. In addition, the uniaxial approximation may not be accurate and variations in the blocking temperature may be caused not only by changes in the main energy barrier but also by variations in the lineshape due to small changes in the landscape of the actual, multivalleyed anisotropy energy. For evident reasons, pure uniaxial symmetry is usually assumed for interpreting the observed phenomena, including differences in blocking temperatures, which may be smaller than the uncertainty on the blocking temperatures themselves, as is deduced from the fits. In the case of a broad distribution of particle sizes the median volume and the average volumes relevant to other techniques may differ notably. The respective T_B values will be difficult to scale if K varies with particle volume and temperature, as in the case of interacting particles. The blocking temperature as measured by Mössbauer spectroscopy is a parameter easy to visualize, but its exploitation may be problematic, especially in the context of a comparison.[294]

In principle, when the distribution of particle volumes is known, the method used for determining the size distribution[292,293] can be applied to the determination of the blocking temperature as a function of the volume.[137] With the same basic assumption that the decomposition of the Mössbauer spectrum into three components (magnetic, perturbed, and superparamagnetic) corresponds to a partition of the τ (V) distribution, $\tau \geq \tau_1$, $\tau_1 < \tau < \tau_2$, and $\tau \leq \tau_2$, from the relative area of the magnetic component at temperature T and the known V distribution, we can deduce the volume V_1 for which $\tau = \tau_1$, that is, $T_B = T$ for $\tau_m = \tau_1$. From the temperature dependence of the magnetic fraction we can therefore determine T_B as a function of V over a significant range of volumes in the sample, provided that τ_1 remains constant. (τ_2 can be similarly exploited, but this is generally of limited interest because it is only concerned with the smallest particles in the sample.) We can expect to maintain τ_1 approximately constant by keeping the parameters of the intermediate component the same for all temperatures and all samples. This indeed assumes that there is no basic change in the anisotropy nor in the hyperfine field distribution. With this mode of operation applied in the case of γ-Fe_2O_3/PVA particles,[137,294] for each sample with negligible interparticle interactions we obtained a linear variation of the type $T_B = aV$, and for each sample of interacting particles the variation presented a curvature at large V. For the spectral decomposition used, τ_m was rather equal to 10^{-8} s. The data were used to scale, for the same average volume, the blocking temperatures as measured by Mössbauer spectroscopy and ac susceptibility experiments over a broad frequency range.[59,137]

Blocking temperature T_B is generally written as $kT_B = E_B/\ln(\tau_m/\tau_0)$. Because T_B depends on $\tau_0(|\mathbf{m}(0)|, \eta_r)$, and E_B, a variation of T_B as measured by Mössbauer spectroscopy can a priori be due to a variation of $\tau_0(|\mathbf{m}(0)|, \eta_r)$ and/or E_B, assuming τ_m constant. Generally, the variation of τ_0 is assumed negligible and a variation of T_B is assigned to a variation of E_B, which may be rather misleading.

Interparticle interactions (see Section E.3) increase the energy barrier, which increases the relaxation time. The additional barrier, $E_{B\text{int}}$, decreases with increasing temperature. The interactions can also damp the motion of the magnetic moment vector.[59,137] An increase in the dimensionless damping factor η_r ($\eta_r \leq 1$) decreases the relaxation time (see Section D.3). Hence, the interactions can increase or decrease the relaxation time depending on which effect, of $E_{B\text{int}}$ or η_r, is prevailing. This is illustrated in Figure F.6.10, which shows a series of Mössbauer spectra at room temperature for 8-nm γ-Fe_2O_3 particles with varying interparticle spacing. When the spacing is sufficiently large, the interactions are negligible and the spectrum (a) can be described in terms of a magnetically split component and an unsplit component with similar areas (median $T_B = 275$ K). As the distance between the particles decreases, the unsplit component grows at the expense of the magnetically split one (b), which nearly disappears (c), then grows (d), and becomes the only visible component (e). The relaxation goes faster and then slower; η_r increases from (a) to (c) and is about the same for (d) as for (c).[59]

Because of the temperature dependence of $E_{B\text{int}}$, the relaxation time

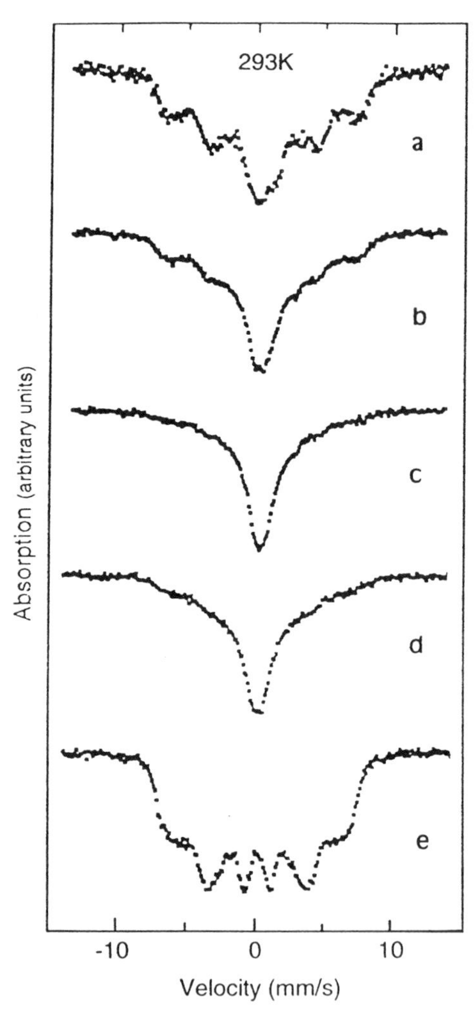

Figure F.6.10. Mössbauer spectra at room temperature of 8-nm γ-Fe_2O_3. Dispersion in a polymer of quasi-non-aggregated particles with a mean center-to-center distance of about (a) 40 nm, (b) 20 nm, and (c) 12 nm, (d) hydrated floculated, and (e) floculate dried at room temperature.[267]

can be decreased by the interactions at high temperature, and increased at low temperature. Therefore, if η_r increases ($\eta_r \leq 1$), we can observe a decrease of the blocking temperature by Mössbauer spectroscopy, and an increase by M_{ZFC} measurements, for instance. Indeed, if η_r does not vary, the blocking temperature will increase for Mössbauer spectroscopy too. These situations have been observed for γ-Fe_2O_3 particles depending on the strength of the interparticle interactions.[59,96,97,134,137,295–299] If the variation of τ_0 is assumed negligible, the decrease of T_B as observed by Mössbauer spectroscopy will be attributed to a decrease of the energy barrier. A model[147,297] was proposed for explaining this feature, and spin glasslike ordering at low temperature was suggested for explaining the increase of T_B observed by M_{ZFC} measurements. For explaining the increase of T_B observed by Mössbauer spectroscopy with further increasing interactions, it was suggested[227,298,299] that the observed blocking was not due to superparamagnetic relaxation but to ordering of the interacting moments. For our γ-Fe_2O_3/PVA samples, the existence of a collective behavior due to the interactions was clearly ruled out by thermoremanence magnetization measurements,[136,138] except perhaps at very low temperature, and the effect of the η_r factor was revealed by ac susceptibility experiments at various frequencies.[59,137]

The Mössbauer spectra are very sensitive to surface phenomena.[223,300] Because such phenomena, like structural defects, may influence the viscosity of the motion of the magnetic moment **m** (and $|\mathbf{m}|$), we can expect a possible role of the damping factor in addition to the energy barrier in determining an increase or a decrease of the blocking temperature in this case too.

F.6.4.5. Magnetic Splitting at $T \ll T_B$. The Mössbauer spectra of fine particle systems at low temperature often show an asymmetrical broadening of the lines (Figs. F.6.5 and F.6.9). Because this can be explained by the size-dependent reduction of the magnetic splitting according to Mørup's model[18,94,285] (see Section F.6.4.3), this model has been largely used for analyzing the spectra at temperatures well below the blocking temperature.

If Eq. (F.6.12) is valid for each particle in the sample, the average observed hyperfine field is given by

$$\langle H_{\mathrm{obs}} \rangle \sim \frac{H_0 \sum_i n_i V_i \overline{\cos \theta_i}}{\sum_i n_i V_i} \qquad \text{(F.6.17)}$$

where n_i is the population of the particles with volume V_i, and H_0 the hyperfine field in the absence of fluctuations assumed independent of the particle size. If the total anisotropy energy per volume unit, K, is independent of the particle volume, the above expression can be developed according to the power expansion of $\overline{\cos \theta}$ (Eq. F.6.14) as a function of various volume averages. In view of the general lack of accuracy in these averages, and the various assumptions in the model, application is usually limited to the first-order approximation given by

$$\langle H_{\mathrm{obs}}\rangle(T) \sim H_0(T)\left(1 - \frac{kT}{2K\langle V\rangle}\right) \tag{F.6.18}$$

where $\langle V\rangle$ is the mean volume. If K is independent of the temperature and if the thermal variation of H_0 is known, the average energy barrier $K\langle V\rangle$ can easily be deduced from the temperature dependence of $\langle H_{\mathrm{obs}}\rangle$. Hence, the analysis of spectra taken at $T \ll T_B$ in terms of a static hyperfine field distribution can provide a simple, most frequently used, way to estimate K if the mean volume is known. In general, however, the thermal variation of H_0 is not known. It is either neglected or taken as that in the corresponding bulk material, although significant differences may result from size and surface effects (see Section F.6.3.2). The K values deduced according to Eq. (F.6.18) are usually found in reasonable agreement with the data deduced by other techniques, but in all cases the uncertainties due to the underlying approximations are generally difficult to estimate; see, for example, Refs. 252, 258, 296, 299, 301, and 302.

We note that if $K = E_B/V$ varies with V, the first-order approximation of Eq. (F.6.17) yields

$$\langle H_{\mathrm{obs}}\rangle(T) \simeq H_0(T)\left(1 - \frac{kT}{2\langle V\rangle}\left\langle\frac{V}{E_B}\right\rangle\right)$$

The average anisotropy energy per volume unit one deduces from the thermal variation of $\langle H_{\mathrm{obs}}\rangle$ is actually the average $\langle V/E_B\rangle^{-1}$. This is not readily exploitable. Such a situation typically occurs when the anisotropy is a combination of, for instance, surface and volume (magnetocrystalline, magnetostatic) anisotropies (see Section C.)

Independent of the assumptions about H_0, the application of Eq. (F.6.18) rests on three basic assumptions. First, the particles are noninteracting. Second, a particle of volume V gives a sextet with Lorentzian lines with width independent of V and T, and third, the position of the lines is determined solely by the value of α (and H_0).

Lineshape effects due to relaxation over the energy barrier may be significant down to low temperature because of the blocking of the smallest particles if the volume distribution is not narrow. Deviations from the ideal width, shape, and position of the lines will also occur if the assumption of fast diffusion is not accurate. The fluctuations around the easy direction have been found faster than about 10^{-10}–10^{-11} s for interacting Fe particles dispersed in alumina ($\eta_r \sim 1$).[130] However, the rate and amplitude of these fluctuations, longitudinal and transverse, depend on α and τ_D, and both the position and shape of the lines should be influenced by these two parameters (see Section F.6.4.2.c).[282] In case of interacting particles, the data should not be exploited insofar as there is no established model to date describing the influence of the interactions when there remain only vibrations in the potential wells and the thermal relaxation is inoperative (very low temperature state). For the intermediate state, when most of the particles are blocked, our statistical model is questionable and certainly requires some adaptations (see Section E.3.3).

For quasi-noninteracting γ-Fe_2O_3/PVA particles[252,294] investigated at $T \lesssim 0.5T_B$, $\langle H_{obs} \rangle$ varies linearly with temperature for all investigated particle sizes, from ca. 3–10 nm, and Mørup's model seems verified. However, the $\langle H_{obs} \rangle$ value extrapolated to zero temperature, $H_0(0)$, slightly decreases with decreasing particle size, and the whole set of data $\langle H_{obs} \rangle(T)/H_0(0)$ scale as $T/\langle V \rangle^{1/3}$. This is not yet well understood and may result from separate or combined features such as special kind of anisotropy, temperature and volume dependence of the hyperfine field, or lineshape effects related to the sample-dependent value of the ratio $K/M_{nr}(0)$ (α/τ_D). For interacting particles, the temperature dependence of $\langle H_{obs} \rangle$ is not linear. The interaction effect, which is negligible, as expected, at 4.2 K, and only weak above, presents a change of regime at some relatively low temperature, ca. 20 K for 7-nm particles with a median blocking temperature of 150–200 K. In the high-temperature regime, the variation of the average hyperfine field due to the interactions resembles the variation of the blocking temperature (see above), suggesting that dynamic features influence the lineshape down to ca. 20 K for the interacting particles, and possibly below for the noninteracting particles as a result of the smaller damping factor.[59,137] The profile of the experimental spectrum indeed changes gradually, and it is difficult to find out if the observed broadening is an effect of the sole position, or both the position and the shape of the lines of the subcomponents. We tend to believe that the application of Eq. (F.6.18), especially with the intention of comparing data, should be restricted to samples with the same magnetization and damping factor.

F.6.5. Influence of a Large Applied Field

In the presence of a large external magnetic field, H_{app}, there is only one energy well (see Section F.3.2). If $KV \ll M_{nr}VH_{app}$, the direction of minimum energy coincides with the direction of the applied field. If the fluctuations of **m** in the energy well are fast compared to the nuclear

Larmor precession, the induced field at the nucleus, H_{ind}, is the average hyperfine field[18,94,303] as deduced from Eq. (F.3.1), that is,

$$H_{ind} \sim H_0 \mathscr{L}\left(\frac{M_{nr} V H_{app}}{kT}\right) \tag{F.6.19}$$

where H_0 is the hyperfine field at the nucleus and $\mathscr{L}(\cdot)$ is the Langevin function. The induced field is parallel or antiparallel to the external field, and the total magnetic field observed by the nucleus is given by

$$\mathbf{H}_{obs} = \mathbf{H}_{ind} + \mathbf{H}_{app} \tag{F.6.20}$$

The relative intensities of the Mössbauer absorption lines will depend on the angle between the direction of the observed magnetic field and the direction of propagation of the γ-rays (see Ap.5).

For $\beta = M_{nr} V H_{app} / kT > 2$, we may use the high-field approximation of the Langevin function in Eq. (F.6.19), and find

$$H_{ind} = |\mathbf{H}_{obs} - \mathbf{H}_{app}| \sim H_0 \left(1 - \frac{kT}{M_{nr} V H_{app}}\right) \tag{F.6.21}$$

which can also be deduced from Eq. (F.6.16). For a volume distribution with H_0 and M_{nr} independent of the volume, the average induced hyperfine field is given by

$$H_{ind} = |\mathbf{H}_{obs} - \mathbf{H}_{app}| \sim H_0 \left(1 - \frac{kT}{M_{nr} \langle V \rangle H_{app}}\right) \tag{F.6.22}$$

where $\langle V \rangle$ is the mean volume.

Equation (F.6.21), which is valid for $\alpha \ll \beta$ and $\beta > 2$, shows that a magnetic hyperfine splitting can be restored by an external field at temperatures at which the magnetic splitting collapses in zero applied field ($T > T_B$). This unambiguously distinguishes the superparamagnetic state from the paramagnetic state.

According to Eq. (F.6.22), if the variation of $M_{nr}(T, H_{app})$ with applied field is negligible, a plot of $|\mathbf{H}_{obs} - \mathbf{H}_{app}|$ as a function $1/H_{app}$ gives a straight line with slope $H_0 kT / M_{nr} \langle V \rangle$ and intercept H_0. If the nonrelaxing magnetization at temperature T is known, the average volume $\langle V \rangle$ can be determined. This method for particle size determination has been widely used.[18,39,223,224,304–306]

As an example Figure F.6.11 shows Mössbauer spectra of α-Fe particles on a carbon support in various applied fields at 80 and 300 K.[304] At both temperatures the spectra show the single line of superparamag-

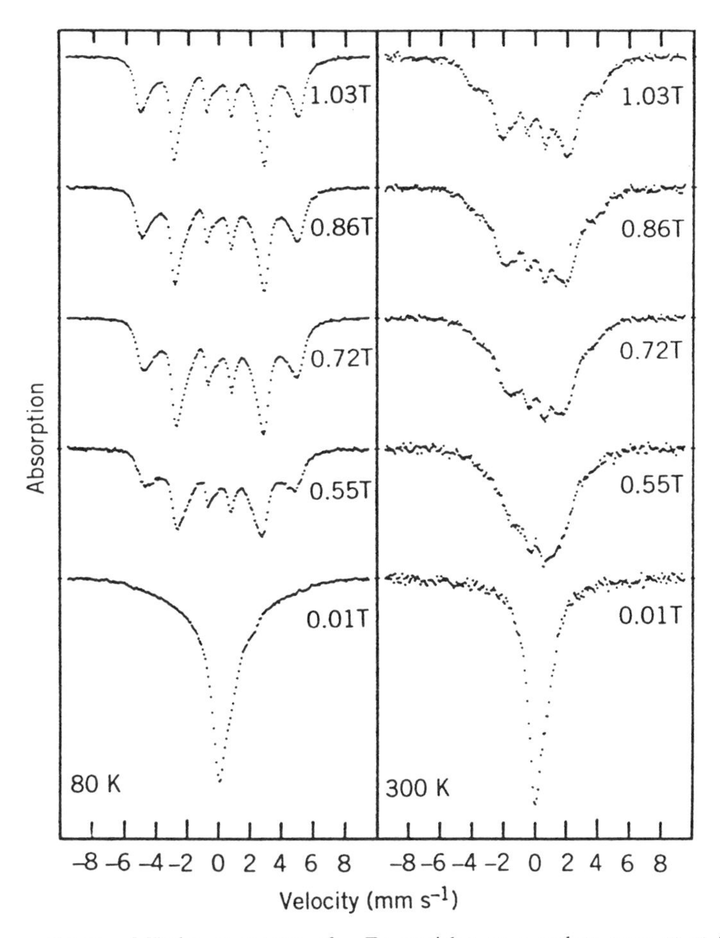

Figure F.6.11. Mössbauer spectra of α-Fe particles on a carbon support at 80 and 300 K obtained in different magnetic fields applied perpendicularly to the γ-ray direction. (Reproduced with permission from Ref. 304.)

netic bcc-iron in zero applied field. In applied field the spectra show a magnetic hyperfine splitting which increases with the strength of the field. Figure F.6.12 shows the induced hyperfine field as a function of the reciprocal of the applied field. At both temperatures a linear dependence is obtained in accordance with Eq. (F.6.22). From the average magnetic moment determined from the slope, an average particle diameter of 2.5 nm was deduced in each case. The hyperfine splittings extrapolated to $H_{\text{app}}^{-1} = 0$ are close to the bulk values at both 80 and 300 K, showing that

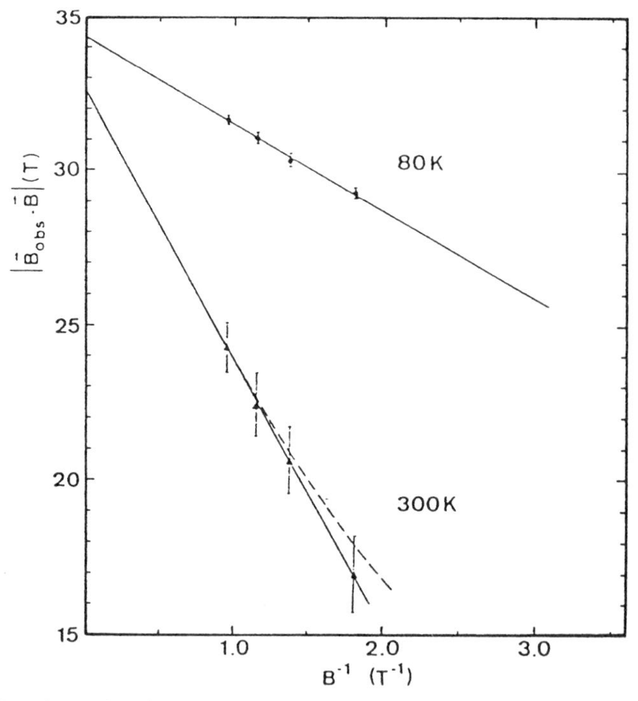

Figure F.6.12. Induced magnetic hyperfine field of the spectra shown in Fig. F.6.11, as a function of the reciprocal applied magnetic field. (Reproduced with permission from Ref. 304.)

the magnetic and electronic properties are similar to those of bulk material apart from the superparmagnetic relaxation.

If the magnetic anisotropy is not very small compared to the Zeeman energy, the induced magnetic hyperfine splitting will depend on both the magnitude of the applied field and the angle between the easy direction and the applied field. A distribution in magnetic hyperfine fields will be present in spectra of samples of randomly oriented particles. In the case when the induced field is antiparallel to the external field, the difference in magnetic splitting for particles with the easy direction parallel and perpendicular to $\mathbf{H}_{app}$ is given by[303]

$$\Delta H_{obs} = H_0 \beta^{-1} \left[\frac{3\alpha}{\beta} - 2\left(\frac{\alpha}{\beta}\right)^2 \right] \tag{F.6.23}$$

The width of the distribution will be of the same order of magnitude. The effects of the distribution in particle sizes will lead to additional broadening. It has been shown[303] that for a sample of identical particles in random orientation, the average induced field is still given by an equation equivalent to Eq. (F.6.21), and that the linear approximation remains a good approximation for quite large values of α, for $\beta \gg 1$. Therefore, the particle size estimated by use of Eq. (F.6.22) is not significantly affected by the magnetic anisotropy.

If the limit of fast fluctuations is not accurate, we can expect dynamic features influencing the lineshape. The situation will be more complicated than for the zero field case discussed in Section F.6.4.2.d, because of the polarization effect on the relative intensities of the absorption lines.

F.6.6. Influence of a Medium or Weak Applied Field

When an external magnetic field is applied along the uniaxial direction, the magnetic energy is given by $E(\theta) = KV \sin^2\theta - M_{nr}H_{app}V \cos\theta$ [Eq. (D.4)] where θ is the angle between **m** and the easy direction. If $h = M_{nr}H_{app}/2K < 1$, there are two unequal energy minima (see Sections B.2 and D.1). The energy barriers, E_{B+} for the passage from the lower minimum ($\theta = 0$) to the upper minimum ($\theta = \pi$), and E_{B-} for the reverse direction, are given by $E_{B\pm} = KV(1 \pm h)^2$.

In the experimental conditions of Mössbauer spectroscopy, the system is in a stationary state. Therefore, as discussed in Section D.3.3, the only appreciable longitudinal modes are the ones associated with $\lambda_{1\pm}$ with $\lambda_1 = \lambda_{1+} + \lambda_{1-}$ where λ_1 is the smallest nonvanishing eigenvalue of Brown's equation. In contrast with the magnetic measurements, which only depend on the net (average) relaxation process along the direction of the applied field, the Mössbauer spectrum is the sum of two subspectra corresponding each to relaxation in one particular direction. The description of the relaxation process in terms of a blocked fraction and a relaxing fraction, which is pertinent for interpreting magnetic measurements (see Section D.1), is not appropriate for Mössbauer spectroscopy.

In the simple discrete model of relaxation between the electronic states $|+)$ and $|-)$ corresponding to $\theta = 0$ and $\theta = \pi$, respectively, the Mössbauer spectrum is determined by the two transition probabilities, $p_+(|+)\rightarrow|-))$ and $p_-(|-)\rightarrow|+))$, and the electronic state populations, ρ_+ and ρ_-. The transition probabilities are given by $p_\pm = \tau_\pm^{-1} = \lambda_{1\pm}/\tau_D$ [Eqs. (D.27) and (D-29)], and the populations are $\rho_\pm = \tau_\pm/(\tau_+ + \tau_-) = p_\mp/(p_+ + p_-)$. Theoretical spectra can easily be calculated.[271,273] With the same notations as in Section F.6.4.2, the Mössbauer spectrum may be expressed by

$$\Phi_{abs}(\omega) = -\mathrm{Im}[\rho_+\Phi_+(\omega) + \rho_-\Phi_-(\omega)] \qquad \text{(F.6.24)}$$

with
$$\Phi_{\pm}(\omega) = \sum_{j=1}^{3} c_j \frac{\varphi_0(\omega) \pm i(p_- - p_+)\varphi_1(\omega)^2}{1 + (p_- - p_+)^2 \varphi_1(\omega)^2}$$

$$\varphi_0(\omega) = \frac{\omega_1}{\omega_1 \omega_2 - \omega_{\mathrm{hf},j}^2}$$

$$\varphi_1(\omega) = \frac{\omega_{\mathrm{hf},j}}{\omega_1 \omega_2 - \omega_{\mathrm{hf},j}^2}$$

$$\omega_1 = \omega + i(p_+ + p_- + \tfrac{1}{2}\Gamma)$$

$$\omega_2 = \omega + \tfrac{1}{2} i\Gamma$$

Most significant differences with respect to the symmetrical case are obtained when τ_+ and τ_- are in the Mössbauer window with τ_+/τ_- significantly greater than 1.[307] The magnetic aspect of the spectrum becomes more marked as h increases since τ_+ and ρ_+ increase whereas τ_- and ρ_- decrease. Two sets of spectra calculated for two values of the parameter $A = 2\pi^{-1/2}\alpha/\tau_D$ and $\alpha = 3$ are shown in Figure F.6.13. One can notice the dual appearance of the spectra for $h = 0.10$.

In actual samples, the easy directions are in random positions with respect to the direction of the applied field. This means that $p_{\pm}$ depend on the angle between H_{app} and the easy axis (see Section D.3.3). In addition, Eq. (F.6.24) must be adapted as the relaxation takes place between θ_1 and θ_2 different from 0 and π. This results in a distribution of relaxation times, which combines with that due to the volume distribution. One may therefore expect[308] a complex evolution of the spectra with temperature, from the six-line pattern to the doublet (singlet), involving mainly a distribution of reduced hyperfine fields dependent on the value of h. To our knowledge, no experimental study by Mössbauer spectroscopy of the relaxation in medium or weak field for ferromagnetic particles has been reported to date.

As noted by Rancourt,[291,307] the nonsymmetrical double potential well can arise, not only from an external magnetic field but also from many other cases, such as structural defects, chemical interaction between the particle and its supporting medium, surface effects, and so on. In all of these cases, the relaxation of **m** can be described using the same model. As discussed in Section F.6.4.3, this model can also describe the effect of the interactions between the particles, but only locally and at a given time.

F.6.7. Concluding Remarks

Mössbauer spectroscopy is quite a unique technique in the present context mainly for two reasons. First, it offers the possibility of investigat-

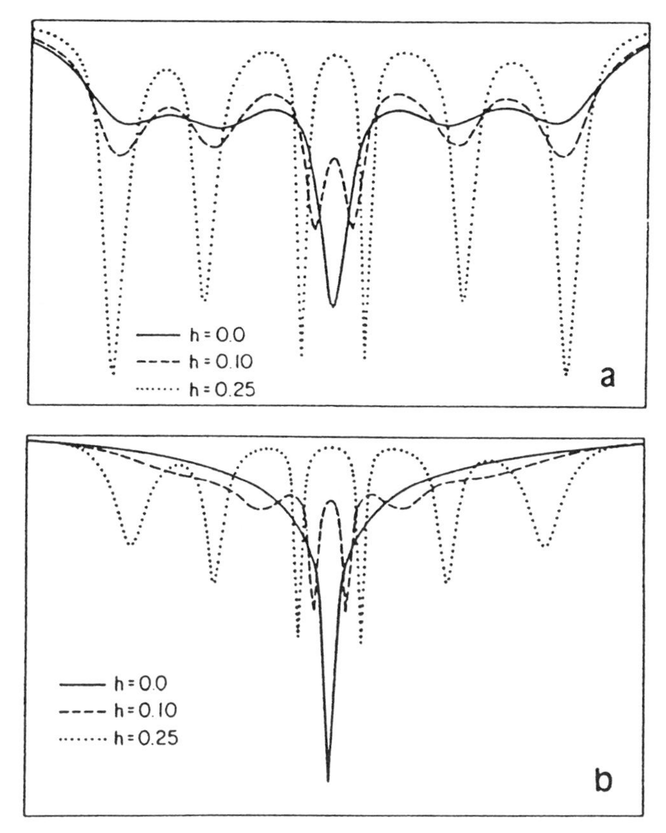

Figure F.6.13. Theoretical Mössbauer spectra in the discrete two-level relaxation model for uniaxial symmetry, with unequal probabilities of the presence of the magnetization vector along the two directions. Calculations for three values of the asymmetry parameter, h, at $\alpha = 3$ with (a) $A = 10^{10}$ Hz and (b) 4×10^{10} Hz (see text). The horizontal scale is from -9.22 to $+10.78$ mm/s, and the static Mössbauer parameters are those of the B site in magnetite at room temperature. (Reproduced with permission from Ref. 307.)

ing the surface properties, and second, it presents some specific advantages for studying the superparamagnetic relaxation due to the relatively short time scale and the local, isotropic character of the measurement. Small variations in the properties of the particles can manifest themselves as rather large changes in the Mössbauer spectrum. Hence, Mössbauer spectroscopy is a very sensitive tool for studies of small particle systems. However, the data are often difficult to interpret in detail, especially in the case of ferromagnetic particles, because difficulties arising from the materials and difficulties inherent in the technique accumulate. The major specific problems are related to the existence of a range of measuring times, and to the requirement of an appropriate lineshape model for deducing any numerical information. Mössbauer spectroscopy is, in principle, sensitive to all relaxation times and subsidiary energy minima may play a significant role. This may influence the distribution in energy barriers as deduced using the assumption of pure uniaxial symmetry. The phenomena as observed by Mössbauer spectroscopy scarcely have a unique interpretation, thus the need for being very cautious when proposing one in particular.

F.7. Ferromagnetic Resonance

It is well known that ferromagnetic resonance (FMR) is a sensitive tool to study the local fields. In fine-particle systems, as the different contributions to the total anisotropy (magnetocrystalline, magnetostatic, surface, shape, interparticle interactions) induce local fields, FMR experiments allow one to obtain information on the anisotropy energy and on the distribution of easy axes.

The theory of ferromagnetic resonance in an assembly of single-domain particles uses, as its basis, the independent grain model, once proposed for the description of FMR in polycrystals.[309–311] Ferromagnetic resonance spectra are described evaluating the dynamic susceptibility $\chi = \delta M/\delta H$ of a single magnetic grain, assuming a fixed direction and strength of the anisotropy field H_a, and then by carrying out the average of χ over the distribution of H_a characteristic of a given sample, thus getting the observable quantity. The model was developed to account for thermal fluctuations of the magnetic moments in superparamagnetic grains[84,312,313] and for the orientational mobility of particles in magnetic fluids (ferrofluids).[84,85]

In an FMR experiment a constant, large, magnetizing field is applied, and a Larmor precession of the particle moments is induced by a small, high-frequency field, perpendicular to H. In the customary FMR technique the spectrometer frequency is fixed (e.g., at 9 and 34 GHz) and the dynamic susceptibility is recorded as a function of the strength and the direction of the external field.

The resonant frequency condition for an isotropic superparamagnet is

$$\omega = \gamma H \tag{F.7.1}$$

On the other hand, for anisotropic magnetic particles, assuming that the anisotropy energy E_a is much smaller than the magnetic energy due to the external field ($E_a = M_{nr} V H_a \ll M_{nr} V H$, where M_{nr} is the nonrelaxing magnetization $M_{nr} = |\mathbf{m}|/V$), the resonance condition is[85]

$$\omega = \gamma \left\{ H_{res} + H_a \left[\frac{\mathscr{L}_2(\xi)}{\mathscr{L}(\xi)} \right] P_2(\psi) \right\} \tag{F.7.2}$$

where ψ is the angle between the anisotropy axis and the external field; $\xi = M_{nr}VH/kT$; $\mathscr{L}(\xi)$ is the Langevin function; $\mathscr{L}_2(\xi) = 1 - 3\mathscr{L}(\xi)$ is the next member of the Langevin function family, defined by the general expression $\mathscr{L}_j = \langle P_j \rangle_0$, with P_j being the Legendre polynomials.

Equation (F.7.2) shows that thermal fluctuations lead to a decrease of the effective anisotropy field (in addition to any actual temperature dependence of the anisotropy constant), which can be written[85]

$$H_{a\text{eff}} = \frac{H_a \mathscr{L}_2(\xi)}{\mathscr{L}(\xi)} \tag{F.7.3}$$

It follows from the asymptotics for $\xi < 1$ that with increasing temperature $H_{a\text{eff}}$ decreases, being $H_{a\text{eff}} \propto \xi \propto 1/T$. Therefore, according to Eq. (F.7.2) the effect of increasing temperature on FMR spectra is that the resonance frequency approaches the value $\omega = \gamma H$, inherent to the isotropic superparamagnet.

For an actual fine-particle system, with volume and easy axis distribution, the dependence of the resonance condition upon the direction of anisotropy axes provides an asymmetric lineshape at low temperature, accompanied by a shift of the resonant field. With increasing temperature, the width reduces and the line transforms to a Lorentzian shape with an isotropic resonance field $H_{res} = \omega/\gamma$, as thermal fluctuations (with $\tau \ll \tau_L$, the Larmor precession) produce an averaging to zero of the anisotropy fields.

The above described temperature evolution of the lineshape is the result of three different mechanisms, having different temperature dependences. In addition to the spreadout of anisotropy axes relative to the external field, another mechanism of inhomogeneous broadening does exist, due to the fact that the instantaneous resonance condition depends on the relative orientation (described by the angle θ) of the magnetic moment of the particle and its anisotropy axis[87,314]:

$$\omega = \gamma[H_{res} + H_a P_2(\psi)\cos\theta] \tag{F.7.4}$$

This mechanism is of a dynamic origin. Finally, in the limit of $\xi < 1$, thermal fluctuations, increasing the rate of precession damping, are expected to yield an homogeneous broadening of the line.

The existence of interparticle interactions leads to an additional term to the local field. For dipolar interactions $H_d \cong \mu/d^3 = \phi_0 M_{nr}$, where d is the interparticle distance and ϕ_0 is the volume fraction of particles. This also yields a contribution to linewidth $\Delta_d H = \phi_0 M_s$ (in the absence of magnetic moments fluctuations).

FMR experiments performed on fine particles dispersed in a nonmagnetic matrix (e.g., Fe, Co, Ni in polymer matrices,[315] α-Fe_2O_3 in Al_2O_3,[316] silica supported Ni particles[311]) show a progressive broadening of the absorption line, accompanied by the shift of the resonance field to lower values, as the temperature decreases.

For some samples (e.g., Fe particles in a polymer matrix,[315] magnesioferrite particles in MgO,[312] and for iron-containing microclusters in borate and silicate glasses[317]) a coexistence of the broad line with a narrow line near $g = 2$ is observed (Fig. F.7.1). With decreasing temperature the broad line grows, the width increasing and the shape becoming more and more distorted, and the narrow line tends to vanish (Fig. F.7.2).

In Ref. 312 the influence of a particles size distribution on the FMR lineshape and width and on their temperature dependence is satisfactorily explained by a model of independent superparamagnetic grains. The two components are attributed to the fractions (changing with temperature) of blocked particles (responsible for the broad line) and superparamagnetic particles (responsible for the narrow line) present in the sample at each temperature. The linewidth is found to increase with decreasing particle size. However, for particles below a given size (corresponding to the characteristic size for superparamagnetism) the linewidth decreases abruptly.[312]

In magnetic suspensions, in addition to the thermal fluctuations of the particle moment, the particle itself moves with respect to the liquid carrier (translational and rotational Brownian diffusion). This leads to the change of the orientational distribution of the particle anisotropy axes under the influence of the applied field.

The theory of ferromagnetic resonance in magnetic suspensions was developed by Raikher and Stepanov,[85] who extended the model for superparamagnetic particles accounting both for thermal fluctuations of magnetic moments and for orientational mobility of particles. The authors calculated the dependence of the equilibrium distribution function for the particle anisotropy axes on the external field. In the limiting cases of $\xi \to 0$ or σ $(= KV/kT) \to 0$ (isotropic particles) the distribution becomes isotropic, whereas for both ξ and σ very large, the distribution function is condensed around the external field direction, along which it shows a sharp peak. They calculated the dynamic susceptibility $\bar{\chi}$ integrating over the contributions $\chi^+(n)$ and over the anisotropy axes distribution function $f(n, H)$, where n is the number density of the particles[85]:

$$\bar{\chi} = \int n\chi^+ f(n, H)\, dn \tag{F.7.5}$$

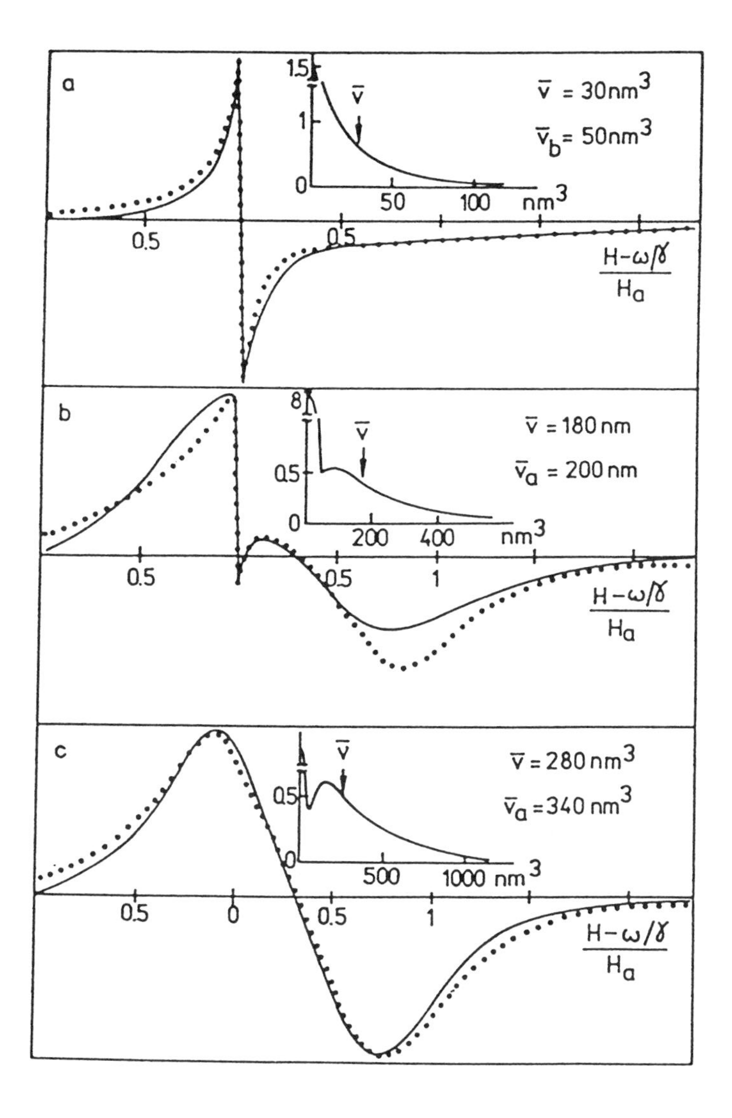

Figure F.7.1. Changes in the lineshape of normalized FMR spectra on magnesioferrite particles in MgO (H parallel to [100] direction of the MgO single crystal for different annealing times at $T = 1073$ K: (a) 5 min, (b) 10 min, (c) 15 min. Dotted lines represent simulations of each line with various particle size distributions; $\bar{v}_a$, $\bar{v}_b$, and $\bar{v}$ represent the mean volumes obtained from anisotropy field measurements, magnetization measurements, and those used for simulations, respectively. (Reproduced with permission from Ref. 312.)

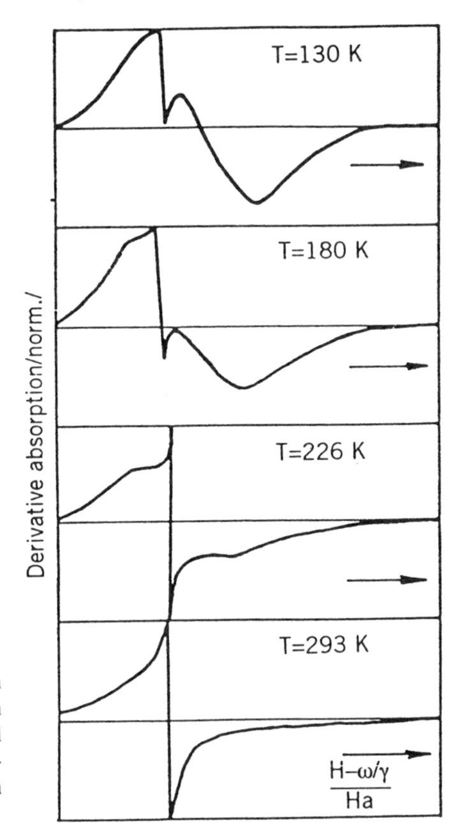

Figure F.7.2. Temperature changes in the lineshape of normalized FMR spectra on magnesioferrite particles in MgO (*H* parallel to [100] direction of the MgO single crystal). (Reproduced with permission from Ref. 312.)

The results of the evaluation of $d\chi''/dH$ for different values of ξ in a dilute magnetic fluid are reported in Figure F.7.3. For small ξ values (Fig. F.7.3f) the effects of anisotropy are averaged to zero by thermal fluctuations, and it results in a Lorentzian line with $H_{res} = \omega/\gamma$, characteristic of an isotropic superparamagnet. For high ξ values (Fig. F.7.3a) the orientational distribution of easy axes yields an asymmetric line shifted from $g = 2$. The two components coexist for intermediate ξ values.

Ferromagnetic resonance experiments were recently performed on diluted γ-Fe_2O_3 ferrofluids as a function of temperature and particle size ($4\ \text{nm} < \phi < 10\ \text{nm}$).[314] The resonance behavior changes drastically with decreasing particle size (Fig. F.7.4): for the sample consisting of large particles (sample 2) a wide Gaussian-like line, shifted from $g = 2$ is obtained; for the sample consisting of sufficiently small particles a narrow line at $g = 2$ is observed. Samples with intermediate particle size exhibit FMR spectra with more complex structure. It can be described by a

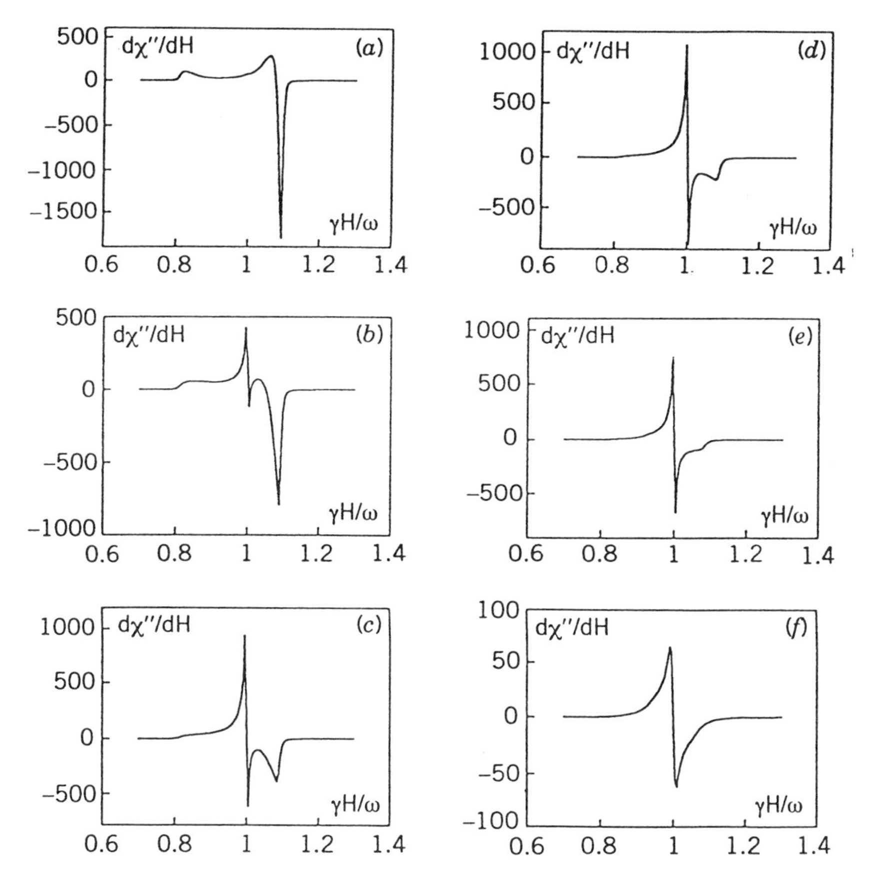

Figure F.7.3. Numerical calculation of $d\chi''/dH$ as a function of the dimensionless field $\gamma H/\omega$ for a randomly oriented assembly of particles; α (damping parameter) = 0.001; $\gamma H_a/\omega = 0.1$; $\xi_0 = 10$, 5, 3, 2, 1, and 0.2, respectively, for curves a, b, c, d, e, and f. (Reproduced with permission from Ref. 314.)

combination of the two contributions, the balance depending on the size distribution function. The g value of the narrow line is $\cong 2$, whatever the sample. For the broad line H_{res} decreases slightly with increasing particle size. As the temperature decreases, the intensity of the narrow line is diminished relative to the broad line. For the intermediate temperature range, spectra display the same complex structure (presence of both signals) as for intermediate particle size. With decreasing T a reduction of the resonance field H_{res} and an increase of the linewidth of the broad component is observed. The narrow line does not shift, but just broadens slightly.

In Ref. 314 experiments were also performed after cooling the sample

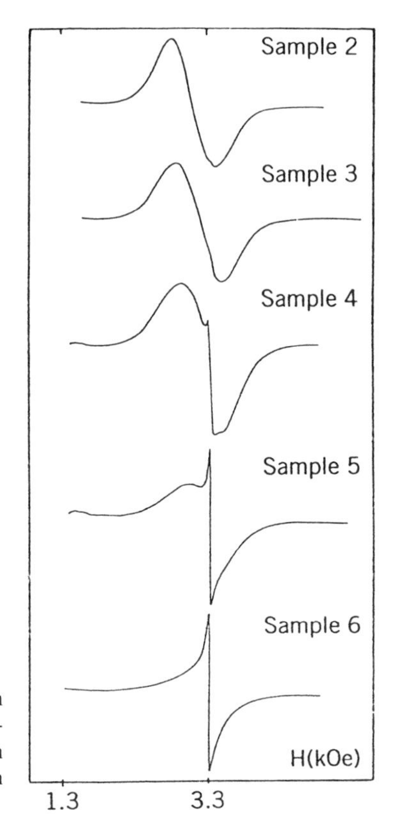

Figure F.7.4. FMR ($f = 9.2$ GHz) at room temperature for a series γ-Fe_2O_3 ferrofluids samples (2–6) with decreasing particle diameter (from 10 to 5 nm). (Reproduced with permission from Ref. 314.)

in a magnetic field (in a plane perpendicular to the radiofrequency field), allowing the anisotropy axes to freeze in a favored direction. Resonance spectra were recorded as a function of the angle between the external resonance field and the cooling field. The two components of the spectrum exhibit different behavior with angular variations. The narrow component is not sensitive to angular variations, neither to the cooling process. The large component, on the contrary, shifts to higher resonance fields when the external field is turned away from the direction of the cooling field. These results confirm that the broad line concerns particles that experience their own anisotropy field, whereas the narrow line concerns particles that do not feel any anisotropy field.[314] The relative intensity of both components as a function of ξ and their distinct behaviors with angular variations fit qualitatively the experimental spectra. However, in the high-temperature limit, the homogeneous broadening predicted by the theory is not observed. Moreover, the theoretical asymmetric shape of the large signal is not found in the experimental spectra, whereas a wide almost symmetric line is observed.

In conclusion, FMR experiments represent a useful tool for studying the dynamical behavior of fine magnetic particles and for obtaining information on the anisotropy energy and on the orientational distribution of easy axes. The proposed theoretical models are able to fit qualitatively the experimental spectra. However some important aspects, strongly affecting the thermal fluctuations of particle moments, have not yet been accounted for, for example, the effect of dipolar interactions in dynamical conditions (the dipolar local fields fluctuate in time in a system of nonidentical particles) and the effect of surface layers. Moreover, some models are valid only if $H_a \ll H$ and cannot be applied when H is small.

F.8. Neutron Experiments

As for X-rays, the neutron technique may give some information about the particle sizes and local ordering, but it also provides an insight about spin correlations and spin fluctuations. We can mainly distinguish two kinds of experiments. The first one, namely the neutron diffraction corresponds to an integration over all the energies of the outcoming neutrons. It thus measures the Fourier transform of the instantaneous spatial correlations. The neutron diffraction may be performed at large angles, for instance, a determination of the particle size may be obtained from the broadening of the Bragg peaks. At small angles, the SANS technique (small-angle neutron scattering) gives much more detailed information. We will discuss this experiment in Section F.8.1. The second one, called inelastic neutron scattering, involves an energy analysis of the neutron cross section, and therefore measures the Fourier transform of the time-dependent correlations. This will be discussed in Section F.8.2. In Section F.8.3, we will report other possible experiments and finally draw some conclusions in Section F.8.4.

F.8.1. SANS Experiments

The neutron cross section is the sum of a nuclear and of a magnetic contribution. If the particle density is high enough, and for quasimonodispersed particles, the nuclear contribution involves a one-particle term and an interparticle term. This nuclear cross section is analogous to the X-rays cross section, and the main results concern the mean particle diameter and the mean distance between neighboring particles. The magnetic contribution involves spin correlations within the particle (the magnetic particle form factor is the Fourier transform of the magnetic density inside the particle), and eventually interparticle spin correlations.

As an example, we show in Figure F.8.1 the Q dependence of the

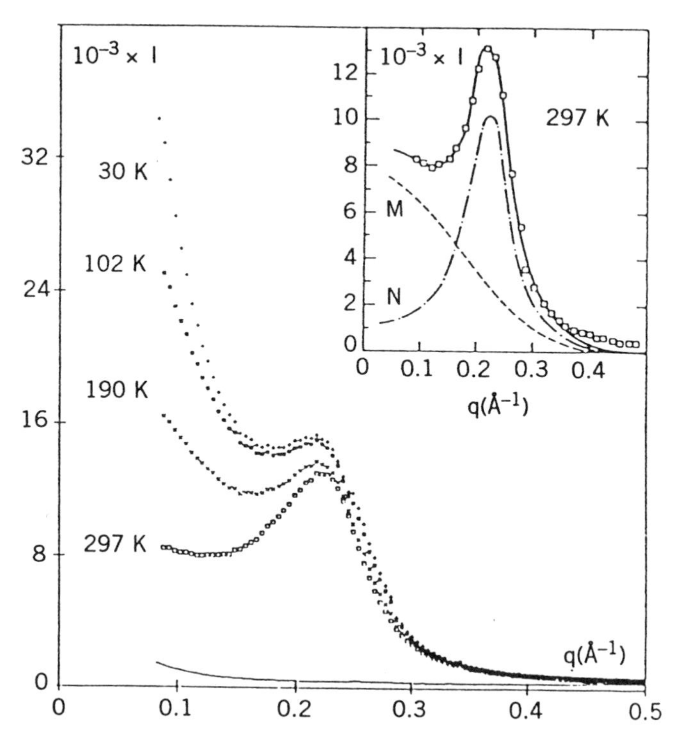

Figure F.8.1. Q dependence of the intensity in neutron counts for fine Fe particles embedded in an Al_2O_3 matrix for several temperatures. In the insert, a fit of the data at 297 K. (Reproduced with permission from Ref. 128.)

intensity for Fe particles embedded in an alumina Al_2O_3 matrix.[128,129] The interference peak at about 0.2 $Å^{-1}$, corresponds to a liquid-like short-range order between particles of very similar sizes. This is the nuclear term analogous to the X-ray one. From this term, one can deduce a mean particle diameter of 20 Å, and a mean distance between neighboring particles of 35 Å. The magnetic contribution, clearly observed in the small Q range, strongly increases with decreasing temperature. From a change in its Q dependence, it is possible to separate the one-particle term, which is dominant down to 100 K, from the interparticle term arising below 100 K. The temperature dependence of the one-particle term corresponds to an increase of the particle magnetization, which may be due to a progressive alignment of spins at the surface of the particle.[132] The interparticle term corresponds to ferromagnetic correlations between the total spins of neighboring particles, induced by dipolar effects. SANS experiments on Fe particles embedded in an SiO_2 matrix performed in a more restricted Q range give similar results, although the interpretation is slightly different.[318] We note that SANS experiments allow the determination of the nonrelaxing magnetization $M_{nr} = |\mathbf{m}|/V$, where **m** and V are the magnetic moment and the volume of the particle, respectively, without applied field and outside any relaxation effect, which is not the case for magnetization measurements (see Sections F.2 and F.3). However, when **m** relaxation is effective, the measured M_{nr} value may be slightly smaller than those corresponding to our M_{nr} definition (**m** is then considered as static; see Section F.3.1), due to the rearrangement of surface spins during rotation (synchronous rotation does not strictly occur, see Section D.3.1).

One may also use a specific property of the neutron scattering, which only measures the spin components perpendicular to the scattering vector **Q**. In applied magnetic field, when using a two-dimensional multidetector, the anisotropy of the intensity in the scattering plane $(\mathbf{Q}, \mathbf{H})$, gives some information about the reorientation of the particles induced by the field H. The intensity in the direction $\mathbf{Q} \| \mathbf{H}$, measures the Fourier transform of the correlations between spin components transverse to the field. By contrast the intensity in the direction $\mathbf{Q} \perp \mathbf{H}$, involves longitudinal together with transverse spin components (see, e.g., Refs. 319 and 320). The results obtained on Fe/Al_2O_3[133] are in good agreement with those expected from the variation of the magnetization under applied field in a superparamagnetic model. In addition, some information is obtained about the effect of the field on the ferromagnetic correlation beween the particles.

Other samples containing particles have been studied by SANS, particularly ferrofluids (see Refs. 320 and 321–324, e.g.), but either particles are too large for showing **m** relaxation or this relaxation is not evidenced.

F.8.2. Inelastic Neutron Scattering

By inelastic neutron scattering, one gets the full $S(Q, \omega)$ function corresponding to the double Fourier transform of the space and time pair correlation function $\langle S_i(0) S_j(t) \rangle$. The characteristic energy scale 0.01–100 meV, which corresponds to the time scale 10^{-10}–10^{-12} s, is typical of anisotropy energies for the lowest values, and of the exchange energies for the highest ones. A special interest in the neutron technique lies in the possibility of observing the dynamics of the particle in zero field so that it can probe the intrinsic susceptibility.

At present, only one study has been published[129–132] and that concern Fe particles in an Al_2O_3 matrix. Two samples have been studied, which present a different particle mean volume, the ratio being about 4.6 from magnetization measurements. Therefore a temperature shift in the properties may be expected between the two samples.

In the small Q range where the magnetic intensity is enhanced due to the particle form factor, the energy spectra consist in two parts: a central peak (delta function) centered at $\omega = 0$ with intensity C_1 and an inelastic or quasielastic spectrum with an energy integrated intensity C_2, both convoluted with the spectrometer resolution function. The central peak consists of the nuclear scattering (static) and of any magnetic scattering fluctuating with a time larger than $1/2\pi\Gamma_0$ (Γ_0 being the energy resolution). The inelastic spectrum can be described by $S(Q, \omega) = C_2(Q)F(\omega)$ where, in the absence of any specific theory, several phenomenological forms of $F(\omega)$ have been used, depending on the experimental cases: a Gaussian or a Lorentzian function centered at $\omega = 0$ with a characteristic linewidth Γ or a "double" Lorentzian centered at + and $-\omega_0$ (creation and annihilation processes) with a linewidth Γ. We recall that in a paramagnetic state, as long as $\hbar\omega \ll kT$, the intensity C_2 observed by neutrons is readily related to the susceptibility χ by $C_2 = kT\chi(T)$.

For the first sample where the particles have the smallest size, one can distinguish two regimes in temperature. Above 250 K, where the magnetic intensity is all contained in the quasielastic peak (isotropic regime) and below 250 K where a magnetic intensity occurs in the central peak. This latter observation points out the slowing down of some magnetic component. Interestingly, this slowing down occurs whereas the typical energy linewidth Γ of the remaining quasielastic spectrum becomes larger and larger (see Fig. F.8.2). The existence of two magnetic components of clearly separated energy ranges emphasizes the relevance of two characteristic times for the fluctuation of the particle magnetization. Long time fluctuations can be attributed to the local "longitudinal" component fluctuating between energy minima and the fast ones to the transverse fluctuations around the temporary mean orientation in one of these minima. The temperature $T = 250$ K where a magnetic "resolution-limited" intensity occurs, corresponds to the blocking temperature T_B of the longitudinal component for the neutron probe (with the characteristic time $\tau = 10^{-10}$ s). The occurrence of an isotropic regime at high temperature with the susceptibility $\chi(T)$ proportional to $1/T$ (see Fig. F.8.3) is expected from theory.[86] However, the characteristic time τ is predicted to vary inversely with temperature ($\alpha = KV/kT$ is smaller than one, then $\lambda_1 = 2$ and τ is proportional to $1/T$; see Section D.3.2), which is not observed as $\Gamma = \frac{1}{2\pi\tau}$ is almost constant (Fig. F.8.2). It is suggested that the fluctuation time has reached a limit value above about 200 K, which can be related to the anisotropy energy. A good agreement is found for

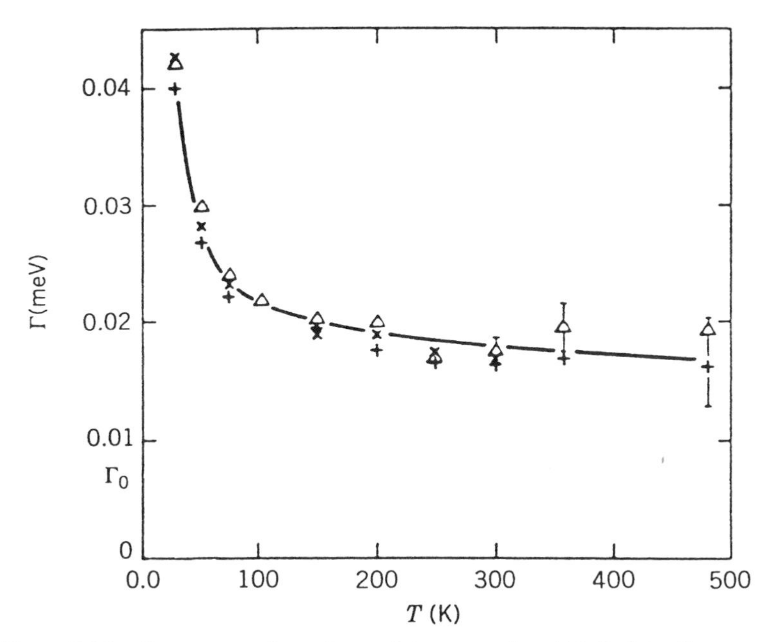

Figure F.8.2. Temperature dependence of the energy linewidth Γ of the C_2 component at different Q values for fine Fe particles embedded in an Al_2O_3 matrix. (Reproduced with permission from Ref. 130.)

$\Gamma = \gamma_0 H^{eff}$ with $H^{eff} = 2K^{eff}/M$ being the anisotropy field. For $T < 250$ K, the fast fluctuations progressively acquire a transverse character, whereas the energy linewidth starts to increase strongly below 100 K. The use of the same relation as above $T = 250$ K, allows one to interpret the strong increase of Γ by including in K^{eff} the interparticle interaction energy (see Section E). We note that the increase of Γ below 100 K is concomitant with the rise of the magnetic intensity observed by SANS in the very low Q range, indicating the growing of ferromagnetic correlations between the particles.[129,130]

For the second sample with a larger particle size,[131] the first regime described above is not really reached because there remains some magnetic intensity in the C_1 component for the highest studied temperature (480 K). Nevertheless, one can see (Fig. F.8.4) that at high temperature, Γ tends to a constant value very close to the asymptotic one measured for the first sample. The second regime described above is observed at the intermediate Q value 0.0775 $Å^{-1}$ for $200 < T < 500$ K and can be explained in the same way.

It is clear that the observed shift in the temperature limits are due to

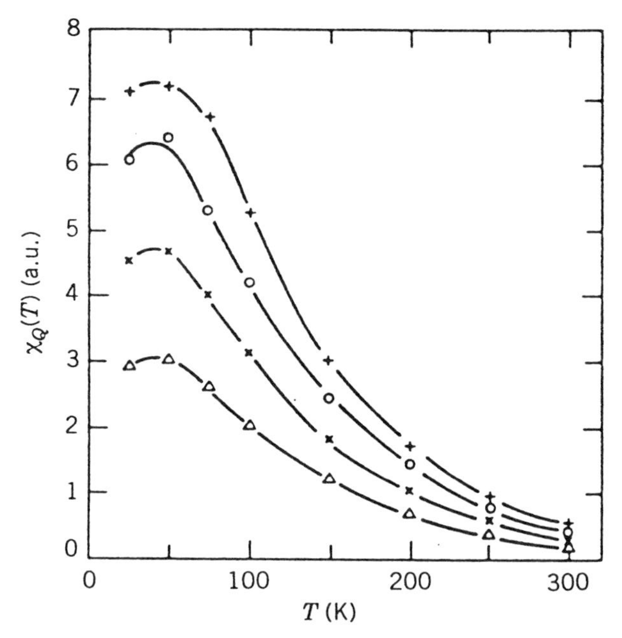

Figure F.8.3. Temperature dependence of the susceptibility $\chi_Q(T)$ of the quasielastic component for fine Fe particles embedded in an Al_2O_3 matrix. (Reproduced with permission from Ref. 130.)

the increase of the mean volume (about four times). Two new regimes occur for $T < 200$–250 K marked by the change between the lineshape of the quasielastic energy spectra (Lorentzian centered at $\omega = 0$ above and at $\omega \neq 0$ below) and the concomitant change of the susceptibility (see Fig. F.8.4). The change of energy spectra toward a weakly inelastic lineshape corresponds to the evolution toward the precession mode. This change may be driven by temperature as well as by interparticle interactions, these latter producing an enhancement of K^{eff}. Therefore down to 200 K, the transverse excitations can be described in terms of local modes, indicating that the interactions are still weak. Finally below 50 K, the energy spectrum turns back toward a quasielastic lineshape with a strong increase of χ with decreasing T (Fig. F.8.4). It seems that the competing interactions have destroyed the local modes. The limits between the regimes described above varies strongly with Q, the shift to higher temperature values corresponding to the smallest Q values, which are weighted by the form factor of the larger particle sizes. This is the obvious consequence of the polydispersity of the particles. However, this strong Q dependence ensures that these transverse fluctuations are still

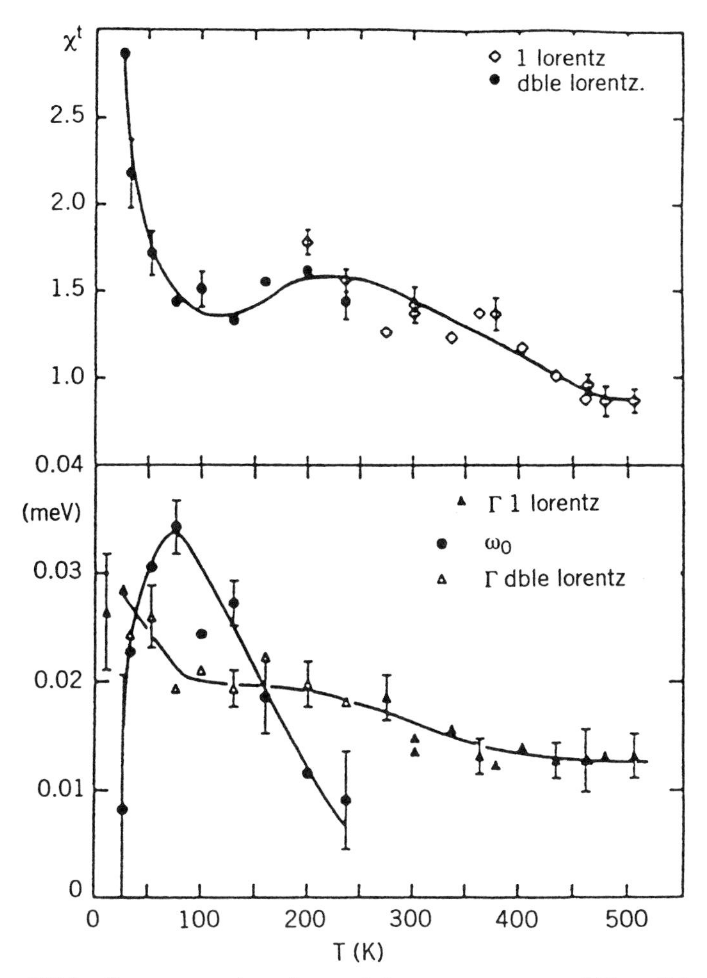

Figure F.8.4. Temperature dependence of the transverse susceptibility $\chi'(Q, \omega = 0)$, the Γ and ω_0 parameters. (Reproduced with permission from Ref. 131.)

incoherent between one particle and its neighbors.[131] This means in our opinion that the collective state for **m** (see Section E.2) does not occur in the studied temperature range (down to 25 K for the second sample, which presents the largest mean volume).

F.8.3. Other Experiments

A neutron depolarization experiment is able to give some information about magnetic inhomogeneities.[324] In such an experiment a polarized neutron beam is transmitted. During transmission, the beam changes its polarization direction **n** and degree of polarization *P*. From a three-dimensional analysis of **n** and *P*, the mean magnetization vector, the magnetic correlation length, and the mean square direction cosines of the local magnetization of the magnetically correlated volume are determined. Ferrofluids have been studied with this technique (see, e.g., Refs. 324 and 325).

Results obtained from polarized neutron reflection measurements have been reported on granular samples consisting of Fe particles in an Al_2O_3 matrix.[326,327] However, this experiment deals with surface studies and is not well adapted to the analysis of fine particle properties. Nevertheless, some features can be deduced.

F.8.4. Conclusion

Neutron experiments lead to very reach features concerning particularly the dynamical properties of fine particles. However, at present, only one particle kind has been studied by inelastic neutron scattering; and the studies of other particle types, other morphologies, would be interesting for the establishment of general behaviors. On the other hand, theoretical development would also be useful for the analysis of the data as well as the understanding of outcomes.

F.9. Some Other Measurements

In this section, we briefly discuss three properties that are measured from particular techniques and have some peculiarities for fine-particle samples, that is, the magnetoresistance, the magnetocaloric effect, and the magneto-optics.

F.9.1. Magnetoresistance

The recent discovery of giant magnetoresistance (GMR) in heterogeneous metallic alloys[25,26] consisting of ferromagnetic grains (e.g., Fe, Co) embedded in an immiscible non magnetic matrix (e.g., Cu, Ag) has determined an enormous increase of interest toward magnetotransport properties and related phenomena in fine magnetic particle systems. The substantial change in the magnetoresistance in such magnetically inhomogeneous media (Fig. F.9.1) has improved the fundamental understanding of the giant magnetoresistance effect, previously observed in a number of antiferromagnetically coupled multilayers (e.g., Fe/Cr).[328] Moreover, GMR in such granular magnetic materials provides large opportunities for potential applications, for example, in magnetoresistive devices (field sensors, magnetoresistive heads for magnetorecording, etc.).

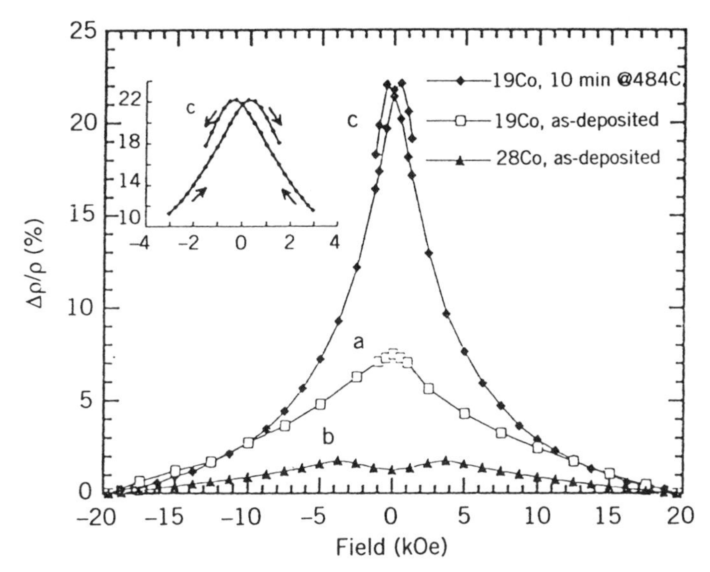

Figure F.9.1. Field dependence of the relative magnetoresistance for three Cu–Co samples. (Reproduced with permission from Ref. 25.)

Giant magnetoresistance comes from the reorientation of the magnetic moments of particles embedded in the nonmagnetic matrix and can be interpreted on the basis of electron spin-dependent scattering[329,330] occurring at the interface of magnetic and nonmagnetic entities.[331] The value of GMR strongly depends on the density and on the size of ferromagnetic particles; their surface-to-volume ratio playing a key role.

Giant magnetoresistance was found in various granular systems: Co–Cu,[25,26] Co–Ag,[332,333] Fe–Cu,[26] Fe–Ni–Ag,[334,335] prepared by a variety of nonequilibrium techniques, such as evaporation,[336] sputtering,[25,26] mechanical alloying,[337] melt spinning, and subsequent thermal annealing of the metastable phase.[338,339]

The interpretation of GMR is basically independent of the superparamagnetism phenomenon. According to the model in Ref. 331, it is expected that the conductivity is minimum for a random alignment of the granule magnetic moments, supposed static. However, the magnetization under applied field is given by the Langevin function, as for fine particles when the anisotropy is neglected (see Section F.3). We note that the Langevin function results from general properties of grains, which do not imply that without field, superparamagnetic properties occur for a measurable time. Magnetic properties are not well established at present. Zero-field-cooled and field-cooled magnetizations show irreversibilities at low field,[25,26,333,334] which could be explained as resulting from superparamagnetic properties, but they are also in agreement with spin glasslike properties (see Section E). However, a common increase of the coercive field when the temperature decreases gives rise to such irreversibilities.[118] For medium concentration of magnetic metal, the magnetization does not saturate under medium applied field.[332,334,335] The lack of precise results at present leads to many unanswered questions such as the correctness of the Langevin function, the variation of the susceptibility at low field, and so forth. In our opinion, superparamagnetism does not seem present in the studied samples due to the large metal concentration and the strong RKKY interparticle interactions. For low concentrations leading to small volumes and weak interparticle interactions, superparamagnetism may be present. Susceptibility in a large frequency range[209] seems in agreement with such properties (see Section F.5) but the entire demonstrations remain to be done.

The influence of the superparamagnetic phenomenon has been clearly evidenced for Ni particles embedded in an SiO_2 matrix.[340] Here, the conduction occurs via tunneling electrons. Without field, below T_B, the magnetic moments **m** of the particles are blocked at random due to the disordered arrangement of particles, and above T_B, **m** relaxes. Then the correlations between spins are weak. Under field, the correlations increase and are maximum near T_B (Fig. F.9.2). For these materials, the magnetoresistance at few thousand Oersted is of the order of 1% and shows a tendency to saturate.

F.9.2. Magnetocaloric Effect

When a material is magnetized by application of a field H, the entropy S_m associated with the magnetic degrees of freedom is changed as the field changes the magnetic order of the material. Under adiabatic condition, the S_m variation must be compensated for by an equal but opposite change in the entropy associated with the lattice, resulting in a change in temperature ΔT of the material, that is, the magnetocaloric effect.

This effect has been used for magnetic refrigeration. Materials presently used fall into two categories: paramagnetic substance (e.g., $Gd_3Ga_5O_{12}$) for use at temperatures up to $\simeq 20$ K, and magnetic materials near the ordering temperature (e.g., Gd or $DyAl_2$), which can operate at temperatures greater than 20 K. However, no material has been found for use at room temperature.

For a paramagnetic material

$$\frac{\Delta T}{\Delta H} = \frac{V}{C_H} \frac{N\mu^2}{3k} \frac{H}{T} \tag{F.9.1}$$

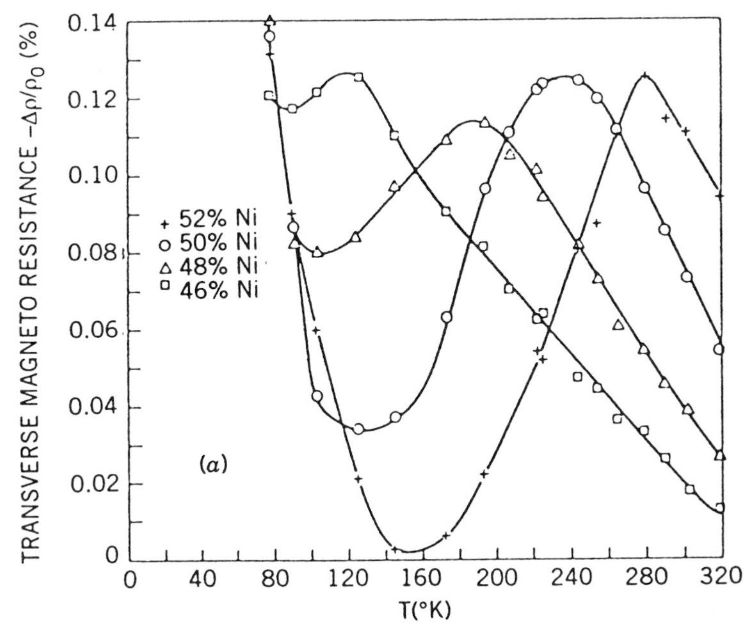

Figure F.9.2. Relative transverse magnetoresistance vs. temperature for four Ni/SiO_2 samples. The field is equal to 260 Oe. (Reproduced with permission from Ref. 340.)

where V is the sample volume, C_H the heat capacity, N the number of spins in the sample, and μ the spin magnetic moment.

For a ferromagnetic material above T_c

$$\frac{\Delta T}{\Delta H} = \frac{V}{C_H} \frac{N\mu^2}{3k} \frac{H}{T} \left(\frac{T}{T - T_c} \right)^2 \tag{F.9.2}$$

It is clear from these formulas that

$$(\Delta T)_{\text{ferro}} \gg (\Delta T)_{\text{para}}$$

For noninteracting particles[24,341,342] in the superparamagnetic state

$$\frac{\Delta T}{\Delta H} = \frac{V}{C_H} \frac{N\mu^2}{3k} \left(\frac{N}{n} \right) \frac{H}{T} \tag{F.9.3}$$

where n is the number of spins of the particle. Due to the order of magnitude,

$$(\Delta T)_{\text{superpara}} \gg (\Delta T)_{\text{ferro}}$$

Finally for interacting fine particles[341,342]:

$$\frac{\Delta T}{\Delta H} = \frac{V}{C_H}\frac{N\mu^2}{3k}\left(\frac{N}{n}\right)\frac{H}{T}\left(\frac{T}{T-\theta_{\rm sp}}\right)^2 \tag{F.9.4}$$

where $\theta_{\rm sp}$ is the superparamagnetic Curie temperature (see Section F.2.3). For interacting particles, $\Delta T/\Delta H$ will be greater than for isolated particles on the condition that $\theta_{\rm sp} > 0$. This condition is achieved only if the sample is thin with the field applied parallel to the surface sample and $\theta_{\rm sp}$ values are rather small (see Section F.2.3).

These formulas indicate that fine-particle materials can be used for magnetic refrigeration for temperatures well in excess of the present maximum of 20 K.[24,341] This is demonstrated by the results obtained on 11%Fe + silica gel nanocomposites (Fig. F.9.3)[343] and $Gd_3Ga_{5-x}Fe_xO_{12}$.[344]

F.9.3. Magneto-optic Properties

Magneto-optical effects are studied in conjunction with their dependence on wavelength, temperature, and applied field. In particular, measurements at specific wavelengths, being associated with transitions of the different atoms of the material, constitute a local probe of the mag-

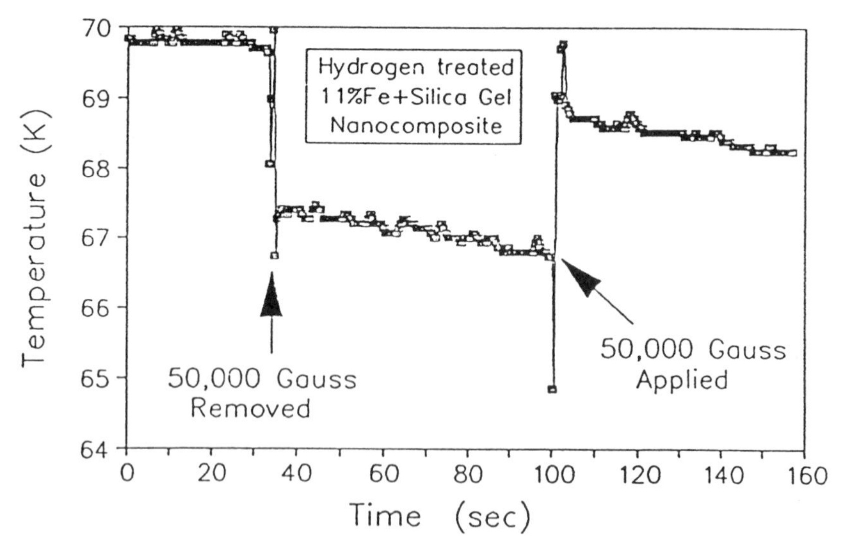

Figure F.9.3. Temperature vs. time for a superparamagnetic 11% Fe + silica gel nanocomposite as a 5T magnetic field was first removed and then applied to the sample. (Reproduced with permission from Ref. 342.)

netization at that position in the crystal. In fine particles, it is possible to distinguish the local magnetic situation of different regions, such as the surface and the core of the particles, when the atomic composition is not uniform throughout them. Basically, the effects are proportional to the local magnetization, such as, for example, for the Faraday rotation, and features can be deduced on its variation with temperature and applied field.[345] Mesurements have been performed on Fe particles embedded in an Al_2O_3 matrix,[345–347] and the main results concern the oxidized iron atoms lying on the particle surface and their magnetic coupling with the particle iron core.

G. QUANTUM TUNNELING OF THE MAGNETIZATION

Macroscopic quantum tunneling, that is, the tunneling of a microscopic variable through the barrier between two minima of the effective potential of a macroscopic system, represents one of the most fascinating phenomena in condensed matter physics. It is well known that quantum phenomena can take place at the macroscopic scale[348] in systems with negligible dissipation (i.e., small interaction of the tunneling variable with the environment), for example, superconductors,[349] one-dimensional metals,[350] and so forth.

In a more recent time it has been predicted that macroscopic quantum tunneling can also be observed in magnetic systems,[351] for example, the tunneling of the magnetization vector of a single-domain particle through its anisotropy energy barrier and the tunneling of the domain wall crossing a larger particle through its pinning energy barrier. These phenomena have been studied both theoretically[351–358] and experimentally.[359–365]

The attention will be focused on macroscopic quantum tunneling in single-domain particles, where the tunneling variable is the magnetization, and the dissipation is weak. Besides the interest in fundamental physics, macroscopic quantum tunneling of magnetization can have in principle practical implications also, for example, in the information storage industry, where it could play a crucial role for low-temperature devices. The use of magnetic memories at low temperature could increase greatly the storage density, but magnetization reversal by quantum tunneling mechanism should place a lower limit in the particle size and in the number of atoms per bit of information.

The possibility of quantum tunneling in fine particles was suggested for the first time by Bean and Livingston,[16] as an explanation of the Weil experimental data,[81] showing that in single-domain nickel particles the transition between different orientations of the magnetic moment did not disappear completely with decreasing temperature to absolute zero. Chudnowsky and Gunther,[351] calculated the probability of tunneling of the magnetization in a single-domain particle, through an energy barrier between easy directions, for several forms of magnetic anisotropy. Many data have been reported on different types of magnetic particles (ferromagnetic and antiferromagnetic, in dispersed systems, in ferrofluids, in magnetic proteins),[359–365] which seem to support the existence of quantum tunneling of the magnetization.

It has been proposed that in the magnetization dynamics of single-domain particles there is a characteristic crossover temperature T^* below which the escape of the magnetization from the metastable states is dominated by quantum barrier transitions, rather than by thermal over barrier activation. Above T^* the escape rate is given by the rate of the thermal transitions, determined by the Boltzman factor, $\Gamma_T \cong \nu \exp(-U/kT)$, where U is the barrier separating two metastable states. In a thermally activated regime it should vanish when the temperature approaches zero.

At $T \cong T^*$ the escape rate begins to depart from the above reported law, tending to a temperature-independent quantum transition rate (Fig. G.1), which has the same form, in the usual WKB (Wentzel–Kramers–Brillouin) approximation for quantum tunneling, as in the case of switching of magnetization via thermal activation:

$$\Gamma_Q \cong \nu \exp(-B) \tag{G.1}$$

where B is the Gamov exponent ($B = U/kT^*$), replacing the exponent U/kT of thermal activation. In the classical approach, $kT^* \cong \hbar\omega_0$, where ω_0 is the frequency of oscillation in the potential well; ω_0 is inversely proportional to the square root of the mass (as for a harmonic oscillator). Then, in the entire temperature range the switching rate of the magnetization vector can be described as $\Gamma = \nu \exp(-U/kT_{esc})$, with $T_{esc} = T$ at $T > T^*$ and $T_{esc} \cong T^*$ at $T < T^*$.

Both U and B are proportional to the volume (or N_s, the number of spin) involved in the tunneling process.[359] Thus, for a given material, T^* depends only on the magnetic field applied, not depending on extensive parameters, like V or N_s.

Let us consider a single-domain particle with uniaxial anisotropy in the presence of an applied field along the easy direction. In a spherical coordinate system the anisotropy energy by volume unit is given by

$$E = (K_\parallel + K_\perp \sin^2\phi)\sin^2\theta - MH\cos\theta \tag{G.2}$$

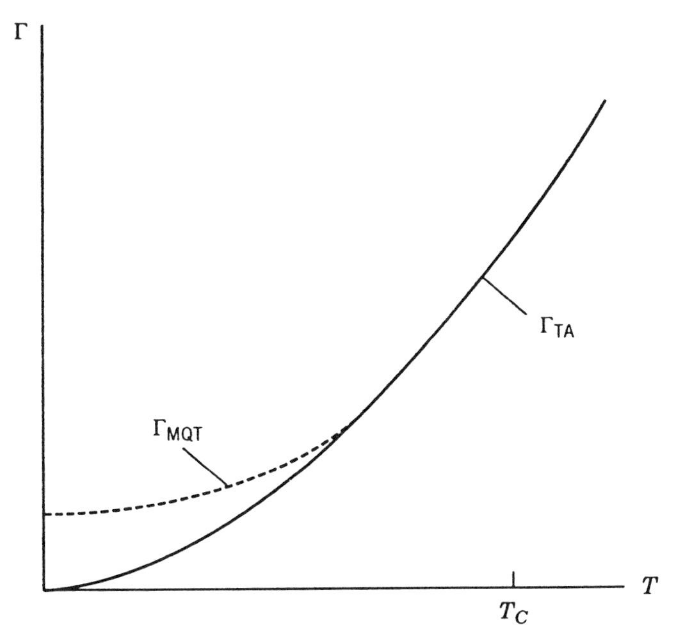

Figure G.1. Switching rate as a function of temperature for both thermal activation (TA) and macroscopic quantum tunneling (MQT). The crossover from the TA regime to the MQT regime takes place at temperature T_c (Reproduced with permission from Ref. 358.)

where $K_{\|}$ and $K_{\perp}$ are the parallel and transverse anisotropy constant, respectively. Analysis reveal that there is an effective mass in the dynamics of a single-domain particle, which is inversely proportional to K. Quantum tunneling of the magnetization cannot take place in the absence of transverse anisotropy, since the exponent B would diverge.

The Gamov exponent B is given by the Euclidean action, as first proposed by Gilbert[366]

$$\frac{S_E}{\hbar} = -\frac{iV}{\hbar} dt\left[\left(\frac{M_0}{\gamma}\right)\phi \cos\theta - E(\theta, \phi)\right] \tag{G.3}$$

evaluated along the instanton trajectory leading M out of the metastable state. This action produces the classical equation for M:

$$\frac{dM}{dT} = -\gamma M \frac{dE}{dM} \tag{G.4}$$

In particular Chudnowsky and Gunther[351] calculated the Euclidean action at absolute zero for fields very close to the critical field H_c ($=2K/M$), above which relaxation is no longer possible, as the actual barrier, $U = KV(1 - H/H_c)^2 = KV\varepsilon^2$, vanishes.

The B exponent was given by[351]

$$B = \frac{8MV}{3\hbar\gamma}\left(\frac{K_\parallel}{K_\perp}\right)^{1/2}\varepsilon^{3/2} \tag{G.5}$$

The crossover temperature between quantum tunneling and thermal activation regime was given, equating the B exponent to U/kT[351]:

$$T^* = \frac{U}{kB} = \frac{3\hbar\gamma(K_\parallel K_\perp)^{1/2}}{8kM}\varepsilon^{1/2} \tag{G.6}$$

In an actual particle assembly the easy axis is in a random direction, and the probability to find it parallel to H is very low. At the present, there are no analytical expressions of the critical field H_c, but $K/M \leq H_c \leq 2K/M$, the lower value corresponding to an angle of $\pi/4$ between H and the easy axis. This implies that for an actual particle assembly T^* is distributed.

From Eq. (G.6) it comes out that in order to make easier the experimental observation of quantum tunneling of the magnetization in small single-domain particles, that is, to have small B (high quantum transition rate) and high T^*, particles with low magnetization and small volume (a radius lower than 5 nm) are needed. Conditions for the observation of quantum tunneling of the magnetization are more favorable in antiferromagnetic particles,[367] with a small uncompensated moment (this can be understood in terms of their effective mass being much smaller than that of ferromagnetic particles). Independent magnetic particles are the best candidate, since interactions lead to an increase of the effective mass of the tunneling object. Moreover, a small $K_\parallel/K_\perp$ ratio and large $K_\parallel K_\perp$ lead to reduce B and to increase T^*, respectively.

T^* is expected to be in the temperature range 0.1–5 K for typical values of the high anisotropy in small single-domain particles, for example, for $CoFe_2O_4$ ($K = 2 \times 10^7$ erg/cm^3) ferrofluid particles ($\langle\phi\rangle \cong$ 5 nm), $T^* \cong 3$ K.[368]

The measurement of the relaxation of the remanent magnetization represents a very good tool to achieve a crossover temperature between thermal activation and quantum tunneling regime. In fine magnetic particle materials, usually characterized by a volume distribution, implying a distribution of energy barriers, the time decay of the remanent magnetization is usually found to follow a logarithmic law, when measurements are performed at low temperature in a not too large time interval:

$$RM(t) = RM(0)\left[1 - S \ln\left(\frac{t}{t_0}\right)\right] \tag{G.7}$$

where S is known as magnetic viscosity, proportional to $kT_{\mathrm{esc}}/\langle U \rangle$; S is usually found to decrease linearly (in a given temperature range) with decreasing T down to the temperature T^* (thermal activation regime), below which it becomes temperature independent (quantum tunneling regime). The transition is expected to be quite sharp in the case of weak dissipation.[369] The crossover temperature T^* between the two regimes is found to depend on the applied magnetic field. The constancy of S at low temperature may be interpreted as evidence of quantum tunneling, revealing the temperature independence of the quantum transitions rate. The experimentally determined T^* agrees well with theoretical predictions.[351]

Many data of this type, supporting quantum tunneling of the magnetization, are reported in the literature, for example, in CrO_2 particles in a magnetic tape (Figs. G.2 and G.3)[370]; Fe_3O_4, FeC, $CoFe_2O_4$ particles in ferrofluids[367]; and antiferromagnetic ferritin proteins.[371] However, as discussed in Section F.4, the proportionality of S with tempera-

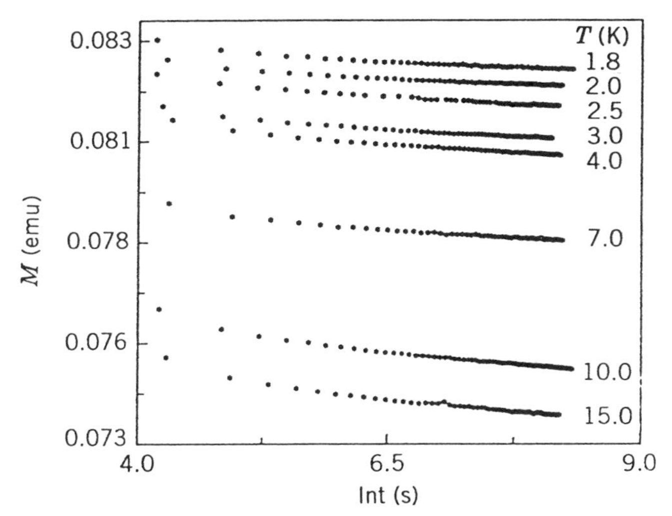

Figure G.2. Time dependence of the remanent magnetization at different temperatures for fine CrO_2 paticles. (Reproduced with permission from Ref. 370.)

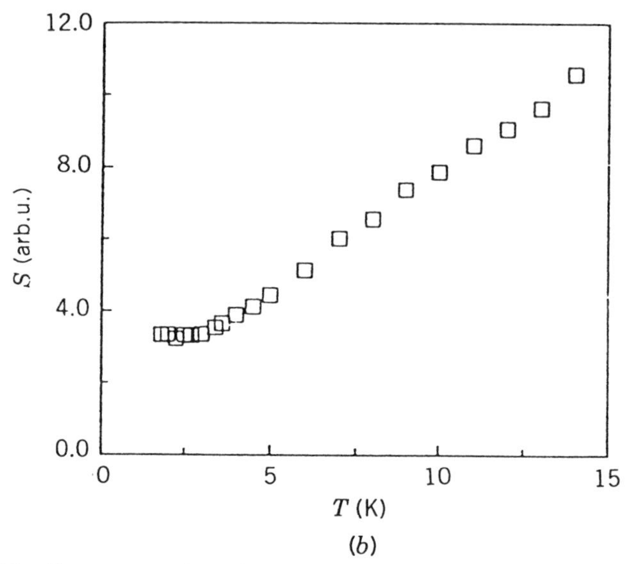

Figure G.3. Temperature dependence of the magnetic viscosity S for fine CrO_2 particles. (Reproduced with permission from Ref. 370.)

ture may be an artefact when the particle volume distribution is large.[90] In this case the S variation mimics the volume distribution. Moreover, for a narrow volume distribution Eq. (G.7) is not valid. These facts imply first the need for carefully analyzing the data in terms of a thermally activated process, by means of a suitable model accounting for the complexity of actual systems, in order to be sure of the deviations from the thermal activation regime.

On the other hand the analysis of magnetic relaxation experiments on noninteracting fine γ-Fe_2O_3 particles dispersed in a polymer revealed the persistence of a thermal activation mechanism down to very low temperatures.[196] In some cases, for example, for experiments on Fe–Al_2O_3,[127] further measurements at lower temperatures are needed in order to have clear evidence of deviations from a thermally activated mechanism.

Some relaxation experiments at different fields gave evidence that the energy barrier is proportional to $1/H$, as reported in Figure G.4 for FeC particles.[365] The effective temperature T^* was reported vs. the measured temperature T by plotting $d(1/H)/d \ln \tau$ vs. T in Figure G.5, where a crossover from thermal activation to temperature independent relaxation is observed at $T \cong 1$ K.

Experiments were reported on Co single-domain particles patterned by

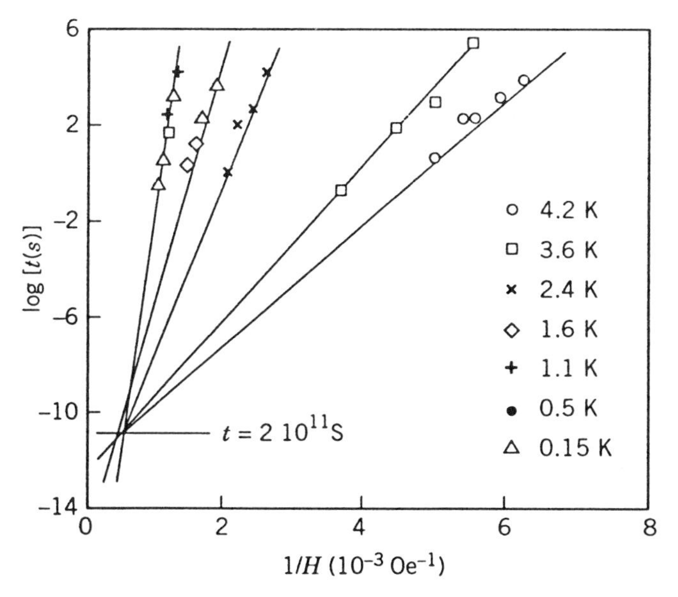

Figure G.4. Plot of $\log_{10}\tau$ vs. $1/H$ for fine FeC particles. (Reproduced with permission from Ref. 365.)

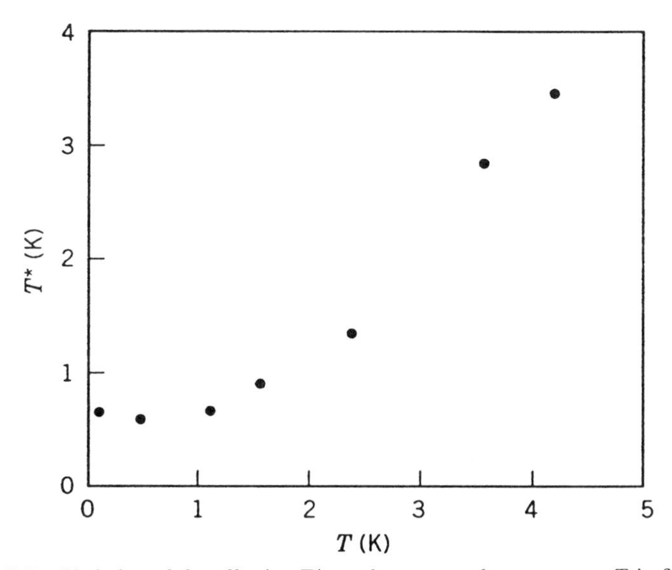

Figure G.5. Variation of the effective T^* vs. the measured temperature T in fine FeC particles. (Reproduced with permission from Ref. 365.)

electron beam lithography, where the angular dependence of the switching field and the temperature dependence of the reversal process were investigated by means of an integrated DC SQUID.[207] The mean value of the switching field was found to become temperature independent below the crossover temperature of about 1 K (Fig. G.6).

Results in favor of the existence of macroscopic quantum tunneling of the magnetization were also reported by means of low-temperature measurements of the frequency-dependent magnetic noise $S(\omega)$ and magnetic susceptibility $\chi(\omega)$.[362,363] In horse-spleen ferritin particles a well-defined resonance below about 200 mK in both $S(\omega)$ and $\chi''(\omega)$ were found (Fig. G.7). The behavior of this resonance as a function of temperature, applied magnetic field, and particle concentration is in qualitative agreement with the theoretical predictions of quantum tunneling in small antiferromagnetic particles.

In conclusion many experiments reported in the literature are in favor of the existence of quantum tunneling of the magnetization in single-domain particles, in agreement with theoretical predictions.

However, both models describing the thermal activation and the quantum tunneling regime do not account for the complexity of actual fine-particle systems (e.g., volume distribution, easy axes in random

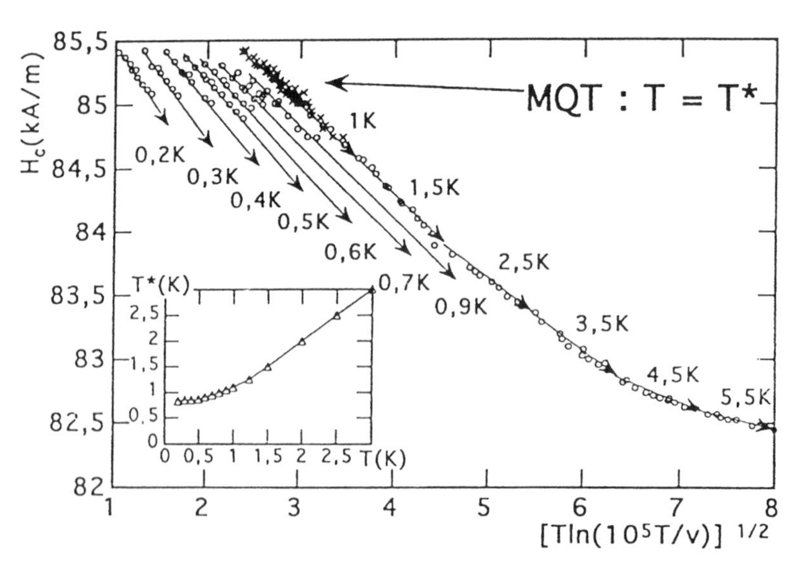

Figure G.6. Scaling, plot of the mean switching field (H_c) for Co particles. Dots are the measured H_c at real temperatures T and crosses are the same values when T is replaced by the effective temperature T^* shown in the inset. (Reproduced with permission from Ref. 207.)

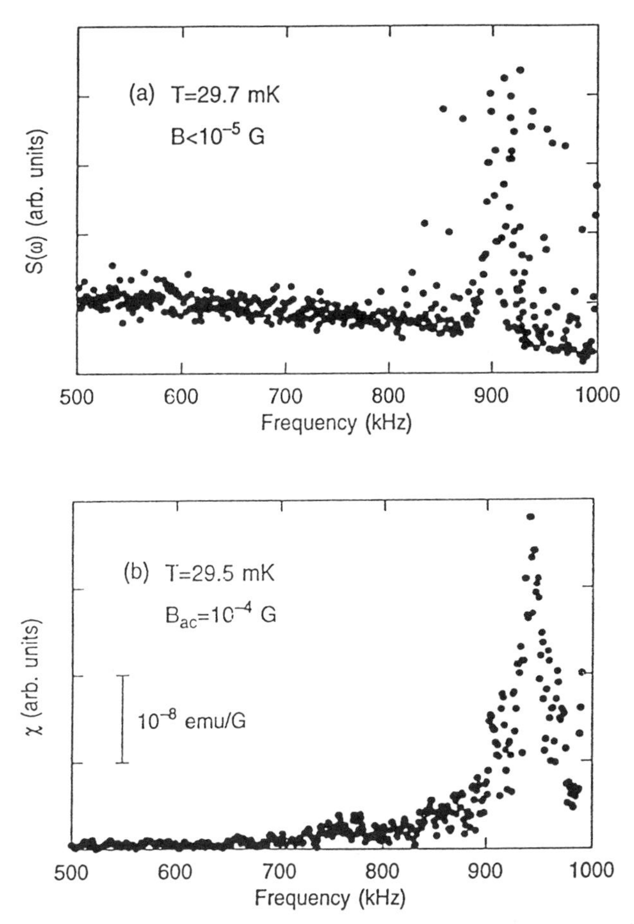

Figure G.7. (a) Magnetic noise spectrum of a 1000:1 diluted solution of magnetic protein (horse-spleen ferritin particles); (b) frequency-dependent magnetic susceptibility of the same sample after cycling the temperature at $T = 4$ K. (Reproduced with permission from Ref. 363.)

position, interparticle interactions, etc.) and do not consider realistic modes of magnetization reversal (i.e., the possibility of modes different from the uniform rotation one). Therefore, it is difficult at the present to derive from the experiments unambiguous evidence of deviations from the thermal activation regime and of the crossover to the quantum tunneling one.

Moreover, for interacting particles an additional complication in the interpretation of the results can derive from the possibility of existence of a collective state of particle magnetic moments at very low temperature (see Section E). A change of properties at the transition temperature can be expected, but it is not known how the magnetization dynamics would be affected.

H. ANTIFERROMAGNETIC PARTICLES

In this section, we discuss the magnetic properties of fine-particle materials for which the bulk presents antiferromagnetic order, mainly when the properties differ from those of ferromagnetic fine particles. In the first section, we describe the magnetic state arising in these particles and we evoke the case of natural oxidized iron samples, which show a disordered magnetic state. We also discuss the validity of the relaxation time calculation performed for ferromagnetic particles. In the second section, results of magnetic measurements are described, pointing out the difference with those obtained with ferromagnetic particles. Mössbauer spectroscopic results are discussed in the third section, and finally some conclusions are given.

H.1. Magnetic State and Relaxation Time

Fine particles of antiferromagnetic materials possess a magnetic moment $|\mathbf{m}|$ not equal to zero. This comes from the fact that the two magnetic sublattices are not exactly compensated due to the small size. Néel[372,373] has suggested that for very small particles, the belonging of a spin to one of the magnetic sublattices is at random. In this case the uncompensated magnetic moment $|\mathbf{m}|$ is equal to $\mu_s\sqrt{N}$ where N is the number of spins in the particle and μ_s is the spin magnetic moment. For larger particles, one can think that the randomness is effective only in the surface, which leads to $|\mathbf{m}| = \mu_s N^{1/3}$. Néel also discussed the case of larger particles with well-defined crystallographic surface planes. Then, $|\mathbf{m}|$ can vary between $\mu_s N^{1/3}$ and $\mu_s N^{2/3}$. It can be equal to zero in a special case.[372,373] Summarizing, we can expect that $|\mathbf{m}| = \mu_s N^p$ with $\frac{1}{3} \leq p \leq \frac{2}{3}$. Experimental results are in agreement with $p \sim \frac{1}{2}$ (see below).

In fact, this discussion is valid for fine particles of materials that show antiferromagnetic order in bulk. However, numerous natural materials mainly corresponding to iron oxides or hydroxides such as goethite, ferrihydrites, or protein cores, show a more or less disordered magnetic state due to a high content of impurities and defects. Depending on the number of defects, the number of nonmagnetic impurities, and the values of exchange interactions, various magnetic states can be observed: (i) antiferromagnetic state with some disorders, for which not too important changes of the properties are observed (increase of the antiferromagnetic susceptibility χ_{AF}, decrease of T_N, e.g.) with regard to perfect materials;

(ii) semidisordered state with spin glasslike properties; and (iii) speromagnetic phase very similar to the spin glass phase.[374] As already discussed in Section E.3.1, we do not know what are the properties of fine particles of materials that show the two latter states for bulk. From the experimental point of view, the difficulties come from the fact that the properties of fine particles resemble those of the spin glasslike state (see Section E.2.4). However, for describing the system in terms of superparamagnetic properties with a blocking temperature, T_B, it is necessary that in a certain range of temperature $T > T_B$, the magnetic state is not paramagnetic until the transition toward this latter state. In our opinion, the only proof at the present for demonstrating superparamagnetic properties is to show the existence of a second transition toward paramagnetism, at $T > T_B$.

Regarding anisotropy energies, the magnetostatic energy is strongly lowered with respect to ferromagnetic particles due to the weakness of $|\mathbf{m}|$. The same trend occurs for magnetic interactions, which can generally be neglected. Therefore, there mainly remains the magnetocrystalline and surface anisotropies. The latter can be of importance especially when the uncompensated moment comes from the surface.

Relaxation of the magnetic moment **m** due to the same order of magnitude for the anisotropy and thermal energies occurs like for ferromagnetic particles.[372,373] In our opinion, the calculations of the relaxation time for ferromagnetic particles (see Section D.3) are valid for antiferromagnetic particles. Basically, the Gilbert equation concerns the spin magnetic moment, extended to the particle magnetic moment (see Section D.3.1), which is equal to $\mu_s N^p$ for antiferromagnetic particles. The only question concerns the synchronous rotation of the spins because one could think that change of a spin direction from one magnetic sublattice to the other could take place during the rotation. However, the exchange interaction energy is much stronger than the anisotropy energy, which partially hinders this kind of change. In addition, we are interested in average properties. This discussion is very similar to that in Section D.3.1 on ferromagnetic particles in connection with the effects of the surface, and the same conclusion can be drawn. Calculations based on synchronous rotation are valid with perhaps some minor change in the pre-exponential factor τ_0.

With regard to ferromagnetic particles the τ_0 factor is strongly lowered because it depends on $|\mathbf{m}|$. This is checked qualitatively by experimental results (see below). In addition, $|\mathbf{m}| = \mu_s N^p$ is an averaged value. This means that for a given volume and a given anisotropy energy, a τ_0 distribution occurs. In fact, as already discussed, only large τ_0 variations are of importance because the relaxation time is mainly stated by the energy barrier through the exponential.

H.2. Magnetic Measurements

We focus our attention on two measurements, the magnetization under high applied field and the dc and ac susceptibility.

For the first measurements, Néel[373,375] has shown that the magnetization is the sum of two components. The first one corresponds to the contribution of the antiferromagnetic network while the second results from the vibrations of the uncompensated moment $|\mathbf{m}|$ in the potential well as described in Section F.3.2 with M_{nr} replaced by $M_{nc} = |\mathbf{m}|/V$. The applied field H_{app} must clearly be larger than $2K_t/M_{nc}$ and kT/VM_{nc}, where K_t is the total anisotropy constant, which leads to high values of H_{app} due to the weakness of M_{nc}. Equations (F.3.6) and (F.3.7) remain valid (with M_{nc} replacing M_{nr}) for the part of the magnetization due to the vibrations of **m**. For antiferromagnetic particles, there is an additional condition,[375] that is, $\chi_{AF}H^2_{app}/2kT$ small compared to the unity, where $\chi_{AF} = \chi_{\perp} - \chi_{\parallel}$ is the difference between the perpendicular and parallel antiferromagnetic susceptibilities. We note that the condition cited above is fulfilled except for very high fields, because of the weakness of χ_{AF}.

However, χ values can be larger than for bulk because the antiferromagnetic network is not perfect (existence of the uncompensated moment, defects in the surface, etc.). Indeed Néel expected an enhancement of χ in a very special case[376] where the superparamagnetic phenomenon is not present, called superantiferromagnetism, but in our knowledge, no demonstration of this state has yet been done (we recall that $|\mathbf{m}| = 0$ and, therefore, the Néel–Brown model is not applicable). In our opinion, the χ_{AF} increase is an evident trend due to the imperfections.

These properties are qualitatively verified for NiO_2, Cr_2O_3, and α-Fe_2O_3 particles,[377] α-Fe_2O_3 particles embedded in an alumina matrix,[378] and ferritin.[270] These results also verify that $|\mathbf{m}|$ is proportional to about $N^{1/2}$.[270,378] We remark that the thermal variation of $|\mathbf{m}|$ can be strongly modified with regard to what is expected from the T_N bulk values, especially if $|\mathbf{m}|$ comes mainly from the surface.

Measurements of dc susceptibility show a broad peak (see, e.g., the results obtained for NiO[148]), probably due mainly to the volume distribution, and which can in principle be modeled as for ferromagnetic particles (see Section F.2), but at present we do not know if the particular magnetic structure, uncompensated moment adding to the antiferromagnetic network, leads to a broadening of the peak. For this experiment, the measuring time is always difficult to define and, therefore, the value of the temperature of the susceptibility maximum, T_{max}, assimilated to the blocking temperature, cannot be used for an accurate determination of the parameters included in the relaxation time law. We note that T_{max} does not appreciably change in low and medium field, which is not the case for ferromagnetic particles.

The magnetization in the superparamagnetic state is also the sum of two components. The first one corresponds to the contribution of the antiferromagnetic network while the second results from the part of the uncompensated moment, at the thermodynamic equilibrium, along the applied field H_{app}. This latter term can be determined as for ferromagnetic particles (see Section F.2.3), and Eqs. (F.2.6) and (F.2.7) are valid, always by replacing M_{nr} by M_{nc}. However, the use of these equations is simpler for antiferromagnetic particles because (i) the effective field is very near to H_{app} due to the weakness of M_{nc} leading to a very weak value for the superparamagnetic Curie temperature θ_{sp}, and (ii) the range of validity of Eq. (F.2.6) is extended until medium H_{app} because the condition is now $VM_{nc}H_{app}/kT \ll 1$.

For easy directions in random orientation, the superparamagnetic susceptibility $\chi_{superpara}$ is therefore given by

$$\chi_{superpara} = \frac{C_{superpara}}{T} + \chi_{AF} \tag{H.1}$$

with

$$C_{superpara} = \frac{\langle M_{nc}^2(T)V^2 \rangle}{3k\langle V \rangle} \tag{H.2}$$

This allows, in principle, the determination of M_{nc}, providing that its thermal variation and the V distribution are known. However, for $p = \frac{1}{2}$, which seems the usual value, $C_{superpara} \sim C_{para}$. In addition, the order of magnitude of χ_{AF} is equal to C_{para}/T_C where T_C is the transition temperature toward the paramagnetic state. This means that a precise determination of M_{nc} will be possible only for $T \ll T_C$ due to the uncertainties on χ_{AF}. To our knowledge, Eqs. (H.1) and (H.2) have never been used for the M_{nc} determination.

For ac susceptibility, the measuring time is well defined, and this experiment is very useful for determining the parameters. Results obtained for ferritin,[270] which also presents a broad peak, show that the model developed for isolated particles is verified with τ_0 clearly smaller than the expected value for ferromagnetic particles, in agreement with the above discussion (see Section H.1). We note that it is the model for isolated particles that has to be checked since the interparticle interactions are weak due to the small $|\mathbf{m}|$ value.

Some other measurements such as partial thermoremanent measurements for CoO[379,380] and NiO,[381] and thermal decay of the remanence for ferritin[210,382] have been performed, but these experiments mainly provide features about the distribution of the energy barriers.

H.3. Mössbauer Spectroscopic Studies

The differences in the properties observed by Mössbauer spectroscopy with respect to ferromagnetic particles (see Section F.6) are due, in zero applied field, to the lowering of the τ_0 factor, and, for in-field experiments, to the small magnitude of the Zeeman energy of the particle compared to the anisotropy energy.

H.3.1. Experiments in Zero Applied Field

τ_0 is normally of the order of 10^{-12}–10^{-13} s,[58,210,270] that is, lower by about 2 orders of magnitude than for ferromagnetic particles, and thus well below the Mössbauer time window. Due to this lowering of τ_0, the temperature (volume) interval required to span the critical window is notably narrowed, and the sextet-to-doublet transition is sharpened.

If the distribution in particle volumes (energy barriers) is large, the fraction of the particles with a relaxation time in the critical range is small at any temperature, and the Mössbauer spectra as a function of the temperature are well described in terms of a sextet-to-doublet (singlet) transition with coexistence over some temperature range (see Section F.6.4.1). Because relaxation lineshape effects are weak, the determination of the median blocking temperature (see Section F.6.4.4) from the equality of the relative areas of the blocked and unblocked components is precise. It is independent of the procedure used for the determination, and the precise estimation of the measuring time is in fact the only difficulty because the value of τ_m depends on the relaxation process and its corresponding model.

Temperature sets of spectra typical of a broad volume distribution are exemplified, for instance, for α-Fe_2O_3,[292,293,383] iron-storage proteins,[210,286,382,384,385] and ferric "hydroxide."[289,386] As an illustration, Figure H.1 shows Mössbauer spectra of human ferritin and hemosiderin.[384] In both cases, one clearly sees a (super)paramagnetic doublet growing at the expense of the six-line pattern as the temperature increases. For hemosiderin, the coexistence range extends practically from 20 to 130 K. Compared to the 4.2-K pattern, the magnetic component shows almost no line shift (apart from that due to the thermal variation of the hyperfine field) nor extra line broadening as the temperature increases, which clearly contrasts with the case of ferromagnetic particles (Fig. F.6.5). In contrast with hemosiderin, the transition

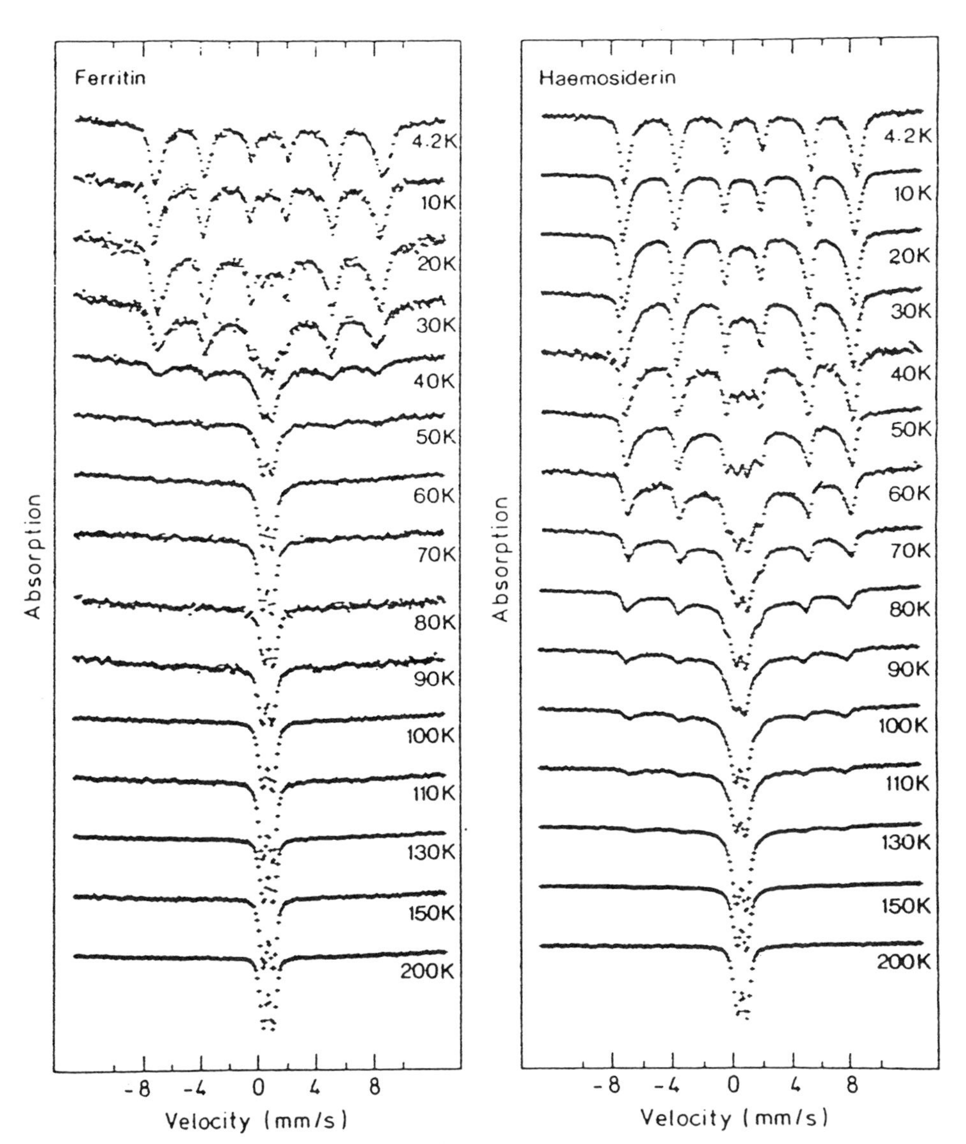

Figure H.1. Mössbauer spectra of human ferritin and hemosiderin at a range of temperatures. (Reproduced with permission from Ref. 384.)

for ferritin (Fig. H.1) is rather sharp and proceeds mostly between 30 and 40 K. The difference in the properties of the two proteins is mainly due to the different structures of the iron core, goethitelike for hemosiderin and ferrihydritelike for ferritin.[385] The line broadening at low temperatures[384,387] is mostly due to a distribution of hyperfine fields, which dominates any effects of the collective magnetic excitations (see Section F.6.4.5).

If the volume distribution is narrow, perturbed spectra will be obtained like for ferromagnetic particles, but over a narrower temperature range. Figure H.2 shows Mössbauer spectra of 2.4- and 3.5-nm α-Fe_2O_3 particles dispersed in alumina.[378] The width of the diameter distribution is about 1 nm. For these samples, the coexistence of the blocked and unblocked components is not observed, and spectra whose shape typically corresponds to relaxation times in the measurement window are prevailing for $30\,K < T < 45\,K$ and $110\,K < T < 200\,K$ for the 2.4- and 3.5-nm particles, respectively. In view of the volume distributions, it can be deduced that these temperature intervals correspond to the crossing of the Mössbauer window by about half of the particles. Consequently, the spectra cannot be analyzed without a suitable lineshape model for superparamagnetic relaxation, which raises the same problems as for ferromagnetic particles (see Section F.6.4.2). In the present cases (Fig. H.2), a distribution of hyperfine fields due to the surface disorder and revealed by some line broadening in the 4.2-K patterns complicates the phenomena.

Like for ferromagnetic particles, the Mössbauer spectra can be sensitive to the state of dispersion/agglomeration of the particles, as observed, for instance, for hematite.[388] It is again clear that the origin of the observed variations cannot reasonably be inferred from the sole Mössbauer features because of the complexity of the phenomena and the many interplaying parameters. This is illustrated by the debate concerning goethite.[94,95,102,389,390]

H.3.2. In-Field Experiments

Since $|\mathbf{m}|$ is smaller by about 2 orders of magnitude than for ferromagnetic particles, the ratio $h = |\mathbf{m}|H_{app}/2K$ is also smaller by about 2 orders of magnitude, for similar values of the applied field, H_{app}, and the anisotropy energy constant, K. Therefore, for antiferromagnetic particles the situation is always that of a weak or medium applied field ($h < 1$) as discussed in Section F.6.6, at least for the field magnitudes classically used, that is, up to about 10 T.

Figure H.3 shows Mössbauer spectra of human ferritin and hemosiderin at 100 and 200 K, respectively, in various applied fields up to 10 T.[391] The considered temperatures are well above the temperatures at

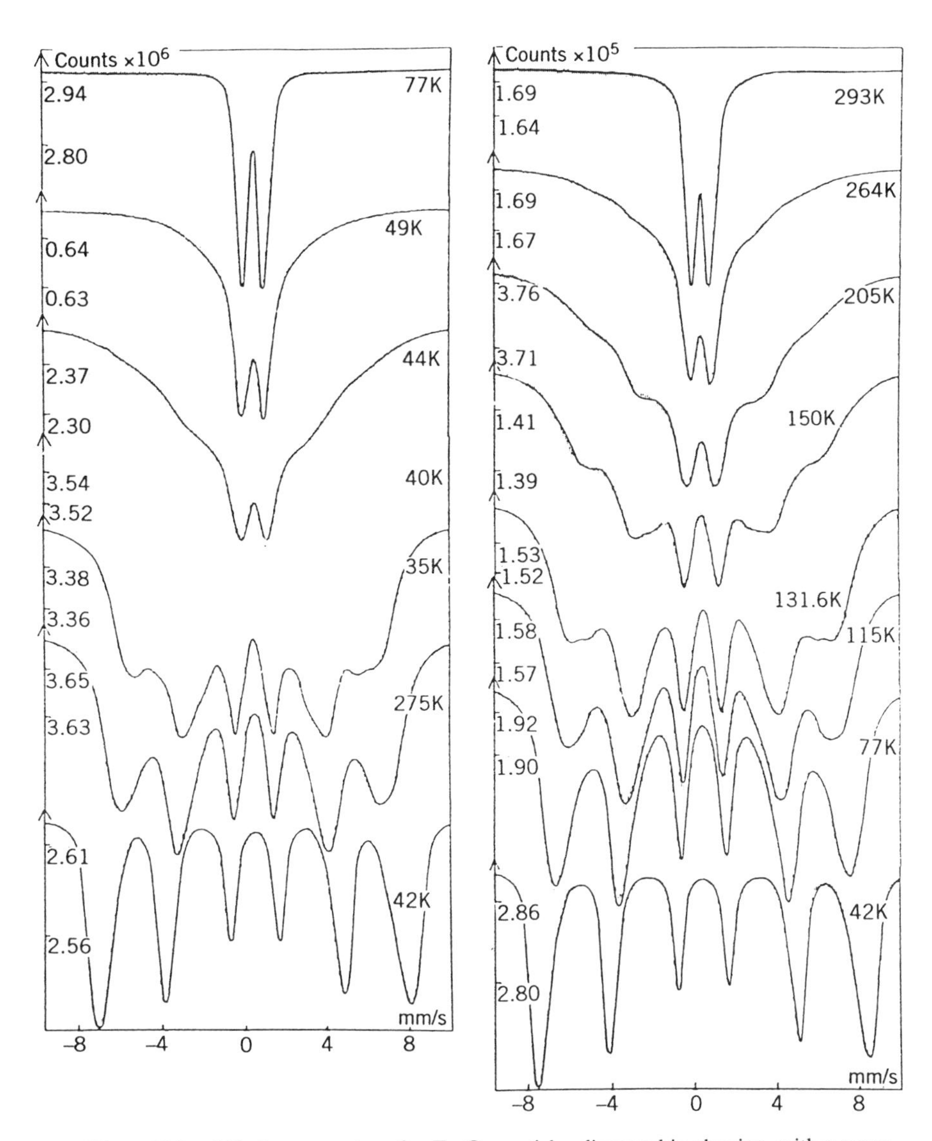

Figure H.2. Mössbauer spectra of α-Fe_2O_3 particles dispersed in alumina, with a mean diameter of 2.4 nm (left) and 3.5 nm (right). (Reproduced with permission from Ref. 378.)

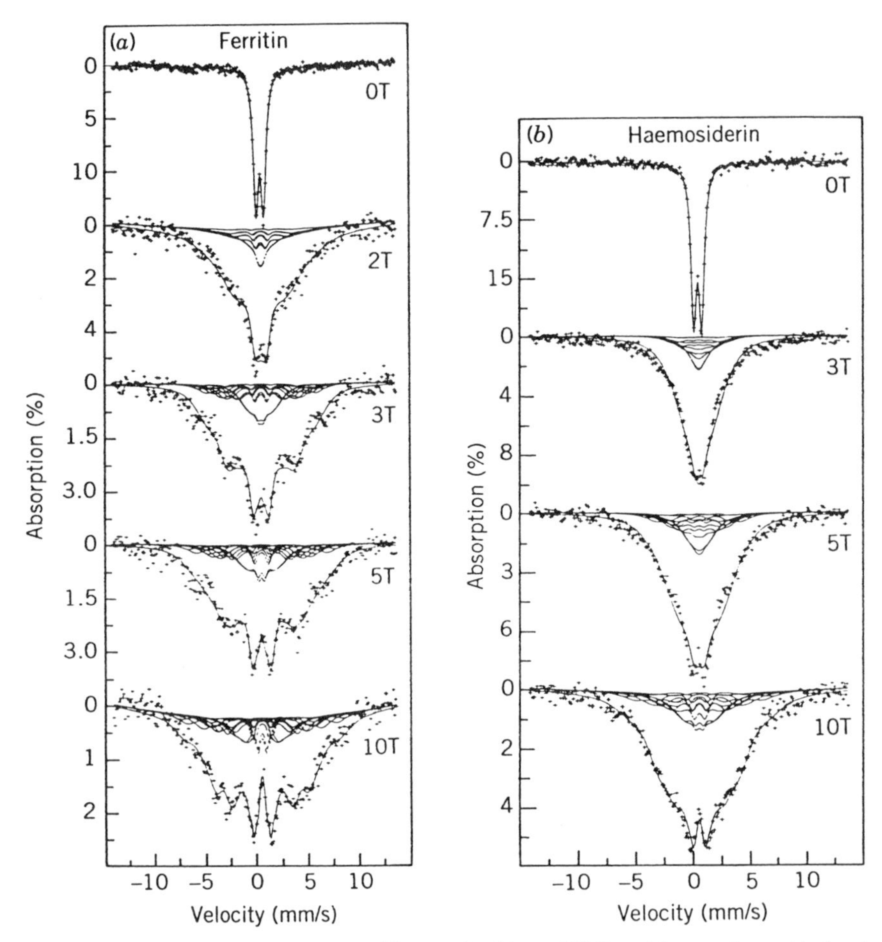

Figure H.3. Mössbauer spectra of human ferritin at 100 K and human hemosiderin at 200 K with magnetic fields up to 10 T applied perpendicularly to the γ-ray beam. (Reproduced with permission from Ref. 391.)

which the sextet component disappears in zero field (Fig. H.1). An applied field (Fig. H.3) restores a magnetic splitting, indicating that the magnetic ordering temperature is above 100 K for ferritin and 200 K for hemosiderin, and confirming the superparamagnetic state in zero field. The in-field spectra were fitted[391] using a model valid for $kT \gg KV$ and based on the calculation of the net magnetization, with a fraction that relaxes (faster than in zero field) and the remaining fraction, which is blocked in the deeper potential well (see Section D.1). This leads to an

$|\mathbf{m}|$ value that decreases with increasing applied field. This is not realistic since, on the contrary, $|\mathbf{m}|$ should slightly increase.

As discussed in Section F.6.6, for Mössbauer spectroscopy it is necessary to consider the two relaxation times τ_+ and τ_-, and the associated populations. Qualitatively, as h (H_{app}) increases, τ_+ increases giving rise first to perturbed spectra and then to sextets with broadened lines, τ_- decreases and always gives rise to a superparamagnetic spectrum but its population decreases. The expressions of τ_+ and τ_- as given by Eq. (D.27) are valid in the case where the applied field is parallel to the easy direction. For actual samples, it is necessary to take into account the random orientation of the particles. To our knowledge, the complete problem has not yet been solved.

H.4. Some Conclusions

An accurate knowledge of the properties of fine antiferromagnetic particles is of great interest because they are present in many natural materials such as clays, laves, protein cores, and so on. From a fundamental point of view several very interesting problems are raised such as the effect of the reduced size on the magnetic state, the effects of defects and impurities in the case of a small size, the possibility of quantum tunneling of the uncompensated moment (see Section G). However, in our opinion, published results do not allow at present a good understanding of the features, due to the lack of systematic studies, of adequate models, but also to the complexity of the materials and the phenomena. For example, theoretical studies on the relaxation process and the calculation of the relaxation time, on the establishment of the magnetic state by simulation methods, on the Mössbauer lineshape under applied fields, as well as experimental studies on well-characterized materials would surely be needful.

I. CONCLUSIONS

In this chapter, we have tried to carry out the restatement of the magnetic properties of fine particles resulting from the relaxation of their magnetic moments from theoretical as well as experimental points of view. We discussed the models, several with some details, that allow one to calculate the relaxation time of the particle magnetic moment, to evaluate the effect of the interparticle interactions, and to interpret the experimental results. We clearly stated our objections, without use of indirect sentences understandable only by specialists, because fine-particle studies can be of interest in various fields, for example, catalysis, biology, mineralogy. Of course, this is our opinion at the present time.

We have pointed out the complexities of the phenomena, which can now be experimentally detected because of the progress in measurement techniques and in materials. These complexities lead to difficulties for an accurate analysis of the experimental results. In particular, we have pointed out that the use of simple expressions, imprecisely stated, for the anisotropy energies and the interparticle magnetic interactions can lead to wrong conclusions, due to the close dependence of the relaxation time on the anisotropic part of the energies. Experimental process too is of fundamental importance for stating the measuring properties. In the same way, comparison without precaution of the results obtained by experiments that do not experience exactly the same properties or do not strictly use the same process can lead to erroneous inferences.

Important progress has been made in the recent last years in the knowledge of the superparamagnetism phenomenon. Numerous imprecisions have been corrected; new models have been developed for explaining new accurate results obtained on well-characterized samples. However, several questions remain to be resolved. Some of them have been pointed out.

Finally, we hope that this chapter will be useful for the establishing of mastered results and the understanding of the phenomena.

ACKNOWLEDGMENTS

We would like to thank Prof. W. T. Coffey, Dr. M. Hennion, and Dr. I. Mirebeau for helpful discussions and comments and F. Gazeau and Dr. R. Perzynski for the communication of new results and calculations. We would also thank our colleagues Dr. F. D'Orazio, Prof. F. Lucari, Prof. M. Godinho, Prof. J. P. Jolivet, Dr. M. Nogues, Dr. A. M. Testa, and our students R. Cherkaoui, Dr. P. Prené, A. Ezzir, L. Spinu, and Dr. J. Maknani who have contributed to our studies on fine particles.

APPENDIX

This appendix provides a summary of the basic formulas giving the parameters of a static Mössbauer spectrum. More detailed information about the theory and applications of Mössbauer spectroscopy can be found in a number of books; for example, see Refs. 392–395.

Ap.1. The Recoil-Free Fraction

When the nucleus of a free atom emits a γ photon, the energy E_γ of the nuclear transition is shared between the γ photon and the recoil energy of the atom, E_R. Conservation of energy and momentum requires that

$$E_R \approx \frac{E_\gamma^2}{2Mc^2} \tag{Ap.1}$$

where M is the mass of the Mössbauer atom. The recoil energy is always much larger than the width of the nuclear level (for a free ^{57}Fe atom, $E_R = 2 \times 10^{-3}$ eV, $\Gamma_N = 4.6 \times 10^{-9}$ eV). Therefore, the emitted γ photon is far too low in energy to be resonantly absorbed by a nucleus of the same type.

If the Mössbauer atom is bound in a solid, the recoil energy may be taken up by the matrix via excitation of lattice vibrations. The recoil energy is then reduced by a factor given by the atom and the solid mass ratio. If the phonon energy is low enough, there will be a finite probability, f, that the emission (absorption) will take place with no creation or annihilation of phonon in the lattice, that is, with no recoil energy loss, and this is the Mössbauer effect. The f factor (recoil-free fraction, Debye–Waller factor, Lamb–Mössbauer factor) is given by

$$f = \exp(-k_\gamma^2 \langle x^2 \rangle) \tag{Ap.2}$$

where $k_\gamma = E_\gamma / \hbar c$ is the magnitude of the wave vector of the γ-ray, and $\langle x^2 \rangle$ is the mean-square vibrational amplitude of the Mössbauer atom in the direction of the γ-ray. The precise form of $\langle x^2 \rangle$ depends on the vibrational properties of the lattice. It can be calculated using, for instance, the Einstein model or the Debye model for the phonon spectrum. In the Debye model, f is given by

$$f = \exp\left\{ -\frac{6E_R}{k\Theta_D} \left[\frac{1}{4} + \frac{T^2}{\Theta_D^2} \int_0^{\Theta_D/T} \frac{x\,dx}{e^x - 1} \right] \right\} \tag{Ap.3}$$

where k is the Boltzmann constant, Θ_D is the Debye temperature, and T is the temperature. Even at zero temperature, the recoil-free fraction depends on E_R and Θ_D. It can be strongly temperature dependent. The rapid decrease at high temperature limits the range over which the Mössbauer effect can be observed.

Ap.2. The Isomer Shift

The electronic charge density at the nucleus is $-e|\psi(0)|^2$ where $-e$ is the electron charge and $\psi(0)$ is the wave function at the origin. This charge density interacts with the charge density of the nucleus. Since the size of the nucleus is different for the excited state and the ground state, the interaction energy is different for these two states. This leads to a shift of the energy of the nuclear transition. In addition, if the chemical environment of the Mössbauer atom is different in the source and in the absorber, there will be a shift (the chemical isomer shift, δ) in the energy of the absorption line observed in Mössbauer spectroscopy given by

$$\delta = C \frac{\Delta R}{R} (|\psi(0)_A|^2 - |\psi(0)|_S^2) \tag{Ap.4}$$

where C is a constant for a given isotope, and $\Delta R/R$ is the relative change of the nuclear radius between the excited state and the ground state ($\Delta R/R < 0$ for ^{57}Fe); A and S stand for "absorber" and "source," respectively.

The isomer shift is proportional to the difference in the s-electron density at the nucleus in the absorber and in the source, and thus yields information on the electronic state and chemical bonding of the Mössbauer atom. The isomer shift is often given relative to a standard material.

Ap.3. The Electric Quadrupole Interaction

A nucleus of spin $I > \frac{1}{2}$ has a quadrupole moment, eQ. If the distribution of the charges surrounding the Mössbauer nucleus is asymmetric, there is an EFG at the nucleus. The interaction with the nuclear quadrupole moment partly lifts the $2I + 1$ degeneracy of the nuclear level.

The Hamiltonian describing the quadrupole interaction has a simple expression only in the principal axis system of the EFG tensor. Choosing the axes so that the principal components are in the order $|V_{zz}| \geq |V_{yy}| \geq |V_{xx}|$, it becomes

$$H_Q = \frac{eQV_{zz}}{4I(2I-1)} [3I_z^2 - \mathbf{I}^2 + \eta(I_+^2 + I_-^2)/2] \tag{Ap.5}$$

where $\mathbf{I}$, I_z, I_+, and I_- are the conventional spin operators and $\eta = (V_{xx} - V_{yy})/V_{zz}$ with $0 \leq \eta \leq 1$. A completely general solution of this Hamiltonian is not possible except if the EFG tensor has uniaxial symmetry ($\eta = 0$) or if $\eta > 0$, for $I = \frac{3}{2}$, since then I_z is a good quantum number. In both cases the energy levels remain twofold degenerated. For $I = \frac{3}{2}$, the energy levels are given by

$$E_Q = \pm \varepsilon (1 + \eta^2/3)^{1/2} \quad \text{with } \varepsilon = eQV_{zz}/4 \tag{Ap.6}$$

with the + and − sign corresponding to $I_z = \pm\frac{3}{2}$ and $\pm\frac{1}{2}$, respectively.

Since the ground state ($I_g = \frac{1}{2}$) of ^{57}Fe is unaffected by the quadrupole interaction, the resultant Mössbauer spectrum consists of two lines. Their separation, $2E_Q$, is called the quadrupole splitting.

The EFG embraces contributions from both the valence electrons of the atom and from the lattice charges. Generally, the valence term is the major contribution unless the ion is an S-state ion such as $Fe^{3+}(^6S)$. In the latter case the EFG will arise solely from charges external to the atom either by direct contribution or by indirect polarization of the electronic shells.

Ap.4. The Magnetic Hyperfine Interaction

A nucleus of spin $I > 0$ has a magnetic moment $\boldsymbol{\mu}$. The Zeeman interaction with a magnetic field $\mathbf{H}_{\text{eff}}$ at the nucleus completely lifts the degeneracy of the nuclear states. The interaction is expressed by

$$H_M = -\boldsymbol{\mu} \cdot \mathbf{H}_{\text{eff}} = -g_I \mu_N \mathbf{I} \cdot \mathbf{H}_{\text{eff}} \qquad \text{(Ap.7)}$$

where μ_N is the nuclear magneton, and g_I the nuclear g factor. If the z axis is taken in the direction of the magnetic field, the resulting Hamiltonian matrix is diagonal. The nuclear state is split into $2I + 1$ equally spaced and nondegenerated substates $|I, I_z\rangle$. The eigenvalues are given by

$$E_M = -g_I \mu_N |\mathbf{H}_{\text{eff}}| I_z \qquad \text{(Ap.8)}$$

The ^{57}Fe excited state ($g_{3/2} = -0.10$) is split into four states and the ground state ($g_{1/2} = 0.18$) into two states. The $\Delta I_z = 0, \pm 1$ selection rules yield six allowed transitions. The resulting Mössbauer spectrum is a symmetric six-line pattern.

When both nuclear quadrupole and Zeeman interactions are present, the Hamiltonian describing the interaction is the sum of Eqs. (Ap.5) and (Ap.7). There is no general solution. When the quadrupole interaction is much weaker than the magnetic interaction, it can be treated as a small perturbation upon the latter. In this case, the $|I = \frac{3}{2}, I_z\rangle$ Zeeman sublevels are shifted by the amount

$$E_Q = (-1)^{|I_z| + 1/2} \varepsilon \tfrac{1}{2}(3 \cos^2\zeta - 1 + \eta \sin^2\zeta \cos 2\xi) \qquad \text{(Ap.9)}$$

where ζ is the angle between $\mathbf{H}_{\text{eff}}$ and the direction of V_{zz}, and ξ is the second angle defining the orientation of the magnetic field with respect to the EFG principal axis system. The resultant Mössbauer spectrum is an asymmetric six-line pattern.

In the presence of an external magnetic field, $\mathbf{H}_{\text{app}}$, the magnetic field seen by the nucleus can be expressed by

$$\mathbf{H}_{\text{eff}} = \mathbf{H}_{\text{app}} + \mathbf{H}_D + \mathbf{H}_L + \mathbf{H}_{\text{el}} \quad \text{(Ap.10)}$$

where $\mathbf{H}_D$ is the demagnetizing field and $\mathbf{H}_L$ is the Lorentz field, but both terms are small in bulk materials; $\mathbf{H}_{\text{el}}$ is the magnetic field due to the surrounding electrons. It has various origins (Fermi contact, electron orbital motion, spin-dipolar interaction, conduction electrons in metal). The effective field at the nucleus in the absence of an applied field is called the hyperfine field. It is typically of the order of 500 kOe in magnetic Fe^{3+} compounds and 300 kOe in metallic materials.

The hyperfine field of a magnetically ordered material is generally proportional to the magnetization. Its temperature dependence will reflect the latter and follows a Brillouin function, becoming zero at the Curie or Néel temperature. In the case where two or more distinct magnetic lattices are present, the Mössbauer spectrum will give the hyperfine field at each individual site, whereas the bulk magnetization is an average effect.

Ap.5. Relative Intensity of the Absorption Lines

The relative areas of the absorption lines are proportional to the probabilities of the corresponding transitions. These probabilities depend on the multipolarity of the nuclear transition, initial and final spin states, and orientation of the quantification axis with respect to the wave vector of the γ-ray.

For ^{57}Fe, the normalized probability $P(I_{z,e}, I_{z,g})$ of transition between substates $|I_e = \frac{3}{2}, I_{z,e}\rangle$ and $|I_g = \frac{1}{2}, I_{z,g}\rangle$ is given by

$$\begin{aligned} P(\pm\tfrac{3}{2}, \pm\tfrac{1}{2}) &= \tfrac{3}{16}(1 + \cos^2\beta) \\ P(\pm\tfrac{1}{2}, \pm\tfrac{1}{2}) &= \tfrac{1}{4}\sin^2\beta \\ P(\pm\tfrac{1}{2}, \mp\tfrac{1}{2}) &= \tfrac{1}{16}(1 + \cos^2\beta) \end{aligned} \quad \text{(Ap.11)}$$

where β is the angle between the wave vector of the γ-ray and the z axis.

The line areas in a Zeeman sextet are in the ratio $3:2p:1:1:2p:3$ with

$$p = \frac{2\sin^2\beta}{1 + \cos^2\beta} \quad \text{(Ap.12)}$$

The second and fifth lines will be absent if $\beta = 0$.

If β is equally distributed over a certain angular range Ω, the total transition probability will be given by

$$P(I_{z,e}, I_{z,g}) = \frac{\int_{\Omega} P \sin \beta \, d\beta}{\int_{\Omega} \sin \beta \, d\beta} \tag{Ap.13}$$

In a powder with no preferred orientation, β is equally distributed between 0 and π ($\beta_{av} = 54.7°$). The lines of a Zeeman sextet are in the intensity ratio 3:2:1:1:2:3, and those of a quadrupole doublet have equal areas. These patterns along with the singlet represent the three basic types of ^{57}Fe Mössbauer spectrum.

REFERENCES

1. L. Néel, *C.R. Acad. Sci. Paris* **228**, 664 (1949); L. Néel, *Ann. Geophys.* **5**, 99 (1949).
2. J. L. Dormann and D. Fiorani, eds., *Magnetic Properties of Fine Particles*, North-Holland, Amsterdam, 1992.
3. G. C. Hadjipanayis and G. A. Prinz, eds., *Science and Technology of Nanostructured Materials*, Plenum Press, New York, 1991.
4. G. C. Hadjipanayis and R. W. Siegel, eds., *Nanophase Materials: Synthesis, Properties, Applications*, Kluwer Acad. Publ., Dordrecht, 1994.
5. J. W. Niemantsverdriet, in Ref. 2, p. 351, and references therein.
6. U. Wagner, R. Gebhard, E. Murad, I. Shimada, and F. E. Wagner, in Ref. 2, p. 339, and references therein.
7. E. Murad, in Ref. 2, p. 339, and references therein.
8. A. Stephenson, in Ref. 2, p. 329, and references therein.
9. D. P. E. Dickson and R. B. Frankel, in Ref. 2, p. 393, and references therein.
10. A. E. Berkowitz, F. T. Parker, F. E. Spada, and D. Margulies, in Ref. 2, p. 309, and references therein.
11. C. Kittel, *Phys. Rev.* **70**, 965 (1946).
12. H. Zijlstra, in *Ferromagnetic Materials*, Vol. 3, E. P. Wohlfarth, ed., North-Holland, Amsterdam, 1982.
13. R. S. Tebble, *Magnetic Domains*, Methuen & Co., London, 1969.
14. D. Givord, Q. Lu, and M. F. Rosignol, in Ref. 3, p. 635.
15. C. P. Bean, *J. Appl. Phys.* **26**, 1381 (1955).
16. C. P. Bean and J. D. Livingstone, *J. Appl. Phys.* **30**, 120S (1959).
17. J. L. Dormann, *Rev. Phys. Appl.* **16**, 275 (1981).
18. S. Mørup, J. Dumesic, and H. Topsøe, in *Applications of Mössbauer Spectroscopy*, R. C. Cohen, ed., vol. 2, Academic Press, New York, 1980, p. 1.
19. D. Fiorani and J. L. Dormann, in *Fundamental Properties of Nanostructured Materials*, D. Fiorani and G. Sberveglieri, eds., World Scientific, 1994, p. 193.
20. A. Aharoni, *IEEE Trans. Magn.* **29**, 2596 (1993), and references therein.
21. R. D. Fredkin and T. R. Koehler, *J. Appl. Phys.* **67**, 5544 (1990).
22. N. A. Usov and S. E. Peschany, *J. Magn. Magn. Mat.* **130**, 275 (1994), and references therein.

23. L. Gunther and B. Barbara, eds., *Quantum Tunneling of Magnetization*, Kluwer Acad. Publ., Dordrecht, 1995.
24. R. D. Shull, R. D. McMichael, J. Swartzendruber, and L. H. Bennet, in Ref. 2, p. 161.
25. A. E. Berkowitz, J. R. Mitchell, M. J. Carey, A. P. Young, S. Zhang, F. E. Spada, F. T. Parker, A. Hutten, and G. Thomas, *Phys. Rev. Lett.* **68**, 3745 (1992).
26. J. Q. Xiao, J. S. Jiang, and C. L. Chien, *Phys. Rev. Lett.* **68**, 3749 (1992).
27. A. Freeman and R. Q. Wu, *J. Magn. Magn. Mat.* **100**, 497 (1991).
28. U. Gradmann, *J. Magn. Magn. Mat.* **100**, 481 (1991).
29. A. Freeman and R. Q. Wu, *J. Magn. Magn. Mat.* **104–107**, 1 (1992).
30. U. Gradmann, in *High Density Digital Recording*, K. H. J. Buschow, ed., Kluwer Acad. Publ., Dordrecht, 1993, p. 315.
31. U. Gradmann in *Handbook of Magnetic Materials*, K. H. J. Buschow, ed., Elsevier Science, Amsterdam, Vol. 7, 1993, p. 1.
32. A. H. Morrish, in Ref. 2, p. 181.
33. A. H. Morrish, K. Haneda, and X. Z. Zhou, in Ref. 4, p. 515.
34. J. L. Dormann, C. Djega-Mariadassou, and J. Jove, *J. Magn. Magn. Mat.* **104–107,** 1567 (1992).
35. C. Djega-Mariadassou, J. L. Dormann, M. Nogues, G. Villers, and S. Sayouri, *IEEE Trans. Magn.* **26**, 1819 (1990).
36. L. Néel, *J. Phys. Radium* **15**, 225 (1954).
37. J. L. Dormann, F. D'Orazio, F. Lucari, L. Spinu, E. Tronc, J. P. Jolivet and D. Fiorani, to be published.
38. A. P. Amulyavichus and I.P. Suzdalev, *Sov. Phys. JETP* **37**, 859 (1973).
39. F. Bødker, S. Mørup, and S. Linderoth, *Phys. Rev. Lett.* **72**, 282 (1994).
40. S. Linderoth, Ll. Balcells, A. Labarta, J. Tejada, P. V. Hendriksen, and S. A. Sethi, *J. Magn. Magn. Mat.* **124**, 269 (1993).
41. A. E. Berkowitz and F. T. Parker, *IEEE Trans. Magn.* **24**, 2871 (1988).
42. B. S. Clausen, S. Mørup and H. Topsoe, *Surf. Sci.* **82**, L589 (1979).
43. S. Mørup, H. Topsøe, and B. S. Clausen, *Phys. Scr.* **25**, 713 (1982).
44. A. Aharoni, *J. Appl. Phys.* **75**, 5891 (1994).
45. J. L. Dormann, L. Bessais, and D. Fiorani, *J. Phys. C* **21**, 2015 (1988).
46. W. F. Brown, Jr., *J. Appl. Phys.* **30**, 130 S (1959).
47. W. F. Brown, Jr., *J. Appl. Phys.* **34**, 1319 (1963).
48. W. F. Brown, Jr., *Phys. Rev.* **130**, 1677 (1963).
49. W. T. Coffey, P. J. Cregg, and Yu. P. Kalmykov, *Adv. Chem. Phys.* **83**, 263 (1993).
50. D. R. Freklin, T. R. Koehler, J. F. Smyth, and S. Schultz, *J. Appl. Phys.* **69**, 5276 (1991).
51. M. E. Schabes and H. N. Bertram, *J. Appl. Phys.* **64**, 1347 (1988).
52. Y. Nakatani, N. Hayashi, and Y. Uesaka, *Jpn. J. Appl. Phys.* **30**, 2503 (1991).
53. D. A. Dimitrov and G. M. Wysin, *Phys. Rev. B* **50**, 3077 (1994).
54. E. C. Stoner and E. P. Wohlfarth, *Phil. Trans. Roy. Soc. A* **240**, 599 (1948).

55. M. Lederman, S. Schultz, and M. Ozaki, *Phys. Rev. Lett.* **73**, 1986 (1994).
56. J. Maddox, *Nature* **371**, 739 (1994).
57. W. Wernsdorfer, K. Hasselbach, A. Benoit, G. Cernicchiaro, D. Mailly, B. Barbara, and L. Thomas, *J. Magn. Magn. Mat.* **151**, 38 (1995).
58. L. Bessais, L. Ben Jaffel, and J. L. Dormann, *Phys. Rev. B* **45**, 7805 (1992).
59. J. L. Dormann, F. D'Orazio, F. Lucari, E. Tronc, P. Prené, J. P. Jolivet, D. Fiorani, R. Cherkaoui, and M. Noguès, *Phys. Rev B* **53**, 14291 (1996).
60. A. Aharoni, *Phys. Rev.* **135**, A447 (1964).
61. L. Bessais, L. Ben Jaffel, and J. L. Dormann, *J. Magn. Magn. Mat.* **104–107**, 1565 (1992).
62. A. Aharoni, *Phys. Rev. B* **46**, 5434 (1992).
63. W. T. Coffey, D. S. F. Crothers, Yu. P. Kalmykov, E.S. Massawe, and J. T. Waldron, *J. Magn. Magn. Mat.* **127**, L254 (1993).
64. W. T. Coffey, D. S. F. Crothers, Yu. P. Kalmykov, E.S. Massawe, and J. T. Waldron, *Phys. Rev. E* **49**, 1869 (1994).
65. W. T. Coffey, P. J. Cregg, D. S. F. Crothers, J. T. Waldron, and A. W. Wickstead, *J. Magn. Magn. Mat.* **131**, L301 (1994).
66. A. Aharoni, *Phys. Rev.* **177**, 793 (1969).
67. W. T. Coffey, D. S. F. Crothers, Yu. P. Kalmykov, and J. T. Waldron, *Phys. Rev. B* **51**, 15947 (1995).
68. W. T. Coffey, D. S. F. Crothers, J. L. Dormann, L. J. Geoghegan, Yu. P. Kalmykov, J. T. Waldron, and A. W. Wickstead, *J. Magn. Magn. Mat.* **145**, L263 (1995); *Phys. Rev. B* **52**, 15951 (1995).
69. H. Pfeiffer, *Phys. Stat. Sol. A* **118**, 295 (1990).
70. H. Pfeiffer, *Phys. Stat. Sol. A* **122**, 377 (1990).
71. J. L. Dormann, D. Fiorani, and M. El Yamani, *Phys. Lett. A* **120**, 95 (1987).
72. I. Eisenstein and A. Aharoni, *Phys. Rev. B* **16**, 1278 (1977); *Physica* **86–88** B, 1429 (1977).
73. D. A. Smith and E. A. Rozario, *J. Magn. Magn. Mat.* **3**, 213 (1976).
74. I. Eisenstein and A. Aharoni, *Phys. Rev. B* **16**, 1285 (1977).
75. J. Korecki, K. Krop, J. Zukrowski, A. Lewicki, and M. Wolny, *Inst. Phys. Conf. Ser.* **39**, 376 (1978).
76. A. M. Afanas'ev, I. P. Suzdalev, M. Ia. Gen, V. I. Goldanskii, V. P. Korneev, and E. A. Manykin, *Sov. Phys. JETP* **31**, 65 (1970).
77. A. M. Afanas'ev, E. A. Manyakin, and E. V. Onishchenko, *Sov. Phys. Solid State* **14**, 2175 (1973).
78. I. Eisenstein and A. Aharoni, *Phys. Rev. B* **5**, 2078 (1976).
79. I. Eisenstein and A. Aharoni, *J. Appl. Phys.* **47**, 321 (1976).
80. A. Aharoni, in Ref, 2, p. 3, and references therein.
81. L. Weil, *J. Chem. Phys.* **51**, 715 (1954).
82. H. B. Braun, *J. Appl. Phys.* **76**, 6310 (1994).
83. D. H. Jones and K. K. P. Srivastava, *J. Magn. Magn. Mat.* **78**, 320 (1989).
84. Yu. L. Raikher and V. I. Stepanov, *Sov. Phys. JETP* **75**, 764 (1992).

85. Yu. L. Raikher and V. I. Stepanov, *Phys. Rev. B* **50**, 6250 (1994).
86. Yu. L. Raikher and M. I. Shliomis, *Sov. Phys. JETP* **40**, 526 (1975).
87. Yu. L. Raikher and V. I. Stepanov, *J. Magn. Magn. Mat.* **149**, 34 (1995); *Phys. Rev. B* **51**, 16428 (1995).
88. J. L. Dormann, *Mat. Sci. Eng. A* **168**, 217 (1993).
89. J. L. Dormann, D. Fiorani, and E. Tronc, in Ref. 4, p. 635.
90. J. L. Dormann and D. Fiorani, *J. Magn. Magn. Mat.* **140–144**, 415 (1995).
91. S. Mørup, *Europhys. Lett.* **28**, 671 (1994).
92. J. P. Bouchard and P. G. Zerah, *Phys. Rev. B* **47**, 9095 (1993).
93. S. Romano, *Phys. Rev. B* **49**, 12287 (1994).
94. S. Mørup, *J. Magn. Magn. Mat.* **37**, 39 (1983).
95. S. Mørup, M. B. Madsen, J. Franck, J. Villadsen, and C. J. W. Koch, *J. Magn. Magn. Mat.* **40**, 163 (1983).
96. P. Prené, E. Tronc, J.P. Jolivet, J. Livage, R. Cherkaoui, M. Noguès, J. L. Dormann, and D. Fiorani, *IEEE Trans. Magn.* **29**, 2658 (1993).
97. R. Cherkaoui, M. Noguès, J. L. Dormann, P. Prené, E. Tronc, J. P. Jolivet, D. Fiorani, and A. M. Testa, *IEEE Trans. Magn.* **30**, 1098 (1994).
98. E. M. Chudnowsky, W. M. Saslow, and R. A. Serota, *Phys. Rev. B* **33**, 251 (1986).
99. E. M. Chudnowsky, *J. Appl. Phys.* **64**, 5770 (1988).
100. R. Ferré, B. Barbara, D. Fruchart, and P. Wolfers, *J. Magn. Magn. Mat.* **140–144**, 385 (1995).
101. G. Parisi, *Phys. Rev. Lett.* **43**, 1754 (1979); *J. Phys. A* **13**, Ll15, 1101, and 1887 (1980).
102. S. Bocquet, R. J. Pollard, and J. D. Cashion, *Phys. Rev. B* **46**, 11657 (1992).
103. J. L. Tholence, *Solid State Com.* **35**, 113 (1980).
104. J. L. Tholence, *Physica B* **126**, 157 (1984).
105. C. Pappa, J. Hammann, J. Jehanno, and C. Jacobini, *J. Phys. C* **18**, 2817 (1985).
106. P. Beauvillain, C. Dupas, J. P. Renard, and P. Veillet, *Phys. Rev. B* **29**, 4086 (1984).
107. E. Vincent, J. Hammann, and M. Alba, *Solid State Com.* **58**, 57 (1986).
108. H. Maletta and W. Flesh, *Phys. Rev. B* **20**, 1245 (1979).
109. D. Fiorani, S. Viticoli, J. L. Dormann, J. L. Tholence, and A. P. Murani, *Phys. Rev. B* **30**, 2276 (1984).
110. S. L. Ginzburg, *Sov. Phys. JETP* **63**, 439 (1986).
111. R. G. Palmer, *Adv. Phys.* **31**, 669 (1982).
112. Ph. Refregier, E. Vincent, J. Hammann, and M. Ocio, *J. Phys. (les Ulis)* **48**, 1533 (1987).
113. K. H. Fisher, *Phys. Stat. Sol. B* **116**, 357 (1983) and **130**, 13 (1985).
114. C. Y. Huang, *J. Magn. Magn. Mat.* **51**, 1 (1985).
115. K. Binder and A. P. Young, *Rev. Mod. Phys.* **58**, 801 (1986).
116. J. L. Dormann and M. Nogues, in *Physic of Magnetic Materials*, W. Gorzkowski, H. K. Lachowicz, and H. Szymczak, eds., World Scientific, Singapore, 1987, p. 531.
117. J. L. Dormann, *Hyp. Int.* **68**, 47 (1991).
118. J. L. Dormann and M. Nogues, *Phase Trans.* **33**, 159 (1991).
119. J. L. Dormann and D. Fiorani, *Proc. of MRS*, Vol. 195, G.D. Cody and T.H. Geballe, eds., Materials Research Society, Pittsburgh, 1990, p. 429 (and references therein).

120. A. Saifi, J. L. Dormann, D. Fiorani, P. Renaudin, and J. Jové, *J. Phys. C* **21**, 5295 (1988).
121. J. Hammann, D. Fiorani, M. El Yamani, and J.L. Dormann, *J. Phys. C* **19**, 6635 (1986).
122. D. Fiorani, in *The Time Domain in Surface and Structural Dynamic*, G. Long and F. Grandjean, eds., Kluwer Acad. Publ., Dordrecht, 1988, p. 391.
123. D. Fiorani, in Ref. 2, p. 135.
124. D. Fiorani, J. L. Dormann, J. L. Tholence, L. Bessais, and G. Villers, *J. Magn. Magn. Mat.* **54–57**, 173 (1986).
125. D. Fiorani, J. L. Tholence, and J. L. Dormann, *J. Phys. C* **19**, 5495 (1986).
126. R. Omari, J. J. Prejean, and J. Souletie, *J. Phys.* (*les Ulis*) **44**, 1069 (1983).
127. D. Fiorani, J. L. Dormann, A. M. Testa, and R. Zysler, in Ref. 4, p. 645.
128. I. Mirebeau, C. Bellouard, M. Hennion, J. L. Dormann, C. Djega-Mariadassou, and M. Tessier, *J. Magn. Magn. Mat.* **104–107**, 1560 (1992).
129. M. Hennion, C. Bellouard, I. Mirebeau, E. Blank, and J. L. Dormann, in Ref. 2, p. 27; C. Bellouard, I. Mirebeau, and M. Hennion, *Phys. Rev.* **53**, 5570 (1996).
130. M. Hennion, C. Bellouard, I. Mirebeau, J. L. Dormann, and M. Nogues, *Europhys. Lett.* **25**, 43 (1994).
131. M. Hennion, C. Bellouard, I. Mirebeau, J. L. Dormann, and R. Ober, *J. Appl. Phys.* **75**, 5900 (1994).
132. C. Bellouard, M. Hennion, and I. Mirebeau, *J. Magn. Magn. Mat.* **140–144**, 357 (1995).
133. C. Bellouard, I. Mirebeau, and M. Hennion, *J. Magn. Magn. Mat.* **140–144**, 431 (1995).
134. P. Prené, E. Tronc, J. P. Jolivet, J. Livage, R. Cherkaoui, M. Nogues, and J. L. Dormann, *Hyp. Int.* **93**, 1409 (1994).
135. M. Godinho, J. L. Dormann, M. Nogues, P. Prené, E. Tronc, and J. P. Jolivet, *J. Magn. Magn. Mat.* **140–144**, 369 (1995).
136. D. Fiorani, A. M. Testa, P. Prené, E. Tronc, J. P. Jolivet, R. Cherkaoui, J. L. Dormann, and M. Nogues, *J. Magn. Magn. Mat.* **140–144**, 395 (1995).
137. E. Tronc, P. Prené, J. P. Jolivet, F. D'Orazio, F. Lucari, D. Fiorani, M. Godinho, R. Cherkaoui, M. Nogues, and J. L. Dormann, *Hyp. Int.* **95**, 129 (1995).
138. E. Tronc, P. Prené, J. P. Jolivet, D. Fiorani, A. M. Testa, R. Cherkaoui, M. Nogues, and J. L. Dormann, *Nanostructured Materials*, **6**, 945 (1995).
139. H. Inoue, N. Kodama, and M. Katsumoto, *J. Magn. Magn. Mat.* **124**, 213 (1993).
140. C. Djega-Mariadassou, J. L. Dormann, L. Bessais, P. Renaudin, E. Agostinelli, and D. Fiorani, *Hyp. Int.* **55**, 933 (1990).
141. M. Olivier, J. O. Strom-Olsen, and Z. Altounian, *Phys. Rev. B* **35**, 333 (1987).
142. S. Shtrikman and E. P. Wohlfarth, *Phys. Lett. A* **85**, 467 (1981).
143. R. W. Chantrell and E. P. Wohlfarth, *J. Magn. Magn. Mat.* **40**, 1 (1983).
144. H. Pfeiffer and R. W. Chantrell, *J. Magn. Magn. Mat.* **120**, 203 (1993).
145. D. Rode, H. N. Bertram, and D. R. Fredkin, *IEEE Trans. Magn.* **23**, 2224 (1987).
146. P. V. Hendriksen, S. Mørup, G. Christiansen, and K. W Jacobsen, in Ref. 3, p. 573.
147. S. Mørup and E. Tronc, *Phys. Rev. Lett.* **72**, 3278 (1994).
148. J. T. Richardson and W. O. Milligan, *Phys. Rev.* **102**, 1289 (1956).
149. J. I. Gittleman, B. Abeles, and S. Bozowski, *Phys. Rev. B* **9**, 3891 (1974).

150. S. H. Liou and C. L. Chien, *J. Appl. Phys.* **63**, 4240 (1988).
151. C. L. Chien, *J. Appl. Phys.* **69**, 5267 (1991).
152. M. El-Hilo, K. O'Grady, J. Popplewell, R. W. Chantrell, and N. Ayoub, *J. Phys. (Les Ulis)* **49**, C8-1835 (1988).
153. R. W. Chantrell, N. Ayoub, and J. Popplewell, *J. Magn. Magn. Mat.* **53**, 199 (1985).
154. M. El-Hilo, K. O'Grady, and R. W. Chantrell, *J. Magn. Magn. Mat.* **114**, 295 (1992).
155. M. El-Hilo, K. O'Grady, and R. W. Chantrell, *J. Magn. Magn. Mat.* **114**, 307 (1992).
156. R. W. Chantrell and E. P. Wohlfarth, *Phys. Stat. Sol. A* **91**, 619 (1985).
157. H. A. Lorentz, *Verhandelingen der koninklijke akademie van wetenschappen, Amsterdam* **18**, 1 (1878).
158. B. K. P. Scaife, *Monographs on the Physics and Chemistry of Materials*, Vol. 45, Oxford Science Publ., Oxford, 1989.
159. L. Onsager, *J. Am. Chem. Soc.* **58**, 1486 (1936).
160. J. L. Dormann, A. Ezzir, and M. Godinho, to be published.
161. M. I. Shliomis, A. F. Psenichnikov, K. I. Morozov, and I. Yu. Shurubor, *J. Magn. Magn. Mat.* **85**, 40 (1990).
162. F. Söffge and E. Schmidbauer, *J. Magn. Magn. Mat.* **24**, 54 (1981).
163. G. Xiao and C.L. Chien, *J. Appl. Phys.* **61**, 3368 (1987).
164. A. Ezzir, J. L. Dormann, M. Noguès, M. Godinho, E. Tronc, and J. P. Jolivet, 2nd Int. Workshop on Fine Particle Magnetism, Bangor, U.K. (1996); J. L. Dormann, A. Ezzir, and M. Godinho, to be published.
165. E. P. Wohlfarth, *J. Magn. Magn. Mat.* **39**, 39 (1983).
166. K. O'Grady and R. W. Chantrell in Ref. 2, p. 93.
167. P. E. Kelly, K. O'Grady, P. I. Mayo and R. W. Chantrell, *IEEE Trans. Magn.* **25**, 3881 (1989).
168. K. O'Grady, M. El-Hilo, and R. W. Chantrell, *IEEE Trans. Magn.* **29**, 2608 (1993).
169. M. El-Hilo and K. O'Grady, *IEEE Trans. Magn.* **26**, 1807 (1990).
170. R. W. Chantrell, K. O'Grady, A. Bradbury, S. W. Charles, and J. Popplewell, *J. Phys. D* **18**, 2505 (1985).
171. R. W. Chantrell, K. O'Grady, A. Bradbury, S. W. Charles, and N. Hopkins, *IEEE Trans. Magn.* **23**, 204 (1987).
172. M. Walker, R. W. Chantrell, K. O'Grady, and S. W. Charles, *J. Phys. (Paris)* **49**, C8-1819 (1988).
173. M. Fearon, R. W. Chantrell, and E. P. Wohlfarth, *J. Magn. Magn. Mat.* **86**, 197 (1990).
174. A. Lyberatos, E. P. Wohlfarth, and R. W. Chantrell, *IEEE Trans. Magn.* **21**, 1277 (1985).
175. R. W. Chantrell, S. R. Hoon, and B. K. Tanner, *J. Magn. Magn. Mat.* **38**, 133 (1983).
176. A. Labarta, O. Iglesias, Ll. Balcells, and F. Badia, *Phys. Rev. B* **48**, 10240 (1993).
177. L. C. Sampaio, Thesis, Grenoble, France, 1995.
178. R. W. Chantrell, M. Fearn, and E. P. Wohlfarth, *Phys. Stat. Sol. A* **97**, 213 (1986).
179. S. Gangopadhyay, G. C. Hadjipanayis, C. M. Sorensen, and K. J. Klabunde, *IEEE Trans. Magn.* **29**, 2619 (1993).

180. J. L. Dormann, D. Fiorani, J. L. Tholence, and C. Sella, *J. Magn. Magn. Mat.* **35**, 117 (1983).
181. P. A. Beck, *Phys. Rev. B* **24**, 2867 (1981).
182. J. Ferré, J. Rajchenbach, and H. Maletta, *J. Appl. Phys.* **52**, 1697 (1981).
183. A. Aharoni and E. P. Wohlfarth, *J. Appl. Phys.* **55**, 1664 (1984).
184. M. El Hilo, K. O'Grady, and R. W. Chantrell, in Ref. 2, p. 45.
185. M. El-Hilo, K. O'Grady, and R. W. Chantrell, *J. Appl. Phys.* **69**, 5133 (1991).
186. M. El-Hilo, K. O'Grady, and R. W. Chantrell, *J. Magn. Magn. Mat.* **140–144**, 359 (1995).
187. L. Néel, *Rev. Mod. Phys.* **25**, 293 (1953).
188. J. F. Liu, S. M. Pan, H. L. Luo, D. L. Hou, and X. F. Nie, *IEEE Trans. Magn.* **26**, 2643 (1990).
189. K. O'Grady, *IEEE Trans. Magn.* **26**, 1870 (1990).
190. G. J. Tomka, P. R. Bissel, K. O'Grady, and R. W. Chantrell, *IEEE Trans. Magn.* **26**, 2655 (1990).
191. A. J. Swartz and W. A. Soffa, *IEEE Trans. Magn.* **26**, 1816 (1990).
192. R. Chamberlin, G. Mozurkewich, and R. Orbach, *Phys. Rev. Lett.* **52**, 867 (1984).
193. A. Khater, J. Ferré, and P. Mayer, *J. Phys. C: Solid State Phys.* **20**, 1857 (1987).
194. O. Iglesias, F. Badia, A. Labarta, and Ll. Balcells, *J. Magn. Magn. Mat.* **140–144**, 399 (1995).
195. J. J. Préjean and J. Souletie, *J. Phys. (Paris)* **41**, 1335 (1980).
196. E. Vincent, J. Hammann, P. Prené, and E. Tronc, *J. Phys. I (Les Ulis)* **4**, 273 (1994).
197. C. Sanchez, J. M. Gonzalez-Miranda, and J. Tejada, *J. Magn. Magn. Mat.* **140–144**, 365 (1995).
198. L. C. Sampaio, C. Paulsen, and B. Barbara, *J. Magn. Magn. Mat.* **140–144**, 391 (1995).
199. R. Hogerbeets, Wei-Li Luo, and R. Orbach, *Phys. Rev. B* **34**, 1719 (1986).
200. T. Jonson, P. Svelindh, and P. Nordlab, *J. Magn. Magn. Mat.* **140–144**, 401 (1995).
201. E. F. Kneller and F. E. Luborsky, *J. Appl. Phys.* **39**, 656 (1963).
202. P. Gaunt, *Phil. Mag.* **17**, 263 (1968).
203. I. Joffe, *J. Phys. C* **2**, 1537 (1969).
204. R. W. Chantrell, J. Popplewell, and S. W. Charles, *J. Magn. Magn. Mat.* **15–18**, 1123 (1980).
205. A. Tsoukatos, H. Wan, G. C. Hadjipanayis, V. Papaefthmiou, A. Kostikas, and A. Simopoulos, *J. Appl. Phys.* **73**, 6967 (1993).
206. S. Gangopadhyay, Y. Yang, G. C. Hadjipanayis, V. Papaefthmiou, C. M. Sorensen, and K. J. Klabunde, *J. Appl. Phys.* **76**, 6319 (1994).
207. W. Wernsdorfer, K. Hasselbach, D. Mailly, B. Barbara, A. Benoit, L. Thomas, and G. Suran, *J. Magn. Magn. Mat.* **145**, 33 (1995).
208. D. Kumar and S. Dattagupta, *J. Phys. C: Solid State Phys.* **16**, 3779 (1983).
209. S. B. Slade, L. Gunther, F. T. Parker, and A. E. Berkowitz, *J. Magn. Magn. Mat.* **140–144**, 661 (1995).
210. D. P. E. Dickson, N. M. K. Reid, C. Hunt, H. D. Williams, M. El-Hilo, and K. O'Grady, *J. Magn. Magn. Mat.* **125**, 345 (1993).
211. J. L. Dormann, P. Gibart, G. Suran, J. L. Tholence, and C. Sella, *J. Magn. Magn. Mat.* **15–18**, 1121 (1980).

212. D. Fiorani, J. L. Tholence, and J. L. Dormann, *Physica* **107**, 643 (1981).

213. D. Fiorani, J. L. Tholence, and J. L. Dormann, *J. Magn. Magn. Mat.* **31–34**, 947 (1983).

214. F. J. Lazaro, J. L. Garcia, V. Schunemann, and A. X. Trautwein, *IEEE Trans. Magn.* **29**, 2652 (1993).

215. J. L. Garcia, F. J. Lazaro, C. Martinez, and A. Corma, *J. Magn. Magn. Mat.* **140–144**, 363 (1995).

216. P. C. Fannin, S. W. Charles, and T. Relihan, *Meas. Sci. Technol.* **4**, 1160 (1993).

217. P. C. Fannin and S. W. Charles, *J. Magn. Magn. Mat.* **136**, 287 (1994).

218. P. C. Fannin and S. W. Charles, *J. Phys. D: Appl. Phys.* **22**, 187 (1988).

219. P. C. Fannin and S. W. Charles, *J. Phys. D: Appl. Phys.* **24**, 76 (1991).

220. P. C. Fannin, *J. Magn. Magn. Mat.* **136**, 49 (1994).

221. P. Debye, *Polar Molecules*, Chemical Catalog Co., New York, 1929.

222. M. I. Shliomis, *Sov. Phys.-Usp.* **17**, 53 (1974).

223. S. Mørup, in *Mössbauer Spectroscopy Applied to Inorganic Chemistry*, G. J. Long ed., vol. 2, Plenum Press, New York, 1987, p. 89.

224. H. Topsøe, J. A. Dumesic, and S. Mørup, in *Applications of Mössbauer Spectroscopy*, R. L. Cohen, ed., vol. 2, Academic Press, New York, 1980, p. 55.

225. K. Haneda and A. H. Morrish, *Phase Transitions* **24–26**, 661 (1990).

226. P. J. Picone, K. Haneda, and A. H. Morrish, *J. Phys. C* **15**, 317 (1982).

227. J. S. van Wieringen, *Phys. Lett.* **26A**, 370 (1968).

228. S. Mørup, C. A. Oxborrow, P. V. Hendriksen, M. S. Pedersen, M. Hanson, and C. Johansson, *J. Magn. Magn. Mat.* **140–144**, 409 (1995).

229. G. von Eynatten and H. E. Bömmel, *Appl. Phys.* **14**, 415 (1977).

230. G. von Eynatten, J. Horst, K. Dransfeld, and H. E. Bömmel, *Hyp. Int.* **29**, 1311 (1986).

231. M. Hayashi, I. Tamura, Y. Fukano, and S. Kanemaki, *Phys. Lett.* **77A**, 332 (1980); *Surf. Sci.* **106**, 453 (1981).

232. M. Hayashi, I. Tamura, Y. Fukano, S. Kanemaki, and Y. Fujiyo, *J. Phys. C* **13**, 681 (1980).

233. A. H. Morrish and R. J. Pollard, *Adv. Ceram.* **16**, 393 (1986).

234. P. V. Hendriksen, S. Mørup, and S. Linderoth, *Hyp. Int.* **70**, 1079 (1992).

235. A. M. van der Kraan, *Phys. Stat. Sol. A* **18**, 215 (1973).

236. J. E. Knudsen and S. Mørup, *J. Phys.* (*Paris*) **41**, C1-155 (1980).

237. P. H. Christensen and S. Mørup, *J. Magn. Magn. Mat.* **35**, 130 (1983).

238. T. Shinjo, *Surf. Sci. Rev.* **12**, 51 (1991).

239. J. Zubrowski, G. Liu, H. Fritzsche, and U. Gradmann, *J. Magn. Magn. Mat.* **145**, 57 (1995).

240. B. S. Clausen, S. Mørup, and H. Topsøe, *Surf. Sci.* **106**, 438 (1981).

241. R. Birringer, H. Gleiter, H. P. Klein, and P. Marquardt, *Phys. Lett. A* **102**, 365 (1984).

242. T. Furubayashi, I. Nakatani, and N. Saegusa, *J. Phys. Soc. Jpn.* **56**, 1855 (1987).

243. F. Bødkert, S. Mørup, C. A. Oxborrow, S. Linderoth, M. B. Madsen, and J. Niemantsverdriet, *J. Phys. Condens. Mat.* **4**, 6555 (1992).
244. C. Djega-Mariadassou and J. L. Dormann, in Ref. 2, p. 191.
245. F. T. Parker, M. W. Foster, D. T. Margulies, and A. E. Berkowitz, *Phys. Rev. B* **47**, 7885 (1993).
246. I. Horio, X. Z. Zhou, and A. H. Morrish, *J. Magn. Magn. Mat.* **118**, 279 (1993).
247. A. Ochi, K. Watanabe, M. Kiyama, T. Shinjo, Y. Bando, and T. Takada, *J. Phys. Soc. Jpn.* **50**, 2777 (1981).
248. T. Okada, H. Sekizawa, F. Ambe, S. Ambe, and T. Yamadaya, *J. Magn. Magn. Mat.* **31–34**, 903 (1983).
249. T. Shinjo, M. Kiyama, N. Sugita, K. Watanabe, and T. Takada, *J. Magn. Magn. Mat.* **35**, 133 (1983).
250. A. Yamamoto, T. Honmyo, N. Hosoito, M. Kiyama, and T. Shinjo, *Nucl. Instr. Meth. Phys. Res. B* **76**, 202 (1993).
251. A. Yamamoto, T. Honmyo, M. Kiyama, and T. Shinjo, *J. Phys. Soc. Jpn.* **63**, 176 (1994).
252. P. Prené, E. Tronc, J. P. Jolivet, and J. L. Dormann, *Proceedings on the International Conference on the Applications of the Mössbauer Effect*, ed. by *Italian Phys. Soc.* **50**, 485 (1996).
253. J. M. D. Coey, *Phys. Rev. Lett.* **27**, 1140 (1971); J. M. D. Coey and D. Khalafalla, *Phys. Stat. Sol. A* **11**, 229 (1972).
254. A. H. Morrish, K. Haneda, and P. J. Schurer, *J. Phys. (Paris)* **37**, C6-301 (1976).
255. A. H. Morrish and K. Haneda, *J. Phys. (Paris)* **41**, C1-171 (1980).
256. P. M. de Bakker, E. de Grave, R. E. Vandenberghe, and L. W. Bowen, *Hyp. Int.* **54**, 493 (1990).
257. P. V. Hendriksen, F. Bødker, S. Linderoth, S. Wells, and S. Mørup, *J. Phys. Condens. Mat.* **6**, 3081 (1994); P. V. Hendriksen, S. Linderoth, C. A. Oxborrow, and S. Mørup, *J. Phys. Condens. Mat.* **6**, 3091 (1994).
258. S. Linderoth, P. V. Hendriksen, F. Bødker, S. Wells, K. Davies, S. W. Charles, and S. Mørup, *J. Appl. Phys.* **75**, 6583 (1994).
259. F. T. Parker and A. E. Berkowitz, *Phys. Rev. B* **44**, 7437 (1991).
260. Q. A. Pankhurst and P. J. Pollard, *Phys. Rev. Lett.* **67**, 248 (1991).
261. K. Haneda and A. H. Morrish, *J. Appl. Phys.* **63**, 4258 (1988).
262. A. H. Morrish and K. Haneda, *J. Appl. Phys.* **52**, 2496 (1981).
263. A. E. Berkowitz, J. A. Lahut, I. S. Jacobs, L. M. Levinson, and D. W. Forester, *Phys. Rev. Lett.* **34**, 594 (1975).
264. A. H. Morrish and K. Haneda, *IEEE Trans. Magn.* **25**, 2597 (1989).
265. K. Haneda, H. Kojima, A. H. Morrish, P. J. Picone and K. Wakai, *J. Appl. Phys.* **53**, 2686 (1982).
266. K. Haneda and A. H. Morrish, *Surf. Sci.* **77**, 584 (1978).
267. P. Prené, Thesis, Univ. Pierre et Marie Curie, Paris, 1995; unpublished results.
268. J. L. Dormann and M. Noguès, *J. Phys. Condens. Mat.* **2**, 1223 (1990).
269. Sho-Chen Lin and J. Phillips, *J. Appl. Phys.* **58**, 1943 (1985).
270. S. H. Kilcoyne and R. Cywinski, *J. Magn. Magn. Mat.* **140–144**, 1466 (1995).
271. H. H. Wickman, M. P. Klein and D. A. Shirley, *Phys. Rev.* **152**, 345 (1966).
272. M. Blume, *Phys. Rev.* **174**, 351 (1968).

273. M. Blume and J. A. Tjon, *Phys. Rev.* **165**, 446 (1968).
274. S. Dattagupta and M. Blume, *Phys. Rev. B* **10**, 446 (1974).
275. A. M. Afanas'ev and E. V. Onishchenko, *Sov. Phys. JETP* **43**, 322 (1976).
276. G. N. Belozerskii and Y. T. Pavlyukhin, *Sov. Phys. JETP* **43**, 371 (1976).
277. G. N. Belozerskii, *Phys. Stat. Sol. A* **46**, 131 (1978).
278. D. H. Jones and K. K. P. Srivastava, *Phys. Rev. B* **34**, 7542 (1986).
279. G. N. Belozerskii and S. Simonyan, *J. Phys. (Paris)* **40**, C2-237 (1979).
280. G. N. Belozerskii and B. S. Pavlov, *Sov. Phys. Solid State* **25**, 974 (1983).
281. A. M. Afanas'ev, O. A. Yakovleva, and V. E. Sedov, *Hyp. Int.* **12**, 211 (1982).
282. V. E. Sedov, *Hyp. Int.* **56**, 1491 (1990).
283. F. Harmann-Boutron, *Ann. Phys.* **9**, 285 (1975).
284. A. M. Afanas'ev, in Ref. 2, p. 13.
285. S. Mørup and H. Topsøe, *Appl. Phys.* **47**, 63 (1976).
286. N. M. K. Reid, D. P. E. Dickson, and D. H. Jones, *Hyp. Int.* **56**, 1487 (1990).
287. S. Mørup, H. Topsøe, and J. Lipka, *J. Phys. (Paris)* **37**, C6-287 (1976).
288. A. Tari, J. Popplewell, S. W. Charles, D. St. P. Bunbury, and K. M. Alves, *J. Appl. Phys.* **54**, 3351 (1983).
289. E. Yu. Tsymbal, F. Kh. Chibirova, and I. G. Kostyuchenko, *J. Magn. Magn. Mat.* **136**, 197 (1994).
290. J. L. Dormann, C. Djega-Mariadassou, and P. Renaudin, *Hyp. Int.* **56**, 1683 (1990).
291. D. G. Rancourt, *Hyp. Int.* **40**, 183 (1988).
292. W. Kündig, H. Bömmel, G. Constabaris, and R. H. Lindquist, *Phys. Rev.* **142**, 327 (1966).
293. B. Ganguly, F. E. Huggins, K. R. P. M. Rao, and G. P. Huffman, *J. Catal.* **142**, 552 (1993).
294. E. Tronc, *Nuov. Cimento D* **18**, 163 (1996).
295. E. Tronc and J. P. Jolivet, in Ref. 2, p. 199.
296. E. Tronc and J. P. Jolivet, *J. Phys. (Les Ulis)* **49**, C8-1823 (1988).
297. J. Z. Jiang, S. Mørup, T. Jonsson, and P. Svedlindh, *Proceedings on the International Conference on the Applications of the Mössbauer Effect, Rimini, 1995* (in press).
298. P. V. Hendriksen, C. A. Oxborrow, S. Linderoth, S. Mørup, M. Hanson, C. Johansson, F. Bødker, K. Davies, S. W. Charles, and S. Wells, *Nucl. Instr. Meth. B* **76**, 138 (1993).
299. S. Mørup, F. Bødker, P. V. Hendriksen, and S. Linderoth, *Phys. Rev. B* **52**, 287 (1995).
300. E. Tronc and J. P. Jolivet, *Hyp. Int.* **28**, 525 (1986).
301. P. V. Hendriksen, G. Christiansen and S. Mørup, *J. Magn. Magn. Mat.* **132**, 207 (1994).
302. O. Jarjayes and P. Auric, *J. Magn. Magn. Mat.* **138**, 115 (1994).
303. S. Mørup, P. H. Christensen and B. J. Clausen, *J. Magn. Magn. Mat.* **68**, 160 (1987).
304. P. H. Christensen, S. Mørup, and J. W. Niemantsverdriet, *J. Phys. Chem.* **89**, 4898 (1985).
305. S. Mørup, B. J. Christensen, J. van Wonterghem, M. B. Madsen, S. W. Charles, and S. Wells, *J. Magn. Magn. Mat.* **67**, 249 (1987).

306. S. Mørup and S. Linderoth, in Ref. 4, p. 595.
307. D. G. Rancourt and J. M. Daniels, *Phys. Rev. B* **29**, 2410 (1984).
308. J. L. Dormann and L. Bessais, *Hyp. Int.* **70**, 1109 (1992).
309. C. A. Morrison and N. Karayianis, *J. Appl. Phys.* **29**, 339 (1958); E. Schlöman and J. R. Zeender, *J. Appl. Phys.* **29**, 341 (1958).
310. E. P. Valstyn, J. P. Hanton, and A. H. Morris, *Phys. Rev.* **128**, 2078 (1962).
311. V. K. Sharma and A. Baiker, *J. Chem. Phys.* **75**, 5596 (1981).
312. J. Dubowik and J. Baszynski, *J. Magn. Magn. Mat.* **59**, 161 (1986).
313. R. S. de Biasi and T. C. Devezas, *J. Appl. Phys.* **49**, 2466 (1978).
314. F. Gazeau, E. Dubois, J. C. Bacri, F. Gendron, R. Perzynski, Yu. L. Raikher, and V. I. Stepanov, *2nd Int. Workshop on Fine Particle Magnetism*, Bangor, U.K. (1996); F. Gazeau, J. C. Bacri, F. Gendron, R. Perzynski, Yu. L. Raikher, and V. I. Stepanov, to be published.
315. G. A. Petrakovskii, V. P. Piskorskii, V. M. Sosnin, and I. D. Kosobudskii, *Sov. Phys. Solid. Stat.* **25**, 1876 (1983).
316. R. Zysler, D. Fiorani, J. L. Dormann, and A. M. Testa, *J. Magn. Magn. Mat.* **133**, 171 (1994).
317. R. Berger, J. Kliava, E. Yahiaoui, J. C. Bissey, P. K. Zinsoun, and P. Beziade, *J. Non Crys. Solids* **180**, 151 (1995).
318. R. Childress, C. L. Chien, J. J. Rhyne, and R. W. Erwin, *J. Magn. Magn. Mat.* **104–107**, 1585 (1992).
319. R. Pynn, *Rev. Sci. Instrum.* **55**, 837 (1984).
320. R. Pynn, in Ref. 2, p. 287.
321. R. Rosman, J. J. M. Janssen, and M. Th. Rekveldt, *J. Appl. Phys.* **67**, 3072 (1990).
322. D. J. Cebula, S. W. Charles, and J. Popplewell, *J. Phys (Paris)* **44**, 207 (1983).
323. J. B. Hayter, R. Pynn, S. W. Charles, A. T. Skjeltorp, J. Trewhella, G. Stubbs, and P. Timmens, *Phys. Rev. Lett.* **62**, 1667 (1989).
324. M. Th. Rekvelt, in Ref. 2, p. 297.
325. R. Rosman and M. Th. Rekveldt, *J. Magn. Magn. Mat.* **95**, 1319 (1991).
326. R. Felici, D. Fiorani, and J. L. Dormann, in Ref. 2, p. 205.
327. R. Felici, D. Fiorani, J. L. Dormann, J. Penfold, and R. C. Ward, *Solid State Com.* **76**, 989 (1990).
328. M. N. Baibich, J. M. Broto, A. Fert, F. van Dau Nguyen, F. Petroff, P. Etienne, G. Creuzet, A. Friederich, and J. Chazales, *Phys. Rev. Lett.* **61**, 2472 (1988).
329. R. E. Camley and J. Barnas, *Phys. Rev. Lett.* **63**, 664 (1989).
330. P. M. Levy, S. Zhang, and A. Fert, *Phys. Rev. Lett.* **65**, 1643 (1990).
331. S. Zhang and P. M. Levy, *J. Appl. Phys.* **73**, 5315 (1993).
332. J. A. Barnard, A. Waknis, M. Tan, E. Hafyek, M. R. Parker, and M. L. Watson *J. Magn. Magn. Mat.* **114**, L230 (1992).
333. J. M. Carey, A. P. Young, A. Starr, D. Rao, and A. Berkowitz, *Appl. Phys. Lett.* **61**, 2935 (1992).
334. A. Tsoukatos, A. H. Wan, G. C. Hadjipanayis, and K. M. Unruth, *J. Appl. Phys.* **73**, 5509 (1993).
335. M. L. Watson, J. A. Barnard, S. Hossain, and M. R. Parker, *J. Appl. Phys.* **73**, 5506 (1993).

336. E. Kneller, *J. Appl. Phys.* **33**, 1355 (1963).
337. C. Gente, O. Ohering, and R. Burman, *Phys. Rev. B* **48**, 13244 (1993).
338. M. E. Ghannami, C. G. Polo, G. Rivero, and A. Hernando, *Europhys. Lett.* **26**, 701 (1994).
339. R. H. Yu, X. X. Zhang, J. Tejada, M. Knobel, P. Tiberto, and P. Allia, *J. Appl. Phys.* **78**, 392 (1995).
340. Y. Goldstein and J. I. Gittleman, *Solid State Com.* **9**, 1197 (1971).
341. R. D. Mc Michael, R. D. Shull, L. J. Swartzendruber, L. H. Bennet, and R. E. Watson, *J. Magn. Magn. Mat.* **111**, 29 (1992).
342. R. D. Shull, R. D. Mc Michael, L. J. Swartzendruber, and L. H. Bennet, in *Proceed. of the 6th Inter. Cryocoolers Conf.*, G. Green and M. Knox, eds., David Taylor Res. Center Publ., Annapolis, MD, 1991, p. 231.
343. R. D. Shull, *I.E.E.E. Trans. Magn.* **29**, 2614 (1993).
344. R. D. Mc Michael., J. J. Riter, and R. D. Shull, *J. Appl. Phys.* **73**, 6946 (1993).
345. F. Lucari, F. D'Orazio, J. L. Dormann, and D. Fiorani, in Ref. 2, p. 255.
346. F. D'Orazio, J. L. Dormann, D. Fiorani, and F. Lucari, in *Fundamental and Applicative Aspects of Disordered Magnetism*, P. Allia, D. Fiorani, and L. Lanotte, eds., World Scientific, Singapore, 1989, p. 183.
347. J. L. Dormann, D. Fiorani, F. Giammaria, and F. Lucari, *J. Appl. Phys.* **69**, 5130 (1991).
348. A. O. Caldeira and A. J. Legget, *Phys. Rev. Lett.* **46**, 211 (1981).
349. R. F. Voss and R. A. Webb, *Phys. Rev. Lett.* **47**, 265 (1981).
350. R. E. Thorne, J. H. Miller, W. G. Lyons, J. W. Lyding, and J. R. Tucker, *Phys. Rev. Lett.* **55**, 1006 (1985).
351. E. M. Chudnowsky and L. Gunther, *Phys. Rev. Lett.* **60**, 661 (1988); *Phys. Rev. B* **37**, 9455 (1988).
352. L. Gunther, *Phys. World* **3** (12), 28 (1990).
353. A. J. Legget, in *Directions in Condensed Matter Physics*, G. Grinstein and G. Mazenko, eds., World Scientific, Singapore, 1986, p. 188.
354. B. Barbara and L. Gunther, eds., *Quantum Tunneling of the Magnetization*, Kluwer Publ., 1995.
355. I. Klik and L. Gunther, *J. Appl. Phys.* **67**, 4505 (1990).
356. G. Tatara and H. Fukuyama, *Phys. Rev. Lett.* **72**, 772 (1994).
357. L. Gunther and B. Barbara, *Phys. Rev. B* **49**, 3926 (1994).
358. L. Gunther in Ref. 2, p. 213.
359. J. Tejada, X. X. Zhang, and E. M. Chudnowsky, *Phys. Rev. B* **47**, 14977 (1993).
360. J. I. Arnaudas, A. del Moral, C. de la Fuente, and P. A. J. de Groot, *Phys. Rev. B* **47**, 11924 (1993).
361. C. Paulsen, L. C. Sampaio, B. Barbara, R. Tucoulou-Tachoueres, D. Fruchart, A. Marchand, J. L. Tholence, and M. Uehara, *Europhys. Lett.* **19**, 643 (1992).
362. D. D. Awschalom, M. A. McCord, and G. Grinstein, *Phys. Rev. Lett.* **65**, 783 (1990).
363. D. D. Awschalom, J. F. Smith, G. Grinstein, D. P. Di Vincenzo, and D. Loss, *Phys. Rev. Lett.* **68**, 3092 (1992).
364. J. Tejada, X. X. Zhang, Ll. Balcells, C. Ferraté, J. M. Ruiz, F. Badia, O. Iglesias, and B. Barbara, in Ref. 2, p. 225.

365. B. Barbara, C. Paulsen, L. C. Sampaio, M. Uehara, F. Fruchard, J. L. Tholence, A. Marchand, J. Tejada, and S. Linderoth in Ref. 2, p. 235; B. Barbara, W. Wernsdorfer, L. C. Sampaio, J. G. Park, C. Paulsen, M. A. Novak, R. Ferré, D. Mailly, R. Sessoli, A. Caneschi, K. Hasselbach, A. Benoit, and L. Thomas, *J. Magn. Magn. Mat.* **140–144**, 1825 (1995).
366. T. L. Gilbert, *Phys. Rev.* **100**, 1243 (1955).
367. B. Barbara and E. M. Chudnowsky, *Phys. Lett. A* **145**, 205 (1990).
368. J. Tejada, Ll. Balcells, S. Linderoth, R. Perzynski, B. Rigau, B. Barbara, and J. C. Bacri, *J. Appl. Phys.* **73**, 6952 (1993).
369. H. Grabert, P. Olschowski, and U. Weiss, *Phys. Rev. B* **32**, 3348 (1985); *Phys. Rev. B* **36**, 1931 (1987).
370. X. X. Zhang and J. Tejada, *J. Magn. Magn. Mat.* **129**, L109 (1994); *J. Appl. Phys.* **75**, 567 (1994).
371. J. Tejada and X. X. Zhang, *J. Phys.: Condens. Matt.* **6**, 263 (1994).
372. L. Néel, *C.R. Acad. Sci. Paris* **252**, 4075 (1961).
373. L. Néel, *J. Phys. Soc. Jpn.* **17**, B-1, 676 (1962).
374. J. M. D. Coey, *J. Appl. Phys.* **49**, 1646 (1978).
375. L. Néel, *C.R. Acad. Sci. Paris* **253**, 9 (1961).
376. L. Néel, *C.R. Acad. Sci. Paris*, **253**, 203 (1961).
377. J. Cohen, K. M. Creer, P. Pauthenet, and K. Srivastava, *J. Phys. Soc. Jpn.* **17**, B-l, 685 (1962).
378. J. L. Dormann, Ji Ren Cui, and C. Sella, *J. Appl. Phys.* **57**, 4283 (1985).
379. P. Mollard, M. Figlarz, and F. Vincent, *C. R. Acad. Sci. Paris* **269**, 448 (1969).
380. P. Mollard, F. de Bergevin, P. Germi, F. Vincent, and M. Figlarz, *J. Phys. (les Ulis)* **32**, C1-1041 (1971).
381. P. Mollard, M. Briane, P. Germi, and F. Fievet, *J. Magn. Magn. Mat.* **7**, 63 (1978).
382. D. P. E. Dickson, in Ref. 4, p. 729.
383. A. M. van der Kraan, *J. Phys. (Paris)* **32**, C1-1034 (1971).
384. S. H. Bell, M. P. Weir, D. P. E. Dickson, J. F. Gibson, G. A. Sharp, and T. J. Peters, *Biochem. Biophys. Acta* **787**, 227 (1984).
385. T. G. St. Pierre, J. Webb, and S. Mann, in *Biomineralization Chemical and Biochemical Perspectives*, S. Mann, J. Webb, and R. J. P. Williams, eds., VCH, Weinkeim (Germany), 1989, p. 295.
386. B. Rodmacq, *J. Phys. Chem. Solids* **45**, 1119 (1984).
387. D. P. E. Dickson, D. H. Jones, and F. Keay, *Hyp. Int.* **41**, 471 (1988).
388. M. A. Polykarpov, I. V. Trushin, and S. S. Yakimov, *J. Magn. Magn. Mat.* **116**, 372 (1992).
389. C. J. W. Koch, M. B. Madsen, and S. Mørup, *Hyp. Int.* **28**, 549 (1986).
390. C. J. W. Koch, M. B. Madsen, S. Mørup, G. Christiansen, L. Gerward, and J. Villadsen, *Clays Clay Miner.* **34**, 17 (1986).
391. T. G. St. Pierre, D. H. Jones, and D. P. E. Dickson, *J. Magn. Magn. Mat.* **69**, 276 (1987).
392. N. N. Greenwood and T. C. Gibb, *Mössbauer Spectroscopy*, Chapman and Hall, London, 1971.
393. D. P. E. Dickson and F. J Berry, eds., *Mössbauer Spectroscopy*, Cambridge University Press, Cambridge, 1986.

394. G. J. Long, ed., *Mössbauer Spectroscopy Applied to Inorganic Chemistry*, Plenum Press, New York, Vol. 1, 1984; Vol. 2, 1987; G. J. Long and F. Grandjean, eds., *Mössbauer Spectroscopy Applied to Inorganic Chemistry*, Plenum Press, New York, Vol. 3, 1989.

395. G. J. Long and F. Grandjean, eds., *Mössbauer Spectroscopy Applied to Magnetism and Materials Science*, Plenum Press, New York, Vol. 1, 1993.

COMPLEX SYSTEMS: EQUILIBRIUM CONFIGURATIONS OF N EQUAL CHARGES ON A SPHERE ($2 \leq N \leq 112$)

T. ERBER

Department of Physics and Department of Mathematics, Illinois Institute of Technology, Chicago, Illinois 60616

G. M. HOCKNEY

Theoretical Physics Department, Fermi National Accelerator Laboratory, P.O. Box 500, Batavia, Illinois 60510

CONTENTS

Advances in Chemical Physics, Volume XCVIII, Edited by I. Prigogine and Stuart A. Rice.
ISBN 0-471-16285-X

1. INTRODUCTION

Symmetry and stability criteria are useful for describing charge configurations in a great variety of situations ranging from J. J. Thomson's original plum pudding model of the atom to current investigations of carbon and indium fullerene cages.[1–5] In particular, the O(4) symmetry associated with the Coulomb interaction underlies both the standard Bohr–Pauli level structure of the elements as well as the nested charge rings of the old plum pudding model.[6–8] This robust symmetry constraint enabled Thomson to establish the first quantitative connections between recurrences in the patterns of charge distributions and the periodicities of Mendeleyev's chemical table. The most striking recent success of symmetries in charge configurations is the discovery that C_{60} can exist in a stable form resembling a truncated icosahedron.[9] However, since this is the last but one of the 13 Archimedean polyhedra, there are no further regular structures of this kind that can serve as templates for more complex chemical cages. One method of extending the inventory of geometric figures is to use computers to search for the static equilibrium states of N equal point charges on the surface of a sphere. In contrast to the plum pudding or "jellium" model, where Thomson[2] and Föppl[10] started with the presumption that the equilibrium states would be a series of symmetric nested rings, locally stable solutions of the surface Coulomb problem can be obtained without imposing any *a priori* constraints of symmetry or other types of structural regularities. For small values of N, the results confirm the intuitive expectation that the charge configurations are symmetric and unique. They are also extremely robust because for the special values $N_L = 2–6$, 12 the equilibrium configurations remain invariant if the Coulomb law r^{-2} is replaced by the limiting form r^{-n}, $n \to \infty$.[11] This "ultrarepulsive" interaction is the basis of the biological *Tammes* problem of finding arrangements of N points on the surface of a sphere with the largest possible minimum distance between any pair.[12–15] Since exact solutions of the Tammes problem are known for the set $N_T^{ex} = 2–12, 24$; this invariance also yields optimum configurations for the surface Coulomb problem for the particular values $N_C^{ex} = 2–6$, 12. Of course, these geometric solutions coincide with the computer-generated patterns. If the mutual charge repulsions are described by logarithmic interactions rather than a power law, the corresponding equilibrium solutions for $N = 2$–6, 12 are again given by the Coulomb set N_C^{ex}.[16] Similar configurations—except for a few changes in length scales—appear in the jellium model.[10] All of these equivalences suggest that in cooperative systems with few degrees of freedom symmetry principles alone may be sufficient to determine the character of the equilibrium states. However, when $N > 6$, the sets of equilibrium configurations for these four different force laws lose their resemblance. These divergences illustrate the symmetry breaking effects associated with the emergence of new levels of complexity in larger systems.

In the range $50 \leq N \leq 112$, the surface Coulomb problem has at least 1945 locally stable solutions. These configurations may be classified with the help of several measures based on geometric and energy criteria. Specifically, for any particular value of N, there are a total of $N(N-1)/2$ angles between the $\mathbf{r}_i$ vectors that specify the locations of the charges on the surface of the sphere. A simple measure of the geometric regularity of a charge distribution is then given by the *angular diversity ratio* (%)

$$D_a(N) \equiv 100 \frac{\text{number of distinct angles}}{N(N-1)/2} \tag{1.1}$$

Clearly, large values of D_a (percentages exceeding 96% occur frequently when $N > 50$) indicate irregular configurations that cannot be identified with any of the 123 standard types of convex polyhedra.[17,18] This irregularity also implies that the vertices, or charge positions, of these Coulomb states cannot be interchanged by means of any of the usual rotational symmetry operations. Nevertheless, lack of congruence in vertex separations or edge lengths does not exclude the persistence of other kinds of order. A quantitative measure of the difference between random and geometrically irregular distributions of N points on the surface of a sphere is given by the dipole moment or center of charge,[19,20] that is,

$$\mathbf{d}(N) = \sum_{i=1}^{N} \mathbf{r}_i \tag{1.2a}$$

In particular, for a unit sphere, where $|\mathbf{r}_i| = 1$, the average value of the dipole moment of a random configuration of N unit charges increases with N,

$$\langle |\mathbf{d}(N)| \rangle_{\text{Ran}} = \left\langle \left| \sum_{i=1}^{N} \mathbf{r}_i \right| \right\rangle_{\text{Ran}} = \left(\frac{8}{3\pi} N \right)^{1/2} \tag{1.2b}$$

On the other hand, the dipole moments of all the equilibrium Coulomb states, for $N \leq 112$, are bounded by 10^{-2}, and typically fall in the range $10^{-5} \lesssim |\mathbf{d}(N)| \lesssim 10^{-3}$. Obviously, this is an orders-of-magnitude reduction from the random values. The regularities of the Coulomb states are even more apparent in cases where the angular diversity ratios are small, say $D_a \leq 10\%$. The computer searches show that there are at least 23 geometrically ordered configurations of this kind for a series of N values between $24 \leq N \leq 112$. None of these patterns match the Archimedean polyhedra. For instance, there are four semiregular Archimedean polyhedra with 24 vertices; and in fact one of them, the snub cube, resembles the ordered Coulomb state with 24 charges because both configurations have 38 faces, 60 edges, and occur in enantiomeric forms. However, all edges of the snub cube have equal length and subtend an angle of 43.68° at the center of the sphere, whereas the 60 edges of the Coulomb configuration are split into three sets with approximately equal lengths: 24 subtending an angle of 42.07°, 24 with an angle of 45.04°, and 12 with an angle of 45.71°. Additional comparisons for other sets of states show that this symmetry breaking is pervasive: There is a general trend away from strict geometric regularity in larger systems.

The emergence of complexity is also reflected in several physical effects. For example the electrostatic interaction energy of N unit charges, $E(N)$, can be represented as the sum of the partial energies associated with the individual charges, $E_i(N)$; that is,

$$E(N) = \sum_{i=1}^{N} E_i(N) \qquad E_i(N) = \frac{1}{2} \sum_{j \neq i}^{N} |\mathbf{r}_i - \mathbf{r}_j|^{-1} \tag{1.3}$$

This energy sharing is completely symmetric for the equilibrium states of the surface Coulomb problem in small systems; that is, $E_i(N) = E(N)/N$ for $N < 5$. However, when $N = 5$, the equilibrium arrangement is a triangular bipyramid with three charges positioned at the vertices of an equilateral triangle around a great circle, specifically the equator, and the other two charges at the north and south poles. Since the distances between pairs of equatorial charges exceed the distance from the equator to either pole, Eq. (1.3) implies that each of the two polar charges has a slightly greater partial energy than the equatorial charges. This energy splitting tends to increase for larger values of N; until at $N = 59$ the state with the greatest capture basin, or statistical weight, is so asymmetric that all of the charges have different partial energies. Beyond this point irregular states with angular diversities at the maximum value $D_a = 100\%$; [cf. Eq. (1.1)], and a complete splitting of all partial energies occur with increasing frequency.

The transition from symmetry to asymmetry also appears in a shift of the center of charge [Eq. (1.2a)]. For all $N < 11$, the equilibrium configurations of the surface Coulomb problem are sufficiently regular so that the center of charge coincides with the center of the sphere. This situation is analogous to the absence of permanent electric dipole moments in symmetric atomic and molecular charge distributions.[21] But parity arguments alone cannot exclude the existence of dipole moments in static situations. In the surface Coulomb problem this symmetry is broken at $N = 11$, where the equilibrium pattern consists of an irregular equatorial pentagon and two tilted isosceles triangles in the northern and southern hemispheres.[22] This state has a moment given by $|\mathbf{d}(11)| \cong 0.0132$; which implies the existence of an intrinsic pattern "direction," as well as a nonvanishing electric field at the center of the sphere. Another kind of dipole symmetry breaking appears when the charge interactions are varied. For instance, if the Coulomb law is replaced by an $|\mathbf{r}_i - \mathbf{r}_j|^{-1}$ force, the dipole moments of all of the corresponding equilibrium configurations vanish identically.[16]

A common feature of all three spherical surface problems—associated with the $|\mathbf{r}_i - \mathbf{r}_j|^{-n}$, $n = 1$, 2 and ∞ (Tammes) interactions—is the occurrence of enantiomeric states beginning at $N = 15$. This division marks another threshold of structural complexity. For example, if computer searches for the equilibrium states of the surface Coulomb problem are started at 10^4 random initial positions of 15 points, the trials will lead with about 50–50% probability to two geometrically distinct terminal configurations, $\mathbb{C}_G^{\mathrm{L}}(15)$ and $\mathbb{C}_G^{\mathrm{R}}(15)$, having precisely the same energy. These pairs of states are labeled left (L) and right (R) because they can be transformed into each other by an improper isometry consisting of a rotation combined with a reflection in a plane perpendicular to the axis of rotation.[13] It is intuitively plausible that there should not be any statistical bias favoring either the L or R states if they are derived from a random mix of initial states by a symmetric process. But in computer simulations the L and R labels may be regarded as a deterministic binary code that can be incorporated into the pseudorandom number algorithms that specify the initial states; and this information can create a preference. Specifically, if $\mathbb{C}_{\mathrm{Ran}}(15)$ denotes a computer-generated initial state of 15 charges, and M is an energy minimizing algorithm, then it can be shown that the mappings $M[\mathbb{C}_{\mathrm{Ran}}^{\mathrm{L,R}}(15)] \rightarrow \mathbb{C}_G^{\mathrm{L,R}}(15)$ induce a correspondence between the L and R enantiomers of the equilibrium configuration and two disjoint sets of initial states, $\{\mathbb{C}^{L}_{Ran}(15)\}$ and $\{\mathbb{C}^{R}_{Ran}(15)\}$. These sets of initial states are also enantiomeric because they occur in L and R variants—each pair related by an improper isometry, and degenerate in energy. In general, the points that make up the initial states are distributed uniformly over the surface of the sphere by sets of pseudorandom number generators. The chirality of the $N = 15$ states then implies that the initial angular coordinates of the charges—and the corresponding sets of pseudorandom numbers—can be labeled by a binary L and R alphabet. By choosing appropriate sequences of states it is therefore possible to construct any desired string or "message" composed of L's and R's. This information, in turn, may be encoded in the pseudorandom number generators by algorithms that retrodict any given sequence.[23] The net effect is that either ground state, $\mathbb{C}^{L}_{G}(15)$ or $\mathbb{C}^{R}_{G}(15)$, can be generated by deterministic means although the initial charge configurations are a racemic mix of L and R enantiomers. This method of choice bypasses some of the controversial issues of biological stereochemistry.[24,25]

The equilibrium states of the surface Coulomb problem exhibit many other types of structural transitions. It almost seems as if the addition of every new charge leads to another level of complexity. Basically, this diversity is due to the long range of the Coulomb force: the stable N-body configurations are the result of all $N(N-1)/2$ charge interactions and not just nearest neighbor forces. Similarly, the domain structures and hysteresis of magnetic Ewing arrays arise from the long reach of multipole forces.[26] Finding the stationary states of these cooperative systems by analytical means is generally very difficult. "Greedy" algorithms that search for global extremals by piecing together a series of local "best" choices can go astray even in simpler packing and covering problems.[27] For instance, the arrangement of N congruent spheres whose convex hull has the smallest volume is a straight line or sausage for all $N \leq 56$; but for larger aggregates of spheres the optimum packings have entirely different shapes.[28] In a similar vein, the Tammes problem is equivalent to finding the maximum density—or fraction of covered area—when N congruent spherical caps are packed on the surface of a sphere. Since any cap can touch at most five other caps, this appears to be a nearest neighbor problem with simple contact forces.[29] But the global constraint that all the caps must fit together on the surface of the sphere, in a not necessarily rigid packing, makes this a hard problem. The geometric methods used to construct exact solutions for the set $N^{ex}_{T} = 2$–12, 24 cannot be extrapolated to algorithms valid for arbitrary N. The best results available for $N \leq 90$ have been obtained by computer searches that simulate the nonoverlapping caps with an ultrarepulsive $|\mathbf{r}_i - \mathbf{r}_j|^{-n}$, $n = 1, 310, 720$ potential.[30] The surface Coulomb problem is still more complicated because both self-consistent boundary conditions and long-range forces determine the extremals. Exact results for this situation are sparse: Topological lower bounds for the number of equilibrium states are known only for $N < 4$,[31] and local stability has been verified for only a few symmetric ring patterns.[32] Computer studies of this problem are complicated by the existence of many metastable states separated by very small energy differences. In the range $N \leq 112$, this requires double precision computations, high statistics searches starting from many random initial configurations, and numerical stability checks. But even with these precautions some states may be missed; and for large N, roundoff errors affect the correspondence between analytical and numerical stability criteria. These ambiguities are also implicit in computer simulations of the formation of ionic "crystals" in electromagnetic traps,[33,34] and the relation of protein structures to amino acid sequences.[35–37]

Prior work on the surface Coulomb problem, and computer results extending to $N = 65$, are discussed in Refs. 38 and 39. The values of the ground-state energies have meanwhile been confirmed by several independent calculations.[16,40–42] The Coulomb configurations have a number of practical applications: These include problems in structural chemistry,[43,44] the design of multibeam laser implosion drives, and the optimum placement of communication satellites. Comprehensive summaries of related packing and covering problems—with applications to error-free data transmission—are given in Ref. 45. Some quantum mechanical extensions are discussed in Refs. 46–48.

A. Contents

In Section 2.A we set up the surface Coulomb problem for N equal point charges and derive a simple relation between the partial energies associated with the individual charges and the dipole moments of the equilibrium states. The computer algorithms and conventions for orienting the charge configurations are described in Section 2.B. Tabulations of the results for the range $2 \leq N \leq 112$ are given in Appendix B. Trends in the number of locally stable states M, found by the computer searches, are summarized in Section 3.A. The results indicate an exponential increase in the number of states, that is, $M \sim \exp\{0.05N\}$, for $N \geq 50$. Energy relations for the random initial states, ground states, metastable states, and the partial energy distributions within states are discussed in Section 3.B. The ground-state energies can be represented by a semiempirical expression of the form $E(N) \simeq 0.5N^2 - 0.55N^{3/2}$ over the entire range $6 < N \leq 112$. Geometric properties of the equilibrium configurations are considered in Section 4: These include the distributions of dipole moments and chiral states in Sections 4.A and 4.D. Measures of order, such as the angular diversity ratios, and comparisons with Tammes configurations and regular polyhedra are summarized in Sections 4.B, 4.C, and 4.E.1. Some general conjectures concerning locally stable states of complex systems are discussed in Section 5. The corresponding analytical and numerical stability criteria are reviewed in Appendix A.

2. THE SURFACE COULOMB PROBLEM

A. Analytic Formulation

The set of N unit vectors $\{\mathbf{r}_i,\ 1 \le i \le N\}$ describes the position of N point charges constrained to lie on the surface of a unit sphere. If all charges are equal, the corresponding dimensionless Coulomb energy is

$$E(N) = \sum_{i=1}^{N} \sum_{j>i}^{N} |\mathbf{r}_i - \mathbf{r}_j|^{-1} \tag{2.1}$$

The static equilibrium configurations of this system are specified by the requirement that the total force $\mathbf{F}_i$ acting on the ith charge is parallel to $\mathbf{r}_i$. This condition implies

$$\mathbf{F}_i = \sum_{\substack{j=1 \\ j\neq i}}^{N} \frac{\mathbf{r}_i - \mathbf{r}_j}{|\mathbf{r}_i - \mathbf{r}_j|^3} = E_i(N)\mathbf{r}_i \tag{2.2}$$

where $E_i(N)$ is the partial energy associated with the ith charge; cf. (1.3). The equilibrium states of the surface Coulomb problem are special cases of the central configurations of the (nonrelativistic) gravitational N-body problem.[49–51] Clearly, the total force on the sphere vanishes because the double sum is odd under an interchange of indices:

$$\sum_{i=1}^{N} \mathbf{F}_i = \sum_{i=1}^{N} \sum_{\substack{j=1 \\ j\neq i}}^{N} \frac{\mathbf{r}_i - \mathbf{r}_j}{|\mathbf{r}_i - \mathbf{r}_j|^3} = \sum_{i=1}^{N} E_i(N)\mathbf{r}_i = 0 \tag{2.3}$$

If all the partial energies are equal, that is, $E_i(N) = E(N)/N$, Eq. (2.3) implies that the corresponding dipole moments also vanish; cf. (1.2a):

$$\mathbf{d}(N) = \sum_{i=1}^{N} \mathbf{r}_i = 0 \tag{2.4a}$$

But this is only a sufficient condition. There are many equilibrium configurations for which

$$\sum_{i=1}^{N} E_i(N)\mathbf{r}_i = \sum_{i=1}^{N} \mathbf{r}_i = 0 \tag{2.4b}$$

even though $E_i(N) \neq E_j(N)$ for at least one pair of indices. If the interaction energies of the charges are logarithmic, Eq. (2.2) is replaced by

$$\mathbf{F}_i = \sum_{\substack{j=1 \\ j\neq i}}^{N} \frac{\mathbf{r}_i - \mathbf{r}_j}{|\mathbf{r}_i - \mathbf{r}_j|^2} = \tfrac{1}{2}(N-1)\mathbf{r}_i \tag{2.5}$$

This expression shows that all the equilibrium forces have the same magnitude and—in analogy with (2.4b)—the corresponding dipole moments vanish identically.[16] These constraints indicate that the equilibrium configurations of the surface logarithm problem generally tend to be more regular than the equilibrium states of the surface Coulomb problem. In both cases the equilibrium coordinates $\mathbf{r}_i$ satisfy sets of linear relations, such as (2.3) and (2.4b), which are vectorial generalizations of cryptographic knapsack problems: These are known to be computationally difficult, or NP-hard.[52]

The locally stable equilibrium configurations of the surface Coulomb problem satisfy the additional constraint that the associated energies are local minima. Specifically, if the charge positions are described by spherical coordinates—the co-latitudes $0 \leq \phi_i \leq \pi$, and longitudes $-\pi \leq \theta_i \leq \pi$—then the Coulomb energy (2.1) is $E(\phi_i, \theta_i)$, $1 \leq i \leq N$; and the equilibrium condition (2.2) is equivalent to

$$\frac{\partial E}{\partial \phi_i} = \frac{\partial E}{\partial \theta_i} = 0 \qquad 1 \leq i \leq N \tag{2.6a}$$

If Ω_κ, $1 \leq \kappa \leq 2N \leftrightarrow \phi_1, \ldots, \phi_N, \theta_1, \ldots, \theta_N$, then a sufficient condition for the local stability of the solutions of (2.6) is that the associated Hessian matrix

$$\mathcal{H}(\Omega_\kappa, \Omega_\mu) = \frac{\partial^2 E}{\partial \Omega_\kappa \, \partial \Omega_\mu} \qquad 1 \leq \kappa, \mu \leq 2N \tag{2.6b}$$

is positive definite. See Appendix A. Physically, this simply means that tangential restoring forces, that is, $\mathbf{F}_i^{\text{rest}} \cdot \mathbf{r}_i = 0$, counter small displacements from equilibrium. In potential theory these locally stable configurations are known as Fekete points, and some asymptotic estimates of the rate of approach to the limit of continuous charge distributions are available.[53,54] In Section 3.B these methods are used to construct an expression for the ground-state energy $E(N)$.

Both in the Coulomb and dipole problems analytic solutions of the equilibrium equations (2.6a) and evaluation of the associated Hessians (2.6b) becomes tedious for as few as four interacting objects.[32,55] At present, the only practical way of surveying the locally stable states of the Coulomb systems for larger values of N is to use computers to find energy minima. However, since the number of minima appears to grow exponentially with N, the energy surface $E(\phi_1, \theta_1; \ldots; \phi_N, \theta_N)$ becomes progressively more convoluted, and for $N > O(10^2)$ has many small hills and valleys. This leads to fundamental difficulties in mapping out the topography of the energy surfaces: It is necessary to distinguish genuine physical features such as minute ridges or clefts arising from the competition among the $N(N-1)/2$ charge interactions from numerical artifacts such as corrugations due to roundoff or truncation errors. Furthermore, even high statistics computer searches can miss some minima with small capture basins or special symmetries. The net result is that computer trials can both over- and underestimate the actual number of locally stable states. Analytic and numerical stability criteria for multidimensional energy surfaces are discussed in more detail in Appendix A.

B. Computer Algorithms

Most of the numerical work was carried out with the ACPMAPS supercomputer at Fermilab. This is a parallel processing machine utilizing 600 double precision nodes. The computer searches for the locally stable states of the surface Coulomb problem were started from sets of points randomly distributed over the surface of the sphere—specifically, 10^4 random starts for every value of N in the range $2 \leq N \leq 64$; 2000 starts for each successive N in the interval $65 \leq N \leq 108$, 111; and 1000 starts for $N = 109$, 110, and 112. The initial charge configurations were described by sets of spherical coordinates $\mathbf{r}_i(\phi_i, \theta_i)$, where each angle is represented by a 24-bit, or 7-decimal, pseudorandom number normalized to yield a uniform spherical distribution.[19,20] The equilibrium states were found by allowing the points to move in the direction of the forces acting on them subject to the constraint of remaining on the surface of the sphere. The steepest descent method of iterating the map $\mathbf{r}_i \rightarrow \mathbf{r}'_i = (\mathbf{r}_i + \gamma \mathbf{F}_i)/|\mathbf{r}_i + \gamma \mathbf{F}_i|$, with γ chosen to maximize convergence, was used for this problem by Claxton and Benson.[43] In the limit $\gamma \rightarrow \infty$, the update formula reduces to $\mathbf{r}_i \rightarrow \mathbf{r}'_i = \mathbf{F}_i/|\mathbf{F}_i|$, which is an overrelaxed update step with good convergence. If this step is so large that the $\{\mathbf{r}'_i\}$ configuration has a higher energy than the $\{\mathbf{r}_i\}$ state, γ is automatically adjusted downward for that step until the energy does decrease. The iterations are terminated when the energies stabilize within the machine precision of one part in 2^{-48} (~14.4 decimals). Since these computations involve the cancellation of large forces, it is essential to use at least 48-bit precision. Conjugate-gradient methods do not improve this technique because of the highly convoluted structure of the energy surface.

To compare the geometric properties of the equilibrium states, it is useful to rotate the configurations into a standard set of orientations. According to (1.3) the N charges of a locally stable state may be labeled by their partial energies. Suppose that these are ordered in a non-decreasing sequence, that is,

$$E_1(N) \leq E_2(N) \leq E_3(N) \leq \cdots \leq E_N(N) \tag{2.7}$$

As a first step in orienting, pick a charge with the lowest partial energy—if $E_1(N) = E_2(N)$, and so forth, this won't be a unique choice!—and rotate the configuration so that this charge is placed at the north pole, $\theta = \phi = 0$. Consider next the set of charges with the second lowest partial energies: for instance, E_3, E_4, E_5, if (2.7) has the special form

$$E_1 = E_2 < E_3 = E_4 = E_5 < E_6, \ldots, \leq E_N \tag{2.8}$$

Find the (not necessarily unique) charge in this set closest to the north pole and rotate the entire configuration so that this second charge is at zero longitude, $\theta = 0$. If the second charge happens to be at the south pole, repeat the process with another charge from the set with the third lowest partial energies. This scheme is adequate because the orientations are unique for irregular configurations, and the ambiguities are irrelevant for comparing symmetric configurations

The numerical reproducibility of the computations can be checked by comparing the results obtained from minimizing runs starting at different random initial configurations. For instance, for $N = 84$, the reproducibilities of some of the typical values that describe the characteristics of the configurations—in this case the chiral states with the largest capture basin—are

Total energy, Eq. (1.3):		3103. 478 717 096	13 digits	(2.9a)
Lowest partial energy, Eq. (1.3):		36. 885 477	8 digits	(2.9b)
Typical angular coordinates (rad.):	θ	0. 039 852 25	7 digits	(2.9c)
	ϕ	0. 010 146 18	7 digits	

The disparity in significant digits between the total and partial energies is not due to statistical fluctuations in roundoff errors. Rather, it indicates that the computer runs end in a multiplicity of shallow stability valleys that merge into the local energy minima. The relation of these *eigenmodes* to the Hessian stability criterion, Eq. (2.6b), is discussed in Appendix A. The basic numerical consequence is that the slight variations of the individual charge positions and energies compensate in such a way that the total energies of the equilibrium configurations are reproducible with a gain of five additional significant digits.

3. LOCALLY STABLE STATES OF THE SURFACE COULOMB PROBLEM

A. Variation of the Number of States with the Particle Number N

The computer trials show that when there are only a few interacting charges—that is, N is in the range $2 \leq N \leq 14$—the energy minimizing algorithm leads to a unique terminal energy $E(N)$ for every value of N. If the associated charge configurations are rotated into a standard orientation by means of the conventions established in Section 2.B, then the resulting geometrical patterns $\mathbb{C}(N)$ are also unique. A new level of complexity appears at $N = 15$. In this case all the computer searches still converge to a unique final energy value $E(15) = 80.670\,244\,11$; but the associated charge configurations are split into a pair of enantiomeric states: Out of a total of 10^4 randomized initial configurations 4958, or 50%, of the energy minimizing sequences terminate in a charge pattern $\mathbb{C}^L(15)$, which is the chiral transform of another pattern $\mathbb{C}^R(15)$ reached in the other 5042 energy minimizations.

Three distinct terminal configurations appear when $N = 16$. As indicated in Table VIII in Appendix B, 75.7% of the 10^4 minimizing runs end at an energy of $E_1(16) = 92.911\,655\,30$. The frequency of occurrence of this state, or *capture basin*, is in turn almost evenly divided (37.7 and 38.0%) between two enantiomeric configurations $\mathbb{C}_1^L(16)$ and $\mathbb{C}_1^R(16)$. The remaining 24.3% of the computer searches end at a locally stable state with a slightly higher energy, $E_2(16) = 92.920\,353\,96$. The associated charge configuration $\mathbb{C}_2(16)$ is a symmetric set of four rings outlining a series of four relatively rotated squares with a charge at every corner. Figures 1(a)–1(d) show these configurations in detail.

A summary of the multiplicities of the states $M(N)$ for all N in the range $2 \leq N \leq 112$ is given in Table I. As indicated in column 2 of Table I, $M(15) = 2$ and $M(16) = 3$ because every chiral configuration is counted as a separate state. Columns 3, 6, 9, and 11 also list the cumulative

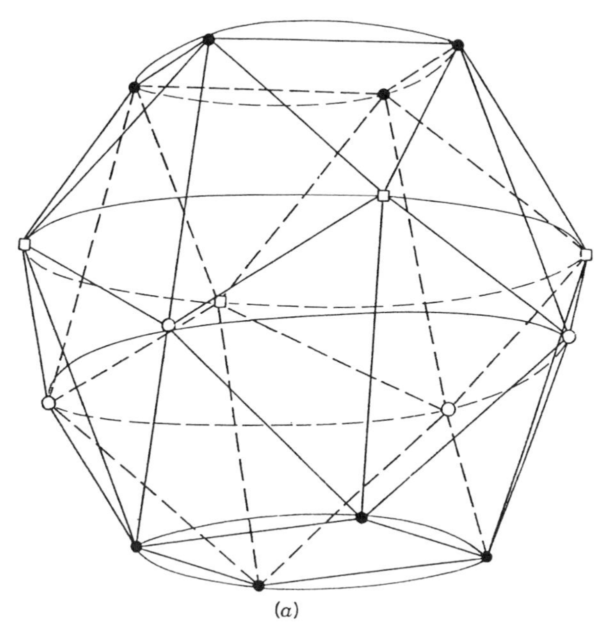

Figure 1(a). Charge configuration for $N = 16$ (metastable state). This is a perspective view showing the charges arranged in a highly symmetric pattern of four rings with four charges in each ring. These rings are symmetrically positioned with respect to the equator at 11.342° north and south latitude, and 51.684° north and south latitude. The auxiliary lines show the associated Coulomb polyhedron: This figure has 26 faces and 40 edges.

number of states

$$M_C(N) = \sum_{j=2}^{N} M(j) \tag{3.1}$$

The graph in Figure 2 shows that $M_C(N)$ increases at an exponential rate with N. In particular, if we assume that

$$M(N) = Ae^{\nu N} \tag{3.2a}$$

then (3.1) implies

$$M_C(N) = A(e^{\nu N} - e^{\nu})/(1 - e^{-\nu}) \tag{3.2b}$$

A Newton–Raphson optimization shows that for $70 \leq N \leq 112$, Eq. (3.2b) provides an excellent fit of the data with

$$A \simeq 0.382 \quad \text{and} \quad \nu \simeq 0.0497 \tag{3.2c}$$

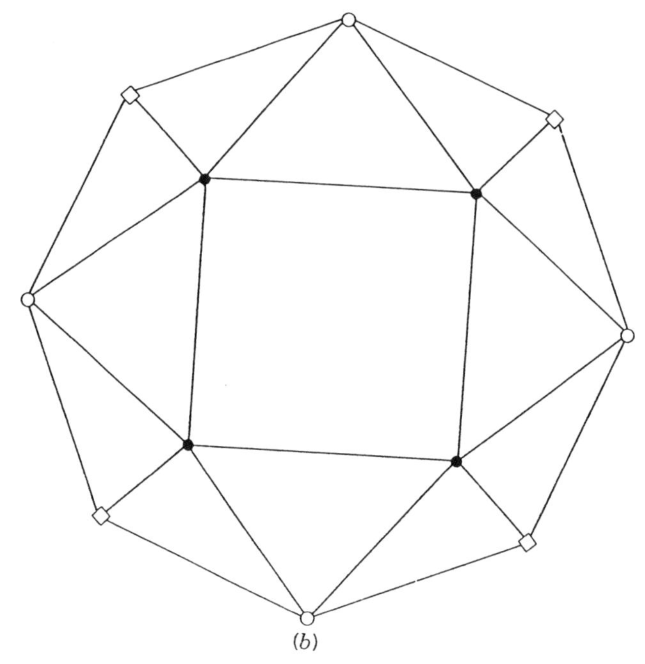

Figure 1(b). Charge configuration for $N = 16$ (metastable state). This is a plan view of the four-ring structure looking down from the north pole. The squares and open and filled circles indicate corresponding points on Figures 1(a) and 1(b).

An exponential growth of the multiplicities of states is also observed in two-dimensional arrays of pivoted magnets. Extensive experiments with $n \times n$, $2 \leq n \leq 6$ systems, initially stirred by fluctuating magnetic fields, and then allowed to settle into locally stable configurations, show that the number of distinct patterns $M^m(N)$ is of the order of

$$M^m(N) \simeq 1.3e^{0.19N} \tag{3.3}$$

where $N = n \times n$ is the number of magnets.[55,56] Figure 2 shows that the multiplicity of the magnetic states grows much more rapidly than the multiplicity of the surface Coulumb states. This trend is plausible because the magnets are coupled by a vector interaction that generates complex domain structures.

There are several other N-body systems that exhibit an exponential growth of $M(N)$ with $\nu \sim 0.07$ and 0.16.[57] In these statistical models the

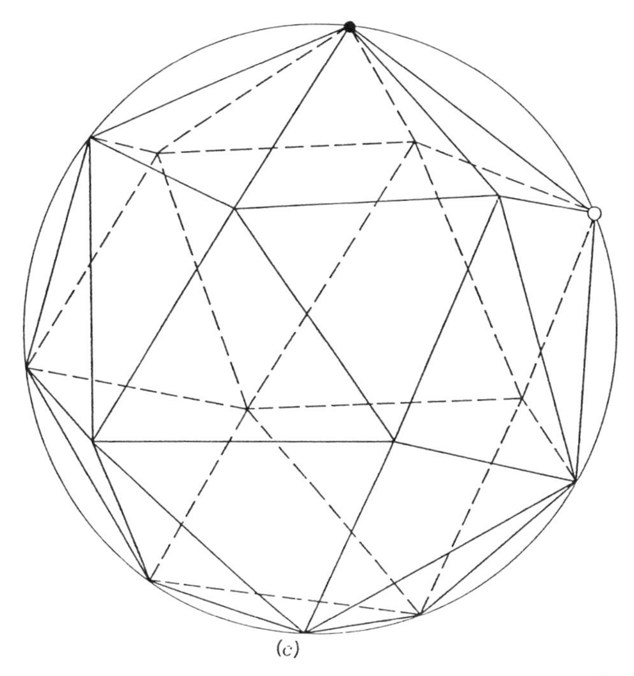

Figure 1(c). Charge configuration for $N = 16$ (ground state). This is another perspective view similar to that shown in Figure 1(a). The ground state is less symmetric than the metastable state. In fact, as shown in Figure 1(d), this configuration can exist in two enantiomorphic variants.

index ν is identified with a *maximum configurational entropy*, that is,

$$\nu = \lim_{N \to \infty} \frac{1}{N} \ln[M(N)] \tag{3.4}$$

If these results are combined with the trends of the surface Coulomb problem and the magnetic arrays, it is plausible to conjecture that in general the number of locally stable states of N-body cooperative systems increases exponentially with N. This conjecture has several practical consequences: If the exponential growth in the number of metastable states of the surface Coulomb problem continues to increase at the rates indicated in (3.2a) and (3.2c), then the numerical simulation of large systems $N > O(10^3)$ involves severe problems. For instance, the energy manifold describing the Coulomb interaction of 2000 charges constrained to the surface of a sphere would have about 5×10^{42} locally stable minima. Implementing numerical optimization or search algorithms and

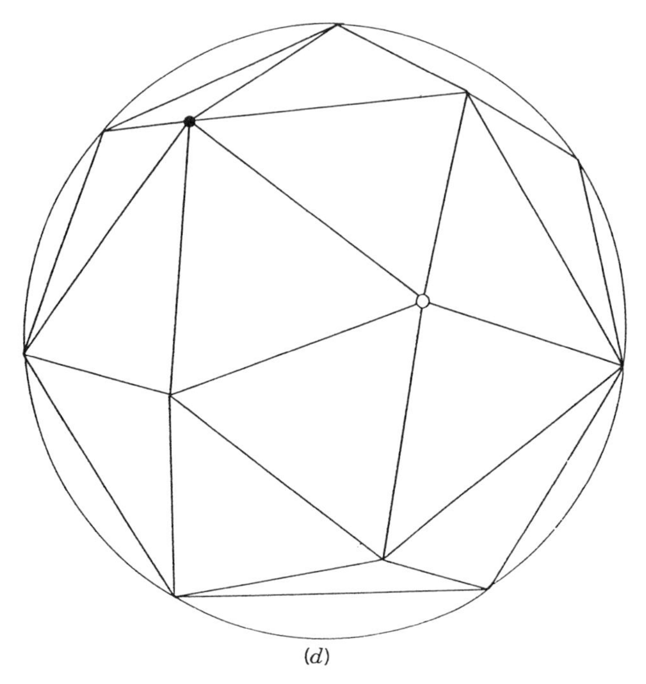

Figure 1(d). Charge configuration for $N = 16$ (ground state). The relative orientation of Figures 1(c) and 1(d) can be inferred by comparing the positions of the filled and open circles on the two diagrams. The enantiomorphic character of this configuration can be verified by copying it on a transparency, flipping the transparency over, and checking that the obverse figure cannot be rotated into coincidence with the original diagram.

testing for stability on such an intricately corrugated energy landscape would strain current computing resources beyond their limits.

B. Energy Distributions

The electrostatic energy of the N-particle surface Coulomb problem, Eq. (2.1), is given explicitly by

$$E(\phi_1, \theta_1; \ldots ; \phi_N, \theta_N) = \frac{1}{2} \sum_{i=1}^{N} \sum_{j>i}^{N} \{\sin \phi_i \sin \phi_j \sin^2[\tfrac{1}{2}(\theta_i - \theta_j)] + \sin^2[\tfrac{1}{2}(\phi_i - \phi_j)]\}^{-1/2} \quad (3.5)$$

where $\phi_i \in [0, \pi]$ and $\theta_i \in [-\pi, \pi]$ are the spherical coordinates of the ith charge.

Geometrically, $E(\phi_1, \ldots, \theta_N)$ corresponds to a surface in a $2N + 1$ dimensional space. The highest peaks on this energy "landscape" are

TABLE I
Variation of the Number of States $M(N)$ with the Particle Number N

N	$M(N)$	$M_c(N)$ [a]	N	$M(N)$	$M_c(N)$ [a]	N	$M(N)$	$M_c(N)$ [a]	N	$M(N)$	$M_c(N)$ [a]
2	1		29	2		57	9		85	19	505
3	1		30	3	43	58	18		86	46	
4	1		31	1		59	9		87	39	
5	1		32	2		60	11	200	88	32	
6	1		33	1		61	13		89	37	
7	1		34	2		62	6		90	44	703
8	1		35	5	54	63	4		91	37	
9	1		36	2		64	10		92	49	
10	1	9	37	3		65	6	239	93	41	
11	1		38	2		66	4		94	55	
12	1		39	4		67	2		95	35	920
13	1		40	6	71	68	9		96	41	
14	1		41	3		69	9		97	21	
15	2	15	42	7		70	13	276	98	37	
16	3		43	1		71	7		99	24	
17	1		44	1		72	10		100	52	1095
18	1		45	3	86	73	10		101	82	
19	1		46	8		74	22		102	87	
20	1	22	47	10		75	6	331	103	52	
21	2		48	3		76	12		104	56	
22	2		49	2		77	9		105	70	1442
23	2		50	1 [b]	110	78	7		106	93	
24	2		51	3		79	7		107	86	
25	1	31	52	8		80	10	376	108	75	
26	2		53	3		81	19		109	86	
27	3		54	10		82	30		110	93	1875
28	2		55	11	145	83	31		111	88	
			56	8		84	30		112	91	2054

[a] Cumulative number of states, Eq. (3.1).
[b] $M(N) > 1$ for $N > 50$.

generated by configurations where some of the charges are close together. The median range of heights is associated with randomly distributed sets of coordinates—such as those used as the starting configurations for the computer searches. The lowest points of the valleys and craters correspond to locally stable configurations of the surface Coulomb problem. As indicated by (3.2a) and (3.2c), the number of these local minima increases at an exponential rate with N. Geometrical comparisons show that for a given value of $N \gg 1$, the charge configurations associated with these minima all tend to be quite different. Nevertheless, the relative energy variations between the lowest and highest local minima are less than 0.006% even for the largest multiplicities of states, that is, $M(112) \simeq 0.382e^{5.56} \simeq 100$.

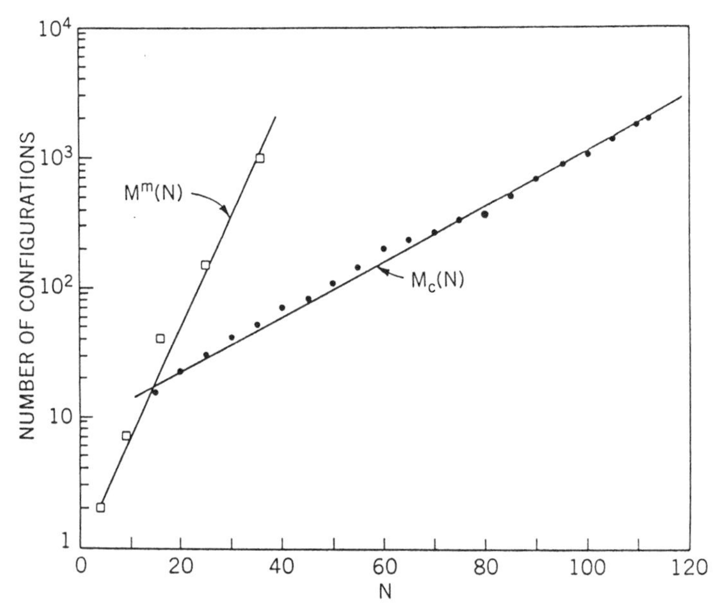

Figure 2. Variation of the number of states with the number of particles. The $M_c(N)$ points are plots of the data given in Table I. The fitted line represents the exponential function in Eqs. (3.2b) and (c). $M^m(N)$ show the corresponding data for magnetic dipole arrays; cf. (3.3) and Refs. 55 and 56.

B.1. Energies of Random Initial Configurations

Let $E_j^{\mathrm{Ran}}(N)$, $1 \leq j \leq p$, denote the energies of a set of random distributions of N charges on the surface of a unit sphere, where a total of $p \gg 1$ configurations are generated. Then ergodic arguments and rigorous results of potential theory[54] both show that the average energy of the set of random states is given by

$$\langle E^{\mathrm{Ran}}(N) \rangle = \lim_{p \to \infty} \frac{1}{p} \sum_{j=1}^{p} E_j^{\mathrm{Ran}}(N) = \frac{N^2}{2} \tag{3.6}$$

where $N^2/2$ is the Coulomb energy of a continuous uniform spherical surface charge distribution with total charge N. Figure 3 and Table II show some of the results obtained from computer simulations with $p = 10^5$, and N varying throughout the range $6 \leq N \leq 100$. The overall agreement is good although the computer-generated averages $\langle E^{\mathrm{Ran}}(N) \rangle$ tend to exceed the theoretical values $N^2/2$ by about 6%. This bias is also evident in the asymmetric distribution of the maximum and minimum energy values about the mean displayed in columns 3, 4, and 5 of Table

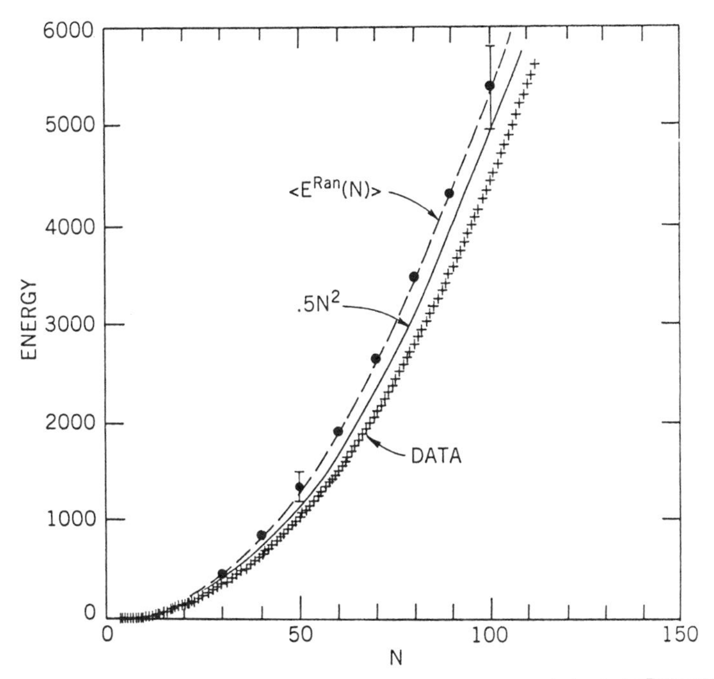

Figure 3. Energies of the surface Coulomb states. The dashed $\langle E^{\mathrm{Ran}}(N)\rangle$ curve represents the average energies of randomly chosen initial states (cf. Table II). The crosses show the minimum energies found by computer searches (cf. Table VIII). The semi-empirical formula (3.8) matches the data points more accurately than can be shown on this graph.

TABLE II
Electrostatic Energies and Dipole Moments of Random Spherical Charge Distributions

N	$N^2/2$ [a]	$\langle E^{\mathrm{Ran}}(N)\rangle$ [a]	σ [b]	$\mathrm{Max}\{E^{\mathrm{Ran}}(N)\}$	$\mathrm{Min}\{E^{\mathrm{Ran}}(N)\}$	R [c]
10	50	47.30	7.76	99.11	34.35	1.116
20	200	197.85	13.32	270.65	164.82	1.092
30	450	457.74	78.76	2594	395.85	1.107
40	800	835.63	198.4	3716	721.6	1.204
50	1250	1317.2	288.9	4416	1143	1.092
64	2048	2182.0	446.5	6229	1920	1.050
80	3200	3428.9	630.3	7824	3047	—
100	5000	5392.4	864.1	10 886	4851	—

[a] Equation (3.6).
[b] Standard deviation.
[c] Equation (3.7).

II. The underlying reason is that random selections of angular coordinates include charge clusters,[58] and these configurations boost the energy values in (3.5).

The computer simulations of the random charge configurations can also be checked by calculating their dipole moments, Eq. (1.2a). In an independent series of trials the *random walk* result $\langle|\mathbf{d}(N)|\rangle_{\mathrm{Ran}} \simeq 0.9213N^{1/2}$, cited in (1.2b), was verified by generating 100 random configurations for every value of N in the range $3 \leq N \leq 64$. Finally, by combining (1.2b) and (3.6) in the invariant ratio

$$R \equiv \frac{\langle|\mathbf{d}(N)|\rangle_{\mathrm{Ran}}}{\langle E^{\mathrm{Ran}}(N)\rangle^{1/4}} = \left[\frac{2^{7/2}}{3\pi}\right]^{1/2} = 1.095\,637 \tag{3.7}$$

it is possible to cross-check the consistency of the energy and dipole moment simulations. The numbers listed in the last column of Table II yield an average ratio of $R = 1.110$, which is within 1.3% of the theoretical value.

B.2. Minimum Energy States

Let $E_1(N)$ denote the lowest energy states of the N-body surface Coulomb problem found by computer searches. A complete set of values, ranging from $E_1(3) = 3^{1/2} = 1.732\ldots$, to $E_1(112) = 5618.044\,882\,33$, is listed in column 4 of Table VIII in Appendix B. In the absence of rigorous analytical bounds, we cannot exclude the existence of other configurations with even lower energies. The sequence of crosses in Figure 3 shows the variation of E_1 with N in graphical form. On this coarse energy scale $E_1(N)$ is a smooth monotonic function: The simple expression

$$E_{1F}(N) = 0.5N^2 - 0.5513N^{3/2} \tag{3.8}$$

fits the data with error bounds of 0.1% at $N = 20$ and 0.01% at $N = 112$. Using $E_{1F}(N)$ as a smooth baseline, it is possible to construct scatter plots of the energy differences $E_{1F}(N) - E_1(N)$ on an enlarged scale. However, searches for systematic deviations resembling the energy peaks associated with atomic clusters[59] or analogues of Thomas–Fermi oscillations[60] have not led to any conclusive results.[16,40] See also Ref. 69.

The functional form of $E_{1F}(N)$ has two physical interpretations[39]: (i) $N^2/2$ is the electrostatic energy of a uniform surface charge density—with total charge N—on a unit sphere. To recover the energy of a distribution of N point charges, it is necessary to subtract the self-energies of a set of N uniformly charged spherical caps centered on these points. For $N \gg 1$, it is plausible to approximate the caps by disks. Since the energy of an infinitely thin disk of charge with radius a is $E_D = 2\pi^2\sigma^2 a^3$ {0.4244}, where σ is the charge density;[61,62] the total self-energy correction is of the order of NE_D where $\sigma = (\pi a^2)^{-1}$. For simplicity, suppose that all the disks have the same radius. Then the crudest measure of the total area covered by the N disks is the surface area of a unit sphere, that is, $N\pi a^2 = 4\pi$. Consequently, the self-energy correction is approximately given by

$$NE_D \simeq 0.4244 N^{3/2} \tag{3.9}$$

which accounts for the second term in (3.8). More elaborate estimates that improve the agreement with the empirical coefficient 0.5513 are outlined in Ref. 40. (ii) Equation (3.6) shows directly that $N^2/2$ can also be identified with the average energy of a set of N unit charges randomly distributed over the surface of a unit sphere. In this case, the $O(-N^{3/2})$ term represents the correlation energies of the ordered Coulomb equilibrium states.

B.3. Energies of Metastable States

The most striking feature of the metastable states is that their energies are closely bunched just above the minimum energy states. This trend begins with the first metastable state at $N = 16$: As indicated in column 4 of Table VIII in Appendix B, the energy difference $\Delta E(16)$ between the two states is

$$\begin{aligned}\Delta E(16) &= E_2(16) - E_1(16)\\ &= 92.920\,353\,96 - 92.911\,655\,30\\ &= 0.008\,698\,66\end{aligned} \tag{3.10}$$

and this implies $\Delta E(16)/E_1(16) \simeq 9.36 \times 10^{-5}$. Figures 1(a) and 1(c) show that this small relative energy difference is not reflected in any geometric similarities between these two states. At the other extreme, for $N = 112$, the computer searches lead to 60 locally stable states with distinct energy values—31 of these states occur in enantiomorphic pairs. In this case it is convenient to describe the level spacings by the average energy difference $\langle \Delta E(112) \rangle$, that is,

$$\begin{aligned}\langle \Delta E(112) \rangle &= [E_{60}(112) - E_1(112)]/59\\ &= [5618.419\,481\,31 - 5618.\,044\,882\,23]/59\\ &= 0.006\,349\,14\end{aligned} \tag{3.11}$$

which indicates that the relative spacings are of the order $\langle \Delta E(112) \rangle / E_1(112) \simeq 1.13 \times 10^{-6}$.

In general, $\langle \Delta E(N) \rangle = [E_n(N) - E_1(N)]/(n-1)$, for N charges, where $n(>1)$ denotes the number of distinct energy levels. Table III shows the trends in level spacings for 18 values of N ranging from "small" to "large." Since $\langle \Delta E(N) \rangle / E_1(N) \sim 10^{-6}$ for $N > 100$, computer searches for the lowest energy states in complex systems of this type require high precision. In fact this energy scale is so fine that neither the empirical fit (3.8) nor its graph on Figure 3 can discriminate between the ground and metastable states.

It is also interesting to display the distribution of the density of states. Table VIII in Appendix B shows that for $N = 112$ there are 60 states with energies spread between 5618.044 and 5618.419. If these states were distributed uniformly, there would be about 8 states per bin for bins of width 0.05. With this particular choice of bin width, the first bin covers the energy interval 5618.044–5618.094 but according to Table VIII contains only two states. The second bin extends from 5618.094 to 5618.144 and contains no states, and so forth. Similarly, for $N = 111$, the first bin of width 0.05 spans the interval 5515.293–5515.343 and contains only the ground state, and so forth. The histogram in Figure 4 shows the combined statistics for $N = 111$ and 112—a total of 112 states. Clearly the level distribution is not uniform. There is a dip, or *level repulsion*, in the energy bin just above the ground state; a pronounced maximum in the middle of the range; and an eventual decrease in the density of the highest levels. This density profile formally resembles the Wigner distribution of the energy level spacings of large *random* Hamiltonian systems.[63]

Figure 5 shows a semilog plot of the density of states weighted by the probability of occurrence. It is a straightforward matter to include this additional information. Specifically, for $N = 112$, Table VIII shows that the two states falling into the first energy bin between 5618.044 and

TABLE III
Variation of the Average Energy Level Spacing $\langle \Delta E(N) \rangle$ with the Number of Charges N

N	16	21	22	27	30	32
n	2	2	2	2	2	2
$\langle \Delta E(N) \rangle$	0.008 70	0.000 29	0.020 42	0.006 99	0.000 45	0.207 12
N	55	56	57	58	59	60
n	6	4	5	10	5	6
$\langle \Delta E(N) \rangle$	0.005 49	0.051 22	0.022 16	0.013 08	0.004 36	0.030 07
N	107	108	109	110	111	112
n	52	47	56	59	52	60
$\langle \Delta E(N) \rangle$	0.007 38	0.007 37	0.004 94	0.004 01	0.007 16	0.006 35

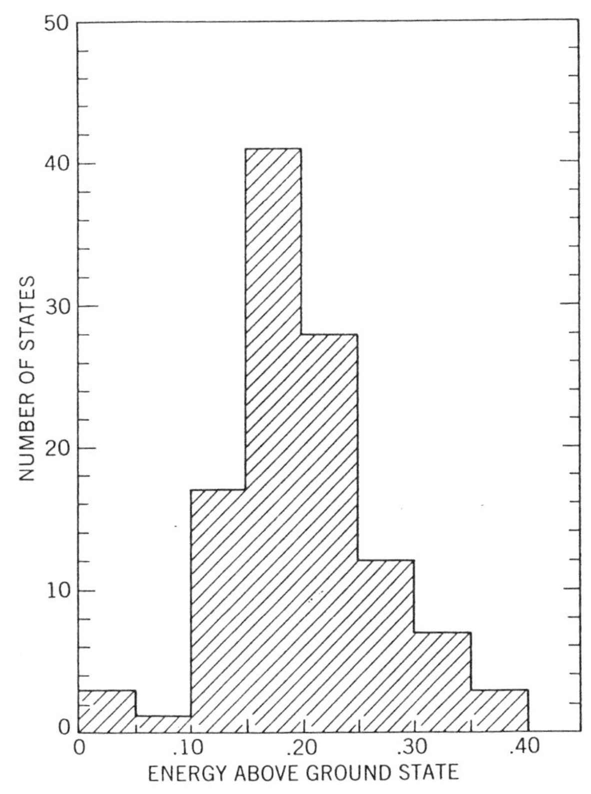

Figure 4. Density of states. The histogram combines level statistics for $N = 111$ and 112 compiled from Table VIII. In both cases, the zero of energy is taken to be the ground-state energy.

5618.094 appeared 620 times in 1000 computer searches starting from different random configurations. On average, therefore, their relative probability of occurrence is 62%. Similiarly, for $N = 111$, the state in the first energy bin occurred in 48% of the computer trials. The combined average for these three states therefore is 55%; and this is the value indicated for the first bin in Figure 5. The rest of the histogram can be obtained by similar means.

The most conspicuous difference between the two histograms in Figures 4 and 5 is that the maximum of the probability density occurs near the minimum energy states. In general, this implies that for values of $N \gtrsim 100$ there is about a 95% probability that a computer search will end at an energy level within 0.003% of the ground states. But it is difficult to improve this precision. In the range $100 \leq N \leq 112$, the average probabili-

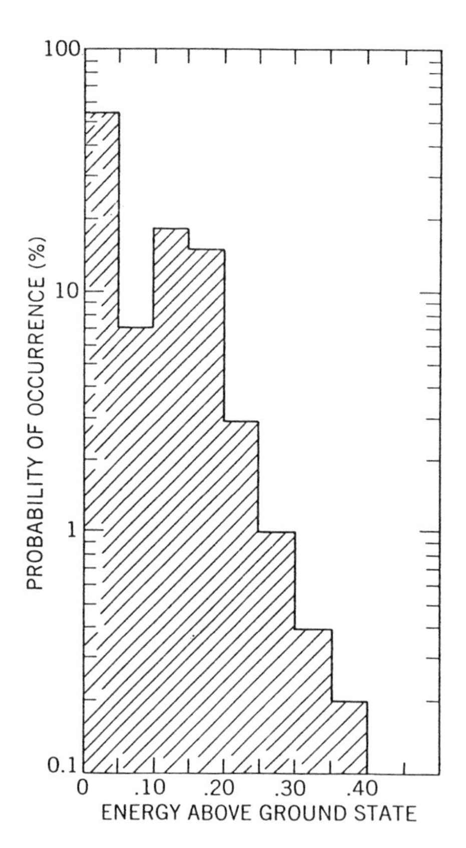

Figure 5. Density of states weighted by the probability of occurrence. This is a semilogarithmic plot that shows the qualitative shift in the histogram of Figure 4 when the probabilities of occurrence are taken into account.

ty that a computer minimization will actually reach the minimum energy state is only 35%. Of course, this result depends on the choice of minimizing algorithm. Nevertheless, similar statistical behavior occurs in the distribution of patterns in magnetic cooperative arrays.[56] All of these systems display the same basic trend: As the number of interacting objects increases, the statistical weight of the ground state decreases.

The survey of metastable states summarized in Table VIII is based on a total of about 7×10^5 computer trials. Rare states, with probabilities of occurrence as low as 0.01% are found for $N = 21$, 30, 42, 48, 58, and 61. Possibly there are additional states with still smaller capture basins. Certainly it is plausible that for $N = 112$ some states on the high-energy tail of the histogram in Figure 4 have been missed due to limited statistics (only 1000 energy minimizing searches). But the essential observation is that none of the numerical trials—for any value of N—has yet turned up any trace of isolated energy levels; that is, single levels separated by large "band gaps" ($\gg \langle \Delta E(N) \rangle$) from the cluster of states above the ground state. It remains to be seen whether this trend continues for still larger values of N.

B.4. Energies of Individual Charges

The total electrostatic energy of a locally stable state of N charges can be represented as the sum of the partial energies associated with the individual charges. These partial energies have two interesting properties: (1) The variation of the individual charge energies *within* a configuration is generally much larger than the variation of the total energy *between* configurations. And (2), since the energy apportioned to a charge is simply the sum of the inverse distances to all the other charges, the variation of the individual energies is a measure of the geometric regularity of the configurations. Figure 6 illustrates some of these energy relations. Specifically, let $E_m(N)$ denote the total energy of the mth state

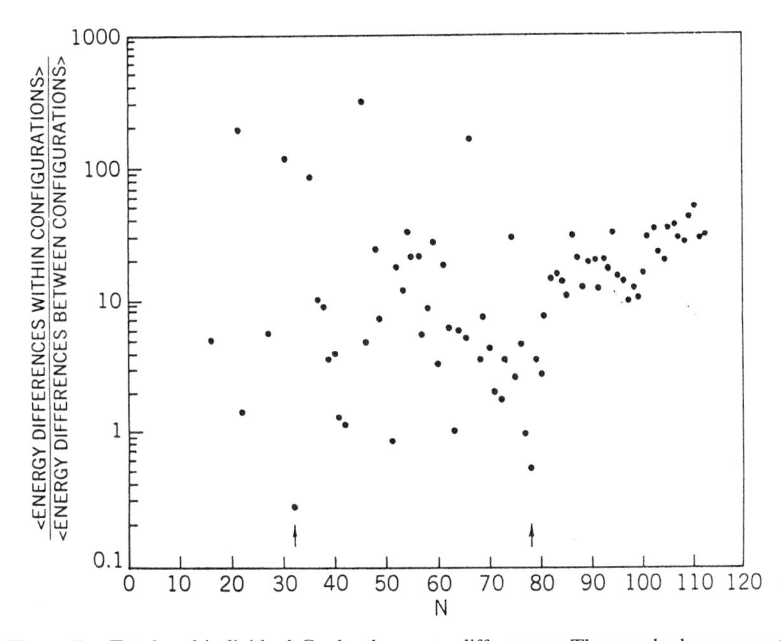

Figure 6. Total and individual Coulomb energy differences. The graph shows a scatter plot of the ratio $R(N)$ defined in Eq. (3.15) for all values of N in the range $16 \leq N \leq 112$. The only values of N for which the average energy differences *within* configurations are smaller than the average total energy differences *between* configurations are $N = 32$, 51, 77, and 78. The two smallest values of R—$R(32) = 0.26$ and $R(78) = 0.51$—are highlighted by the arrows. The maximum value is $R(45) = 305$.

of N charges. Then a slight extension of (1.3) shows that

$$E_m(N) = \sum_{i=1}^{N} E_{m,i}(N) \qquad E_{m,i}(N) = \frac{1}{2}\sum_{j\neq i}^{N} |\mathbf{r}_i - \mathbf{r}_j|^{-1} \tag{3.12}$$

where $E_{m,i}(N)$ is the partial energy of the ith charge in the mth state of N objects. The scatter plot in Figure 6 begins at $N = 16$. This entry corresponds to the following array of total and partial energies; cf. (3.10):

$$\begin{array}{ll}
N = 16;\ \text{ground state} & N = 16;\ \text{metastable state} \\
\text{Figs. 1(c) and 1(d)} & \text{Figs. 1(a) and 1(b)} \\
E_1(16) = 92.911\,655\,30 & E_2(16) = 92.920\,353\,96 \\
E_{1,1}(16) = 5.762\,143\,2 & E_{2,1}(16) = 5.793\,787\,0 \\
\downarrow & \downarrow \\
E_{1,4}(16) = 5.762\,143\,2 & E_{2,8}(16) = 5.793\,787\,0 \\
E_{1,5}(16) = 5.821\,923\,5 & E_{2,9}(16) = 5.821\,257\,1 \\
\downarrow & \downarrow \\
E_{1,16}(16) = 5.821\,923\,5 & E_{2,16}(16) = 5.821\,257\,1
\end{array} \tag{3.13}$$

The spread of partial energies in the ground state is $E_{1,16}(16) - E_{1,1}(16) = 0.059\,780\,3$; and in the metastable state $E_{2,16}(16) - E_{2,1}(16) = 0.027\,470\,1$. Consequently, the average maximum energy variation *within* these configurations is 0.043 625, whereas the total energy difference *between* the configurations is only $E_2(16) - E_1(16) = 0.008\,698\,66$—smaller by a factor of 5. This disparity is also reflected in the individual charge energies: Twelve charges in the ground state, $E_{1,5}(16), \ldots, E_{1,16}(16)$, have greater energies than any of the charges in the metastable state!

In the general case, when there are n distinct energy levels associated with N charges, the average maximum variation of partial energies within the configurations $\langle \Delta E_{\text{part}}(N) \rangle$ is given by

$$\langle \Delta E_{\text{part}}(N) \rangle = \frac{1}{n}\sum_{m=1}^{n} \{E_{m,N}(N) - E_{m,1}(N)\} \tag{3.14}$$

Table IV shows that this energy spread is a slowly increasing function of N. The differences in partial and total energies can be combined in the

TABLE IV
Variation of the Partial Energy Differences $\langle \Delta E_{\text{part}}(N) \rangle$ Within Configurations

N	16	21	22	32	55	60	111	112
$\langle \Delta E_{\text{part}}(N) \rangle$ [a]	0.043	0.055	0.030	0.054	0.117	0.098	0.211	0.208
$R(N)$ [b]	5.01	189	1.44	0.26	21.3	3.19	29.5	32.7

[a] Equation (3.14).
[b] Equation (3.15).

ratio

$$R(N) = \frac{\langle \Delta E_{\text{part}}(N) \rangle}{\langle \Delta E(N) \rangle} \sim \frac{\text{energy differences within configurations}}{\text{energy differences between configurations}} \tag{3.15}$$

which is the ordinate of the scatter plot in Figure 6. Some representative values are also listed in Table IV. Clearly most of the points in Figure 6 fall into the band between $5 < R(N) < 50$. This demonstrates that the scale of total energy differences between successive metastable states is much finer than the variation of the individual charge energies. A complementary pattern is exhibited by the stabilities: Equations (2.9a) and (2.9b) show that the numerical reproducibilities of the total energies of the configurations generally exceed the reproducibilities of the partial energies by 5 orders of magnitude.

The contrast between individual and collective energies is also illustrated by the following example: Suppose that the partial energy of a charge has the value 36.935 241. Then it is easy to verify from the computer results that this charge cannot be a constituent of any locally stable state with either $N \leq 83$ or $N \geq 85$; it must belong to one of the 30 configurations with $N = 84$. However, there is no finer scale of energy rankings to help in locating this charge. Every one of the 30 states is comprised of sets of 84 partial energies that straddle the value 36.935 241. Consequently all of these states have to be examined in detail before it can be established that 36.935 241 corresponds to $E_{10,37}(84)$—the partial energy of the 37th charge in the 10th equilibrium state of 84 objects. This assignment is unique because all 84 partial energies in the 10th state are different, and $E_{10,37}(84) \neq E_{m,i}(84)$ for all $1 \leq i \leq 84$ and $m \neq 10$. The only remaining ambiguity is geometric: as indicated in Table VIII, $E_{10}(84)$ has two enantiomeric configurations.

Equation (3.12) shows that the partial energy of a charge is proportional to the sum of its inverse distances to all the other charges. This implies that highly symmetric equilibrium configurations that "look alike" from every charge or vertex have unique partial energies, that is, $E_{m,i}(N) = E_m(N)/N$ for all $1 \leq i \leq N$. Indeed, this is the case for three of the Platonic solids, the tetrahedron, octahedron (dipyramid), and icosahedron, whose vertices are the equilibrium positions of the surface Coulomb problem for $N = 4$, 6, and 12 respectively. The partial energies are also unique for $N = 8$ and 24, even though these configurations are not included among the standard semiregular (Archimedean) polyhedra. Clearly, less symmetric charge distributions will have a greater variety of reciprocal distances, and this dispersion can be used as a measure of geometric irregularity analogous to the angular diversity ratio (1.1): If $n_e(N, m)$ denotes the number of distinct partial charge energies that occur in the mth state of N objects, then the corresponding *energy diversity ratio* (%) is given by

$$D_e(N, m) = 100 \frac{n_e(N, m)}{N} \tag{3.16}$$

In the range $2 \leq N \leq 112$, the computer trials yield 1142 equilibrium states with distinct energies; 912 of these states occur in enantiomorphic pairs; cf. Table I. The associated energy diversity ratios are listed in column 9 of Table VIII in Appendix B and displayed graphically in Figure 7. Two trends are evident: (i) $D_e(N, m)$ is a slowly increasing function of N. The first configuration that is so irregular that all of its partial charge energies are different occurs at $N = 35$; that is, $D_e(35, 4) = 100\%$. By the time N reaches 102, 34 out of a total of 54 locally stable states have energy diversity ratios in excess of 95%. This is another confirmation of the basic trend that increasing complexity is correlated with greater geometric irregularity. (ii) Figure 7 also shows that the energy diversity ratios tend to cluster in a series of bands near $\frac{1}{6}$, $\frac{1}{4}$, $\frac{1}{2}$, $\frac{3}{4}$, and 1. It is plausible that this regularity is connected with a deeper symmetry of the surface Coulumb problem.

4. GEOMETRIC PROPERTIES OF THE SURFACE COULOMB STATES

The locally stable solutions of the N-charge surface Coulomb problem are constrained solely by spherical boundary conditions and the $O(4)$ symmetry of the Coulomb interaction. The exponential growth of the multiplicity of solutions—$M(N) \sim e^{0.05N}$, Eq. (3.2a)—shows that these restrictions are compatible with a great variety of geometric structures. Only in the simplest systems is there an overlap with the criteria of strict regularity that underlie the classical theories of polygons and polyhedra.[13] For

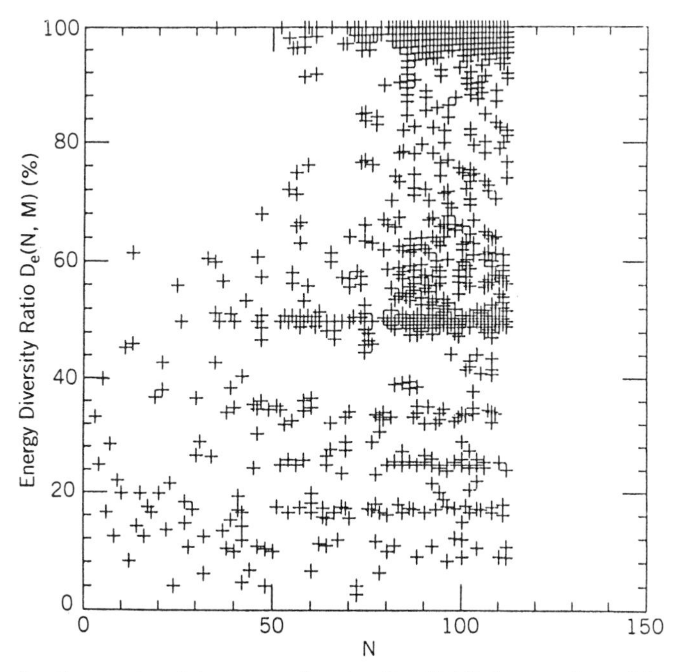

Figure 7. Scatter plot of the energy diversity, Eq. (3.16). Large values of this ratio are correlated with irregular charge configurations. The bands in the vicinity of $D_e \sim 100\%$ and 50% indicate a marked statistical preference for these values.

instance, the Coulomb solution for $N = 3$ corresponds to an equilateral triangle inscribed in a great circle: This is the simplest example of a *regular polygon*, that is, a plane polygon with equal interior angles and equal sides. Similarly, *regular polyhedra* are bounded by congruent regular polygons and have congruent vertices. Only the solutions for $N = 4$ (tetrahedron), $N = 6$ (dipyramid), and $N = 12$ (icosahedron) share this high degree of symmetry. The other Platonic solids, the cube with 8 vertices and the dodecahedron with 20 vertices, do not correspond to solutions of either the surface Coulomb or Tammes problems. The *semiregular polyhedra* are also bounded by regular polygons with congruent vertices and edges, but the polygons do not all have to be congruent to each other. This class of objects includes the 13 Archimedean polyhedra as well as infinite sets of semiregular prisms and antiprisms. None of the surface Coulomb configurations match any of these semiregular polyhedra. In particular, the well-known "bucky ball," or truncated icosahedron, associated with C_{60} is not a solution of either the Tammes or surface Coulomb problems for $N = 60$.

Every Archimedean polyhedron has a dual formed by joining a point that is above the center of each face of the polyhedron to equivalent points above all the neighboring faces. The lines connecting these points are constrained to intersect the edges of the original polyhedron. The resulting *duals of the semiregular polyhedra* have congruent faces but none of these faces are regular polygons. These duals are also less symmetric than the Archimedean figures because not all of their vertices lie on a single sphere; consequently none of the dual polyhedra coincide with any of the solutions of the surface Coulomb problem.[64–66] However, there is an interesting "near miss" for $N = 32$. The *pentakis dodecahedron* is a convex polyhedron with 32 vertices, 90 edges, and 60 faces composed of congruent isosceles triangles: This object is the dual of the truncated icosahedron, which has 60 vertices and 32 faces. The two types of edges of the pentakis dodecahedron intercept angles of

$$\sin^{-1}(\tfrac{2}{3}) = 0.729\,727\,656$$

and (4.1)

$$\tfrac{1}{2}[\pi - \sin^{-1}(\tfrac{2}{3}) - \tan^{-1}(2)] = 0.652\,358\,139$$

as seen from the center of symmetry, that is, the origin of the intersphere.[65,66] These values agree to within six significant figures with the corresponding angles of the minimum energy Coulomb configuration for $N = 32$ (see the entries on lines 13 and 14 of Table V). A pictorial comparison of the pentakis polyhedron and the Coulomb configuration would show that they are essentially identical. But pentakis breaks strict spherical symmetry because its 32 vertices are distributed over two concentric spheres whose diameters differ by 2.58%. Consequently, the ratio of the two edge lengths of the pentakis dodecahedron, 1.127 322, deviates by 0.77% from the corresponding edge ratio, 1.118 600, of the Coulomb solution. In this instance, the surface Coulomb problem actually leads to a more symmetric "dual" partner of an Archimedean polyhedron than the original construction of pentakis by Catalan in 1862.[67] Moreover, the minimum energy solution for $N = 32$ is not only geometrically regular, but it is also robust: In the range $12 < N \leq 65$, it is the only equilibrium configuration common to both the Coulomb and logarithmic interactions.[16]

In addition to the 5 Platonic solids and 26 Archimedean polyhedra and their duals, there are only 92 other convex polyhedra whose faces are entirely composed of regular polygons—generally not all of the same kind.[17,18] These objects are geometrically irregular or *nonuniform* in the sense that there are no symmetry operations that transform a particular

TABLE V
Regular Coulomb Polyhedra

N [a]	E [b]	f [c]	e/ℓ [d]	n_e [e]	D_a^{nn} (%) [f]	Nearest neighbor angles (degrees)	Multiplicity [g]
12	49.165	20	30/1	1	3.3	63.4349 [i]	30
16* [h]	92.911	28	42/4	2	9.5	48.9362	6
						52.5452	12
						54.6580	12
						61.8004	12
16	92.920	26	40/4	2	10.0	50.1269	8
						52.0044	8
						54.2578	16
						63.0252	8
24*	223.347	38	60/3	1	5.0	42.0653	24
						45.0400	24
						45.7102	12
32	412.261	60	90/2	2	2.2	37.3773	60
						41.8103	30
72*	2255.001	140	210/4	2	1.9	24.4917	60
						24.9262	30
						25.4334	60
						28.2068	60

[a] Number of charges or vertices.
[b] Coulomb energy, Eq. (3.5).
[c] Number of faces, Eq. (4.6).
[d] Number of edges/distinct edge lengths, Eq. (4.7).
[e] Number of distinct partial energies, Eq. (3.16).
[f] Diversity ratio, Eq. (4.7).
[g] Number of times this angle appears.
[h] Enantiomeric states.
[i] $2\sin^{-1}[\frac{1}{2}(2-2/5^{1/2})^{1/2}]$.

vertex into each of the other vertices in turn. Twenty-four of these nonuniform polyhedra may be inscribed in a sphere.[66] By comparing the corresponding numbers of vertices and faces, it is easy to verify that none of these 24 objects match any of the surface Coulomb equilibrium configurations. In summary, therefore, out of a total of 2054 surface Coulomb states and 123 convex polyhedra derived from classical geometry, there are only three configurations common to both sets. This number is also an upper bound because further extensions of the Coulomb problem to larger systems with $N > 112$ cannot yield any additional matches. These results show that the locally stable states of complex cooperative systems of this kind tend to have symmetries that differ from those that characterize the regular polyhedral configurations of classical geometry.

A. Dipole Moments

The distribution of the dipole moments of the surface Coulomb states can be used to answer two basic questions: (1) Are the configurations for large values of N so irregular that they are approximately equivalent to random networks of points on a sphere? And furthermore, (2) do these networks approach some kind of universal asymptotic statistical distribution that is independent of the laws of repulsion that act between the individual charges? To settle these issues, it is convenient to recall from Eq. (1.2b) that the average value of the dipole moment of a random configuration of N unit changes on a sphere is an increasing function of N, that is, $\langle |\mathbf{d}(N)| \rangle_{\text{Ran}} \sim N^{1/2}$. As indicated in connection with Eq. (3.7), the applicability of this random walk result to the Coulomb problem can be confirmed by computer trials. In particular then for $N = 100$, the expectation value of the dipole moment of a random distribution is quite large, $\langle |\mathbf{d}(100)| \rangle_{\text{Ran}} \simeq 9.2$; whereas the entries in column 5 of Table VIII show that $0 \leq |\mathbf{d}(100)| \leq 0.0037$ for all 52 of the Coulomb states found by computer searches. This upper bound indicates that the metastable state with the highest energy and nearly maximal angular diversity (see below) for $N = 100$ has a dipole moment that is about 4×10^{-4} smaller than that expected for a random configuration. Figure 8 shows that this trend of small dipole moments prevails for all the Coulomb configurations in the range $N \leq 112$. The logarithmic ordinate scale of the graph extends down to 10^{-6}, which is near the limit of numerical accuracy for large systems, $N \sim O(100)$. Table VIII shows that states with vanishing dipole moments are quite common for small values of N, but tend to become less frequent as N approaches 100. Nevertheless, they do not disappear entirely: The (ground) state with the largest capture basin for $N = 112$ apparently has a vanishing moment. These results clearly show that the charge distributions of the surface Coulomb configurations have intrinsic regularities that persist despite the lack of the congruences or symmetries associated with the polyhedra of classical geometry.

There are systematic variations of the dipole moments that depend on the strength of the force acting between the charges. According to Eq. (2.5), if the interaction is logarithmic, or "soft," all locally stable configurations have vanishing dipole moments.[16] At the other extreme, the "hard" Tammes potential, $|\mathbf{r}_i - \mathbf{r}_j|^{-n}$, $n \to \infty$, leads to states with sizable moments. Spot checks of some of the Tammes configurations found by Kottwitz's computer searches[30] yield moments larger than unity.

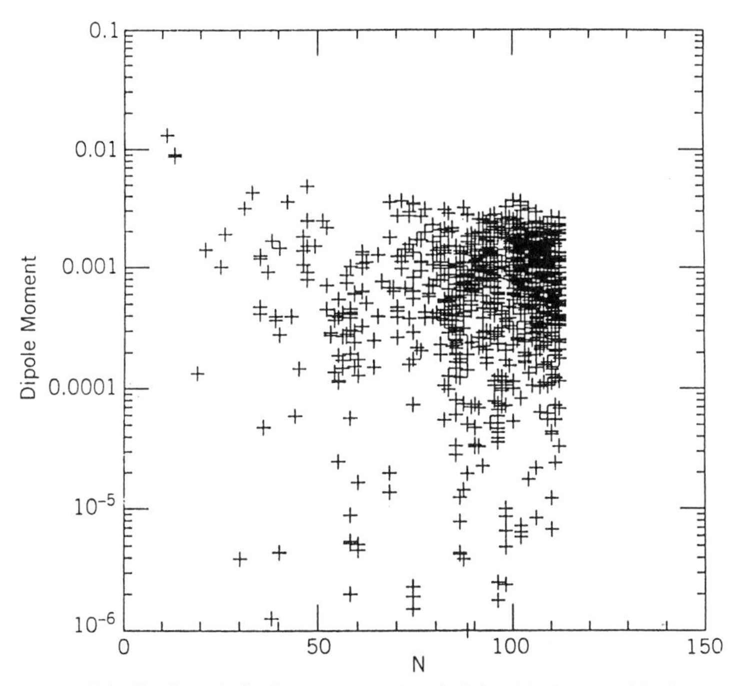

Figure 8. Distribution of dipole moments, Eq. (1.2a). This is a graphical summary of all the dipole moments listed in column 5 of Table VIII.

All the available information can be summarized as follows:

Force Law	*Size of Dipole Moment*	*Source of Result*
$\lvert \mathbf{r}_i - \mathbf{r}_j \rvert^{-1}$	0	analytical identity, Eq. (2.5)
$\lvert \mathbf{r}_i - \mathbf{r}_j \rvert^{-2}$	$0–10^{-2}$	computer trials ($N \leq 112$)
$\lvert \mathbf{r}_i - \mathbf{r}_j \rvert^{-n}, \quad n \gg 1$	$O(1)$	computer trials ($N \leq 90$)
Random	$(8N/3\pi)^{1/2}$	combinatorial lemma, Eq. (1.2b)

(4.2)

Obviously, in the range $2 \leq N < O(100)$, there is no tendency for a convergence of the dipole moments associated with the logarithmic, Coulomb, or Tammes interactions. This diversity suggests the conjecture that for large values of N different force laws lead to distinct asymptotic distributions of spherical charge networks. Comparisons of trends in the Tammes and Coulomb angles (see Section 4.E.1) also support this surmise.

B. Distributions of Angles

Another measure of the regularity of the surface Coulomb configurations is the *angular diversity ratio* introduced in Eq. (1.1). This has a simple basis: If $\mathbf{r}_i$ and $\mathbf{r}_j$ specify the locations of two charges on the surface of a sphere with unit radius, then the set of $N(N-1)/2$ angles, $\psi_{ij} = \cos^{-1}(\mathbf{r}_i \cdot \mathbf{r}_j)$, where $\psi_{ij} \leq 180°$, $1 \leq i, j \leq N$, $i \neq j$, describes the geometry of the charge distribution. The degeneracy of this set is a measure of the symmetry of the configuration. For instance, if 5 points are distributed arbitrarily over the surface of a sphere, there will generally be $5 \times 4/2 = 10$ distinct angles between pairs of points. However, in the case of the surface Coulomb problem, the unique equilibrium arrangement of 5 charges is a triangular dipyramid—one charge at the north pole, another at the south pole, and the remaining three charges equally spaced around the equator. Obviously, only three distinct angles appear between any pair of charges in this highly symmetric configuration: 180° occurs once, 120° occurs three times, and 90° occurs six times. The corresponding angular diversity ratio therefore has the low value of

$$D_a(N) = 100\,\frac{\text{number of distinct angles}}{N(N-1)/2} \rightarrow 100\,\frac{3}{10} = 30\% \qquad (4.3)$$

Similarly, the clustering of the irregular $N = 11$ and 13 configurations around the highly symmetric icosahedron at $N = 12$ is immediately apparent from the D_a fluctuations, without the need for any graphical comparisons; viz.

N	$D_a(N)$	$D_e(N)$	$\lvert\mathbf{d}(N)\rvert$
11	36.4%	45.5%	0.0132
12	4.5%	8.3%	0
13	37.2%	46.2%	0.0088

(4.4)

This array shows that all three indices of regularity—the angular diversity ratio D_a, the energy diversity ratio D_e [Eq. (3.16)], and the dipole moment $\lvert\mathbf{d}\rvert$—yield consistent results. These correlations also appear in the detailed list of values in columns 5, 9 and 10 of Table VIII in Appendix B, as well as in the graphical summaries in Figures 9 and 10. In particular, the parallel increase of both the angular and energy diversity ratios confirms once again the general conjecture that increasing complexity tends to be associated with decreasing symmetry. For instance, the first configuration that is so irregular that all of its vertices are inequivalent ($D_e = 100\%$) and most of its edges have different lengths

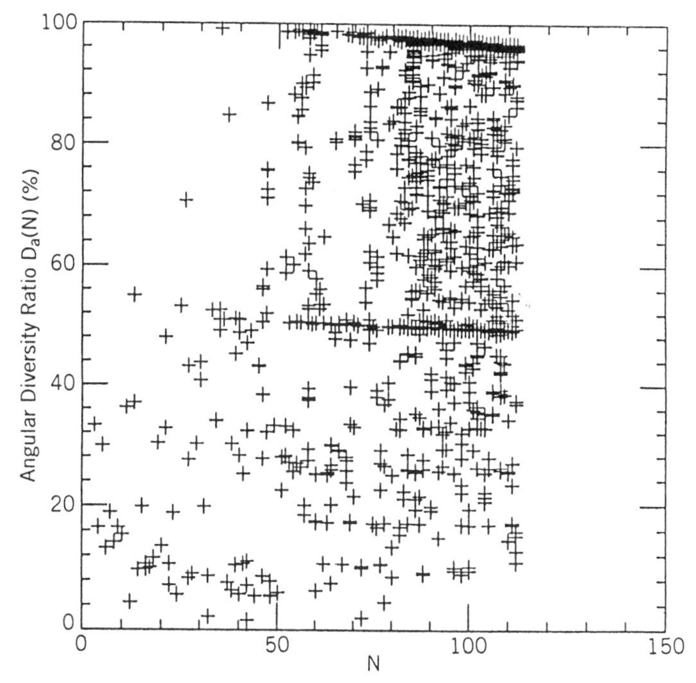

Figure 9. Scatter plot of the angular diversity, Eq. (4.3). Large values of this ratio indicate irregular charge configurations; small values are correlated with symmetric polyhedra; cf. Table V. The bands at 50 and 95% are the result of statistical preferences analogous to those in Figure 7.

($D_a = 99.2\%$) occurs at $N = 35$. Figure 9 shows the development of this trend in graphical form. At $N = 102$, 30 out of a total of 54 locally stable states have energy and angular diversity ratios in excess of 95%. These irregularities are pervasive for $N \sim O(100)$.

The distribution of values in the sets of angles ψ_{ij} is also useful for comparing the structures of different charge configurations belonging to the same value of N. Since the data in Section 3.B.3 show that the energies of all of these locally stable states are very nearly the same—within 0.007% for $N = 102$—it is possible that some of these states also have geometrical resemblances. Well-known examples of sets of complex configurations with common "backbones" and minor "peripheral" variations include the tautomers and conformers of structural chemistry. However, every one of the surface Coulomb states with nonidentical energies appears to have a distinct structure. For instance, at $N = 102$, there are 87 configurations (cf. Table I) each of which is described by a ψ_{ij} set with 5151 angles. Comparisons show that there are 33 sets that occur

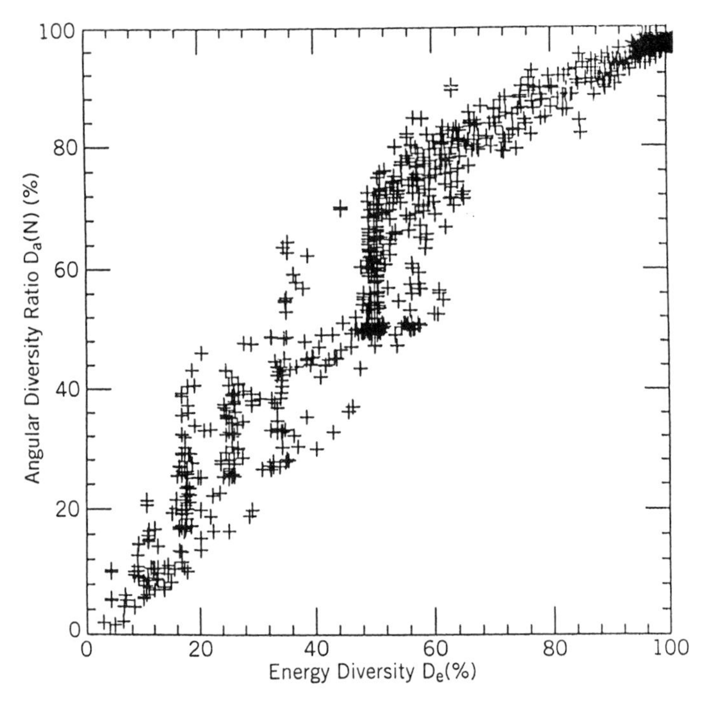

Figure 10. Correlation of angular diversity and energy diversity. Symmetric configurations cluster near the origin, D_e and $D_a < 20\%$; irregular configurations near $D_e \approx D_a \sim 100\%$. There is also a statistical cumulation around $D_e \approx D_a \sim 50\%$. The plot indicates that D_e and D_a always yield consistent measures of regularity.

twice: Each matching pair has the same energy and is geometrically related by an improper isometry—evidently these are just the enantiomeric configurations. Apart from these degeneracies, there are then a total of $87 - 66/2 = 54$ different states. Further comparisons of the associated angular sets, ψ_{ij}^k, $1 \leq k \leq 54$, show that the *maximum* fraction of coincidences among any pair of these sets is bounded by 9%. Computer surveys for all N in the range $50 < N \leq 112$, where multiple states become more frequent, indicate that this overlap estimate is actually a general result; that is, if $V(N)$ denotes the fraction of common angles, then

$$V(N) = \frac{\#(\psi_{ij}^k \cap \psi_{ij}^l)}{N(N-1)/2} \leq 0.09 \tag{4.5}$$

where $k \neq l$, and the set intersections exclude enantiomeric pairs. The low value of this overlap ratio shows that it is implausible that configurations with nonidentical energies share any major structural features such as common backbones.

The overlap bound in (4.5) is based on very conservative angle matching criteria. When $N \lesssim 100$, the precision of the angular coordinates of the individual charges in rare states can decrease to about one part in 10^5. This is degraded further by the computation of the interparticle angle sets ψ_{ij}. Finally, the coarseness of the matching may be relaxed even more to ensure that all the enantiomeric states are correctly paired up. Consequently, the actual values of the overlap ratios $V(N)$ may be significantly smaller than the bound shown in (4.5). For example, at $N = 84$, all 16 states with distinct energies are sufficiently irregular so that the positioning conventions of Section 2.B yield unique orientations. Under these circumstances, the charge coordinates of all of these states—which are known to seven figures, (2.9c)—can be compared directly. Extensive spot checks have failed to turn up even one matching charge position, apart from the common fixed point at the north pole. It seems, therefore, that the exponential increase in the number of states for larger values of N (>50) is accompanied by a tremendous proliferation of distinct geometric structures.

C. Coulomb Polyhedra: Regular Configurations

The coexistence of order and disorder in the geometric structure of the surface Coulomb states is illustrated in Figure 11. This diagram shows the equilibrium configuration of 19 charges on the surface of a sphere. The apparent symmetry of this arrangement is highlighted by the auxiliary polyhedron whose vertices coincide with the charge positions. The faces and edges of this polyhedron can be constructed with the help of some computer graphics: Given N (>3) points on the surface of the sphere, the set of all combinations of 3 points determines a maximum of $N(N-1)(N-2)/6$ planes. Associated with each plane and triple of points—located by the unit vectors $\mathbf{r}_j$, $j = \alpha, \beta, \gamma$—is another vector $\mathbf{r}_c$ extending from the center of the sphere to the plane and perpendicular to it. Since the plane and sphere intersect in a circle ($C_{\alpha\beta\gamma}$) all the scalar products $\mathbf{r}_c \cdot \mathbf{r}_j$ are equal. Suppose now that $\mathbf{r}_k$ ranges over the positions of all the charges *not* included in the $\mathbf{r}_j$ triplet—that is, the set $\{\mathbf{r}_\ell\}_1^N \backslash \mathbf{r}_a$, $\mathbf{r}_\beta$, $\mathbf{r}_\gamma$—and furthermore that $\mathbf{r}_k \cdot \mathbf{r}_c \leq \mathbf{r}_j \cdot \mathbf{r}_c$; then the plane containing the charges α, β, γ is a face of the polyhedron. Geometrically, this inequality simply means that the spherical cap bounded by $C_{\alpha\beta\gamma}$ contains no other charges. In cases where two or more charge triplets determine coincident planes, the associated polyhedron face is bounded by four or more vertices. Figure 11 includes an example of this situation. The end result of this construction is that the Coulomb polyhedron for $N = 19$ has a total of

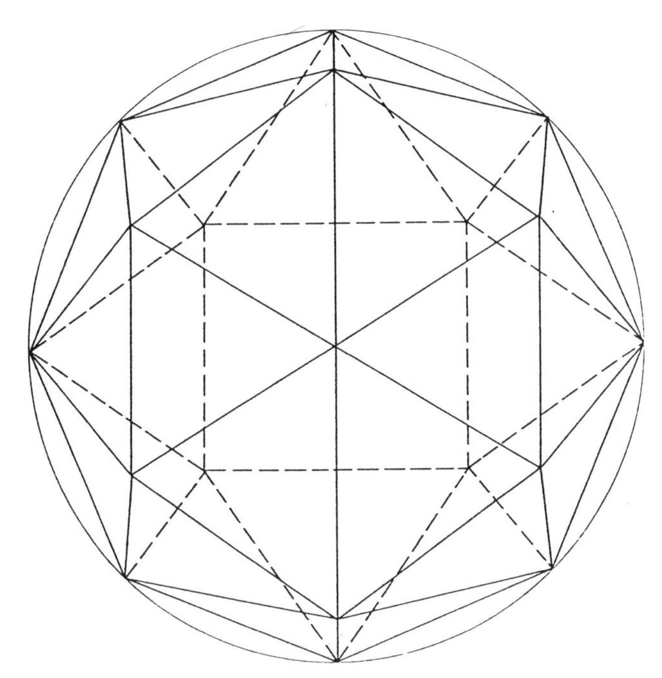

Figure 11. Surface Coulomb configuration for $N = 19$. The symmetry of this arrangement is partly illusory. The 50 edges are composed of 10 groups of 4 congruent edges and 5 groups of 2 congruent edges. The polygonal faces are too irregular to fit into the standard scheme of polyhedra.[17,18]

33 faces. The corresponding number of edges (e) then follows from Euler's formula

$$N + f - 2 = e$$

or (4.6)

$$19 + 33 - 2 = 50$$

Column 11 of Table VIII in Appendix B lists the number of faces (f) of the Coulomb polyhedra for all configurations in the range $4 \leq N \leq 112$.

The symmetries of the Coulomb polyhedron in Figure 11 are reflected in the low values of the energy diversity, $D_e = \frac{7}{19} \simeq 36.8\%$ [Eq. (3.16)], and the angular diversity, $D_a = \frac{52}{171} \simeq 30.4\%$ [Eq. (4.3)]. In particular—apart from the charge at the north pole with the least partial energy—all the other 18 charges occur in pairs: each partner with the same partial energy and latitude, but the two charges differing by 180° in longitude.

This symmetric pattern has a small but nonvanishing dipole moment, $|\mathbf{d}(19)| = 0.000\,135$, pointing toward the north pole. The contrasting irregular features of this polyhedron arise from a lack of congruence among the edge lengths. No more than four edges have equal lengths. In fact, the 50 edges are composed of 10 groups of 4 congruent edges and 5 groups of 2 congruent edges. Consequently, the polygonal faces in Figure 11 are too irregular to fit into the standard set of nonuniform polyhedra.[17,18]

A useful measure of the degree of congruence in the Coulomb polyhedra is the ratio of the number of distinct edge lengths to the total number of edges. Since the edge lengths are determined by the central angles between the corresponding vertices, this congruence measure is equivalent to a *nearest neighbor angular diversity ratio* analogous to (4.3), viz.

$$D_a^{\text{nn}}(N) = 100\,\frac{\text{number of distinct edge lengths } (\ell)}{\text{total number of edges } (e)} \tag{4.7}$$

Whereas the angular diversity D_a is a global index of the variety of all possible angles between charges, D_a^{nn} is a strictly local measure that takes into account only the diversity of angles between adjacent charges. In the case of the $N = 19$ polyhedron, both the local and global measures of regularity yield nearly the same result

$$D_a^{\text{nn}}(19) = 100 \times \tfrac{15}{50} = 30.0\% \leftrightarrow 30.4\% \simeq 100 \times \tfrac{52}{171} = D_a(19) \tag{4.8}$$

Computer surveys of all the other Coulomb polyhedra with N vertices in the range $4 < N \leq 112$ show a similar equivalence. If this trend extends beyond $N \sim 112$, it would simplify the identification of regular charge patterns: estimates of $D_a^{\text{nn}}(N)$ for $N \gg 1$ require at most the comparison of $3N$ nearest neighbor angles.

Although the dominant geometric trend of the Coulomb states is one of increasing irregularity for larger values of N, the sporadic appearance of small percentages among the diversity ratios listed in columns 9 and 10 of Table VIII shows that some ordered patterns persist up to the limits of the computer explorations. The distribution of these special states is indicated graphically by the set of points in the 0–20% bands in Figures 7 and 9. Quantitative information concerning the most regular configurations is summarized in Table V on p. 525. For reference, the entries in the first line recapitulate the data for the icosahedron ($N = 12$)—the largest Platonic solid whose vertices coincide with the solutions of the logarithmic, Coulomb, and Tammes problems. Comparisons with the indices for

$N = 16$, 24, 32 and 72 show that these new polyhedra are also highly symmetric. The two $N = 16$ configurations are depicted in Figures 1(a)–1(d): They illustrate the interesting point that the lowest energy state is not necessarily the most symmetric. Table VIII shows that this situation recurs at several other values of N; for example, the most symmetric $N = 82$ pattern is ranked eighth in order of increasing energy and has an extremely low probability of occurrence. The $N = 24$ Coulomb polyhedron resembles the snub cube, one of the semiregular Archimedean solids. However, the Coulomb interactions distort the symmetry of the classical polyhedron: Whereas the snub cube has 32 triangular and 6 square faces, all with equal edges, the faces of the Coulomb polyhedron include 24 scalene triangles.[41] The $N = 32$ situation corresponds to the "near miss" of the pentakis dodecahedron discussed previously in connection with Eq. (4.1). In this case the Coulomb polyhedron is slightly more symmetric than its classical counterpart. The lowest energy Coulomb state for $N = 72$ is also conspicuously symmetric. All faces of this polyhedron are triangular. There is no resemblance to the aspherical $N = 72$ "fullerene" cage containing 12 pentagons and 26 hexagons.[68] $N = 122$, 132 may also be regular.[69]

The entries in Table V do not continue beyond $N = 72$ because the more complex symmetric polyhedra contain at least 11 different nearest neighbor angles. Nevertheless, the ordered patterns stand out clearly among the increasing variety of irregular polyhedra. For example, at $N = 112$, there are at least 60 locally stable states with distinct energies. The first, second, and tenth levels are clearly different because their nearest neighbor ratios D_a^{nn} [Eq. (4.7)] are 10.5, 8.2, and 24.1%, respectively; all the other states have angular diversities exceeding 45%. The marked regularity of the second level is also apparent from the small number of partial charge energies—equivalent to 10 types of polyhedron vertices—and the symmetric grouping of the 330 nearest neighbor angles: These occur in 26 sets of 12 equal angles, and a residual set of 18 angles, also all alike. Unraveling the complex order of these large polyhedra is a challenging problem in "physical" geometry.

D. Enantiomorphic Configurations

A set of points on the sphere may be transformed by *isometries* or *congruence mappings* that preserve the distances between all pairs of points. All isometries, in turn, can be built up from three basic types of transformations.[70] (i) rotations about an axis, (ii) mirror reflections in a plane, and (iii) parallel displacements of all points. If the mappings are restricted to a fixed sphere, parallel displacements play no role, and the congruence transformations reduce to *proper isometries* or (rigid body) rotations and *rotatory reflections* composed of a reflection and a rotation whose axis is perpendicular to the mirror.[13,71] *Central inversions*, in which the coordinates of all points are reflected in the origin of the sphere, that is, $\mathbf{r} \rightarrow -\mathbf{r}$, are special cases of rotatory reflections in which the rotation is a half-turn.

If a pattern $\mathbb{C}_i$ of identical charges on the surface of a sphere is sufficiently irregular—though not necessarily random—then the only isometric mapping, $I: \mathbb{C}_i \rightarrow \mathbb{C}_f$, that yields a final configuration $\mathbb{C}_f$ identical to the initial state is the identity transformation. In contrast, highly symmetric configurations such as the icosahedron are invariant under a great variety of isometric transformations, for example, the composite group $A_5 \times C_i$.[14] The set of solutions of the surface Coulomb, logarithmic, and Tammes problems interpolates between these two extremes: In all three cases larger values of N are associated with less symmetric point groups.[16,30,41,42] However, as emphasized in connection with the dipole moments in Section 4.A, even Coulomb states whose only isometric symmetry is the identity transformation have ordered structures.

When $N \geq 50$, the surface Coulomb states tend to cluster in pairs, each with the same sequence of partial energies, equal total energy, and nearly equal probability of occurrence. Suppose that $\mathbb{C}^L(N)$ and $\mathbb{C}^R(N)$ denote such a pair of states. Since the orientation conventions established in Section 2.B automatically include rotational degeneracies, it remains to check whether these states are related by an improper isometry. In practice, this mirror symmetry can be verified by picking a state, say $\mathbb{C}^{\mathrm{L}}(N)$, and reflecting it in an arbitrary plane through the center of the sphere. The resulting configuration is then rotated so that the charge with the lowest partial energy is positioned at the north pole, $\theta = \phi = 0$, and the charge with the next lowest partial energy is at zero longitude, $\theta = 0$. If all the partial energies are different, this orientation is unique, and the final configuration will coincide with $\mathbb{C}^{\mathrm{R}}(N)$. In case there is a degeneracy in the partial energies, some auxiliary comparisons may be required.

The distinctions between proper and improper isometries can be illustrated with two simple examples: Figure 1(b) is a plan view of the symmetric four-ring structure of the $N = 16$ metastable Coulomb solution, $\mathbb{C}_2(16)$. Obviously, this pattern is invariant under 90° rotations and reflections—if the rings are copied on a transparency, and the transparency is flipped over, the reversed image will coincide with the original pattern. This symmetry is broken by the greater complexity of the two $N = 16$ ground states. If Figure 1(d) is copied, the image on the flipped transparency cannot be rotated into coincidence with the original pattern, but it will match the other ground-state configuration. In general, any configuration that cannot be brought into coincidence with its mirror image by rotations is *chiral* or *enantiomorphic*. Hence the familiar example of right (R) and left (L) handedness suggests the notation $\mathbb{C}_1^R(16)$ and $\mathbb{C}_1^L(16)$ for the two $N = 16$ chiral ground states. But for arbitrary patterns—in fact, even the simple perspective view in Figure 1(c)—there are no obvious pictorial cues of handedness, or a "screw sense," and chirality has to be checked by other means such as exhaustive computer comparisons.[72]

The asterisks in column 3 of Table VIII mark the enantiomeric states of the surface Coulomb problem. Comparisons show that $N = 15$ is the common threshold for the appearance of chiral configurations in the surface Coulomb, logarithmic, and Tammes problems.[14,16,39] Furthermore, in the range $15 \leq N \leq 65$, the ground states of the logarithmic potential are chiral if and only if the ground states of the associated surface Coulomb problem are chiral.[16] However, the results for $N = 15$, 16, 19, 21, and so forth show that there is no such one-to-one correspondence between the ground states of the surface Coulomb and Tammes problems.[30]

There are interesting connections between chirality, "chaos," symmetry breaking, and cryptography in the surface Coulomb problem. Let $M[\mathbb{C}_{\text{Ran}}^R(15)] \rightarrow \mathbb{C}_1^R(15)$ represent the mapping of a randomly chosen initial state of 15 charges, $\mathbb{C}_{\text{Ran}}^R(15)$, to one of the pair of chiral ground states, $\mathbb{C}_1^R(15)$ by means of an energy minimizing algorithm M. Suppose further that the initial configuration is sufficiently irregular so that it can be verified that $\mathbb{C}_{\text{Ran}}^R(15)$ is indeed a chiral state with a mirror image $\mathbb{C}_{\text{Ran}}^L(15)$. Then it can be shown that the minimizing algorithm of Section 2.B, as implemented on a computer, preserves chirality. (An analytic analogue is discussed in Ref. 73.) This leads to an array of parallel mappings that can be extended to include many initial states:

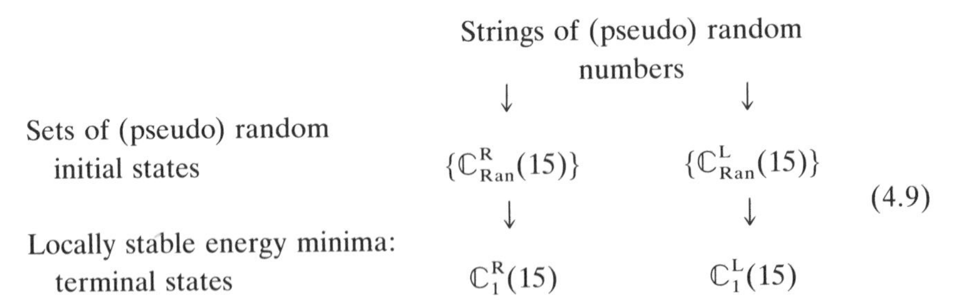

This diagram shows that the net effect of the chirality preserving map M is to transfer the L and R labels from the ground states up to the level of the random initial states, and to split these into two corresponding subsets $\{\mathbb{C}_{\text{Ran}}^R(15)\}$ and $\{\mathbb{C}_{\text{Ran}}^L(15)\}$. Since the initial configurations are on the average distributed uniformly over the surface of the sphere, it is plausible that slight changes in the angular coordinates of the charges in any particular state $\mathbb{C}^{\mathrm{R}}_{\mathrm{Ran}}(15)$ can transform it into a $\mathbb{C}^{\mathrm{L}}_{\mathrm{Ran}}(15)$ state, and vice versa. Consequently the end result of an energy minimization can be sensitively affected by slight perturbations of the initial conditions: This mix of randomized states and unstable evolution is a basic characteristic of "chaotic" dynamics.[74]

Chiral symmetry breaking can occur in a variety of ways. For instance, varying the index n in the power law $|\mathbf{r}_i - \mathbf{r}_j|^{-n}$ can induce transitions between chiral and nonchiral states. The simplest illustration is provided by $N = 16$. In this case the ultrarepulsive Tammes potential $|\mathbf{r}_i - \mathbf{r}_j|^{-n}$, $n \rightarrow \infty$, can be approximated by choosing $n = 1\,310\,720$.[15,30] Both geometrical arguments[75] and computer trials then show that the $N = 16$ Tammes solution is a symmetric four-ring structure closely resembling the pattern in Figures 1(a) and 1(b). (The latitudes of the rings are $\pm 13.632°$ and $\pm 51.490°$ in the Tammes case, and $\pm 11.342°$ and $\pm 51.684°$ in the Coulomb case.) But the lowest energy solution for the surface Coulomb problem is quite different: It is split into a pair of chiral states one of which is shown in Figures 1(c) and 1(d). Evidently then, as the potential index n decreases from 1 310 720 to 1, there must be at least one threshold where chiral states appear.

The chiral L and R indices are equivalent to a binary alphabet. In principle, therefore, it is possible to construct any desired string or "message" with an appropriate series of $\mathbb{C}^{\mathrm{R}}_1(15)$ and $\mathbb{C}^{\mathrm{L}}_1(15)$ configurations. But as (4.9) shows, each ground-state configuration can be enciphered in an enormous number of ways by the mappings $M[\{\mathbb{C}^{\mathrm{R,L}}_{\mathrm{Ran}}(15)\}] \rightarrow \mathbb{C}^{\mathrm{R,L}}_1(15)$. For instance, on a double precision computer, the number of initial states with a particular chirality can easily exceed 10^{10}. The element of ambiguity or concealment than lies in the assignment of a specific L or R label to any one of these random initial states. Although it is easy to verify that a particular state is chiral, the spatial arrangement of charges is usually too complex to exhibit an obvious handedness—it is necessary to go through an explicit energy minimizing sequence leading to either $\mathbb{C}^{\mathrm{R}}_1(15)$ or $\mathbb{C}^{\mathrm{L}}_1(15)$ in order to identify whether an initial state is L or R.

The strings of random numbers in the top line of (4.9) refer to the angular positions of the charges in the initial configurations. In particular, if the latitudes and longitudes of the charges are specified to an accuracy of 12 decimals, then the configurations $\mathbb{C}^{\mathrm{R,L}}_{\mathrm{Ran}}(15)$ can be represented by strings of $15 \times 2 \times 12$ nominally random digits, $\{d_j\}_1^{360}$, $d_j = 0, 1, \ldots, 9$. The security of this "chiral-energy" encipherment therefore relies both on the algorithmic complexity of the mapping M and the tremendous redundancy of the correspondence

$$\mathbb{C}^{R,L}_{Ran}(15) \leftrightarrow \{d_j\}_1^{360} \rightarrow \text{R or L} \tag{4.10}$$

In analogy with other schemes involving "trap-door" or "one-way" functions,[76] Eq. (4.10) is hard to invert because the reversion is a *set-valued* function that associates an entire set with a particular input.[77]

In practice, the charge coordinates of the initial configurations are derived from deterministic pseudorandom number generators. The complete sequence of the chiral-energy encipherment is therefore a combination of (4.9) and (4.10), that is,

$$\begin{aligned} &\text{Pseudorandom} \\ &\quad \text{number generator} \rightarrow \{d_j\}_1^{360} \leftrightarrow \mathbb{C}^{R,L}_{Ran}(15)\text{: } M[\mathbb{C}^{R,L}_{Ran}(15)] \rightarrow \mathbb{C}^{R,L}_1(15) \\ &\qquad \rightarrow \text{R or L} \end{aligned} \tag{4.11}$$

Since the number generators can be programmed to produce any sequence, Eq. (4.11) is a slow but feasible means of encipherment.

The concealed propagation of order through pseudorandom numbers and geometric complexity also adds a novel twist to the problem of chiral bias. This concerns the observation that naturally occurring proteins are almost exclusively composed of chiral amino acids of the L variety.[24,25] Although these compounds are far more complex than the surface Coulomb states, the basic production mechanisms are presumed to be similar in both cases: The underlying idealization is that a uniform statistical mix of initial states evolves toward equilibrium in a symmetric pair of potential wells whose minima correspond to states of opposite chirality. Since processes of this kind always lead to a racemic mix of final states, the observed "handedness" of the biosphere is usually attributed to a critical fluctuation ("spontaneous" symmetry breaking) or a fundamental chiral force (e.g., β *decay*) that introduces an asymmetry in the potential wells.[24,25] Equation (4.11) indicates still another possibility: The final chirality is actually predetermined by a set of algorithmic instructions at a nongeometric level. It is certainly feasible to generate long strings of pseudorandom numbers that will consistently produce L-handed initial configurations.[23] The appearance of a geometrically unbiased mix of initial states is therefore an illusion—the L die has already been cast before the game begins.

The binary code of chirality disappears when (4.11) is rewritten for 14 charges. The essential difference in this case is that the ground state is not enantiomorphic even though the pseudorandom initial configurations may be chiral, that is,

$$\text{Pseudorandom number generator} \rightarrow \{d_j\}_1^{336} \leftrightarrow \mathbb{C}_{\text{Ran}}^{\text{R,L}}(14)\text{: } M[\mathbb{C}_{\text{Ran}}^{\text{R,L}}(14)] \rightarrow \mathbb{C}_1(14) \tag{4.12}$$

The transition from (4.12) to (4.11) illustrates another threshold of structural complexity. When there are 15 charges represented by 30 blocks of 12 digit numbers—as in Eq. (4.10)—each string of 360 digits specifies a unique dichotomic variable, an L or an R. However, if the strings are parsed differently—as in Eq. (4.12)—they are too simple to generate the chiral alphabet. By this means the threshold of a geometric property is expressed as a minimum complexity requirement for a coding algorithm.

E. Influence of Force Laws on Charge Distributions

E.1. Coulomb Angles and Tammes Angles

The Tammes problem is equivalent to finding the largest angular diameter $\Theta_T(N)$ of N congruent caps that can be packed on the surface of a sphere without overlapping.[12–14] Column 7 of Table VIII lists the optimum values of $\Theta_T(N)$ obtained by Kottwitz[30] and Tarnai[78] for $3 \le N \le 100$. Clearly, $\Theta_T(N)$ is a (not strictly) decreasing function of N, with an asymptotic dependence $\Theta_T(N) \simeq (8\pi/3^{1/2}N)^{1/2}$, for $N \gg 1$. There is an analogous angle for the surface Coulomb problem $\Theta_c(N)$ determined by the minimum angular separation between neighboring charges in a locally stable configuration.[39] Several examples are contained in column 7 of Table V: $\Theta_c(16) = 48.9362°$, $\Theta_c(24) = 42.0653°$, $\Theta_c(32) = 37.3773°$, and so forth. A comprehensive survey is given in column 6 of Table VIII. Since the optimization in the surface Coulomb problem is carried out with respect to total energy rather than nearest neighbor separations, the two sets of angles are related by $\Theta_T(N) > \Theta_c(N)$ when $N > 6$, $N \ne 12$. $\Theta_c(N)$ is a nonmonotonic but generally decreasing function of N with an asymptotic estimate resembling the Tammes result; $\Theta_c(N) \sim (4\pi/N)^{1/2}$, for $N \gg 1$. If this estimate were accurate to leading order in N, then the relative difference between the two sets of angles would approach a constant value for large N,

$$[\Theta_T(N) - \Theta_c(N)]/\Theta_T(N) \rightarrow 1 - 3^{1/4}/2^{1/2} \sim 0.07 \qquad N \gg 1 \tag{4.13}$$

Figure 12 shows this relative difference in graphical form when $\Theta_c(N)$ is averaged over all locally stable states belonging to a given value of N.

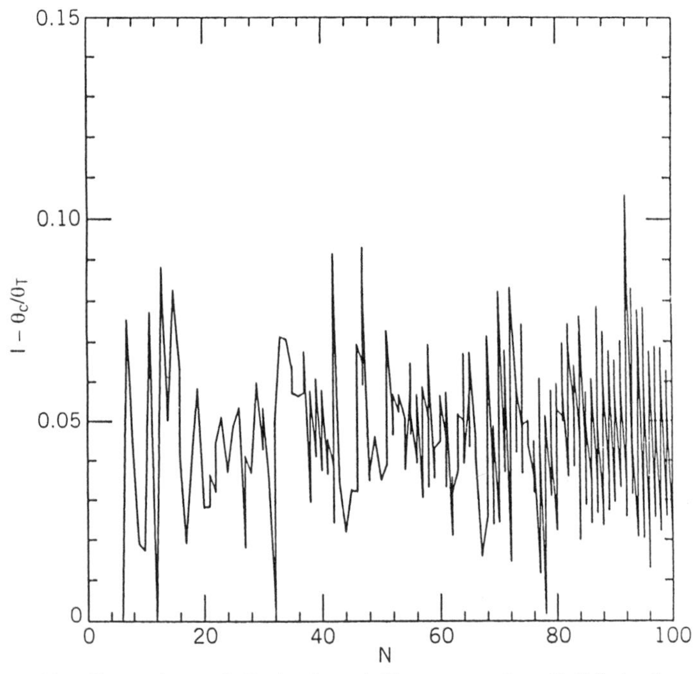

Figure 12. Comparison of Coulomb and Tammes angles. $\Theta_T(N)$ is the optimum angular diameter of a spherical cap in the Tammes packing problem. $\Theta_c(N)$ is the minimum angular separation between adjacent charges in the spherical Coulomb problem. The graph shows the relative difference between the two sets of angles, Eq. (4.13).

Despite the prominent fluctuations, the overall trend is roughly consistent with (4.13).

The basic purpose of these comparisons is to see whether the configurations of points have some kind of asymptotic regularity for large values of N that is insensitive to the precise nature of the underlying interactions. If the trends in Figure 12 can be extrapolated beyond $N \sim 112$, then it would be an indication that the local equilibrium states of the Coulomb law $|\mathbf{r}_i - \mathbf{r}_j|^{-2}$ and the Tammes interaction $|\mathbf{r}_i - \mathbf{r}_j|^{-1\,310\,720}$ retain a distinct character even for arbitrarily large values of N.

E.2. The Largest "Hole" Angle

A useful complement to the Coulomb angle $\Theta_c(N)$, which measures the minimum separation between charges, is the hole angle $\Theta_H(N)$, which is the angular diameter of the largest spherical cap containing *no* charges in its interior. The set of these empty regions was obtained previously in Section 4.C as an aid in the construction of the Coulomb polyhedra. Column 8 of Table VIII lists approximate values of the hole angle $\Theta_H(N)$

for all surface Coulomb configurations in the range $4 \le N \le 112$. As expected, for regular configurations such as the icosahedron the ratio of the hole and Coulomb angles is close to 1—$\Theta_H(12)/\Theta_c(12) \simeq 1.18$—but shows a larger disparity for irregular states—$\Theta_H(13)/\Theta_c(13) \simeq 1.34$. For larger values of N, the minimum ratio (1.10) occurs at $N = 44$, and the maximum ratio (1.32) at $N = 92$. There is no evidence for conspicuous empty regions that may be analogues of the interstices in complex molecules.

In the Tammes problem, the densest packing of congruent caps on a sphere does not necessarily lead to rigid configurations. In fact, Kottwitz's solutions show a slowly increasing trend in the number of caps free to "rattle" for larger values of N.[30] These partially empty regions can also be characterized by a set of hole angles. Specifically, if a Tammes configuration is represented by the center points of a set of caps, the corresponding *minimum* hole angle $\psi_H(N)$ is the angular diameter of the *smallest* circular region that does not contain any of these points in its interior. It is easy to show that

$$\Psi_H(N) = 2\sin^{-1}\left\{\frac{2}{3^{1/2}}\sin\left[\frac{\Theta_T(N)}{2}\right]\right\} \tag{4.14}$$

where $\Theta_T(N)$ is the standard Tammes angle (Sec. E.1). Numerical comparisons of the Coulomb and Tammes hole angles for the respective ground states lead to the inequality $\Theta_H(N) > \Psi_H(N)$ for $12 < N \le 90$. If this relation were valid for all $N > 12$, it would indicate still another optimum property of the Tammes configurations.

E.3. Edges and Faces of Coulomb Polyhedra

The conjecture that the surface Coulomb and Tammes configurations remain distinct for large values of N is also supported by the statistical behavior of the number of faces and edges of the associated polyhedra. As in Section 4.C, let e and f denote the total number of edges and faces of a convex polyhedron whose N vertices coincide with the charge positions. Then if $\langle\eta(N)\rangle$ is the average number of edges that meet at a vertex, Euler's theorem (4.6) shows that

$$\langle\eta(N)\rangle \equiv \frac{2e}{N} = 2\left(1 - \frac{2}{N} + \frac{f}{N}\right) \tag{4.15}$$

In the special case that all the faces of a polyhedron are triangular, $3f = 2e$, and (4.6) implies

$$f = 2(N-2) \quad \text{and} \quad e = 3(N-2) \tag{4.16a}$$

TABLE VI
Values of the Edge/Vertex Ratio $\langle\eta(N)\rangle$ for the Coulomb and Tammes Polyhedra

N	12	32	71	72	109	112
Surface Coulomb[a]	5	5.625	5.775	5.833	5.853	5.893
Triangular tessellation[b]	5	5.625	5.831	5.833	5.890	5.893
Tammes[c]	5	4.125	3.887	4.083	—	$\leq$4.9988[d]

[a] Equation (4.15) and Table VIII.
[b] Equation (4.16b).
[c] Ref. 30.
[d] Equation (4.17).

These relations yield the sharper constraint

$$\langle\eta(N)\rangle_{\text{tri}} = 6 - \frac{12}{N} \tag{4.16b}$$

Table VI displays the trends in the numerical values of the "edge/vertex" ratio $\langle\eta(N)\rangle$ for the ground states of the surface Coulomb and Tammes problems. The middle row of the table lists the corresponding values for polyhedra whose faces consist solely of triangles.

Obviously, for large values of N, most of the Coulomb polyhedra have triangular faces. However, the "hard" Tammes potential $(\sim|\mathbf{r}_i - \mathbf{r}_j|^{-1\,310\,720})$ can generate more complex polyhedra because of inherent geometric constraints. Specifically, since in any packing of congruent caps on the surface of a sphere any cap can have at most 5 neighbors, Eq. (4.16b) and Table VI show that the icosahedron is the only figure with a triangular covering whose associated caps have the maximum possible mutual contact.[29] For $N \gg 12$, the edge/vertex ratios of the Coulomb polyhedra tend toward the triangular limit (4.16b), that is, $\langle\eta\rangle_{\text{tri}} \rightarrow 6$; whereas the Tammes ratios cannot exceed 5. In fact a rigorous sharper bound is available:

Theorem.[79] There is an $r_0 > 0$ such that for any packing of congruent caps of radius $r \leq r_0$ on the sphere of unit radius, the average number of neighbors of the caps in the packing is at most

$$\frac{4204}{841} \simeq 4.99881\ldots \geq \langle\eta(N > [2/r_0]^2)\rangle \tag{4.17}$$

Further trials suggest that for any congruent cap packing on the sphere—not necessarily the densest packing—the maximum possible value of $\langle\eta(N \rightarrow \infty)\rangle$ is 4.4.[79]

E.4. Minimal Properties of Coulomb Energies

The computer results for $N = 112$ illustrate three general trends in the distributions of the Coulomb states: (i) For every $N \gg 1$ there are many locally stable states—at least 60 for $N = 112$. (ii) These states are nearly degenerate in energy—for $N = 112$, Eq. (3.11) shows that the relative energy spacings are $\sim 10^{-6}$. (iii) And, finally, angular comparisons indicate that the geometrical configurations of all of these states are quite different. These features impose interesting constraints on the energy "landscapes" of the Coulomb states. The data in Table II and Figure 3 show that the average value of the energy E for arbitrary choices of the angles $\phi_1, \ldots, \phi_N$; $\theta_1, \ldots, \theta_N$ in Eq. (3.5) is $\sim N^2/2$. In other words, most of the energy surface lies in the highlands. Furthermore, it is obvious, since the number of arbitrary (initial) states is far larger than the number of terminal equilibrium states, that these highlands surround a few valleys whose lowest points lie at a depth $\sim 0.55N^{3/2}$ below the average height of the landscape; cf. Eq. (3.8). What is not at all obvious is that according to (ii) all of the valley bottoms are situated at nearly the same depth, even though (iii) indicates that these minima are widely dispersed over the landscape. The essential implication is that the energy surface is bounded from below by a *single* hyperplane that is effectively tangent to *every* local minimum. This hyperplane is also tangent to every rigid Tammes configuration in the range $2 \leq N \leq 87$. Specifically, if $\mathbb{C}_T(N)$ denotes the optimum Tammes solution for N points; and each point is assigned a unit charge, Eqs. (2.1) and (3.5) can be used to compute a Coulomb energy $E_{\mathrm{CT}}(N)$ for this configuration. Table VII shows some representative comparisons of the ground-state Coulomb energies $E_1(N)$ and Coulomb–Tammes energies $E_{\mathrm{CT}}(N)$ for various values of N.

Since the Coulomb problem yields the minimum energy for the $|\mathbf{r}_i - \mathbf{r}_j|^{-1}$ potential, whereas the Tammes problem optimizes interparticle distances, it is evident that $E_{\mathrm{CT}}(N) > E_1(N)$, except for $N = 2-6$, 12. But there is no *a priori* reason for the pervasive near equality $E_{\mathrm{CT}}(N) \approx E_1(N)$ displayed in Table VII. In fact, significant differences might be expected to arise from the tremendous disparity between the effective Tammes potential, $|\mathbf{r}_i - \mathbf{r}_j|^{-1\,310\,720}$ and the Coulomb law, as well as the general lack of resemblance of the Tammes and Coulomb equilibrium configurations. For instance, in the simplest case $N = 7$, the surface Coulomb distribution consists of five points spaced equally around the equator, with the remaining two points at the poles (pentagonal dipyramid), while the points of the Tammes solution are given by the vertices of an equilateral triangle at 43.476 677° south latitude, another equilateral triangle, rotated by 60°, at 12.130 450° north latitude, and the

TABLE VII
Comparisons of Ground-State Coulomb Energies $E_1(N)$ and Coulomb–Tammes Energies $E_{CT}(N)$

N	$E_1(N)$	$E_{CT}(N)$	$[E_1 - E_{CT}]/E_{CT}$
2–6	—	—	0
7	14.452 977	14.461 864	6.15×10^{-4}
8	19.675 288	19.725 173	2.53×10^{-3}
12	49.165 253	—	0
13	58.853 231	58.909 592	9.57×10^{-4}
16	92.911 655	92.951 183	4.25×10^{-4}
32	412.261 274	412.376 77[a]	2.80×10^{-4}
78	2662.046 474	2662.677	2.37×10^{-4}
79	2733.248 357	2734.540	4.72×10^{-4}
80	2805.355 876	2805.908	1.97×10^{-4}
84	3103.465 124	3104.142	2.18×10^{-4}
87	3337.000 750	3337.978	2.93×10^{-4}

[a] Values for $N \geq 32$ from Ref. 30.

remaining point at the north pole. Computer trials also confirm that if the $N = 7$ Tammes solution is chosen as the initial configuration of a surface Coulomb minimization, the algorithm of Section 2.B will eventually converge to the pentagonal dipyramid solution. Despite these qualitative distinctions, the Coulomb energies associated with these two configurations differ by less than 0.062%.

Although these insensitive energy variations may be accidental, they are certainly not isolated accidents. Extensive numerical evidence from the analysis of magnetic cooperative systems[55,56] and computer simulations of vortex arrays[80] also yield examples of complex systems with a variety of metastable states all of which are very nearly degenerate in energy. The most prominent example of this type is the electrostatic interaction energy of ionic crystals. Madelung, in his initial computations,[81] already emphasized the severe requirements of accuracy necessary for discriminating between different kinds of lattices. A canonical example is the 0.857% difference in Madelung constants between the sodium chloride and cesium chloride structures.[82] After a long series of evolutionary developments, a useful description of these structures is finally available; but this requires a quantum mechanical density functional formalism and implementation on "supercomputers".[83,84] Even these sophisticated methods have not resolved the inverse problem: Given a complex gradient system, what are the characteristics of the configurations that are nearly degenerate in energy with the ground state? Or, more informally, why is it that the energy surfaces of some complex gradient systems have multiple valley bottoms with very little height variation?

5. CONJECTURES CONCERNING THE STABLE EQUILIBRIUM CONFIGURATIONS OF COMPLEX GRADIENT SYSTEMS

The locally stable equilibrium states of the surface Coulomb problem share many of the characteristics of planar magnetic dipole configurations.[55,56] Similar trends are exhibited by other systems such as arrays of vortex pattern;[80,85] the *jellium* model, or its equivalent, the three-dimensional spherical Coulomb problem;[10] and sets of floating magnets interacting with an external magnetic field.[86] These common features suggest several general conjectures concerning the locally stable states of complex gradient systems.[87]

In the simplest cases, when there are only a few identical interacting objects, general arguments of balance and symmetry show that the equilibrium configurations are regular polygons or polyhedra whose form is essentially independent of the detailed nature of the forces.[11] The underlying assumption is that the potential energy of these systems can be derived from the superposition of identical pairwise interactions. Although these results agree with observations, the steps from arguments or conjectures to rigorous assertions are incomplete even for the smallest gradient systems consisting of only three to six objects. Analytical methods involve tedious computations,[10,32,55,88,89] and topological estimates of the number of critical points, including stable equilibria, are just being developed.[90]

For the spherical charge systems, the influence of the force laws becomes dominant in the transition from six to seven objects. Specifically, when $N \leq 6$, all of the force laws, ranging from the soft logarithmic interaction to the hard Tammes potential, generate identical equilibrium patterns; whereas for $N \geq 7$ ($\neq 12, 32$), all of the ground states appear to be markedly different.[43] This sensitivity to the form of the interactions also appears in magnetic arrays: As the number of interacting objects increases, the organization of the domain structures shifts from the control of the strong dipole interactions to the weaker octupole forces.[26] It is plausible to conjecture that this sensitivity is a general attribute of cooperative gradient systems when all of the $N(N-1)/2$ interparticle forces are taken into account. A more complicated analogue is the folding of protein molecules under the influence of nominally weak secondary and tertiary forces.[35,36,91]

The diminishing importance of strict geometric symmetry as an organizing principle in complex systems is also illustrated in a different but related problem concerning the positions of N points in a unit square arranged so that the minimum distance d_N between any pair is as large as possible. Evidently, this is a two-dimensional version of the Tammes problem in which a set of N congruent circles with diameter d_N is packed into a square with side $1 + d_N$. The successive panels in Figure 13, taken from Ref. 28, show that beginning at $N = 10$, the optimum configurations in this problem also tend to be asymmetric. Clearly, this trend parallels the evolution of the surface Coulomb states, where for increasing values of N, computer graphics and other indices such as the dipole moments (1.2a), and energy and angular diversity ratios (3.16), (4.3), (4.7), show that the equilibrium configurations became more irregular. When $N > 12$, there is no overlap with any of the 123 convex regular polyhedra,[17,18] and indeed many states are so disordered that their only invariant isometry is the identity transformation. Since this progression from "broken" sym-

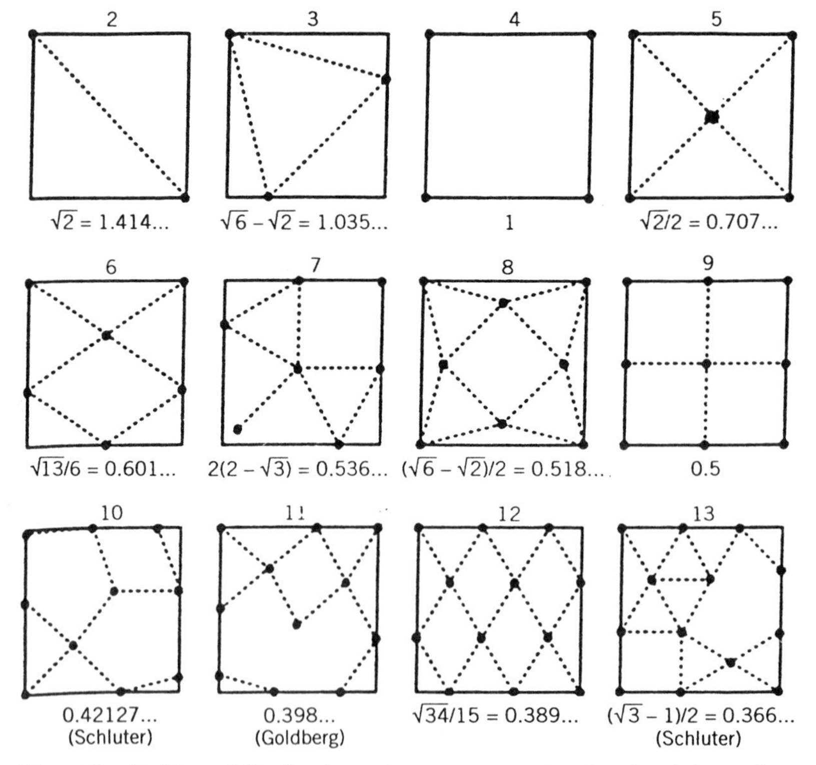

Figure 13. Positions of N points in a unit square arranged so that the minimum distance between any pair is as large as possible. The configurations for $N \leq 9$ are known to be optimal. This diagram is reproduced from Ref. 28 with permission of Springer Verlag. See also Ref. 92.

metries to "fragmented" symmetries appears in other N-body problems, it may be a general attribute of many complex systems. Nevertheless, it still remains unclear what characteristics distinguish force laws that lead to complicated equilibrium states from those that generate extended regular lattices.

As emphasized in Ref. 91, *diversity* is a prerequisite for complexity: In general, complex systems should exhibit many significantly different states. A typical chemical illustration of the proliferation due to simple combinatorics is the estimate that "$C_{167}H_{336}$ is the smallest alkane with more realizable isomers than the observed universe has 'particles' $\sim 10^{80}$."[93] Similar arguments can be applied to sets of coupled nonlinear oscillators to determine their maximum configurational entropy (3.4). The results support the general surmise that the number of local minima of the energy landscapes of N-body cooperative systems increases at an exponential rate with N.[57,94] Direct confirmation of these trends is displayed in Figure 2. The magnetic data are derived from extensive experimental observations of the stable configurations of planar dipole arrays:[56] The graph showing the increase in the number of surface Coulomb states is based on the summary of computer results given in Tables I and VIII. In the currently accessible ranges of N, the evidence for exponential growth is consistent and convincing.

Angular comparisons among the sets of surface Coulomb configurations for fixed values of N ($\gg 1$) show very little overlap (4.5). This is a counter-example to the presumption that the structural properties associated with different local minima are fairly similar;[36] but is in accord with the general diversity conjecture which asserts that complex systems should appear in many *significantly different* states. This diversity is also connected with the irregularity of configurations and their sensitive dependence on the nature of the underlying interactions. If, for instance, contrary to the observations, the surface Coulomb configurations were all in the form of regular polyhedra, then there could not be any more than 123 different equilibrium states.[17] Analogous constraints of symmetry limit the total number of possible crystallographic space groups to 230 distinct types.[13] In this obvious sense, significant diversity is not compatible with symmetry.

For large values of N, it is possible that the trend toward increasing diversity eventually merges into a statistically regular sequence of patterns that is effectively insensitive to the details of the interparticle forces. However, all the available evidence from the spherical charge systems points in the opposite direction. The stable configurations can be characterized by their dipole moments, or centers of charge [cf. (1.2a) and Fig. 8]; the nearest neighbor angles (4.7); and the average number of edges that meet at the vertices of the associated polyhedra (4.15). Comparisons of these indices for the 'soft' logarithmic interaction, the Coulomb law, and the 'hard' Tammes repulsion, given in Eq. (4.2), Figure 12, and Table VI, show that they are all different. Moreover, in the range N up to $O(100)$, there is no indication of any trend toward confluence. Of course, it remains an open question whether there are any qualitative changes for still larger values of N.

Planar magnetic dipole arrays exhibit a different type of structural stability. Experimental observations and computer checks show that the domain patterns generated by pivoted dipoles are *insensitive* to variations of the individual magnetic moments and perturbations of the underlying lattices. The net result is that the domain structures are robust under changes of scale, but vulnerable to qualitative shifts in the strength of multipolarities.[26] These examples indicate that various levels of structural stability and instability can coexist in complex systems.

This duality also appears in the mingling of order and disorder in complex systems. For instance, most of the surface Coulomb configurations seem to be highly irregular, but their small dipole moments (4.2), and the band structures in Figures 7 and 9, clearly show that the charge distributions are far from random. The most striking element of order is the uniformity of the energies displayed in Table III. These correlations imply that the corresponding Coulomb energy landscapes are bounded from below by a single hyperplane that is effectively tangent to every local minimum (cf. Section 4.E.4 and Ref. 95). Table VII shows that this *same* "flat bottom" underlies the Tammes landscapes when the energy minima of the $|\mathbf{r}_i - \mathbf{r}_j|^{-1\,310\,720}$ interactions are rescaled to the Coulomb values. Furthermore, the occurrence of "rattlers"—or nonrigid configurations—in more complex Tammes solutions indicates that the associated minima actually lie in flat valley bottoms that are nearly tangent to this bounding plane.[30] Similar uniformities of the energy minima occur in magnetic arrays even though the corresponding domain patterns are quite distinct. All of these observations run counter to the expectation that the energy landscapes of complex systems have a random—possibly Gaussian—distribution of local minima as a function of "altitude."[91] But this does not necessarily imply a contradiction. Combinatorial arguments can lead both to narrow as well as widely scattered distributions of pseudorandom variables. The near degeneracy of the local energy minima in many cooperative systems may be connected with the sharply peaked distribution of the zeros of random polynomials.[96]

Another basic characteristic of complex systems is *contingency*, or more precisely, the history dependence of their evolution.[91] In the special case of gradient systems, all continuous quasi-static changes can be described in terms of trajectories on the corresponding energy landscapes. The situation that is most frequently considered is the transition from some arbitrarily chosen set of initial states, scattered throughout the highlands of the landscape, down to the local energy minima in the valley bottoms. Since these steepest descents are not equilibrium processes, and the energy minima are often nearly degenerate, Boltzmann statistics cannot be used to infer the occupation probabilities of the terminal states. An obvious alternative is to assume that the occupation probabilities are proportional to the size of the drainage or capture basins that surround each local minimum. Experiments and computer trials show that this size is determined by the lowest neighboring mountain passes in the "slow cooling" limit, and by the mountain crest lines or watershed basins, in the "fast quench" limit.[55] In either case, the topography of the energy landscapes controls the occupation probabilities of the various minima.

All the available evidence indicates that in many cooperative systems the occupation probabilities are highly nonuniform even if there are many local minima that are nearly degenerate in energy. A typical example is shown in Figure 5. In first approximation, the decrease from the peak follows a steep power law, which then trails off into a plateau. This type of empirical Pareto or Zipf distribution has long been familiar in demographics (ranking of cities by population) and economics (apportionment of wealth).[97] Occupation probabilities that are concentrated in a small number of states are also connected with search problems such as the Levinthal paradox:[35,91] Namely, how does a protein find a global optimum without an unreasonably long global search? In the case of the gradient systems, the explanation is simply that a few states are favored because they are fed by the largest drainage basins on the energy landscapes. Although this picture is consistent with computer simulations, it has to be interpreted with caution because minimizing algorithms—such as the procedure described in Section 2.B—do not necessarily correspond to physical processes. This is illustrated by the discussions in Refs. 98–100, which show that discrepancies can arise from differences in the computing schemes without any import on the basic physics.

APPENDIX A. MINIMA OF COMPLEX ENERGY LANDSCAPES

The surface Coulomb problem is a special type of gradient system with constraints.[87] Its extremals are determined by the simultaneous solutions of

$$\frac{\partial E}{\partial \phi_i} = 0 \qquad \frac{\partial E}{\partial \theta_j} = 0 \qquad \forall i, j \tag{A.1}$$

where E is the energy given in (3.5). There are also infinite energy maxima at cusps due to coincident charges. The extremals are often referred to as critical, stationary, or equilibrium points. The simplest locally stable minima are a subset of extremal points $p(\phi_i, \theta_j)$ whose associated Hessian matrices $\mathscr{H}[p(\phi_i, \theta_j)]$ are positive definite;[107,108] *pace rotational degeneracies*. Since these Hessians are arrays of all second-order partial derivatives of the energy—including terms such as $\partial^2 E/\partial\phi_i\, \partial\phi_j$, $\partial^2 E/\partial\phi_i\, \partial\theta_j$, and $\partial^2 E/\partial\theta_i\, \partial\theta_j$—it is convenient to introduce a single symbol Ω_κ, $1 \leq \kappa \leq 2N$, that ranges over all the angles $\phi_1, \ldots, \phi_N$, $\theta_1, \ldots, \theta_N$. The Hessian at p then is

$$\mathscr{H}[p] = \left.\frac{\partial^2 E}{\partial\Omega_\kappa\, \partial\Omega_\mu}\right|_p \quad 1 \leq \kappa,\ \mu \leq 2N \tag{A.2}$$

A necessary and sufficient condition for a real symmetric matrix such as $\mathscr{H}[p]$ to be positive definite is that all of its eigenvalues are positive.

Figure 14(a) is a plan view of part of an energy landscape containing three locally stable minima at E_m, E_{m+1}, and E_{m-1}. For N charges, this landscape is actually a surface in a $2N+1$ dimensional space spanned by $\Omega_1, \ldots, \Omega_{2N}$, and E. Figure 14(b) is an elevation showing the altitude or energy variations along a steepest descent path (path 1) connecting E_m and E_{m+1}. As indicated in Figure 14(a) there may be several mountain pass routes between adjacent valleys. The saddle points at $SP^{(1)}, \ldots, SP^{(5)}$, and so forth are, of course, also extremals satisfying (A.1). These points are distinguished by the property that the associated Hessians $\mathscr{H}[SP]$ have at least one negative eigenvalue: specifically, if λ denotes the number of negative eigenvalues, then $0 < \lambda < 2N$.

Figure 14(a) illustrates some of the technical problems that can occur in mapping complex energy landscapes by means of computers. For instance, the convolutions of path 3 indicate that some minima may be difficult to reach from random initial points located in the energy highlands. The large iteration values listed in column 2 of Table VIII for $N = 7$, 13, 19, 36, and so forth are probably due to such labyrinthine obstacles. But these results have to be interpreted with caution because the *convolution of paths* has a dual significance: (1) *Analytically* it refers to steepest descent paths constrained to wind through highly corrugated landscapes and (2) *numerically* it corresponds to complicated patterns of steps generated by computer algorithms seeking lower ground on rugged terrain. Path 4 illustrates these distinctions in a complementary setting. Suppose for the moment that E_m were *not* a strict minimum because—as shown in Figures 14(a) and 14(b)—there is another winding narrow defile leading from E_m to a lower energy minimum at E_{m-1}. If this track passes

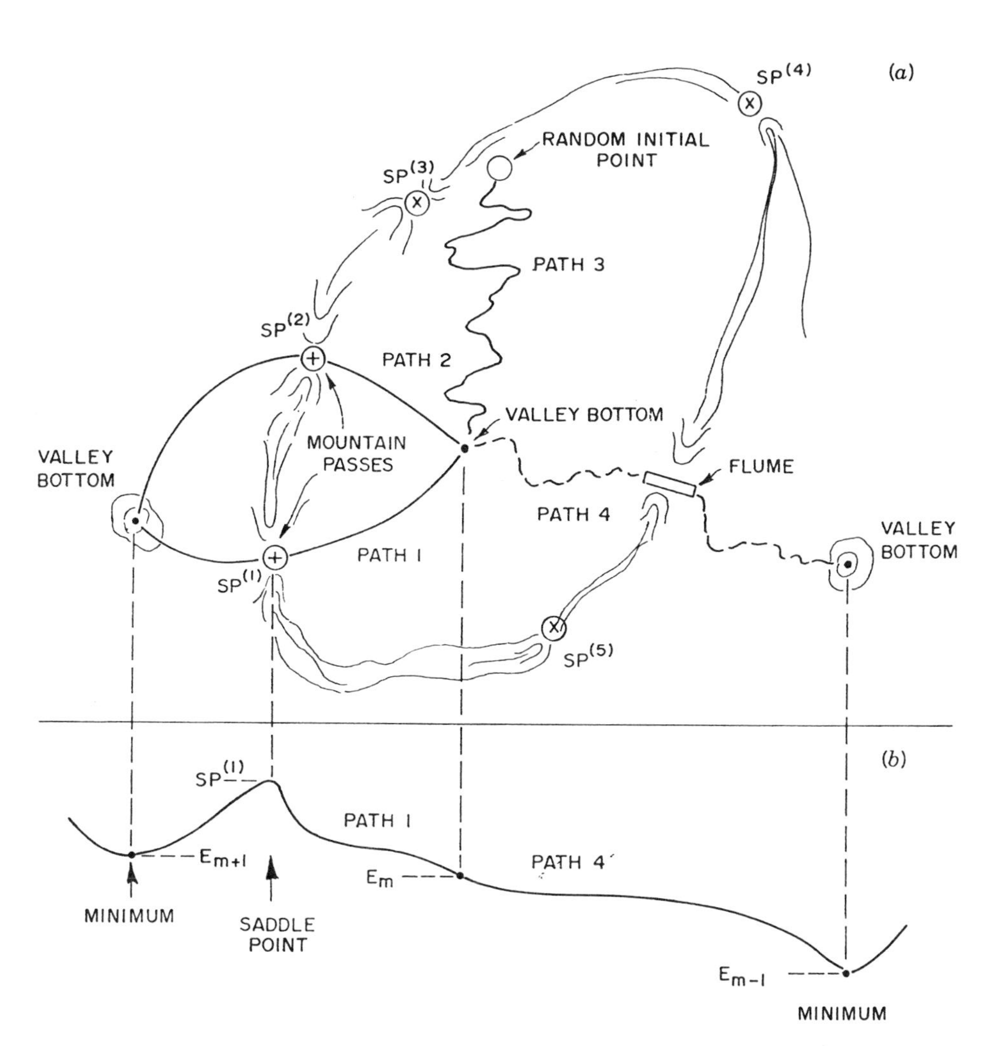

Figure 14(a) and (b). Energy landscape of a gradient system. The plan view in (a) shows three minima and a mountain range. Some of the corresponding heights are indicated in the elevation (b). Paths 1 and 2 are two possible gradient routes linking E_m and E_{m+1}. If $E(\mathrm{SP}^{(1)}) < E(\mathrm{SP}^{(2)})$, then path 1 is the minimax route.[102] The valley bottom at E_m is a locally stable minimum if we omit path 4. But if path 4 is joined to E_{m-1} through a narrow exit, then E_m is numerically stable only relative to a coarse search grid. The positions of the saddle points $\mathrm{SP}^{(1)}, \ldots, \mathrm{SP}^{(5)}$, indicate the extent of the capture basin surrounding E_m.

through the mountains via a flume rather than a saddle point, the energy can decrease monotonically between E_m and E_{m-1}. In principle, the existence of such a narrow exit from E_m presents no difficulties for an analytical description. It simply means that the associated Hessian would show that E_m is a shallow saddle point with respect to one direction. But numerically it would be very difficult to detect the existence of this escape route if the scale of topographic variations of the energy surface were comparable to the roundoff errors of the search routines. Similar computational problems can occur for minima with small capture basins. If we revert to the assumption that $\mathscr{H}[E_m]$ is positive definite, standard Morse theory shows that E_m is an isolated minimum.[101,102] This means that E_m is located in the interior of an open neighborhood all of whose points have energies exceeding E_m. But this theorem is of little practical use in cases where the neighborhood is so small that it falls below the threshold of resolution of numerical surveys.

All the surface Coulomb states listed in Table VIII were screened for numerical robustness with respect to roundoff. Generally, N-charge equilibrium configurations are described by $2N$ angular coordinates with a resolution of 10 decimal digits. Every coordinate was successively truncated to 6, and then to 3 decimal digits. All of these sets of truncated coordinates were taken as the starting configurations of new energy minimizing searches. Numerical stability was then verified in every instance by checking that these minimizations led back to the original equilibrium configurations. Nevertheless, despite these precautions, numerical methods can both under- and overestimate the actual number of minima. As indicated previously, states may be missed because they are concealed by tortuous approaches or have minute capture basins. And states may be counted as locally stable minima because narrow escape routes such as path 4 on Figure 14(a) can be overlooked by numerical surveys.

The correspondences between the analytical and numerical descriptions of multivariable gradient systems tend to be even more complicated in situations where the Hessian matrices are singular at critical points. Figures 14(c) and 14(d) indicate some of the topographic complexities that can appear on the energy landscapes. In particular, the rippled stalagmite in Figure 14(c) is a schematic representation of the cumulation of critical points around a nonisolated singularity. Typical one-dimensional potentials illustrating such a clustering of sequences of maxima and minima around the origin are $\mathsf{U}(q) = e^{-1/q^2}\cos(1/q)$, and $e^{-1/q^2}\sin^2(1/q)$.[101–104] Although the surface Coulomb potential (3.5) clearly does not contain any factors resembling terms such as $\sin(1/q)$, the exponential increase of extremals (3.2a) on a $(2N+1)$-dimensional surface is bound

(*c*)

Figure 14(c). Topography of the energy surface in the vicinity of a Hessian singularity. This rippled stalagmite is a schematic representation of the cumulation of local maxima and/or minima around a nonisolated Hessian singularity.

to produce a crowding that is numerically equivalent to a clustering. Figure 14(d) shows two extremal *lines* in the form of intersecting valley bottoms. If the crossing point is taken as the origin of a set of local coordinates, say $\alpha, \beta, \ldots$, then potential expressions such as $\mathsf{U} \sim \alpha^2\beta^2$ will generate this kind of topography. Generally, the lower the rank r of the Hessian, that is, $r < 2N$, or equivalently, the larger the dimensionality of the *nullity* ($=2N - r$), the "flatter" will be the associated valley bottom on the energy surface.[101] The rattling or labile states found by Kottwitz[30] at $N = 19, 20$, and so forth for the extremely repulsive Tammes potential $|\mathbf{r}_i - \mathbf{r}_j|^{-1\,310\,720}$ are examples of such extended minima. Similiar trends appear in the surface Coulomb problem. For instance, the 13-digit reproducibility of the total energy in Eq. (2.9a) represents the sharply defined level of a valley bottom on the energy landscape—that is, $\liminf E[\mathbb{C}_2(84)] = 3\,103.478\ldots$. But the much lower precision of the individual charge coordinates (2.9b,c) reflects the influence of shallow grooves (or eigenmodes) surrounding the minimum. This quasi-degenerate behavior is also connected with the poor performance of conjugate-gradient methods in the Coulomb problem.

It is plausible that for increasing values of N, the energy landscapes of

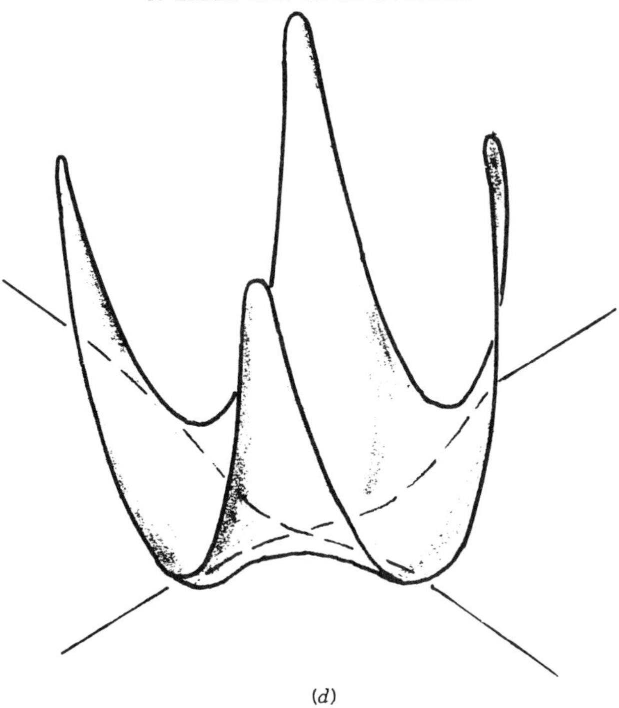

(*d*)

Figure 14(d). Topography of the energy surface in the vicinity of a Hessian singularity. The fluted landscape surrounds valleys whose minima are networks of lines rather than isolated points. In higher dimensions these lines correspond to areas, that is, *flat* valley bottoms.

the surface Coulomb problem include a greater proportion of complex features such as those shown in Figures 14(c) and 14(d). Analytically, this incidence of singular Hessians means that more of the critical point behavior falls outside the scope of Morse theory.[102] The corresponding numerical description of singular or close-to-singular $2N \times 2N$ matrices with $N > 100$, then also requires greater computational effort:[105,106] First, it is necessary to evaluate all of the $N(2N+1)$-independent matrix elements in (A.2) at the relevant critical points. And then, since roundoff errors make it impractical to check directly whether $\det|\mathscr{H}[p]|$ vanishes, the proximity of singularities has to be detected by sensitivity analyses. One criterion of this type is the condition number of a matrix, which is also proportional to the inverse "distance" to a singularity.[105] Standard software packages are available for implementing these diagnostics, but it remains to be seen whether they will be of any significant help in extending the surface Coulomb analysis to more complex systems.

APPENDIX B: SUMMARY OF RESULTS

Table VIII contains a survey of the numerically robust states of the surface Coulomb problem extending up to 112 charges. The first column lists the number of charges N. The next column shows the *average* number of iterations required to reach an equilibrium state. (Note that every charge is simultaneously moved at each iterative step.) The frequency of occurrence, or capture basin, of each state is indicated in the third column. The asterisks mark enantiometric states. Column 4 lists the dimensionless energy (1.3) of each state. The center of charge, or magnitude of the dipole moment (1.2a), of every configuration is given in column 5. Column 6 shows the minimum angular separation (radians) between pairs of points of the surface Coulomb states. A corresponding set of values for the Tammes problem is listed in column 7. The hole angles, given in column 8, approximate the angular diameters of the largest spherical caps containing *no* charges in their interior; cf. Section 4.E.2. Columns 9 and 10 list the energy and angular diversity ratios defined in (3.1) and (1.1), respectively. Finally, column 11 indicates the number of faces of the polyhedra associated with each configuration; cf. Section 4.C.

ACKNOWLEDGMENTS

We are grateful to many colleagues and students for information and assistance. We would particularly like to thank Prof. B. Bernstein, J. Cheevers, Prof. R. Filler, Prof. A. Florian, Prof. M. J. Frank, Dr. D. R. Gavelek, Prof. N. W. Johnson, Dr. D. A. Kottwitz, Prof. P. Lykos, Prof. S. Miller, Prof. E. T. Olsen, Prof. A. Sklar, Prof. T. Tarnai, and R. K. Bernett-Zasadzinski. This work was supported in part by the Research Corporation and AFOSR. Fermilab is operated by Universities Research Association Inc. under contract DE-AC02-76CHO3000 with the U.S. Department of Energy.

TABLE VIII
Equilibrium Configurations for the Surface Coulomb Problem

N	Average iterations	Frequency %	Coulomb energy	Dipole moment	Coulomb angle	Tammes angle	Hole angle	% Energy diversity	% Angular diversity	Faces N_f
3	14	100.00	1.73205081	0	2.094395	2.094395	3.1416	33.3	33.3	
4	16	100.00	3.67423461	0	1.910633	1.910633	2.4619	25.0	16.7	4
5	78	100.00	6.47469149	0	1.570796	1.570796	2.2143	40.0	30.0	6
6	42	100.00	9.98528137	0	1.570796	1.570796	1.9106	16.7	13.3	8
7	2160	100.00	14.45297741	0	1.256637	1.359080	1.7812	28.6	19.0	10
8	182	100.00	19.67528786	0	1.251299	1.306527	1.6161	12.5	14.3	10
9	280	100.00	25.75998653	0	1.207589	1.230959	1.4835	22.2	16.7	14
10	442	100.00	32.71694946	0	1.134387	1.154480	1.4666	20.0	15.6	16
11	499	100.00	40.59645051	0.013220	1.021708	1.107149	1.3455	45.5	36.4	18
12	76	100.00	49.16525306	C	1.107149	1.107149	1.3047	8.3	4.5	20
13	1280	100.00	58.85323061	0.008820	0.913103	0.997224	1.2268	46.2	37.2	22
14	255	100.00	69.30636330	0	0.922687	0.971635	1.2060	14.3	9.9	24
15	437	100.00*	80.67024411	0	0.859136	0.936506	1.1307	20.0	20.0	26
16	293	75.66*	92.91165530	0	0.854098	0.911837	1.0730	12.5	10.0	28
	394	24.34	92.92035396	0	0.874880		1.0938	12.5	10.8	26
17	677	100.00	106.05040483	0	0.874550	0.891694	1.0637	17.6	10.3	30
18	502	100.00	120.08446745	0	0.829632	0.864927	1.0517	16.7	11.8	32
19	9109	100.00	135.08946756	0.000135	0.783822	0.832381	0.9934	36.8	30.4	33
20	662	100.00	150.88156833	0	0.804480	0.827828	0.9716	20.0	13.7	36
21	3976	99.99	167.64162240	0.001406	0.773536	0.796101	0.9348	42.9	32.9	38
	847	0.01	167.64183186	0.001425	0.767113		0.9334	38.1	48.1	36
22	541	97.30	185.28753615	0	0.755763	0.780863	0.9350	13.6	7.4	40
	1442	2.70	185.30795160	0	0.746305		0.9498	13.6	10.8	40
23	442	100.00*	203.93019066	0	0.723982	0.762883	0.9213	21.7	19.0	42
24	440	100.00*	223.34707405	0	0.734178	0.762548	0.9018	4.2	5.8	38
25	7560	100.00	243.81276030	0.001021	0.691333	0.726658	0.8593	56.0	53.3	45

26	2292	100.00*	265.13332632	0.001919	0.677923	0.716242	0.8531	50.0	70.8	48
27	662	99.97	287.30261503	0	0.697089	0.709958	0.8397	14.8	8.5	50
	402	0.03*	287.30961176	0	0.680734		0.8422	18.5	27.6	50
28	529	100.00*	310.49154236	0	0.660149	0.686877	0.8309	10.7	9.3	52
29	3133	100.00*	334.63443992	0	0.635147	0.675681	0.8073	17.2	30.3	54
30	2470	99.99*	359.60394590	0	0.644763	0.673647	0.7921	26.7	40.9	56
	1290	0.01	359.60439833	0.000004	0.637774		0.7886	36.7	43.9	54
31	390	100.00	385.53083806	0.003205	0.634831	0.658161	0.7744	29.0	20.0	58
32	239	97.93	412.26127465	0	0.652358	0.654066	0.7920	6.2	2.2	60
	673	2.07	412.46839720	0	0.621525		0.7699	12.5	8.9	58
33	7818	100.00	440.20405745	0.004356	0.588174	0.632761	0.7621	60.6	52.7	61
34	1591	100.00*	468.90485328	0	0.580730	0.624964	0.7448	26.5	34.0	64
35	3873	81.76*	498.56987249	0.000419	0.577711	0.616423	0.7290	51.4	51.1	66
	961	0.02	498.56991740	0.000480	0.576922		0.7288	42.9	49.2	64
	5649	18.20	498.57345404	0.001266	0.581227		0.7257	60.0	52.8	66
	988	0.02	498.57346232	0.001202	0.581086		0.7250	100.0	99.2	66
36	19168	100.00*	529.12240842	0.000048	0.579503	0.614177	0.7203	50.0	51.1	68
37	1846	19.91	560.61888773	0	0.564307	0.600784	0.7294	13.5	7.8	70
	3543	80.09*	560.62797306	0.000925	0.558252		0.7033	56.8	84.8	70
38	640	43.87	593.03850357	0.000001	0.580086	0.597787	0.6967	10.5	6.5	72
	1119	56.13	593.04894354	0.001687	0.563323		0.6999	34.2	30.3	72
39	472	65.27	626.38900902	0	0.559429	0.584494	0.6991	15.4	10.7	74
	5256	30.18*	626.44095841	0.000399	0.547982		0.6869	51.3	51.3	74
	4013	4.55	626.44096635	0.000371	0.547924		0.6866	38.5	45.3	72
40	753	66.94	660.67527883	0	0.557045	0.578722	0.6905	10.0	5.9	76
	7442	23.54*	660.72530410	0.000004	0.551384		0.6812	35.0	48.8	76
	959	0.14*	660.72530735	0.000282	0.550872		0.6810	50.0	51.2	76
	743	9.38	660.74121431	0.001465	0.545251		0.6826	35.0	28.3	76
41	476	93.51	695.91674434	0	0.550264	0.571230	0.6802	17.1	10.9	78
	1466	6.49*	695.97869944	0	0.545295		0.6778	19.5	25.2	78
42	488	98.08	732.07810754	0	0.545324	0.567343	0.6760	11.9	7.3	80
	1953	0.96*	732.15182672	0	0.540498		0.6650	16.7	32.4	80

TABLE VIII (*Continued*)

N	Average iterations	Frequency %	Coulomb energy	Dipole moment	Coulomb angle	Tammes angle	Hole angle	% Energy diversity	% Angular diversity	Faces N_f
	4625	0.92*	732.19816736	0.003638	0.529379		0.6624	40.5	47.2	78
	255	0.03	732.25624103	0	0.553574		0.6754	4.8	1.6	80
	761	0.01	732.45155921	0	0.515500		0.6713	14.3	11.3	80
43	7239	100.00	769.19084646	0.000400	0.538723	0.559980	0.6600	53.5	49.2	82
44	3606	100.00	807.17426309	0.000060	0.545548	0.558216	0.6486	6.8	5.6	78
45	6635	99.89*	846.18840106	0	0.527214	0.546691	0.6419	24.4	43.2	86
	908	0.11	846.18864878	0.000147	0.527383		0.6382	35.6	43.4	86
46	729	11.24*	886.16711364	0	0.519938	0.540339	0.6408	10.9	8.8	88
	1355	44.66	886.17021602	0.001066	0.504669		0.6341	34.8	27.8	86
	1821	26.25*	886.17143242	0.001395	0.509081		0.6336	50.0	50.7	88
	1561	17.23*	886.17710517	0.001838	0.509578		0.6336	60.9	56.2	88
	10912	0.62	886.25028042	0	0.500098		0.6415	30.4	38.6	88
47	1443	55.74*	927.05927068	0.002483	0.502432	0.537244	0.6250	57.4	59.6	89
	4315	31.32*	927.06226967	0.002536	0.487302		0.6272	68.1	86.8	89
	12603	8.81	927.07222457	0.004938	0.505445		0.6233	36.2	32.3	90
	1816	0.10	927.07227174	0.000921	0.505369		0.6245	46.8	52.2	88
	2067	3.69*	927.08823351	0.000803	0.505048		0.6280	51.1	75.9	90
	1942	0.34*	927.14108835	0.001525	0.491380		0.6325	48.9	72.6	90
48	1061	99.99*	968.71345534	0	0.518182	0.536912	0.6282	4.2	5.7	86
	538	0.01	968.71550891	0	0.517172		0.6258	10.4	8.0	90
49	1112	100.00*	1011.55718265	0.001529	0.495439	0.522265	0.6184	34.7	33.3	94
50	827	100.00	1055.18231473	0	0.501108	0.519287	0.6084	10.0	6.1	96
51	1502	98.56*	1099.81929032	0	0.491578	0.512456	0.5977	17.6	22.6	98
	1521	1.44	1099.94023114	0.002506	0.474308		0.5988	35.3	28.1	96
52	661	53.50*	1145.41896432	0.000457	0.482931	0.509545	0.5940	34.6	33.3	100
	1208	29.81*	1145.42198063	0	0.485412		0.6015	25.0	28.1	100
	1436	9.84*	1145.43570898	0.000720	0.484667		0.5978	50.0	59.0	100
	3641	6.85*	1145.43759698	0.002189	0.480104		0.5944	100.0	98.8	100
53	7869	69.19	1191.92229042	0.000279	0.473629	0.502897	0.5881	32.1	27.2	99
	1028	30.81*	1191.93158471	0.000293	0.471423		0.5868	50.9	50.6	102

54	1846	80.27*	1239.36147473	0.000138	0.471755	0.501205	0.5856	50.0	60.3	104
	2205	4.08*	1239.36525530	0	0.475519		0.5864	16.7	25.6	104
	3846	7.26*	1239.37119227	0.000371	0.474284		0.5871	72.2	88.3	104
	1033	0.02*	1239.37125018	0.000405	0.472860		0.5868	98.1	98.7	103
	1355	8.37*	1239.37320071	0	0.478224		0.5850	25.9	32.6	104
55	1080	26.82*	1287.77272078	0.000392	0.464521	0.493279	0.5777	50.9	50.8	106
	1657	19.71*	1287.77702746	0.000114	0.461470		0.5780	50.9	50.8	106
	2337	16.37	1287.77726081	0.000118	0.470319		0.5826	32.7	27.1	104
	2953	14.65*	1287.78870934	0.000025	0.466465		0.5762	56.4	80.5	106
	1476	21.03*	1287.78905724	0.000191	0.464988		0.5806	96.4	99.1	106
	3478	1.42*	1287.80015929	0.000551	0.467767		0.5767	58.2	84.6	106
56	2231	10.82*	1337.09494528	0	0.465704	0.491276	0.5757	25.0	26.4	108
	2502	49.91*	1337.09534827	0.000174	0.464412		0.5770	50.0	50.6	108
	2674	39.17*	1337.09872742	0.000275	0.465148		0.5744	96.4	98.7	108
	4680	0.08*	1337.24862285	0.000149	0.457559		0.5749	75.0	90.1	108
57	1886	90.37*	1387.38322925	0	0.466045	0.485667	0.5719	17.5	18.6	110
	2511	3.05	1387.42008235	0.000753	0.453763		0.5682	66.7	79.6	110
	4016	1.84*	1387.43037248	0.000285	0.453468		0.5700	63.2	89.4	110
	2107	4.28*	1387.43113006	0.000273	0.452877		0.5713	50.9	70.0	110
	1494	0.46*	1387.47189278	0.000870	0.452564		0.5678	49.1	62.1	110
58	1941	23.51*	1438.61825064	0	0.456495	0.480923	0.5680	25.9	29.4	112
	2656	17.69*	1438.62550858	0.000058	0.455362		0.5630	50.0	50.6	112
	2362	5.43*	1438.62628995	0	0.454155		0.5633	25.9	37.6	112
	2260	25.42*	1438.62722515	0.000308	0.456654		0.5598	53.4	74.6	112
	5281	4.71*	1438.63370800	0.000002	0.454605		0.5638	50.0	52.9	110
	3329	0.01	1438.63374161	0.000009	0.454111		0.5640	91.4	94.7	111
	1550	22.64*	1438.63810500	0.000198	0.452373		0.5643	100.0	98.7	112
	2411	0.39*	1438.64735982	0.001029	0.453854		0.5648	36.2	59.3	112
	1839	0.11*	1438.67209913	0	0.464794		0.5593	25.9	38.1	112
	1345	0.04	1438.73596329	0.000388	0.449459		0.5648	100.0	97.6	112
59	2127	27.53*	1490.77333528	0.000154	0.456757	0.478133	0.5593	55.9	73.9	114
	2747	61.39*	1490.77438608	0.000623	0.456849		0.5570	100.0	98.7	114
	7687	3.58*	1490.78475584	0.000245	0.457361		0.5604	76.3	91.4	114
	898	0.01	1490.78478568	0.000301	0.456913		0.5601	100.0	98.5	114
	1646	7.49*	1490.79077309	0.000608	0.453876		0.5581	50.8	50.4	114
60	893	26.04*	1543.83040098	0	0.453046	0.474604	0.5566	16.7	17.5	116
	883	69.38*	1543.83509960	0.000130	0.452967		0.5562	50.0	50.6	116
	1946	3.71*	1543.84153514	0.000177	0.452851		0.5544	36.7	58.0	116

TABLE VIII (*Continued*)

N	Average iterations	Frequency %	Coulomb energy	Dipole moment	Coulomb angle	Tammes angle	Hole angle	% Energy diversity	% Angular diversity	Faces N_f
	1189	0.36	1543.86465762	0.000017	0.447471		0.5556	6.7	6.4	116
	6212	0.40*	1543.96947231	0	0.451689		0.5608	20.0	25.3	114
	1561	0.11*	1543.98073384	0.000005	0.447795		0.5526	18.3	17.5	116
61	1907	62.11*	1597.94183020	0.001091	0.443169	0.469011	0.5507	98.4	98.7	118
	1174	11.83*	1597.95155534	0.000648	0.442474		0.5530	49.2	53.0	118
	7882	13.29*	1597.95512785	0.001364	0.444069		0.5483	100.0	98.1	117
	1516	0.03*	1597.95514972	0.001289	0.445279		0.5504	50.8	53.7	118
	2744	10.45*	1597.97036059	0.000634	0.445656		0.5503	100.0	98.5	118
	9463	0.03*	1597.97266027	0.000330	0.449018		0.5530	91.8	96.4	118
	4070	2.26*	1597.98080362	0.001003	0.437864		0.5492	100.0	98.7	117
62	2785	26.00*	1652.90940990	0	0.451689	0.465714	0.5517	11.3	10.8	120
	2546	61.95*	1652.92859368	0.001117	0.444933		0.5483	50.0	64.9	120
	1477	12.05*	1652.94201427	0.000513	0.446840		0.5481	51.6	50.6	120
63	1922	99.71*	1708.87968150	0	0.440812	0.462284	0.5457	15.9	25.5	122
	946	0.29*	1709.00838502	0	0.434248		0.5429	17.5	17.4	122
64	1695	85.57*	1765.80257793	0	0.434936	0.457888	0.5388	25.0	26.7	124
	2255	3.24	1765.81619775	0	0.435611		0.5371	15.6	20.3	120
	2552	8.02*	1765.82032129	0.000253	0.430896		0.5371	50.0	50.4	124
	1100	1.05	1765.87533511	0	0.434467		0.5443	10.9	7.6	122
	2022	1.19*	1765.89790410	0.000152	0.427295		0.5380	50.0	50.4	124
	1520	0.93*	1765.91167428	0	0.439880		0.5380	26.6	30.2	124
65	5625	92.95*	1823.66796026	0.000400	0.428072	0.455004	0.5335	60.0	81.0	126
	6714	1.50*	1823.69459614	0	0.434582		0.5358	32.3	48.9	120
	2478	5.55*	1823.71802820	0.001283	0.423836		0.5321	100.0	98.8	126
66	12273	32.40*	1882.44152535	0.000776	0.432235	0.452868	0.5312	50.0	50.5	128
	2446	20.05*	1882.44209276	0	0.434503		0.5293	16.7	29.0	128
67	1017	100.00*	1942.12270041	0	0.431572	0.448270	0.5294	11.9	10.8	130
68	2004	70.25*	2002.87470175	0	0.426435	0.444372	0.5221	23.5	28.1	132
	2244	16.05*	2002.88294764	0.000014	0.422194		0.5246	17.6	23.8	132
	8889	13.40	2002.89000272	0.000688	0.418399		0.5199	57.4	51.1	131
	3192	0.20*	2002.98206733	0.001807	0.413769		0.5236	50.0	50.3	132
	2505	0.10*	2003.06105788	0.003608	0.406507		0.5100	97.1	98.3	132

69	1114	5.55*	2064.53348323	0	0.421261	0.442131	0.5215	17.4	17.6	134
	8757	32.35*	2064.53606623	0	0.425080		0.5202	27.5	39.9	134
	1891	53.25*	2064.53944940	0.000688	0.419259		0.5187	97.1	98.7	134
	3182	8.00	2064.55491294	0.000685	0.420971		0.5161	33.3	33.6	133
	5525	0.85*	2064.60448326	0.000643	0.414273		0.5186	100.0	98.1	134
70	2604	61.15	2127.10090155	0	0.423952	0.439315	0.5142	15.7	21.7	132
	3377	9.55*	2127.11628069	0.000268	0.415831		0.5129	34.3	33.1	136
	2768	13.05*	2127.11814054	0.001250	0.416111		0.5138	34.3	33.1	136
	4194	8.65*	2127.12667460	0.000393	0.421508		0.5146	55.7	81.6	136
	4200	5.05*	2127.13628990	0.000474	0.422922		0.5129	57.1	75.7	136
	1733	2.40*	2127.21103001	0.000442	0.418839		0.5121	50.0	50.8	136
	4131	0.15*	2127.30372335	0.002756	0.398821		0.5072	64.3	81.0	134
71	4140	95.50*	2190.64990643	0.001257	0.415434	0.436121	0.5115	50.7	50.3	136
	3787	3.05*	2190.69381272	0.003693	0.410524		0.5026	98.6	97.9	138
	1874	1.40*	2190.76494784	0.001128	0.405700		0.5096	100.0	98.2	138
	8180	0.05	2190.88650071	0.000984	0.402219		0.5042	50.7	50.1	138
72	852	82.95*	2255.00119098	0	0.427461	0.435049	0.5140	2.8	2.0	140
	3747	7.45*	2255.13111676	0	0.423770		0.5082	4.2	10.1	134
	3986	5.45	2255.26271483	0.000678	0.404687		0.5073	58.3	70.3	137
	2762	3.30*	2255.29359234	0.001255	0.399868		0.5053	100.0	98.4	140
	12174	0.80*	2255.31431104	0.001425	0.403415		0.5038	100.0	98.1	140
	1842	0.05	2255.39323492	0.000681	0.397730		0.5035	98.6	97.7	140
73	3432	90.35*	2320.63388375	0.001573	0.398115	0.428544	0.5076	56.2	77.6	142
	4995	8.90*	2320.67535341	0.000383	0.405192		0.5059	98.6	98.0	142
	1384	0.35*	2320.79425065	0.000160	0.399849		0.5016	50.7	54.1	142
	8113	0.20*	2320.79742109	0.002966	0.403042		0.5019	84.9	95.4	142
	2687	0.20*	2320.82910544	0.002754	0.398270		0.4931	97.3	98.7	142
74	3039	28.60*	2387.07298184	0.000642	0.400828	0.426225	0.5006	52.7	70.9	144
	15994	20.05*	2387.07516717	0.000177	0.401819		0.4999	48.6	50.4	144
	2482	24.05*	2387.07993444	0.001345	0.402612		0.4980	97.3	97.8	144
	5762	20.50*	2387.08728748	0.000559	0.403742		0.5015	100.0	97.9	143
	7775	0.60*	2387.09418960	0.000834	0.399111		0.5009	52.7	62.5	144
	2728	0.55*	2387.10121441	0	0.400112		0.5000	50.0	49.6	144
	7289	0.30*	2387.10460931	0.000002	0.405705		0.5005	45.9	47.2	144
	4049	3.40*	2387.10711171	0.000295	0.402163		0.4994	50.0	50.3	144
	3038	0.40*	2387.11556802	0.003553	0.391427		0.5006	52.7	69.8	144
	5016	0.50*	2387.11557541	0.003445	0.389949		0.4999	83.8	88.7	144

TABLE VIII (*Continued*)

N	Average iterations	Frequency %	Coulomb energy	Dipole moment	Coulomb angle	Tammes angle	Hole angle	% Energy diversity	% Angular diversity	Faces N_f
	3594	1.00*	2387.12809098	0.000074	0.400626		0.4994	66.2	83.9	144
75	5316	97.15*	2454.36968904	0	0.396826	0.424145	0.4999	17.3	17.3	146
	5583	0.65*	2454.47518023	0.001730	0.397887		0.4961	96.0	98.1	146
	3823	2.20*	2454.48282529	0.000221	0.399361		0.4942	48.0	50.0	146
76	2160	72.65*	2522.67487184	0.000943	0.399431	0.420968	0.4932	50.0	57.9	148
	7668	3.85	2522.70043604	0.001113	0.395642		0.4932	63.2	80.9	148
	5015	18.40*	2522.71740900	0.001136	0.397212		0.4910	98.7	97.4	146
	3483	3.00*	2522.73513045	0	0.400992		0.4954	28.9	39.3	148
	1999	0.70*	2522.79463635	0.001703	0.400674		0.4927	50.0	61.8	148
	6714	1.20	2522.79646997	0.002723	0.396414		0.4921	76.3	89.2	146
	2258	0.15	2522.83098418	0	0.400978		0.4913	17.1	16.7	148
	18057	0.05	2522.91577137	0.000209	0.396677		0.4931	46.1	49.4	148
77	1272	96.30*	2591.85015235	0	0.406420	0.418807	0.4947	11.7	10.8	150
	10481	0.30*	2592.01257632	0.003137	0.386295		0.4874	83.1	92.6	149
	4061	2.30	2592.01416288	0.001999	0.402892		0.4913	23.4	22.7	145
	1733	0.95*	2592.06613681	0.000433	0.392480		0.4886	98.7	97.7	150
	2255	0.10	2592.15817266	0	0.392444		0.4920	18.2	29.2	150
	9326	0.05	2592.16547594	0.000385	0.392832		0.4912	33.8	27.1	149
78	1192	0.80	2662.04647457	0	0.408867	0.417675	0.4876	6.4	4.6	152
	1944	93.90*	2662.04721329	0	0.404565		0.4877	16.7	17.3	152
	4909	3.05	2662.12291358	0.000584	0.397320		0.4857	33.3	37.0	151
	3831	1.45	2662.22689315	0.000615	0.394945		0.4863	30.8	26.6	149
	3430	0.80*	2662.27142693	0.000723	0.388632		0.4835	98.7	98.1	151
79	2625	92.60	2733.24835748	0.000703	0.395073	0.412316	0.4856	62.0	67.1	153
	3171	6.70	2733.27367184	0.001057	0.392687		0.4839	34.2	40.6	154
	4669	0.05	2733.43501631	0.001652	0.390014		0.4810	67.1	83.6	154
	3123	0.45	2733.44491374	0.000395	0.390023		0.4838	32.9	37.9	152
	5193	0.10	2733.45513921	0.002295	0.384218		0.4842	89.9	95.4	154
	2334	0.10*	2733.47618113	0.001284	0.390059		0.4815	50.6	50.1	154
80	2287	74.25	2805.35587598	0	0.397557	0.410912	0.4851	10.0	8.7	154
	6104	15.85*	2805.43731924	0.000424	0.390855		0.4820	66.3	84.0	156
	3538	4.05	2805.47175377	0	0.389448		0.4875	16.3	13.5	154
	6259	3.40	2805.47441738	0	0.385954		0.4840	25.0	25.1	152
	6135	1.75*	2805.52242977	0.001283	0.382592		0.4803	100.0	97.6	156
	2302	0.55*	2805.53769852	0.001393	0.385740		0.4814	50.0	50.1	156
	4869	0.15	2805.67140367	0.000801	0.385097		0.4786	58.8	65.0	156

81	5873	48.70*	2878.52282967	0.000194	0.382079	0.407493	0.4769	70.4	86.4	158
	8035	27.50*	2878.52853268	0.000381	0.379170		0.4779	97.5	97.5	158
	3140	7.35*	2878.54384730	0.000234	0.377999		0.4801	55.6	78.6	158
	7977	10.20*	2878.58533588	0.000788	0.374717		0.4776	100.0	97.7	157
	2963	0.80*	2878.59244543	0.001113	0.381752		0.4758	55.6	68.9	158
	8548	1.80*	2878.59516266	0.001047	0.382537		0.4795	98.8	98.0	158
	3832	0.75*	2878.59851237	0.000990	0.380933		0.4757	97.5	98.0	158
	3642	2.70*	2878.61621814	0.002079	0.380754		0.4755	97.5	97.5	158
	1584	0.05	2878.65259291	0.001830	0.379521		0.4768	33.3	32.9	158
	3544	0.15*	2878.70501519	0.001470	0.379058		0.4745	100.0	98.1	158
82	4515	27.75*	2952.56967529	0	0.387567	0.404787	0.4784	25.6	32.7	160
	5593	13.05*	2952.57368611	0	0.384226		0.4762	39.0	44.2	160
	3318	16.45*	2952.57478472	0.000107	0.375075		0.4746	48.8	49.9	160
	4092	15.45*	2952.58504951	0.000366	0.376074		0.4734	100.0	97.2	160
	3511	9.95*	2952.59404559	0.000869	0.378304		0.4757	98.8	97.9	159
	3995	7.40*	2952.59952750	0.001395	0.380152		0.4719	100.0	97.4	160
	2468	0.90*	2952.60156975	0.000424	0.382119		0.4771	50.0	70.9	160
	1581	0.10*	2952.60827268	0	0.382946		0.4784	11.0	16.7	160
	7724	3.70*	2952.61233304	0.000451	0.372213		0.4732	98.8	97.9	160
	9095	1.90	2952.61405693	0.000530	0.375110		0.4762	65.9	81.2	158
	6667	1.85*	2952.62580274	0.002129	0.376405		0.4762	100.0	97.6	160
	4510	0.60	2952.62681055	0.000549	0.381781		0.4729	74.4	86.3	160
	3051	0.10*	2952.64690930	0.001319	0.378503		0.4746	53.7	80.1	160
	2492	0.25*	2952.68656376	0.000516	0.377377		0.4738	50.0	50.3	160
	2569	0.40*	2952.70171196	0.000055	0.376233		0.4717	50.0	50.0	160
	4097	0.15*	2952.73357764	0.003106	0.374029		0.4697	100.0	97.7	160
83	5438	24.20*	3027.52848893	0.000340	0.377795	0.402874	0.4758	73.5	85.4	162
	2968	17.85*	3027.54137715	0.000128	0.377682		0.4734	50.6	50.0	162
	9602	19.80*	3027.55862902	0.000875	0.376526		0.4723	90.4	92.3	160
	3747	9.00*	3027.59245704	0.000099	0.379116		0.4705	54.2	73.1	162
	2898	9.10*	3027.59319407	0	0.383514		0.4716	18.1	22.4	162
	4424	5.25*	3027.59924209	0.000530	0.376939		0.4713	96.4	97.9	162
	2679	3.30*	3027.60050001	0.000521	0.374436		0.4729	50.6	50.2	162
	6300	2.10	3027.61937175	0.000291	0.376050		0.4712	78.3	87.7	158
	7075	1.50	3027.62808194	0.000669	0.380499		0.4677	57.8	67.3	162

TABLE VIII (*Continued*)

N	Average iterations	Frequency %	Coulomb energy	Dipole moment	Coulomb angle	Tammes angle	Hole angle	% Energy diversity	% Angular diversity	Faces N_f
	4879	1.95*	3027.63179386	0.001527	0.373643		0.4724	97.6	97.9	162
	4567	1.35*	3027.63256198	0.000883	0.377717		0.4733	98.8	97.8	162
	4299	3.20*	3027.63425680	0.000991	0.375605		0.4709	100.0	98.0	162
	3594	0.75*	3027.64213174	0.001039	0.375153		0.4684	98.8	97.8	162
	10202	0.25*	3027.66414535	0.001978	0.378210		0.4710	98.8	97.5	162
	5174	0.25*	3027.66441132	0.000407	0.376792		0.4723	59.0	82.0	162
	3453	0.05	3027.68231381	0.001979	0.376238		0.4708	100.0	98.1	162
	6021	0.10*	3027.71099742	0.002902	0.373357		0.4653	97.6	97.4	162
84	8308	11.45*	3103.46512444	0.000407	0.375467	0.402329	0.4708	58.3	77.8	164
	2243	34.10*	3103.47871710	0	0.385740		0.4708	16.7	17.3	164
	2843	8.10*	3103.48273357	0	0.379225		0.4708	25.0	26.0	164
	5600	14.35*	3103.49243447	0.000587	0.375862		0.4676	98.8	97.4	164
	4610	4.95*	3103.49295073	0.000480	0.377327		0.4682	97.6	97.4	164
	5784	0.55	3103.49438022	0	0.383778		0.4705	27.4	28.5	164
	3960	3.25*	3103.50450535	0.000226	0.375240		0.4708	61.9	80.9	164
	2724	1.95*	3103.50942982	0.000373	0.376401		0.4690	50.0	49.8	164
	2598	8.65	3103.51025966	0.000479	0.375186		0.4702	58.3	69.6	164
	5379	6.85*	3103.51055938	0.000569	0.376813		0.4668	97.6	97.6	163
	5520	3.15*	3103.51244522	0.000544	0.378262		0.4697	50.0	50.1	164
	5423	1.00*	3103.54674529	0	0.379786		0.4686	39.3	45.4	164
	6238	0.45*	3103.55004326	0.000317	0.376672		0.4667	78.6	87.9	164
	3600	0.30*	3103.56428084	0.000369	0.374846		0.4679	57.1	79.9	164
	2897	0.45*	3103.59823623	0.002032	0.373768		0.4648	98.8	97.8	164
	4248	0.45*	3103.63951517	0.002169	0.363683		0.4642	97.6	97.8	163
85	3124	18.70*	3180.36144294	0.000417	0.375204	0.397563	0.4671	50.6	50.1	166
	6119	12.10*	3180.37847263	0.001008	0.374286		0.4665	50.6	49.9	166
	5850	13.00*	3180.38006415	0.001031	0.371294		0.4671	97.6	97.4	166
	3822	23.45*	3180.38254415	0.000826	0.372080		0.4650	98.8	97.9	166
	4525	7.25*	3180.38926265	0.001080	0.379221		0.4660	49.4	54.6	166
	10567	9.90*	3180.38946622	0.000324	0.376140		0.4670	49.4	69.8	166
	4834	7.40*	3180.39289716	0.000731	0.368077		0.4653	100.0	97.4	166
	5573	4.20*	3180.41240577	0.000241	0.373119		0.4642	100.0	97.4	165
	5820	2.40*	3180.42055571	0.001020	0.378541		0.4631	100.0	97.5	166
	2271	0.35	3180.49260922	0.001515	0.371106		0.4640	54.1	54.9	166

86	2290	15.25*	3258.21160571	0.001379	0.375622	0.395742	0.4631	50.0	56.3	168
	2897	5.05*	3258.21366308	0.000013	0.377280		0.4660	17.4	21.2	168
	7523	7.80*	3258.21995934	0.000165	0.368917		0.4612	48.8	50.0	168
	8512	6.90*	3258.22282967	0.000124	0.371239		0.4626	47.7	49.9	168
	3746	18.25*	3258.22413239	0.000179	0.371405		0.4624	54.7	77.2	168
	6651	3.75*	3258.22424493	0	0.374381		0.4651	39.5	45.6	168
	2182	13.40*	3258.23636072	0.000842	0.373502		0.4596	96.5	97.2	168
	2683	2.90*	3258.24051336	0.001038	0.372867		0.4616	59.3	73.1	168
	2495	7.60*	3258.24155989	0.001298	0.372814		0.4631	97.7	97.6	168
	3846	1.80*	3258.24178774	0.000004	0.372126		0.4624	48.8	49.8	168
	3935	2.90*	3258.25311944	0.000726	0.370914		0.4596	53.5	74.2	168
	2552	0.30*	3258.26358795	0	0.379686		0.4657	25.6	25.9	168
	5394	3.40*	3258.26882630	0.000727	0.370793		0.4584	98.8	98.0	168
	3750	1.70*	3258.27038695	0.000655	0.367979		0.4594	100.0	97.6	168
	4339	1.50*	3258.27058422	0.000564	0.367010		0.4597	96.5	97.7	168
	5915	1.05*	3258.27683885	0.000745	0.372503		0.4593	98.8	97.0	167
	3497	0.40*	3258.28163525	0.000569	0.372030		0.4629	64.0	83.2	168
	5225	0.20*	3258.28558850	0.000690	0.374568		0.4596	62.8	78.9	168
	9177	1.05*	3258.28614166	0.000563	0.369638		0.4603	98.8	97.8	166
	2981	2.30*	3258.28694076	0.000500	0.372375		0.4637	97.7	97.9	168
	3000	1.25*	3258.30567932	0.000833	0.369457		0.4617	98.8	97.3	168
	4136	0.65*	3258.32740180	0.000280	0.373141		0.4601	62.8	78.2	168
	8679	0.15*	3258.33606379	0.000876	0.365530		0.4608	98.8	97.3	168
87	1836	10.50*	3337.00075002	0.000755	0.374486	0.393513	0.4585	49.4	49.9	170
	3117	37.75*	3337.00264299	0.000858	0.370122		0.4576	97.7	97.6	170
	2725	14.85*	3337.01479783	0.000957	0.370999		0.4576	98.9	96.9	170
	2033	9.85*	3337.02117824	0.001104	0.371262		0.4624	49.4	50.0	170
	3487	2.80	3337.02362435	0.000711	0.371217		0.4562	55.2	51.2	166
	3393	5.95*	3337.02502927	0.000768	0.369781		0.4603	62.1	77.2	168
	2965	0.80*	3337.03446331	0.000075	0.375119		0.4602	57.5	74.9	170
	4297	5.50*	3337.04552047	0.000594	0.371494		0.4607	50.6	49.9	170
	1772	0.15	3337.05086065	0	0.375040		0.4614	17.2	21.6	170
	2351	0.80*	3337.05185933	0.000857	0.371193		0.4594	50.6	61.9	170
	3236	3.55*	3337.05440265	0.000767	0.371009		0.4583	97.7	97.5	170
	3587	1.75*	3337.06733548	0.000644	0.367472		0.4579	97.7	97.2	169
	3510	3.10*	3337.07025805	0.000394	0.368212		0.4596	98.9	97.5	170
	2789	0.90	3337.07431278	0.001329	0.376930		0.4605	33.3	34.8	168
	20000	0.15*	3337.08185385	0.000014	0.374393		0.4580	49.4	48.8	170
	18681	0.35*	3337.08619885	0.000651	0.367931		0.4554	94.3	96.6	170

TABLE VIII (*Continued*)

N	Average iterations	Frequency %	Coulomb energy	Dipole moment	Coulomb angle	Tammes angle	Hole angle	% Energy diversity	% Angular diversity	Faces N_f
	6771	0.10*	3337.08860953	0	0.372228		0.4586	17.2	17.2	170
	2499	0.35*	3337.12378055	0.000641	0.368926		0.4627	50.6	49.9	170
	5071	0.35*	3337.13487826	0.000496	0.367306		0.4591	72.4	85.6	168
	9751	0.10	3337.14268447	0.000316	0.366066		0.4590	56.3	56.2	168
	4413	0.05	3337.18362893	0.003243	0.356901		0.4547	82.8	86.2	167
	7587	0.15*	3337.19679142	0.001123	0.368213		0.4536	74.7	88.1	170
88	1490	30.65*	3416.72019676	0	0.374994	0.392138	0.4607	25.0	27.6	172
	2017	15.35*	3416.73289041	0.000020	0.371882		0.4581	48.9	53.0	172
	1842	5.10*	3416.73291048	0.000144	0.373766		0.4573	34.1	33.2	172
	1679	1.05*	3416.73691630	0	0.376714		0.4565	9.1	9.5	172
	2637	3.25*	3416.74783597	0.000051	0.372089		0.4574	34.1	32.9	172
	4292	11.40*	3416.77258601	0.000771	0.370835		0.4578	100.0	97.0	172
	4805	13.95*	3416.77625791	0.000889	0.370321		0.4564	100.0	97.4	172
	6811	9.10*	3416.77838595	0.000001	0.372959		0.4565	38.6	62.4	168
	5142	7.15*	3416.78202265	0.000900	0.367094		0.4572	98.9	97.1	172
	2957	0.40*	3416.81629359	0.000686	0.367167		0.4552	51.1	65.4	172
	3871	1.15*	3416.82016056	0.000867	0.366852		0.4557	97.7	97.4	172
	7510	0.50*	3416.82931948	0.001497	0.366057		0.4552	96.6	97.7	172
	3374	0.55*	3416.84374846	0.000239	0.368103		0.4593	50.0	50.0	172
	2082	0.20*	3416.86252342	0.000195	0.368193		0.4587	47.7	60.5	172
	2861	0.05	3416.91071105	0.002782	0.357957		0.4540	98.9	97.4	172
	9667	0.05	3416.93710995	0.001651	0.364377		0.4481	98.9	97.3	172
	2792	0.05	3416.94511277	0.002845	0.362872		0.4533	100.0	97.6	172
	4248	0.05	3416.96926530	0.001848	0.362298		0.4548	95.5	97.5	172
89	4297	40.30*	3497.43901863	0.000071	0.369698	0.389498	0.4563	50.6	49.8	174
	3948	20.10*	3497.45922155	0.000444	0.369944		0.4531	98.9	97.1	174
	5382	0.95*	3497.48136472	0.000246	0.369170		0.4540	48.3	49.8	174
	2481	4.45*	3497.48778781	0.000712	0.370412		0.4555	52.8	65.0	174
	7108	9.05*	3497.49123325	0.000647	0.365935		0.4536	97.8	97.3	173
	2363	0.05	3497.49124830	0.000668	0.365652		0.4537	97.8	97.3	172
	4174	3.65*	3497.49643601	0.000644	0.372884		0.4546	57.3	77.1	174
	4072	10.20*	3497.49781554	0.000599	0.364690		0.4544	100.0	97.0	174
	4161	1.75*	3497.49806433	0.001277	0.368111		0.4525	62.9	76.0	174
	2991	2.40*	3497.50964066	0.000835	0.365823		0.4557	50.6	58.9	174
	3473	1.45*	3497.51520013	0.000788	0.366995		0.4519	97.8	97.4	174
	4100	0.05	3497.52108788	0.001096	0.372382		0.4553	59.6	75.2	174

	3082	1.05*	3497.52321530	0.000495	0.367655		0.4546	49.4	62.4	174
	3566	0.90*	3497.55221987	0.000726	0.365735		0.4536	50.6	49.7	174
	3138	2.10*	3497.55978555	0.000533	0.365594		0.4527	97.8	97.1	174
	4114	0.25*	3497.56016553	0.000631	0.367741		0.4533	64.0	74.4	174
	1704	0.10	3497.56437230	0.000599	0.370195		0.4558	57.3	50.9	174
	3126	0.45*	3497.58896492	0.001931	0.362087		0.4555	98.9	97.3	174
	7504	0.25*	3497.60394456	0.001580	0.357530		0.4557	97.8	97.0	174
	8698	0.50*	3497.61268734	0.001836	0.358466		0.4471	98.9	97.3	173
90	1776	24.90*	3579.09122272	0	0.370538	0.386661	0.4557	16.7	19.3	176
	11403	18.45*	3579.12767705	0.000074	0.365325		0.4518	85.6	94.0	172
	2338	12.25*	3579.12846222	0.000048	0.367046		0.4532	50.0	50.0	176
	11803	1.50*	3579.17005864	0.000623	0.365565		0.4535	87.8	94.1	174
	4637	11.00*	3579.17029217	0.000974	0.363114		0.4515	95.6	97.2	176
	3823	5.75*	3579.17065197	0.000939	0.365783		0.4503	98.9	97.2	176
	4174	1.75*	3579.17673789	0	0.366697		0.4524	26.7	39.3	176
	19207	6.20*	3579.18145257	0.000033	0.365676		0.4506	94.4	96.6	176
	4266	0.85	3579.18243002	0.001140	0.369312		0.4529	72.2	81.3	176
	4477	3.40*	3579.19150205	0.001428	0.364526		0.4529	98.9	97.5	176
	3473	0.60	3579.19355887	0.001843	0.357642		0.4458	57.8	56.9	174
	1978	2.00*	3579.19477433	0.000826	0.363556		0.4531	50.0	58.4	176
	3027	2.00*	3579.19786368	0.000662	0.362691		0.4499	97.8	97.6	176
	3343	1.15*	3579.19863938	0.001108	0.367189		0.4512	98.9	97.6	176
	8549	0.50	3579.19907772	0.001254	0.368101		0.4536	62.2	77.7	174
	4497	2.15*	3579.19964436	0.000946	0.362293		0.4525	100.0	97.2	176
	4882	1.20*	3579.20381518	0.000667	0.359699		0.4516	98.9	97.5	175
	7147	0.15*	3579.20634325	0.000822	0.361884		0.4516	98.9	97.1	176
	2276	1.45*	3579.21087514	0.000841	0.359267		0.4520	100.0	96.8	176
	1822	0.25	3579.22252725	0.000739	0.363807		0.4560	32.2	33.2	176
	5565	0.60*	3579.22894703	0.001375	0.356951		0.4518	98.9	97.2	175
	2469	0.10*	3579.22985701	0	0.367525		0.4542	25.6	42.2	176
	3014	0.10	3579.24035346	0.000663	0.358755		0.4489	50.0	65.5	176
	3825	0.05	3579.24117875	0.001173	0.361032		0.4523	100.0	97.8	176
	3552	1.55*	3579.24720647	0.001563	0.360988		0.4516	97.8	97.4	175
	4372	0.10	3579.31345102	0.000960	0.359610		0.4551	55.6	50.9	176
91	1813	51.60*	3661.71369932	0.000033	0.368345	0.381338	0.4517	50.5	60.5	178
	3471	14.75*	3661.73060424	0.000070	0.365807		0.4498	61.5	78.0	178
	2577	10.75*	3661.74414545	0.000223	0.365919		0.4499	50.5	66.3	178

TABLE VIII (*Continued*)

N	Average iterations	Frequency %	Coulomb energy	Dipole moment	Coulomb angle	Tammes angle	Hole angle	% Energy diversity	% Angular diversity	Faces N_f
	2363	9.05*	3661.79159223	0.000942	0.363419		0.4496	97.8	97.3	178
	6112	3.15*	3661.79263713	0.001204	0.362612		0.4503	50.5	49.9	178
	3103	2.25*	3661.80392012	0.000980	0.363667		0.4527	52.7	68.4	178
	2633	2.35	3661.80949650	0.000972	0.362586		0.4513	54.9	51.1	178
	2082	0.60*	3661.81797984	0.000267	0.368668		0.4513	50.5	56.4	178
	5209	1.75*	3661.82455480	0.000941	0.364284		0.4471	97.8	97.6	178
	2479	0.55	3661.83475823	0.000815	0.363118		0.4500	56.0	57.7	178
	4778	0.45*	3661.84030385	0.001125	0.354871		0.4489	100.0	97.7	177
	5567	0.25*	3661.85032596	0.000715	0.361957		0.4509	70.3	84.7	178
	8358	1.30*	3661.86007289	0.001042	0.358145		0.4491	98.9	97.6	178
	14052	0.30*	3661.88538438	0.002588	0.354657		0.4450	95.6	97.5	178
	1236	0.05	3661.88539500	0.002570	0.355246		0.4449	97.8	97.1	178
	13450	0.20*	3661.89681239	0.001013	0.355467		0.4486	98.9	97.2	177
	4817	0.05	3661.90331648	0.001947	0.355836		0.4452	96.7	97.3	178
	3194	0.15*	3661.90669487	0.001062	0.359267		0.4494	96.7	97.4	178
	20000	0.20*	3661.90686604	0.001758	0.358115		0.4429	98.9	97.1	175
	4433	0.20*	3661.94247551	0.002006	0.355772		0.4492	97.8	97.1	176
	3379	0.05	3662.00817185	0.000034	0.358704		0.4476	35.2	62.9	178
92	2256	28.10*	3745.29163624	0	0.366970	0.380107	0.4512	25.0	25.7	180
	1946	13.60*	3745.32218334	0.000023	0.365232		0.4504	48.9	50.0	180
	1968	9.40*	3745.32555284	0.000227	0.363114		0.4490	48.9	60.5	180
	1949	6.90*	3745.33860835	0.000466	0.363444		0.4485	50.0	50.1	180
	3287	6.70*	3745.36351634	0.001062	0.360649		0.4486	97.8	97.0	180
	2576	0.15	3745.36772048	0	0.369486		0.4487	10.9	15.0	180
	7273	6.05*	3745.38100894	0.001837	0.353498		0.4411	97.8	97.2	180
	4231	7.80*	3745.38177221	0.001112	0.354172		0.4443	94.6	96.9	180
	2949	2.30*	3745.38212626	0.000809	0.361606		0.4477	50.0	49.9	180
	4784	0.95*	3745.38297242	0.000764	0.362623		0.4479	98.9	97.6	180
	9572	0.70*	3745.38713069	0	0.370293		0.4483	21.7	33.3	180
	8056	4.55*	3745.38930840	0.001010	0.355401		0.4435	98.9	96.8	180
	3404	3.10*	3745.38943470	0.000940	0.362068		0.4468	97.8	96.8	180
	2972	2.65*	3745.39152813	0.000684	0.363450		0.4477	52.2	69.5	180
	3519	4.05*	3745.39598792	0.000796	0.359341		0.4484	97.8	97.4	180
	3955	0.35*	3745.40796449	0.000293	0.358002		0.4496	51.1	73.1	180
	3589	0.60	3745.40999059	0.000263	0.365047		0.4488	57.6	50.8	180
	5133	0.25*	3745.41663889	0.000909	0.362572		0.4472	67.4	79.9	178
	4972	0.15	3745.42597609	0.000464	0.357381		0.4482	78.3	88.6	180

	6210	0.25*	3745.43377765	0.000385	0.361479		0.4483	50.0	49.7	180
	4037	0.55*	3745.46354936	0.001120	0.357536		0.4462	100.0	97.5	180
	3062	0.05	3745.48063056	0.000593	0.359309		0.4470	64.1	76.6	180
	4430	0.20*	3745.48989730	0.002360	0.352056		0.4424	100.0	97.4	178
	18325	0.20	3745.50315713	0.001691	0.359301		0.4414	82.6	88.6	178
	3312	0.10*	3745.50321893	0.001051	0.350216		0.4475	97.8	96.9	180
	3344	0.10*	3745.51089154	0.002534	0.346900		0.4460	98.9	97.6	179
	3552	0.05	3745.52379435	0.001989	0.350982		0.4456	100.0	97.6	180
	2109	0.05	3745.57536192	0.002598	0.339922		0.4488	50.0	49.9	180
93	3182	44.65*	3829.84433842	0.000213	0.362167	0.375872	0.4477	59.1	70.7	182
	6639	15.50	3829.85868994	0.000164	0.360397		0.4435	60.2	69.1	180
	2508	4.00	3829.90619684	0.000523	0.363138		0.4431	32.3	26.6	182
	8282	8.65*	3829.91210881	0.000825	0.353932		0.4468	98.9	97.1	181
	2572	5.10*	3829.91995003	0.001253	0.359324		0.4432	98.9	97.4	182
	6742	1.50*	3829.92076446	0	0.362784		0.4463	17.2	29.4	182
	2225	3.80*	3829.93282075	0.001393	0.364055		0.4450	49.5	54.3	182
	4030	4.40*	3829.93683327	0.001717	0.356058		0.4443	98.9	97.0	180
	2890	2.80*	3829.93726481	0.001055	0.352693		0.4446	96.8	97.2	182
	6625	1.40	3829.94200079	0.001630	0.356002		0.4422	63.4	70.5	180
	2733	1.05*	3829.94230874	0.001116	0.351249		0.4452	97.8	97.0	182
	2617	1.05*	3829.94773193	0.001301	0.357394		0.4432	98.9	97.5	182
	10082	0.60*	3829.96290203	0.000183	0.356474		0.4458	86.0	91.2	180
	10392	0.65*	3829.96620757	0.001237	0.358821		0.4401	94.6	95.0	180
	2971	0.95*	3829.97421741	0.000776	0.356428		0.4452	98.9	97.4	182
	7000	0.70*	3829.97815077	0.002412	0.351322		0.4397	97.8	96.8	182
	2336	0.40*	3829.98564216	0.000300	0.360589		0.4460	50.5	65.7	182
	4924	0.20*	3829.98611334	0.001385	0.361367		0.4440	97.8	96.9	182
	3938	1.40*	3829.99329087	0.001490	0.347258		0.4447	98.9	96.8	182
	5630	0.10	3830.00626820	0.001435	0.353843		0.4460	97.8	97.0	181
	5166	0.05	3830.02249121	0.001789	0.355793		0.4441	50.5	49.8	182
	5158	0.10	3830.02481672	0.001418	0.344687		0.4453	96.8	97.3	182
	2722	0.05	3830.06290088	0.001280	0.354053		0.4415	100.0	97.0	182
	7179	0.05	3830.07937624	0.001795	0.351081		0.4444	97.8	97.5	181
	6637	0.10	3830.12098837	0.000302	0.359076		0.4490	53.8	47.4	182
94	4917	43.50*	3915.30926962	0	0.365679	0.373440	0.4471	24.5	29.4	184
	3249	9.90	3915.33763030	0.000260	0.364033		0.4425	66.0	77.0	184

TABLE VIII (*Continued*)

N	Average iterations	Frequency %	Coulomb energy	Dipole moment	Coulomb angle	Tammes angle	Hole angle	% Energy diversity	% Angular diversity	Faces N_f
	1878	9.10*	3915.35216101	0.000505	0.361205		0.4448	34.0	39.6	184
	2290	5.20*	3915.41514225	0.000467	0.357315		0.4413	48.9	50.4	184
	4017	8.70*	3915.42460992	0.000958	0.359197		0.4413	97.9	97.3	184
	2373	4.15*	3915.42890409	0.000798	0.354464		0.4409	98.9	97.2	184
	5381	7.05*	3915.43908332	0.000703	0.350277		0.4422	98.9	96.8	184
	7796	2.05*	3915.44315884	0.001099	0.360016		0.4426	57.4	74.6	182
	4799	1.25*	3915.44481487	0.001392	0.355902		0.4447	97.9	97.3	184
	4409	1.10*	3915.45027523	0.000835	0.351244		0.4400	98.9	97.1	183
	5114	0.70	3915.45159808	0.002106	0.355880		0.4335	56.4	61.0	184
	5434	0.60	3915.45281248	0.000827	0.358784		0.4403	62.8	71.7	184
	2856	1.10*	3915.45967717	0.000052	0.360993		0.4452	50.0	70.6	184
	10741	0.30	3915.46280175	0.000456	0.355575		0.4437	98.9	97.2	182
	5489	0.10	3915.47416587	0.001327	0.354867		0.4423	95.7	95.3	184
	2324	0.05	3915.47460951	0.000953	0.358904		0.4439	48.9	64.3	184
	4134	0.95*	3915.47657801	0.001302	0.353930		0.4432	100.0	97.3	184
	1632	0.05	3915.47973753	0.000431	0.361005		0.4449	33.0	44.8	184
	4300	0.50*	3915.49216438	0.001251	0.351729		0.4397	97.9	97.0	184
	2955	0.20*	3915.49266383	0.002485	0.354183		0.4424	98.9	97.3	184
	3738	0.25	3915.49293708	0.001363	0.353588		0.4434	97.9	97.1	184
	2233	0.70*	3915.49372578	0.002351	0.354794		0.4394	97.9	97.2	184
	14961	0.40*	3915.49662960	0.000960	0.356566		0.4405	91.5	93.3	184
	3617	0.25	3915.49763862	0.000298	0.355783		0.4390	58.5	78.3	184
	9377	0.10*	3915.49815180	0.001574	0.351081		0.4420	96.8	97.5	184
	2446	0.25*	3915.50141193	0.001366	0.352030		0.4415	100.0	97.3	184
	5043	0.15	3915.50394770	0.001829	0.348852		0.4372	76.6	82.1	184
	3588	0.10	3915.50617610	0.001480	0.356503		0.4420	48.9	49.8	184
	7130	0.15	3915.51252701	0.001202	0.356538		0.4405	100.0	97.3	184
	4173	0.05	3915.51407010	0.000501	0.353108		0.4423	98.9	97.4	184
	3005	0.40*	3915.52008036	0.002364	0.351132		0.4370	96.8	96.8	184
	2754	0.10*	3915.52356293	0.002761	0.352597		0.4446	50.0	61.5	184
	6624	0.05	3915.52886015	0	0.358211		0.4464	20.2	46.2	180
	7251	0.10*	3915.54319693	0.002329	0.351277		0.4371	96.8	97.4	184
	3839	0.10	3915.54720228	0.002455	0.344401		0.4399	100.0	97.1	184

95	3146	56.95*	4001.77167557	0.000117	0.361479	0.369935	0.4400	53.7	65.9	186
	2965	23.90*	4001.80663427	0.000408	0.358976		0.4408	51.6	63.2	186
	4554	7.75*	4001.81894117	0.000682	0.358154		0.4405	64.2	81.5	184
	8736	3.70	4001.84343013	0.000106	0.357665		0.4396	80.0	86.6	186
	4092	0.55*	4001.87348103	0	0 362345		0.4438	18.9	40.7	186
	8716	3.85*	4001.88717904	0.002060	0.349877		0.4386	96.8	96.8	185
	5071	0.60*	4001.90696584	0.000871	0.356407		0.4393	97.9	96.6	186
	3155	0.25	4001.92060379	0.000499	0.355199		0.4404	49.5	49.8	186
	5270	0.30*	4001.92245595	0.001013	0.351863		0.4419	66.3	84.4	186
	5163	0.30*	4001.93659630	0.000770	0.356533		0.4394	100.0	96.8	184
	4763	0.65*	4001.94231614	0.001961	0.348660		0.4423	34.7	54.7	186
	3161	0.15*	4001.95525848	0.000518	0.350565		0.4428	49.5	49.7	186
	2788	0.20*	4001.96789955	0.001285	0.348385		0.4398	50.5	49.6	186
	10409	0.10	4001.96821304	0.002227	0.354032		0.4420	50.5	49.9	186
	15400	0.15*	4001.96821587	0.002195	0.352959		0.4417	49.5	49.8	185
	5496	0.15	4001.97185179	0.000973	0.354439		0.4399	71.6	79.4	184
	2724	0.15	4001.97930328	0.000284	0.354455		0.4389	49.5	59.6	186
	3067	0.05	4002.00545579	0.000861	0.355418		0.4399	100.0	96.9	186
	3528	0.05	4002.00636613	0.000831	0.356324		0.4411	100.0	95.4	186
	2763	0.10*	4002.00958028	0.001681	0.348189		0.4399	97.9	96.5	186
	5173	0.05	4002.01386735	0.002842	0.340982		0.4366	82.1	86.1	185
	10444	0.05	4002.01893501	0.000079	0.347362		0.4417	49.5	49.7	186
96	10339	56.20*	4089.15401006	0.000036	0.361049	0.368544	0.4377	49.0	49.5	186
	7508	1.90*	4089.15425120	0.000060	0.360714		0.4377	84.4	91.0	184
	15016	1.15*	4089.18961607	0.000003	0.363750		0.4378	24.0	25.4	188
	4043	6.45*	4089.19508088	0	0.359195		0.4404	17.7	37.4	188
	2621	9.15*	4089.22051843	0.000236	0.355515		0.4376	53.1	66.1	188
	7348	8.10*	4089.25754163	0.000513	0.357001		0.4368	95.8	96.7	187
	2015	0.55*	4089.25810401	0	0.357872		0.4385	8.3	9.7	188
	20000	0.95*	4089.27994849	0.000625	0.349075		0.4331	100.0	96.6	188
	3156	1.60*	4089.29813692	0.001006	0.346557		0.4395	60.4	76.4	186
	3560	1.65*	4089.31409559	0.000818	0.351814		0.4376	100.0	96.9	188
	4514	0.25	4089.32601764	0.002184	0.345804		0.4369	71.9	82.0	188
	4944	0.40*	4089.32884800	0.000975	0.349171		0.4378	99.0	96.8	188
	3423	0.90*	4089.32914181	0.000298	0.351588		0.4371	99.0	96.7	188
	3059	0.90*	4089.32934173	0.000128	0.355321		0.4382	52.1	73.5	188
	3245	0.70*	4089.32937644	0.002078	0.343830		0.4334	97.9	96.4	188
	10229	0.45*	4089.33178663	0.001262	0.355442		0.4363	100.0	97.5	187
	4143	0.20*	4089.36465058	0.000271	0.354412		0.4385	65.6	80.1	188
	7069	0.05	4089.39134612	0.000601	0.347028		0.4384	99.0	97.1	188

TABLE VIII (*Continued*)

N	Average iterations	Frequency %	Coulomb energy	Dipole moment	Coulomb angle	Tammes angle	Hole angle	% Energy diversity	% Angular diversity	Faces N_f
	2039	0.05	4089.39260339	0.000238	0.353831		0.4388	49.0	53.4	188
	12704	0.05	4089.41039115	0.001791	0.354552		0.4381	50.0	47.5	188
	2010	0.05	4089.41837532	0.002204	0.347503		0.4313	100.0	97.1	188
	10089	0.30*	4089.41857907	0.000560	0.349882		0.4403	77.1	89.9	188
	5646	0.05	4089.42100567	0.001137	0.349877		0.4399	67.7	79.6	188
	2419	0.05	4089.47265364	0.002235	0.343688		0.4338	100.0	97.6	188
97	10458	90.20*	4177.53359963	0.000096	0.356912	0.366315	0.4353	69.1	81.3	190
	6950	5.60*	4177.59058909	0.000319	0.354241		0.4345	96.9	96.8	188
	5341	1.95*	4177.61875516	0.000443	0.353788		0.4376	44.3	70.0	190
	2406	0.95*	4177.64356494	0.000414	0.354661		0.4338	49.5	49.8	190
	5597	0.10*	4177.66799753	0.002142	0.345708		0.4305	97.9	96.6	190
	2035	0.40*	4177.71819311	0.000072	0.352609		0.4360	49.5	49.8	190
	5244	0.15*	4177.72270181	0.001956	0.343084		0.4312	97.9	97.2	190
	4572	0.10*	4177.73376330	0.001032	0.350770		0.4323	99.0	97.0	190
	3275	0.10	4177.74816006	0.001981	0.341206		0.4315	96.9	97.2	190
	3661	0.15*	4177.75166330	0.001118	0.345083		0.4379	49.5	49.8	190
	8617	0.05	4177.77558985	0.002416	0.341406		0.4296	99.0	97.0	189
	2697	0.05	4177.83805800	0.002598	0.346954		0.4332	95.9	97.0	190
98	3420	51.95*	4266.82246416	0.000113	0.356424	0.365017	0.4355	50.0	49.7	192
	2812	10.85*	4266.82617987	0.000010	0.356512		0.4370	17.3	23.5	192
	2157	8.35*	4266.83001454	0.000005	0.355677		0.4342	25.5	32.1	192
	2514	15.45*	4266.83424531	0.000073	0.356879		0.4339	50.0	49.7	192
	16766	3.45*	4266.83546709	0.000175	0.356828		0.4333	50.0	54.8	192
	1891	0.15	4266.86496975	0.000002	0.354672		0.4349	12.2	9.2	192
	4714	5.20*	4266.88905193	0.000361	0.353096		0.4323	65.3	80.8	192
	1540	0.70	4266.90771087	0.000107	0.356083		0.4323	32.7	27.9	192
	3776	0.30*	4266.94010421	0	0.352888		0.4360	17.3	17.1	192
	5620	0.30*	4266.99141130	0.001695	0.348640		0.4346	50.0	49.5	192
	6272	1.75*	4266.99290589	0.001817	0.346274		0.4323	99.0	97.0	192
	2555	0.40*	4267.02331924	0.000211	0.352952		0.4336	52.0	73.0	192
	3884	0.20*	4267.03957642	0.001741	0.340284		0.4304	99.0	96.6	192
	3170	0.05	4267.04219229	0.002046	0.341199		0.4311	100.0	97.3	192
	3763	0.20*	4267.04360487	0.001124	0.346337		0.4335	100.0	96.5	191
	7524	0.10	4267.05612322	0.002518	0.344810		0.4295	98.0	96.5	192
	3776	0.15*	4267.05691148	0.001886	0.340080		0.4321	100.0	97.2	192
	2887	0.05	4267.07313307	0.003289	0.346711		0.4294	50.0	49.4	192
	3007	0.20*	4267.09421053	0.000376	0.347308		0.4317	99.0	97.0	192
	3245	0.10*	4267.11281169	0.001422	0.341775		0.4319	100.0	96.6	191
	6130	0.05	4267.11376752	0.001001	0.347395		0.4337	87.8	92.4	192
	6586	0.05	4267.13819960	0.001122	0.341793		0.4346	99.0	97.2	192

99	14180	81.95*	4357.13916314	0.000157	0.354032	0.363478	0.4333	49.5	49.6	194
	8978	8.30*	4357.17268162	0.000332	0.352303		0.4324	97.0	97.1	194
	5779	3.30	4357.18495891	0.000852	0.351978		0.4312	58.6	63.6	194
	2252	1.05*	4357.20285542	0.000387	0.353588		0.4313	33.3	42.5	194
	3721	0.20	4357.25433676	0.001464	0.346708		0.4319	54.5	50.5	194
	4163	1.75	4357.25694427	0.000799	0.350033		0.4292	57.6	57.0	190
	5514	0.20	4357.26204992	0.000926	0.344357		0.4317	75.8	83.8	194
	5939	1.65	4357.26865095	0.000480	0.343308		0.4317	57.6	70.7	192
	3646	0.45*	4357.27634050	0.000940	0.344401		0.4300	98.0	96.7	194
	4815	0.25	4357.32511953	0.001487	0.347475		0.4319	47.5	43.6	194
	3889	0.25*	4357.32893422	0.002696	0.345178		0.4286	97.0	96.8	193
	3882	0.20	4357.33023200	0.001586	0.348309		0.4324	56.6	50.4	192
	2741	0.05	4357.35477108	0.002329	0.342488		0.4284	100.0	96.6	194
	4303	0.10*	4357.35789854	0.001819	0.342169		0.4291	99.0	96.9	191
	4149	0.15*	4357.37036331	0.001641	0.340640		0.4285	99.0	96.8	194
	11928	0.05	4357.37577617	0.002043	0.344761		0.4226	92.9	93.6	192
	2776	0.10	4357.46779855	0.000954	0.343514		0.4281	100.0	96.9	194
100	6632	43.75*	4448.35063434	0	0.354240	0.361806	0.4329	15.0	19.6	196
	4451	22.35	4448.41042065	0.000548	0.351242		0.4308	63.0	71.9	196
	1985	1.95*	4448.42088460	0	0.356163		0.4333	9.0	10.4	196
	2461	1.80*	4448.43456411	0.000232	0.351233		0.4302	49.0	50.7	196
	2746	1.10*	4448.43664953	0.000797	0.348810		0.4293	34.0	33.0	196
	12634	6.80*	4448.44720777	0.000610	0.346536		0.4332	50.0	49.6	194
	7183	2.00*	4448.44834304	0.000115	0.350778		0.4294	95.0	94.8	196
	2729	3.05*	4448.46648310	0.000681	0.346444		0.4282	100.0	96.7	196
	3908	1.75*	4448.46871243	0	0.349249		0.4302	25.0	40.8	196
	3536	4.15*	4448.47370541	0.000100	0.346810		0.4289	50.0	62.4	196
	1910	1.30*	4448.47440810	0.000596	0.349473		0.4293	50.0	49.6	196
	7079	4.75*	4448.47538457	0.000627	0.346445		0.4283	98.0	96.5	196
	4940	0.25	4448.47902256	0.001645	0.348765		0.4280	29.0	37.5	196
	3579	0.55*	4448.48018390	0.000054	0.346106		0.4303	50.0	68.9	196
	5121	0.85*	4448.48730379	0.000677	0.348757		0.4296	49.0	54.3	196
	5750	0.75*	4448.48960197	0.000466	0.349570		0.4292	49.0	59.6	196

TABLE VIII (*Continued*)

N	Average iterations	Frequency %	Coulomb energy	Dipole moment	Coulomb angle	Tammes angle	Hole angle	% Energy diversity	% Angular diversity	Faces N_f
	3614	1.05*	4448.49667946	0.000878	0.343913		0.4281	100.0	96.8	196
	3366	0.45*	4448.50686302	0.000312	0.346927		0.4302	50.0	49.8	196
	3438	0.10	4448.54040487	0.000940	0.342441		0.4260	62.0	74.0	196
	5421	0.05	4448.56301774	0.001107	0.339114		0.4288	99.0	96.0	196
	12096	0.20*	4448.57164151	0.001349	0.339049		0.4285	97.0	96.9	195
	2441	0.05	4448.57479340	0.002504	0.338709		0.4298	63.0	73.5	196
	3806	0.05	4448.57642589	0.001049	0.340203		0.4290	66.0	81.2	196
	3372	0.05	4448.59743687	0.002539	0.341857		0.4276	75.0	86.6	196
	3653	0.15	4448.61021879	0.001183	0.342874		0.4253	99.0	96.7	195
	10105	0.10	4448.62161144	0.002121	0.332642		0.4260	98.0	96.5	195
	7394	0.05	4448.62525087	0.001769	0.338185		0.4251	97.0	96.8	196
	3762	0.15*	4448.62928026	0.001209	0.334653		0.4252	98.0	97.1	196
	3665	0.05	4448.63317178	0.001823	0.339986		0.4263	96.0	97.2	196
	5249	0.10	4448.66828993	0.000779	0.342168		0.4260	99.0	96.9	195
	3530	0.05	4448.66909039	0.001739	0.336776		0.4265	100.0	96.9	196
	9861	0.05	4448.67654431	0.001400	0.339756		0.4267	99.0	96.6	196
	2745	0.05	4448.70899783	0.001896	0.335485		0.4258	99.0	96.8	195
	2709	0.05	4448.78255856	0.003722	0.334545		0.4228	99.0	97.1	196
101	3628	40.75*	4540.59005170	0	0.349259		0.4287	17.8	36.2	198
	4832	13.30*	4540.65878548	0.000762	0.346173		0.4284	96.0	96.5	198
	3452	4.10*	4540.67532223	0.000754	0.345065		0.4279	99.0	96.8	197
	4611	2.00*	4540.67666960	0.000736	0.343724		0.4260	99.0	97.6	197
	4588	0.70	4540.67798465	0.000728	0.348626		0.4288	41.6	44.1	198
	14862	5.00*	4540.68233258	0.001700	0.344000		0.4264	90.1	90.9	198
	7956	4.25*	4540.68572575	0.001179	0.343152		0.4262	98.0	96.8	198
	8304	1.90*	4540.68669238	0.001057	0.343349		0.4262	97.0	96.5	196
	5241	2.50*	4540.69461564	0.001797	0.340756		0.4262	98.0	96.3	196
	6561	1.85*	4540.69602516	0.001118	0.343969		0.4280	56.4	74.8	198
	2065	0.70*	4540.70249096	0.000820	0.346209		0.4289	50.5	59.2	198
	7108	1.35*	4540.70405116	0.001311	0.341813		0.4280	96.0	96.5	198
	3798	2.20*	4540.70410867	0.001225	0.341380		0.4286	100.0	97.0	197
	12931	2.35*	4540.70507628	0.001346	0.338717		0.4265	99.0	96.5	198
	4280	1.55*	4540.70634842	0.001777	0.343050		0.4252	96.0	96.8	198
	8735	3.30*	4540.70663798	0.000750	0.342249		0.4278	100.0	97.0	197
	2487	1.50	4540.70701619	0.001192	0.339750		0.4290	56.4	50.2	198
	9679	0.90*	4540.70780726	0.001562	0.340204		0.4269	97.0	96.5	198
	10634	1.60	4540.71126420	0.000589	0.343480		0.4293	57.4	57.8	198

	4236	0.65*	4540.71328571	0.000828	0.341285	0.4269	97.0	96.6	198
	4249	0.55*	4540.71537082	0.000576	0.344120	0.4274	94.1	97.4	198
	6848	0.10	4540.71609656	0.001220	0.344720	0.4291	97.0	95.0	195
	5481	0.85*	4540.72616655	0.002221	0.342467	0.4291	98.0	96.3	198
	15245	0.65*	4540.73175426	0.001360	0.338741	0.4262	100.0	96.6	198
	2186	0.45*	4540.73643501	0.001409	0.337389	0.4262	98.0	96.6	198
	5746	0.30*	4540.74219207	0.001395	0.344564	0.4259	98.0	96.4	198
	7194	0.20*	4540.74368177	0.001127	0.344476	0.4265	99.0	97.0	198
	18839	0.95*	4540.74470982	0.000265	0.344598	0.4248	99.0	96.8	197
	4328	0.40*	4540.75252547	0.001231	0.341404	0.4268	99.0	96.9	198
	4120	0.75*	4540.75307454	0.001371	0.341351	0.4261	97.0	96.8	198
	2570	0.10	4540.75459130	0.001067	0.345703	0.4309	33.7	43.1	198
	6314	0.05	4540.76551282	0.001480	0.340515	0.4267	97.0	96.9	198
	3930	0.10	4540.76683798	0.001361	0.339038	0.4270	99.0	97.0	198
	3814	0.30*	4540.76853869	0.002324	0.338765	0.4258	98.0	96.9	198
	2860	0.25*	4540.77391959	0.000312	0.339712	0.4250	100.0	96.8	198
	3322	0.25*	4540.77601345	0.000804	0.344874	0.4302	63.4	81.5	198
	6089	0.10*	4540.78271386	0.001149	0.342709	0.4232	100.0	96.8	198
	7166	0.05	4540.78294548	0.001212	0.342083	0.4271	99.0	97.0	198
	5832	0.20*	4540.78977525	0.001613	0.338516	0.4284	98.0	96.6	198
	2420	0.05	4540.79799071	0.000830	0.345848	0.4295	50.5	70.9	198
	8432	0.05	4540.79800707	0.001583	0.339397	0.4284	96.0	96.8	198
	2620	0.15*	4540.79856369	0.002600	0.335211	0.4210	98.0	96.5	198
	3291	0.25*	4540.82275811	0.001659	0.333337	0.4228	98.0	96.5	198
	7546	0.05	4540.82951697	0.001935	0.341129	0.4245	50.5	49.6	198
	2910	0.05	4540.84548695	0.001689	0.340170	0.4242	99.0	97.0	198
	4638	0.05	4540.85872567	0.002410	0.332157	0.4220	72.3	80.7	197
	3721	0.10*	4540.86780715	0.002192	0.339999	0.4241	100.0	97.0	197
	5498	0.05	4540.92415697	0.002169	0.336749	0.4216	100.0	96.8	198
102	5562	44.85*	4633.73656590	0	0.349762	0.4283	16.7	35.7	200
	3487	0.05	4633.74993810	0	0.349998	0.4274	27.5	34.7	200
	3007	11.35*	4633.83689270	0.000860	0.340877	0.4238	96.1	96.4	200
	15595	1.00	4633.83812704	0.000007	0.345146	0.4229	50.0	47.3	200
	3245	0.25*	4633.83829706	0	0.345712	0.4233	43.1	45.4	196
	3073	2.00*	4633.85199367	0.001859	0.339422	0.4222	97.1	96.7	199
	7064	4.60*	4633.85207449	0.001664	0.338472	0.4227	96.1	96.8	200

TABLE VIII (*Continued*)

N	Average iterations	Frequency %	Coulomb energy	Dipole moment	Coulomb angle	Tammes angle	Hole angle	% Energy diversity	% Angular diversity	Faces N_f
	8993	6.95*	4633.85227866	0.001764	0.338007		0.4233	98.0	97.1	200
	6035	2.60*	4633.85754458	0.000667	0.342413		0.4233	98.0	96.6	200
	3015	0.65*	4633.85931485	0	0.349061		0.4287	24.5	35.9	200
	4434	1.85*	4633.86290383	0.001930	0.337926		0.4228	97.1	96.8	200
	2642	1.75*	4633.86313546	0.000083	0.346016		0.4283	49.0	68.6	200
	3879	2.20*	4633.86358398	0.000677	0.341600		0.4243	98.0	97.3	199
	6259	1.60*	4633.86605294	0.000947	0.343398		0.4259	50.0	49.6	200
	4070	0.05	4633.86666295	0.002084	0.341072		0.4241	98.0	96.4	200
	4424	1.45*	4633.86951464	0.001037	0.337419		0.4256	99.0	96.9	200
	3587	1.85*	4633.87171213	0.001306	0.337476		0.4235	96.1	96.9	199
	5480	0.25*	4633.87266256	0.000809	0.341949		0.4253	70.6	83.9	200
	3411	0.20	4633.87295600	0.001868	0.340708		0.4241	58.8	65.9	199
	7853	0.85*	4633.88106601	0.001077	0.343487		0.4232	98.0	96.7	199
	3133	1.25*	4633.88152617	0.001448	0.341558		0.4256	97.1	96.8	200
	8247	0.20*	4633.88475300	0.000811	0.342239		0.4259	90.2	94.1	200
	5877	0.95*	4633.88698437	0.002330	0.338992		0.4177	97.1	97.0	200
	4837	0.55*	4633.88887689	0.002066	0.342432		0.4252	100.0	97.0	199
	7660	0.05	4633.88888748	0.001135	0.341990		0.4234	89.2	90.1	200
	11195	1.25*	4633.89082450	0.001926	0.345677		0.4258	82.4	91.8	200
	3824	1.60*	4633.89636032	0.000616	0.340606		0.4269	97.1	96.6	200
	3019	0.15*	4633.89906101	0.000999	0.339183		0.4262	99.0	96.3	200
	6732	0.55*	4633.90529547	0.001556	0.339756		0.4265	98.0	96.8	199
	11486	0.20*	4633.91542780	0.000970	0.339636		0.4251	95.1	96.8	199
	20000	0.20*	4633.91548667	0.000818	0.340036		0.4267	96.1	96.7	199
	4593	0.05	4633.92151679	0.001515	0.344350		0.4266	73.5	82.7	200
	1868	0.20*	4633.92154073	0.000656	0.339689		0.4276	49.0	63.4	200
	6057	0.50*	4633.92219304	0.000503	0.339727		0.4216	98.0	97.0	199
	2862	0.05	4633.92448524	0.001445	0.344620		0.4267	50.0	49.9	200
	5120	0.25*	4633.92642546	0.000502	0.340915		0.4245	73.5	82.8	200
	2953	0.05	4633.92675171	0.001141	0.340939		0.4264	64.7	72.3	200
	10611	0.10	4633.92796901	0.001733	0.337318		0.4248	100.0	96.3	200
	3246	0.10*	4633.92901937	0.000508	0.339559		0.4273	49.0	66.8	200
	3379	0.50*	4633.93247439	0.002091	0.335457		0.4194	98.0	96.5	200
	5847	0.40*	4633.93302261	0.000713	0.341365		0.4248	98.0	97.2	199
	5340	0.30*	4633.93360282	0.002304	0.337316		0.4194	99.0	96.4	200
	2327	0.05	4633.94030983	0.000414	0.339619		0.4268	50.0	68.5	200
	16801	0.05	4633.94882459	0.001482	0.341347		0.4261	95.1	94.7	200
	3737	0.05	4633.95962334	0.001724	0.338635		0.4247	99.0	96.6	199

	2344	0.10	4633.97006396	0.003630	0.335170	0.4208	96.1	96.8	200
	2291	0.05	4633.97445084	0	0.340043	0.4271	16.7	26.2	200
	3492	0.20	4633.97457151	0.003064	0.340149	0.4224	99.0	97.0	200
	3758	0.10*	4633.97895241	0.001189	0.340785	0.4238	99.0	97.0	200
	5067	0.05	4633.99094034	0.001769	0.336441	0.4266	75.5	84.5	200
	2409	0.05	4634.00951995	0.000648	0.334067	0.4243	99.0	96.6	198
	8370	0.05	4634.03801133	0.001009	0.339310	0.4184	88.2	91.5	200
	7070	0.05	4634.05539599	0.000962	0.336984	0.4266	99.0	96.8	199
103	4758	57.60*	4727.83661684	0.000201	0.347443	0.4248	50.5	49.4	202
	5130	30.65*	4727.87233500	0.000171	0.347829	0.4240	37.9	57.0	202
	4393	1.65*	4727.96451737	0.000794	0.344944	0.4225	50.5	49.7	202
	2460	1.05*	4727.97788313	0.001201	0.342853	0.4250	50.5	58.2	202
	19089	1.50*	4727.97860150	0.001467	0.344326	0.4216	97.1	96.9	200
	3267	0.05	4727.97861093	0.001488	0.344607	0.4215	96.1	96.2	198
	4423	0.55*	4727.98878377	0.000450	0.342837	0.4228	50.5	49.8	202
	6067	1.20*	4728.00679250	0.000358	0.341955	0.4187	34.0	33.1	202
	20000	0.05	4728.01191498	0.001942	0.337083	0.4239	96.1	96.1	201
	2721	0.95*	4728.01375563	0.000245	0.342511	0.4203	61.2	78.7	202
	3746	0.05	4728.02087435	0.000696	0.337122	0.4223	99.0	96.6	202
	4778	0.80*	4728.02169914	0.000188	0.341271	0.4224	67.0	81.2	202
	12345	1.00*	4728.02343366	0.000763	0.338137	0.4206	99.0	96.5	199
	4557	0.30*	4728.03114014	0.001001	0.340910	0.4218	99.0	96.4	201
	7863	0.15	4728.03201156	0.001489	0.339532	0.4229	74.8	88.5	202
	8450	0.25	4728.04710707	0.000571	0.339761	0.4244	99.0	96.5	200
	2126	0.20*	4728.04851625	0.000489	0.338829	0.4239	50.5	49.7	202
	4988	0.05	4728.05487686	0.001705	0.336514	0.4206	98.1	96.8	202
	10123	0.25*	4728.05733810	0.001016	0.334814	0.4243	100.0	96.7	201
	3005	0.35	4728.05869705	0.000466	0.338715	0.4210	98.1	96.3	202
	4538	0.10	4728.06415117	0.000165	0.340939	0.4237	99.0	96.1	201
	5276	0.15*	4728.06466480	0.001595	0.340494	0.4201	97.1	96.4	201
	3991	0.35*	4728.06784859	0.000830	0.342047	0.4236	98.1	96.6	202
	2580	0.10*	4728.07516653	0.001187	0.334958	0.4208	99.0	96.1	202
	20000	0.05	4728.08206930	0.000490	0.333983	0.4233	98.1	96.3	202
	3408	0.05	4728.09569659	0.001283	0.339427	0.4199	98.1	96.7	202
	3675	0.10*	4728.09795720	0.001141	0.341058	0.4216	99.0	96.6	200
	9033	0.05	4728.10261840	0.000583	0.340430	0.4231	99.0	96.8	201

TABLE VIII (*Continued*)

N	Average iterations	Frequency %	Coulomb energy	Dipole moment	Coulomb angle	Tammes angle	Hole angle	% Energy diversity	% Angular diversity	Faces N_f
	2406	0.10*	4728.11348291	0.001322	0.333478		0.4201	100.0	97.0	202
	5666	0.05	4728.12048167	0.002155	0.333740		0.4231	98.1	97.2	202
	6609	0.05	4728.12107616	0.000920	0.334107		0.4208	100.0	96.6	202
	8637	0.05	4728.12232759	0.000282	0.333524		0.4224	99.0	96.7	201
	2781	0.05	4728.13299914	0.000778	0.335127		0.4228	99.0	96.4	201
	8292	0.05	4728.13599495	0.001045	0.336212		0.4248	79.6	91.7	200
104	4969	41.95*	4822.87652275	0	0.348307		0.4218	10.6	20.8	204
	7229	14.55*	4822.88031733	0	0.346364		0.4209	44.2	46.7	204
	2758	25.80*	4822.92698786	0	0.347422		0.4203	25.0	35.4	204
	3828	3.85	4822.92747363	0	0.347076		0.4237	22.1	22.3	204
	3636	2.60*	4823.01967103	0.000018	0.342357		0.4208	50.0	49.7	204
	6692	6.15*	4823.04488010	0.000611	0.337480		0.4206	97.1	96.2	204
	20000	0.05	4823.05579007	0.001039	0.338858		0.4203	98.1	96.2	202
	3056	0.35*	4823.06318322	0	0.342268		0.4232	26.0	25.4	204
	2087	0.05	4823.07366060	0	0.341980		0.4232	17.3	20.7	204
	5029	0.60*	4823.09919587	0.000697	0.335411		0.4212	98.1	96.6	204
	5319	0.35*	4823.10629624	0.002171	0.329398		0.4182	98.1	96.3	204
	4304	0.55*	4823.11156855	0.002170	0.329781		0.4172	100.0	96.3	204
	3026	0.10*	4823.12721353	0.000192	0.338367		0.4197	50.0	49.5	204
	4532	0.15	4823.12827998	0.000772	0.336981		0.4227	51.9	60.9	202
	4767	0.40*	4823.12955754	0.001213	0.333409		0.4184	98.1	97.0	203
	9419	0.10*	4823.13066751	0.000540	0.328185		0.4221	87.5	88.5	204
	3901	0.10*	4823.13066844	0.001652	0.338055		0.4190	97.1	96.5	204
	11709	0.60*	4823.13238924	0.001000	0.337818		0.4224	100.0	96.6	203
	6272	0.05	4823.13350534	0.001217	0.337101		0.4208	89.4	92.1	202
	4061	0.05	4823.13488062	0.001539	0.333034		0.4188	99.0	96.7	204
	3012	0.20*	4823.13507703	0.000475	0.333717		0.4176	99.0	96.7	204
	3064	0.05	4823.13617395	0.001688	0.334602		0.4170	97.1	96.8	204
	3466	0.15*	4823.13730803	0.001596	0.338416		0.4203	98.1	96.5	204
	7260	0.20	4823.14146505	0.001785	0.333094		0.4154	97.1	96.8	203
	13077	0.15*	4823.14729381	0.000635	0.337112		0.4211	97.1	96.7	204
	6746	0.20*	4823.15027704	0.001212	0.334667		0.4164	95.2	96.3	204
	7936	0.05	4823.15115646	0.000135	0.339055		0.4232	90.4	91.4	200
	4215	0.05	4823.15217494	0.001290	0.336176		0.4195	98.1	96.5	204
	5927	0.05	4823.17391107	0.002039	0.336342		0.4224	99.0	96.8	204
	3175	0.05	4823.17760089	0.001383	0.334108		0.4192	96.2	96.6	203
	5523	0.05	4823.18486438	0.001766	0.336363		0.4188	99.0	96.5	204
	6462	0.05	4823.18812069	0.000634	0.335494		0.4214	98.1	96.7	204

	9175	0.05	4823.18900353	0.001468	0.330288	0.4167	99.0	96.7	204
	3692	0.05	4823.19463495	0.001094	0.334986	0.4207	99.0	96.9	204
	12115	0.05	4823.20788707	0.002344	0.335637	0.4190	98.1	96.8	204
	2228	0.05	4823.21788086	0.002422	0.330432	0.4159	98.1	96.5	204
	17153	0.05	4823.23730313	0.003195	0.332397	0.4162	97.1	95.1	202
	4526	0.05	4823.24022881	0.001672	0.333170	0.4193	97.1	96.8	203
105	2104	2.55*	4919.00063762	0	0.346306	0.4185	17.1	17.0	206
	2463	2.15*	4919.01224999	0	0.346599	0.4215	17.1	27.9	206
	5422	26.50*	4919.01355167	0.000471	0.336634	0.4184	98.1	96.3	205
	9992	14.65*	4919.01702056	0.000106	0.339367	0.4180	90.5	91.8	206
	2745	9.05*	4919.01858666	0.000507	0.338347	0.4196	99.0	96.4	206
	3874	10.30*	4919.02415335	0.000756	0.338330	0.4187	97.1	96.5	205
	3280	12.35*	4919.02530716	0.000470	0.339674	0.4164	97.1	96.4	206
	4750	7.60*	4919.03708362	0.000829	0.337467	0.4196	100.0	96.5	206
	4402	2.45*	4919.03754388	0.001051	0.338603	0.4168	98.1	96.4	206
	2216	1.95*	4919.03831509	0.000450	0.339523	0.4192	50.5	49.8	206
	4698	2.95*	4919.05619732	0.001023	0.339728	0.4168	98.1	96.8	205
	4239	1.80*	4919.05909379	0.000654	0.337724	0.4173	99.0	96.2	206
	16383	0.05	4919.07988863	0.000255	0.337926	0.4191	97.1	94.2	206
	2718	0.55*	4919.09887396	0.001257	0.341413	0.4208	50.5	54.8	206
	4044	0.95*	4919.10387341	0.000342	0.338604	0.4169	97.1	96.6	205
	2956	0.05	4919.10445765	0.000456	0.338484	0.4163	99.0	96.6	206
	8912	0.20*	4919.12311186	0.001457	0.336513	0.4166	100.0	96.1	205
	4259	0.15	4919.12706371	0.001050	0.330145	0.4192	50.5	49.5	204
	5006	0.05	4919.14620223	0.000880	0.338121	0.4191	93.3	93.8	206
	9658	0.75*	4919.14879307	0.001846	0.332323	0.4152	98.1	96.3	205
	6642	0.60*	4919.15663690	0.000601	0.337815	0.4213	98.1	96.6	204
	3839	0.20*	4919.16085510	0.001798	0.330379	0.4138	99.0	96.2	206
	3024	0.25*	4919.16162258	0.000287	0.335135	0.4198	100.0	96.5	206
	2632	0.10*	4919.16264803	0.002048	0.326137	0.4157	97.1	96.4	206
	4901	0.05	4919.16797734	0.000320	0.333943	0.4189	97.1	96.4	206
	2529	0.10*	4919.17541968	0.000905	0.336405	0.4199	49.5	63.1	206
	3677	0.10*	4919.18685060	0.001853	0.328175	0.4172	98.1	96.6	206
	4095	0.15*	4919.18927498	0.001499	0.331345	0.4155	100.0	96.5	205
	1962	0.20*	4919.19282572	0.001262	0.333217	0.4162	99.0	96.2	206

TABLE VIII (*Continued*)

N	Average iterations	Frequency %	Coulomb energy	Dipole moment	Coulomb angle	Tammes angle	Hole angle	% Energy diversity	% Angular diversity	Faces N_f
	3521	0.20*	4919.19429654	0.001220	0.332252		0.4186	98.1	96.4	206
	2124	0.15*	4919.19715296	0.001917	0.333341		0.4156	97.1	96.0	206
	4491	0.05	4919.19777251	0.001666	0.330154		0.4150	100.0	96.8	206
	2833	0.05	4919.19869612	0.001254	0.336963		0.4152	96.2	96.9	206
	3281	0.05	4919.20125556	0.001417	0.332326		0.4192	49.5	49.5	206
	4004	0.25*	4919.20167685	0.000890	0.335103		0.4188	98.1	96.6	206
	6984	0.10*	4919.20325721	0.001997	0.337011		0.4161	99.0	96.4	206
	3840	0.05	4919.20670267	0.000698	0.335856		0.4222	41.0	49.2	206
	2901	0.05	4919.22001554	0.001541	0.327261		0.4166	99.0	96.5	206
	3749	0.05	4919.22297231	0.000699	0.333374		0.4180	97.1	96.6	206
	12141	0.05	4919.22499337	0.000292	0.334094		0.4196	88.6	92.0	206
	5105	0.05	4919.22789000	0.001635	0.334827		0.4179	100.0	96.7	205
	6839	0.05	4919.24138309	0.001518	0.331023		0.4169	98.1	96.2	206
106	2997	32.40*	5015.98459571	0.000022	0.343101		0.4175	24.5	35.6	208
	2627	5.45*	5016.02907073	0.000263	0.333412		0.4164	96.2	96.2	208
	10952	24.95*	5016.03188558	0.000193	0.337740		0.4143	98.1	96.4	207
	3372	2.25*	5016.03997847	0	0.339986		0.4194	25.5	39.1	208
	2599	4.55*	5016.04404567	0.000528	0.333778		0.4160	100.0	96.4	208
	3209	0.45	5016.05148374	0.000313	0.339617		0.4158	55.7	51.2	207
	4968	5.45*	5016.05372399	0.000585	0.336943		0.4148	100.0	96.7	208
	4851	4.00*	5016.05739622	0.000488	0.336990		0.4176	60.4	79.8	208
	2628	1.20*	5016.06003956	0.000349	0.337665		0.4195	49.1	54.1	208
	6312	1.45*	5016.07238095	0.000674	0.332479		0.4161	50.0	49.7	208
	3470	0.45*	5016.07555956	0.000299	0.331744		0.4153	65.1	71.6	208
	4485	1.95*	5016.07687796	0.000420	0.331698		0.4173	98.1	96.4	207
	3885	2.50*	5016.08222957	0.001289	0.334367		0.4145	99.1	96.0	208
	6255	0.85*	5016.08328859	0.000991	0.339068		0.4150	73.6	83.1	208
	3981	0.25*	5016.08683971	0.000392	0.336611		0.4177	48.1	55.6	208
	7247	1.05*	5016.09129480	0.001014	0.335959		0.4166	100.0	96.2	207
	6096	1.90*	5016.09595341	0.000249	0.332510		0.4151	96.2	96.7	208
	3895	0.80*	5016.09732472	0.000868	0.335870		0.4170	97.2	96.1	208
	7804	0.50*	5016.09771716	0.001325	0.339396		0.4169	99.1	95.7	208
	3442	0.15*	5016.09812109	0.001338	0.338685		0.4171	98.1	96.4	208
	5184	2.35*	5016.09948825	0.001462	0.339661		0.4151	99.1	95.9	207
	2951	0.15	5016.09996656	0.000397	0.333024		0.4159	56.6	72.6	208
	5305	0.25*	5016.10952568	0.001362	0.332964		0.4167	97.2	97.1	208
	4110	0.25*	5016.11020446	0.001487	0.328548		0.4166	96.2	96.4	207

	4882	0.05	5016.11122491	0.001140	0.336392	0.4157	97.2	96.8	208
	5701	0.20*	5016.11256548	0.001232	0.331912	0.4150	98.1	96.7	208
	4422	0.95*	5016.11293661	0.000681	0.335466	0.4135	98.1	96.4	208
	2977	0.25*	5016.11667164	0.000996	0.332268	0.4145	99.1	96.1	208
	2387	0.10*	5016.12312979	0.000379	0.335243	0.4181	49.1	63.0	208
	2086	0.05	5016.12511371	0.000471	0.332627	0.4175	50.9	65.4	208
	4720	0.25*	5016.12937726	0.001574	0.330080	0.4153	98.1	96.0	208
	4646	0.10	5016.13428403	0.000382	0.333062	0.4171	84.9	84.4	208
	3083	0.25*	5016.14095562	0.002017	0.332506	0.4101	98.1	96.5	208
	3184	0.60*	5016.14460498	0.002464	0.328331	0.4112	99.1	96.4	208
	2980	0.05	5016.14845404	0.000150	0.337118	0.4170	49.1	49.8	208
	11392	0.05	5016.16609934	0.001934	0.334814	0.4126	97.2	96.4	208
	3611	0.10	5016.16730770	0.002037	0.332123	0.4085	98.1	96.4	208
	4686	0.05	5016.16837348	0.000332	0.332373	0.4152	99.1	96.4	208
	8827	0.05	5016.18438729	0.002968	0.329205	0.4135	99.1	96.4	207
	4004	0.10*	5016.18719575	0.002143	0.328745	0.4159	97.2	96.4	208
	14167	0.05	5016.19056300	0.001870	0.328538	0.4137	98.1	96.8	207
	3686	0.05	5016.19208897	0.000690	0.331646	0.4176	99.1	97.0	207
	4299	0.10*	5016.19602441	0.001489	0.330527	0.4126	98.1	96.4	207
	6259	0.05	5016.19721054	0.001927	0.329449	0.4141	95.3	97.0	208
	14737	0.05	5016.19746973	0.001256	0.334776	0.4130	91.5	94.7	206
	7877	0.05	5016.19825750	0.001862	0.329210	0.4135	97.2	96.0	207
	2427	0.05	5016.20760384	0.000499	0.331291	0.4183	49.1	49.9	208
	4805	0.05	5016.20994256	0.001107	0.329374	0.4128	99.1	96.4	208
	3669	0.05	5016.21724073	0.001073	0.331695	0.4125	98.1	96.8	208
	4699	0.05	5016.21800419	0.001003	0.333949	0.4136	98.1	96.4	208
	11792	0.05	5016.21929950	0.001644	0.327098	0.4140	100.0	96.4	208
	3374	0.05	5016.22112056	0.001616	0.329402	0.4126	96.2	96.4	207
	4237	0.05	5016.22124328	0.002021	0.327434	0.4137	98.1	95.8	208
	9289	0.05	5016.22278968	0.001946	0.328857	0.4142	98.1	96.1	208
	11011	0.05	5016.22618835	0.002488	0.332864	0.4141	80.2	89.7	206
	5031	0.10*	5016.22778255	0.002422	0.326437	0.4144	98.1	96.2	206
	6423	0.05	5016.22820174	0.000565	0.329783	0.4132	49.1	49.7	208
	3497	0.05	5016.23013185	0.000363	0.330685	0.4195	49.1	49.7	208
	4103	0.05	5016.24242732	0.000668	0.331721	0.4187	99.1	96.5	208
	4817	0.05	5016.30307750	0.001566	0.328122	0.4130	98.1	96.8	208
	8157	0.05	5016.32705914	0.001171	0.332073	0.4163	99.1	96.5	208

TABLE VIII (*Continued*)

N	Average iterations	Frequency %	Coulomb energy	Dipole moment	Coulomb angle	Tammes angle	Hole angle	% Energy diversity	% Angular diversity	Faces N_f
107	3645	12.95*	5113.95354772	0.000064	0.337317		0.4155	50.5	49.5	210
	4563	8.80*	5113.97385010	0.000213	0.338380		0.4129	64.5	78.4	210
	5804	18.35*	5113.98085775	0.000220	0.337460		0.4154	49.5	49.6	210
	3397	1.15*	5113.98698336	0.000341	0.332633		0.4136	49.5	51.9	210
	4519	7.50*	5113.99261825	0.000811	0.333919		0.4144	99.1	96.2	210
	7598	16.40*	5114.00057902	0.000104	0.337051		0.4125	99.1	95.9	209
	3945	1.40*	5114.00134457	0.000326	0.335514		0.4121	50.5	64.3	210
	3065	3.10*	5114.00268858	0.001195	0.332411		0.4159	50.5	53.5	210
	5265	2.90*	5114.01281908	0.001369	0.333985		0.4137	95.3	96.5	209
	3716	2.00*	5114.01951688	0.001297	0.332430		0.4121	99.1	96.0	210
	6380	6.85*	5114.02098442	0.000817	0.333393		0.4164	99.1	96.2	209
	4322	2.85*	5114.02380411	0.001380	0.336739		0.4144	99.1	96.0	209
	2474	0.60*	5114.02683190	0.000267	0.331019		0.4136	50.5	49.6	210
	4003	1.55*	5114.02775097	0.001630	0.332551		0.4135	100.0	96.2	209
	8022	3.10*	5114.02844932	0.001632	0.329639		0.4125	94.4	96.0	208
	3271	0.15*	5114.03333531	0.001155	0.333825		0.4143	56.1	68.4	210
	16455	0.05	5114.03350687	0.000649	0.334338		0.4155	100.0	96.5	209
	4797	1.65*	5114.03406517	0.000540	0.334241		0.4122	99.1	96.1	209
	2125	0.90*	5114.03792376	0.000572	0.327212		0.4158	48.6	53.4	210
	4015	2.45*	5114.04005697	0.001811	0.330313		0.4130	100.0	96.0	210
	4512	0.70*	5114.04345820	0.001234	0.331829		0.4144	100.0	96.3	208
	3875	0.50*	5114.05197947	0.001333	0.336017		0.4145	98.1	96.5	210
	3902	0.25*	5114.05831959	0.002016	0.330082		0.4110	100.0	96.0	210
	4397	0.05	5114.06301739	0.002210	0.332700		0.4110	98.1	95.9	210
	4209	0.25*	5114.06764051	0.001190	0.334763		0.4152	96.3	96.0	209
	3623	0.15	5114.06999407	0.001784	0.334388		0.4143	98.1	96.4	210
	3348	0.45*	5114.07068401	0.000548	0.335563		0.4154	98.1	96.1	210
	5775	0.20*	5114.08279997	0.000553	0.333070		0.4147	98.1	96.3	210
	2395	0.25*	5114.08602239	0.000416	0.335598		0.4156	50.5	67.1	210
	2902	0.05	5114.09328058	0.000535	0.330003		0.4154	61.7	83.4	210
	6678	0.10	5114.09357220	0.001560	0.329440		0.4111	98.1	96.1	210
	2179	0.25*	5114.09869247	0.000228	0.332794		0.4136	99.1	96.1	210
	2757	0.15	5114.10110776	0.001261	0.334056		0.4151	49.5	49.7	210
	3281	0.10	5114.11514557	0.000317	0.330099		0.4152	99.1	96.6	210
	4771	0.10*	5114.11571613	0.000618	0.331224		0.4135	99.1	96.0	208
	5861	0.25*	5114.11577237	0.001645	0.334536		0.4138	98.1	96.8	210

	7484	0.05	5114.11647970	0.001003	0.331746	0.4154	99.1	96.8	210
	2645	0.25*	5114.11703802	0.000855	0.329999	0.4132	98.1	96.5	210
	2111	0.10*	5114.11962949	0.000313	0.331565	0.4175	34.6	48.7	210
	3530	0.15	5114.12171504	0.001247	0.330743	0.4136	57.0	69.2	209
	3606	0.05	5114.12234959	0.001297	0.332845	0.4149	64.5	82.8	210
	3462	0.20*	5114.12269861	0.000495	0.332245	0.4168	96.3	96.8	210
	3989	0.05	5114.14155186	0.001916	0.328905	0.4120	99.1	96.8	210
	5100	0.05	5114.14320671	0.001803	0.323100	0.4111	97.2	96.4	210
	8618	0.10*	5114.14567219	0.000800	0.330746	0.4154	98.1	96.5	208
	13181	0.05	5114.14676496	0.000923	0.328624	0.4092	56.1	50.0	209
	6087	0.10*	5114.15773096	0.000635	0.332440	0.4150	99.1	96.8	210
	5350	0.10	5114.16353781	0.000816	0.329894	0.4161	100.0	96.4	208
	4990	0.05	5114.18531290	0.000503	0.329894	0.4149	98.1	96.3	210
	4146	0.10	5114.21543765	0.001346	0.329707	0.4127	72.0	79.2	210
	4711	0.05	5114.25760386	0.002341	0.327919	0.4091	98.1	96.5	210
	3608	0.05	5114.33023734	0.001604	0.323896	0.4128	100.0	96.2	210
108	3661	30.40*	5212.81350783	0.000432	0.337313	0.4129	50.0	49.7	212
	6951	41.20*	5212.81758312	0.000436	0.336759	0.4121	98.1	95.6	212
	2481	0.10	5212.81762475	0.000486	0.337182	0.4121	97.2	95.9	211
	3439	0.75*	5212.81890852	0	0.340819	0.4147	16.7	39.4	212
	3702	5.85*	5212.82103064	0	0.340402	0.4123	40.7	42.2	212
	3386	8.55*	5212.87259090	0.000230	0.336660	0.4113	50.0	51.2	212
	1979	0.40*	5212.94775343	0	0.335144	0.4160	17.6	26.3	212
	2350	0.70*	5212.95468499	0.000352	0.336837	0.4118	32.4	38.5	212
	4050	2.60*	5212.96203966	0.000396	0.326349	0.4137	98.1	96.7	212
	3572	0.75*	5212.96252713	0.000113	0.334889	0.4144	50.0	70.0	212
	6248	0.65*	5212.96595200	0.001095	0.334867	0.4148	48.1	55.2	212
	3190	0.25*	5212.97031079	0.000543	0.335739	0.4149	49.1	64.7	212
	4952	0.65*	5212.97252200	0.001043	0.335022	0.4114	98.1	96.1	212
	8333	0.45*	5212.97645040	0.000748	0.329659	0.4111	97.2	96.5	211
	4485	1.00*	5212.97811922	0.000938	0.328288	0.4129	97.2	96.1	212
	5110	0.25*	5212.97878645	0.001098	0.336460	0.4116	50.0	66.8	212
	3052	0.75*	5212.98362560	0.000278	0.333295	0.4154	50.9	49.7	212
	9097	0.40*	5212.98609793	0.001111	0.334672	0.4114	98.1	96.4	209
	6320	0.30*	5212.98695079	0.000827	0.327804	0.4117	99.1	96.2	212
	6920	0.70*	5212.98857439	0.001051	0.330021	0.4123	99.1	95.9	212
	8180	0.45*	5212.99775092	0.001462	0.328015	0.4142	100.0	96.2	212

TABLE VIII (*Continued*)

N	Average iterations	Frequency %	Coulomb energy	Dipole moment	Coulomb angle	Tammes angle	Hole angle	% Energy diversity	% Angular diversity	Faces N_f
	5471	0.20*	5213.00379554	0.000874	0.333383		0.4109	97.2	96.1	212
	3178	0.10	5213.00591463	0.000798	0.332016		0.4121	99.1	96.2	212
	5375	0.10	5213.00739612	0.001465	0.331220		0.4139	56.5	60.3	212
	5622	0.10	5213.01171880	0.001636	0.326744		0.4088	99.1	96.5	212
	3429	0.10	5213.01501059	0.001314	0.334884		0.4149	52.8	73.6	212
	4443	0.10*	5213.01541423	0.000158	0.325609		0.4121	59.3	75.0	212
	6962	0.55*	5213.01951579	0.001878	0.323555		0.4092	98.1	96.6	212
	4338	0.05	5213.01993758	0.002125	0.327208		0.4106	99.1	95.7	212
	7898	0.05	5213.02597286	0.001409	0.328084		0.4065	99.1	96.2	212
	2604	0.25*	5213.02687860	0.000706	0.328136		0.4127	100.0	96.3	212
	2856	0.05	5213.03439974	0.000551	0.331577		0.4135	63.9	82.6	212
	5244	0.10	5213.03628821	0.002347	0.326050		0.4062	99.1	96.0	212
	3240	0.15*	5213.04548392	0.002385	0.327668		0.4107	96.3	96.0	212
	3861	0.10*	5213.05063949	0.000983	0.327036		0.4137	97.2	95.7	210
	4009	0.10	5213.05096591	0.000159	0.325736		0.4120	50.0	49.7	212
	9144	0.05	5213.05122718	0.001912	0.332580		0.4138	50.0	54.2	212
	4683	0.20*	5213.05637717	0.002119	0.327328		0.4099	99.1	96.5	211
	3254	0.05	5213.06426316	0.001990	0.329165		0.4106	100.0	95.9	211
	4030	0.05	5213.06735088	0.001686	0.328932		0.4110	98.1	96.3	211
	6307	0.10*	5213.07431452	0	0.335165		0.4152	41.7	44.2	208
	4634	0.05	5213.07692279	0.000617	0.331603		0.4108	60.2	81.0	212
	8018	0.05	5213.07915908	0.001605	0.324787		0.4124	98.1	96.1	211
	5727	0.05	5213.08168306	0.001294	0.326452		0.4114	97.2	96.4	211
	6338	0.05	5213.08363709	0.001622	0.324108		0.4067	100.0	96.1	212
	4314	0.05	5213.12705126	0.001661	0.326574		0.4126	100.0	96.4	211
	3189	0.05	5213.15265756	0.001140	0.324681		0.4102	100.0	96.1	212
109	4800	45.50*	5312.73507994	0.000647	0.333401		0.4112	49.5	49.4	212
	2272	3.20*	5312.75464830	0.000247	0.337508		0.4127	50.5	61.2	214
	4050	3.60*	5312.77396021	0.000484	0.331108		0.4107	99.1	96.1	212
	5262	7.40*	5312.77508749	0.000056	0.334593		0.4095	57.8	65.4	214
	5075	1.30*	5312.77632903	0.000524	0.333564		0.4115	99.1	96.1	214
	2893	6.00*	5312.77865178	0.000524	0.331816		0.4119	95.4	96.1	213
	5852	3.20*	5312.79249977	0.001129	0.330098		0.4107	99.1	96.0	214
	4419	1.60*	5312.79715528	0.000062	0.330997		0.4117	99.1	96.1	214
	4662	1.40*	5312.79856954	0.001076	0.330162		0.4115	98.2	96.4	214
	4416	3.20*	5312.79985513	0.000895	0.331708		0.4107	99.1	96.5	214

	3451	3.80*	5312.80010079	0.000412	0.329964	0.4111	100.0	96.2	214
	13361	0.80	5312.80175514	0.000394	0.332414	0.4110	50.5	49.8	214
	4147	0.40	5312.80437836	0.000397	0.334958	0.4112	50.5	61.0	214
	3161	2.40*	5312.80648273	0.000629	0.329112	0.4112	99.1	96.7	214
	6775	0.30	5312.80971877	0.000106	0.331442	0.4093	70.6	83.4	214
	4992	0.20	5312.82165401	0.000167	0.330655	0.4117	50.5	49.4	214
	3811	1.90*	5312.82350767	0.000628	0.330873	0.4102	99.1	96.3	214
	4087	0.30	5312.82575643	0.000988	0.328546	0.4109	100.0	96.6	213
	2564	0.30*	5312.82582086	0.000885	0.336168	0.4131	33.9	43.9	214
	6412	0.20*	5312.82608204	0.001235	0.333333	0.4099	61.5	80.6	214
	8362	0.60*	5312.82835406	0.000833	0.333385	0.4109	83.5	88.6	214
	5370	0.40*	5312.82948191	0.000293	0.330721	0.4099	89.0	90.9	214
	3202	1.00*	5312.83631357	0.000486	0.331409	0.4103	98.2	96.0	214
	2929	0.40*	5312.83766190	0.000776	0.329988	0.4103	100.0	96.0	213
	3995	1.00*	5312.84056515	0.000108	0.332912	0.4126	50.5	72.2	214
	20000	0.30*	5312.84203408	0.001080	0.332610	0.4107	98.2	94.7	212
	5967	0.10	5312.84330444	0.000879	0.327770	0.4096	99.1	96.7	214
	4305	0.50*	5312.84536551	0.000439	0.333114	0.4109	100.0	96.0	214
	9002	0.70*	5312.84841327	0.000767	0.331461	0.4120	99.1	96.0	213
	4892	0.20	5312.85353630	0.000640	0.327640	0.4107	98.2	95.8	213
	6414	0.50*	5312.85358451	0.001019	0.325733	0.4093	100.0	96.7	214
	10183	0.10	5312.86062859	0.000288	0.328071	0.4114	98.2	96.3	214
	5292	0.80*	5312.86551212	0.000846	0.324200	0.4103	100.0	96.0	214
	5083	0.50*	5312.86634833	0.002036	0.325374	0.4060	99.1	96.0	214
	2869	0.10	5312.87443286	0.001662	0.321002	0.4113	99.1	96.3	214
	7426	0.20	5312.87619264	0.000242	0.328733	0.4072	100.0	96.4	214
	6910	0.80*	5312.88224787	0.000651	0.331100	0.4119	98.2	95.9	213
	3911	0.50*	5312.89066290	0.001968	0.326534	0.4082	99.1	96.5	214
	11181	0.20*	5312.89655576	0.001539	0.325941	0.4040	100.0	96.1	213
	12194	0.40	5312.90655106	0.001695	0.325138	0.4096	100.0	96.2	214
	9720	0.20*	5312.90689816	0.002307	0.320748	0.4044	99.1	96.0	214
	5266	0.20	5312.91325624	0.001631	0.322474	0.4083	97.2	96.1	213
	3468	0.10	5312.92460605	0.001494	0.324379	0.4076	97.2	96.3	214
	5906	0.10	5312.93676853	0.002367	0.325234	0.4076	50.5	49.5	214
	3479	0.10	5312.93983337	0.001353	0.324586	0.4098	100.0	96.4	214
	4062	0.20*	5312.94003492	0.001370	0.324357	0.4096	98.2	96.8	213
	4772	0.10	5312.94198517	0.001991	0.324873	0.4069	50.5	49.6	214
	3281	0.20	5312.94221283	0.001501	0.323345	0.4058	99.1	96.0	214

TABLE VIII (*Continued*)

N	Average iterations	Frequency %	Coulomb energy	Dipole moment	Coulomb angle	Tammes angle	Hole angle	% Energy diversity	% Angular diversity	Faces N_f
	6146	0.10	5312.94365738	0.001400	0.327313		0.4047	49.5	49.3	214
	4195	0.10	5312.95862195	0.000633	0.325645		0.4105	98.2	96.7	214
	6815	0.10	5312.95956019	0.001653	0.326406		0.4058	99.1	95.8	214
	6813	0.10	5312.96329655	0.001730	0.329108		0.4091	99.1	96.4	212
	6107	0.10	5312.99107321	0.001512	0.326766		0.4095	99.1	96.5	212
	5307	0.10	5312.99439478	0.001535	0.325601		0.4037	100.0	96.4	214
	2998	0.10	5313.00685905	0.002025	0.323621		0.4076	99.1	96.2	214
110	3883	16.70*	5413.54929420	0.000012	0.339912		0.4106	9.1	14.6	216
	2794	17.20*	5413.56758230	0.000659	0.334684		0.4115	49.1	53.0	216
	3903	7.70*	5413.59868690	0.000112	0.333575		0.4104	50.9	72.1	216
	2368	7.10*	5413.61155043	0.000547	0.329299		0.4088	100.0	96.3	216
	2322	0.70*	5413.61560102	0	0.333605		0.4104	25.5	25.8	216
	7304	2.40*	5413.62179208	0.000783	0.329568		0.4083	97.3	96.0	216
	3392	3.80*	5413.62905287	0.000740	0.327782		0.4093	98.2	96.2	216
	3596	3.30*	5413.63183228	0.000145	0.328092		0.4092	49.1	65.7	216
	4016	6.30*	5413.63712239	0.001885	0.324388		0.4045	97.3	96.1	216
	2889	0.30*	5413.63881652	0.000152	0.331131		0.4093	54.5	71.0	216
	7111	3.70*	5413.63946560	0.000402	0.326498		0.4093	100.0	96.3	216
	6406	1.20*	5413.64231129	0.000347	0.331186		0.4089	95.5	96.2	215
	6329	3.20*	5413.64881901	0.000336	0.329843		0.4085	100.0	96.2	216
	5969	1.10*	5413.64907064	0.000397	0.328765		0.4110	50.0	49.6	216
	2700	0.40*	5413.64996374	0.000367	0.329199		0.4078	98.2	96.5	216
	4849	1.40*	5413.65048130	0.000540	0.328199		0.4088	99.1	96.5	215
	2900	2.60*	5413.65157268	0.000674	0.325866		0.4080	98.2	96.1	216
	5428	0.90*	5413.65238277	0.000368	0.327661		0.4075	97.3	96.5	216
	4014	0.80*	5413.65499335	0.000042	0.332282		0.4072	50.0	49.4	216
	4447	0.50*	5413.65724954	0.000376	0.328621		0.4065	100.0	95.7	216
	4247	0.40*	5413.65903088	0.000419	0.327953		0.4067	100.0	95.9	214
	6327	1.00*	5413.66012127	0.000435	0.327757		0.4085	100.0	95.9	214
	5988	2.10*	5413.66137962	0.000866	0.328801		0.4096	99.1	96.3	216
	2714	1.10*	5413.66159998	0.000272	0.330180		0.4086	98.2	95.9	216
	3682	0.50*	5413.66242354	0.000131	0.330004		0.4100	99.1	96.1	215
	5346	2.10*	5413.66350044	0.000474	0.329736		0.4082	99.1	96.2	215
	3986	3.20*	5413.66453234	0.001187	0.325963		0.4088	100.0	96.3	216
	4726	0.20	5413.66747167	0.000468	0.327746		0.4060	99.1	95.7	215
	4974	0.10	5413.66800668	0.000928	0.329748		0.4082	95.5	93.9	216

	3853	0.10	5413.66804358	0.000972	0.328783	0.4084	99.1	96.1	216
	14320	0.20	5413.66821468	0.000555	0.329686	0.4089	100.0	96.5	214
	6487	0.20	5413.67106097	0.000515	0.329227	0.4054	98.2	96.5	215
	9344	0.20*	5413.67225890	0.000741	0.327790	0.4082	96.4	96.6	215
	2919	0.50*	5413.67945550	0.000293	0.328632	0.4065	100.0	96.1	216
	7672	0.30	5413.68392561	0.000467	0.329572	0.4077	99.1	96.3	215
	4081	1.20*	5413.68395005	0.000991	0.331016	0.4085	99.1	96.1	216
	2916	0.30	5413.68431490	0.000542	0.327983	0.4089	98.2	95.7	216
	7199	0.70*	5413.68677176	0.000919	0.329573	0.4104	82.7	89.6	214
	11354	0.10	5413.68934422	0.000044	0.327158	0.4085	93.6	92.8	214
	5280	0.30	5413.69424335	0.000713	0.330287	0.4085	97.3	96.1	216
	13110	0.10	5413.70114673	0.000750	0.326517	0.4074	97.3	96.0	215
	5351	0.20	5413.70292809	0	0.332070	0.4096	25.5	25.4	216
	2422	0.20*	5413.70491131	0.000461	0.325462	0.4099	98.2	96.1	216
	3462	0.50*	5413.71547656	0.001773	0.322695	0.4044	97.3	96.3	216
	4859	0.10	5413.71762705	0.000510	0.327216	0.4078	57.3	73.7	216
	4486	0.10	5413.71919780	0.000259	0.327419	0.4076	79.1	85.0	214
	4136	0.90*	5413.72313200	0.000563	0.328159	0.4097	98.2	96.6	214
	2530	0.10	5413.72818239	0.002019	0.328043	0.4100	100.0	96.0	216
	3532	0.10	5413.73561358	0.002328	0.326255	0.4080	100.0	95.7	216
	6384	0.10	5413.73788035	0.000295	0.326627	0.4092	98.2	96.1	215
	12108	0.10	5413.74943853	0.000635	0.328731	0.4094	98.2	96.4	216
	9027	0.20	5413.74947669	0.001175	0.322815	0.4032	100.0	95.7	216
	3283	0.30*	5413.75213359	0.001228	0.323427	0.4100	100.0	96.6	216
	3914	0.10	5413.75692110	0.001260	0.319915	0.4059	97.3	96.7	216
	5163	0.10	5413.76229840	0.002057	0.324118	0.4016	97.3	96.0	215
	8233	0.10	5413.76322607	0.001136	0.326166	0.4040	95.5	96.2	216
	3977	0.10	5413.77236829	0.001364	0.324254	0.4097	98.2	96.2	216
	5643	0.10	5413.78197893	0.001673	0.323928	0.4041	99.1	96.3	216
111	5788	47.95*	5515.29321459	0	0.336066	0.4107	16.2	17.3	218
	5081	14.10*	5515.37822022	0.000378	0.332049	0.4088	50.5	54.6	218
	2979	0.90*	5515.39964901	0.000998	0.329809	0.4097	50.5	49.5	218
	2964	2.35*	5515.40973823	0.000628	0.331042	0.4082	98.2	96.0	218
	2523	0.10	5515.42133925	0	0.332921	0.4085	18.0	23.5	215
	4886	5.20*	5515.42556838	0.000645	0.325343	0.4065	96.4	96.1	218
	6086	3.35*	5515.42695246	0.000606	0.329950	0.4076	99.1	96.2	218
	7732	2.10*	5515.42750020	0.000248	0.330319	0.4074	59.5	77.8	218

TABLE VIII (*Continued*)

N	Average iterations	Frequency %	Coulomb energy	Dipole moment	Coulomb angle	Tammes angle	Hole angle	% Energy diversity	% Angular diversity	Faces N_f
	2205	1.00*	5515.43691472	0.000666	0.329852		0.4070	95.5	95.5	218
	3585	2.45*	5515.43714361	0.001040	0.330017		0.4059	99.1	96.1	217
	6490	1.45*	5515.43879234	0.002212	0.327013		0.4043	94.6	95.8	218
	2729	0.25*	5515.43969989	0	0.331704		0.4101	16.2	27.0	218
	6357	1.30*	5515.44354698	0.001224	0.327275		0.4080	98.2	96.0	218
	4072	0.90*	5515.44744430	0.000347	0.329144		0.4077	53.2	74.4	218
	6392	1.10*	5515.45213408	0.000776	0.327113		0.4068	97.3	96.3	218
	7213	0.50*	5515.45333861	0.000317	0.328230		0.4059	99.1	96.0	215
	18992	2.70*	5515.45408736	0.000481	0.329046		0.4062	97.3	96.1	217
	3953	0.70*	5515.45472014	0.000870	0.330363		0.4067	99.1	96.0	217
	3313	0.15*	5515.45608011	0.000192	0.331467		0.4088	54.1	72.9	218
	5892	1.10*	5515.45771562	0.000024	0.332937		0.4112	80.2	89.9	215
	5502	0.10	5515.45884084	0.000437	0.330178		0.4073	61.3	76.9	218
	6263	0.95*	5515.46471350	0.000517	0.330129		0.4077	99.1	95.8	217
	4885	0.20*	5515.46721864	0.000055	0.330393		0.4058	50.5	49.4	218
	6650	0.30*	5515.46995593	0.000213	0.333295		0.4103	50.5	52.0	218
	4239	0.15*	5515.47277914	0.000122	0.332090		0.4057	58.6	73.2	218
	3971	0.25*	5515.47650805	0.000206	0.324838		0.4074	97.3	96.1	217
	4346	0.90*	5515.47784442	0.000331	0.325787		0.4092	98.2	96.1	217
	4325	0.85	5515.47803364	0.000367	0.326291		0.4080	56.8	50.3	218
	7784	0.45*	5515.47806647	0.000742	0.326214		0.4071	97.3	96.1	218
	4993	0.15*	5515.47920645	0.000596	0.327959		0.4062	98.2	96.1	218
	4101	0.75*	5515.47931863	0.001835	0.323489		0.4034	93.7	95.5	218
	3342	0.35*	5515.48021626	0.000125	0.326311		0.4079	50.5	65.1	218
	8738	0.30*	5515.48051908	0.000541	0.325241		0.4058	99.1	96.0	218
	5358	0.65*	5515.48439202	0.000529	0.325913		0.4069	99.1	96.8	218
	4124	0.05	5515.49412689	0.001576	0.323115		0.4011	98.2	96.3	218
	7601	0.05	5515.49885869	0.000344	0.329428		0.4079	99.1	96.6	218
	5253	0.70*	5515.50969953	0.001109	0.323321		0.4080	99.1	96.2	216
	3604	0.20*	5515.51281749	0.000073	0.329731		0.4067	50.5	49.6	218
	4363	0.15	5515.51619137	0.000191	0.325759		0.4059	97.3	96.3	217
	6225	0.25*	5515.52023040	0.001617	0.324555		0.4016	98.2	96.2	218
	3995	0.10	5515.52288453	0.000771	0.326751		0.4068	99.1	96.2	218
	4127	0.05	5515.53088257	0.001108	0.325726		0.4046	99.1	96.0	218
	3678	0.10	5515.53848192	0.000444	0.322840		0.4058	100.0	95.8	218
	7577	0.10	5515.54254505	0.000663	0.325932		0.4090	97.3	96.4	218
	7690	0.15	5515.54940105	0.001348	0.329487		0.4085	50.5	49.5	218

	5260	0.15*	5515.55299866	0.000515	0.322224	0.4064	100.0	96.2	217
	6676	0.35*	5515.55307500	0.000762	0.328647	0.4080	97.3	96.1	218
	4062	0.10	5515.56968339	0.001502	0.318141	0.4069	100.0	95.7	218
	4482	0.10*	5515.57176579	0.000889	0.325678	0.4070	100.0	96.2	217
	4643	0.05	5515.59452162	0.001952	0.318722	0.3999	99.1	95.9	217
	4238	0.05	5515.65828199	0.002024	0.320332	0.4046	98.2	96.7	217
112	3352	43.60*	5618.04488233	0	0.337747	0.4078	10.7	15.3	220
	4197	18.40*	5618.05824444	0	0.335550	0.4095	8.9	12.8	220
	6522	1.70	5618.16281873	0.001122	0.327323	0.4038	56.3	51.3	218
	8122	0.40*	5618.17022822	0.000033	0.325734	0.4037	76.8	87.5	220
	5483	6.20*	5618.17975389	0.001003	0.328821	0.4039	96.4	96.1	220
	2741	1.00*	5618.18139715	0.000508	0.319826	0.4080	49.1	49.7	220
	3771	2.40*	5618.18692576	0.000391	0.328264	0.4046	97.3	96.0	220
	3849	3.40*	5618.19123408	0.000376	0.324163	0.4056	99.1	96.0	219
	3606	1.50*	5618.19340191	0.000331	0.327719	0.4071	97.3	96.0	220
	2737	0.50*	5618.19617462	0	0.330723	0.4089	24.1	37.0	220
	3729	0.50	5618.19669597	0.000908	0.329067	0.4058	74.1	79.7	218
	3031	1.40*	5618.19682071	0.000594	0.326403	0.4066	97.3	96.3	220
	6419	0.70*	5618.19738268	0.000257	0.326579	0.4043	98.2	96.0	219
	5227	1.40*	5618.19784373	0.000363	0.329961	0.4043	98.2	95.8	219
	6874	2.10*	5618.19876503	0.001005	0.327286	0.4084	99.1	96.0	219
	10697	0.60*	5618.19996075	0.000532	0.325018	0.4069	98.2	96.4	219
	6892	0.80*	5618.20008193	0.000528	0.324998	0.4066	99.1	96.0	219
	5447	0.40*	5618.20403595	0.000831	0.327900	0.4046	97.3	96.3	220
	5129	0.80*	5618.20575583	0.000116	0.329203	0.4054	96.4	95.7	219
	7540	0.20	5618.20591159	0.000800	0.328497	0.4044	99.1	95.9	220
	3502	1.50*	5618.20836091	0.000181	0.327292	0.4043	98.2	96.0	220
	6680	0.40*	5618.20886000	0.000667	0.323681	0.4070	98.2	96.2	220
	5141	1.00*	5618.20936383	0.000675	0.327058	0.4055	95.5	96.2	220
	4532	0.80*	5618.21058012	0.000213	0.326902	0.4055	100.0	96.1	220
	6421	0.30	5618.22022487	0.000069	0.325049	0.4086	82.1	86.3	220
	5133	1.30*	5618.23448912	0.001165	0.321334	0.4066	97.3	96.4	220
	3418	0.50*	5618.24074945	0.000495	0.322422	0.4053	99.1	96.4	220
	5871	0.10	5618.24899109	0.000678	0.324842	0.4070	97.3	96.7	220
	2495	0.50*	5618.24925481	0.000256	0.327010	0.4065	98.2	96.0	220
	4381	0.60*	5618.25368022	0.001349	0.322400	0.4040	96.4	95.9	219
	11615	0.30	5618.25621803	0.001191	0.316907	0.4049	81.3	88.1	219

TABLE VIII (*Continued*)

N	Average iterations	Frequency %	Coulomb energy	Dipole moment	Coulomb angle	Tammes angle	Hole angle	% Energy diversity	% Angular diversity	Faces N_f
	7119	0.30*	5618.25646911	0.000242	0.324503		0.4033	99.1	96.3	220
	2887	0.30	5618.25870781	0.001925	0.324573		0.4022	98.2	96.1	220
	3491	0.10	5618.27042281	0.001689	0.320620		0.4041	97.3	96.3	220
	11459	0.20	5618.27104971	0.002336	0.322996		0.3987	98.2	95.9	220
	5254	0.10	5618.27405701	0.001196	0.325236		0.4048	99.1	96.0	219
	4306	0.10	5618.27590295	0.000480	0.322857		0.4060	98.2	96.4	220
	5897	0.10	5618.27669984	0.000342	0.324160		0.4041	96.4	95.9	220
	3449	0.40*	5618.27802049	0.000681	0.325544		0.4059	98.2	95.8	219
	5045	0.10	5618.28223927	0.000720	0.328029		0.4060	49.1	49.6	220
	5368	0.20	5618.28245930	0.001964	0.325069		0.3981	98.2	96.2	219
	8640	0.20*	5618.28263752	0.000412	0.325306		0.4072	96.4	96.0	219
	3632	0.30*	5618.28434640	0.001703	0.322639		0.4011	100.0	96.1	220
	12996	0.10	5618.28804357	0.001351	0.324640		0.4071	98.2	95.8	219
	9407	0.10	5618.29354208	0.002647	0.318589		0.4028	97.3	95.9	220
	5354	0.30	5618.29659227	0.001327	0.324433		0.4024	98.2	95.8	220
	3864	0.10	5618.30033892	0.000400	0.322764		0.4054	51.8	61.0	218
	3396	0.20	5618.30235974	0.000911	0.324960		0.4058	100.0	96.3	219
	13423	0.10	5618.30367493	0.001183	0.317356		0.4028	98.2	96.4	220
	3425	0.20*	5618.30931247	0.001975	0.323746		0.4019	97.3	95.9	220
	7187	0.10	5618.32187570	0.001985	0.319207		0.4001	100.0	96.0	220
	4210	0.10	5618.34027361	0.000826	0.321902		0.4069	97.3	95.7	220
	3095	0.10	5618.34424060	0.000384	0.323462		0.4069	50.0	63.9	220
	5022	0.10	5618.34604924	0.001145	0.322203		0.4050	98.2	96.1	220
	5121	0.10	5618.34947946	0.000791	0.319789		0.4053	97.3	96.6	219
	8154	0.10	5618.36426242	0.001736	0.314963		0.3995	99.1	96.3	220
	18663	0.20*	5618.37527701	0.000542	0.324870		0.4073	91.1	93.8	217
	4829	0.20	5618.39446210	0.001252	0.323779		0.4042	97.3	96.0	220
	4277	0.10	5618.41155067	0.002074	0.321258		0.4014	98.2	96.1	220
	5520	0.10	5618.41948131	0.002252	0.317741		0.4041	99.1	96.3	219

REFERENCES

1. J. J. Thomson, *Phil. Mag.* **44** (ser. 5), 293–316 (1897).
2. J. J. Thomson, *Phil. Mag.* **7** (ser. 6), 237–265 (1904).
3. J. J. Thomson, *The Electron in Chemistry*, Franklin Institute, Philadelphia, 1923.
4. D. S. Bethune, R. D. Johnson, J. R. Salem, M. S. de Vries, and C. S. Yannoni, *Nature* **366**(6451), 123–128 (1993).
5. S. C. Sevov and J. D. Corbett, *Science* **262**(5135), 880–883 (1993).
6. H. D. I. Abarbanel, in *Studies in Mathematical Physics—Essays in Honor of Valentine Bargmann*, E. H. Lieb, B. Simon, and A. S. Wightman, eds., Princeton University Press, Princeton, 1976, pp. 3–18.
7. N. Bohr, *Zeit. Physik* **9**, 1–67 (1922).
8. W. Pauli, *Zeit. Physik* **31**, 765–783 (1925).
9. H. W. Kroto, J. R. Heath, S. C. O'Brien, R. F. Curl, and R. E. Smalley, *Nature* **318**, 162–164 (1985).
10. L. Föppl, *J. Reine Angew. Math.* **141**, 251–302 (1912).
11. J. Leech, *Math. Gazette* **41**, 81–90 (1957).
12. P. M. L. Tammes, *Recueil des Travaux Botaniques Néerlandais* **27**, 1–84 (1930).
13. L. Fejes-Toth, *Regular Figures*, Macmillan, New York, 1964.
14. H. S. M. Coxeter, *Trans. N.Y. Acad. Sci.* **24** (Series II), 320–331 (1961–1962).
15. D. E. Lazić, V. Šenk, and V. Šeškar, *Bull. Appl. Math. Techn. U. Budapest* **479** (1986).
16. B. Bergersen, D. Boal, and P. Palffy-Muhoray, *J. Phys. A* **27**(7), 2579–2586 (1993).
17. N. W. Johnson, *Can. J. Math.* **18**, 169–200 (1966).
18. V. A. Zalgaller, *Convex Polyhedra with Regular Faces*, Vol. 2, Seminars in Mathematics, V. A. Steklov Mathematical Institute, Leningrad, Consultants Bureau, New York, 1969.
19. N. I. Fisher, T. Lewis, and B. J. J. Embleton, *Statistical Analysis of Spherical Data*, Cambridge University Press, Cambridge, 1987.
20. G. S. Watson, *Statistics on Spheres*, Wiley-Interscience, New York, 1983.
21. E. Wigner, *Zeit. Physik* **43**, 624–652 (1927).
22. H. Cohn, *Math. Tables Other Aids to Comp.* **10**, 117–120 (1956).
23. D. Gavelek and T. Erber, *J. Comput. Phys.* **101**, 25–50 (1992).
24. D. K. Kondepudi and G. W. Nelson, *Nature* **314**(6010), 438–441 (1985).
25. V.A. Avetisov, V. I. Goldanskii, and V. V. Kuz'min, *Phys. Today* **44**, 33–41 (1991).
26. T. Erber, H. G. Latal, and R. P. Olenick, *J. Appl. Phys.* **52**, 1944–1946 (1981).
27. A. Beck, *Am. Math. Monthly* **97**, 289–294 (1990).
28. H. T. Croft, K. J. Falconer, and R. K. Guy, *Unsolved Problems in Geometry*, Springer, New York, 1991.
29. R. M. Robinson, *Math. Ann.* **179**, 296–318 (1969).
30. D. A. Kottwitz, *Acta Cryst.* **A47**, 158–165 (1991).
31. R. F. Brown and J. H. White, *Indiana Univ. Math. J.* **30**, 501–512 (1981).
32. L. Föppl, Stabile Anordnungen Von Elektronen Im Atom, Göttinger Dissertation, G. Reimer, Berlin (1912).

33. R. W. Hasse and J. P. Schiffer, *Ann. Phys. (N.Y.)* **203**, 419–448 (1990).
34. S. L. Gilbert, J. J. Bollinger, and D. J. Wineland, *Phys. Rev. Lett.* **60**, 2022–2025 (1988).
35. H. S. Chan and K. A. Dill, *Phys. Today* **46**(2), 24–32 (1993).
36. J. D. Honeycutt and D. Thirumalai, *Biopolymers* **32**, 695–709 (1992).
37. P. L. Privalov, *Adv. Prot. Chem.* **35**, 1–104 (1982).
38. T. Erber and G. M. Hockney, *Am. Math. Soc. Abstr.* **6**, 487 (1985).
39. T. Erber and G. M. Hockney, *J. Phys. A* **24**, L1369–L1377 (1991).
40. L. Glasser and A. G. Every, *J. Phys. A* **25**, 2473–2482 (1992).
41. J. R. Edmundson, *Acta Cryst. A* **48**, 60–69 (1992).
42. J. R. Edmundson, *Acta Cryst. A* **49**, 648–654 (1993).
43. T. A. Claxton and G. C. Benson, *Can. J. Chem.* **44**, 157–163 (1966).
44. T. W. Melnyk, O. Knop, and W. R. Smith, *Can. J. Chem.* **55**, 1745–1761 (1977).
45. J. H. Conway and N. J. A. Sloane, *Sphere Packings, Lattices and Groups*, 2nd ed., Springer, New York, 1993.
46. G. S. Ezra and R. S. Berry, *Phys. Rev. A* **25**, 1513–1527 (1982).
47. A. Martin, J.-M. Richard, and T. T. Wu, *Stability of Three Unit-Charge Systems*, TH-6227, CERN, 1991.
48. E. H. Lieb and H.-T. Yau, *Phys. Rev. Lett.* **61**, 1695–1697 (1988).
49. F. Pacella, *Arch. Rat. Mech. Anal.* **97**, 59–74 (1987).
50. C. K. McCord, "Planar Central Configuration Estimates in the N-Body Problem," preprint, 1994.
51. J. Merkel, private communication, 1994.
52. E. F. Brickell and A.M. Odlyzko, in *Contemporary Cryptology*, G. J. Simmons, ed., IEEE Press, Piscataway, N. J. 1991, pp. 501–540.
53. P. Sjögren, *Arkiv Mat.* **11**, 147–151 (1973).
54. J. Korevaar, "Problems of Equilibrium on the Sphere and Electrostatic Fields," University of Amsterdam, Report 76-03, 1975.
55. T. Erber, H. G. Latal, and B. N. Harmon, in *Advances in Chemical Physics*, S. Rice and I. Prigogine, eds., Vol. 20, Wiley, New York, 1971, pp. 71–134.
56. T. Erber, G.R. Marousek, and G. K. Forsberg, *Acta Phys. Austriaca* **30**, 271–294 (1969).
57. J. Vollmer and W. Breymann, *Phys. Rev. B* **47**(18), 11767–11773 (1993).
58. W. Feller, *An Introduction to Probability Theory and Its Applications*, Vol. II, Wiley, New York, 1971.
59. G. Benedek, T. P. Martin, and G. Pachioni, eds., *Elemental and Molecular Clusters*, Springer Series in Material Sciences, Vol. VI, Springer, Berlin, 1988.
60. B. G. Englert, *Semiclassical Theory of Atoms*, Lecture Notes in Physics 300, Springer, Berlin, 1988.
61. H. Buchholz, *Elektrische und Magnetische Potentialfelder*, Springer, Berlin, 1957.
62. D. R. Gavelek, unpublished, 1984.
63. C. E. Porter, *Statistical Theories of Spectra: Fluctuations*, Academic, New York, 1965.
64. P. Pearce and S. Pearce, *Polyhedra Primer*, Van Nostrand Reinhold, New York, 1978.

65. R. E. Williams, *Handbook of Structure Part I: Polyhedra and Spheres*, Res. Commun. 75, Douglas Advanced Reasearch Laboratories, 1968.
66. N. W. Johnson, private communication.
67. E.C. Catalan, *J. École Impériale Polytechnique* **24**, 1–71 (1865).
68. J. R. Heath, *Spectroscopy* **5**, 36–43 (1990).
69. J. R. Morris, D. M. Deaven, and K. M. Ho, *Phys. Rev.* **B53**(4), R1740–R1743 (1996).
70. M. Hamermesh, *Group Theory and Its Application to Physical Problems*, Addison-Wesley, Reading, MA, 1962.
71. H. S. M. Coxeter, *Introduction to Geometry*, Wiley, New York, 1961.
72. K. Mislow, *Introduction to Stereochemistry*, W.A. Benjamin, New York, 1965.
73. F. J. Almgren, Jr., and E.H. Lieb, *Ann. Math.* **128**, 483–530 (1988).
74. T. Erber and A. Sklar, "Macroscopic Irreversibility as a Manifestation of Micro-Instabilities" in *Modern Developments in Thermodynamics*, B. Gal-Or, ed., Wiley, New York, and Israel Universities Press, Jerusalem, 1974, pp. 281–301.
75. K. Schütte and B. L. van der Waerden, *Math. Ann.* **123**, 96–124 (1951).
76. G. J. Simmons, ed., *Contemporary Cryptology*, IEEE Press, Piscataway, NJ, 1992.
77. J. Radon, *Sitz. K. Akad. Wiss. Wien, Math–Naturw. Kl. II a*, **122**(7), 1295–1438 (1913).
78. T. Tarnai, private communication, 1991.
79. K. Bezdek, R. Connelly, and G. Kertesz, in *Intuitive Geometry*, K. Böröczky and G. Fejes Toth, eds., Colloq. Math. Soc. Janos Bolyai, Vol. 48, North-Holland, Amsterdam, 1987, pp. 37–52.
80. L.J. Campbell and R. Ziff, A Catalog of Two-Dimensional Vortex Patterns, LA-7384, Los Alamos (unpublished), 1979.
81. E. Madelung, *Phys. Zeitschr.* **19**, 524–532 (1918)
82. C. Kittel, *Introduction to Solid State Physics*, Wiley, New York, 1961.
83. R. O. Jones and O. Gunnarsson, *Rev. Mod. Phys.* 61, 689–746 (1989).
84. Y. Y. Ye, C. T. Chan, K. M. Ho, and B. N. Harmon, *Int. Jr. Supercomputer Appl.* **4**, 111–121 (1990).
85. G. A. Williams and R. E. Packard, *Phys. Rev. Lett.* **33**, 280–283 (1974).
86. A. M. Mayer, *Sci. Am. Suppl.* **V**(129), 2045–2047 (1878).
87. C. Lanczos, *The Variational Principles of Mechanics*, University of Toronto Press, Toronto, 1966.
88. K. F. Sundman, *Acta Math.* **36**, 105–179 (1913).
89. Z. Xia, *Ann. Math.* **135**, 411–468 (1992).
90. R. C. Howison and C. K. McCord, "Equilibria of Spherical Charge Distributions," preprint, 1994.
91. H. Frauenfelder and P. G. Wolynes, *Phys. Today* **47**(2), 58–64 (1994).
92. R. Peikert, *El. Math.* **49**, 16–26 (1994).
93. R. E. Davies and P. J. Freyd, *J. Chem. Ed.* **66**(4), 278–281 (1989).
94. F. H. Stillinger and T. A Weber, *Science* **225**(4666), 983–989 (1984).
95. E. A. Rakhmanov, E. B. Saff, and Y. M. Zhou, *Math. Res. Lett.* **1**, 647–662 (1994).
96. A. Edelman and E. Kostlan, *Bull. Am. Math. Soc.* **32**(1), 1–37 (1995).

97. V. Pareto, *Manuale di Economia Politica*, Società Editrice Libraria, Milan, 1909.
98. E. L. Altschuler, T. J. Williams, E. R. Ratner, F. Dowla, and F. Wooten, *Phys. Rev. Lett.* **72**(17), 2671–2674 (1994).
99. T. Erber and G. M. Hockney, *Phys. Rev. Lett.* **74**(8), 1482 (1995).
100. E. L. Altschuler, T. J. Williams, E. R. Ratner, F. Dowla, and F. Wooten, *Phys. Rev. Lett.* **74**(8), 1483 (1995).
101. J. Milnor, *Morse Theory*, Annals of Math. Studies, Vol. 51, Princeton University Press, Princeton, 1963.
102. J. Mawhin and M. Willem, *Critical Point Theory and Hamiltonian Systems*, Springer, New York, 1989.
103. P. Painlevé, *Compt. Rend.* **138**, 1555–1557 (1904).
104. A. Wintner, *The Analytical Foundations of Celestial Mechanics*, Princeton University Press, Princeton, 1941.
105. G. E. Forsythe, M. A. Malcolm, and C. B. Moler, *Computer Methods for Mathematical Computations*, Prentice-Hall, Englewood Cliffs, NJ, 1977.
106. R. L. Burden and J. D. Fares, *Numerical Analysis*, P.W.S. Kent, Boston, 1989.
107. T. M. Apostol, *Mathematical Analysis*, Addison-Wesley, Reading, MA, 1957.
108. R. Bellman, *Introduction to Matrix Analysis*, McGraw-Hill, New York, 1960.

AUTHOR INDEX

Numbers in parentheses are reference numbers and indicate that the author's work is referred to although his name is not mentioned in the text. Numbers in *italic* show the pages on which the complete references are listed.

SUBJECT INDEX